LUMINOUS CHEMICAL VAPOR DEPOSITION
and
INTERFACE ENGINEERING

ADDITIONAL VOLUMES IN PREPARATION

LUMINOUS CHEMICAL VAPOR DEPOSITION and INTERFACE ENGINEERING

Hirotsugu Yasuda

University of Missouri–Columbia
Columbia, Missouri, U.S.A.

CRC Press
Taylor & Francis Group
Boca Raton London New York

CRC Press is an imprint of the
Taylor & Francis Group, an **informa** business

First published 2005 by Marcel Dekker, Inc.

Published 2019 CRC Press
Taylor & Francis Group
6000 Broken Sound Parkway NW, Suite 300
Boca Raton, FL 33487-2742

First issued in paperback 2020

ISBN 13: 978-0-367-57817-6 (pbk)
ISBN 13: 978-0-8247-5788-5 (hbk)

Visit the Taylor & Francis Web site at
http://www.taylorandfrancis.com

and the CRC Press Web site at
http://www.crcpress.com

Preface

In 1985, the author wrote a book titled *Plasma Polymerization*, Academic Press. This book is an expanded version of that publication, but the title is changed according to the progress made in understanding the phenomena and important developments of interface engineering. *Plasma Polymerization* has two arguable terms, i.e., "plasma" and "polymerization." Is the glow discharge used for "plasma polymerization" in the state of "plasma"? Are the chemical reactions that occur in glow discharge "polymerizations"? The answers to these questions are yes and no depending on the definition and the meaning of these terms.

If *plasma* is not an adequate term to describe low-pressure glow discharge, what term can be used to describe glow discharge? Similarly, if polymerization is not a proper term to describe the material formation that takes place in the glow discharge, what term represents the process better than polymerization? This book explains why *luminous chemical vapor deposition* (LCVD) is chosen to replace *plasma polymerization*.

Upon publication of the first book on *plasma polymerization*, the author had a well established concept that plasma polymerization is not an ionization driven chemical process, but a chemical process driven by the dissociation of molecules caused by the impact of low energy electrons—but could not find the decisive phenomenon to support the concept. After beginning to write the follow up of the first book on *plasma polymerization*, on the basis of the hypothesis, the author decided to present a picture of a DC glow discharge, of which the distribution profile of electron temperature and electron density were available.

Argon (Ar) glow discharge showed a good matching of the glow and the electron temperature profile. However, when trimethylsilane was used under the otherwise identical setup and conditions, it was astonishing to find that the cathode that should be in the dark space was actually in the midst of glow; i.e., the cathode glow was found in the glow discharge of the organic compound, but not in the glow discharge of argon.

About a year prior to the discovery of the cathode glow, it was recognized that the deposition kinetics of DC cathodic polymerization (deposition on the cathode) was completely different from that of deposition on the substrate floating in glow discharge. In order to describe the difference, it was explained by designating the cathodic polymerization as *dark polymerization* assuming that the polymerization took place in the cathode dark space. This was obviously a mistake, because it was intuitively assumed that the cathode was in the dark space. It has never occurred to researchers that such a glow should be examined, because it was well known that the cathode is in the dark space based on the detailed studies of inert gas discharges.

In order to correct the mistake in designating *dark polymerization*, a manuscript with photographs of cathode glows, to point out that the glow discharge of an organic gas is totally different from the discharge of an inert gas, and to explain the mechanisms of material formation in glow discharge, was submitted for publication within a journal. However, the manuscript was rejected because "the dark space of DC glow discharge is well known."

LCVD has more profound implications than the means of depositing nano film on the surface of material. The surface is the boundary of a condensed matter, particularly of a solid. This means that something else exists beyond the boundary including nothing in the case of high vacuum. Thus, what we conceive as a surface is in reality an interface. In a strict sense, surface properties, which are interfacial properties under a specific set of conditions of surface analysis, cannot be described without specifying the contacting phase of an interface.

Interface engineering requires the understanding of "surface" or "interface." In an effort to understand the surface state, low-pressure glow discharge has a significant role. The view on the fundamentals of surface and interface is presented in this book based on the knowledge gained by applying low-pressure plasma technology to the study of surface and interface. LCVD is actually an interfacial phenomenon at the interface of substrate and the luminous gas phase that make contact with the substrate.

Because of the system-dependent nature of the process and of the very wide range of characteristic properties of the depositions obtainable from a monomer (starting gas), a comprehensive summary of literature on *plasma polymerization* is nearly impossible to attain. In this book, it is intended to discuss general views of the forest rather than microscopically analyzing the details of leaves without knowing where they come from. The attention is focused only on *plasma polymerization*, which can be utilized in *low-pressure plasma interface engineering*—since the first book, *Plasma Polymerization*, was published in 1985—following the line of thought developed at the Camille Dreyfus Laboratory, Research Triangle Institute; Thin Film Institute, Material Research Center, University of Missouri-Rolla; and Center for Surface Science and Plasma Technology, University of Missouri-Columbia.

The views presented in this book are not the consensus views of publications in the literature, but are strictly based on the author's best effort and knowledge at the time of writing. When one considers that a subject is well understood, the progress in pursuit of the truth ends. The author hopes this book will provide enough impetus for pursuit of the truth in the ever-expanding inter-disciplinary domains of luminous chemical vapor deposition and interface engineering.

The author is deeply indebted to his colleagues for creating valuable information through dedicated hard work and discussions. Special thanks are due to his family, especially his wife Gerda, for their understanding, sacrifice and encouragement, which made the writing of this book possible.

Hirotsugu Yasuda

Contents

Part II. Operation of LCVD and LCVT

Part III. Fundamentals of Surface and Interface

Part IV. Interface Engineering

1

Introduction

Since the publication of *Plasma Polymerization* in 1985 [1], applications of ultrathin film, e.g., 10–200 nm, by means of plasma polymerization have reached the stage in which such processes are being carried out on an industrial scale. An example of such processes could be seen in the inner-surface coating of plastic bottles by plasma polymerization of acetylene, ACTIS process developed by SIDEL (France), by which approximately 150-nm-thick barrier coating is applied in 3 s by a plasma polymerization process, with total handling time of a bottle in 7 s [2]. Photographs of equipment shown in Figures 1.1–1.3 could aid in the recognition of the current state of plasma polymerization or luminous chemical vapor deposition (LCVD) process, by which the author prefers to describe the process according to the knowledge gained since the publication of *Plasma Polymerization*.

Figure 1.3 clearly demonstrates the luminous gas phase created under the influence of microwave energy coupled to the acetylene (gas) contained in the bottle. This luminous gas phase has been traditionally described in terms such as low-pressure plasma, low-temperature plasma, nonequilibrium plasma, glow discharge plasma, and so forth. The process that utilizes such a luminous vapor phase has been described as plasma polymerization, plasma-assisted CVD (PACVD), plasma-enhanced CVD (PECVD), plasma CVD (PCVD), and so forth.

However, the luminous gas phase created in low pressure under the influence of some kinds of electromagnetic power seems to be so far deviated from the definition of *plasma state* that the use of the term *plasma* could be misleading. Furthermore, the luminous gas phase deserves its own identity. Plasma state has a characteristic glow (luminous gas phase); however, whether the luminous gas phase created by an electrical discharge is in plasma state is a serious question. A very recent discovery of molecular dissociation glow in glow discharge of organic molecules strongly suggests that major body of luminous gas phase dealt in LCVD is not in the plasma state [3,4], as described in Chapters 3 and 4.

Any chemical reaction that yields polymeric material can be considered polymerization. However, polymerization in the conventional sense, i.e., yielding high enough molecular weight materials, does not occur in the low-pressure gas phase (without a heterogeneous catalyst). With a heterogeneous catalyst, polymerization is not a gas phase reaction. Therefore, the process of material deposition from luminous gas phase in the low-pressure domain might be better represented by the term *luminous chemical vapor deposition* (LCVD). Plasma polymerization and LCVD (terms explained in Chapter 2) are used synonymously in this book, and the former

20 station machine
- continuous rotary kinematics
- 10 000 bph
- bottles < 0,6 litre
- Energy: 100 kW (est.)
- Bottles handled by the neck
  at infeed and exit
  (base handling optional)
- 39 m2

Figure 1.1 Pictorial view of ACTIS coating station, courtesy of Sidel.

Basic Process Conditions

- Inner coating
- Monomer: C_2H_2
- Power: Microwave 2.45 GHz
- Process under vacuum (< 1 mbar)
- Deposition time : 3 s
- Total cycle time : 7.2 s

Figure 1.2 Microwave powered plasma polymerization unit, courtesy of Sidel.

is preferentially used dealing some aspects of material formation, especially the kinetics of growth reactions and of deposition.

The ACTIS process described above is a typical example of low-pressure plasma polymerization or LCVD, which is an ultimate green process with no effluent in the practical sense. Microwave plasma is used for plasma polymerization of acetylene. ACTIS process, as an example of LCVD, has an ideal combination of unique advantages in (1) very high reaction yield (monomer to coating), (2) no effluent from the process, (3) no reactor wall contamination because the reactor wall is the substrate surface, and (4) very short reaction time. However, whether such an ideal LCVD process is an industrially viable practice is a totally different issue,

Figure 1.3   Pictorial view of the luminous gas phase created in a bottle, courtesy of Sidel.

which involves many other factors beyond the processing of LCVD. The list of advantageous features is not necessarily an endorsement of such a process for an industrial application.

Plasma polymerization has a built-in unique feature of forming strong bonds between the coating layer and the substrate polymer, which is difficult to obtain by other conventional coating processes; however, this advantageous feature cannot be obtained automatically, and it is necessary to tailor the operation for such a purpose. ACTIS process, mentioned above, is considered as a passive plasma polymerization coating according to the concept of interface engineering, which will be explored in this book. While passive plasma coating on a polymeric substrate works well in certain cases, the change at the interface between the substrate and the coating is rather abrupt, which is not an ideal interface that could be created by LCVD interface engineering.

When a passive plasma coating process is applied to a smooth surface of nonpolymeric substrate, such as metal, ceramics, and glass, insufficient level of adhesion of the plasma coating to the substrate becomes the major drawback, which hampers the successful application of plasma coatings. LCVD interface engineering can eliminate this problem by smearing out the discrete boundary or by creating a graded interphase within the thickness range of 10–100 nm. Well-executed low-pressure plasma interface engineering not only eliminates the drawback but also creates the ideal interphase in between two completely different types of materials that could not be attained by other means.

Plasma polymers have a wide spectrum with respect to their characteristics. At one end, the material is an extremely tight three-dimensional network of elements (mainly C, Si, or mixture of C and Si) that have terminal groups of H or F (halogens). Those plasma polymers are insoluble and infusible, with very little segmental mobility and hence no chemical reactivity. This kind of plasma polymers can be designated as type A plasma polymers.

On the other end of the spectrum, plasma-formed materials are oligomeric and could contain various functional groups rather than terminal H or F. Those materials are often soluble, chemically reactive depending on the nature of functional groups, but lack cohesive strength. This kind of plasma polymers can be designated as type B plasma polymers.

In the middle of a spectrum, the gain in one feature is attained on sacrifice of another feature. Therefore, one must choose a plasma polymerization process, including type of reactor, reaction conditions, and type of monomer (starting gas or vapor), aimed at a specific type of plasma polymer; i.e., type A or type B, suitable for an application.

It is a great dream of practitioners of plasma polymerization to form a type A plasma polymer with desired functional groups on the surface. However, it is a dream because of the lack of understanding of basic principles involved in plasma polymerization. A simple plasma polymerization process cannot form a tight network with functional groups. This is where the interface engineering by means of LCVD becomes important.

For LCVD interface engineering of composite materials, it is considered that the tight three-dimensional network is necessary because of the required structural integrity, i.e., type A plasma polymers are essential for LCVD interface engineering. Low-pressure plasma polymerization is considered ideal for such applications because of the mechanisms of polymerization or material formation.

The proper execution of system approach interface engineering (SAIE) requires the understanding of fundamentals of the interface. Conversely, utilization of nanofilms of unique plasma polymers provided tools to investigate the fundamentals of interfaces, which were nearly impossible to examine otherwise. The use of LCVD in the interface engineering and the basic investigation of interface worked hand in hand to broaden the knowledge in this specific topical area.

In order to provide better understanding of SAIE, this book is divided into four major sections. Part 1 provides basics of LCVD (plasma treatment and polymerization). Part 2 describes various modes of LCVD operation. The knowledge on the surface and the interface gained by utilizing low-pressure plasma technology is described in Part 3, which covers the fundamentals of surface and interface. Part 4 presents some examples of SAIE, which were achieved by virtue of LCVD.

A significant portion of SAIE attempts involves metallic substrate, which is difficult to be treated with microwave plasma. While the inner surface of a plastic bottle (about 1 liter or less) is an ideal surface to coat by microwave plasma, which propagates on the surface, a larger or smaller bottle makes the process less effective and less feasible. The substrates that are the target of SAIE are generally much more complex in size and shape than the inner surface of a plastic bottle used in ACTIS. Consequently, most attempts made so far involve low-pressure plasma polymerization by direct current (DC) to the radio frequency range power source. Dealing with electrically conducting substrates, DC discharge, which has not been in the major stream of plasma polymerization research and development, was found to be very useful and practical in SAIE.

A best process to achieve a goal, as indicated by the above examples, is a combination of many factors including the type of electrical discharge, chemical and physical properties of the substrate, size and shape of the substrate, and ultimately the objective to be accomplished by the process. Therefore, it is dangerous to prematurely determine the merit of a seemingly successful operation. After all, so many industrial scale attempts indicate that an excellent technological success does not necessarily translate to an industrially successful (profitable) operation.

SAIE by low-pressure plasma processes must cope with practical demands:

1. Relatively large total surface area should be treated in a unit process, regardless of whether it is carried out in a batch mode or in a continuous in-line mode operation.
2. Coating should be applied uniformly regardless of the size and shape of substrate.
3. Process should be operated in a sustainable manner with minimal interruptions for maintenance, e.g., for a month.
4. There should be high enough, e.g., >10%, conversion yield of the deposition onto the substrate.
5. There should be minimal effluent and unwanted coating on nonsubstrate surfaces.

Above all of these requirements, SAIE must produce products that are superior to the conventional products. In other words, low-pressure plasma SAIE is not an alternative process; it should be a new approach to create superior composite materials that could not be obtained by other means, which is of utmost importance with respect to the use of LCVD. It is often mentioned that plasma polymerization was successfully used in the surface modification but that a conventional, more economical, wet chemical process later replaced it. Such an attempt to use LCVD process based only on the laboratory curiosity is an absolutely wrong approach. This aspect is explored in Chapter 12.

SAIE applied to the corrosion protection of a metal can be used to illustrate the overall features of the approach, which are described in Part 4. The factors to be considered in a corrosion protection system include (1) the surface state of the substrate on which a protective coating system is applied, (2) interfacial factors, (3) the bulk characteristics of the plasma polymerization coating, and (4) the surface state characteristics of plasma polymerization coating on which a primer coating is applied. The *surface state* onto which a plasma polymer coating will be applied depends on the preceding processing step, which must be tailored for a particular desired outcome. The term surface state describes the properties of materials in the top surface region that are significantly different from those in the bulk phase of the same material [6], which is described in Chapter 24.

SAIE corrosion protection emphasizes the fact that the corrosion protection of a metal depends on the overall corrosion protective behavior of an entire system. If a plasma polymerization coating is changed, all factors must be optimized to yield the best result that can be attributed to the change. The essence of interface engineering lies in the tailoring of surfaces to facilitate the equilibration of surface states of different materials.

Low-temperature plasma processes, such as gas plasma treatment and plasma polymerization, have unique advantages in that active (depositing) species strongly interact with the surface of the substrate and modify the surface state. An ultrathin layer of plasma polymer, e.g., thickness less than 50 nm, can be viewed as a new surface state because such a thin layer does not develop a characteristic bulk phase. Contact electrification measurements indicate that surface electrons are still influenced by those from the substrate, up to a film thickness of roughly 20 nm, as described in Chapter 24. Thus, LCVD (low-pressure plasma surface modification and polymerization) could be considered as a means to create an entirely new surface state grafted onto the surface state of a substrate.

A nanofilm of a plasma polymer could be used as an interface modifier, which joints two distinctively different materials, or could be used as the top layer of a material, which would become a part of new interface when the material is placed in a different environment such as biological tissues, bloodstream, etc. (SAIE in biomedical materials is described in Chapter 35).

REFERENCES

1. Yasuda, H. *Plasma Polymerization*; Academic Press: San Diego, 1985.
2. Courtesy of Sidel, Le Havre, France.
3. Hirotsugu Yasuda; Qingsong Yu Plasma Chem. Plasma Proc. **2004**, *24*, 325.
4. Yu, Q.S.; Huang, C.; Yasuda, H.K. J. Polym. Sci.; Polym. Chem. Ed. **2004**, *42*, 1042.
5. Reddy, C.M.; Yu, Q.S.; Moffitt, C.E.; Wieliczka, D.M.; Johnson, R.; Deffeyes, J.E.; Yasuda, H.K. Corrosion **2000**, *56* (8), 819.
6. Yasuda, H.; Charlson, E.J.; Charlson, E.M.; Yasuda, T.; Miyama, M.; Okuno, T. Langmuir **1991**, *7*, 2394.

2
Domain of Luminous Chemical Vapor Deposition

1. TERMINOLOGY

The material deposition that occurs in the low-pressure electrical discharge has been discussed under various terminologies such as plasma polymerization (PP), plasma-enhanced chemical vapor deposition (PECVD), plasma-assisted chemical vapor deposition (PACVD), plasma chemical vapor deposition (PCVD), and so forth [1]. However, none of these terminologies seems to represent the phenomenon adequately. The "plasma" aspect in the low-pressure discharge is remote, although it plays a key role in creating the environment from which material deposition occurs to the extent that no chemical reaction occurs without the "plasma." In this sense, PECVD and PACVD could be out of the context in many cases in which nothing happens without plasma. In such cases, PP or PCVD would describe the phenomenon better. If the substrate was not heated substantially above the ambient temperature, the use of PECVD or PACVD should be avoided.

The common denominator factor that has not been emphasized but deserves its own identity is the "luminous gas phase" from which the material deposition occurs. The key issues are how the luminous gas phase is created in the low-pressure electrical discharge and how chemically reactive species are created in the luminous gas phase. In this chapter we focus on the domain of CVD that functions only under the influence of the luminous gas phase by using the term *luminous chemical vapor deposition* (LCVD).

In order to find the domain of LCVD, it is necessary to compare various vacuum deposition processes: chemical vapor deposition (CVD), physical vapor deposition (PVD), plasma chemical vapor deposition (PCVD), plasma-assisted CVD (PACVD), plasma-enhanced CVD (PECVD), and plasma polymerization (PP). All of these terms refer to methods or processes that yield the deposition of materials in a thin-film form in vacuum. There is no clear definition for these terms that can be used to separate processes that are represented by these terminologies. All involve the starting material in vapor phase and the product in the solid state.

PVD such as evaporation deposition and sputter deposition of a metal also share the common denominator aspects that the starting material, in the crucial stage of material deposition, is in the vapor phase and the product in a solid state, if one ignores the process of creating the vapor of the starting material. Some processes use plasma to create gas phase material to deposit.

The major difference between PVD and CVD is the change of the chemical nature of material in the vapor phase and in the solid state. PVD has no change of chemical nature of material used. For instance, gold vapor deposits as a solid-state gold in PVD. In contrast to this situation, the material deposits in CVD are different from the starting material with respect to the chemical composition and/or chemical structure. In other words, chemical reactions are crucially important factors of the material deposition. According to the definitions just described, there is no distinction among PP, PACVD, PECVD, PCVD, and CVD. Therefore, the further distinction of these terms must rely on the processing factors that are related to the reaction mechanisms.

In a general CVD process, the substrate is heated to a substantially high temperature, e.g., higher than 300°C, but the reactor itself is generally not heated. Therefore, the activation of the starting material in vapor phase is done by the thermal energy provided by the substrate surface. Here, the important factor is (1) thermal activation and (2) the creation of the activated chemically reactive species both occur at the surface of the substrate.

In hot-wire CVD (HWCVD), hot wires are used to initiate the reaction and the substrate is kept in lower temperature. In this case the thermal activation occurs in a spatially separated location, and the substrate is the deposition surface. In such a process the chemical activation of vapor and the deposition of materials are spatially separated, whereas in the ordinary CVD both processes occur in the same place.

2. PARYLENE POLYMERIZATION AND PLASMA POLYMERIZATION

The process known as parylene coating, which is described in Chapter 5, is a typical thermally activated CVD, in which the activation and the deposition are spatially separated, i.e., *A/D decoupled*, as is the case of HWCVD. In parylene coating, a cracking chamber is generally used instead of hot wires. Parylene coating is a typical (thermal) CVD (not LCVD) as far as the processing is concerned. However, parylene polymerization has a very unique distinction from the rest of CVD that the very specific chemical structure is created from a well-defined chemical structure designed to perform parylene polymerization process. Parylene coating process is schematically depicted in Figure 2.1.

In a plasma polymerization, the substrate is generally not heated, nor is the vapor heated. The chemical activation is done by the interaction of gas phase molecules with plasma (luminous gas) or by the generation of plasma of the starting material. In other words, activation of the starting material occurs in the vapor (plasma) phase, and the substrate is merely the collector of the product unless the substrate is used as an electrode.

PP is a CVD process in which the chemically reactive species are created by plasma. Terms such as plasma-assisted CVD (PACVD) and plasma-enhanced CVD (PECVD) are inadequate to describe PP. Unless the substrate temperature is raised

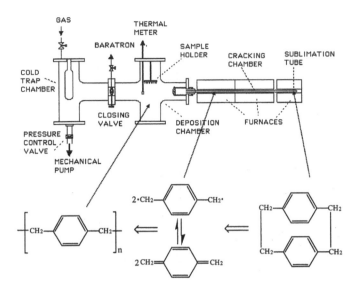

Figure 2.1 Schematic representation of parylene polymerization.

substantially so that (1) surface thermal activation of the starting material becomes significant and (2) the substrate surface temperature is above the ceiling temperature of plasma polymerization (see Chapter 5). If the second requirement was met, the process is indeed PACVD or PECVD. It appears that if the first requirement is met, the second requirement is also met. Thus, the substrate temperature more or less determines whether a process is PP or PACVD or PECVD. PACVD or PECVD should proceed without plasma because in order to enhance or assist, there must be an independent process that can be enhanced or assisted by other processes. If nothing occurs when glow discharge is switched off, the process is a PP or PCVD. PP or PCVD is, therefore, not PACVD or a PECVD; however, these terms are frequently used to describe essentially the identical process as PP.

The spatially decoupled activation and deactivation can be also seen in a mode of PP known as low-pressure cascade arc torch (LPCAT) polymerization), which is described in Chapter 16. The activation of a carrier gas (e.g., argon) occurs in a cascade arc generator, and the chemical activation of a monomer or a treatment gas takes place near the injection point of the argon torch in the deposition chamber. The material deposition (deactivation) occurs in the deposition chamber. This is the same situation as the HWCVD, except that the mode of activation is different.

LPCAT is a luminous CVD process with decoupled ionization process, i.e., the chemically reactive species are created by the neutral species–impact dissociation of molecules, but not by ion–impact or electron–impact dissociations. Because of this aspect, LPCAT provides important information pertinent to the nature of the creation of chemically reactive species in LCVD and will be discussed in some detail in the following chapters.

Some important distinctive factors for these vacuum deposition processes, which have common denominator aspects, are tabulated in Table 2.1. From the operational viewpoint, these processes can be compared as shown in Table 2.2.

Table 2.1 Comparison of Vacuum Deposition Processes

Process	Activation	Deposition (deactivation)	T_S, T_V, and ΔE for deposition
Plasma polymerization	Vapor phase by plasma	Substrate in the same chamber	$T_S < T_V$ $\Delta E < 0$
LPCAT polymerization	Vapor phase in a separate chamber by plasma	Substrate in a separate chamber	$T_S < T_V$ $\Delta E < 0$
Parylene polymerization	Active species is created in a separate chamber by thermal cracking	Substrate in a separate deposition chamber	$T_S < T_V$ $\Delta E < 0$
CVD	Substrate surface is the thermal activation site	Substrate (in the same chamber)	$T_S > T_V$ $\Delta E > 0$
HWCVD	Active species is created in a separate section of the same reactor by thermal cracking	Substrate in the same chamber	$T_S < T_V$ $\Delta E < 0$
PECVD or PACVD	Plasma and substrate surface	Substrate surface in the same chamber	$T_S > T_V$ $\Delta E > 0$

T_S, temperature of substrate surface; T_V, temperature of vapor phase; $\Delta E < 0$, deposition rate decreases with increasing substrate temperature.
Plasma polymerization is a vapor phase–activated (by plasma) A/D-coupled CVD. Parylene polymerization is a vapor phase–activated (by thermal process) A/D-decoupled CVD. Cascade arc torch (CAT) polymerization is a vapor phase–activated (by plasma) A/D-decoupled CVD. General CVD is a surface-activated (by thermal process) A/D-coupled CVD. HWCVD is a hot wire activated A/D-decoupled CVD. Coupled and decoupled refer only to the spatial separation.

Table 2.2 Comparison of Operational Factors of Vacuum Deposition Processes

Process	Energy input	Phase of activation	Activation/ deactivation[a]
Plasma polymerization or PCVD	Electrical discharge	Gas	Coupled
Parylene polymerization	Thermal energy	Gas	Decoupled
CVD	Thermal energy	Substrate surface	Coupled
HWCVD	Thermal energy	Wire surface	Decoupled
LPCAT polymerization	Electrical discharge	Gas	Decoupled
PECVD or PACVD	Electrical discharge and heat	Gas and substrate surface	Coupled

[a]Coupled means that both processes occur in the same space; decoupled means that both processes occur in separate spaces.

3. PLASMA-INDUCED POLYMERIZATION

Plasma can be utilized in the polymerization of monomer liquid. In this case, no substrate is employed, and monomers are typically organic compounds with an olefinic double bond (monomer for chain growth polymerizations). In a typical case, the vapor phase of a monomer liquid in a sealed tube is used to create plasma. The duration of plasma is generally very short (on the order of a few seconds). After plasma exposure, the tube is shaken to mix plasma-induced reactive species with the monomer and is kept at a constant temperature (polymerization temperature) for a prolonged period.

Plasma-induced species act as initiator of polymerization. Polymerization characteristics and properties of polymers formed by plasma-induced polymerization strongly resemble those of the thermal polymerization of the corresponding monomer [2–12]. Results indicate that plasma-induced polymerization is a free radical addition polymerization initiated by difunctional free radicals created by plasma. The molecular weight of polymer increases with the polymerization time, which is distinctively different from the initiator-initiated free radical addition polymerization.

After a long reaction time, polymers with exceptionally high molecular weight can be synthesized by plasma-induced polymerization. Since only brief contact with luminous gas phase is involved, plasma-induced polymerization is not considered to be LCVD. However, it is important to recognize that the luminous gas phase can produce chemically reactive species that trigger conventional free radical addition polymerization. This mode of material formation could occur in LCVD depending on the processing conditions of LCVD, e.g., if the substrate surface is cooled to the extent that causes the condensation of monomer vapor.

4. PLASMA SURFACE MODIFICATION OF POLYMERS

Exposing a polymer surface to various kinds of plasmas can modify polymer surfaces. Plasmas of argon, oxygen, hydrogen, and air are frequently used in plasma surface modification of polymers. Plasmas of non-polymer-forming gases are used in this process, and a plasma and/or plasma-induced reactive species interact with the polymer substrate placed in a reactor. This is in contrast to plasma (state) polymerization in which plasma mainly interacts with monomer vapor.

No deposition of materials occurs in most cases; however, the deposition of plasma polymer could occur depending on the nature of substrate polymer. Such a deposition of materials can be viewed as PP of organic vapors, which emanated from the substrate, by the interaction with plasma. Because the major player is the luminous gas phase, the surface treatment is included in this book under the term luminous chemical vapor treatment (LCVT).

Comparing the terms plasma chemical vapor deposition and luminous chemical vapor deposition, the difference exists in the meaning of plasma and luminous gas and its implications to the nature of chemical reactions that occur in the gas phase. Without referring the details of the difference, however, the process could be described either plasma polymerization (plasma CVD) or luminous CVD in all practical purposes.

The terms luminous chemical vapor deposition and plasma polymerization are used synonymously in this book. Dealing with mechanism of reactions that lead to formation of solid deposition, PP is used according to the traditional use of the term. When dealing with the formation of reactive species and other operation and processing aspects, LCVD is preferentially used.

REFERENCES

1. Yasuda, H. Plasma polymerization and plasma modification of polymer surfaces. In *New Methods of Polymer Synthesis*; Ebdon J.R., Eastmond, G.C., Eds.; Blackie: London, 1995; Vol. 2, 161–196.
2. Osada, Y.; Bell, A.T.; Shen, M. J. Polym. Sci., Polym. Lett. Ed. **1978**, *16*, 309.
3. Osada, Y.; Shen, M.; Bell, A.T. ACS Symp. Ser. **1979**, *108*, 253.
4. Simionescu, B.C.; Leanca, M.; Ananiescu, C.; Simionescu, C.I. Polym. Bull. **1980**, *3*, 437.
5. Osada, Y.; Iriyama, Y.; Takase, M. Kobunshi Ronbunshu **1981**, *38*, 629.
6. Johnson, R.D.; Osada, Y.; Bell, A.T.; Shen, M. Macromolecules **1981**, *14*, 118.
7. Simionescu, B.C.; Leanca, M.; Ioan, S.; Simionescu, C.I. Polym. Bull. **1981**, *4*, 415.
8. Osada, Y.; Takase, M.; Iriyama, Y. Polym. J. **1983**, *15*, 81.
9. Kuzuya, M.; Kawaguchi, T.; Daikyo, T.; Okuda, T. J. Polym. Sci., Polym. Lett. Ed. **1983**, *21*, 515.
10. Osada, Y.; Takase, M. J. Polym. Sci., Polym. Lett. Ed. **1983**, *21*, 643.
11. Akovali, G.; Orhan, B. J. Polym. Sci., Polym. Chem. Ed. **1984**, *22*, 3351.
12. Kuzuya, M.; Kawaguchi, T.; Daikyo, T.; Okuda, T. J. Polym. Sci., Polym. Chem. Ed. **1985**, *23*, 77.

3
Luminous Gas Phase

1. LUMINOUS COLUMNS IN DC DISCHARGE OF ARGON

Glow discharge, particularly the location of luminous columns in the system, has been often explained by direct current (DC) glow discharge of an inert gas, e.g., argon [1–4], as shown in Figure 3.1. The major electrical field exists near the cathode (cathode region), and a secondary electron emitted from cathode surface gains energy in the electrical field while traveling toward the anode. In this acceleration of electron, the kinetic energy of the electron traveling in the cathode region is a function of the distance traveled from the cathode. When the electron gains enough kinetic energy to ionize the gas, the luminous gas phase recognized as *negative glow* develops. In the cathode dark space, electrons have not gained sufficient energy to cause the creation of luminous gas phase and thus the space remains dark. Beyond negative glow, it has been postulated that the positive charge density and the negative charge density are nearly equal and the gas phase maintains the electrical neutrality, which fits the criteria of *plasma*.

However, the glow discharge of Ar in a luminous chemical vapor deposition (LCVD) reactor, e.g., a pair of parallel electrodes placed in a bell jar–type reactor (about 75 linter), is quite different from those originally used in a long tube. Consequently, the identification of glows does not match the glow that occurs under LCVD conditions. Characteristic glows in DC discharge of Ar in an LCVD reactor are shown schematically in Figure 3.2.

The negative glow extends nearly to the anode at low pressure. At higher pressure (under the same power input), the cathode dark space and negative glow are pulled toward the cathode and leave dark space in front of anode. The size and location of glows are dependent on the system pressure p and the distance between the cathode and the anode d. The parameter given by pd determines the glow characteristics, including the breakdown voltage of discharge [5]. Figure 3.3 shows the change of the location, the intensity, and the size (volume) of glow as functions of pressure and discharge power. When the discharge power is increased, negative glow approaches the cathode, and the intensity and the volume of negative glow increases. At a higher pressure, negative glow approaches the cathode, and the intensity increases but the volume decreases (the anode dark space expands) at a given discharge power. With inclusion of the discharge power W, the overall glow characteristics are governed by W and pd.

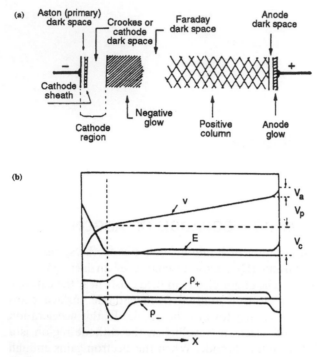

Figure 3.1 Schematic representation of DC glow discharge: X, distance from cathode: V, Potential: E, electrical field; r_+, positive charge density; r_-, negative charge density; the shaded areas are luminous. Adapted from Ref. [4].

These characteristics of glow in an LCVD reactor cast some serious questions regarding the nature of glow and the domain of plasma in a reactor. It is certain that one cannot intuitively assume that the luminous gas phase (glow) in glow discharge is plasma, while plasma has characteristic glow.

2. ELECTRON TEMPERATURE AND ELECTRON DENSITY

The distribution of electron temperature (energy of electrons) and the density of electrons in a DC argon glow discharge in an LCVD (plasma polymerization) reactor [6] are shown in Figures 3.4 and 3.5, respectively. The electron temperature rises as an electron is accelerated in the electric field (cathode dark space). In the cathode region where the acceleration of electron takes place, the electron density is relatively small. When the electron temperature reaches the maximal level as a function of the distance from the cathode, T_e drops as the electron impact ionization of Ar atom occurs, which creates an additional electron. In this region, the electron density starts to increase as more electrons are created and are pulled toward the anode. Photon-emitting neutrals of Ar are also created in this region, forming luminous gas phase recognized as negative glow.

The positively charged argon ions are pulled toward the negatively charged cathode surface and cause the emission of secondary electrons, which

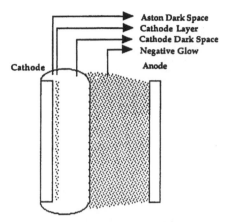

(a): System pressure < 6.66 Pascal (50 mTorr)

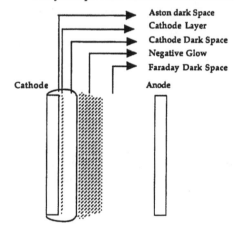

(b): System Pressure > 13.33 Pascal (100 mTorr)

Figure 3.2 Schematic view of nonmagnetron DC glow discharge of argon in the interelectrode space, interelectrode distance: 102 mm; the luminous regions are shown shaded.

will be accelerated while traveling in the cathode fall region. Beyond the cathode fall region, little electric field exists and no significant acceleration of electrons occurs.

The electron temperature increases as a function of the distance traveled from the cathode surface as shown in Figure 3.4. Accordingly, the location of the glow with respect to the distance from the cathode surface is the important parameter that indicates the energy of electrons required to cause the luminous gas phase. The main luminous gas phase of Ar corresponding to the distance for the maximal electron temperature is observed in Figure 3.4 and discussed in the following sections. Thus, the presence and the location of the luminous gas phase can be explained well by the mechanisms of ionization of Ar. The nature of glow matches well with the distribution of electron temperature shown in Figure 3.4.

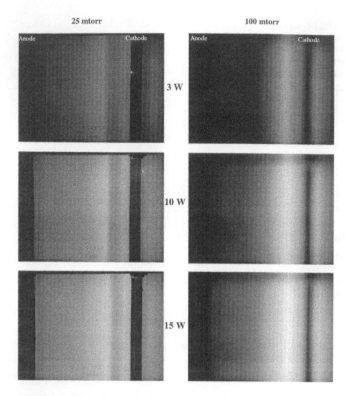

Figure 3.3 Change of the intensity and location of luminous gas phase depending on the discharge power and the system pressure of Ar DC discharge. Left column: 25 mtorr, right column: 100 mtorr. Top row: 3 W, middle row: 10 W, bottom row: 15 W.

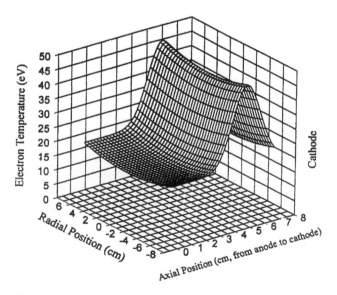

Figure 3.4 Distribution profile of electron temperature in an argon DC glow discharge in a plasma polymerization reactor.

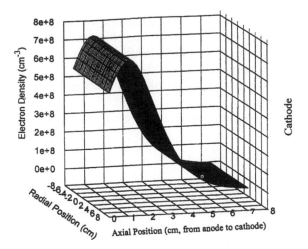

Figure 3.5 Distribution profile of electron density in an argon DC glow discharge in a plasma polymerization reactor.

3. PLASMA AND LUMINOUS GAS

In an ionized gas created by electric discharge, the formation of charges and the diminishing of charges occur independently and simultaneously. In a steady state, the number of electrons, n_e, and the number of ions, n_i, are determined by the balance between these two processes. In an ionized gas, however, there is a strong tendency for the density of negative charge and the density of positive charge to become equal. Because of this tendency, in certain types of ionized gas the system as a whole remains electrically neutral. Such an ionized gas is called *plasma*. The following sections examine the key aspect of maintaining electrical neutrality according to the illustration by Hatta [3].

As shown in Figure 3.1, in a low-pressure glow discharge of an inert gas, the electric potential in the majority of space between the electrodes, except in the vicinity of the cathode, is nearly constant. If we take the distance along the axis of a tube as x,

$$dV/dx = \text{const} \qquad d^2V/dx^2 = 0 \tag{3.1}$$

The space charge density ρ is related to d^2V/dx^2 by

$$d^2V/dx^2 = -4\pi\rho \tag{3.2}$$

Therefore, if $d^2V/dx^2 = 0$, we obtain $\rho = 0$. The space charge density originates from the difference in density of positive charge ρ_+ and that of negative charge ρ_-:

$$\rho = \rho_+ - \rho_-$$

Hence, if $\rho = 0$,

$$\rho_+ = \rho_- \tag{3.3}$$

which is the criterion for plasma.

During the processes of charge formation and recombination of charges, q_+ and q_-, always appear or disappear as a pair. However, once charged species are formed, electrons and ions behave independently. The velocity of thermal motion and acceleration in an electric field are completely different for electrons and ions. Therefore, maintaining of electrical neutrality is not an obvious phenomenon and requires further examination.

Let us consider a simple case in which no negative ions exist and all positive ions are monovalent:

$$\rho_- = qn_e \quad \text{and} \quad \rho_+ = qn_i$$

Hence, if $\rho = 0$, $n_e = n_i$.

Let us further consider a simplified model of plasma in which the properties of plasma vary only in direction x and are uniform in directions y and z. Assuming that ions stay still and that only electrons move within a given time span, we can examine the fluctuation of charge density created in a plasma as shown in Figure 3.6. In the space shown by $2d$, too many electrons have accumulated, and the negative charge density $q\,\Delta n_e$ is created. In this region, $\Delta n_e = n_e - n_i$, but beyond this region, $n_e = n_i$ and

$$d^2V/dx^2 = 4\pi qn_e \tag{3.4}$$

and with conditions $V = 0$ at $x = 0$, and $dV/dx = 0$, we obtain

$$V = 2\pi q\,\Delta n_e x^2 \tag{3.5}$$

Therefore, the electric potential V_d at $x = d$ is given by

$$V_d = 2\pi q\,\Delta n_e d^2 \tag{3.6}$$

The average kinetic energy of the x component of the velocity of electrons is $1/2\ kT_e$. We shall roughly assume that all electrons have this kinetic energy. All electrons in the space shown by width $2d$ are accelerated by the electric potential V_d

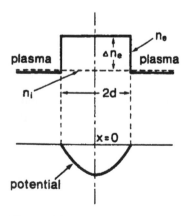

Figure 3.6 Potential well for calculation of Debye length.

and tend to move out of the space. Electrons in the outer space, however, try to come into the space by kinetic energy $1/2\,kT_e$. Under these conditions, the following two situations can be considered:

1. If $1/2\,kT_e < |qV_d|$, the electrons outside of the space do not have enough energy to overcome the electric potential V_d and consequently cannot come into the space.
2. If $1/2\,kT_e > |qV_d|$, the electrons outside of the space can penetrate the space. Therefore, in the first case, the excess electrons simply decrease and electric neutrality is regained, but in the second case, the negative charge density cannot be dissipated and remains in the space.

Accordingly, the borderline case of

$$\tfrac{1}{2}kT_e = |qV_d| = 2\pi q^2 d^2 \Delta n_e \tag{3.7}$$

becomes the critical condition for whether or not the electric neutrality can be regained. The distance given by

$$2d = 2\left(\frac{kT_e}{4\pi q^2 \Delta n_e}\right)^{\tfrac{1}{2}} \tag{3.8}$$

is the maximal distance in which such a space charge can exist and can be used to define electric neutrality, hence the term plasma.

The value of $\Delta n_e/n_e$ can be considered a parameter that indicates the degree of deviation from neutrality. For the limiting case of $\Delta n_e/n_e = 1$, one can assign the value of d as h_0. Thus defined, h_0 is a constant that is determined by the characteristics of plasma and given by

$$h_0 = (kTe/4\pi q^2 n_e)^{\tfrac{1}{2}} \tag{3.9}$$

By virtue of Eq. (3.8), we obtain

$$\Delta n_e/n_e = (h_0/d)^2 \tag{3.10}$$

This equation indicates that if the value of d is 10 times greater than the value of h_0, then Δn_e of 1% can be tolerated to maintain electric neutrality, and if d is 100 times h_0, then Δn_e of 0.01% can be tolerated. In other words, if $d \gg h_0$, an ionized gas can be considered plasma, and if $d \approx h_0$, the deviation from electric neutrality is significantly great, and the ionized gas cannot be considered plasma. Therefore, h_0 can be used as a measure of the criterion of plasma and is generally called *Debye length* according to the originator of the concept. The value of h_0 is given by

$$h_0(\text{cm}) = 6.90[T_e(\text{K})/n_e(\text{cm}^{-3})]^{\tfrac{1}{2}} \tag{3.11}$$

The values of h_0 for $T = 10^4$K, $n_e = 10^8$, 10^{10}, and 10^{12} cm^{-3}, are 6.9×10^{-2}, 6.9×10^{-3}, and 6.9×10^{-4} cm, respectively, indicating that the value of h_0 is small.

Figure 3.7 depicts distribution of electron temperature, electron density, and Debye length, calculated according to Eq. (3.11), in glow discharge of Ar in the LCVD reactor, which is shown in Figures 3.4 and 3.5. The application of Debye length, which is for the equilibrium plasma, to nonequilibrium plasma is probably a meaningless practice; however, it raises a question of what "nonequilibrium plasma" means. Nonequilibrium plasma could possibly mean the luminous gas phase that is not in plasma state.

There is no direct indication where glow exists according to T_e and n_e; that is, T_e and n_e can be measured both in dark space and in glow. The calculated Debye length decreases nearly linearly with the distance from the cathode covering the dark space and luminous gas phase, i.e., the value alone does not indicate where is plasma.

The sharp increase of electron density, roughly 3 cm away from the cathode, can be taken as a clear indication that beyond this point there can be no electrical neutrality, i.e., it is impossible to accumulate large number of positively charged ions near the anode. The luminous gas phase in this space cannot be considered as plasma. Thus the domain of the luminous gas phase extends beyond the domain of plasma or the state that is close to the plasma state. The space in which T_e and N_e

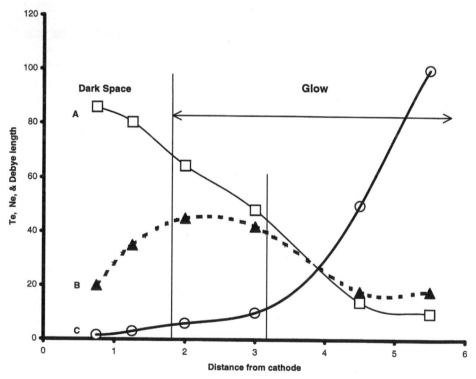

Figure 3.7 Distribution of electron temperature, electron density and Debye length in DC discharge of Ar in an LCVD reactor; A: Debye length, cm $\times 10^2$; B: electron temperature, eV; C: electron density, cm$^{-3} \times 10^{-7}$.

remain nearly constant can be viewed as a more or less uniform layer of luminous gas phase, and the remainder of space can be divided into many more or less uniform layers throughout the glow region of the discharge. Thus, a glow discharge exists as an onion layers structure of luminous gas phase.

Plasma has characteristic glow, and "glow" associated with glow discharge of gases and vapors is often treated as plasma without recognition that the definition of plasma has to be changed to do so. For instance, the concentration of ions in glow discharge has not been measured simultaneously and independently of the measurement of electron density and calculated based on the assumption of the electroneutrality from data of electron density. Figure 3.7 clearly indicates that the luminous gas phase created by glow discharge of Ar cannot be treated as plasma. If we take a broad meaning of plasma, which is synonymous with luminous gas phase, it becomes necessary to examine physical parameters that control the luminous gas phase. Parameters such as electron temperature, electron density, and associated phenomena such as ionization might not be major parameters that control the luminous gas phase. This situation becomes quite clear when we examine the luminous gas phase created in the glow discharge of organic molecules.

4. GLOW DISCHARGE OF ARGON AND OF ACETYLENE

When an organic vapor rather than an inert gas is used in the same discharge reactor, a nearly completely different phenomenon occurs, in which deposition of material is an aspect. Deposition of material constitutes the foundation of LCVD. In an LCVD environment, the composition of the gas phase changes continuously as deposition proceeds. This difference could be further illustrated by examples for glow discharge of argon and of acetylene.

In a closed system, e.g., 50 mtorr of argon or acetylene, the discharge is completely different in the following way. The inception of glow discharge of argon does not change the system pressure. The glow discharge of argon in a closed system can be maintained indefinitely because argon is not consumed by the process, and diagnostic measurements such as electron temperature measurement and emission spectroscopy could be carried out in such a glow discharge.

In contrast to this situation, the glow discharge of acetylene in a closed system extinguishes in a few seconds to few minutes depending on the size of the tube and the system pressure. This is because acetylene forms deposit and coats the wall of the reactor. In this process of LCVD (plasma polymerization) of acetylene, very little hydrogen or any gaseous species is created, and the LCVD of acetylene acts as a vacuum pump. When the system pressure decreases beyond a certain threshold value, the discharge cannot be maintained.

The similar situation can also be seen in a flow system discharge. Again, argon discharge can be maintained indefinitely under a steady-state flow of argon at a given pressure. This is not the case with acetylene glow discharge. When the flow rate is relatively low, e.g., 1 sccm, and there is no adjustment to the pressure change in a relatively small reactor (e.g., less than 1 liter), acetylene is consumed (by plasma polymerization) faster than it is replenished by the flow of acetylene. Consequently, the system pressure decreases and the glow discharge extinguishes. This phenomenon is, of course, dependent on the size of reactor and flow rate.

Since acetylene is fed into the system continuously at a given flow rate, the system pressure increases as soon as the glow discharge extinguishes. As the system pressure increases to that at which the breakdown of the gas phase could occur under the applied voltage, glow discharge is ignited again, but the reignited glow discharge follows the same path of the first discharge. As a consequence of these processes, the glow discharge occurs as a self-pulsating intermittent discharge.

Polymer-forming species created in the cathode region of a DC discharge deposit mainly on the cathode surface and form a solid film by reacting with the substrate surface or each other. However, in order to maintain glow discharge, electrons must be emitted from the cathode surface. This means that electrons must come out of the deposited plasma polymer when the cathode surface is sufficiently covered by the plasma polymer.

The emission of electrons from the surface of dielectric materials is a well-known phenomena. In the case of plasma polymers, it can be understood that the surface state electrons, which are responsible for the contact electrification of polymer surfaces (static charges), could be emitted as the electrons to sustain the glow discharge (see Chapter 24). This raises a fundamental question on what is the first step to create a DC glow discharge. It is generally conceived that the primary electrons are first accelerated under the electrical field and ionize gases when electrons gain enough energy to ionize gases. The ions created by the ionization are accelerated toward the cathode. The bombardment of ions causes the emission of the secondary electrons, which follow the same step of the primary electrons to sustain the discharge. Now the most serious question is where the primary electrons come from? There has not been a reasonable and satisfactory answer to this question.

A nanofilm of plasma polymer (up to about 100 nm) has sufficient electrical conductance as evidenced by the fact that an LCVD-coated metal plate can be coated by the electrolytic deposition of paint (E coating), i.e., plasma polymer–coated metals can be used as the cathode of the electrolytic deposition of paint (see Chapter 31). Thus, the plasma polymer layer remains in the same electrical potential of the cathode (within a limited thickness) and the work function for the secondary electron emission does not increase significantly. When the thickness of plasma polymer deposition increases beyond a certain value, the coated metal becomes eventually insulated, and DC discharge cannot be sustained. DC cathodic polymerization is primarily aimed to lay down a nanofilm (10–100 nm) on the metal surface that is used as the cathode (see Chapter 13).

5. NEGATIVE GLOW IN DC DISCHARGE OF ARGON

Figure 3.8 depicts a pictorial view of electrodes arrangement, which was used to examine luminous gas phase in glow discharges. A cold-rolled steel plate was used as the center electrode, which is the cathode in DC discharge. The two counterelectrodes are equipped with magnetic field assembly (magnetron), which can be seen on the backside of the right electrode shown in Figure 3.8a. The details of anode magnetronics are described in Chapter 15. The use of a magnetron is not necessary to investigate the glow near the cathode, but it is easier to identify the negative glow by the magnetron, especially when alternating current (AC)

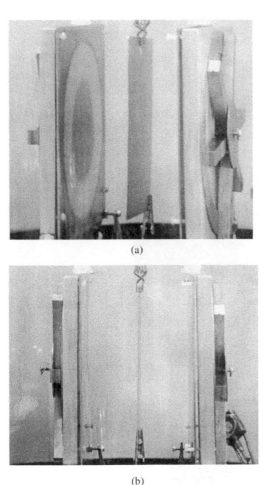

(a)

(b)

Figure 3.8 Pictorial view of electrode assembly: (a) an angled view, (b) side view by which photographs of glow were taken.

discharge is employed, in which the counterelectrode acts as cathode on the cathodic cycle and as anode on the anodic cycle. The same magnetron when used as the cathode creates toroidal glow shown in Figure 3.9. The anode magnetron creates a funnel-shaped negative glow (as seen in Figure 10) and pushes the negative glow toward the center of the cathode. For the examination of the luminous gas phase of Ar or trimethylsilane (TMS), photographs were taken from the side view of the electrode assembly shown in Figure 3.8b, at the distance of approximately 1 m. While the center plate cathode appears as a side view of the plate (thin line), a low-angle view of the anode surface appears on the both counter electrodes.

Figure 3.10 depicts the negative glow of Ar obtained by the electrode setup. It is important to recognize that the energy of electron (electron temperature) increases with the distance from the cathode surface up to the point where the maximal electron temperature is observed as shown in Figure 3.4. Therefore, the

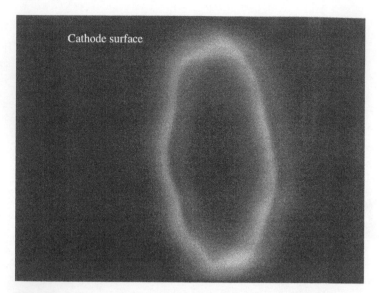

Figure 3.9 Angled view of the toroidal glow on cathode magnetron.

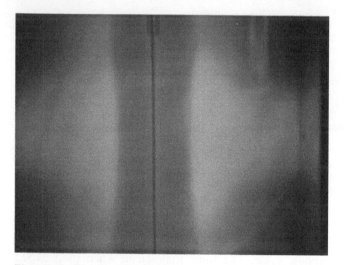

Figure 3.10 Ar DC discharge: a cathode (steel plate) and two magnetron anodes, 1 sccm, 50 mtorr, 5 W; the intensity of Ar glow is more than five times greater than that of TMS.

distance from the cathode surface where glow appears is the indication of what level of electron energy is responsible for the creation of the glow. Based on this point of view, the negative glow appearing in DC glow discharge of Ar can be attributed to the excited species that have the energy in the vicinity of the ionization of Ar, i.e., the negative glow can be considered as the high-energy glow, which is caused by electrons having high enough energy to ionize Ar. Chapman's calculations have suggested that the negative glow in DC discharge is the main region where electron impact ionization of ground state Ar atoms occurs [7]. For the simplicity of discussion, the high-energy glow (negative glow) could be termed *ionization glow* (IG). The distance between the cathode surface and the edge of negative glow roughly match the distance where the maximal electron temperature was observed in Figure 3.4. The cathode in Ar discharge is surrounded by the dark space in a practical sense, and a well-recognizable negative glow exists some distance away from the cathode on both sides of the cathode because two anodes are used against one center cathode plate. In Ar discharge, the primary glow is the IG, and the cathode is in the dark space.

6. CATHODE GLOW IN GLOW DISCHARGE OF TRIMETHYLSILANE

The concept gained by the analysis of inert gas discharge is generally applied to discharge of molecular (organic) gases with the a priori assumption that the same principle applies. However, DC glow discharge of multiatomic organic molecules is significantly different from that of monoatomic gases, and the direct application of knowledge gained from inert gas glow discharge to LCVD could lead to serious misunderstandings of the process.

With organic molecules, the consequence of glow discharge is much more complex than the argon glow discharge shown. In this respect, the DC glow discharge that can be thoroughly characterized is limited to the discharge of inert gases. However, the movement of electrons can be considered to be similar to the case shown for argon glow discharge, and the location of glow in an organic vapor discharge could be correlated to the energy of electrons that causes the formation of the luminous gas phase, i.e., the further away from the cathode surface, the higher energy electrons are involved, and the closer to the cathode surface the lower is the energy.

The author's effort to capture luminous glow by the two-anodes setup led to a surprising discovery of cathode glow of TMS discharge, which has never been recognized [8]. Figure 3.11 depicts the luminous gas phase associated with the cathode in TMS discharge. The intensity of glow in Ar discharge is more than 20 times stronger than that for TMS under the same conditions employed. The film speed, aperture, and shatter speed were adjusted to yield the comparable pictures shown.

In a strong contrast to the situation for Ar glow discharge, the cathode in TMS discharge is surrounded by the glow, which is weaker than the negative glow observed in Ar discharge but is the primary glow of the system. The primary glow literally touches the cathode surface in TMS discharge. This glow can be termed *cathode glow* for the ease of discussion. The cathode glow, according to the above-mentioned principle, is a low-energy glow caused by the impact of electrons with very low energy. This glow could be designated as molecular *dissociation glow* (DG).

It is also important to recognize that a very faint ionization glow exists away from the cathode (approximately the same distance between the cathode where the negative glow was observed with Ar discharge) separated by a dark space between the cathode glow and the negative glow. The intensity of the ionization

TMS, Flow
5 s

Figure 3.11 TMS glow discharge: a cathode (steel plate) and two magnetron anodes, 1 sccm, 50 mtorr, 5 W.

glow increases with the reaction time due to the increase of H_2 in the gas phase as depicted in Figure 3.12, which compares the dissociation glow and ionization glow with and without anode magnetrons. The presence of the dissociation glow is not due to the influence of the anode magnetron. The major differences are seen in the shape of the ionization glow and in the thickness of the dissociation glow. The influence of anode magnetronics is described in Chapter 15.

The presence of cathode glow, as the primary glow, means that the major reactions take place at the cathode surface in DC discharge of TMS but not in the negative glow. This implies that the chemically reactive species are not created by electron impact ionization of TMS molecules, which should occur at the fringe of the negative glow, but by the energy transfer reaction with low-energy electrons.

7. DISSOCIATION OF MOLECULE VS. IONIZATION

The dissociation of a molecule by collision with an accelerated electron is an essential process for creating luminous gas phase of an organic compound. How the dissociation of an organic molecule is related to the ionization of the molecule is a critically important issue. Here we must recognize the difference between the ionization of atoms and that of molecules, particularly relatively complex organic molecules. With inert gas atoms, the ionization can occur only by the elimination of an electron from an electron orbital, and the process requires a relatively high energy (e.g., 13–25 eV for inert gas atoms). The formation of excited species of inert gas atoms that are responsible for luminous gas phase also requires high-energy electrons. Therefore, the main glow in DC discharge of Ar (negative glow) appears at the same distance from the cathode surface where the maximal electron temperature is observed.

In the discharge of organic molecules, the ionization is not an accurate picture of the step that creates the luminous gas phase. First of all, the ionization energy

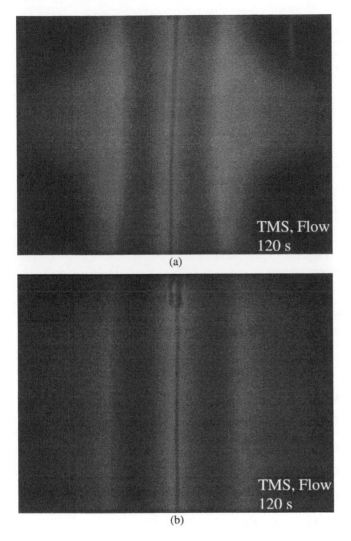

Figure 3.12 Comparison of dissociation and ionization glows with and without anode magnetron; (a) with anode magnetrons, and (b) without anode magnetrons.

of greater than 10 eV is far above the bond energies of primary bonds involved in organic compounds. Typical bond energies are given in Table 3.1. The dissociation energy, metastable energy, and ionization energy for noble gases and diatomic gases are compared in Table 3.2. The low-energy electrons and/or off-centered collisions that cannot ionize molecules can break bonds in organic molecules or create excited species that can trigger chemical reactions. These reactions are absent in the ionization of Ar.

Thus, the ionization of an organic molecule is far more complex than the ionization of Ar, and one can easily estimate the extent of the dissociation reactions by comparing the energies necessary for the dissociation reactions and the ionization energy of a molecule. The following examples provide some indication of the

Table 3.1 Bond Dissociation Energy (eV)

Bond	Dissociation energy
C–C	3.61
C=C	6.35
π bond in C=C	2.74
C–H	4.30
C–N	3.17
C≡N	9.26
C–O	3.74
C=O	7.78
C–F	5.35
C–Cl	3.52
N–H	4.04
O–H	4.83
O–O	1.52

energies involved in the dissociation reactions that occur in the glow discharge of organic molecules (the enthalpies of reactions are given in units of electron volts):

$$\Delta H \text{ (eV)}$$

$$e^- + C_2H_4 \quad \rightarrow C_2H_2 + H_2 + e^- \quad\quad 1.8$$

$$\rightarrow C_2H_2 + 2H\bullet + e^- \quad\quad 6.3$$

$$e^- + C_2H_3Cl \quad \rightarrow C_2H_2 + HCl + e^- \quad\quad 1.1$$

$$\rightarrow C_2H_2 + H\bullet + Cl\bullet + e^- \quad\quad 5.6$$

$$e^- + C_2H_3F \quad \rightarrow C_2H_2 + HF + e^- \quad\quad 0.8$$

$$\rightarrow C_2H_2 + H\bullet + F\bullet + e^- \quad\quad 6.6$$

Table 3.2 Dissociation Energy, Metastable Energy, and Ionization Energy for Noble and Diatomic Gases (eV)

Gas	Dissociation energy	Metastable energy	Ionization energy
He	—	19.8	24.6
Ne	—	16.6	21.6
Ar	—	11.5	15.8
Kr	—	9.9	14.0
Xe	—	8.32	12.1
H_2	4.5	—	15.6
N_2	9.8	—	15.5
O_2	5.1	—	12.5

Whereas most ionization requires energy greater than 10 eV, the dissociation of a molecule requires much less energy. The removal of small molecules from the original organic molecule, such as the removal of H_2, HCl, HF, etc., requires very little energy in comparison with the ionization energy. The relative ease with which such dissociation reactions occur should depend on the activation energy ΔE associated with each reaction. However, a rough estimate by the magnitude of the heats of reaction (ΔH) seems to provide reasonably accurate comparisons between the dissociation and ionization of a molecule [9].

Ionization is the essential step in sustaining luminous gas phase but is not necessarily the primary step in initiating LCVD reactions. The scission of bonds occurs with a far greater frequency than the formation of ions. It was estimated that the concentration of neutral species in *low-pressure plasma* is usually five to six orders of magnitude higher than that of ions and electrons [8]. In other words, the scission of bonds does not occur as the consequence of the ionization of molecules but rather as the primary step to create luminous gas phase.

A significant difference between glow discharge of an inert gas, such as argon, and that of an organic compound, such as acetylene, is the fundamental step of creating excited or chemically reactive species. The low-energy electron on the slope of increasing energy in the cathode fall region shown in Figure 3.3 could be energetic enough to dissociate an organic molecule (slow-electron impact dissociation). Thus, the direct ionization of the entire molecule is rather an unlikely primary event whereas such a step should be the major and mandatory step, if one applies the fundamental knowledge gained by argon discharge to LCVD without modifications to account for the basic difference.

The characteristics of luminous gas phase in LCVD reactor are dependent on the nature of gas and operational parameters. When the glow is compared at a similar level of input power of LCVD (see Chapter 8), the intensity of glow can be compared. Simple molecular gases, such as N_2, O_2, and CO_2, do not create the dissociation glow; the addition of simple molecular gas to an organic molecule changes the basic characteristics of glow completely. Figure 3.13 shows the effect of

Figure 3.13 Comparison of glows for TMS and (TMS + O2) at the comparable level of energy input, 40 kHz discharge, without mangnetron.

adding oxygen to TMS in the ordinary parallel electrodes reactor with 40 kHz power source. The intensity as well as the expansion of glow volume is completely different depending on the absence or presence of oxygen in the gas. Since O_2 creates only negative glow, the intensity of glow become much stronger and glow expands by the addition of O_2. The mechanisms for the creation of chemically reactive species for these systems are described in Chapter 4.

In recognition of the cathode glow in DC discharge of TMS, the first step of creating DC glow discharge could be illustrated as follows. When electrical voltage is applied between cathode and anode, the surface state electrons are pulled out of the cathode surface and moved toward the anode. In this process, electrons are accelerated under the electrical field in the cathode region. As soon as electrons gain enough energy to dissociate LCVD gas, the dissociation of gas occurs. Some of dissociated species are in excited states and emit photons to dissipate the excess energy. The photon emission causes the luminous gas phase, which is recognized as the cathode glow because the dissociation took place near the cathode surface. Some of dissociated species are non–polymer forming, in most cases H_2, and create the ionization glow, which takes place away from the cathode surface. In the absence of LCVD gas, e.g., glow discharge of Ar, the electrons pulled out of the cathode surface, which are considered as the primary electrons under this scheme, are accelerated to higher electron energy to create the ionization glow.

8. ONION LAYER STRUCTURE OF LUMINOUS GAS PHASE

Regardless of what term could represent the luminous gas phase created by an electrical discharge of gas or vapor, the total volume of luminous gas phase (glow) could be divided into many layers, in which the gas phase could be treated as a more or less uniform luminous gas phase. That is, the total luminous gas phase can be expressed by the onion layer structure, as indicated by Figure 3.7. In such an onion layer structure, only a relatively thin layer could be considered to be in plasma state or a state close to plasma state.

Very important factors in LCVD are (1) the location of the critically important layer, i.e., the dissociation glow, in a glow discharge, and (2) the location of the substrate with respect to the onion layer structure, i.e., in which layer of an onion structure the substrate is placed. The location of the critical layer depends on what kind of discharge system is employed to create a luminous gas phase. In a strict sense, it is impossible to uniformly coat a substrate placed in a fixed position in a reactor, and the relative motion of a substrate to the onion layer structure of luminous gas phase is a mandatory requirement if high uniformity of coating is required.

Adaptability of an LCVD process in an industrial scale operation greatly depends on the nature of the onion structure of the luminous gas phase that could be accommodated in the operation. The change of reactor size inevitably changes the basic onion layer structure of the luminous gas phase, which constitutes the main (often insurmountable) difficulty in the scale-up attempt by increasing the size of reactor. (The scale-up principle is discussed in Chapter 19.)

For the ease of comparison, let us term the most critical layer in a luminous gas phase as "core." The meaning of the core depends on what kind of process with respect to the objective is considered. The reactor parameters, such as the distance between electrodes, system pressure, applied electrical voltage and magnetic enhancement, etc., change the core and, accordingly, the size and the shape of onion layer structure. It should be cautioned here that the photo-emitting species in the luminous gas phase are not necessarily the major chemically reactive species that form the deposition in LCVD as described in detail in Chapter 4.

9. MODE OF DISCHARGE AND ONION LAYER STRUCTURE

9.1. DC

In DC discharge for LCVD, the main core is the DG adhering to the cathode surface, and the anode is out of the onion structure in most cases. In DC discharge of Ar for glow discharge treatment or sputter deposition of the cathode material, the core is the IG, which does not touch the cathode surface.

9.2. Alternating Current

In alternating current discharge for LCVD, up to about 100 kHz, the DG adhering to the electrode surface (in the cathodic cycle) is the core. The discharge system has two cores, and the interelectrode space is filled with two onion structures overlapping in part, which approach each other when the value of W/pd increases.

9.3. RF

In radio frequency discharge, the electrode does not function as electrode in the lower frequency discharge. Consequently, the core of discharge moves away from the electrode surface, i.e., two cores for DG, which are located away from electrodes but near the electrode surface (in the case with internal electrodes). Because DG does not adhere to the electrode surface, the main medium of material deposition shifts to the interelectrode space, which leads to the higher deposition rate observed in the interelectrode space under otherwise identical discharge conditions carried out by different power source (see Chapter 8).

9.4. Microwave

In microwave plasma, electrons adhere to the wall because the microwave energy propagates on surface. In this case, the inner surface of the vacuum chamber becomes the core of the luminous gas phase for LCVD. If the inner surface of a vacuum chamber is the main substrate surface on which LCVD coating is aimed, such as the case of the coating the inner surface of a plastic bottle, the microwave discharge is the most efficient LCVD.

9.5. Atmospheric Pressure Glow Discharge

When the system pressure of such a luminous gas phase is raised, it is generally considered that the electron temperature, T_e, ion temperature, T_i, and temperature of neutral species, T_n approach an equilibrium state (thermal equilibrium), and the plasma becomes hot at round 100 torr. On the other hand, there is a type of glow discharge that is termed *atmospheric-pressure glow discharge*, which is claimed to be atmospheric low-temperature plasma. Obviously, such low-temperature plasma cannot be obtained as an extension of the low-pressure luminous gas phase by increasing the system pressure without creating high-temperature plasma.

This discrepancy is linked, at least in part, to the vague definition and/or the interpretation of the terms *glow discharge* and *plasma*. In an atmospheric glow discharge, the mode of discharge is different from that for low-pressure glow discharge. In the former case, the discharge occurs as filamentary discharge. In contrast to the onion model of low-pressure discharge, the filamentary discharge could be represented by the model structure of celery, i.e., celery bundle structure rather than onion layer structure. The corona discharge is also a filamentary discharge.

While the atmospheric pressure operation might be considered as the major advantage, particularly among people who suffer "vacuum phobia," the atmospheric pressure operation inevitably requires that a higher consumption of gases be involved because the number of gas molecules is proportional to the system pressure. The high consumption of gases and high production of effluent gases more or less necessitate the addition of gas separation/recovery systems, which negates the advantage based on not using a vacuum system. The economical aspects of low-pressure LCVD processes are discussed in Chapter 36. Because of the limitation on the size of glow that can be created by atmospheric glow discharge, it is considered beyond the scope of LCVD processing for the system approach interface engineering dealt with in this book.

9.6. Low-Pressure Cascade Arc Torch

The luminous gas phase created by a special mode of DC discharge recognized as the low-pressure cascade arc torch (LPCAT) provides an especially important case for understanding the fundamental aspects of the luminous gas phase. The luminous gas phase in form of luminous gas jet stream or torch are created by blowing out DC discharge into an expansion chamber in vacuum. The luminous gas jet of Ar mainly consists of photon-emitting excited neutral species of Ar, which is certainly not the plasma of classical definition. The core of LPCAT is the tip of injection nozzle; however, it is not the core of electrical discharge.

The luminous gas phase that can be characterized in the LPCAT very likely exists as remnant of electrical discharge plasma. The luminous gas phase in LPCAT provides an important foundation to elucidate the creation of chemically reactive species in LCVD, which is presented in Chapter 4. The details of LPCAT are given in Chapter 16. LPCAT provides an example that can distinguish the luminous gas phase from plasma, i.e., the luminous gas jet created by LPCAT is not in the state of plasma. In this book, the luminous gas phase is emphasized without being bound by the concept or definition of plasma.

REFERENCES

1. Francis, G. The glow discharge at low pressure. In *Handbuch der Physik*; Flügge, S. Ed.; Springer-Verlag: Berlin, Germany, 1956; Vol. 22, No. 2, 53–208.
2. Howatson, A.M. *An Introduction to Gas Discharges*, 2nd Ed.; Pergamon Press: New York, 1976; 84–91.
3. Hatta, Y. Gas discharge. In *Introduction to Basic Electronics*, 2nd Ed.; Tohoku University Lecture Series; Kindai Kagaku-Sha: Tokyo, 1979; Vol. 4, 60–71.

4. Shakin, M.M. In *Reaction under Plasma Conditions*; Venugopalan, M. Ed.; Wiley: New York, 1971; 298.

5. Wei-Han Tao; Hirotsugu K. Yasuda. Plasma Chem. Plasma Proc. **2002**, *22* (2), 313.

6. Tao, W.H.; Prelas, M.A.; Yasuda, H.K. J. Vac. Sci. Tech. **1996**, *A14* (4), 2113.

7. Chapman, B. *Glow Discharge Processes*; Wiley: New York, 1980; 116–124.

8. Hirotsugu Yasuda; Qingsong Yu Plasma Chem. Plasma Process, **2004**, *24*, 325.

9. Kobayashi, H.; Shen, M.; Bell, A.T. J. Macromol. Sci. Chem. **1974**, *A8*, 1354.

4
Creation of Chemically Reactive Species in Luminous Chemical Vapor Deposition

1. POLYMERIZABLE SPECIES VS. PHOTON-EMITTING SPECIES

The photon-emitting species are vitally important in luminous chemical vapor deposition (LCVD) process, and the location of the luminous gas phase indicates where actions occur within the interelectrode space. On the other hand, whether any particular photon-emitting species is primarily responsible for LCVD is a different issue. As is described in Chapter 5 for the growth and deposition mechanisms, many chemically reactive species, such as various forms of free radicals that do not emit photons, are major reactive species that carry the growth reactions. No single species could be identified as the precursor or chemically reactive species for the process.

The intensity of glow (dissociation glow, DG) in LCVD system is faint, which often necessitates use of a dark room in order to see it, while the glow (ionization glow, IG), of nonpolymerizing gas, such as Ar, under the comparable discharge conditions is often easily visible in a bright room. The faint glow indicates that the probability of a reactive species chemically reacting without emitting photon is high, i.e., most reactive species form polymers without emitting photons.

In LCVD, the creation of chemically reactive species and the reaction of reactive species (material formation) occur within the same space concurrently. Furthermore, the creation of chemically reactive species and the formation of materials are not kinetically coupled. The reactive species are created irrespective of what happened to the previously formed reactive species.

In conventional free radical polymerization, the initiation, propagation, and termination are kinetically coupled. Consequently, the increase of initiation rate increases the overall polymerization rate but reduces the degree of polymerization. In contrast to this situation (kinetically coupled initiation, propagation, and termination), the formation of chemically reactive species is not the initiation of a subsequent polymerization. Under such an activation/deactivation decoupled reaction system, the mechanism for how chemically reactive species are created and how these species react to form solid material deposition cannot be viewed in analogy to polymerization.

The key issue of how the luminous gas phase is created in the low-pressure electrical discharge is described in Chapter 3. In this chapter we focus on how polymerizable species are created in the luminous gas phase as the initial steps of plasma polymerization that function only under the influence of the luminous gas phase without the assistance of other forms of energy input.

2. LCVD ON ELECTRODE AND IN GAS PHASE

The deposition characteristics of trimethylsilane (TMS) in low-pressure DC discharge, in comparison to 40-kHz and 13.5-MHz discharge, revealed that there existed two distinctively different modes of material formation in the luminous gas phase created by low-pressure DC discharge [1,2]. In DC discharge, the majority of deposition (more than 80%) occurs on the cathode surface, which was attributed to the cathodic polymerization, and the remainder of deposition was attributed to the negative glow polymerization, which is essentially the glow discharge polymerization by alternating polarity power sources including radio frequency.

The cathodic polymerization could be considered to take place in the dissociation glow (DG); likewise the glow discharge polymerization occurs in the negative glow (ionization glow, IG). These two types of polymerization have distinctively different deposition kinetics (see Chapter 8), which is helpful in elucidating how reactive species are created in an LCVD environment.

The cathodic polymerization was found to be pressure dependent (with a fixed flow rate) and independent of the flow rate, whereas the negative glow polymerization is pressure independent (at a fixed flow rate) but flow rate dependent (at a fixed pressure). The cathodic polymerization yields a much tighter network, manifested by high refractive indices, than the product of the negative glow polymerization.

These characteristics of DC cathodic polymerization were also found on the electrodes of 40-kHz discharge, but the deposition rate dropped to one-half of the value for DC discharge because an electrode acts as the cathode only one-half of the discharge time. The electrodes in 13.5-MHz discharge, however, did not show any feature of the cathode described above. These observations strongly suggest that examination of the luminous gas phases in those discharges is valuable in elucidating the mechanisms for creating polymerizable species in LCVD.

In plasma polymerization, the species that form polymeric deposition is not the original gas (monomer) but reactive fragments (mainly free radicals) of the original molecule. Although the contribution of the original monomer could be enhanced to some extent by employing special operational conditions that favor the preservation of the original monomer structure, e.g., low power input, off-glow deposition, pulsed discharge, and so forth, those conditions generally yield low molecular weight oligomers, which drastically reduce the practical value particularly as a nanofilm, and the yield of process drops beyond the realm of industrially feasible processing.

The growth mechanism of plasma polymerization cannot be expressed by the growth mechanism that is specific to the functional groups existing in the molecule, i.e., neither chain growth or step growth mechanisms are applicable to plasma polymerization. Polymerizable species are created in glow discharge, and different polymerizable species are created depending on the operational conditions of plasma polymerization. Thus, the conceptual image of polymerizable species in glow discharge is completely divorced from the concept of "monomer" for conventional sense of polymerization.

The complex nature of plasma polymerization was explained by the rapid step growth polymerization (RSGP) in that the recombination of free radicals constitutes the main mechanism to increase the size of molecules, as described in Chapter 5. The presence of many dangling bonds (free radicals trapped in a three-dimensional network), the change of the concentration of dangling bonds as a function of the duty cycle of pulsed discharge, and the dependence of the deposition rate on molecular structures all support the RSGP mechanism. Conversely, these phenomena all together could not be explained reasonably without the principle of RSGP mechanism. This implies that the main polymerizable species are free radicals that are created in the luminous gas phase.

It has been stipulated that the fragmentation of an organic molecule precedes the ionization of the fragmented species and the ionization occurs with the fragmented entities. That is, ionization of the original organic molecules, used as the monomer, is a very unlikely event [3]. However, no phenomenon that directly shows this principle was known. The discovery of the cathode glow in glow discharge of organic gas (TMS), as described in Chapter 3, seems to be the first direct evidence to support the principle that the dissociation or fragmentation of the monomer molecule precedes the ionization. A significant difference between glow discharge of an inert gas such as argon and that of an organic molecule, such as TMS, is the initial step of creating excited and/or chemically reactive species.

3. CHANGE OF GAS PHASE COMPOSITION WITH DISCHARGE TIME

The direct ionization of the entire molecule is rather an unlikely primary event; however, the ionization would occur with fragmented moieties, primarily H, in the case of organic molecules. This situation could be visualized by the change of several representative gas phase species in a closed system glow discharge of TMS shown in Figure 4.1, which was obtained by repeating a closed system discharge for varying discharge time [2]. The analysis of gas remaining in the reactor was carried out after the discharge was terminated. The figure clearly shows that species that could cause the deposition disappear in the early stage of the discharge and that the glow discharge, after a certain discharge time, is maintained by hydrogen liberated by dissociation of the monomer.

TMS increases the number of gaseous species by slow-electron impact dissociation leading to the increase of system pressure of a closed system in spite of the material deposition that occurs at the expense of gas phase molecules. If dissociation of the original TMS molecule did not take place and TMS polymerized as the original molecule, the system pressure should have dropped with discharge time because gas phase molecules decrease due to polymerization and deposition. Figure 4.2 depicts the increase of the system pressure as a function of the discharge time. It can be seen that the system pressure increased considerably in the first 2 min after the onset of glow discharge, indicating the substantial fragmentation of TMS monomers during this stage of LCVD. After 2 min, the rise of the system pressure with discharge time slowed down due to the consumption of polymerizable species. The continuous but

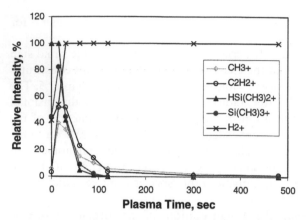

Figure 4.1 Change of gas phase species in a plasma after plasma polymerization of TMS with plasma time; closed system DC discharge, 25 mtorr TMS, 2 panels of Alclad 7075-T6 as electrode, DC 1000 V.

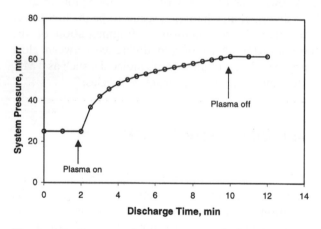

Figure 4.2 Increase of system pressure during closed system plasma polymerization of TMS.

slow rising of the system pressure after 2-min plasma polymerization could be due to the liberation of gaseous species (mainly H_2) from once-deposited material, which is similar to the evolution of gases when a polymer film is exposed to gas plasma.

Before and after the discharge, the system pressure remains constant, indicating that the total number of gases remains constant without electrical discharge. One important aspect that can be deduced from Figures 4.1 and 4.2 is that the composition of gas phase changes continuously with discharge time in an LCVD environment, especially in a closed-system discharge.

4. CHANGE OF DISSOCIATION GLOW AND IONIZATION GLOW WITH REACTION TIME

The faint negative glow observed with the primary dissociation glow of TMS shown in Figure 3.11 intensifies with discharge time, particularly in a closed system but to a lesser extent in a flow system. The increase of the IG with discharge time is depicted in Figure 4.3 for both closed-system and flow system operations. On the inception of glow discharge, where the gas phase is (pure) TMS, the luminous gas phase is mainly the DG, and the IG is barely visible. However, the IG becomes more and more visible as the glow discharge time increases, reflecting the change of gas composition in the luminous gas phase.

The change of luminous gas phase is buffered by the nonluminous gas phase, which surrounds the luminous gas phase, and its influence is dependent on the size of (nonluminous) gas phase and also whether a closed system or a flow system is employed. In a closed-system operation of TMS plasma polymerization, which is shown in the left column of Figure 4.3, the IG becomes the primary glow in 3 min

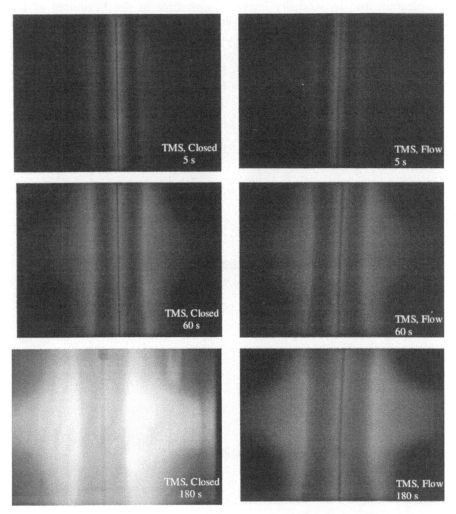

Figure 4.3 Change of the luminous gas phase of trimethylsilane (TMS) with reaction time for cathodic polymerization with two magnetron-anodes, 5 W; left column: closed system from the initial pressure of 50 mtorr, right column: flow system at 50 mtorr, top row: 5 s, middle row: 60 s, bottom row: 180 s.

reaction time, and the DG virtually disappears. At this stage of discharge, the glow appears essentially the same with Ar glow discharge, except that the color is different.

The gas phase analysis data shown in Figure 4.1 indicate that the gas phase at this point was mainly hydrogen (nonpolymerizable fragmentation product of TMS). Thus, 3 min after the inception of glow discharge of TMS in a closed system, the negative glow reached the maximal intensity, but no polymerization occurred simply because all polymerizable species were already consumed. This observation is in accordance with the trend that the intensity of glow is inversely proportional to the ease of polymerization in glow discharge.

In a flow system operation of the same process, shown in the right column of Figure 4.3, the change of gas phase species is compensated by the continuous feeding of TMS. The IG does not become the primary glow, although its presence becomes quite conspicuous, and its intensity increases with time at the beginning of the process until a steady-state luminous gas phase is established. The DG remains nearly the same throughout the operation.

The direct evidence to show that reactive species are created in the dissociation glow rather than in the ionization glow was found in the in situ Optical Emission Spectroscopy (OES) analysis aimed specifically at the dissociation glow and at the ionization glow of TMS DC discharge in a closed system [4]. Figures 4.4 and 4.5

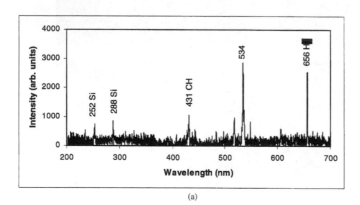

(a)

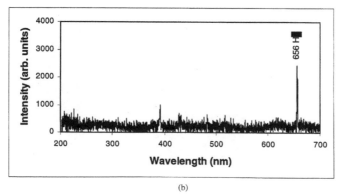

(b)

Figure 4.4 Optical emission spectra (OES) measured from (A) dissociation glow and (B) negative glow in DC glow discharge of trimethylsilane (TMS): flow system, 1 sccm TMS, 50 mtorr, DC power 5 W.

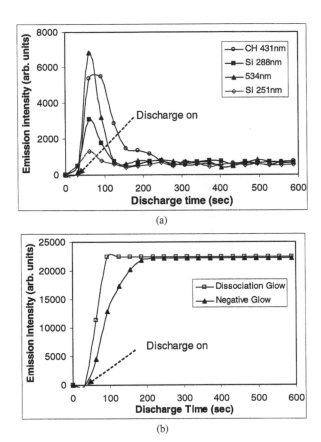

(a)

(b)

Figure 4.5   The time dependence of emission intensity of various photo-emitting species detected by OES in TMS DC glow discharge in a closed reactor: (a) polymerizable species in dissociation glow, (b) Hα emission line at 656 nm; 50 mtorr TMS, DC power 5 W.

shows reactive species in the DG and in the IG, respectively. Every aspect postulated based on ex situ data presented in Figures 4.1 and 4.2 is proven right by these data. The DG and IG of TMS DC glow discharge and their constituent species are summarized in Figure 4.6.

5.   MIXTURE OF POLYMERIZABLE AND NONPOLYMERIZABLE GASES

When the second gas, which cannot deposit material but can participate in material deposition from TMS, is added to a TMS flow system, both the IG and the DG appear at the onset of discharge. Figure 4.7 depicts the initial glow of (TMS 1 sccm + O_2 4 sccm), which clearly shows the DG of TMS and the IG of O_2 in different colors. The shape of glow did not change with discharge time in this case. However, the color of the IG changed from gold-white in the beginning to blue-white in few seconds (about 10 s) and remained essentially the same color during the remainder of 2-min discharge time. The color of the DG didn't change.

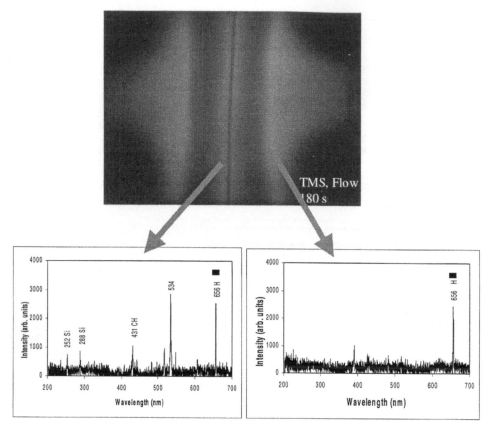

Figure 4.6 Dissociation glow and ionization glow in DC glow discharge of TMS and respective OES species; material forming species are mainly in the dissociation glow and the ionization glow consists of mainly hydrogen species.

The change of the color of IG indicates that the luminous gas phase consisting of TMS and O_2 was established in a few seconds but the cathode glow existed during the entire period of discharge. This indicates that the concurrent processes of the material deposition and the modification of depositing species occur in the luminous gas phase in a flow system.

The chemical modification of depositing species due to the addition of O_2 is evident in the characteristics of the deposit. Plasma polymer from TMS is highly hydrophobic (contact angle of water greater than 95 degrees) and contains a large number of Si base dangling bonds (see Chapter 6). Plasma polymer from the mixture of (1 sccm TMS + 4 sccm O_2) is highly hydrophilic (water contact angle less than 5 degrees) and contains no dangling bond [5,6].

6. INFLUENCE OF FREQUENCY OF POWER SOURCE

The glow discharge initiated by an AC power source can be visualized by an alternating cathode and anode in the DC glow discharge, up to a certain frequency,

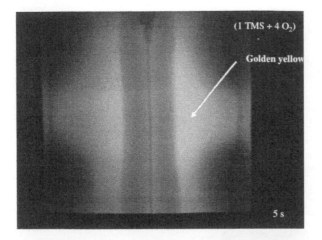

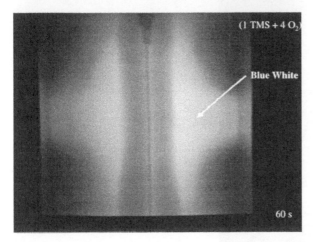

Figure 4.7 The glow of a mixed flow of 1 sccm TMS and 4 sccm O_2; both the cathode glow and the negative glow appear at the onset of glow discharge; top: 5 s after, bottom: 60 s after the onset of glow discharge.

e.g., about 100 kHz. Therefore, the polymer deposition in DC discharge as well as in an AC discharge is a mixture of DC cathodic polymerization that occurs in the cathode glow (DG) and negative glow polymerization that occurs in the negative glow (IG). The deposition rate on the electrode in 40-kHz discharge is one-half of that for DC discharge because an electrode is the cathode only on the cathodic cycle [1].

Figure 4.8 depicts the hybrid aspect of cathode glow and the negative glow in a 40-kHz discharge with the change of the contribution of these two types of glows due to the change of gas phase composition with discharge time. At the onset of discharge, the DG develops while the IG is hardly visible. As the IG becomes visible, two types of IG are clearly visible. One is the ring-shaped toroidal glow, which is characteristic of the cathode magnetron, representing the magnetron toroidal glow on the cathodic cycle. The other is the funnel-shaped negative glow, which is

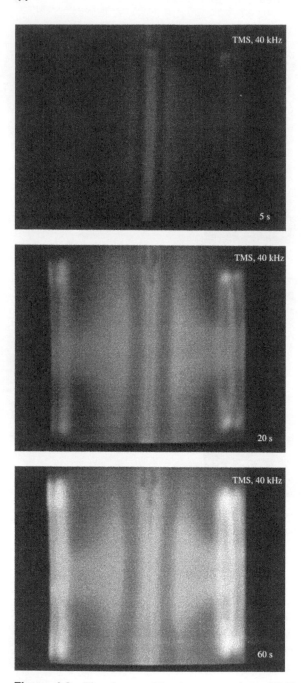

Figure 4.8 The change of luminous gas phase (TMS, 40 kHz) with discharge time.

characteristic of the anode magnetron in DC discharge, representing the IG on the magnetron on the anodic cycle.

At a higher frequency (e.g., more than 100 kHz), an electron can no longer travel the distance to reach the anode before the polarity changes depending on the parameter *pd*. Consequently, oscillation rather than linear motion becomes the major movement of electrons in 13.56-MHz discharges. In this case the collision of an oscillating electron with an atom or a molecule is the principal mechanism of ionization as well as the creation of luminous gas phase, both of which occur away from the electrode surface.

Because of the oscillation of electrons rather than linear motion, the influence of magnetic field imposed by the magnetrons also diminishes. Thus, the predominant role of the cathode observed in DC and AC discharge diminishes in 13.5-MHz discharge. The deposition rate characteristics (on electrode surface), i.e., the system pressure dependence and the refractive index of the deposit, are completely different from those for DC or 40-kHz discharge [1].

Figure 4.9 depicts the DG in radio frequency discharge, with the development of IG with discharge time, in the same experimental setup. The pictures clearly show that the center electrode is surrounded by the dark space, i.e., the cathode glow does not exist in radio frequency discharge. The DG is located away from the surface of electrode in radio frequency discharge. While the intensity and the volume of the IG increase, the center electrode plate remains in the dark space. The time-dependent establishment of the luminous gas phase is clearly evident in the increase of IG intensity.

7. POLYMERIZABLE SPECIES CREATED IN DC CATHODE GLOW

In the DC glow discharge of organic molecules, species that contribute to the deposition of materials in the cathode region are created mainly by the slow-electron impact dissociations in the cathode glow, which is located in the dark space recognized in Ar DC discharge. The plasma polymerization does not occur without this luminous gas phase. The cathode glow, which is the DG appears as touching the cathode surface, is observed only with gases that cause the deposition of materials. Simple non-polymer-forming gases, such as N_2, O_2, and CO_2, create only negative glow (the cathode remain in the dark space). Therefore, the DG could be considered the most important factor of LCVD.

If one applies the same principle for Ar discharge to the glow discharge of organic molecules, the first step of ionization could be written as

$$M + e \rightarrow M^+ + 2e \tag{4.1}$$

The ionization of an organic molecule is the basic principle of mass spectroscopy; however, it should be recognized that mass spectroscopy of a simple molecule such as ethane generally shows multiple ions, covering the fragmented species to partially polymerized species, which indicates that the dissociation of the molecule and some extent of polymerization of fragmented species occurred in the mass spectrometer. The presence of the DG (cathode glow) in DC cathodic polymerization implies that the formation of chemically reactive species via ionization, as depicted by Eq. (4.1), is a very unlikely primary event under the conditions of LCVD.

A monomer consisting of plurals of different elements could be represented by ABH instead of M for the entire molecule. In the case of TMS, A represents silicon,

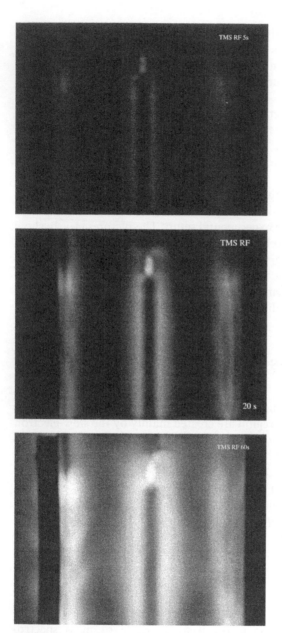

Figure 4.9 The development of the ionization glow with discharge time in TMS RF discharge.

B represents carbon, and H represents hydrogen. The dissociation that occurs in the cathode glow (DG) could be represented by the following energy transfer schemes.

$$\text{ABH} + e \rightarrow \text{A}^* + \text{B}^* + \text{H} + e \tag{4.2}$$

$$\text{A}^* \rightarrow \text{A}\bullet + h\nu_3 \tag{4.3}$$

$$\text{B}^* \rightarrow \text{B}\bullet + h\nu_4 \tag{4.4}$$

where A^* represents Si-based photon-emitting species, including some Si-C species collectively, and B^* represents C-based photon-emitting species collectively. H represents detached hydrogen atom. $A\bullet$ and $B\bullet$ represent chemically reactive but not photon-emitting Si-based and C-based species, respectively. In most cases, these chemically reactive species are free radicals, but this does not preclude the involvement of any other chemically reactive species, such as negative ions of certain atoms with high electronegativity.

Since the dissociation of molecules occurs as the first step, the deposition of Si-based moieties and C-based moieties procedes the growth and deposition steps more or less independently, although the cross-reaction is inevitable. The characteristic deposition rates of Si-based species are roughly seven times greater than those for C-based species [7]. This difference in characteristic deposition rates creates the change of composition of elements in the deposition with reaction time. In a closed system, this difference is significant as the change of the luminous gas phase change shown in the left column of Figure 4.3 indicates. This leads to a graded composition film, when plasma polymerization of TMS is performed in a closed system.

XPS C1s/Si2p ratio steadily increases with the reaction time (film thickness) by a closed-system plasma polymerization, while the ratio more or less stays at a constant level by a flow system cathodic polymerization, as shown in Figure 4.10. Such a graded ultrathin film was found to provide an excellent corrosion protection of aluminum alloy when an organic coating was applied on top of the ultrathin film [8].

The IG (negative glow) in plasma polymerization could be explained by the following reactions, in which high-energy electron is shown by e^*.

$$H + e^* \rightarrow H^+ + 2e \tag{4.5}$$

$$H + e^* \rightarrow H^* + e \tag{4.6}$$

$$H^* \rightarrow H\bullet + h\nu_7 \tag{4.7}$$

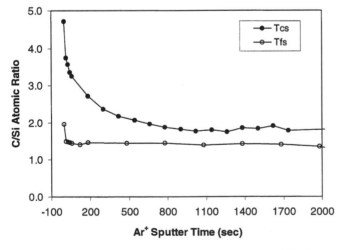

Figure 4.10 C/Si ratios of plasma polymer films of TMS prepared in a flow system reactor (Tfs) and in a closed system reactor (Tcs), as generated by XPS depth profiling.

The electron energy in Eq. (4.5) is much greater than that for Eq. (4.2), which causes the difference in the distance from the cathode surface where the cathode glow and negative glow appear.

The importance of the DG might give impression that those photon-emitting species are the precursor species in plasma polymerization. However, both the photon emission and the chemical reaction of reactive species (deposition of material or chemical modification of surface) are dissipation of excess energy (deactivation), which are exclusive of each other. In other words, if an excited species reacts chemically, it cannot emit photons, and conversely, if an excited species releases the excess energy by emitting photons, it cannot react chemically. If the photon-emitting species dissociates further by releasing the excess energy, its reaction product could react (polymerization).

With plurals of excited species, the fraction or probability α of excited species reacting with molecules or other reactive species determines the extent of photon-emitting species participating in LCVD material formation. The value of α depends on the concentration or the availability of other target species, and hence depends on the operational factors of electrical discharge. Thus, increasing the number of photon-emitting species (e.g., by increasing discharge power) would increase the deposition rate, even if photon-emitting species do not react: chemically. There are three major considerations:

1. Polymerizable species could be photon emitting.
2. Photon-emitting species create polymerizable species (which does not emit photon) via energy transfer reaction.
3. Polymerizable species that do not emit photons are created by the molecular dissociation reaction occurring in the dissociation glow, i.e. A^* and/or B^* in Eq. (4.2) should be replaced by $A\bullet$ and $B\bullet$, in this case.

Because the intensity of photon-emitting species within a given system would positively correlate to the deposition rate [9], there is a trend that the identified photon-emitting species be taken as the precursor structures of the depositing species without confirming the polymerization reactions. However, the most efficient polymerization (with respect to the molecular structure of monomer) is associated with the weakest glow at an equivalent energy input, indicating that a majority of polymerizable species do not emit photons. For instance, the glow of acetylene is much weaker than that of other hydrocarbons when they are compared at the same energy input per mass of monomer.

8. POLYMERIZABLE SPECIES CREATED IN JET STREAM OF LUMINOUS GAS

The existence of the dissociation glow in DC discharge strongly suggests that the creation of chemically reactive species in LCVD involves different mechanism than those in the electron impact ionization. However, in DC discharge, electron impact and ion impact reactions cannot be eliminated. Low-pressure cascade arc torch (LPCAT) provides a unique opportunity to investigate the formation of chemically reactive species with minimal influence of ions and electrons. That is, the creation of chemically reactive species from an organic molecule by the luminous gas phase of Ar, which is a primarily excited neutral species of Ar, can be investigated.

When plasma polymerization is carried out in a flow system, in which glow covers the entire cross-section of reactor with respect to the direction of monomer flow, the monomer molecules coming into the reactor first encounter the luminous gas phase. It is very unlikely that the molecules pass through the luminous gas phase without interacting with it and reach the relatively narrow zone in which IG or DG, located near the electrode surface, occurs. Therefore, the mode of activation that occurs in LPCAT without the influence of ionization is important in terms of the creation of chemically reactive species in LCVD. The creation of reactive species by the luminous gas is the mechanism considered here.

In contrast to the situation in conventional glow discharge by DC to radio frequency power source, LPCAT provides a unique opportunity to examine the role of photon-emitting species (without the influence of accelerating electrons) in the creation of polymerizable species in plasma polymerization [10–12]. The details of LPCAT are given in Chapter 16. In LPCAT, plasma formation (ionization/excitation of Ar) occurs in the cascade arc generator and the major body of luminous gas is blown into an expansion chamber in vacuum. The majority of electrons and ions are captured by anode and cathode, respectively, of the cascade arc generator, and there is no electrical field in the expanding plasma jet. Consequently, the photon-emitting excited neutrals of Ar cause the majority of chemical reactions that occur in the plasma jet.

In cascade arc plasma polymerization, a monomer (or monomers) is introduced in the expansion chamber. Because of an extremely high velocity of gas injected from a small nozzle (e.g., 3 mm in diameter), the second gas injected into the expansion chamber in vacuum cannot migrate into the cascade arc generator. Thus, the activation of Ar in the cascade arc generator and deactivation of the excited neutral species of Ar in the expansion chamber, which activate the monomer introduced in the expansion chamber, are totally decoupled. LPCAT plasma polymerization occurs under such a spatially and temporally decoupled activation/deactivation system.

As soon as the monomer is introduced, the plasma jet shrinks, indicating that the excess energy carried by the photon-emitting species in the plasma jet is consumed. Polymerizable species in the cascade arc plasma polymerization is created by photon-emitting neutrals of Ar, i.e., electronically excited Ar neutrals, and no electron impact ionization is involved. This is the similar situation under consideration for the molecular DG in DC discharge of organic molecules, except that energy of photon-emitting neutrals is much higher than that in the cathode glow.

The ionization of Ar by high-energy electron, e^*, occurs in the cascade arc generator.

$$Ar + e^* \rightarrow Ar^+ + 2e \qquad (4.8)$$

At the same time, the creation of excited neutrals by the high-energy electron e^* impact also occurs in the generator.

$$Ar + e^* \rightarrow Ar^* + e \qquad (4.9)$$

In the expansion chamber without addition of the second gas, excited species Ar^* decay by emitting photons:

$$Ar^* \rightarrow Ar + h\nu_{10} \tag{4.10}$$

The formation and the dissipation of luminous gas phase of organic vapor by excited Ar neutrals in the cascade arc polymerization could be described as follows. A fraction of photon-emitting Ar neutrals (Ar^*), expressed by nA^*, interacts with the monomer (ABH), and causes the dissociation of the monomer yielding fragmented photon-emitting species (A^* and B^*) and hydrogen atom (H) according to the scheme described by Eq. (4.2).

$$nAr^*m(ABH) \rightarrow m(A^* + B^*) + xH + (n - m)Ar^* \tag{4.11}$$

The remaining Ar^* emits photon according to Eq. (4.10). The photon emission by A^* and B^* is described by Eqs. (4.3) and (4.4), respectively. A fraction (β) of fragmented photon-emitting species reacts with the monomer or its fragmented moiety (represented by X) to yield polymerizable species (represented by $AX\bullet$).

$$\beta A^* + X \rightarrow \beta AX\bullet \tag{4.12}$$

In the third case described above, the formation of polymerizable species can be represented by a modified Eq. (4.11) as

$$nAr^* + m(ABH) \rightarrow m(A\bullet + B\bullet) + xH + (n - m)Ar^* \tag{4.13}$$

Clear-cut demonstration for this case, in which polymerizable species that do not emit photons are created by the molecular dissociation according to the energy transfer principle, was found with addition of CF_4 or C_2F_4 in the argon cascade arc torch. The OES spectra of low-temperature cascade arc torch for carrier gas Ar only, and with CF_4 or C_2F_4 addition, are identical except that the intensity of the latter is weaker as shown in Figure 4.11. Low-temperature cascade arc luminous gas jets of CF_4 and C_2F_4 do not show any additional peaks or continuums compared to the Ar plasmas, such as CF_2^+ or the CF lines that were found in the OES spectra ($Ar + CF_4$ and $Ar + C_2F_4$) radio frequency plasmas. None of the species previously reported for CF_4 radio frequency plasma [13], such as CF_2 and CF bands, as well as fluorine atom emissions, exist in the OES spectra of CF_4 or C_2F_4 cascade arc luminous gas jets. In other words, the addition of CF_4 or C_2F_4 quenches the cascade arc torch flame (the excess energy is transferred to the second gas) but does not create such species of the added gas that emits photons. Figure 4.12 depicts OES data for radio frequency discharge, which was obtained by the same reactor but without cascade arc torch, and parallel plate electrodes placed in the expansion chamber created radio frequency discharge. Radio frequency discharge of mixtures of Ar and perfluorocarbon, which consists of the DG and the IG, shows F-containing species as reported.

The finding implies that the photon emission per se is not the essential indication of chemical reaction pertinent to plasma polymerization. The CF_4 and

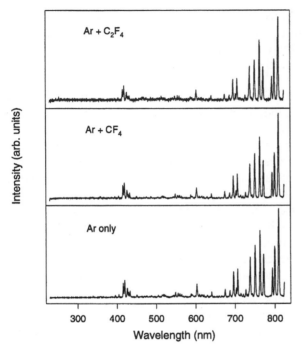

Figure 4.11 Optical emission spectra of cascade arc plasmas, Ar, 2000 sccm, CF$_4$, 20 sccm, C$_2$F$_4$, 20 sccm, 320 W, 0.56 torr.

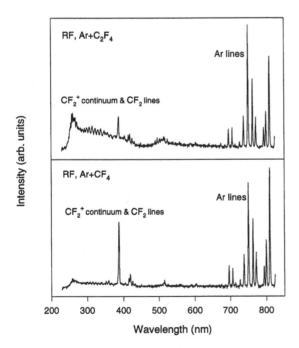

Figure 4.12 Optical emission spectra of RF plasmas, Ar, 2000 sccm, CF$_4$, 20 sccm, C$_2$F$_4$, 20 sccm, 50 W, 0.56 torr.

C_2F_4 plasmas are the well-investigated fluorinated carbon plasma systems by OES diagnostics according to literature with conventional plasma sources. However, without influence of ionization, these cases turned out to be the case 3 described earlier, i.e., the formation of chemically reactive species that do not emit photons by the energy transfer mechanism.

XPS data, on the other hand, showed that the LTCAT treatment of $Ar + CF_4$ and $Ar + C_2F_4$ yielded just as good, if not better, fluorination of PET fibers than radio frequency plasma treatment with these gases [14,15]. These examples clearly demonstrate that polymerizable species in plasma polymerization are not photon-emitting species in most cases. This is in accordance with the growth and deposition mechanism based on free radicals, which account for the presence of large amount of dangling bonds in most plasma polymers.

Most photon-emitting species are high-energy species, i.e., energy above 9–10 eV, and hence considered to be associated with IG rather than DG. The analysis of the cascade arc excitation of the gases introduced in the expansion chamber indicates that there are three possible cases that can be identified as the consequence of the injection of the monomer into the expansion chamber.

1. Immediate quenching of luminous gas jet accompanied with the change of color
2. Immediate quenching of plasma jet but no color change, and
3. No quenching and no color change

The first case is the energy transfer from an electronically excited species to another electronically excited species, which can be represented by

$$Ar^* + M \rightarrow Ar + M^* \tag{4.14}$$

where M represents the monomer molecule or a fragmented moiety of the monomer and M^* represents the electronically excited species of M. The energy transfer reaction represented by Eq. (4.14) is governed by the energy matching between the excited species and the energy of the product species. If the energy is too far above the energy level of the product, the energy transfer that creates a new excited species does not occur. If the energy is too low to electronically excite M, no quenching or color change occurs, which is the second case.

The third case is very important in elucidation of the mechanisms of the creation of polymerizable species. Quenching of the luminous gas jet indicates that the excess energy carried by the electronically excited neutral species of Ar is consumed. No color change or absence of new photon-emitting species indicates that the product of the energy transfer reaction is not an electronically excited species. Consequently, no energy matching rule applies in this case. The dissociation of monomer to yield fragmented moieties could be caused by any form of energy, including thermal energy. In this respect, it could be quite possible that the low-energy electrons (less than 1 eV) found in the luminous gas jet might participate in creating polymerizable species in LPCAT. When methane is added to Ar LPCAT, the number of electrons decreases according to the level of shrinking of the flame, as shown in Figure 4.13.

In any case, dissociation of organic molecules is the main route to create chemically reactive or polymerizable species in LCVD processes. The dissociation of monomer by the luminous gases occurs based on the principle of the energy

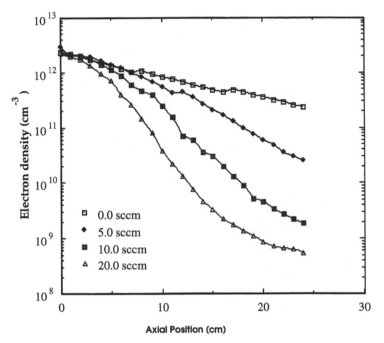

Figure 4.13 Electron density (cm^{-3}) as a function of axial position and methane flow rate in low pressure cascade arc torch (LPCAT); 8.00 A, 2000 sccm argon, and 560 mtorr (75 Pa).

transfer from the excited gas species to the monomer molecule, in which the energy matching between the energy-carrying species and that of the product species plays an important role. The highly energetic species does not produce any lower energy species, and the polymerizable species could be created without losing much intensity of the luminous gas phase. Thus, the electron impact ionization of an organic molecule is not the primary event causing the formation of materials in LCVD. The interaction of the luminous gas phase with monomer causes the formation of materials in LCVD.

In many practical situations of flow system LCVD, monomer is introduced into a reactor, in which a steady-state glow discharge is already established. As an example, in a tubular reactor depicted in Figure 4.14, monomer gas is introduced at one end and pumped out from the other end of a tube. The size and location of glow depends on the energy input level. Since all gas introduced flows through the reactor, the first event with the incoming gas is the interaction with the luminous gas phase. It is highly unlikely that the monomer passes through the luminous gas phase without interacting and reaches the energy input zone where ionization could occur. This situation is also true in nearly all reactors after the transitional period to establish a steady-state glow discharge has passed. Under such a condition, it is clear that reactive species are created by the interaction of luminous gas phase with monomers as described above.

In DC discharge, the situation is quite different. First, the cathodic polymerization is aimed at the short-term polymerization to coat the metal substrate

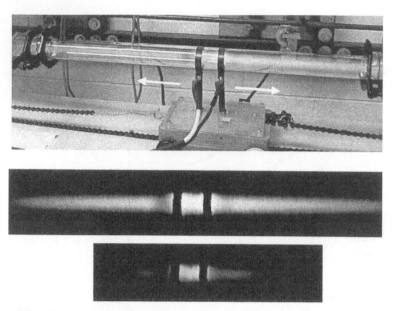

Figure 4.14 Glow in a simple glass tube reactor, in which gas is introduced in the right end and pumped out from the left end of the tube.

used as the cathode, of which the thickness is generally less than 100 nm. DC discharge is created in the presence of monomer in every operation. Second, in DC cathodic polymerization, the cathode is the product, which mandates the batch operation with new cathode (substrate). Details of DC cathodic polymerization are described in Chapter 13.

The glow is highly localized as the cathode glow, of which the volume fraction is very small, and monomer is fed into the DG mainly by monomer diffusion from the surrounding volume within a reactor. Because of this mechanism, the normalized deposition rate as well as the normalized energy input power are different from those in conventional glow discharges as described in Chapter 8. The observation of glow in the early stage of DC glow discharge provided an ideal situation to distinguish the DG and IG. It is reiterated that the DG is observed only with LCVD gases that form solid deposition, and molecular gases such as N_2 and CO_2 do not create the DG.

9. ROLE OF FREE RADICALS IN LUMINOUS CHEMICAL VAPOR DEPOSITION

In the chain growth free radical polymerization of a vinyl monomer (conventional polymerization), the growth reaction is the repeated reaction of a free radical with numbers of monomer molecules. According to the termination by recombination of growing chains, 2 free radicals and 1000 monomer molecules leads to a polymer with the degree of polymerization of 1000. In contrast to this situation, the growth and deposition mechanisms of plasma polymerization as well as of parylene polymerization could be represented by recombinations of 1000 free radicals (some of them are diradicals) to form the three-dimensional network deposit via 1000 kinetic pathlength, which is equivalent of the polymer with a degree of polymerization of 1000. This situation is further complicated by the fact that the formation of free radicals and the recombination of free radicals are not kinetically coupled, which yield numerous dangling bonds (unreacted free radicals trapped in the three-dimensional network) in the plasma polymer. The presence of dangling bonds in nearly all cases is the strong evidence that free radicals are reactive species in plasma polymerization. The details of dangling bonds are described in Chapter 6.

REFERENCES

1. Yasuda, H.K.; Yu, Q.S. J. Vac. Sci. Tech. **2001**, *A19* (3), 773.
2. Yu, Qingsong.; Moffitt, C.E.; Wieliczka, D.M.; Yasuda, Hirotsugu. J. Vac. Sci. Tech. **2001**, *A19* (5), 2163.
3. Yasuda, H. *Plasma Polymerization*; Academic Press: Orlando, FL, 1985; 72–177.
4. Yu, Q.S.; Huang, C.; Yasuda, H.K.; J. Polym. Sci.; Polym. Chem. Ed. **2004**, *42*, 1042.
5. Oldfield, F.F.; Cowan, D.L.; Yasuda, H.K. Plasmas Polym. **2000**, *5* (3/4), 235.
6. Oldfield, F.F.; Cowan, D.L.; Yasuda, H.K. Plasmas Polym. **2001**, *6*, 51.
7. Yu, Q.S.; Yasuda, H.K. J. Polym. Sci., Polym. Chem. Ed. **1999** *37*, 967.
8. Yu, Qingsong.; Moffitt, C.E.; Wieliczka, D.M.; Deffeyes, Joan.; Yasuda, Hirotsugu Prog. Organic Coatings **2002**, *44*, 37.
9. Millard, M.M.; Kay, E. J. Electrochem. Soc. **1982**, *129*, 161.
10. Fusselman, P. Steven; Yasuda, H.K. Plasma Chem. Plasma Proc. **1994**, *14* (3), 251.
11. Fusselman, Steven P.; Yasuda, H.K. Plasma Chem. Plasma Proc. **1994**, *14* (3), 277.
12. Yu, Q.S.; Krentsel, E.; Yasuda, H.K. J. Polym. Sci. A: Polymer Chem. **1998**, *36*, 1583.
13. Coburn, J.W.; Chen, M. J. Appl. Phys. **1980**, *51*, 3134.
14. Yasuda, T.; Okuno, T.; Yasuda, H. J. Polymer Sci., Polym. Chem. Ed. **1994**, *32*, 1829.
15. Krentsel, E.; Fusselman, S.; Yasuda, H.; Yasuda, T.; Miyama, M.; J. Polym. Sci. Polym. Chem. Ed. **1994**, *32*, 1839.

5
Growth and Deposition Mechanisms

1. GROWTH MECHANISMS OF POLYMER FORMATION IN LUMINOUS GAS

Growth mechanism refers to the chemical reaction mechanism of small molecules (monomers) reacting to increase the size of molecules. Chain growth polymerization, step growth polymerization, and the like are growth mechanisms of conventional polymerization. How chemically reactive species are created in glow discharge of organic molecules is described in Chapter 4. Whether or not such reactive species lead to the formation of polymeric material is another issue, which is the topic of growth mechanisms. Chemical species found in gas phase analysis of glow discharge are often termed "precursors" of glow discharge polymerization without examining if such species participate in polymer formation. Without knowing the growth mechanisms of glow discharge polymerization, all precursors could at best, be, building blocks of "black box" polymerization. Luminous chemical vapor deposition (LCVD) does not proceed by black box polymerization.

In discussions of the mechanism of plasma polymerization appearing in the literature, polymerization, particularly the growth mechanism of polymer formation, is dealt with in a somewhat vague manner without any clear distinction between *mechanism of polymerization* and *mechanism of polymer deposition*. For instance, the hypothesis that plasma polymerization occurs via the polymerization of adsorbed monomer on the surface invokes the location of polymer formation rather than mechanism of polymerization; that is, the mechanism of polymerization, whatever that would be, is intuitively or a priori assumed. Nevertheless, such a hypothesis constitutes an important school of thought in dealing with the polymerization mechanism.

Some important but often neglected factors that hamper accurate elucidation of the growth mechanism and the deposition mechanism are associated with the way glow discharge polymerization is practiced. It is generally conceived that glow discharge polymerization starts at the point of the inception of glow discharge, i.e., ignoring what could have taken place in the reactor prior to the time zero. In reality, the glow discharge polymerization starts when the reactor with substrates is evacuated. Factors involved from this point on until the discharge is initiated greatly influence the outcome of glow discharge polymerization. If highly condensable vapor is introduced without glow discharge, the adsorption of the vapor on the substrate surely occurs, and its influence extends far beyond the ordinary reaction time in a flow system reactor, which is in a transient state to establish a steady state of glow discharge polymerization, i.e., the experiment is over before a steady state is established. In such a case, the result is tainted by the effect of adsorbed monomer on the substrate. (Some details of these factors are described in Chapter 12.) It is important to emphasize that the data for LCVD of perfluoro-2-butyltetrahydrofuran (PFBTHF) described below were collected under the true steady-state glow discharge.

Some factors of glow discharge polymerization that are necessary to construct overall growth mechanisms of polymer formation in LCVD are shown as comparisons of plasma, radiation, and parylene polymerizations in Tables 5.1 and 5.2. The organic compounds, which cannot be polymerized by chain growth polymerizations, can be polymerized by plasma polymerization (LCVD), and the deposition rates of these "nonmonomers" are by and large the same with those for "monomers" [1]. Therefore, the chain growth polymerization such as the addition polymerization of vinyl monomers cannot be the major principal growth mechanism of plasma polymerization (material formation in LCVD).

2. POLYMERIZATION IN VACUUM

The entropy of polymerization is negative, i.e., randomly oriented monomer molecules are transformed to a highly ordered chain molecule. In order to have a

Table 5.1 Comparison Plasma, Radiation, and Parylene Polymerizations

Polymerization	Monomer (starting material)	Initiator	Reaction phase
Plasma	No specific functional group is needed	No initiator	Gas phase (gas–solid interface)
Radiation	Monomer structure for addition polymerization	No initiator	Liquid or solid phase
Parylene	Dimer of para-xylene derivatives	No initiator	Gas phase (gas–solid interface)

Table 5.2 Comparison of Kinetic Factors for Plasma, Radiation, and Parylene Polymerizations

Polymerization	Activation	Growing species	Deactivation
Plasma	Dissociation of monomer by energy transfer from excited species	Mono- and diradicals	Recombination in gas phase and at the surface
Radiation	Ionization of monomer	Difunctional ion radicals and difunctional free radicals	Termination of growing chains, of which mechanisms depends on the concentration of impurities
Parylene	Thermal dissociation of cyclic dimer	Difunctional free radicals	Recombination of free radicals

spontaneous chemical reaction to occur, Gibbs free energy, $\Delta F = \Delta H - T\Delta S$, must be negative. Since ΔS for polymerization is negative, ΔH must be negative (exothermic reaction) and its value must be greater than the value of $T\Delta S$. As the temperature of reaction T increases to a critical temperature, the value of $T\Delta S$ reaches the value of ΔH, and ΔF becomes zero. Beyond this temperature, $T_c = \Delta H/\Delta S$, polymerization cannot occur. This critical temperature, above which polymerization cannot proceed, is termed the "ceiling temperature" of polymerization.

Polymerization in gas phase must cope with larger entropy change than polymerization in liquid phase. Therefore, polymerization of gas phase monomers such as olefins is carried out in superatmospheric pressure and/or in the presence of heterogeneous catalyst. Polymerization in gas phase in low pressure (in vacuum) does not occur easily due to the limitation of the ceiling temperature of polymerization, and there are only few cases in which the deposition of polymeric material from gas phase starting material occurs in vacuum. Those main exceptional cases are plasma polymerization and parylene polymerization.

3. INITIATOR AND POLYMERIZABLE SPECIES

Plasma polymerization is initiated via the dissociation of molecules caused by varieties of energetic species in the luminous gas phase as described in Chapter 4. It is important to recognize that the reactive species created in the luminous gas phase are not initiators of plasma polymerization. Some species, e.g., free radicals, could be initiators of some monomers that have specific functional groups under special conditions, e.g., in the off period of pulsed glow discharge and in the nonglow zone of a reactor (remote plasma). In most cases, the reactive species created in luminous gas phase are reactive building blocks of LCVD.

On the other hand, the initiators created by high-energy radiation of monomers initiate radiation polymerization in condensed phases. The initiator that controls the overall radiation polymerization could be cation, anion, or free radical depending on the purity of the monomer. Nevertheless, the concentration of the initiator must be very small and the majority of monomer must remain unaffected; otherwise the formation of sufficiently high molecular weight polymers cannot occur.

In parylene polymerization, the thermal cracking of the dimer (starting material) creates monomeric diradicals. All starting materials are converted to the reactive species, i.e., diradicals. No specific initiator for the chemical structure of starting material is formed. The situation is close to that of plasma polymerization. The comparison of plasma polymerization and radiation polymerization, and the comparison of the two vacuum deposition polymerizations (parylene polymerization and plasma polymerization) enable us to construct an overall view of material formation in the luminous gas phase.

3.1. Comparison of LCVD and Radiation Polymerization

Polymerization initiated by ionizing radiation such as γ rays or high-energy electron beams is somewhat similar to plasma polymerization. Understanding of radiation polymerization is very helpful in elucidating mechanisms of plasma polymerization. In radiation polymerization, no initiator is employed, and the chain-carrying species are created by the ionization of monomer. In this respect, radiation polymerization is similar to plasma polymerization. Particularly if we choose the radiation polymerization by electron beam, we are dealing with essentially identical phenomena of collision of an electron with an organic molecule.

Under irradiation, the energy is transferred to the monomer causing the ionization, i.e., an electron is ejected from the monomer, which leads to a cation of the monomer. The monomers for radiation polymerization are vinyl-type double-bond-containing compounds that are recognized as "monomers" of conventional polymerization. The ionization occurs by ejecting an electron from π bond of olefinic double bond of a monomer molecule, which has the lowest bond energy within the molecule. Consequently, the ionization of monomer by the irradiation leads to a cation radical, in which one end of molecule has a free radical (unpaired electron) and the other end has an excess positive charge (cation).

The electron ejected from the monomer molecule attaches to the double bond of another monomer molecule, which leads to an anion-free radical. In essence, the irradiation produces cation, anion, and free radical, any of which can initiate the monomer unaffected by the irradiation. Which end of the initiator, i.e., cation, anion, or free radical, initiates the polymerization is dependent on the nature of the double bond (electrophilic or electrophobic) and the purity of the monomer [2]. In some monomers in extremely high purity, polymerization proceeds by all three polymerizations, i.e., cationic, anionic, and free radical, as depicted schematically in Figure 5.1.

Radiation-induced polymerization, which generally occurs in liquid or solid phase, is essentially conventional chain growth polymerization of a monomer, which is initiated by the initiators formed by the irradiation of the monomer; i.e., ion radicals. An ion radical (cation radical or anion radical) initiates polymerization by free radical and ionic polymerization of the respective ion. In principle, therefore, radiation polymerization could proceed via free radical polymerization, anionic polymerization, and cationic polymerization of the monomer that created the initiator. However, which polymerization dominates in an actual polymerization depends on the reactivity of double bond and the concentration of impurity because ionic polymerization, particularly cationic polymerization, is extremely sensitive to the trace amount of water and other impurities.

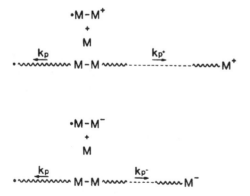

Figure 5.1   General reaction scheme of radiation induced polymerization of a double bond containing monomer M.

Irradiation of monomer vapor does not yield substantial polymerization that is observed in plasma polymerization. In radiation polymerization, the bombardment of electrons creates the initiator for polymerization of the monomer, whereas in plasma polymerization, the bombardment of electrons produces the polymerizable species out of a nonpolymerizable organic molecule as well as of a polymerizable monomer. The polymerizable species could act as an initiator if polymerizable monomers exist, which is not affected by the glow discharge. This situation occurs only in the pulsed discharge of a polymerizable organic molecule in short duty cycle (long resting period).

In radiation polymerization, the concentration of the initiator species is generally low but depends on the dose rate of irradiation. When the dose rate is increased, the concentration of the initiator and the rate of polymerization increase but the degree of polymerization decreases because all initiation, propagation, and termination reactions are coupled. From the viewpoint of the dose rate of irradiation, plasma polymerization corresponds to an extreme case in which an extraordinarily high dose rate (in the context of radiation polymerization) is employed. However, this analogy suggests that the molecular weight of polymer should be very low because the higher initiation rate decreases the degree of polymerization. In other words, at such a high dose rate, only oligomers (or no polymer) can be formed. This aspect also indicates that the growth mechanism of plasma polymerization of styrene should be completely different from that for conventional (ionic or free radical) chain growth polymerization of styrene.

3.2. Comparison of LCVD and Parylene Polymerization

So far as the growth mechanism is concerned, parylene polymerization is the closest kin of plasma polymerization. It is advantageous to go into a little details of parylene polymerization in order to understand how vacuum phase deposition of material occurs starting from a vapor phase material.

The process of parylene polymerization is presented schematically in Figure 5.2 using parylene N, unsubstituted poly(para-xylylene). Parylene dimer is heated until it sublimes. The dimer vapor passes through a high temperature pyrolysis zone where it cracks and becomes monomer vapor, i.e., monomer is created in vacuum. The monomer polymerizes and deposits in the deposition chamber, which is usually at room temperature. Parylene polymerization completed in a vacuum is a process involving no solvents, no curing, and no liquid phase. Its use essentially eliminates concern about the operator's health and safety, air pollution, and waste disposal.

The common denominator features of plasma polymerization (LCVD) and parylene polymerization are as follows:

1. Both polymerizations yield solid-state polymer (in a form of film in most cases) from a gas phase monomer in vacuum.
2. The polymer formed by both processes contains a large amount of trapped free radicals in the solid polymer (dangling bonds).

These two processes are unique in that polymeric films are formed from monomer in vacuum phase, which does not occur via most other types of polymerization. Therefore, there must be common basic principles pertaining to the growth mechanisms of polymerization for these two processes.

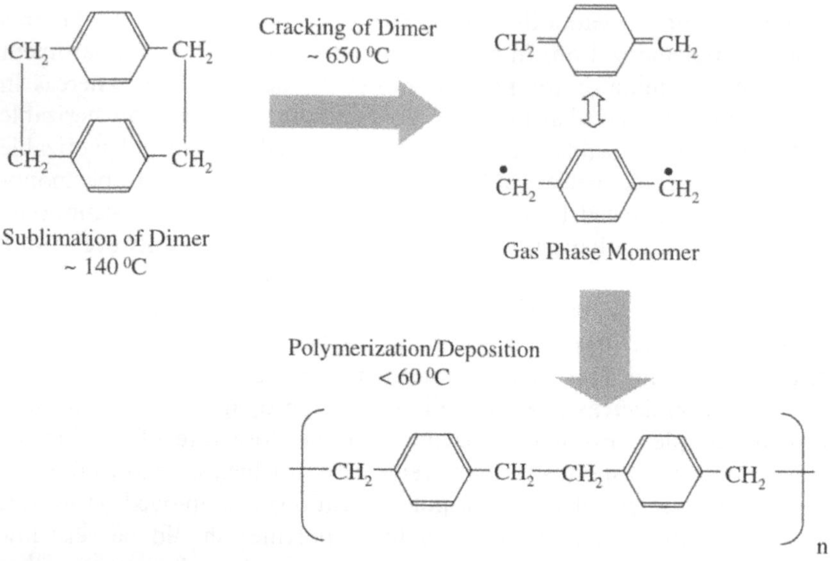

Figure 5.2   Schematic diagram of parylene polymerization steps.

Parylene polymerization proceeds with well-defined chemical species, whereas plasma polymerization proceeds via variety of not-well-defined chemical species, which are created in the luminous gas phase. The reactive species for parylene polymerization is para-xylylene, which has features of (1) difunctional (e.g., diradicals), (2) reactive but relatively stable, and (3) highly selective reactivity (see Fig. 2.1). The exact nature of reactive species involved in glow discharge polymerization is not well known; however, (1) they are not exclusively bifunctional, (2) they are highly reactive, and (3) consequently they have very low selectivity. The difference in the stability or the selectivity of reactive species is reflected in the distinctively different characters of polymer depositions of these two processes.

Highly reactive (unstable) and nonselective species tend to react with any surface on which the species strike and form a polymer deposition with a high level of bonding or adhesion to the surface. Because of this aspect, plasma polymerization tends to form a thin film with a good adhesion with various kinds of substrate materials. Because of nonselective reactivity, the reactive species of plasma polymerization have poor penetration into small cavities such as those of porous structures. Reactive species tend to react with wall material at the entrance of a cavity rather than penetrating into the cavity (which requires that the species not react with the wall at the entrance).

In contrast to the above situations, parylene polymer deposition has very poor adhesion to a smooth surface substrate but can penetrate deep into small cavities. para-Xylylene prefers to react with another para-xylylene or its derivatives. Although it has the feature of difunctional free radical, it is rather stable and does not initiate polymerization of other monomers for conventional free radical polymerization. In spite of numerous attempts, the polymerization of various vinyl monomers initiated by para-xylylene or copolymerization of vinyl monomers with *para*-xylylene has been elusive.

Because of the high selectivity (toward the same kind of species), a para-xylylene molecule (in vacuum phase) can penetrate deep into the cavity without interacting with the wall material and wait for another para-xylylene molecule to arrive. The second molecule reacts immediately without losing the overall reactivity because of being a difunctional free radical. By repeating this step, parylene deposition can penetrate deep into cavities or microvoids existing on the substrate surface, yielding a very good adhesion to porous substrate by virtue of mechanical anchoring. Thus, parylene film deposition can be characterized by (1) no adhesion to a smooth surface and (2) very good adhesion to porous substrate by forming an interpenetrating thin-layer film.

Because of very important similarities and contrasting features, it is highly advantageous to review the basic kinetics of plasma polymerization and of parylene polymerization. Parylene polymerization and plasma polymerization are two major polymerization reactions, which can be viewed as "vacuum deposition polymerization," in which the process of deposition is the key factor for the polymer formation. This principle can be further postulated as *deposition growth polymerization*. In this context, deposition growth polymerization can be considered as the growth mechanism for the vacuum deposition polymerization in a similar manner as step growth polymerization and chain growth polymerization used to characterize the growth mechanisms. In the both polymerizations, free radicals are species that are responsible for the formation of bonds in the depositing materials.

3.2.1. Growth Mechanism of Parylene Polymerization

The following aspects of parylene polymerization [3–9] seem to have important implications in an effort to understand the growth mechanisms of plasma polymerization.

1. The deposition of polymerizable species (the quinoid form and singlet-state monomer) can occur without polymerization. (The deposition of polymerizable species is followed by the polymerization.)
2. In most cases, the deposition and polymerization proceed simultaneously, but under certain circumstances the delayed polymerization of the deposited monomer was observed. When a liquid nitrogen trap was used as the deposition surface, the polymerization of condensed monomer was observed when the temperature of the surface was allowed to rise. The conspicuous vigorous boiling of liquid nitrogen by the heat of polymerization, which caused abrupt change of white deposition to a clear film, was observed.
3. Because of a well-defined chemical structure for the repeating unit, the vacuum-deposited polymer is significantly crystalline. The crystallinity of a film depends on the deposition conditions.
4. The growth step of vacuum deposition polymerization can be conceived as poly-recombination of free radicals. Because para-xylylene (monomer) has the feature of diradical, the recombination does not terminate the propagation reaction (free radical living polymer) and leads to the presence of free spins in the final product (film).
5. Temperature dependence of film formation is negative, i.e., the deposition rate is higher at the lower temperatures.

6. The presence of the ceiling temperature (T_c) for deposition and polymerization is evident. At temperatures above the ceiling temperature, no polymerization and consequently no deposition occur.

7. The polymer deposition (polymerization) is significantly accelerated by the addition of cool inert gas into the deposition chamber. This phenomenon indicates the significance of the reactive species temperature, T_x (in the context similar to that of the electron temperature, T_e). Unless T_x can be cooled down (below T_c), which usually occurs as a consequence of collisions with walls, the polymerization cannot proceed. In other words, the dissipation of the excess energy in some forms is necessary for the deposition growth polymerization to occur.

8. T_x is not the temperature of reaction system and is polydispersed.

4. GROWTH MECHANISMS OF LCVD

Considering these common and contrasting aspects, the rapid step growth polymerization (RSGP) mechanisms has been proposed [10] in which the recombination of reactive species and reactivation of the reaction products play key roles. A schematic presentation of growth mechanisms is presented in Figure 5.3. The actual growth mechanisms involve the participation of the surface or the third body in each reaction shown in the figure. In order to simplify the overall picture of reactions in luminous gas phase, however, the contribution of the surface is not shown in the figure. How these reactions contribute to the deposition of materials is schematically illustrated in Figure 5.4.

Rapid Step-Growth Polymerization (RSGP) Mechanism

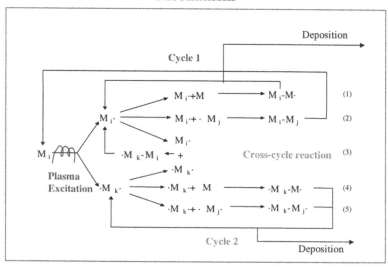

Figure 5.3 Growth and deposition mechanisms in the luminous gas phase.

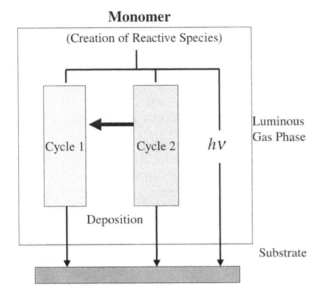

Figure 5.4 Gas phase reactions and material deposition.

In Figure 5.3, the reactive species are represented by free radicals, but any other reactive species could, in principle, contribute to any of the steps. However, the ions with the same charge cannot react each other, and an equal number of positively charged and negatively charged ions are needed for the ions to participate in the growth mechanisms. Furthermore, neutral species in the luminous gas phase outnumber ions roughly 10^6 to 1 [11]. Therefore, the role of ions in the growth mechanisms can be virtually eliminated.

Reactions (1) and (4) are essentially the same as the addition of reactive species to the monomer, which is the same as the initiation and propagation reactions in the free radical chain growth polymerization. However, the kinetic chain length in vacuum is very short, and in a practical sense these reactions can be considered to be stepwise reactions. Cycle I consists of reactions of reactive species with a single reactive site, and cycle II is based on divalent reactive species. Reaction (3) is a cross-cycle reaction from cycle II to cycle I. The growth via cycle I requires the reactivation of the product species, whereas cycle II can proceed without reactivation as long as divalent reactive species or monomers with double bond or triple bond exist.

Any of the species involved will collide with the substrate surface as frequently as the kinetic theory of gases predicts, but only the collisions between reactive species that will lead to the growth of the species are considered in the reaction scheme (deposition growth polymerization). Therefore, any species in any reaction in the scheme can be either in the gas phase or on the surface.

In LCVD, polymerization and deposition are inseparable components of the polymerization–deposition mechanism. None of the reactions considered in Figure 5.3 is a polymerization by itself. While repeating the steps via cycle I or cycle II, the species varying in size, depending on how many cycles has been progressed, deposit on the substrate surface.

The number for how many times the cycles, which are shown in Figure 5.3, repeat before a species deposits can be expressed by the term *kinetic pathlength*. As the kinetic pathlength in gas phase increases, the size of the gaseous species increases, and the saturation vapor pressure of the species decreases, which forces the species to deposit.

4.1. Effect of Substrate Temperature on Growth Reactions

The formation of reactive species from the monomer or from the (nonreactive) products of reaction (2) is essentially a destructive process, i.e., it requires breaking of a bond (e.g., C-H, C-F, or C-C). Consequently, how far these step reactions have progressed before deposition occurs will influence the chemical and physical nature of the polymeric deposit. This situation can be visualized by the temperature dependence of material deposition for LCVD of PFBTHF at different discharge power input, which was obtained by changing the temperature of a thickness monitor placed in a steady-state glow discharge of the monomer only without changing other parameters of glow discharge [12].

In LCVD, the first step to create reactive species is not thermal process, and the temperature (of substrate surface) dependence of overall material deposition merely reflects the temperature dependence of the deposition step. If the deposition rates k at different substrate temperature is plotted against $1/T$, it yields a negative slope, indicating that the rate process cannot express the deposition rate, i.e., the deposition process is more or less an adsorption or a condensation process.

The plasma polymerization reactor used is schematically shown in Figure 5.5. Parallel electrodes, equipped with magnetic enhancement, by a 10-kHz power source created glow discharge of the monomer. A thickness monitor sensor is placed at the projected circumference of electrodes intercepting the midelectrode plane. The electrodes are 13.2 cm in diameter and 6.1 cm apart. The thickness monitor surface is perpendicular to the plane parallel to the electrodes.

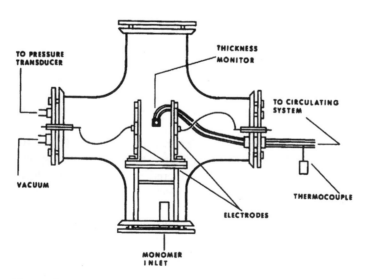

Figure 5.5 Schematic representation of reactor used in the temperature dependence study.

The circulation of a temperature-controlled liquid controlled the temperature of the crystal surface on which the plasma polymer deposits. In order to measure the substrate temperature accurately, two thermocouples are placed in the fluid-circulating tubes (inlet and outlet) just outside of the plasma reactor. The substrate temperature is estimated from the average of the thermocouple readings.

The temperature dependence of plasma polymer deposition is measured by the following procedure. The plasma polymerization of a monomer is investigated by starting at the highest temperature (80°C). After a constant temperature of crystal surface is confirmed, a constant flow of monomer is established. The discharge power is adjusted every 10 min by reading current and voltage until a steady glow discharge is established, and it is checked (and readjusted if necessary) throughout the duration of a run. In such a reactor, the plasma polymerization seems to reach a very stable steady state after the initial transient stage in 30 min to a couple of hours, and once the steady state is obtained very little adjustment of the operational parameter is necessary.

When the deposition rate and the system pressure shown on the recorder are confirmed to be steady, the deposition rate reading and the crystal temperature were recorded. Then changing the thermostat control of the circulating bath, while the plasma polymerization is kept at the steady state, lowered the temperature of the crystal. The deposition rate at the next temperature is read and recorded after steady-state readings are obtained at the new temperature. In this way, the relationship between deposition rate and substrate temperature can be obtained at a set of flow rates and power. A similar procedure is repeated for another set of flow rates and power.

Figure 5.6 depicts the temperature dependence of deposition rate for the plasma polymerization of PFBTHF shown as plots of k versus T. The XPS C 1s spectra of polymers deposited at different temperatures under different energy input levels are shown in Figure 5.7. Table 5.3 depicts the details of XPS C 1s spectra shown in Figure 5.7. The important aspects of the results are as follows:

1. The temperature dependence is negative, indicating that what we observe as the temperature dependence of the polymer deposition is not a reflection of the reaction rate of the growth reactions.
2. The reaction products obtained at different substrate temperatures are not the same, indicating that the kinetic pathlength (how many cycles has progressed) of depositing species changes at different substrate temperatures.

It is important to emphasize that the plasma polymerization was carried out under an identical set of conditions throughout the experiments. The steady state of glow discharge was first established and maintained for the entire series of temperature dependence studies. The only change of temperature of substrate that occurred was the change of cooling liquid temperature for the thickness monitor. The temperature of substrate (a quartz thickness monitor), which is not in the thermal equilibrium with the remaining system but in the dynamic equilibrium with the surrounding gas phase, was allowed to establish the dynamic steady state before the deposition rate was measured. The monomer is the same throughout, and the discharge conditions are identical. Therefore, the chemistry of the luminous gas phase should be identical in each case. The facts found under such conditions, i.e., (1) the products obtained at different temperatures are not the same and (2) the

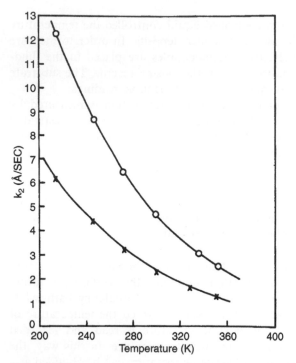

Figure 5.6 Temperature dependence of k_2 for tetrafluoroethylene; flow rate $(cm^3_{STP},/min)$ and power(W): $\bigcirc$: 1.04 sccm and 9.84W, $\times$: 0.54 sccm and 4.89 W.

temperature dependence is negative, indicate that the change of chemical structure of plasma polymers is caused by the change of deposition process, but not by the change of chemistry involved in the luminous gas phase.

 We could understand this situation if we refer to the complex scheme of the RSGP mechanisms presented in Figure 5.3 and also if we recognize that the feature of deposition growth polymerization is built into the RSGP mechanisms. According to the RSGP mechanisms, the deposition of polymer and the growth reactions cannot be clearly separated. In other words, the deposition of material (polymer) occurs neither by the deposition of formed polymers (in gas phase) nor by the discrete process of molecular growth of the deposited monomers.

 At this point, it is important to recognize following:

1. Multiple chemically reactive species are created from a monomer.
2. The kinetic pathlength changes the chemical structure of species.
3. As the size of species increases due to reactions in gas phase, the vapor pressure of the species decreases, enhancing the condensation (deposition of reactive species) on the surface.
4. A kinetic pathlength of probably less than five cycles would yield a large enough molecule from the smallest reactive species created by the dissociation of a monomer that cannot remain in the gas phase.
5. The ambient temperature of the glow in the typical plasma polymerization reactions described in this book is generally in the vicinity of 380–400 K and remains reasonably constant after a steady-state condition has been established.

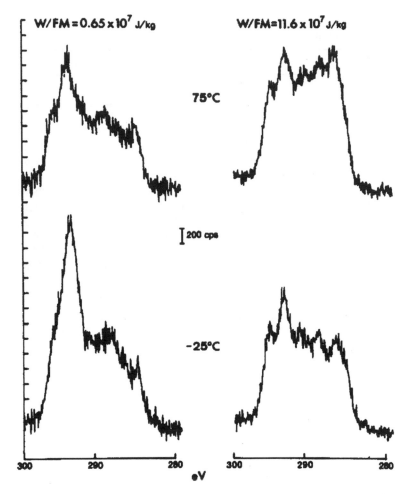

Figure 5.7 ESCA C 1s spectra of plasma polymers of perfluoro-2-butyltetrahydrofuran obtained at different *W/FM* and substrate temperature.

Any species involved in the RSGP mechanisms (Fig. 5.3) can deposit on the substrate surface. Deposition occurs when an impinging particle fails to bounce back from a colliding surface. Such a deposition may result from the loss of kinetic energy or from the formation of a chemical bond with a target molecule or atom. The sticking coefficient, or deposition coefficient, can be defined as the number of particles deposited divided by the total number of impinging particles.

Thus, as defined, the sticking coefficient is a function of the nature of the species involved (mass, kinetic energy, chemical reactivity, etc.) and the surface temperature. What we observe as the temperature of substrate changes is a reflection of the temperature dependence of the sticking coefficients of various species involved in the RSGP mechanisms. The kinetic pathlength plays a key role in the change of the sticking coefficients.

The results of deconvolution of C 1s peaks, together with the assignment of peaks, are shown in Table 5.3. At both levels of *W/FM* it is evident that polymers

Table 5.3 Change in ESCA C 1s Peaks of Plasma Polymers of Perfluoro-2-Butyltetrahydrofuran Due to a Change in Substrate Temperature and W/FM

W/FM $[(J/kg) \times 10^{-7}]$	Substrate temp. (°C)	Area of component peaks (%)				
		1	2	3	4	5
0.65	75	16.0	21.1	19.4	26.6	17.0
	50	16.5	26.8	18.5	22.9	15.2
	25	17.3	27.2	19.5	22.7	13.3
	0	18.5	27.8	14.9	22.1	16.7
	−25	19.1	30.9	16.8	21.4	11.8
11.6	75	14.9	18.1	18.8	19.3	28.9
	50	15.0	18.8	18.3	20.0	28.1
	25	16.6	20.0	17.7	18.8	27.2
	0	16.7	20.2	17.5	19.4	26.2
	−25	17.8	21.0	17.8	18.8	24.6
	−49.5	18.0	21.2	17.1	19.8	24.0

Peak		Approximate peak position (eV)	Approximate peak width (eV)
1	$-CF_3$	295	2.0
2	$-CF_2-$	293	1.9
3	$-\overset{\vert}{\underset{\vert}{C}}-F$	291	2.3
4	$-\overset{\vert}{\underset{\vert}{C}}-\overset{*}{C}F$	288	2.3
5	$\searrow C-H$ and $-\overset{\vert}{\underset{\vert}{C}}-\overset{\vert}{\underset{\vert}{C}}-$	286	2.8

deposited at lower temperature have higher contents of high-fluorine-containing moieties. Peak 1 and peak 2 steadily increase at the expense of peaks 4 and 5 as the substrate temperature decreases. A relatively small change is observed for peak 3. It is also evident that, at a higher level of W/FM, the fluorine-containing moieties decrease and peak 5 increases significantly, indicating the increase of the kinetic pathlength at the same substrate temperature. The change in polymer structure can be attributed not to the change in chemistry involved in the luminous gas phase but to the substrate temperature effect on the plasma polymerization–deposition mechanisms, which can be best explained by the mechanisms described in Figure 5.3.

The activity **a** of a vapor in a system is given by $\mathbf{a} = p/p_0$, where p is the partial pressure of the species and p_0 is the saturation vapor pressure of the species at the specific temperature of the system. As p_0 decreases as a consequence of the increasing size, **a** approaches the unity, where the condensation of the species takes place. In the vapor phase near the substrate, which is at a lower temperature, the value of p_0 is lower. Consequently, at a lower substrate temperature, growing species with shorter kinetic pathlengths deposit.

4.2. Effect of Energy Input on Growth Reactions

The most important factor that influences the properties of plasma polymers from a monomer is the energy input level of plasma polymerization process. The energy input level determines the extent of fragmentation or scrambling of the monomer molecule. The molecular nature of plasma polymerization decreases with the increasing level of energy input. The energy level in the diffused plasma can be manifested by the parameter, W/FM, where W is wattage, F is molar or volume flow rate, and M is molecular weight of monomer. The parameter has units of J/kg, i.e., energy per mass of monomer. The significance of this parameter is described in Chapter 8.

The most significant aspect found in these experimental data is that the same trend found as the consequence of a change in substrate temperature is also found as the consequence of a change in discharge power. At a higher energy input to the luminous gas phase, the kinetic pathlength in the luminous gas phase increases due to the increase of the density of reactive species within a unit of time, which enhances gas–gas reactions in the luminous gas phase. At both substrate temperatures, the polymers deposited at the higher energy input have the lower contents of high-fluorine-containing moieties. Peaks 1 and 2 decrease and peaks 4 and 5 increase as the energy input increases. The data shown in Figure 5.6 and Table 5.3 clearly support the explanation based on the growth mechanism shown in Figure 5.3.

The bicyclic RSGP mechanism shown in Figure 5.3 has an important implication for the interpretation of diagnostic data of the luminous gas phase. Namely, any species identified in the plasma phase are intermediate species of step growth polymerization but not precursors of black box plasma polymerization.

5. DEPOSITION MECHANISM IN LCVD

5.1. Temperature Dependence of Deposition Rate

A chemical reaction is generally associated with positive activation energy, i.e., a reactant must be activated to overcome an energy barrier. The rate constant k of the reaction can be given in terms of the activation energy of reaction ΔE as

$$k = Ae^{-\Delta E/RT} \tag{5.1}$$

where A is the pre-exponential parameter. Therefore, the temperature dependence of the reaction can be generally shown by a plot of $(\ln k)$ versus $(1/T)$, which yields a straight line with a slope $-\Delta E/R$.

The temperature dependence of plasma polymer deposition is generally negative. Some monomers show very little dependence, but it seems that no plasma polymerization system that has positive temperature dependence exists. Consequently, polymer deposition can be prevented if the temperature of the substrate is raised above the ceiling temperature of deposition, which is far above the steady-state ambient temperature of the plasma.

The deposition rate of plasma polymerization depends on many experimental factors of glow discharge. A large number of attempts have been made to correlate the polymer deposition rate with such operational variables as flow rate, discharge power, current density, and system pressure. Although reasonable agreement is found in the various ways of expressing the kinetics of plasma polymerization, the system-dependent aspect of plasma polymerization makes it difficult to examine the general applicability or to confirm the relationship in different systems.

On the basis of the concept of RSGP via bicyclic propagation mechanisms shown in Figure 5.3, Yasuda and Wang [12] examined the polymer deposition rates of four different monomers at substrate temperatures ranging from $-50°C$ to $80°C$ under various combinations of flow rate and discharge wattage. The results of the study seem to reveal a greatly simplified system-dependent aspect of the overall kinetic scheme and its dependence on the structure of monomers. A part of data for the plasma polymerization of PFBTHF shown as plots of k versus T was used in elucidation of growth mechanisms in the preceding section.

The monomers used are as follows:

Monomer	Mol wt	Boiling point ($°C$)
Perfluoro-2-butyltetrahydrofuran	416	107
Tetrafluoroethylene	100	-76
Styrene	104	145
Ethylene	28	-104

As noted before, it took 30 min to 2 h before a steady-state luminous gas phase was established. This means that if one carries out a plasma polymerization for a short time, a steady-state luminous gas phase couldn't be established within the experiment. The details of the non-steady state in the early stage of plasma polymerization are described in Chapter 12.

The deposition rates could be expressed by the following parameters:

k_1, the deposition rate (kg/m^2-s, mg/cm^2- s, etc.).
k_2, the thickness growth rate (m/s, Å/s, etc.). k_2 is related to k_1 by $k_2 = k_1/\rho$, where ρ is the specific weight (kg/m^3) of polymer.
k_0, the normalized deposition rate, which is given by $k_0 = k_1/FM$ (m^{-2} or cm^{-2}), where F is the molar flow rate and M the molecular weight of monomer; i.e., FM is the mass flow rate (k_0 is the polymer conversion ratio per unit area).

The overall polymer conversion ratio is given by k_0 times the total surface area. The temperature dependence of polymer deposition for various monomers (of different molecular weights) under different discharge conditions is best expressed in terms of the normalized deposition rate k_0 rather than deposition rate k_1 or k_2 because the actual deposition rate observed under a set of conditions is dependent on the mass flow rate (see Chapter 8).

In Figure 5.5, values of k_2 (angstroms per second) versus temperature are shown for two sets of flow rate and discharge wattage, which give the same value of composite parameter W/FM (1.28×10^8 J/kg). When the same data are plotted in k_0/ρ (angstroms per kilogram), the two lines shown in Figure 5.5 converge into a single line as shown in Figure 5.8. The data shown in Figure 5.8 also show that the normalized deposition rate is identical for both sets of conditions.

Normalized deposition rates for PFBTHF at various temperatures are shown in Figure 5.9 as plots of $\ln(k_0/\rho)$ versus absolute temperature T, which was found to

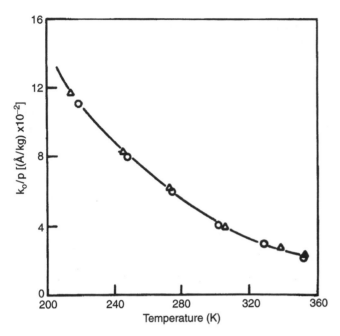

Figure 5.8 Temperature dependence of k_0/ρ for tetrafluoroethylene at the same W/FM $(1.3 \times 108\,\text{J/kg})$ but different flow rate and power: 1.04 sccm and 9.84 W and 0.52 sccm and 4.89 W.

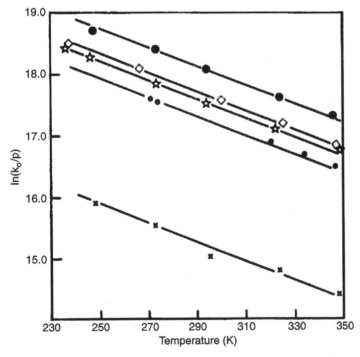

Figure 5.9 Effect of substrate temperature on k_0/ρ for perfluoro-2-butyltetrahydrofuran.

give the best fits by a regression analysis. As shown in these figures, the temperature dependence of k_0 is given by

$$k_0 = Ae^{-bT} \qquad (5.2)$$

where A represents the extrapolated specific deposition rate at $T = 0\,\text{K}$. Thus, the polymer deposition rate can be expressed by two parameters: the extrapolated specific deposition rate at $T = 0$ and the temperature dependence parameter b. The value of b for a monomer was found to be independent of discharge conditions (i.e., flow rate and discharge wattage) and may be considered as a characteristic parameter of a monomer. The values of b for four monomers are shown in Table 5.4

It is important to note that the temperature dependence parameter b is nearly the same for an easily condensable monomer (PFBTHF) and a gas monomer (TFE), and also for styrene and ethylene, but is dependent on the type of monomers (i.e., perfluorocarbons versus hydrocarbons). The fact that the temperature dependence, as well as the normalized deposition rate at a given temperature (e.g., at 273 K and at a given value of W/FM as shown in Table 5.5), is not directly related to the condensability of a monomer clearly indicates that the adsorption of "monomer" plays a small role in plasma polymerization that occurs under steady-state conditions. It should be noted that if a monomer were introduced into a reactor with low temperature substrate without glow discharge, the boiling temperature of the monomer would control the condensation, which constitutes a common mistake in study of glow discharge polymerization of easily condensable monomers.

As already pointed out, the temperature dependence of polymer deposition is not related to the conditions of plasma polymerization (i.e., the flow rate and

Table 5.4   Values of b for Monomers ($k_0 = Ae^{-bt}$)

Monomer	Molecular weight	Boiling point (°C)	b
Ethylene	28	−104	0.0045 ± 0.0003
Styrene	104	145	0.0060 ± 0.0003
Tetrafluoroethylene	100	−76	0.0112 ± 0.002
Perfluoro-2-butyltetrahydrofuran	416	107	0.0143 ± 0.004

Table 5.5   Values of k_0/ρ at $W/FM \sim 1.8 \times 10^8\,\text{J/kg}$ and Substrate Temperature 273 K, $[k_0/\rho = (A/p)e^{-bT}]$

Monomer	k_0/ρ $[(\text{Å/kg}) \times 10^{-7}]$
Ethylene	1.6
Styrene	6.0
Tetrafluoroethylene	10.8
Perfluoro-2-butyltetrahydrofuran	7.0

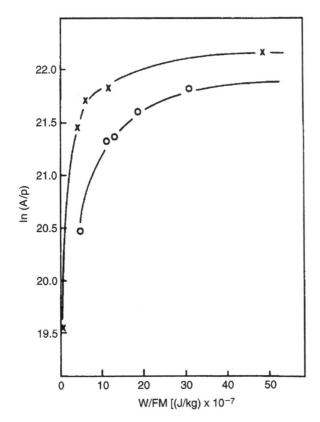

Figure 5.10 Plot of $\ln(A/\mathrm{p})$ versus W/FM; ×: perfluoro-2-butyltetrahydrofuran; ○: tetrafluoroethylene.

discharge power). It is important to examine the effect of plasma polymerization conditions on the value of the pre-exponential factor A in Eq. (5.2).

In Figure 5.10, the values of A for PFBTHF and TFE are plotted against the composite parameter W/FM. A similar plot for ethylene is shown in Figure 5.11. The leveling off of the deposition rate as W/FM increases is clearly seen here. On the basis of the dependence on W/FM, it is possible to distinguish domains of plasma polymerization conditions. Two major domains can be identified: (1) a W/FM-dependent region at lower W/FM, and (2) a W/FM-independent (plateau) region at higher W/FM. The first region is the energy-deficient region and the second is the monomer-deficient region.

5.2. Energy-Deficient Domain

In general LCVD, there are two important domains: (1) energy deficient and (2) monomer deficient. Details of domains are described in Chapter 8. Let us first examine how W/FM would affect the value of A, which is the normalized deposition rate at $T = 0\,\mathrm{K}$, in the energy-deficient region by assuming that A is proportional to $(W/FM)^n$. In Figure 5.12, $\ln(A/\rho)$ is plotted against $\ln(W/FM)$. From the slopes of the lines, the values of n can be estimated. By a regression analysis, the value of n is

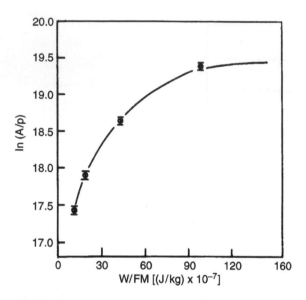

Figure 5.11   Plot of $\ln(A/p)$ versus W/FM for ethylene.

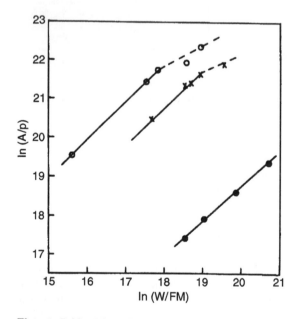

Figure 5.12   Plot of $\ln(A/p)$ versus $\ln(W/FM)$; $\bigcirc$: perfluoro-2-butyltetrahydrofuran; $\times$: Tetrafluoroethylene; $\bullet$: ethylene.

found to be 0.98 for PFBTHF, 0.92 for TFE, and 0.92 for ethylene. Therefore, in the energy-deficient region, A can be given by

$$A = a(W/FM)^n \tag{5.3}$$

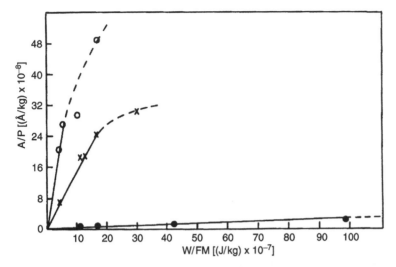

Figure 5.13 Plot of A/p versus W/FM; ○: perfluoro-2-butyltetrahydrofuran; x: tetrafluoro-ethylene; ●: ethylene.

and n is very close to unity. With the first approximation of $n = 1$, the effect of the molecular weight of the monomer can be estimated by plotting A/ρ against W/FM (Fig. 5.13). The value of A/ρ increases nearly linearly with W/FM, and it is mainly dependent on the molecular weight of the monomer:

$$a = cM \tag{5.4}$$

The slopes of the initial linear portion of the plots in Figure 5.13 are plotted against the molecular weights of the monomers in Figure 5.14. Thus, in this region A can be given by

$$A = c(W/F) \tag{5.5}$$

Therefore, in this region the normalized deposition rate can be rewritten as

$$k_0 = c(W/F)e^{-bT} \tag{5.6}$$

and the deposition rate k_1 is given by

$$k_1 = CWMe^{-bT} \tag{5.7}$$

At a given temperature,

$$k_1 = c'WM \tag{5.8}$$

and the observed deposition rate is linearly proportional to the discharge wattage W and the molecular weight of monomer M but is independent of the molar (or volume) flow rate F.

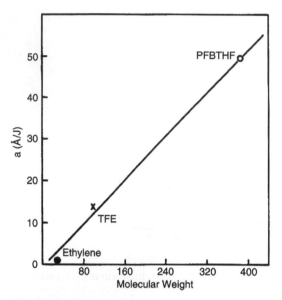

Figure 5.14 Slopes of Fig. 5.13 versus molecular weight; PFBTHF: Perfluoro-2-butyltetrahydrofuran; TFE: tetrafluoroethylene.

5.3. Monomer-Deficient Domain

As shown in Figures 5.10 and 5.11, in the monomer-deficient region the value of A becomes independent of W/FM. Therefore, the normalized deposition rate for this region can be written as

$$k_0 = B_e^{-bT} \tag{5.9}$$

where B is a constant, which depends on the nature of the monomer. The observed deposition rate k_1 in this region can be given by:

$$k_1 = BFMe^{-bT} \tag{5.10}$$

and at a given temperature,

$$k_1 = B'FM \tag{5.11}$$

The deposition rate in this region is linearly proportional to the mass flow rate FM or linearly proportional to the molar flow rate F and the molecular weight of monomer M but is independent of discharge power W.

5.4. Normalized Bond Energy of Molecules

By virtue of Eq. (5.4), the value of B is expected to be larger where M is larger; however, an exact estimate of the value of B is difficult to make because of the transition region from the energy-deficient region to the monomer-deficient region.

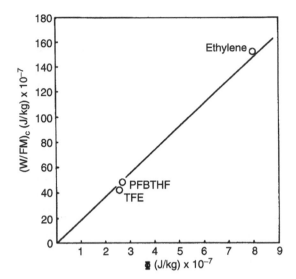

Figure 5.15 Critical value of W/FM versus specific bond energy for each monomer. PFBTHF, perfluorobutyltetrahydrofuran; TFE, tetrafluoroethylene.

An interesting correlation is found, however, between the value of $(W/FM)_c$, the critical W/FM value above which plasma polymerization can be considered in the monomer-deficient region, and the nature of the monomers. In Figure 5.15 values of $(W/FM)_c$, are plotted against values of total bond energy divided by the molecular weight of the corresponding monomer.

Thus, the W/FM necessary to bring the plasma polymerization system into the monomer-deficient region is proportional to the value of the total bond energies divided by molecular weight, which can be considered the *normalized bond energy* of the monomer. This dependence clearly indicates that in the monomer-deficient region nearly all bonds in a monomer are broken, and because of this fact, the additional energy input does not influence the nature of LCVD. The LCVD in this region is a typical "atomic" polymerization because the original monomer structure is nearly totally destroyed and what accounts for the polymer formation are the atoms that comprised the original monomer molecule but are fragmented under the plasma conditions.

Thus, the value of $(W/FM)_c$ above which the pre-exponential term A becomes a constant B, which is characteristic of a monomer, can be given by

$$(W/FM)_c = \alpha\Phi \tag{5.12}$$

where

$$\Phi = \frac{\Sigma(\text{bond energy})}{M} \tag{5.13}$$

is the *normalized bond energy* of the monomer and α the proportionality constant for a given reactor. Values of B/ρ, $(W/FM)_c$, and Φ are given in Table 5.6. From

Table 5.6 Values of B/ρ, $(W/FM)_c$ and Φ for Monomers

Monomer	ln (B/ρ)	B/ρ [(Å/kg) $\times 10^{-8}$]	$(W/FM)_c$ [(J/kg) $\times 10^{-7}$]	$\Phi = \Sigma \, B.E./M$ [(J/kg) $\times 10^{-7}$]
Ethylene	19.55	3.0.9	152	8.0
Tetrafluoroethylene	21.90	32.4	42	2.6
Perfluoro-2-butyltetrahydrofuran	22.15	41.6	48	2.7

Table 5.7 Normalized Bond Energy and Estimated Critical Energy Input

Compound	Bond energy (kJ/mol)	Molecular weight	Bond energy/mass (MJ/kg)	19 times bond energy (GJ/kg)
Hexafluoropropene	3871	150	25.8	0.49
Perfluorobutane	5891	238	24.8	0.47
Hexafluoropropene oxide	4320	166	26.0	0.49
Tetrafluoroethane	3113	102	30.5	0.58
Methane	1652	16	103.3	1.96
Butane	5171	58	89.2	1.69
Trimethylsilane	5190	74	70.1	1.33

these data, the value of α was estimated to be about 19. Thus, in the polymerization system employed, when the energy input in joules per kilogram exceeds about 19 times the normalized bond energy (joules per kilogram), the plasma polymerization becomes a typical atomic polymerization. Normalized bond energy and anticipated critical energy input given by 19 times normalized bond energy for some molecules are shown in Table 5.7. The values indicate that perfluorocarbons have significantly lower critical energy input, implying that LCVD of hydrocarbons and of perfluorocarbons cannot be compared at an arbitrarily selected discharge wattage W.

It is important to recognize that a large molecule has small normalized bond energy. On the other hand, because of the larger mass, at a given discharge wattage the value of W/FM is also small for a large monomer. These trends clearly indicate that the plasma polymerizations of various monomers cannot be compared at an arbitrarily chosen discharge power and/or flow rate.

It is also important to recognize the domain in which a plasma polymerization is carried out under a given set of operational conditions. The value of W/FM alone does not identify whether a plasma polymerization is in the energy-deficient or the monomer-deficient region. A crude estimate of the domain might be made by the parameter $(W/FM)/\alpha\Phi$ if the value of α were known for the reactor. The following conditions can be used for this purpose:

$(W/FM)/\alpha\Phi > 1$ monomer-deficient region

$(W/FM)/\alpha\Phi < 0.5$ energy-deficient region

$1 > (W/FM)\alpha\Phi > 0.5$ transient region

An important implication of growth and deposition mechanisms described in this chapter is that an arbitrarily selected set of discharge conditions, e.g., discharge wattage and flow rate, does not provide a basis for comparing LCVD characteristics of different monomers. Multiple chemically reactive species are created in the luminous gas phase, and those species changes during the resident time in the luminous gas phase. The kinetic pathlength depends on the operational parameters and the design factors of the reactor employed.

REFERENCES

1. Yasuda, H.; Lamaze, C.E. J. Appl. Polym. Sci. **1973**, *17*, 1533.
2. Hayashi, K. Polym. J. **1980**, *12*, 583.
3. Gorham, W.F. J. Polym. Sci. A-1 **1966**, *4*, 3027.
4. Yasuda, H.; Kramer, P.; Sharma, A.K.; Hennecke, E.E. J. Polym. Sci., Polym. Chem. Ed. **1984**, *22*, 475.
5. Gazicki, M.; Surendran, G.; James, W.; Yasuda, H. J. Polym. Sci., Polym. Chem. Ed. **1985**, *23*, 2255.
6. Gazicki, M.; Surendran, G.; James, W.; Yasuda, H. J. Polymer Sci., Polym. Chem. Ed. **1986**, *24*, 215.
7. Gazicki, M.; James, W.J.; Yasuda, H.K. J. Polym. Sci., Polym. Lett. **1986**, *23*, 639.
8. Surendran, G.; Gazicki, M.; James, W.J.; Yasuda, H. J. Polym. Sci., Part A: Polym. Chemistry **1987**, *25*, 1481.
9. Surendran, G.; Gazicki, M.; James, W.J.; Yasuda, H. J. Polym. Sci.: Part A: Polym. Chemistry **1987**, *25*, 2089.
10. Yasuda, H. *Plasma Polymerization*; Academic Press: Orlando, FL, 1985; 44–177.
11. Kobayashi, H.; Bell, A.T.; Shen, M. J. Macromol. Sci. Chem. **1976**, *10*, 491.
12. Yasuda, H.; Wang, C.R. J. Polym. Sci. Polym. Chem. Ed. **1985**, *23*, 87.

6
Dangling Bonds

1. FREE RADICALS AND DANGLING BONDS

A free radical is a chemical structure that contains an unpaired electron. A chemical bond is made from a pair of electrons. The homolytic cession of a chemical bond yields two unpaired electrons (free radicals), and a free radical could be viewed as a half bond, which seems to be the logic of the term *dangling bond*, i.e., dangling half bond. A mobile free radical is usually very reactive, reacting with other chemicals and forming a new chemical bond. For example, a free radical can be added on to a double bond, in which the free radical breaks the π bond of a double bond, creating a σ bond with the moiety that had the free radical and leaving another free radical originated from the broken π bond. This is the principle of free radical chain growth polymerization of vinyl monomers.

The term *free radical* is often used in the context of a reactive intermediate, as in the case of polymerization of vinyl monomers, but the same structure (unpaired electron) can and does exist in a kind of immobilized environment. For example, a bulk-polymerized (monomer and initiator only in the polymerization system) poly(methyl methacrylate) (PMMA) contains an appreciable number of free radicals that can be detected by electron spin resonance (ESR) [1]. When the polymerization system becomes highly viscous toward the end of the bulk polymerization, "gel formation" occurs and immobilizes the growing end of free radical chain growth polymerization, preventing recombination of two free radical ends of growing chains.

"Immobilized" or "trapped" free radicals are often formed in a solid by ionizing radiation such as ultraviolet (UV) light, electron beam, and γ ray. The term *dangling bond* is often used to describe the free radicals created in a solid [2]. Those immobilized free radicals have the same chemical reactivity as the corresponding mobile free radicals, but the net reactivity is severely reduced due to space restriction or lack of mobility. In some cases, free radicals are well protected and remain in the solid nearly indefinitely. The terms *dangling bond* and *trapped free radical* could be considered synonymous in the context of chemistry. However, in this book, dangling bond is used to describe the free radicals trapped in the tight three-dimensional network of plasma polymers (LCVD product) to distinguish them from the free radicals trapped in polymers that are exposed to glow discharge or polymers, which are used as the substrate for plasma polymer deposition. The latter could be termed *polymer free radical*. According to this notation, the free radicals trapped in PMMA are polymer free radicals. ESR signals for these two types of free radicals are significantly different, as seen in the following figures.

ESR signals of dangling bonds are broad single-line signals without hyperstructure [3] as depicted in Figure 6.1, whereas the polymer free radicals have hyperstructures that are characteristic of the polymer structure as depicted in Figure 6.2 for synthetic polymers and Figure 6.3 for natural polymers, which were

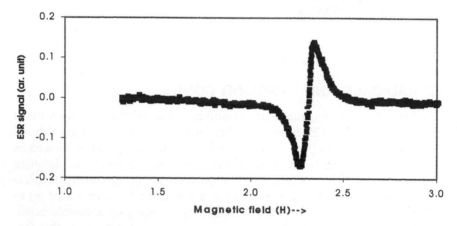

Figure 6.1   ESR signal of dangling bond of LCVD coating of trimethylsilane (TMS).

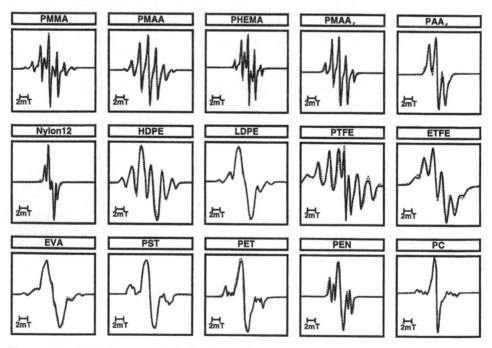

Figure 6.2   ESR spectra of the luminous gas phase induced free radicals in synthetic polymers, courtesy of Dr. M. Kuzuya.

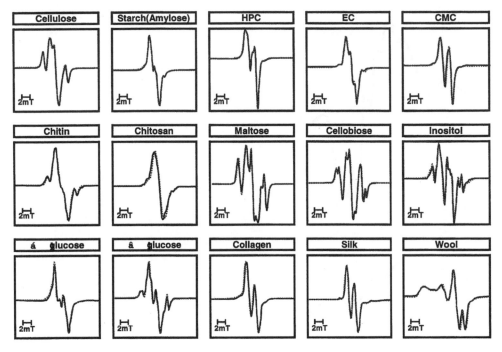

Figure 6.3 ESR spectra of the luminous gas induced radicals in natural polymers, courtesy of Dr. M. Kuzuya.

exposed to argon (Ar) glow discharge [4,5]. Irradiating polymers by Ar glow discharge in sealed tubes created the free radicals shown in Figures 6.2 and 6.3.

The hyperstructure of ESR signal indicates the degree of interaction of free radical spin with hydrogen atoms in the neighboring medium. The broad single line without hyperstructure indicates that the free radical is surrounded by a tight network of main atoms, e.g., C or Si, without H nearby.

In luminous chemical vapor deposition (LCVD), irradiation with photons at various energy levels and the deposition of materials that contain free radicals occur concurrently. The balance of these processes is closely related to the chemical structure of the monomer and its characteristic behavior in the luminous gas phase. The investigation of this balance by ESR provides valuable information pertinent to the growth and the deposition mechanisms in LCVD. The details of this aspect are described in Chapter 7, which deals the chemical structure of the monomer (starting material for LCVD).

By means of ESR spectroscopy, Morosoff et al. studied the free radicals in plasma polymers deposited on a glass rod [6]. The free spin signals observed with a plasma polymer–coated glass rod consist of the free spin signal of the glass and that of the plasma polymer. A typical ESR signal observed with a glass rod exposed to N_2 plasma (nonpolymer-forming plasma) together with the background signal observed with an untreated glass rod is shown in Figure 6.4.

ESR signals [7] observed with LCVD-coated glass rods are shown in Figures 6.5, and 6.6. By removing plasma polymer coating from the surface of a glass rod, it is possible to quantitatively examine both the free spins in the plasma

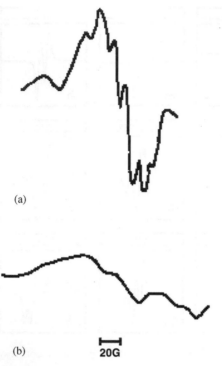

(a)

(b)   20G

Figure 6.4 (a) ESR spectrum of radicals formed by exposure of a glass tube to N_2 plasma at an initial N_2 pressure of $12 \mu m$ Hg, 30 W power, for 5 min with subsequent exposure of the tube to air. (b) "Background" from untreated glass tube. Relative ordinate scale: 1 : 1.

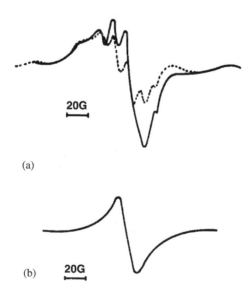

20G

(a)

(b)   20G

Figure 6.5 ESR signal of plasma polymer of cyclohexane. (a) Polymer signal superimposed on glass signal. (b) Polymer signal after subtracting glass signal.

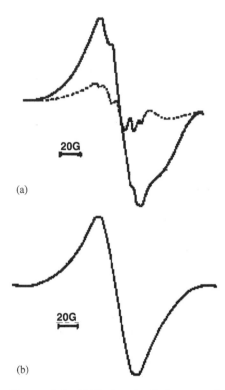

(a)

(b)

Figure 6.6 ESR signal of plasma polymer of tetrafluoroethylene (by pulsed discharge). (a) Polymer signal superimposed on glass signal. (b) Polymer signal after subtracting glass signal.

polymer and the free spins in the substrate glass (by subtracting the glass signal from the composite signal observed with the coated glass rod). The coating signal is the same as the dangling bond signal shown in Figure 6.1. ESR signals of the substrate glass rod are due to the irradiation effect of the luminous gas phase of the monomer. The coating shown in Figure 6.6 was prepared with a pulsed radio frequency glow discharge, which reduces the intensity of glow significantly. The contribution of glass signal in this figure is significantly lower than that shown in Figure 6.5.

2. FREE RADICALS IN PLASMA POLYMER AND FREE RADICALS IN SUBSTRATE

The ESR signals observed with a coating could in principle be due to the deposition of unreacted free radicals and also to the free radicals formed by the irradiation effect of the luminous gas phase on the as-deposited coating. One can examine this situation by changing the type of monomers to be used.

If the ESR signal of the dangling bonds was caused by the irradiation of as-deposited materials during the process of LCVD, the polymer signal should be proportional to the glass signal because the polymer signal should come from the irradiation effect of the luminous gas phase from which the coating deposits. If the dangling bond is due mainly to the unreacted free radicals in the depositing species, on the other hand, one cannot expect direct proportionality between the dangling bond signal and the glass signal.

The experimental results showed not only that there is no direct proportionality between the dangling bond signal and the glass signal, but also that there is an inverse proportionality between them, i.e., the higher the dangling bond signal, the lower is the glass signal. This means that the dangling bond signal is due mainly to the unreacted free radicals in the depositing species.

The data of Morosoff et al. [6] obtained with 4-picoline, ethylene, and acetylene, used as pure monomer and also in combination with gases that are nonpolymerizable by themselves but can copolymerize with other monomers, are shown in Table 6.1. The change in the steady-state system pressure due to glow discharge is expressed in the table by $\delta = p_g/p_m$, where p_g is the steady-state system pressure in a glow discharge and p_m is that observed before a glow discharge is initiated. The value of $\delta > 1$ means that the total number of gas molecules increases under the environment of luminous gas phase, which in most cases increases the intensity of glow.

It is important to note that in no system studied (Table 6.1) did the glass signal increase with time. The free spins in the glass are evidently generated in the first few minutes (i.e., in a time period smaller than 20 min), and the coating deposited on it prevents the further formation of free spins in the substrate.

The concept that the free spins in the glass are created at an early stage and the material deposited on the surface protects the glass from further UV irradiation preventing the further formation of glass free spins is also supported by the dependence of the glass signal on the rate of polymer deposition. The decrease in the total pressure (16 mtorr, 4-picoline; 16 mtorr, N_2) leads to a slower rate of polymer deposition and an increase in the glass free spins; the increase in the total pressure (49 mtorr, 4-picoline; 49 mtorr, N_2) shows the opposite effect (Table 6.1).

Assuming that photon irradiation is the sole cause of the formation of free spins in the glass substrate, the intensity of a monochromatic light penetrating through the plasma polymer deposit can be expressed by Beer's law. The number of free spins in the substrate (glass rod) can be correlated with the deposition rate of polymer using the following assumptions:

1. The intensity of the light after passing through a polymer film of thickness L can be given by

$$I = I_0 A e^{-aL} \tag{6.1}$$

    where I_0 is the intensity of incident light, and A and a are proportionality constants.
2. The number of free spins, S, is proportional to the exposure time,

$$dS = kI\,dt \tag{6.2}$$

    where k is the quantum yield of the spin formation.

Table 6.1 Unpaired Spins Detected in Deposited Film and Substrate After Glow Discharge Treatments for 1 Hour or Less at 30 W Power

Components and p_m (mtorr) of components	Total p_m (mtorr)	Duration of glow discharge treatment (min)	Yield of polymer film (mg/cm²)	Spin concentration in polymer film [(spins/g) $\times 10^{-19}$]	Linewidths of ESR first-derivative signal (G)	Spins near surface of glass substrate [(spins/cm²) $\times 10^{-15}$]	$\delta = p_g/p_m$	p_g (mtorr)	Code on Fig. 6.10
4-Picoline (30), N_2 (30)	60	20	0.05	0.23	16	—	—	—	—
		40	0.11	0.20	17	—	—	—	B
		60	0.18	0.18	16	0.4	0.15	9	—
N_2 (30), 4-picoline (20)	50	20	0.03	0.20	17	0.5	0.14	7	—
		40	0.06	0.14	15	0.5	0.11	6	A
		60	0.09	0.11	15	0.5	0.11	6	C
4-Picoline (16), N_2 (16)	32	60	0.03	0.17	17	0.75	0.12	3.8	—
4-Picoline (50),	50	60	0.10	0.20	16	0.4	0.16	8	D
4-Picoline (49), N_2 (49)	98	20	0.16	0.21	15	0.1	0.61	60	E
4-Picoline (25), N_2 (25), H_2O (25) (10)	60	20	0.05	0.19	14	0.7	0.33	20	—
		40	0.06	0.21	16	0.7	0.33	20	F
		60	0.14	0.24	19	0.8	0.33	20	—
N_2 (30), ethylene oxide (30)	60	20	0.008	—	—	2.5	1.0	60	—
		40	0.016	—	—	2.5	1.0	60	—
		60	0.05	0.41	17	2.8	1.0	60	—
N_2 (20), acetylene (30) H_2O (10)	60	20	0.03	—	—	3.1	0.32	19	—
		40	0.06	—	—	3.0	0.32	19	—
		60	0.07	0.18	16	3.3	0.32	19	—

3. The thickness of polymer deposition is proportional to the deposition time (i.e., $L = rt$, where r is the thickness growth rate constant).

Then, the total number of free spins, S, after the plasma polymerization time t has elapsed can be given by

$$S = (kI_0A/ar)(I - e^{-art}) \tag{6.3}$$

This relationship indicates that (1) S is proportional to the intensity I_0, (2) S is inversely proportional to the rate of thickness growth r, and (3) S approaches a constant value as the deposition time increases. Namely, when art is very large, Eq. (6.3) reduces to

$$S_\infty = kI_0A/ar \tag{6.4}$$

indicating that after a certain thickness of coating is built up, no radiation reaches the substrate.

In the case of a polychromatic radiation source with a constant spectral composition, it would be expected that the quantity S_i would be expressed by a relationship somewhat more complex than Eq. (6.5), but it could be shown that S_∞ can be expressed as

$$S_\infty = (I_0/r)C \tag{6.5}$$

where I_0 is the total intensity of the light emitted by the polymer-forming luminous gas phase and C is a constant. To relate this expression to the measured quantities, we may use the empirical relationship that I_0 increases with the system pressure p_g in a glow discharge in the range of low values of p_g. Assuming that I_0 is directly proportional to p_g, we may write

$$S_\infty = C'p_g \tag{6.6}$$

The quantity S_∞ is that given for glass signals in Table 6.1. None of the glass signals changes as a function of the posttreatment time (no decay with time). A plot of the number of free spins induced in the glass rod against the quantity p_g/r is shown in Figure 6.7 for all 4-picoline systems described in Table 6.1.

Point E in Figure 6.8 represents the case in which p_g is so high that we may expect that I_0 is no longer proportional to p_g due to the insufficient energy input in a fixed-wattage experiment scheme. Thus, Eq. (6.6) describes reasonably well the generation of free spins in the glass substrate used in LCVD.

The relationship shown above is exactly what we would expect from the growth and the deposition mechanism described in Chapter 5. The quantity of dangling bonds is determined largely by the chemical structure of the monomer [2,7]. The larger the contribution of the cycle II, which is based on bifunctional free radicals (see Fig. 5.3), the larger is the quantity of dangling bonds in the product solid. On the other hand, the luminous gas phase, which yields more dangling bonds, emits fewer photons according to the creation of chemically reactive species described in Chapter 4, and hence causes less glass signal.

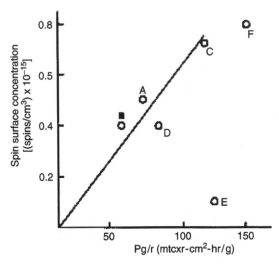

Figure 6.7 Spin surface concentration of glass spins obtained after plasma treatment with all 4-picoline systems given in Table 6.1 plotted against p_g/r; p_g is the total pressure during plasma treatment and r the rate of film deposition.

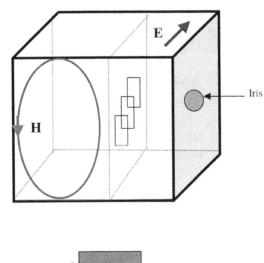

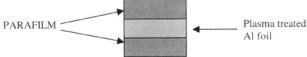

Figure 6.8 Orientation of the Al foils sample in the TE 102 microwave cavity of the ESR spectrometer and the composition of the sample as a sandwich of the Al foils and the PARAFILM layers.

When a polymer substrate is employed the situation becomes much more complex because the formation of free radicals in the polymer matrix by photon irradiation depends on the chemical structure of the polymer and the energy level of irradiation. The adhesion of LCVD coating is generally stronger to a polymer substrate than to a glass rod, which makes it more difficult, often impossible, to separate the coating from the substrate. ESR signal of an LCVD-coated polymer substrate is a complex signal of dangling bonds and *polymer free radicals* in the substrate polymer, but these two kinds of signals cannot be easily separated or identified. The complexity could be visualized by overlapping the ESR signal shown in Figure 6.1 with any ESR signal shown in Figures 6.2 and 6.3.

3. DANGLING BONDS IN PLASMA POLYMER

When a gas that has more than one backbone-forming element such as C and Si, e.g., H Si(CH$_3$)$_3$, is employed in LCVD, the dangling bonds could be, in principle, on either or both of these elements. In order to distinguish ESR signals for these two kinds of dangling bonds, i.e., Si based and C based dangling bonds, it is necessary to use the substrate material, which does not form free radical when it is exposed to the luminous gas phase. Glass, quartz, and polymers cannot be used for this purpose.

The ESR signal shown in Figure 6.1 was obtained by depositing cathodic polymerization of trimethylsilane (TMS) on an aluminum foil used as the cathode and following a specially developed method of sample preparation [3]. The principle of this method is schematically depicted in Figure 6.8. Aluminum does not form free radical on exposure to the luminous gas phase; consequently, the ESR signal of the coating only can be obtained. The aluminum substrate technique enabled us to distinguish C-dangling bonds and Si-dangling bonds.

3.1. Comparison Between TMS and CH$_4$ Plasma Deposition Signals

Since the TMS monomer gas contains both carbon and silicon, the TMS signal could originate from either carbon or silicon radicals. A comparison of the TMS signals with carbon-based signals obtained by methane plasma deposition disclosed some interesting features.

First, under the same condition, the DC cathodic plasma polymerization of TMS produced a broad single line ($\Delta H_{pp} \sim 15\,\text{G}$) but CH$_4$ deposition gave a narrow line ($\Delta H_{pp} \sim 6\,\text{G}$) (Fig. 6.9). The TMS signal with its unique shape and linewidth is clearly very different from the CH$_4$ signal (Fig. 6.10). Analysis of signal decay showed that TMS signal decayed by 40% in 420 min whereas methane signal showed only 10% decay in 4320 min. The stability of the methane signal is a key feature of a graphite-type radical. The features of CH$_4$ signal are also similar to those radicals in amorphous carbon (a microcrystalline form of graphite) films obtained by the microwave plasma chemical vapor deposition of CH$_4$/H$_2$ mixture [8].

Second, in contrast to the TMS systems where identical signals can be obtained using the DC or the 40-kHz discharge, the deposition of CH$_4$ by these two methods produced two different signals (Fig. 6.11). While the DC cathodic produced a narrow signal ($\Delta H_{pp} \sim 6\,\text{G}$) (Fig. 6.11b) with very little decay (10% decay in 4320 min), the 40-kHz discharge produced a broad signal ($\Delta H_{pp} \sim 12\,\text{G}$) (Fig. 6.11a) with very fast decay (about 80% in 20 min). This fast-decaying signal has the spectral feature that is similar to those radicals in hydrocarbon films obtained by

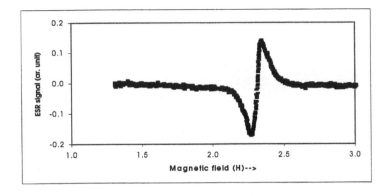

(a)

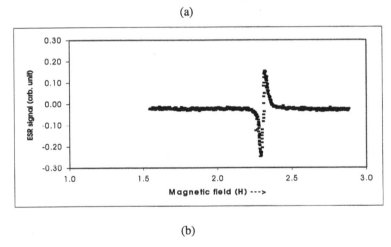

(b)

Figure 6.9 Comparison between the TMS and the CH_4 signals obtained by the DC cathodic plasma polymerization method (5 W, 50 mtorr, 3 min); (a) TMS plasma (b) CH_4 plasma.

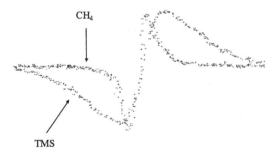

Figure 6.10 A picture of the overlap of the ESR signals from TMS and CH_4 plasma polymerization; DC cathodic plasma polymerization (5 W, 50 mtorr, 3 min).

the radio frequency plasma discharge of unsaturated hydrocarbons [9]. These results demonstrate that the TMS signal is very different from carbon-based signals. Table 6.2 summarizes the decay rates and refractive indices of film prepared by these methods.

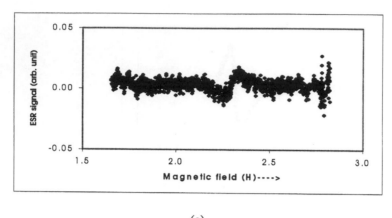

(a)

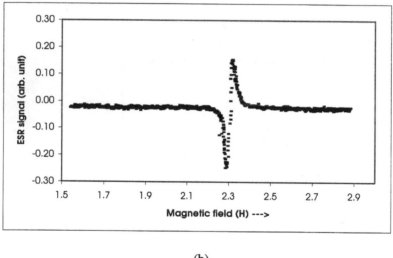

(b)

Figure 6.11 Effect of plasma deposition mode on the ESR signal of CH_4 obtained by (a) the AF discharge method (115 W, 100 mtorr, 90 min) (b) the DC cathodic method (5 W, 50 mtorr, 3 min). Field sweep settings as in (2a).

Table 6.2 ESR Signal Decay and Refractive Index Measurements as a Function of Plasma Type and Deposition Method

Plasma	Method	Type of radical	ESR signal decay	Refractive index
TMS	DC cathodic[a]	Silicon	40% in 420 min	2.04
TMS	40-kHz discharge[b]	Silicon	40% in 150 min	1.55
CH_4	DC cathodic[a]	Graphitic carbon	10% in 4320 min	2.40
CH_4	40-kHz discharge[b]	Hydrocarbon	80% in 20 min	1.67

[a]5 W, 50 mtorr, 1 min.
[b]115 W, 0.5 sccm, 100 mtorr, 15 min.

3.2. Effect of the Substrate Material on the TMS Signal

When the substrate polyethylene (PE) is placed on the cathode, unlike Al, it will not act as a part of the cathode, and the film produced is identical to that prepared by the 40-kHz discharge where the substrate is floating in the luminous gas. For comparison, both the substrates Al foils and PE fibers were placed in the reactor and plasma coated at the same time. Results showed that signals from TMS LCVD on PE are very different from those of TMS LCVD on Al as shown in Figure 6.12. Unlike the broad ESR line observed when Al was used as the substrate, hyperfine structures were observed with use of the substrate PE.

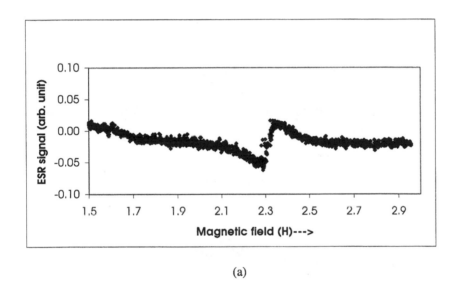

(a)

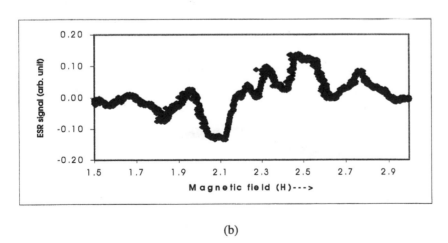

(b)

Figure 6.12 Variation of the ESR signals as a function of substrates; TMS plasma polymerization, AF discharge (115 W, 100 mtorr, 15 min) deposited on (a) Al foils (b) PE fibers.

While the ESR signal decay showed only a decrease in intensity when Al was used as the substrate, the complex spectrum observed when PE was used as the substrate continued to change its pattern. A decrease in the spectral intensity and sharpening of the central ESR line were observed (Fig. 6.13). In addition, long-term air exposure (7 days) of the coated fibers resulted in a major change, where only the central line was observed. Similar signals were also observed due to CH_4 plasma deposition, Ar plasma treatment, or γ irradiation (with an approximate dose of 2.5 Mrad) of the PE fibers. These complex spectra together with the change of pattern are very similar to the reported carbon-based signals observed from PE powder treated with Ar plasma [4]. Radicals generated by the Ar plasma treatment are hydrocarbon-based signals derived from the substrate PE. Kuzuya and

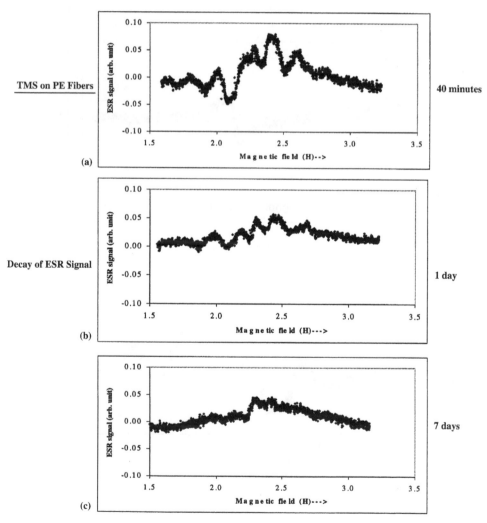

Figure 6.13 Effect of air exposure on the ESR signals obtained by TMS plasma coated PE fibers using the AF glow discharge (115 W, 100 mtorr, 15 min); (a) 40 min after air exposure (b) one day after air exposure (c) 7 days after air exposure.

colleagues assigned these PE radicals as the overlap of a sextet spectrum due to midchain alkyl radicals (-CH$_2$·CHCH$_2$-), a septet spectrum due to midchain allylic radicals (-CH$_2$·CHCH=CHCH$_2$-), and a broad line due to immobilized dangling-bond sites.

TMS deposition on PE showed only substrate signals with no detectable TMS signal (Fig. 6.12b). The absence of the TMS signal in this system could be due to the fast reaction of TMS radicals with the surface radicals generated from PE. The more likely explanation is that the number of free radicals in the plasma polymer layer is too small in comparison with the free radicals created in the bulk of the substrate, PE. What we see in Figure 6.13 is the decay of PE *polymer free radicals*, which were created by the luminous gas of TMS. With substantial decay of the PE free radicals, TMS dangling bonds, which decay much slower, became discernible.

3.3. Identification of the TMS Signal

The TMS dangling bond signal represents a large number of free radicals with concentration in the order of $c \sim 1.3 \times 10^{14}$ spins/cm^2 in a typical 50-nm layer. The signal provides only a few diagnostic clues for a microscopic interpretation. A single line, centered near the free electron g value ($g \approx 2.003$), with no suggestion of underlying fine or hyperfine structure (Fig. 6.1) was observed. The most unusual feature is the large linewidth ($\Delta H_{pp} \sim 15$ G). It may be reasonable to assume that the TMS signal is a composite of more than one signal, but there was no observable change in the line shape as the signal decayed with time.

The spin orbit broadening is given by

$$\Delta H/H \sim \Delta g/g \sim \lambda_{(so)}/\delta E$$

For carbon, the spin orbit constant $\lambda_{(so)}$ is very small and the contribution to line broadening is normally negligible. But the significantly larger $\lambda_{(so)}$ leads to substantial g shift in silicon. The interactions between the spin system and its environment are also directly related to $\lambda_{(so)}$. Both intrinsic g shifts and lifetime broadening effects are less than 15 G for carbon, but 15 G is a typical broadening for silicon at the X band. Broadening is, therefore, more apparent in silicon than in carbon radicals [10].

However, one cause of significant line broadening for carbon radicals is the anisotropic hyperfine in immobilized molecules, such as radicals produced in frozen hydrocarbon glasses [11] or radicals trapped in polymer matrix [12]. These large hyperfine effects also show resolved structure. None of these objections apply to the interpretation of our ESR signal as a dangling silicon bond. The absence of hyperfine structure is a strong argument in favor of the model. Also on the experimental side, the comparisons of the TMS signals with the dangling bond signal of methane plasma polymer, which is due to carbon-based radicals with graphitic nature, clearly show that they are very different (Fig. 6.10).

Furthermore, the broad TMS signal is similar to the reported silicon-dangling bond centers observed from silane plasma deposition [13,14]. In addition, a well-studied class of paramagnetic silicon defects, the P_b centers [15,16], has precisely the g anisotropy ($\delta g \approx 0.006$) required to account for the width of the TMS signal. The overall effect of including all these P_b defects together would be to

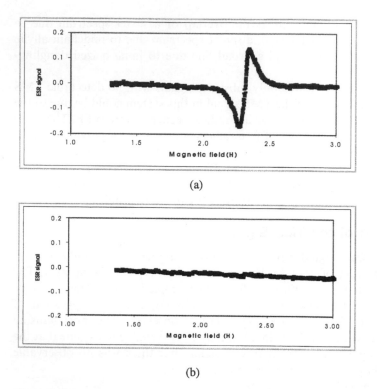

(a)

(b)

Figure 6.14 Effect of oxygen as a mixture to the TMS plasma; (a) ESR signal from TMS plasma coating of Al foils, and (b) ESR signal from TMS + O2 (1 scm + 4 scm) plasma coating of Al foils.

produce a nearly isotropic line about 15 G wide. Since the P_b centers are known to be associated with the Si-SiO$_2$ interface, this is in agreement with the system of Si on AlO$_X$. Therefore, the uniqueness of the TMS signal with its shape, linewidth, and absence of hyperfine structure is consistent with a highly localized silicon-based dangling bond.

When oxygen is mixed with TMS, the deposition of LCVD changes from amorphous silicon-type networks to SiO$_x$, ceramic-type materials that showed no dangling bond as depicted in Figure 6.14. The dangling bond can be virtually eliminated from the deposition by converting the nature of deposition to ceramic-type materials; however, the intensity of glow increases significantly as shown in Figure 3.13, and its influence on the substrate, particularly of polymers, should not be overlooked in practical applications.

4. DECAY OF THE TMS SIGNAL

4.1. Decay Kinetics

When a plasma polymer is exposed to ambient air, oxygen reacts with the free radicals residing at the surface or free radicals that have access to oxygen. Oxygen is an effective scavenger of free radicals, converting a free radical to a peroxy radical and then further to a carbonyl or carboxyl moieties. Because of this mechanism, nearly all plasma polymers of hydrocarbons, which do not contain oxygen atoms, show oxygen atoms at the surface. XPS analysis reveals the presence of approximately 20 at.% of oxygen at the surface of nearly all plasma polymers derived from various hydrocarbons.

If a plasma polymer structure is not very tight, eventually all trapped free radicals will be quenched by reacting with oxygen or by recombining them, although this might take several minutes to several days. If the structure is very tight, oxygen molecules cannot reach the dangling bond, and consequently the dangling bonds remain in the film for a long time. The decay of free radicals in a plasma polymer due to the first exposure to air is generally very fast but does not involve the change of free radical signal; the intensity only decreases, indicating that the free spins disappear during this process. The reaction of oxygen with surface free radicals is mainly responsible for the fast decay as mentioned above. This stage of decay slows down within a few minutes to an hour, in general, and is followed by the second stage of slow decay.

ESR signals of two identical TMS plasma–coated samples are compared in Figure 6.15, of which one sample was stored in air for 3 days, and the other sample

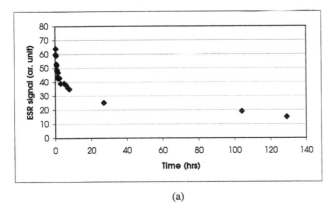

(a)

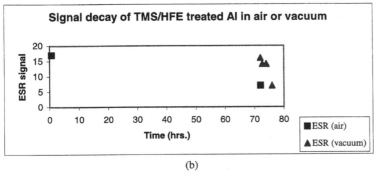

(b)

Figure 6.15   Effect of air exposure on the ESR signal obtained by the DC cathodic plasma deposition (5 W, 50 mtorr, 3 min) of TMS (a) ESR signal decay with storage time in air (b) ESR signal intensity from two identical samples; one sample was held in vacuum for 72 h showing no decay in vacuum and rapid decay when exposed to air, the other identical sample was exposed to air for 72 h showing decay with time.

was held in a vacuum for the same length of time. A comparison of the ESR signals from the two samples showed that after 72 h the signal intensity from the vacuum sample was essentially the same as that of the initial signal obtained for the sample exposed to air (72 h ago). In addition, as soon as the protected sample was exposed to air its ESR signal fell very sharply, and after only a few hours the two samples showed nearly identical free radical concentrations. These results demonstrate that the ESR signal of dangling bonds does not decay in vacuum (or the decay rate is orders of magnitude slower), and that the decay represents a reaction of the dangling bonds with ambient oxygen.

The overall decay curve of ESR signal with time after a sample is exposed to ambient air encompasses two major types of decay. The initial decay immediately after the sample is exposed to ambient air is usually very fast, starting with the first-order decay and deviating from the first decay as the decay rate slows down. Thus, the entire decay curve cannot be expressed by one decay-kinetic mechanism.

The slower decay follows second-order kinetics if the permanent residual concentration is subtracted from the signal intensity as shown in Figure 6.16. Insofar as the decay kinetics is concerned, the decrease of the ESR signal at this stage is due to the recombination of dangling bonds. However, it cannot be a straightforward recombination of dangling bonds because no decay occurs in a vacuum or in an inert gas environment.

Without speculating about the details of the mechanism, this decay rate in the second (slow) stage can be used as a measure of the mobility of segments in the network structure on which the dangling bonds reside. In other words, the slower decay may indicate that the network has a tighter structure than that showing a faster decay rate. The rate of ESR signal decay in the slower stage seems to have significant implications in practical application of LCVD materials.

Decay kinetics from TMS/HFE treated Al. foil

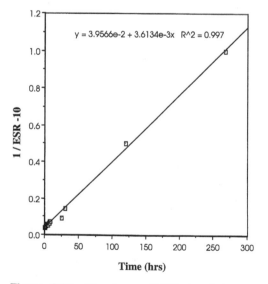

Figure 6.16   The decay of ESR signal shown as the second order kinetics.

4.2. Decay Rate of Cathodic Plasma Polymer and Glow Discharge Polymer

The DC cathodic discharge deposition of TMS (5 W, 50 mtorr, 1 min) gave an identical ESR signal with the same intensity as one produced by the 40-kHz discharge (115 W, 100 mtorr, 15 min). However, these two signals showed a great difference in their decay rates. TMS deposition by the DC cathodic polymerization showed the ESR signal decay by 40% in 420 min. In comparison, when the deposition was carried out by the 40-kHz glow discharge, 40% decay occurred in only 150 min. In addition, differences were observed in the refractive index of these films.

The DC cathodic polymerization yields much higher refractive indices than the 40-kHz glow discharge (Table 6.2). Films deposited by the DC cathodic polymerization have a tighter network than to the deposition by the 40-kHz discharge [17]. Since signal decay in the second stage, which is the second-order decay, represents recombination of dangling bonds, tighter network will restrict the motion of the matrix segments and the signal will show a slower decay. These results demonstrate the correlation between the decay rate of the ESR signals and the film characteristics.

4.3. Sequential LCVD Processes

Describing a sequential LCVD process, the first process is followed by the second process separated by / sign. For example, LCVD of CH_4 is followed by glow discharge treatment of O_2, and the sequential process is expressed by CH_4/O_2. If O_2 is mixed with CH_4, it is expressed by LCVD of $(CH_4 + O_2)$ by placing the gas mixture in parentheses.

4.3.1 TMS/HFE

Postdeposition plasma modifications to the plasma polymer of TMS have been seen to greatly improve bonding to various primers and paints [18–20]. One particular system has been observed to have tremendous adhesion between plasma-coated Al alloy panels and paint applied to them. This system involves cathodic DC plasma deposition of a roughly 50-nm primary plasma polymer film from TMS onto a properly pretreated alloy substrate, followed by the deposition of an extremely thin fluorocarbon film by DC cathodic deposition of hexafluoroethane (HFE). It was the "superadhesion" aspect of this particular system that triggered the series of ESR studies [3,21].

It is important to note that HFE is not a monomer of general plasma polymerization because it does not polymerize in the absence of hydrogen in a reactor system [22]. When it is applied to a TMS plasma polymer surface, however, hydrogen is abstracted from the surface and forms a very thin layer of plasma polymer. It is essentially a self-terminating deposition process leading to an extremely thin layer of HFE plasma polymer. XPS analysis indicated that the thickness of the F-containing layer in the TMS/HFE system is less than a few nanometers [23].

It is also important to recognize that HFE plasma is a good etching agent for silicon, and the substrate (plasma polymerized TMS) contains silicon. Consequently, the deposition of HFE plasma polymers on the surface might partially etch the loose

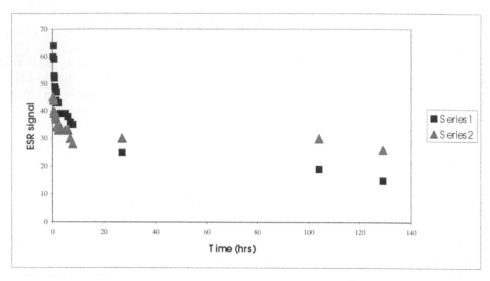

Figure 6.17 Difference in the ESR signal and its decay rate of TMS (Series 1) and TMS/ HFE (Series 2) plasma polymer coatings on Al foil.

TMS plasma polymer layer at the top of deposition. The plasma polymerization may increase the free radical concentration, whereas the latter may decrease it. ESR signals from both TMS and TMS/HFE are shown in Figure 6.17 as a function of time (decay). Figure 6.17 shows the following two important effects of HFE plasma treatment of plasma polymer of TMS: (1) a significant decrease in the ESR signal and (2) a significant decrease in the decay rate in a longer time scale range.

There is a consistent reduction in the free radical signal from the TMS layer when an HFE layer is deposited on top. It should be noted that Si free spins could exist in the network structure of the plasma polymer of TMS and also on the oligomeric moieties, which are loosely embedded in or adherent to the network structure. The magnitude of the reduction is about 20% of the total, or about 2.6×10^{13} spins/cm^2. This number may represent nearly all of the silicon-dangling bonds in the upper 20% of the TMS plasma polymer film. If this 10-nm-thick transition region (20% of TMS layer) contained 2.6×10^{13} cross-links/cm^2, as these numbers suggest, it would likely indicate the removal of loosely kept oligomeric structures deposited at the conclusion of the LCVD deposition of TMS. This restructuring would yield exceptionally strong surface structure for the HFE-derived film by converting a weak boundary (the oligomeric deposits) to a stronger surface structure for the bonding of primer coatings. Namely, the "superadhesion," at least in part, is due to the strong cohesive strength of the composite plasma coating; conversely, the superadhesion cannot be achieved without the strong cohesive strength of the plasma polymer layer.

Plasma deposition on a hydrogen-containing surface from HFE is a self-terminating plasma polymerization process. When the initial plasma polymer surface, which contains hydrogen atoms, is sufficiently covered by the plasma polymer of HFE, the deposition ceases because the supply of hydrogen diminishes.

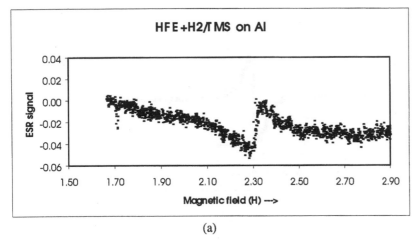

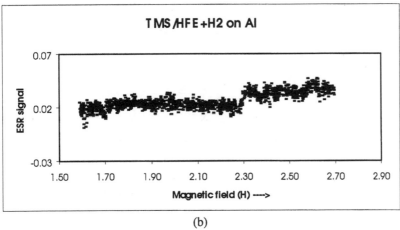

Figure 6.18 Comparison of the ESR signals from (a) (HFE + H$_2$)/TMS and (b) TMS/
(HFE + H$_2$).

Consequently, it is extremely difficult to detect ESR signals from the HFE plasma
polymer.

4.3.2 TMS/(HFE + H$_2$)

LCVD of (HFE + H$_2$) provides a mean to investigate the etching effects described
above because this system is not self-terminating and enables us to deposit a thicker
layer. Figure 6.18 compares ESR signals in plasma polymer of (HFE + H$_2$)/TMS
(Fig. 6.18a) and of TMS/(HFE + H$_2$), (Fig. 6.18b). The ESR signal of the TMS film
almost completely disappeared due to the second deposition of plasma polymer of
(HFE + H$_2$) as shown in Figure 6.18b. There is still no ESR signal attributable to
(HFE + H$_2$) plasma polymer.

4.3.3 TMS/CH$_4$

The ESR signal from TMS/CH$_4$ is compared with the TMS signal in Figure 6.19.
Analysis of the TMS/CH$_4$ showed that it is a composite signal of TMS and CH$_4$ film

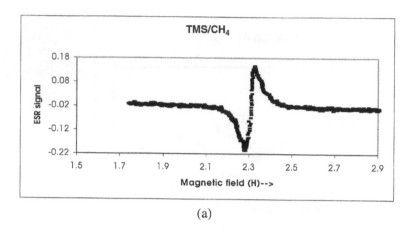

(a)

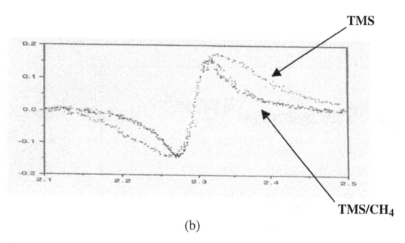

(b)

Figure 6.19 (a) ESR signals from TMS/CH$_4$ (b) picture of the overlap of TMS and TMS/CH$_4$ signals.

signals, which is different from the TMS signal and also from the CH$_4$ signal. This composite signal showed that the TMS portion of the signal was larger than the signal from TMS alone. The second plasma polymerization, of CH$_4$, increases the ESR signal intensity dramatically. The large increase of the ESR signal intensity from that portion due to the TMS component, as well as its small percentage of decay over 24 h (see TMS/CH$_4$ in Fig. 6.20a), suggests that the second layer of CH$_4$ plasma polymer acts as a barrier to oxygen diffusion and prevents reaction of oxygen with dangling bonds.

4.4. Glow Discharge Treatment of TMS LCVD Film

ESR results for the TMS/HFE system, together with the adhesion characteristics of the system, strongly suggest that the etching of oligomers, or the conversion of oligomers in the plasma-polymerized TMS to more stable polymeric networks [24],

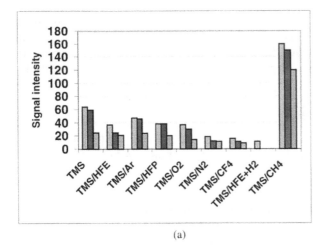

(a)

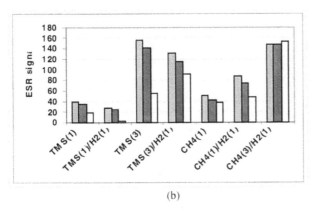

(b)

Figure 6.20 (a) Comparison of the ESR signal decay after 0.3 and 24 h from the plasma coating of 1 min TMS followed by 1 min of plasma treatment of O_2, H_2, Ar, HFE and the plasma coating of CH_4; (b) Comparison of the effect of H_2 plasma treatment on plasma polymer of TMS and of CH_4, where the number in parenthesis indicates the treatment time in minute.

seems to be an important factor that accounts for the superadhesion found with this system. It is likely that a second plasma treatment with nonpolymerizing gases might produce a similar effect observed with the TMS/HFE system. From this point of view, various plasma posttreatments of plasma-polymerized TMS were investigated. The ESR signal intensity and its decay characteristics for various combinations of sequential treatments are depicted in Figure 6.20, in which the specific effects, in relation to the ESR signal from the plain TMS film, with respect to the increase or decrease of ESR signal and of the decay characteristics shown by bar graphs are detailed. The first bar in each bar graph (in Fig. 6.20) is the first ESR signal, which was obtained roughly 20 min after the completion of LCVD due to the sample preparation, and the second and the third bars represent ESR signal obtained 0.3 h and 24 h, respectively, after the first measurement.

TMS/Ar, TMS/O$_2$, TMS/N$_2$, and TMS/H$_2$ all showed the general trends that (1) the plasma treatment of plasma polymer of TMS decreases ESR signal and (2) the decay of ESR signal is slower than that of untreated plasma polymer of TMS. The effect of H$_2$ plasma treatment of plasma polymer of TMS is compared with that of CH$_4$ in Figure 6.20b.

4.5.   LCVD of Mixed Gases

The ESR signal intensity and its decay characteristics for (TMS + H$_2$), (TMS + Ar), and (TMS + O$_2$) are compared with those for TMS in Figure 6.21.

4.5.1 (TMS + H$_2$)

A drastic change occurred when hydrogen was added to TMS in the plasma polymerization process. As a result, the initial ESR signal intensity increased but no significant decay was observed over a 24-h period.

4.5.2 (TMS + Ar)

Addition of Ar to the TMS plasma also increases the ESR signal significantly from that for TMS alone. The relative decay rate decreased but not as much as when H$_2$ was mixed with TMS.

4.5.3 (TMS + O$_2$)

Results show that when a sufficient amount of oxygen was added to the TMS plasma, the ESR signal was virtually extinguished as shown in Figure 6.14. This indicates that the plasma polymer of (TMS + O$_2$) is a completely different material from the plasma polymer of TMS. Oxygen mixed with TMS reacts with TMS during the process of LCVD yielding more inorganic character of the structure, which does not contain Si-dangling bonds. This is a completely different process from O$_2$ plasma treatment of plasma polymer of TMS, in which the O$_2$ plasma reacts with Si-dangling bonds only in the near-surface region.

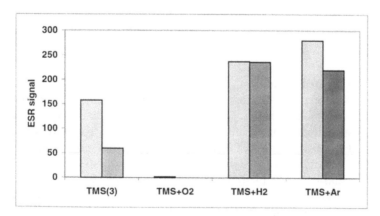

Figure 6.21 Comparison between the ESR signal intensities of TMS, (TMS + O$_2$), (TMS + H$_2$) and (TMS + Ar); LCVD time was 3 min, and signal intensities were measured 0.3 h and 24 h after air exposure.

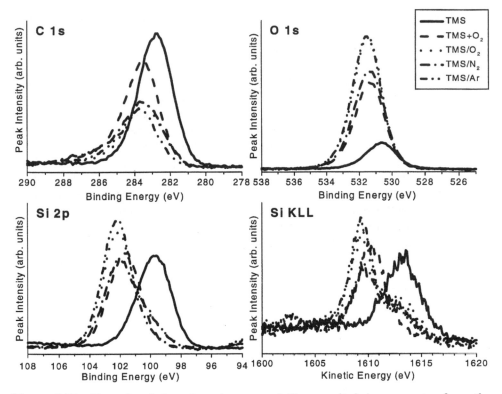

Figure 6.22 Normal emission photoelectron and X-ray excited Auger spectra from the surface of various films, including a plain TMS film, a mixed TMS + O$_2$ deposited film, and several TMS films with postdeposition plasma treatments.

XPS data support this assessment of the effects. Figure 6.22 shows photoelectron spectra from the surface of several films at normal emission (90° take-off angle). It is immediately clear that the postdeposition treatments and combined gas deposition yield different surface chemistries from the as-deposited, pure TMS film.

All of the postdeposition treatments, even treatment with the more inert gas plasmas, leave the surface of the films in a silica-like state. This is thought to be due to the creation of free radicals in the luminous gas phase, which are quickly oxidized upon exposure to the atmosphere prior to analysis.

The relation of the oxidized silicon peak and the separation from the untreated TMS film peak and those of other samples, coupled by the modified Auger parameter [25], indicate that the higher binding energy Si peak is likely due to the formation of SiO$_2$. This is particularly distinct on the O$_2$ plasma–treated film, with the other more inert plasma treatments having some suboxide components [26]. A value of 1711.6 for the modified Auger parameter of the oxygen plasma–treated film is quite consistent with those reported for various forms of silica-type bonding [27]. Then using a shift of about 1.2 eV for all of the peaks, it is seen that the high binding energy peak in the C 1s spectra is now at 284.8 eV, which is a standard position for the graphitic and hydrocarbon bonding generally used for charge neutralization.

There is some contribution at even higher binding energy, appearing around 286–287 eV in the uncorrected spectra, which is consistent with some C-O bonding due to the plasma treatment. The charging shift of the lower binding energy C 1s peak from the untreated, plain TMS film then makes it lie in a region consistent with Si-C or Si-CH$_x$ bonding. This and the associated position of the Auger peak then yield a value of 1713.0 for the modified Auger parameter, just slightly lower than that observed on pure SiC. The point of this lengthy argument is that the treatment with oxygen plasma modifies both the silicon and carbon bonding, forming Si-O and C-C structures from the Si-CH$_x$ structure of the original film.

The incorporation of oxygen with TMS in the deposition process also yields a distinct modification of the local bonding structure, beyond the surface effects observed in the postdeposition treatments. Figure 6.23 shows spectra from the same films deep beneath the modified surfaces, after inert ion sputtering, with the addition of a second spectra set from the TMS film at an advanced stage of aging. It now becomes apparent that while the treated films maintain a high degree of Si-C bonding in the bulk, the film formed with the addition of O$_2$ in the plasma does not. It has the carbon bound in C-C structures while the silicon is fairly exclusively bound to oxygen, similar to the surface of the treated films.

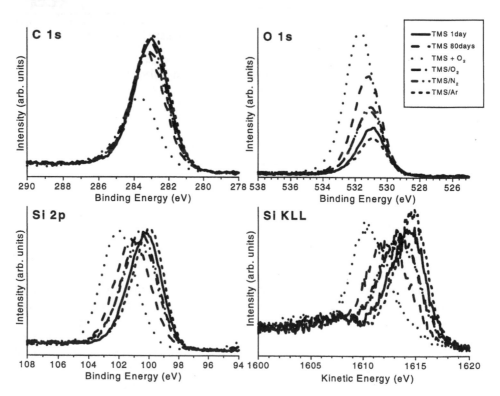

Figure 6.23 Normal emission photoelectron and X-ray excited Auger spectra taken from the bulk of various films, after the surface regions have been removed with substantial inert ion sputtering.

The level of oxidation is not as complete as on the surface of the O_2-treated film, with the peak positioned at slightly lower binding energy and some additional suboxide tailing on the low binding energy side; however, some level of oxidation is visible throughout. The level of Si oxidation in some of the plasma-treated films appears somewhat misleading due to the fact that they were not all analyzed immediately after formation, having varying lengths of exposure to atmosphere. The trend toward higher binding energy Si 2p and lower kinetic energy Si KLL peaks, however, is consistent with the longer exposures, as evidenced by the spectra from the same plain TMS film at two different ages.

Figure 6.24 complements these arguments. It shows spectra taken as a function of take-off angle, which indicates that the modification is restricted to a thin region on the surface of the film. At smaller take-off angles, a shallower volume of the material is sampled. It is clear that the underlying structure from the bulk of the TMS film is present in both of the Si spectra when sampling at the higher take-off angles, which looks deeper into the film. Although not as markedly obvious, the carbon spectra also support the points discussed earlier. The spectra taken at higher take-off angles (deeper) show that the component associated with Si-CH_x bonding resides below the C-C structure, essentially disappearing at the lower take-off angles that sample the outermost, treated region of the film. As mentioned earlier,

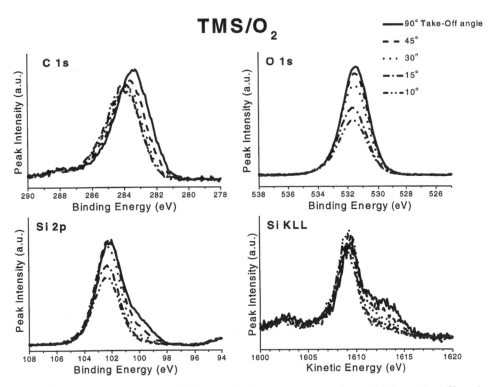

Figure 6.24 Photoelectron and X-ray excited Auger spectra taken at various take-off angles from the surface of the TMS film treated with an O_2 plasma, showing how the contribution from the underlying bulk TMS film disappears at low take-off angle, revealing the layered structure left after the plasma treatment.

some C-O bonding is associated with the small component on the high binding energy side at about 287–288 eV in this figure.

Using a simple model, a rough estimate of the modified layer thickness can be obtained. Since atomic densities are not known for these materials, an approximation based on XPS sensitivity factors gives a plausible value. Using elemental ratios calculated from the 10-degree take-off angle for the outer region and values from deep in the bulk after sputtering for the bulk values, an approximate ratio of 1.36 is found for the number of silicon atoms in the inner versus outer regions. Likewise, effective attenuation lengths (EALs) of Si 2p electrons ejected by Mg-Kα excitation for these materials are not well known, but again approximate values could be used. Using the average value (2.7 nm) from a recent review by Powell and Jablonski [28] for SiO_2 as the outer layer EAL and a rough guess of 2.5 nm for the bulk film isn't unreasonable.

The value for close-packed crystalline Si is about 2.3 [29], and above 3.0 nm for traditional polymers [30]. Strohmeir's method [31] for calculating the thickness from the normal emission spectrum then yields:

$$d = 2.7 * \ln\left[(1.36)\left(\frac{2.5}{2.7}\right)\left(\frac{I_t}{I_b}\right) + 1 \right]$$

where I_t and I_b are the intensities of the top and bulk contributions of the normal emission spectrum, respectively. The intensities are calculated from curve fitting the normal emission Si 2p peak with two gaussian-lorentzian peaks (30% lorentzian for the 80-eV pass energy), one for the surface oxide and one for the bulk. This gives a thickness d of roughly 6.3 nm, with a large error from the approximations. So it is seen that the thickness of the treated layer is roughly 10–15% of the thickness of the film, when discussing films on the order of 50 nm thick, depending on whether any sputtering/ablation of the original TMS film occurred during the postdeposition plasma treatment. Depth profiling of the treated films also indicates that a substantial decrease in carbon content, as compared to the bulk composition or the untreated TMS film, takes place in this layer.

XPS analysis supports the ESR data and yields insight into the mechanisms responsible for the changes in free radical concentrations in both the treated films and the mixture, and points to the overall mechanism responsible for the changes in the pure film with exposure to atmosphere. The ESR data showed that the incorporation of oxygen in the plasma polymerization process quenched the free radical signal. This directly correlates to the XPS data showing that in this combined deposition the silicon is now dominantly bonded with oxygen and the carbon no longer has the binding energy characteristic of Si-C type bonds. The postdeposition treatment with O_2 directly oxidizes the surface, again converting the local bonding in a similar manner as the mixture with O_2 in the deposition, but only in the outer region of the film. Likewise, treatment with the other more inert plasmas causes the surface region to oxidize upon exposure to atmosphere, likely due to the breaking of relevant bonds and the creation of free radicals during treatment. This new silicated/oxidized structure on the surface is then void of the free radicals that contribute to the pure TMS ESR signal and may have some inhibiting effect on the diffusion of oxygen into the bulk.

5. EX SITU ESTIMATION OF DANGLING BONDS AND POLYMER FREE RADICALS

Free radicals react most efficiently with other free radicals, and the chemicals that have relatively stable free radicals, such as 2,2-diphenyl-1-pricylhydrazyl (DPPH), could be used, in principle, to quantitatively estimate the amount of free radicals (DPPH method). The peroxide formed on the surface that derived from free radical could be quantitatively analyzed by the determination of iodine liberated from KI solution, which could be used to calculate the amount of the original free radical from that peroxide was derived (iodine method).

It is tempting to use such relatively simple wet chemical methods to determine the amount of free radical on plasma polymers and on polymers treated with glow discharge. However, these methods have serious limitations when applied to the dangling bonds in plasma polymers or polymer free radicals in polymers treated with glow discharge. The most serious limitation is the accessibility of the chemical to the free radicals to be analyzed. Another serious limitation is the specificity of chemical reactions.

The dangling bonds and polymer free radicals (on the surface of glow discharge treated polymer) capture molecular O_2 or H_2O, producing hydroxyl and carbonyl groups as schematically shown:

The surface of plasma polymer and the surface of polymer treated with glow discharge are exposed to ambient oxygen, the accessible free radicals are already reacted with oxygen, and free radicals that couldn't react with oxygen certainly could not be accessed by much large DPPH molecule in solution. In this sense, the analysis of peroxide by iodine method might make more sense, if the analysis could be performed at the right time duration, in which peroxide has not progressed too far in the cascading transformation depicted above. However, a more serious limitation was that the specificity of the both methods was not high enough for quantitative determination. For instance, the aluminum surface, which doesn't show any ESR signal, consumes both chemicals. Chemical cleaning of aluminum alloy surfaces, such as alkaline cleaning and deoxidizing treatment, was found to leave large amount of DPPH scavengers, which could be removed by argon or oxygen plasma treatment, implying that the reactivity of DPPH is not exclusive to free radicals. The DPPH method should be viewed as an analytical tool to detect scavengers of DPPH (rather than free radicals) existing on the surface region accessible to the DPPH solution. The iodine method should likewise be viewed as an analytical tool for iodine-liberating species existing on the surface region. Plasma treatment (O_2 or Ar) of the aluminum surface did not remove the iodine-liberating species and in some cases increased the amount.

Both methods might be useful to assess the overall reactivity of surfaces and could be used in many areas where the surface reactivity is an important factor.

REFERENCES

1. Oldfield, F.F.; Yasuda, H.K. J. Biomed. Mater. Res. **1999**, *44*, 436–445.
2. Elliot, S.R. *Physics of Amorphous Materials*; Wiley: New York, 1990; 339–380.
3. Oldfield, F.F.; Cowan, D.L.; Yasuda, H.K. Plasmas Polym. **2000**, *5* (3/4), 235–253.
4. Kuzuya, M.; Niwa, J.; Ito, H. Macromolecules **1993**, *26*, 1990.
5. Kuzuya, M.; Kondo, S.; Sugito, M. Macromolecules, **1998**, *31* (10), 3230.
6. Morosoff, N.; Crist, B.; Bumgarner, M.; Hsu, T.; Yasuda, H.K. J. Macromol. Sci. Chem. **1976**, *A10*, 451.
7. Yasuda, H.; Hsu, T.S. J. Polym. Sci., Polym. Chem. Ed. **1977**, *15*, 2411.
8. Watanabe, I.; Sugata, K. Jpn. J. Appl. Phys. **1988**, *27*, 1808.
9. Kuzuya, M.; Ishikawa, M.; Noguchi, A.; Ito, H.; Kamiya, K.; Kawaguchi, T. J. Mater. Chem. **1991**, *1* (3), 387.
10. Atkins, P.W.; Symons, M.C.R. *The Structure of Inorganic Radicals: An Application of Electron Spin Resonance to the Study of Molecular Structure*; Elsevier: New York, 1967; 9–33, 208–218.
11. Ingram, D.J.E. *Free Radicals as Studied by Electron Spin Resonance*; Academic Press: Butterworth, London. 1958; 171–181.
12. Oldfield, F.F.; Yasuda, H.K. J. Biomed. Mater. Res. **1999**, *44*, 436–445.
13. Inokuma, T.; He, L.; Kurata, Y.; Hasegawa, S. J. Electrochem. Soc. **1995**, *142* (7), 2346.
14. Hari, P.; Taylor, P.C.; Finger, F. *Amorphous Silicon Technology. Materials Research Society Symposium Proceedings*, **1996**, *420*, 491.
15. Caplan, P.J.; Poindexter, E.H. J. Appl. Phys. **1979**, *50* (9), 5847.
16. Poindexter, E.H.; Caplan, P.J.; Deal, B.E.; Razouk, R.R. J. Appl. Phys. **1981**, *52*, 879.
17. Miyama, M.; Yasuda, H.K. J. Appl. Poly. Sci. **1998**, *70*, 237.
18. Reddy, C.M.; Yu, Q.S.; Moffitt, C.E.; Wieliczka, D.M.; Johnson, R.; Deffeyes, J.E.; Yasuda, H.K. Corrosion **2000**, *56* (8), 819–831.
19. Yu, Q.S.; Reddy, C.M.; Moffitt, C.E.; Wieliczka, D.M.; Johnson, R.; Deffeyes, J.E.; Yasuda, H.K.; Corrosion **2000**, *56* (9), 887–900.
20. Moffitt, C.E.; Reddy, C.M.; Yu, Q.S.; Wieliczka, D.M.; Johnson, R.; Deffeyes, J.E.; Yasuda, H.K.; Corrosion **2000**, 56 (10), 1032–1045.
21. Oldfield, F.F.; Cowan, D.L.; Yasuda, H.K. ESR study of trimethyl silane plasma polymer. Part II: Effect of consecutive treatments and mixed gases. Plasmas Polym. **2001**, *6*, 51–69.
22. Masuoka, T.; Yasuda, H.; J. Polym Sci., Polym. Chem. Ed. **1982**, *20*, 2633.
23. Moffitt, C.E.; Reddy, C.M.; Yu, Q.S.; Wieliczka, D.M.; Yasuda, H.K. Appl. Surf. Sci. **2000**, *161* (3–4), 481–496.
24. Yasuda, Hirotsugu.; Yasuda, Takeshi. J. Polym. Sci., Part A: Polym. Chem. 2000, *38*, 943.
25. Wagner, C.D.; Six, H.A.; Jansen, W.T.; Taylor, J.A. Appl. Surf. Sci. **1981**, *9*, 203.
26. Waddington, S.D. In *Practical Surface Analysis*, 2nd Ed.; *Auger and X-ray Photoelectron Spectroscopy*, Briggs, D., Seah, M.P., Eds.; Wiley: New York, 1990; Vol. 1, 587–594.
27. Himpsel, F.J.; McFeely, F.R.; Taleb-Ibrahimi, A.; Yarmoff, J.A.; Hollinger, G. Phys. Rev. B **1988**, *38*, 6084.

28. Wagner, C.D. In *Practical Surface Analysis*, 2nd Ed.; *Auger and X-ray Photoelectron Spectroscopy*, Briggs, D., Seah, M.P., Eds.; Wiley: New York, 1990; Vol. 1, 602.
29. Powell, C.J.; Joblonski, A. J. Phys. Chem. Ref. Data **1999**, *28*, 19.
30. Briggs, D. In *Practical Surface Analysis*, 2nd Ed.; *Auger and X-ray Photoelectron Spectroscopy*, Briggs, D., Seah, M.P., Eds.; Wiley: New York, 1990; Vol. 1, 443.
31. Strohmeier, B.R. Surf. Interface Anal. **1990**, *15*, 51.

7
Chemical Structures of Organic Compounds for Luminous Chemical Vapor Deposition

1. HYDROCARBONS AND THEIR DERIVATIVES

1.1. Chemical Structure

Because of the unique growth mechanism of material formation, the monomer for plasma polymerization (luminous chemical vapor deposition, LCVD) does not require specific chemical structure. The monomer for the free radical chain growth polymerization, e.g., vinyl polymerization, requires an olefinic double bond or a triple bond. For instance, styrene is a monomer but ethylbenzene is not. In LCVD, both styrene and ethylbenzene polymerize, and their deposition rates are by and large the same. Table 7.1 shows the comparison of deposition rate of vinyl compounds and corresponding saturated vinyl compounds.

The material formation in the luminous gas phase (plasma polymerization) is less specific to the chemical structure of molecules. Benzene, which is a nonpolymerizable solvent in the free radical chain growth polymerization, polymerizes readily in the luminous gas phase. Benzene not only polymerizes, but its rate of deposition is nearly equivalent to that of acetylene, i.e., a benzene molecule is equivalent to three molecules of acetylene in the luminous gas phase.

However, there are indications that some structures are favorable for the material deposition more than others. According to the growth and the deposition mechanisms described in Chapter 5, the most important feature of organic molecules, which determines the ease of material deposition in LCVD, is the presence or absence of chemical structures that split to form diradicals by the energy transfer reaction with electrons or excited species in the luminous gas phase. The double bonds, triple bonds, and cyclic structures are the major chemical structures that create the reactive species that could proceed via cycle II of the growth and the deposition mechanisms (Figure 5.3). Without these structures, molecules depend on hydrogen abstraction or the detachment of simple stable species such as HF and HCl to create free radicals.

Oxygen in aliphatic molecular structure is readily liberated under the environment of luminous gas; the liberated oxygen acts as a radical scavenger and retards the material deposition (poisoning effect). Oxygen in a cyclic structure, on the other

Table 7.1 Comparison of Deposition Rates for Vinyl Monomers and Corresponding Saturated Compounds

Vinyl compounds	k^a	Saturated vinyl compounds	k^a
(pyridine)–CH=CH$_2$	7.59	(pyridine)–CH$_2$–CH$_3$	4.72
(phenyl)–C(CH$_3$)=CH$_2$	5.33	(phenyl)–CH(CH$_3$)–CH$_3$	4.05
(phenyl)–CH=CH$_2$	5.65	(phenyl)–CH$_2$–CH$_3$	4.52
H$_3$C–(pyridine, N=)–CH=CH$_2$	7.65	H$_3$C–(pyridine, N=)–CH$_2$–CH$_3$	7.38
(piperidinone ring)N–CH=CH$_2$	7.55	(piperidinone ring)N–CH$_2$–CH$_3$	3.76
$H_2C{=}CH{-}C{\equiv}N$	5.71	$H_3C{-}CH_2{-}C{\equiv}N$	4.49
$H_2C{=}C{\big\langle}^{Cl}_{Cl}$	5.47	$H_3C{-}CH{\big\langle}^{Cl}_{Cl}$	2.98
$H_2C{=}CH{-}CH_2{-}NH_2$	2.86	$H_3C{-}CH_2{-}CH_2{-}NH_2$	2.52

^{a}Parameter k is expressed in units of $cm^{-2} \times 10^4$; $r = kF_w$, where r is the rate of polymer deposition (g/cm^2 min) and F_w the weight-based monomer flow rate (g/min).

hand, acts as the initiation site to induce a diradical by detaching itself from a cyclic ring. This aspect could be seen in the absence of O 1s XPS signal in the LCVD deposition from perfluoro-2-butyltetrahydrofuran, which was used in the deposition rate investigation described in Chapter 5. Figure 7.1 depicts the decrease of O 1s signal as the energy input parameter, W/FM, increases and the deposition rate also increases [1]. (Details about these parameters are given in Chapter 8.) Table 7.2 summarizes the effects of chemical structures in LCVD.

Hydrogen detachment seems to be a very important parameter that is related to the mechanism for the material deposition in LCVD (plasma polymerization). As shown in Figure 4.2, the pressure of closed systems, that contain a known amount of organic molecules subjected to glow discharge levels off at a certain pressure after sufficient discharge time. Some hydrocarbon increases the pressure and some decreases the pressure. Some reaches the plateau value quickly and some takes longer. Yasuda et al. [2] examined the gas phase of a *closed system* after a known amount of a hydrocarbon was subjected to radio frequency glow discharge in a relatively small (about 0.5 liter) tubular reactor. According to the results, nearly all hydrocarbons were converted to polymers, with the yield varying from 85% to more than 99% in a relatively short time under the conditions used, and the gas phase after the polymerization (excluding unreacted organic vapor, which is 0–15% of the monomer depending on the polymer yield) consisted mainly of hydrogen.

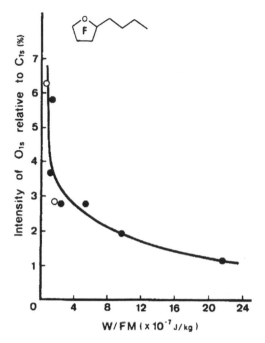

Figure 7.1 Dependence of oxygen content in the plasma polymer of perfluoro-2-butyltetrahydrofuran on W/FM. Adapted from Ref. 1.

The hydrogen production expressed as the hydrogen yield per monomer molecule (number of hydrogen molecules evolved when a monomer participates in polymer formation) increases with the increasing number of hydrogen atoms in a hydrocarbon, as depicted in Figure 7.2. In order to distinguish the role of double bond, triple bond, cyclic structure, and aromatic structure, the hydrogen yield is plotted against the structure parameter, which is given by the ratio of (number of hydrogen atoms in a molecule)/(number of structures), in Figure 7.2. For instance, in the case of cyclohexene, the total number of hydrogen, 10, is divided by two structures (i.e., one cyclic structure and one double bond).

There are clear separations of curves depending on the types of monomer structure, which indicate the contribution of chemical structure in a molecule. There is also a strikingly regular dependence of the hydrogen yield on the number of hydrogen atoms in a molecule (within a group). This smooth and regular dependence strongly indicates that every C–H bond in organic molecules has an equal probability for hydrogen detachment and participation in cycle I reaction described by Figure 5.3.

Hydrogen detachment can also be correlated with the deficiency of hydrogen in the deposition compared to the corresponding monomer. It has been reported by a number of investigators that the H/C ratio in a plasma polymer is significantly lower than the corresponding H/C ratio in the monomer. Typical data for the change in H/C ratio on plasma polymerization are shown in Table 7.3 [3]. The actual value of H/C in a plasma polymer is highly dependent on the conditions of polymerization, and therefore the absolute values shown in Table 7.2 should not be taken as values specific to any particular monomer in the general sense. It is important to note that

Table 7.2 Classification of Organic Compounds in Luminous Chemical Vapor Deposit

Type	Chemical structure	LCVD reaction characteristics	Characteristic features of LCVD deposits
I	Aromatic, heteroaromatic, triple bond	Polymerize readily with little hydrogen production and low photon emission	High concentration of dangling bonds and unsaturation
II	Double bond, cyclic structure	Polymerize with moderate hydrogen production and moderate photon emission	Moderate level of dangling bonds and unsaturation
III	Saturated hydrocarbons	Polymerize with high yield of hydrogen and high level of photon emission	Low concentration of dangling bonds and unsaturation
IV	Oxygen (in aliphatic structure)	Low deposition rate, high rate of by-product gases, high level of photon emission (poisoning effect)	Very sensitive to the energy input level of LCVD. Often the oxygen containing groups are absent in polymers
V	Oxygen (in cyclic structure including epoxides)	Oxygen atom is preferentially removed yielding diradical for polymerization	Oxygen is absent in polymers

the three are clearly separated lines (in Fig. 7.2) representing the corresponding types of chemical structures—type I (bottom line), type II (middle line), and type III (top line)—shown in Table 7.2.

1.2. Effect of Chemical Structure Investigated by Pulsed Discharge

Because of the chemical structure–insensitive (nonspecific) nature of plasma polymerization illustrated above, the structure of the monomer appears to have relatively little influence on the polymerization characteristics as well as on the characteristics of plasma polymers. This is largely true in the context of selective reactivity due to the chemical structure of monomers in conventional polymerization (i.e., monomer vs. nonmonomer). However, the influence of monomer structure, as classified by five types in Table 7.2, is actually accentuated when the operational conditions are varied. In this context, therefore, the chemical structure of a monomer is a key factor in its deposition characteristics and also in determining the properties of the deposition.

It is important to reiterate here that chemical structure of a species that deposits on a substrate depends on the kinetic pathlength described in Chapter 5, which is greatly influenced by the energy input to the luminous gas phase and the temperature of the substrate as shown in Figure 5.6. In other words, a monomer with a certain structure does not yield a deposition with a specific chemical structure

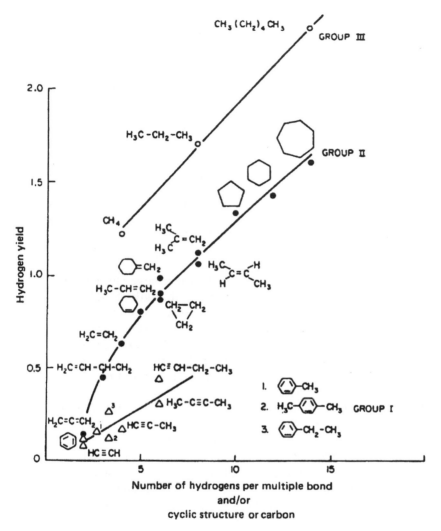

Figure 7.2 Number of hydrogen molecules evolved per molecule of starting material when hydrocarbons polymerize (hydrogen yield) as a function of chemical structure.

that can be predicted by the chemical structure of the monomer. Furthermore, the characteristics of LCVD deposition greatly depend on the kinetic pathlength also. A high-energy input lengthens the kinetic pathlength in the luminous gas phase because more reactive species are created in the luminous gas phase in a given time; it yields a practically useful tight network system but loses the identity of chemical structure of the monomer. A low-energy input shortens the kinetic pathlength and preserves the identity of chemical structure of the monomer, although the deposition could be useless in practical application.

The change of polymerization characteristics and consequential properties of polymers, which can be recognized to be most sensitive to the chemical structure of monomer, may be seen in the influence of pulsed discharge [4]. The pulsed discharge changes the balance between the contribution of cycle 1 and cycle 2

Table 7.3 Hydrogen/Carbon Ratios in Plasma Polymers and Corresponding Monomers

Monomer	H/C (monomer)	H/C (polymer)	H/C (polymer) / H/C (monomer)
Acetylene	1.0	0.95	0.95
Ethylene	2.0	1.49	0.75
Propylene	2.0	1.40	0.70
Isobutylene	2.0	1.44	0.72
cis-2-Butene	2.0	1.34	0.67
Butadiene	1.50	1.33	0.88
Methane	4.00	2.40	0.60
Ethane	3.00	1.55	0.52
Propane	2.67	1.58	0.59

Adapted from Ref. 3.

without changing specific energy input. The pulsed discharge shortens the kinetic pathlength of cycle 1 but could increase the kinetic pathlength of cycle 2 depending on the chemical structure of the monomer and the system pressure (concentration of the monomer during the off time). However, within the duration of the on time, the energy input level is still a very important parameter for LCVD by pulsed discharge.

The effect of pulsed discharge on plasma polymerization may be viewed as the analogue of the rotating sector in photoinitiated polymerization. The ratio r of off time t_2 to on time t_1, $r = t_2/t_1$, is expected to influence the polymerization rate depending on the relative time scale of t_2 to the lifetime of free radicals in free radical addition polymerization of a monomer. This technique was used to estimate the average lifetime of free radicals in the polymerization.

The rate of vinyl polymerization (free radical chain growth polymerization) decreases when a pulsed initiation is employed. The decrease of polymerization rate can be correlated to the duty cycle r of the initiation. It decreases to a rate that is in the range $1/(1+r)$, (if $t_2 \gg$ average lifetime of free radicals), to $1/(1+r)^{1/2}$ (when t_2 is close to the average lifetime) [5]. Contrary to this expectation, pulsed glow discharge exhibits rather abnormal effects of the pulse, which can be explained best by consideration of the growth and deposition mechanisms of LCVD depicted in Figure 5.3.

Some monomers show a more or less anticipated decrease in polymer deposition rates based on the concept that a pulsed discharge decreases the initiation rate, but some monomers show dramatically increase deposition rates. The most significant effect of pulsed discharge, however, can be seen in the concentration of dangling bonds, which reflects the unique mechanisms of material formation in LCVD.

Two important aspects of the growth mechanisms (Fig. 5.3) are as follows: (1) cycle 2 leads to "free radical living polymer" and (2) the cross-cycle reaction shown by equation (3) in Figure 5.3 is the terminator of the living polymer radicals during the on period of a pulsed discharge. In other words, during the on time some cycle 2 reactions are diverted to cycle 1. Because of the balance of the contribution of cycle 1 and cycle 2, the concentration of dangling bonds can be correlated to the monomer types shown in Table 7.2. The greater the contribution of cycle 2 in the overall growth and deposition mechanisms, the more dangling bonds are found in the deposition.

The effects of pulsed discharges may be understood by examining the roles of cycle 1 and cycle 2 for monomers in each type. For type I monomers, cycle 2 dominates in the growth and deposition mechanisms. Type II monomers polymerize via both cycle 1 and cycle 2. Type III monomers polymerize mainly via cycle 1. Oxygen in type IV monomers tends to be liberated in glow discharge and blocks cycle 2 and, to a lesser extent, cycle 1. This blocking effect is termed the "poisoning effect" of oxygen in type IV monomers. Oxygen in a type V molecule acts as the site to create a diradical, like a π bond in type I molecules, and type V monomers polymerize mainly via cycle 2.

The influence of pulsed discharge for some typical monomers, investigated by using a pulsing cycle of 100 ms "on" and 900 ms "off", is shown in Table 7.4. The most striking effect of pulsed discharges is that the conspicuous increase of dangling bonds in the deposits of some monomers. Such dramatic increases of dangling bond concentration are seen for monomers for which the contribution of cycle 1 and cycle 2 are nearly equal. If cycle 1 or cycle 2 dominates, the cross-cycle reaction has little influence on the overall effect and hence the conspicuous effect of pulsed discharge is not seen.

The amount of free radicals in a glass substrate is largely due to UV irradiation of an LCVD system. The glass free radical decreases in the pulsed glow discharge in all cases, which indicates that the interaction of the luminous gas with the substrate decreases by pulsing plasma. This is in accordance with the decreased intensity of glow by pulsed discharge.

The most pronounced increase in dangling bonds due to the pulsed discharge is found for LCVD of ethylene, for which cycle 1, based on the detachment of hydrogen, and cycle 2, based on the opening of the double bond, have nearly equal roles in the continuous discharge polymerization. Consequently, a significant portion of difunctional species is diverted to cycle 1 (monofunctional species) by the cross-cycle reaction during the on period of pulsed discharge and in the continuous discharge.

During the off period, however, the cycle 1 reactions as well as the cross-cycle reaction cease, and during this period difunctional species created are largely conserved by the continued reactions with (unexcited) monomers in cycle 2. As a net effect of the pulsed glow discharge, the contribution of cycle 2 to the overall plasma polymerization increases despite the rather long resting period involved in the pulsed glow discharge. Thus, the increase in the contribution of cycle 2 results in an increase in dangling bonds.

The increase of deposition rate by pulsed discharge occurs when the fragmentation of monomer, particularly oxygen-containing monomer to cause the poisoning effect, is great in the continuous discharge. The pulsed discharge reduces the contribution of the poisoning effect, and the increase of deposition rate is observed.

Acetylene ($HC{\equiv}CH$) and benzene are very similar in their LCVD characteristics. Both compounds form plasma polymers with the least amount of hydrogen production (type I monomer), and their characteristics of copolymerization with N_2 and/or H_2O are nearly identical if we consider that one molecule of benzene is equivalent to three molecules of acetylene. Analysis of the gas phase in both closed and flow systems are given in Tables 7.5 and 7.6.

Table 7.4 Effect of Pulsing on Plasma Polymerization of Monomers (I)

Monomer	Types of reaction involved	Cycles in plasma polymerization	Pressure change, δ	Deposition rate	Dangling bonds	Glass free radicals	cos θ (H_2O)
Tetramethyldisiloxane	Detachment of H	1	−9	−47	−90	0	−0.28
	Detachment of Me	1					
Propionic acid	Detachment of H	1	−18	−110	0	−81	0.53
	Detachment of CO_2H	1 & poisoning					
Acrylic acid	Detachment of H	1	−29	+120	+140	−100	0.49
	Detachment of CO_2H	1 & poisoning					
	Opening double bond	2					
Cyclohexane	Detachment of H	1	21	−90	−100	−100	−0.09
	Ring opening	2					
Ethylene	Detachment of H	1	19	+2	+970	−79	−0.09
	Opening double bond	2					
Acetylene	(Detachment of H)	1	200	−23	+81	0	−0.01
	Opening triple bond	2					
Benzene	Detachment of H	1	150	−8	−50	0	−0.76
	Aromatic ring opening	2					
Styrene	Detachment of H	1	87	−16	−86	0	−0.09
	Aromatic ring opening	2					
	Opening double bond	2					

Monomer	Chemical Structure	Number					
Vinyl acetate	Detachment of H	1	−5	−48	−21	−71	−0.03
	Detachment of OCOMe	1					
	Opening double bond	2					
Ethylene oxide	Detachment of H	1	−5	−7	−33	−76	−0.09
	Ring opening	2 & poisoning					
Tetrafluoroethylene	Detachment of F	1 & ablation	−11	+106	−35	−84	−0.01
	Opening double bond	2					
Hexafluorobenzene	Detachment of F	1 & ablation	50	−22	−27	0	−0.05
	Aromatic ring opening	2					
Vinyl fluoride	Detachment of H	1 & ablation	47	−45	+480	−100	−0.07
	Detachment of F	1					
	Opening double bond	2					
Vinylidene fluoride	Detachment of H	1 & ablation	−6	−47	+100	−64	−0.36
	Detachment of F	1					
	Opening double bond	2					

Number, in column 4–8, represents the percent change due to pulsing from the corresponding value for the continuous plasma. $\delta = p_g/p_m$, where p_g is pressure of glow discharge and p_m is pressure of the monomer flow before discharge. Glass free radical is due to UV radiation during plasma polymerization. $\cos\theta$ (H_2O) is the value of cosine of the contact angle of water; negative value means that the plasma polymer becomes more hydrophobic, and a positive value indicates that plasma polymer becomes more hydrophilic by pulsed glow discharge.

Table 7.5 Effect of Pulsed Radio Frequency on Acetylene and Benzene in a Closed System

Polymerization parameter	Acetylene		Benzene	
	Continuous	Pulsed	Continuous	Pulsed
$t^{1/2}$ (s)	1.0	5.0	2.5	18.5
Monomer type parameter γ^a	0.17	0.16	0.15	0.15
Fraction of residual vapor x^b	0.023	0.021	0.014	0.010
Hydrogen yield y^c	0.15	0.14	0.13	0.14
Polymer yield $Z = 1 - x$	0.98	0.98	0.99	0.99

[a]Monomer type parameter $\gamma = p_\infty/p_0$, where p_0 is initial monomer pressure in a closed system and p_∞ the pressure after glow discharge.
[b]Fraction of residual vapor $x = (p_\infty - p_{H2})/p_0$ where p_{H2} is the pressure after the cold finger is surrounded by liquid N_2.
[c]Hydrogen yield $y = p_{H2}/p_0$.

Table 7.6 Vapor Phase Analysis in a Flow System at Steady State

Parameter	Acetylene		Benzene	
	Continuous	Pulsed	Continuous	Pulsed
Total vapor pressure at steady state in glow discharge (μm Hg)	5.1	10.7	5.8	12.4
Vapor pressure after condensation with liquid N_2 (H_2 pressure) (μm Hg)	3.4	4.8	3.8	6.3
H_2 (%)	68	45	66	51
Condensable vapor (%)	32	55	34	49

The most remarkable difference in these data is found in the dissipation rates of monomer given by $t_{1/2}$ indicating that acetylene polymerizes (disappears from the gas phase) about 2.5 times faster than benzene. This difference is not very surprising if we recognize that we are measuring the rate of disappearance of the number of gas molecules. As far as the effect of pulsed radio frequency discharge (on the gas phase kinetic data) is concerned, these data indicate that the effect is nearly the same for acetylene and benzene.

In contrast, the concentration of ESR spins for dangling bonds showed noteworthy differences between acetylene and benzene. Spin concentrations of acetylene, benzene, styrene, and ethylene in continuous and pulsed discharges are compared in Table 7.7. The pulsed radio frequency discharge showed the most pronounced distinction of chemical structure of these monomers in the dangling bonds. When pulsed discharge is employed, the concentration of the dangling bonds changes from that in continuous discharge. The ratio of (dangling bonds by pulsed discharge)/(dangling bonds by continuous discharge) for acetylene is 1.8, and the corresponding ratios for benzene and styrene are 0.5 and 0.14, respectively. Thus, an increase in dangling bonds by the use of pulsed discharge is observed with acetylene and ethylene, and a decrease is observed with benzene and styrene.

Table 7.7 Change of Spin Concentration Due to Pulsed Discharge, $C_s^0 \times 10^{-19}$ (spins/cm^3)

Compound	Continuous	Pulsed	Ratio
Acetylene	8.6	15.6	1.81
Benzene	3.2	1.6	0.50
Styrene	3.8	0.54	0.14
Ethylene	1.4	14.5	10.36

It is important to note that the first three monomers are type I monomers, and their behaviors in plasma polymerization, except under pulsed plasma conditions, are very similar. These conspicuous differences observed as a consequence of the use of pulsed discharge may be attributed to the behavior of the aromatic structure under the conditions of LCVD.

Although numerous kinds of reactions could occur in the luminous gas phase, as far as the dissipation of vapor phase molecules (LCVD deposition) is concerned, one benzene molecule behaves as three acetylene molecules. Consequently, the final polymers formed (from acetylene and benzene) under the condition of relatively high W/FM are very similar. The transport characteristics of ultrathin films of plasma polymers and copolymers (with N_2 and/or H_2O) of acetylene and benzene are nearly identical.

However, with respect to the kinetics of polymer formation, benzene is one step behind acetylene, and in its original form (without dissociation of the molecule in the luminous gas) lacks reactions (1) and (4) of Figure 5.3. Pulsed radio frequency discharge highlights the difference in a dramatic way. Whereas the olefinic double bonds or triple bonds could conserve the overall concentration of radicals that have built up during the on period by the addition onto these bonds during the off period of the pulsed discharge, the conjugated double bonds in the aromatic structure could not react with free radical and could not conserve free radicals created in the on period. In other words, the diradicals can be conserved by the reaction with the monomer during the off period in the case of acetylene and partially in the case of styrene. However, in the case of aromatic compounds, cycle 2 reactions cease during the off period because benzene is not a monomer of free radical chain growth polymerization.

Styrene can be viewed as a monomer that combines the effects of a double bond and an aromatic structure. In a continuous discharge, the dominating factor is that of the phenyl group (styrene is a group I monomer), but not the double bond, because a phenyl group could be equivalent of three triple bonds in acetylene. A large concentration of dangling bond is due to the phenyl group. With a pulsed radio frequency discharge, the dissociation of the phenyl group ceases during the off period, but the contribution of the double bond is much smaller than that of the phenyl group (equivalent of three triple bonds of acetylene), which leads to the net loss.

The large increase found with ethylene is due largely to the low value of trapped free radical concentration by the continuous discharge owing to the dissipation of free radicals via the cross-cycle reaction (3) in Figure 5.3. During the off period of pulsed discharge, cycle 1 essentially ceases, and the free radicals created during the on period are conserved by the addition to the monomer during the off period leading to the highest increase in the dangling bonds, 10.4 times over the continuous discharge, due to the pulsed discharge.

Hydrocarbons that show a conspicuous increase of free radicals in LCVD upon the application of a pulsed radio frequency discharge can be categorized as compounds having a double or triple bond but not an aromatic structure. It is important to note that even with those monomers the deposition rates generally decrease with the pulsed discharge, although the reduction is not as great as one might expect on the basis of the duty cycle of the pulsed discharge.

Notable exceptions to this observation on deposition rates are found for acrylic acid and tetrafluoroethylene. In order to visualize the overall effect of a pulsed discharge, one should refer to the data given in the following tables: polymerization parameter in Table 7.8a, pressure parameters in Table 7.8b, deposition rates of polymers in Table 7.9, characteristics of ESR spin signals in Table 7.10, and contact angles of water in Table 7.11.

The LCVD of acetylene and allene are characterized by a high concentration of dangling bond and very low glass spin signals, which are typical features of type I monomers. When these monomers are copolymerized with H_2O, the characteristics of the plasma on the polymer free radicals change from those of type I to those of type III (i.e., low dangling bonds and high glass signals) as depicted in Table 7.12.

Because of the poisoning effect of oxygen-containing groups, which is caused by the detachment of the groups from the depositing species, acrylic acid shows a lower deposition rate than one would expect from the molecular weight of the monomer, and the plasma polymer is rather hydrophobic, as seen in Table 7.11. In a pulsed glow discharge, the poisoning effect is reduced due to the low duty cycle of the pulse, but once formed the radicals can be conserved by the characteristically high polymerization rate of acrylic acid via double bond during the off period of the pulse. Consequently, the net effects of a pulsed radio frequency discharge on acrylic acid are the following:

1. Remarkable increase in the polymer deposition rate (Table 7.9)
2. Increase in the concentration of dangling bonds (Table 7.13)
3. Decrease in the substrate glass signals (Table 7.13)
4. Significant increase in the wettability by water of the plasma polymer (Table 7.11)

For monomers with growth mechanisms in which cycle 2 mechanisms have a minor role, a pulsed glow discharge generally decreases the concentration of dangling bond, but the reduction is much less than one might expect from the duty cycle of the pulsed discharge. The reduction of dangling bond in such cases can be explained by the more complete coupling of free radicals before new free radicals are created by the subsequent on cycle of the pulsed glow discharge.

2. PERFLUOROCARBONS

2.1. Characteristics of Perfluorocarbons in Discharge

Perfluorocarbons are plasma-etching gases used in the microelectronics industry, and are less likely to polymerize in glow discharge than the hydrocarbon counterparts, unless a double bond or a cyclic structure exists in the molecule. Some factors that

Table 7.8a Hydrocarbon Polymerization Parameters

Compound	Structure	$t_{1/2}$ (s)	Monomer-type parameter γ	Fraction of residual vapor x	Hydrogen yield y (ratio)	Polymer yield $(1-x)$
Methane	CH_4	—	1.22	0.017	1.22	—
Propane	$CH_3CH_2CH_3$	~3	1.75	0.055	1.70	0.946
n-Hexane	$CH_3(CH_2)_4CH_3$	12	2.62	0.285	2.33	0.715
n-Octane	$CH_3(CH_2)_6CH_3$	7	2.75	0.133	2.61	0.867
Ethylene	$H_2C{=}CH_2$	0.75	0.647	0.020	0.628	0.980
Propene	$CH_3CH{=}CH_2$	2.25	0.920	0.040	0.880	0.960
trans-2-Butene	$CH_3CH{=}CHCH_3$	2.75	1.16	0.038	1.12	0.962
Isobutylene	$(CH_3)_2C{=}CH_2$	3	1.11	0.048	1.07	0.955
1,3-Butadiene	$H_2C{=}CHCH{=}CH_2$	3.75	0.471	0.019	0.452	0.981
Allene	$H_2C{=}C{=}CH_2$	3	0.135	0.007	0.128	0.994
Acetylene	$HC{\equiv}CH$	1.1	0.103	0.011	0.091	0.989
Methylacetylene	$HC{\equiv}CCH_2$	2.25	0.179	0.008	0.171	0.992
Dimethylacetylene	$CH_2C{\equiv}CCH_3$	6	0.326	0.020	0.307	0.980
Ethylacetylene	$CH_3CH_2C{\equiv}CH$	5.5	0.462	0.148	0.447	0.985
Cyclopropane	△	0.5	0.907	0.033	0.873	0.96
Cyclopentane	⬠	2.5	1.41	0.081	1.33	0.92
Cyclohexane	⬡	3	1.50	0.043	1.43	0.95
Cycloheptane	⬡	3.5	1.69	0.083	1.61	0.91
Cyclohexene	⬡	2	0.833	0.033	0.800	0.96
Methylenecyclohexane	⬡=CH_2	—	1.01	0.026	0.979	0.97
Benzene	⬡	<2	0.110	0.003	0.107	0.99
Toluene	⬡–CH_3	3	0.174	0.001	0.172	9.90
p-Xylene	H_3C–⬡–CH_3	4	0.133	0.000	0.133	1.00
Ethylbenzene	⬡–CH_2CH_3	6	0.298	0.020	0.278	0.98
Styrene	⬡–$CH{=}CH_2$	4	0.105	0.017	0.088	0.98

Table 7.8b Pressure Parameter δ of Various Monomers in Continuous and Pulsed Radio Frequency Discharges

Monomer	δ^a		
	Continuous	Pulsed	Change[b]
C_2H_2	0.10	0.30	0.2 (200)
C_6H_6	0.13	0.33	0.2 (150)
C_6F_6	0.10	0.15	0.05 (50)
Styrene	0.15	0.28	0.13 (87)
C_2H_4	0.63	0.75	0.12 (19)
C_2F_4	0.65	0.58	−0.07 (−11)
Cyclohexane	1.03	1.25	0.22 (21)
Ethylene oxide	1.45	1.38	−0.07 (−5)
Acrylic acid	2.00	1.43	−0.57 (−29)
Propionic acid	2.30	1.88	−0.42 (−18)
Vinyl acetate	2.25	2.13	−0.12 (−5)
Methyl acrylate	2.25	1.93	−0.32 (−14)
Hexamethyldisilane	1.50	1.20	−0.30 (−20)
Tetramethyldisiloxane	1.15	1.05	−0.10 (−9)
Hexamethyldisiloxane	1.50	1.15	−0.35 (−23)
Divinyltetramethyldisiloxane	0.73	0.75	0.02 (3)

[a]$\delta = p_g/p_m$ where p_g is the pressure of a steady-state flow in glow discharge and p_m the pressure of monomer before the discharge.
[b]Changes are based on values of continuous discharge. Numbers in parentheses are percentages.

Table 7.9 Deposition Rates of Various Monomers in Continuous and Pulsed Radio Frequency Discharges

Monomer	Deposition rate $\times 10^8$ (g/cm^2 min)		
	Continuous	Pulsed	Change[a]
C_2H_2	31	24	−7 (−23)
C_6H_6	110	101	−9 (−8)
C_6F_6	190	149	−41 (−22)
Styrene	173	145	−28 (−16)
C_2H_4	42	43	1 (2)
C_2F_4	18	37	19 (110)
Cyclohexane	92	9	−83 (−90)
Ethylene oxide	15	14	−1 (−7)
Acrylic acid	28	61	33 (120)
Propionic acid	7	15	8 (110)
Vinyl acetate	31	16	−15 (−48)
Methyl acrylate	32	33	1 (3)
Hexamethyldisilane	251	65	−186 (−74)
Tetramethyldisiloxane	191	102	−89 (−47)
Hexamethyldisiloxane	233	43	−190 (−82)
Divinyltetramethyldisiloxane	641	277	−364 (−57)

[a]Changes are based on values of continuous discharge. Numbers in parentheses are percentages.

Table 7.10 Characteristics of Spin Signals

Monomer	Normalized peak height (a.u.)[a]		Peak width (G)		Half-life of initial decay (min)	
	Continuous	Pulsed	Continuous	Pulsed	Continuous	Pulsed
C_2H_2	20.6	56.5	15.7	12.7	60	90
C_6H_6	4.9	2.7	19.6	18.6	32	15
C_6F_6	2.2	2.3	44.1	37.2	60	64
Styrene	5.8	1.5	19.6	14.7	30	55
C_2H_4	2.3	20.1	18.6	20.6	17	18
C_2F_4	7.4	2.4	52.9	45.1	40	25
Cyclohexane	1.6	—[b]	17.6	—[b]	36	—[b]
Ethylene oxide	1.2	1.1	18.6	16.7	65	40
Acrylic acid	1.6	3.5	16.7	17.6	83	115
Propionic acid	1.9	2.4	17.6	15.7	13	10
Vinyl acetate	0.7	0.7	18.6	16.7	26	25
Methyl acrylate	0.5	0.3	18.6	17.6	100	110
Hexamethyldisilane	11.3	4.1	5.9	5.9	15	14
Tetramethyldisiloxane	5.2	1.1	7.5	5.1	28	20
Hexamethyldisiloxane	3.6	—[b]	5.9	—[b]	20	—[b]
Divinyltetra-methyldisiloxane	1.6	0.5	7.5	7.5	19	21

[a]Arbitrary units.
[b]Signal not detected.

Table 7.11 Contact Angles of Water with Plasma Polymers

Monomer	$\cos \theta$ (H$_2$O)		
	Continuous	Pulsed	Change
C_2H_2	0.82	0.81	−0.01
C_6H_6	0.94	0.18	−0.76
C_6F_6	0.05	0.01	−0.04
Styrene	0.29	0.20	−0.09
C_2H_4	0.15	0.06	−0.09
C_2F_4	−0.22	−0.23	−0.01
Cyclohexane	0.01	−0.08	−0.09
Ethylene oxide	0.44	0.34	−0.10
Acrylic acid	0.50	0.99	0.49
Propionic acid	0.46	0.99	0.53
Vinyl acetate	0.39	0.36	−0.03
Methyl acrylate	0.39	0.37	−0.02
Hexamethyldisilane	−0.04	−0.12	−0.08
Tetramethyldisiloxane	0.23	−0.05	−0.28
Hexamethyldisiloxane	−0.18	−0.29	−0.11
Divinyltetramethyldisiloxane	0.03	0.03	0

Table 7.12 Unpaired Spins Detected in Polymer Films and Substrates After Glow Discharge Treatments for 1 Hour at 30 W Power

Components and p_m (mtorr) of components	Total p_m (mtorr)	Rate of polymer deposition (mg/cm²·h)	Spin concentration in polymer film [(spins/g) × 10^{-19}]	Linewidth of ESR first-derivative signal (G)	Spin near surface of glass substrate [(spins/cm²) × 10^{-15}]	$\delta = p_g/p_m$	p_g (mtorr)
Acetylene (81)	81	0.09	4.8	16	—	0.10	8
Acetylene (60)	60	0.04	7.4	18	—	0.12	7
Acetylene (40)	40	0.02	6.5	16	0.6	0.10	4
N$_2$ (30), acetylene (30)	60	0.02	3.3	16	0.3	0.08	5
Acetylene (30), H$_2$O (20)	50	0.05	—	—	3.8	0.79	40
N$_2$ (20), acetylene (30), H$_2$O (10)	60	0.07	0.18	16	2.7	0.37	22
CO (20), acetylene (30)	50	0.19	4.1	16		0.10	5
CO (20), acetylene (30), H$_2$O (15)	65	0.14	0.29	12	2.1	1.3	84
Allene (40)	40	0.05	2.0	20	1.1	0.20	8

Table 7.13 ESR Spin Concentration in Plasma Polymers and Glass Substrates

Monomer	$C_g^0 \times 10^{-19}$ (spins/cm^3) Continuous	Pulsed	Change[a]	$C_g \times 10^{-15}$ (spins/cm^2) Continuous	Pulsed	Change[a]
C_2H_2	8.6	15.6	7 (81)	0	0	0
C_6H_6	3.2	1.6	−1.6 (−50)	0	0	0
C_6F_6	7.4	5.4	−2.0 (−27)	0	0	0
Styrene	3.8	0.54	−3.26 (−86)	0	0	0
C_2H_4	1.36	14.5	13.1 (970)	4.0	0.85	−3.15 (−79)
C_2F_4	13.0	8.4	−4.6 (−35)	11.2	1.8	−9.4 (−84)
Cyclohexane	0.84	0	−0.84 (−100)	1.1	0	−1.1 (−100)
Ethylene oxide	0.75	0.5	−0.25 (−33)	6.6	1.6	−5.0 (−76)
Acrylic acid	0.76	1.85	1.09 (140)	4.4	0	−4.4 (−100)
Propionic acid	1.0	1.0	0 (0)	6.3	1.6	−4.7 (−75)
Vinyl acetate	0.42	0.33	−0.09 (−21)	6.1	1.8	−4.3 (−71)
Methyl acrylate	0.31	0.15	−0.16 (−52)	6.4	1.5	−4.9 (−77)
Hexamethyldisilane	0.5	0.24	−0.26 (−52)	0	0	0
Tetramethyldisiloxane	0.49	0.05	−0.44 (−90)	0	0	0
Hexamethyldisiloxane	0.21	0	−0.21 (−100)	0	0	0
Divinyltetra-methyldisiloxane	0.15	0.05	−0.10 (−67)	0	0	0

[a]Changes are based on values of continuous discharge. Numbers in parentheses are percentages.

distinguish perfluorocarbons from hydrocarbons are as follows: (1) the C–F bond is stronger than the C–H bond; (2) the C–F bond is stronger than the C–C bond; (3) F is a highly electronegative element; and (4) the normalized bond energy, which is given by $\sum$ (bond energy)$/M$, is smaller than those for hydrocarbons, and consequently more sensitive to the level of energy input (see Chapter 5).

The strong C–F bond accounts for the chemical stability of polytetrafluoroethylene (Teflon), and it is generally considered that making a Teflon surface hydrophilic by plasma treatment is difficult. On the other hand, F is highly electronegative and sensitive to luminous gas phase. When a Teflon surface is exposed to a jet stream of excited Ar neutral species in low-pressure cascade arc torch (LPCAT), it renders a surface-dynamically stable hydrophilic surface after a short exposure time. This indicates that the behavior of perfluorocarbons in glow discharge cannot be explained solely by comparison of bond energies. The electronegative F could form a negative ion by attaching an electron; consequently, completely different reactions that were not involved in glow discharge of hydrocarbons could have significant role in perfluorocarbon discharge.

2.2. Electronegativity of Fluorine

A study involving an electron probe method revealed that a considerable quantity of negative ions exist in the plasma of tetrafluoroethylene (TFE) [6]. The negative ions are found only with atoms that have high electronegativity (i.e., halogens and oxygen); consequently, in the glow discharge of most monomers without those elements, the presence and the role of negative ions can be virtually neglected. The negative ions formed by an electric discharge are generally very reactive, and it is difficult to detect negative ions quantitatively because ordinary plasma diagnostic methods, such as positive ion mass spectrometry are not effective in the detection of negative ions. Thus, it is not surprising that the role of negative ions has been overlooked in the postulations of the plasma polymerization of perfluorocarbons.

The plasma probe method can be used to determine the electron temperature and the electron density of plasma. The quantitative value of the method has been an academic point of argument among investigators; however, the qualitative indication of the presence of negative ions obtainable by the method seems to be very valuable for the interpretation of the plasma polymerization of perfluorocarbons.

A plot of the probe current and voltage for most plasma in which the number of positive ions is equal to the number of electrons (i.e., no negative ion exists) shows a typical S-shaped curve as depicted in Figure 7.3. Probe measurements carried out in plasma of most organic compounds show this kind of probe current–voltage relationship. However, when the probe measurement is applied to TFE, a conspicuously different probe current–voltage curve is obtained, as shown in Figure 7.4. This change of shape from type I (Fig. 7.3) to type II (Fig. 7.4) has been attributed to the presence of negative ions. Type II curves are obtained with plasma of gases that contain fluorine. O_2 plasma also shows the type II probe current–voltage curve.

When argon or hydrogen is mixed with TFE, the probe current–voltage relationship changes from type II to type I, or vice versa depending on the mole fraction of TFE in the gas mixture. Figure 7.5 depicts the change in electron temperature and number of positive ions, which is electron density, with identification of types as a function of mole fraction of TFE in a TFE–Ar system. Figure 7.6 shows a similar change in the TFE–H_2 system. Thus, plasma probe measurements indicate that a considerable quantity of negative ions exist in the plasma polymerization system of TFE.

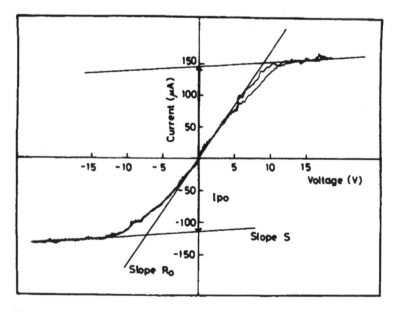

Figure 7.3   Probe current–probe voltage diagram for plasma of ethylene.

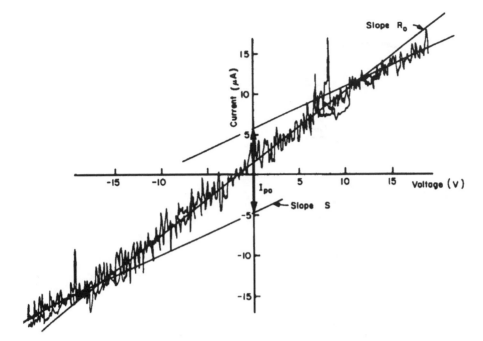

Figure 7.4 Probe current–probe voltage diagram for plasma of tetrafluoroethylene.

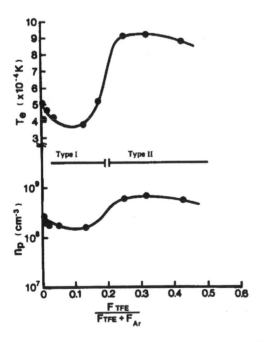

Figure 7.5 Change in electron temperature T_e and number of positive ions, n_p, and the type I–type II transition observed as a function of the mole fraction of tetrafluoroethylene (TFE) in a tetrafluoroethylene/argon system.

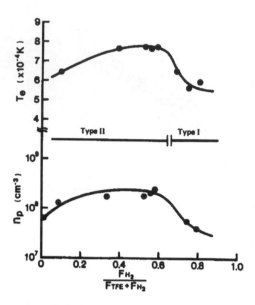

Figure 7.6 Change in T_e and n_p and the type I–type II transition observed as a function of the mole fraction of H_2 in an H_2/tetrafluoroethylene system.

In Figures 7.5 and 7.6 (positive) ion density is plotted; however, the number is equivalent to the electron density because the electrical neutrality is intuitively assumed based on the concept of plasma state. The electron density and the electron temperature generally follow the rule that the product of both is constant, which represents the total energy in the system. In both figures, however, the increase of electron temperature is accompanied by the increase of the electron density when the system change from type I to type II. This transition seems to indicate that when negative ions exist in the system, the electron density and the electron temperature both increase, indicating that the system absorbs the electrical energy more efficiently. This postulation agrees with the trends that perfluorocarbons should be treated with significantly less energy input based on W/FM than that for hydrocarbon counterparts.

If there are numerous negative ions in the plasma of TFE, then we cannot ignore the presence of F^- because fluorine is one of the most electronegative elements. Negative ion F^- can be formed by electron attachment on the collision of an electron and a fluorine atom, which is detached from the monomer molecule on the scission of a C–F bond. The F^- would react somewhat rapidly with a positive ion, which is formed by the dissociation ionization of a double bond, which rejects an electron from π bond of the double bond:

$$F_2C = CF_2 \rightarrow \cdot F_2C - \overset{+}{C} F_2 + e^-$$

$$\cdot F_2C - \overset{+}{C} F_2 + : F^- \rightarrow [\cdot F_2C - \overset{F:^-}{\underset{+}{C}F_2}] \rightarrow \cdot F_2C - CF_3$$

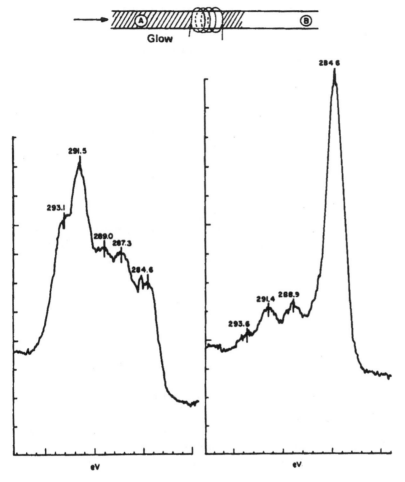

Figure 7.7 Dependence of the ESCA C_{1s} peaks of glow discharge polymers of tetrafluoroethylene on discharge conditions and location of polymer deposition; polymer deposition occurred at two locations in the reactor shown in the inset: (A) before the radio frequency coil and (B) after the radio frequency coil, discharge power 1.9×10^7 J/kg. Adapted from Ref. 7.

Other reactions also probably lead to the formation of $-CF_3$ in the polymer, but for the simplicity of discussion, let us consider that the $-CF_3$ formation step, regardless of the reaction mechanism, occurs on the inception of glow discharge and examine how the content of $-CF_3$ varies with the location of polymer deposition within a reactor.

Figure 7.7 depicts type of plasma polymer of TFE depending on the location in a small tube reactor [7]. In the tubular reactor shown, the formation of F^- would occur at the upstream side of the reactor, where the monomer flow makes contact with the luminous gas phase of TFE. Then, the $-CF_3$ could be used as a labeled species or an indicator of the change in the chemical nature of the polymer due to the kinetic pathlength of a growing species. The XPS data obtained with polymers

Table 7.14   Summary of ESCA Data for Plasma Polymer of Tetrafluoroethylene

Distance (cm)	Peak area,[a] (counts eV) $\times 10^{-4}$				Elemental ratio			Peak height ratio (C 1s 291.5 eV/ C 1s 284.6 eV)
	O 1s	F 1s	C 1s	Al2	O/C	F/C	Al/C	
$F = 5.6\,\mathrm{cm^3_{STP}/min}$, 8 W, $1.9 \times 10^7\,\mathrm{J/kg}$								
−18	0.135	4.01	2.65	0	0.051	1.51	0	1.75
0	0.098	3.71	2.65	0	0.037	1.40	0	1.52
9	0.108	3.78	2.58	0	0.042	1.47	0	1.41
27	0.961	2.18	1.92	0.664	0.50	1.14	0.35	0.261
36	1.33	1.38	1.73	0.913	0.77	0.80	0.53	0.184
$F = 0.56\,\mathrm{cm^3_{STP}/min}$, 32 W, $7.7 \times 10^8\,\mathrm{J/kg}$								
−36	1.18	1.58	1.41	1.09	0.84	1.12	0.77	0.052
−18	1.16	1.38	1.38	1.29	0.84	1.00	0.94	0.020
0	1.09	1.39	1.49	0.95	0.73	0.94	0.64	0.0098
18	1.29	1.21	1.38	1.24	0.94	0.88	0.90	0.023
36	1.50	0.73	1.57	1.01	0.96	0.46	0.64	0.013

[a]Corrected peak area using photoelectric cross-sections relative to C 1s.

deposited at various locations are shown in Table 7.14. The peak height ratio (C 1s, 291.5 eV)/(C 1s, 284.6 eV) serves as an indicator. Our true interest is in the C 1s peak at 293.5 eV for $-CF_3$. Unfortunately, the data did not include the C 1s peak height or the (C 1s, 293.5 eV)/(C 1s, 284.6 eV) ratio. However, because of a general trend in which the peak height of C 1s at 293.5 eV is roughly proportional to that of C 1s at 291.5 eV in the plasma polymers of fluorine-containing monomers, the (C 1s, 291.5 eV)/(C 1s, 284.6 eV) ratio may be used in place of the (C 1s, 293.5 eV)/(C 1s, 284.6 eV) ratio.

As can be seen from the data for the case of lower W/FM ($1.9 \times 10^7\,\mathrm{J/kg}$) shown in the upper half of Table 7.14, the ratio decreases along the length of the reactor. Thus, it appears that $-CF_3$ is introduced where the activation of monomer molecules occurs (at the tip of the glow, which contacts the monomer flow).

When the input energy is increased to $7.7 \times 10^8\,\mathrm{J/kg}$, the size of the glow zone expands (pushing the tip of glow far beyond the sample collecting zone) and, perhaps more importantly, the energy density reaches a level where the ablation effect becomes significant. Consequently, the plasma polymer becomes fluorine poor and the ratio decreases drastically, as seen in the lower half of Table 7.14. At low energy input of $1.9 \times 10^7\,\mathrm{J/kg}$, the XPS C 1s peaks show a higher content of $-CF_3$ and $-CF_2$ in the plasma polymer deposited in the glow region but a very small content of $-CF_3$ and $-CF_2$ in the polymer deposited in the nonglow downstream side of the reactor (Fig. 7.7). At a higher energy input of $7.7 \times 10^8\,\mathrm{J/kg}$, the glow expands, the samples are farther away from the tip of glow, and XPS C 1s peaks show little content of F-containing C moieties, as shown in Figure 7.8.

The same parameter can be used to demonstrate the effect of the frequency of the electric field on the structure of the polymer depositions and also the structural difference between polymers on the electrode and on the substrate. As shown

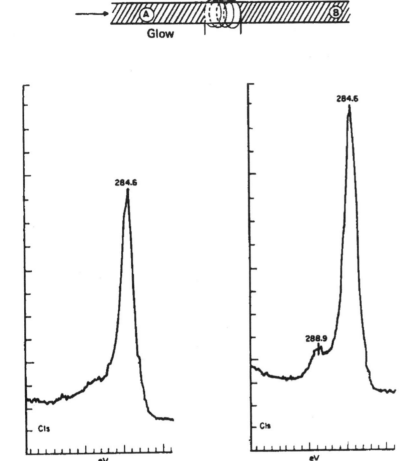

Figure 7.8 ESCA C_{1s} peaks of glow discharge polymers of tetrafluoroethylene in the same reactor shown in Fig. 6.41, but at the higher discharge power level of 7.7×10^8 J/kg. Adapted from Ref. 7.

in Table 7.15, more $-CF_3$-rich polymers are obtained with the radio frequency, and the parameter decreases as the frequency decreases. The data also indicate that more $-CF_3$ groups are found in the polymer deposit on the substrate than on the electrode in the low-frequency discharge (about 2.3 times at 60-Hz discharge). This ratio (shown in the last column of Table 7.15) gradually decreases with increasing frequency; in the radio frequency discharge (13.56 MHz) the situation is reversed, and the ratio drops to less than 0.8. These trends are in accordance with the location of dissociation glow described in Chapter 4 and the principle that the activation starts at the tip of the dissociation glow.

2.3. Chemical Reactivity of F in Luminous Gas

The deposition rates of perfluorocarbons are influence by the presence of H atoms in the luminous gas phase. If a polymer substrate is used, H atoms in the polymer

Table 7.15   Effect of the Frequency of an Electric Field on the Peak Height Ratio (C 1s, 291.5 eV)/(C 1s, 284.6 eV)

Frequency	Code[a]	Peak height ratio		
		Electrode	Substrate	Substrate/electrode
13.5 MHz	0.10	2.44	2.05	0.84
	0.100	3.57	0.15	0.04
10 kHz	0.19	1.14	1.61	1.41
	4.19	1.16	1.61	1.39
	0.71	0.12	0.15	1.25
	4.71	0.10	0.19	1.90
60 Hz	0.50	0.54	1.20	2.22
	4.50	0.44	1.06	2.41

[a]Code: The first number indicates the distance (centimeters) from the center of the electrode; the second number indicates the discharge wattage (watts) for 13.56-MHz radio frequency and the discharge current (milliamperes) for 10 kHz and 60 Hz.

are abstracted by the luminous gas of perfluorocarbons by forming HF and promote the deposition. Using glass plates as substrate and eliminating all polymers in the reactor system permits examination of the behavior of perfluorocarbons in absence of the influence of H atoms. Results obtained by such experiments indicate that the deposition of perfluorocarbons is dictated by the etching capability of species created in glow discharge [8].

One commonly held belief about fluorocarbon plasmas is that unsaturated (and cyclic) fluorocarbons are easier to polymerize than saturated fluorocarbons. Although it appears there is little room for argument in this statement, the difference probably stems not from the different type of structure but from the ratio of F atoms to C atoms. Based on the concept of "atomic polymerization" in plasmas, nearly all the bonds in a monomer can be broken and as the building block of a plasma polymer is generally smaller than that of the starting material, including atoms that constituted the starting material, there is no significant difference between "saturated" and "unsaturated" particularly at high W/FM domain. In consideration of etching capability of F in luminous gas phase, the polymerization characteristics, such as whether a fluorocarbon is polymer forming or not, might entirely depend on the ratio of F to C at least in continuous discharge at relatively high W/FM. Then the ratio of F/C in the monomer could be the factor that controls the deposition from perfluorocarbon discharge. Figure 7.9 depicts plots of the normalized deposition rates versus F/C ratio.

As F/C ratio falls below 2.5, the deposition rate abruptly increases. The deposition rate of a C_8F_{18} is as high as that of a C_2F_4, which is about 10 times higher than that of a CF_4. This contrast is remarkable in comparison with the hydrocarbon plasma described in Chapter 5, in which the normalized deposition rate of n-hydrocarbons at a fixed W/FM is almost independent regardless of its structure. This result clearly indicates that saturated perfluorocarbons cannot be classified as "polymer forming" or "nonpolymer forming" based on their structures, i.e., unsaturated or saturated. The difference between CF_4 and C_8F_{1s} is not due to the

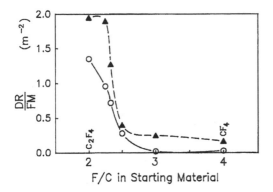

Figure 7.9 The effect of F/C ratio in monomer on the normalized deposition rates of saturated perfluorocarbons; W/FM: 0.3 GJ/kg for the lower curve with open circle, 1.6 GJ/kg for the upper curve with solid triangle.

type of structure but rather to the F/C ratio, i.e., 4 and 2.25, respectively. The primary step for successful polymerization is the elimination of excess fluorine. It is reasonable to conclude that the polymerization is easier if the starting fluorocarbon has fewer fluorine atoms per carbon atom.

In general, nonpolymer-forming plasmas, especially inert plasmas, can readily sputter electrodes. In polymer-forming plasmas, sputtering does not readily occur, unless a very high power is supplied, because most of the plasma energy is consumed for polymerization. For example, in Ar plasma, which causes only physical reactions, the sputtering rate is almost proportional to the plasma power supply, while a high power ($W/FM \sim 5$ GJ/kg) is required to cause sputtering in CH_4 plasma, which forms a polymer [9]. The sputtering phenomenon can be an indicator of whether a gas is polymer forming or not in glow discharge.

After exposure to saturated fluorocarbon discharge, the glass slides were analyzed for aluminum, which is the electrode material, using XPS. There was no aluminum peak following all discharge exposures at a low W/FM. However, when a high W/FM was applied, as shown in Figure 7.10, which depicts the atomic percentage of aluminum on the surfaces versus F/C, some aluminum was found on the substrate exposed to CF_4 and C_2F_6 plasmas. This, along with data shown in Figure 7.9, makes it clear that CF_4 and C_2F_6 act as etching gas and do not form polymer in the absence of H atoms in the system.

In sputter coating with inert plasmas such as Ar, physically sputtered metal from the electrodes usually deposits on the substrate and is evidenced by its metallic color. However, when chemically reactive fluorocarbons are used, it is possible that the metal from the electrodes reacts and forms metal fluorides. As Dilks and Kay demonstrated with C_3F_8 plasma, a small amount of metal fluoride was found incorporated in the plasma polymer [10,11]. Figure 7.11 shows the F 1s spectra of a substrate exposed to CF_4, C_2F_6, and C_8F_{18} glow discharges at a high W/FM. For surfaces exposed to CF_4, and C_2F_6 plasmas, two peaks are observed: the one at the higher binding energy is due to the fluorine bonded to carbon and the other to the fluoride (ion). For surfaces exposed to C_4F_{10}, C_6F_{14}, C_8F_{18}, and C_2F_4 plasmas, no broad peak is present in Figure 7.11c (C_8F_{18}). This indicates that most of the F atoms are contained in the plasma polymer. Considering the amount of

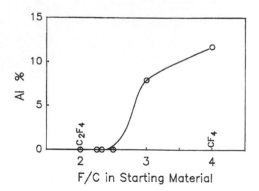

Figure 7.10 Aluminum percent on the surface of glass slide estimated from ESCA spectra after a series of saturated perfluorocarbon plasma exposures versus F/C ratio of the monomer; $W/FM = 1.6\,GJ/kg$.

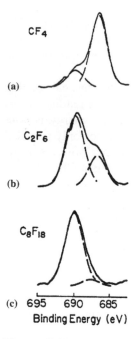

Figure 7.11 ESCA F 1s spectra on the surface of glass slide after exposure to (a) CF_4, (b) C_2F_6, and (c) C_8F_{18}; $W/FM = 1.6\,GJ/kg$.

sputtered aluminum, the peak at the lower binding energy can be attributed to aluminum fluoride and possibly to a small amount of sodium fluoride, in which the sodium is from the glass substrates. As described in the previous section, the fluorine anion seems to play an important role in fluorocarbon plasma. When a metal is present in the system, a metal fluoride could form preferentially unless conditions are in favor of polymerization. These results strongly suggest:

1. Perfluorocarbons are easily broken down, even to atoms, by glow discharge due to the high electronegativity of F.

2. The liberated F, in form of atom or negative ion, dominates the chemistry that occurs in the luminous gas phase.
3. Because of these factors, chemical structure, e.g., the presence of double bond or cyclic structure, has relatively minor role, at least in the continuous discharge, and the total number of F atoms contained in a molecule, which can be most adequately represented by the ratio F/C, becomes the dominant factor.

The combination of these three factors could explain why hexafluorobenzene ($F/C = 1.0$) is the fastest polymerizing monomer among hydrocarbons and perfluorocarbons, and the effect of pulsed radio frequency discharge is significantly different in hydrocarbons and perfluorocarbons.

In general, the concentrations of dangling bonds in plasma polymers of fluorine-containing monomers are higher than those of other organic compounds. However, the substrate free spin concentrations observed with glass for both fluorine-containing monomers and hydrocarbons are by and large the same. The effect of a pulsed radio frequency discharge is generally a reduction in the dangling bonds.

The facts found with pulsed radio frequency discharge that (1) the largest increase in dangling bonds is observed with ethylene and fluorohydrocarbons (e.g., vinyl fluoride and vinylidene fluoride), (2) the dangling bonds in tetrafluoroethylene decrease, and (3) dangling bonds in most perfluorocarbons decrease support a significant difference between the plasma polymerization of hydrocarbons and that of perfluorocarbons summarized above.

3. INCORPORATION OF NONPOLYMERIZABLE GASES IN PLASMA POLYMERS

Another important and unique feature of plasma polymerization is the incorporation of gases that do not form polymer or solid deposits in plasma by themselves during polymer formation of organic molecules in plasma. This incorporation of gases is plasma copolymerization and not the trapping of gas molecules in plasma polymers.

There is a significant difference between copolymerization and codeposition. An example of codeposition is the simultaneous deposition of parylene or a plasma polymer and an evaporated metal, in which each component can be deposited regardless of whether or not the other component is being deposited. Plasma polymerization of a mixture of two hydrocarbons, e.g., CH_4 and C_2H_4, is essentially codeposition of the respective plasma polymers. In contrast, plasma polymerization of gases occurs only in the presence of polymer-forming plasma. This is similar to the copolymerization of maleic anhydride, which does not polymerize, with other vinyl monomers.

An example of plasma copolymerization of gases is the incorporation of N_2 in the plasma polymer of styrene. N_2 mixed with styrene was consumed in plasma polymerization [12]. In a closed-system experiment, pressure measurement is a very useful tool for investigating plasma polymerization, particularly when the monomer used does not produce gaseous by-products. The pressure changes observed in a closed-system plasma reactor with mixtures of N_2 and styrene are shown in

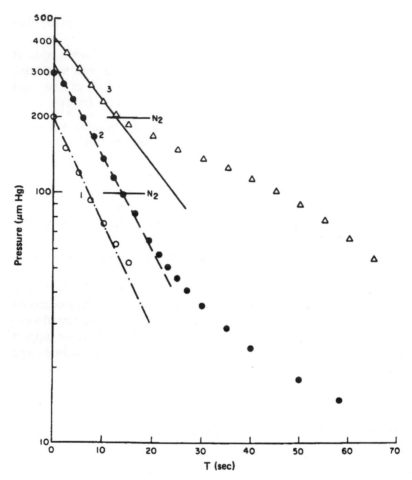

Figure 7.12 Pressure change in a closed system containing (1) 200 μm Hg styrene, (2) 200 μm Hg styrene and 100 μm Hg N_2, and (3) 200 μm Hg styrene and 200 μm Hg N_2, with time of glow discharge at 60 W.

Figure 7.12. The partial pressure of N_2 in each experiment is shown as a horizontal line crossing the pressure–decay curve. As can be seen, the system pressure decreases beyond the partial pressure of N_2 in a mixture, indicating that N_2 is incorporated into the plasma polymer (and thus disappears from the gas phase). Such a copolymerization of an unusual monomer has been observed for N_2, CO, and H_2O and is particularly efficient with monomers containing triple bond(s), double bond(s), or aromatic structure(s) (type I and type II monomers).

Figure 7.13 depicts the change of flow pressure due to glow discharge copolymerization of acetylene and H_2O. Acetylene is a typical type I monomer that reduces the pressure of a flow system. The addition of H_2O gradually changes the pressure change to increase above the system pressure before glow discharge is initiated [13]. The increase of system pressure is typically observed with type III and type IV monomers (Table 7.2). Comparison of Figure 7.13, for H_2O, and Figure 7.12, for N_2, clearly shows the characteristic difference of these two types of comonomers.

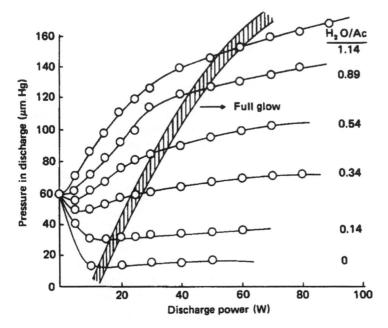

Figure 7.13 Dependence of flow pressure in the discharge of acetylene/H$_2$O mixtures on discharge power.

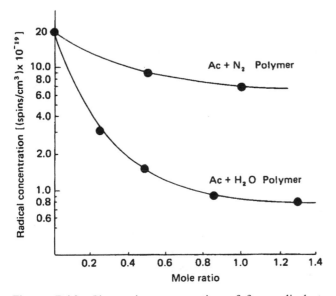

Figure 7.14 Change in concentration of free radicals trapped in plasma polymers by copolymerization of acetylene with N$_2$ or H$_2$O.

Figure 7.14 depicts the influence of copolymerization of N$_2$ and H$_2$O on the dangling bonds in plasma polymers.

The incorporation of N$_2$, CO, and H$_2$O in the plasma polymer of acetylene is evident from the results of elemental analysis as well as from infrared (IR) spectra

Table 7.16 Elemental Analysis of Glow Discharge Polymers of Acetylene with N_2, CO, and H_2O

Monomer (pressure, $\mu g\,Hg$)	C (%)	H (%)	N (%)	O (%)	Empirical formula	Color
Acetylene (50)	79.5	5.4	—	15.1	$C_2H_{1.6}O_{0.3}$	Dark brown
Acetylene/N_2 (50:33)	64.0	5.8	16.7	13.5	$C_2H_2N_{0.5}O_{0.3}$	Dark brown
Acetylene/H_2O (40:20)	66.5	7.6	—	25.9	$C_2H_{2.7}O_{0.4}$	Off-white
Acetylene/N_2/H_2O (30:20:15)	53.2	6.5	15.7	24.6	$C_2H_{2.9}N_{0.5}O_{0.7}$	Brown
Acetylene/CO (30:20)	82.6	6.9	—	10.5	$C_2H_{1.8}O_{0.2}$	Dark brown
Acetylene/CO/H_2O (30:20:15)	72.0	8.4	—	19.6	$C_2H_{2.5}O_{0.4}$	Light brown

obtained with the copolymers. Table 7.16 shows the results of elemental analysis of plasma polymers of acetylene copolymerized with N_2, CO, and H_2O. The complex nature of plasma polymers makes the precise interpretation of their IR spectra difficult. However, much useful information concerning the general nature of polymers and the general trends in the addition of unusual comonomer can be obtained, especially when other data, such as dangling bond concentrations and elemental analysis, are available. The IR spectra of the six plasma polymers are shown in Figures 7.15 and 7.16, and a summary is presented in Table 7.17. Data on the dangling bond concentrations from the ESR measurements can be found in Table 7.18.

In summary, the addition of H_2O, N_2, CO, or various combinations of these unusual comonomers to a plasma polymerization of acetylene produces chemically distinct polymers. The copolymerization of H_2O reduces the quantity of dangling bond in a remarkable manner, as shown in Figure 7.14, and enhances the stability of the polymers.

Nitrogen and/or H_2O are active only in the luminous gas phase. The incorporation or copolymerization of N_2 and H_2O must involve the activated species of these comonomers, which means that reactions that occur in the luminous gas phase are essential to plasma polymerization.

As far as the influence of added gas on the growth mechanism is concerned, it is important to recognize the following points:

1. Nitrogen and carbon monoxide have similar electron structures and evidently participate in the chemical reactions of cycle 2. Consequently, the copolymerization of these gases does not decrease the concentration of free radicals trapped in the plasma polymers.
2. Water is a blocking agent of cycle 2, and excessive H_2O added to a monomer inhibits plasma polymerization.

Because H_2O acts as an efficient modifier of the growth mechanism, which shifts the major growth path (if it is applicable) from cycle 2 to cycle 1, the addition of H_2O decreases the concentration of dangling bonds in the plasma polymers.

The long-term stability of a plasma polymer seems to be related to the concentration of dangling bonds and the tightness of the network which dangling

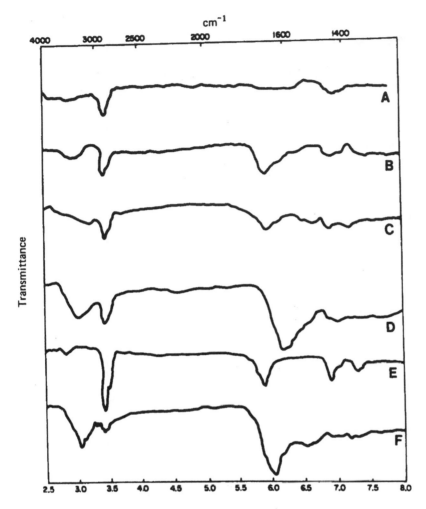

Figure 7.15 Infrared spectra ($4000-1250\,\mathrm{cm}^{-1}$) of plasma copolymers of acetylene: A, acetylene; B, acetylene/H_2O; C, acetylene/CO; D, acetylene/N_2; E, acetylene/CO/H_2O; F, acetylene/N_2/H_2O.

bonds reside. Figures 7.17 and 7.18 depict the change of IR spectrum with post–plasma polymerization time for plasma polymer of acetylene/N_2 and acetylene/H_2O, respectively. While the former shows gradual but appreciable change in a 15-month period, the latter shows virtually no difference in the same period.

4. DERIVATIVES OF SILANE AND SILOXANE

Silane (SiH_4) has been extensively used in the microelectronics industry in essentially the same process with LCVD. In LCVD dealing with relatively larger substrates, however, derivatives of silane are preferentially used because derivatives are more stable and easy to handle. SiH_4 reacts with oxygen in an explosive manner and requires special precautions in handling of the gas. The specific reactivity of Si with

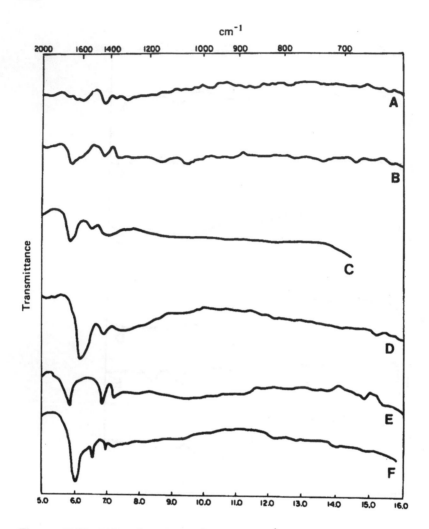

Figure 7.16 Infrared spectra $(2000–625\,cm^{-1})$ of plasma copolymers of acetylene: A, acetylene; B, acetylene/H_2O; C, acetylene/CO; D, acetylene/N_2; E, acetylene/CO/H_2O; F, acetylene/N_2/H_2O.

oxygen makes the major difference from C-based compounds described above. O atoms in the structure of monomers act as retarder or inhibitor for plasma polymerization of C-based compounds, and compounds having O in their structure are reluctant to polymerize. In contrast to O in C-based compounds, O in siloxanes is stable and does not detach from Si, and the O/Si ratio remains nearly constant in plasma polymers of siloxanes whereas C/Si changes depending on the conditions of glow discharge. Furthermore, O_2 acts as a comonomer in LCVD of silane and siloxane derivatives, and changes the characteristics of plasma polymers to the characteristics of inorganic or ceramic-type materials gradually according to the mole ratio of O_2 added to the mixture.

Derivatives of silane and siloxanes contain Si- and C-based moieties. In LCVD environment, the dissociation of monomer causes the separation of Si-based moieties

Table 7.17 Infrared Absorption Characteristics of Electrodeless Glow Discharge Polymers

Absorption region (cm^{-1})	Source	Acetylene	Acetylene/ H$_2$O	Acetylene/ CO	Acetylene/ N$_2$	Acetylene/ CO/H$_2$O	Acetylene/ N$_2$/H$_2$O
					Monomer system		
1370–1380	C–H symmetric bend, methyl			W		M	
1325–1440	C–C aldehyde			W		M	
1430–1470	C–H asymmetric bend, methyl	W	M	W	W	M	
1445–1485	C–H asymmetric bend, methylene	W	W	W	W	M	
1490–1580	N–H bend, secondary amine				W		W
1515–1570	N–H bend, secondary amide				S		
1560–1640	N–H bend, primary amine						
1630–1680	C=O stretch, secondary amide						S
1630–1670	C=O stretch, tertiary amide						S
1665–1685	C=O stretch, α,β-unsaturated ketone		S	M		S	S
1680–1705	C=O stretch, α,β-unsaturated aldehyde		S	M		S	S
1705–1725	C=O stretch, saturated ketone		M	W		S	
1710–1740	C=O stretch, saturated aldehyde					M	
2843–2863	C–H symmetrical stretch, methylene	M	S	S	S	S	M
2916–2936	C–H asymmetrical stretch, methylene	S	S	S	S	S	M
2862–2882	C–H symmetrical stretch, methyl	S	S	S	S	S	M
2952–2972	C–H asymmetrical stretch, methyl	S	M	S	S	S	M
3070–3100	N–H stretch, secondary amide bonded NH, cis or trans						W
3140–3180	N–H stretch, secondary amide bonded NH, cis						M
3270–3370	N–H stretch, secondary amide bonded NH, trans						S
3310–3350	N–H stretch, dialkylamine				S		
3400–3500	N–H stretch, primary amine				M		
3200–3600	O–H stretch, bonded hydroxyl	S	M^b	W	M	W^b	M

^{a}Given as peaks: strong (S), medium (M), and weak (W).
^{b}Possibly carbonyl overtone.

Table 7.18 Dangling Bonds in Plasma Polymers of Acetylene
with H_2O, N_2, and CO

Monomer (pressure, $\mu m\,Hg$)	Spin concentration $[(spins/cm^3) \times 10^{-18}]$
Acetylene	280
Acetylene/N_2 (30:30)	180
Acetylene/N_2/H_2O (30:10:20)	9
(30:20:10)	9
Acetylene/H_2O (30:20)	0
Acetylene/CO (30:20)	217
Acetylene/CO/H_2O (30:20:15)	1.5

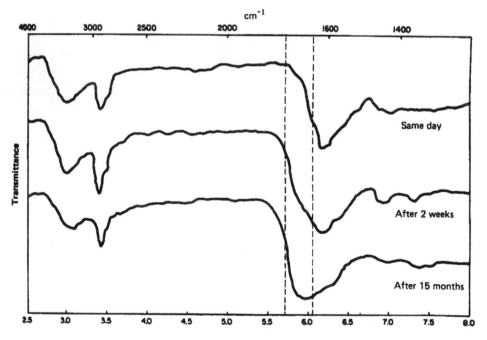

Figure 7.17 Infrared spectra of the plasma polymer of acetylene and N_2 taken at various times after the polymerization.

and C-based moieties, and the rate by which these two types of moieties are incorporated in the deposition is dependent on the deposition rate of each moiety. The plasma polymerization of two basically different moieties proceeds more or less independently and concurrently but leading to a more or less homogeneously mixed phase. The specific rates by which these moieties polymerize are different, which causes the variation of C/Si ratio depending on conditions of LCVD operation. The normalized deposition rates for Si-based moieties are roughly seven times greater

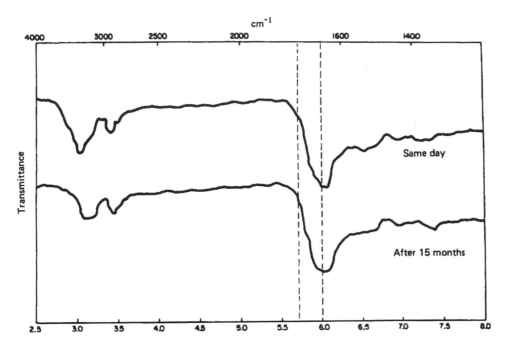

Figure 7.18 Infrared spectra of the plasma polymer of acetylene, N_2, and H_2O taken at various times after the polymerization.

than that for C-based moieties. Details of these aspects are described in various chapters dealing with specific aspects.

REFERENCES

1. Clark, D.T.; Abrahman, M.Z. J. Polym. Sci., Polym. Chem. Ed. **1982**, *20*, 691.
2. Yasuda, H.; Bumgarner, M.O.; Hillman, J.J. J. Appl. Polym. Sci. **1975**, *19*, 531.
3. Kobayashi, H.; Bell, A.T.; Shen, M. Macromolecules **1974**, *7*, 277.
4. Yasuda, H.; Hsu, T. J. Polym. Sci., Polym. Chem. Ed. **1977**, *15*, 81.
5. Burnett, G.M.; Melville, H.W. Proc. R. Soc. London **1947**, *A189*, 456.
6. Yanagihara, K.; Yasuda, H. J. Polym. Sci., Polym. Chem. Ed. **1982**, *20*, 1833.
7. Yasuda, H.; Morosoff, N.; Brandt, E.S.; Reilley, C.N. J. Appl. Polym. Sci. **1979**, *23*, 1003.
8. Iriyama, Yu; Yasuda, H. J. Polym. Sci., Polym. Chem. Ed. **1992**, *30*, 1731.
9. Yasuda, H. *Plasma Polymerization*; Academic Press: Orlando, FL, 1985.
10. Dilks, A.; Kay, E. ACS Symp. Ser. **1979**, *80*, 195.
11. Kay, E.; Dilks, A. J. Vac. Sci. Technol. **1981**, *18*, 1.
12. Yasuda, H.; Lamaze, C.E. J. Appl. Polym. Sci. **1973**, *17*, 1519.
13. Yasuda, H.; Hirotsu, T. J. Polym. Sci., Polym. Chem. Ed. **1977**, *15*, 2749.

8
Deposition Kinetics

1. MASS BALANCE FOR DEPOSITION IN A FLOW SYSTEM

The following parameters are considered *per unit time*.

Total mass of monomer introduced into the system: $W_1 = FM$
Total mass of deposition: W_2
Total mass exits from the reactor: W_3
Monomer–polymer conversion ratio: $Y_p = W_2/W_1$
Total surface area: S

$$W_2 = \oint_S k_1 ds$$

$$S = \oint ds$$

$$k_0 = \frac{k_1}{W_1} = \frac{k_1}{FM}$$

$$\bar{k}_0 = \frac{\oint_S K_0 ds}{\oint ds}$$

$$= \frac{\bar{k}_1}{FM}$$

$$\bar{k}_1 = \frac{\oint_S k_1 ds}{\oint ds}$$

where k_1 is the local mass deposition rate (kg/m^2 s)
$\bar{k}_1$ is the average deposition rate
k_0 is the normalized deposition rate (1/m^2)
$\bar{k}_0$ is the average normalized deposition rate
The unit of normalized deposition rate is m^{-2}, and normalized deposition rate decreases with the total surface on which deposition occurs. This aspect can be conceived as "loading factor" of luminous chemical vapor deposition (LCVD).

Mass balance in a reactor (flow system) can be established as

$$W_1 = W_2 + W_3$$

The material formation in LCVD generally requires the production of gaseous by-products, which do not form deposition, in order to create new chemical bonds for the material formation. For instance, LCVD of saturated hydrocarbons requires hydrogen abstraction in the dissociation glow. In presence of double and triple bonds, the hydrogen production becomes very small.

Polymer yield, which is given by $Y_p = W_2/W_1$, cannot be unity because of the gas formation for which gas yield can be defined by $Y_g = W_3/W_1$. The value of Y_p and Y_g can be determined by measurement of pressure of a closed system, which is subjected to electrical discharge. If thus obtained Y_p can be considered as a physicochemical parameter that is characteristic to a specific monomer. The average specific deposition rate can be expressed by

$$\bar{k}_0 = Y_p W_1 \left(\frac{1}{S}\right)$$

where the first term, Y_p, is the monomer characteristic, and the second term, mass flow rate, is an operational parameter, and the third term $(1/S)$ is a factor that depends on the reactor design and the loading factor of operation. In other words, the average normalized deposition rate or the average deposition rate obtained by plasma polymerization is a function of the monomer, operational parameter, and the design factor of reactor employed:

Deposition rate $= F$ (monomer characteristics, operational parameter, design factor)

Accordingly, the deposition rate, or the deposition characteristic in general, cannot be dealt with as a sole function of the monomer or operational parameters such as flow rate and discharge power. This aspect is recognized as the *system-dependent* nature of LCVD.

Since the dissociation glow can be considered to be the major medium in which polymerizable species are created, the location of the dissociation glow, i.e., whether on the electrode surface or in the gas phase, has the most significant influence on where most of the LCVD occurs. The deposition of plasma polymer could be divided into the following major categories: (1) the deposition that occurs to the substrate placed in the luminous gas phase (deposition G) and (2) the deposition onto the electrode surface (deposition E). The partition between deposition G and deposition E is an important factor in practical use of LCVD that depends on the mode of operation.

2. DEPOSITION IN LUMINOUS GAS PHASE (DEPOSITION G)

2.1. Normalized Energy Input Parameter to Luminous Gas Phase

In LCVD, activation (formation of the reactive species) and deactivation (deposition of materials) occur in the same gas phase because the power input is directly applied to monomer gases and the material formation occurs mainly in the same gas phase.

The activation of organic molecules in a flow system operation of LCVD starts at the boundary of glow, i.e., boundary of luminous gas phase and gas phase monomer. The supply of the monomer into the luminous gas phase is a crucial factor because the monomer is consumed in the glow by depositing polymers and the numbers of polymer-forming species in the glow decreases.

The most important factor that influences the properties of plasma polymers from a monomer is the energy input level of plasma polymerization process. The energy input level determines the extent of dissociation and the extent of scrambling atoms in the monomer molecule. The retention of molecular structure of the original monomer decreases with the increasing level of energy input. The energy level in the luminous gas phase can be manifested by the parameter, W/FM, where W is the discharge wattage, F is molar or volume flow rate, and M is molecular weight of monomer [1]. The parameter has units of J/kg, i.e., energy per mass (of monomer). The numerical value of W/FM in J/kg can be calculated from W in watts, F in sccm, and M in g/mole by

$$[W/FM \text{ in J/kg}] = W/FM \times (1.34 \times 10^9)$$

It is important to recognize that W in watts is the energy input into the electrical discharge system, whereas W/FM is the energy input into the luminous gas phase in which LCVD (plasma polymerization) occurs. This subtle but very important difference could be visualized by the analogy in how a 10-W light bulb can be used as a heat source in a closed system. A 10-W light bulb consumes 10 W of electrical energy. In this case, 10 W means the energy input into the electrical circuit of the bulb.

The influence of the bulb to the surrounding medium, on the other hand, depends on the conditions of the environment. The bulb placed in a room hardly influences the temperature of the room. The same bulb is placed in a small box used as an incubator could be the sole heat source to control the temperature of the incubator. If the same bulb is kept on a hand, it could cause a burn. In these cases the energy input to the bulb, 10 W, has no meaning by itself with respect to the thermal effect to the system in which the bulb is placed. In order to see the effect of the light bulb to the surrounding system, it is necessary to divide the energy input to the bulb by the total mass of air surrounding the bulb. If the bulb is placed in a flow system, the total mass of air is given by the mass flow rate of air.

In the same analogy, W is the energy input to the electrical circuit of the glow discharge generator, and W/FM, in J/kg of monomer, represents the energy input to the luminous gas phase in which plasma polymerization occurs. It has been well established that the plasma polymerization that occurs in the luminous gas phase is primarily controlled by a composite power parameter, W/FM [2].

2.2. Domains of LCVD Operation

When the deposition rate is measured at a fixed flow rate and varying power input W, a line showing the dependence on W is obtained. However, when the same experiment is carried out at a different flow rate, another line is obtained. Figure 8.1

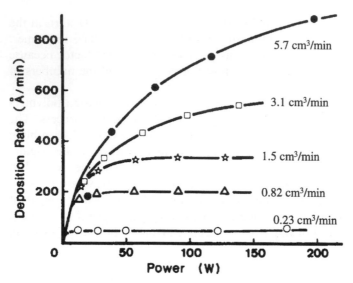

Figure 8.1 Dependence of deposition rate of plasma polymer of tetramethyldisiloxane on discharge wattage at various monomer flow rates (cm^3/min).

depicts the interrelated effect of W and F on the deposition rate. At lower flow rates, the deposition rate reaches a plateau value and becomes wattage independent. The critical value of W, at which the leveling off occurs, increases with increasing flow rate.

When the same experiments were carried out as a function of flow rate at various fixed input power W, the deposition rates could be plotted as a function of flow rate as depicted in Figure 8.2. Similar saturation effects are also seen in this graph as a function of flow rate. At a low flow rate, however, the decline of deposition rate, rather than staying at the plateau value, is observed. The same data shown in Figures 8.1 and 8.2 can be presented in one graph as a function of W/FM as shown in Figure 8.3.

Based on W/FM, the material formation in LCVD can be divided into two regimes: an energy-deficient regime and a monomer-deficient regime. In the energy-deficient domain, ample monomer is available but the power input rate is insufficient. In this domain, the deposition rate increases with the power input. In the monomer-deficient (power-saturated) domain, sufficient discharge power is available but the monomer feed-in rate is the determining factor for the deposition.

The decline of deposition rate with increasing flow rate seen in Figure 8.2 is due to the fact that the increase of flow rate at a fixed W decreases the value of W/FM. In the low-W case, the decrease of W/FM crosses the borderline of two domains, i.e., from the energy-deficient domain to the monomer-deficient domain. It is important to note that these two domains cannot be identified based simply on the value of operational parameters. The domain can be identified only by the dependence of the deposition rate on operational parameters W/FM as depicted in Figure 8.3. The critical value of W/FM at which the domain changes is dependent on the nature of organic molecules as described in Chapter 5.

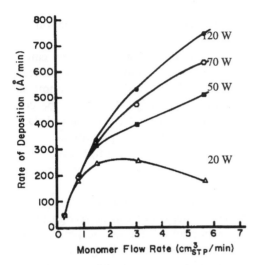

Figure 8.2 Dependence of deposition rate of the plasma polymer of tetramethyldisiloxane on flow rate of the monomer.

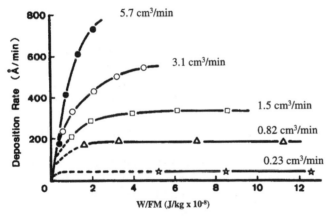

Figure 8.3 Dependence of deposition rate of the plasma polymer of tetramethyldisiloxane on W/FM at various monomer flow rate.

2.3. Normalized Deposition Rate

Most experiments start from the power-deficient domain, where the deposition rate (D.R.), which can be k_1 or $\bar{k}_1$, can be expressed by the following expression:

$$\text{D.R.} = \alpha W \tag{8.1}$$

Dividing both sides of the equation by FM, one obtains

$$\text{D.R.}/FM = \text{N.D.R.} = \alpha W/FM \tag{8.2}$$

$$\text{N.D.R.} = \alpha(W/FM) \tag{8.3}$$

where D.R. is the mass deposition rate, and N.D.R. is the normalized deposition rate (k_0 or $\bar{k}_0$).

The normalized deposition rate is the only form of deposition rate that can be used to compare deposition characteristics of different monomers with different chemical structures and molecular weights under different discharge conditions (flow rate, system pressure, and discharge power). Similarly, W/FM can be considered as the normalized power input. When only one monomer is employed, D.R. can be used to establish the dependency of deposition rate on operational parameters. Even in such a simple case, D.R. cannot be expressed by a simple function of W or F, and its relationship to those parameters varies depending on the domain of plasma polymerization.

As the power input is increased (at a given flow rate), the domain of plasma polymerization approaches the monomer-deficient one, which can be recognized by the asymptotical approach of D.R. value to a horizontal line as the power input increases. In the monomer-deficient domain, the deposition rate (plateau value) increases as the flow rate is increased and shows a linear dependence on the monomer feed-in rate at a given discharge power and the system pressure (Fig. 8.2), i.e.,

$$D.R. = \beta(FM) \tag{8.4}$$

This relationship is valid only in the monomer-deficient domain. The further increase of the flow rate (FM) will eventually decrease the deposition rate as the domain of plasma polymerization changes to the energy-deficient domain. Figure 8.2 includes data points in such a decreasing part in Equation (8.3). An increase in flow rate (at a given discharge power) has the same effect as decreasing the discharge power (at a given flow rate); conversely, an increase of discharge power has the same effect as decreasing the flow rate.

Figure 8.4 illustrates how well the thickness growth rate, GR/FM, in 15 kHz and 13.5 MHz LCVD of methane and n-butane, can be expressed as a function of the composite input parameter W/FM [3]. It is important to recognize that regardless of the mass of monomer, flow rate, and discharge wattage, a single line fits all data obtained in 15-kHz or 13.5-MHz LCVD of hydrocarbons employed, in which the deposition occurs on an electrically floating conductor or on a dielectric substrate placed in the glow.

Thus, the material formation in the luminous gas phase (deposition G), which is given in the form of normalized deposition rate (D.R./FM), can be controlled by the composite parameter W/FM (normalized energy input parameter), which represents the energy per unit mass of gas, J/kg. Because of the system-dependent nature of LCVD, W/FM is not an absolute parameter and varies depending on the design factor of the reactor. The value of W/FM in a reactor might not be reproduced in a different reactor; however, the dependency remains the same for all deposition G.

The results in Figure 8.4 indicate that the deposition rate by radio frequency discharge is significantly greater than that by 15-kHz discharge. However, this is not a general trend applicable to all cases. The deposition rate on substrate placed in gas phase depends on the nature and the location of the dissociation glow, which vary depending on the nature of the monomer. For instance, higher deposition rates were

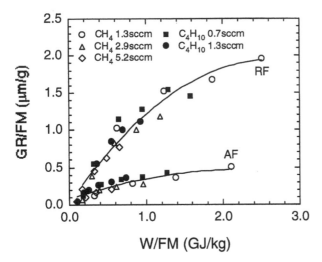

Figure 8.4 Dependence of GR/FM on W/FM for 15 kHz (AF) and 13.56 MHz (RF) magnetron discharge; flow rate; methane 1.3, 2.9, 5.2 sccm, n-butane 0.7, 1.3 sccm.

found for some fluorine-containing monomers for 15-kHz discharge than for 13.5-MHz discharge in the same experimental setup.

The product of deposition rate and deposition time determines the film thickness. Hence, $(W/FM)t$ is an important practical parameter to control the thickness of the deposition. In many practical applications in which the actual thickness of deposition is extremely difficult to measure, the overall functional character of LCVD process itself can be controlled by this parameter.

Figure 8.5 depicts the comparison of efficiency of Ar and H_2 plasma treatment compared as a function of the composite parameter $(W/FM)t$. Figure 8.6 depicts the comparison of efficiency of CH_4 and C_2H_4 LCVD. These figures clearly demonstrate the importance of the composite parameter $(W/FM)t$ in practical applications [4].

3. DEPOSITION ON ELECTRODE SURFACE (DEPOSITION E)

3.1. Deposition on Cathode in DC Discharge

When the equation for *plasma polymerization* [Eq. (8.2)] is applied to express the thickness growth rate of the material that deposits on the cathode (*cathodic polymerization*), it becomes quite clear that the deposition kinetics for the cathodic polymerization is quite different. There is a clear dependence of the deposition rate on W/FM, but no universal curve could be obtained. In other words, the relationship given by Eq. (8.2) does not apply to cathodic polymerization. The best universal dependency for cathodic polymerization was found between D.R./M (not D.R./FM) and the current density (I/S), where I is the discharge current and S is the area of cathode surface [5]. Figure 8.7 depicts this relationship for all cathodic polymerization data, which were obtained in the same study, covering experimental parameters such as flow rate, size of cathode, and mass of hydrocarbon monomers but at a fixed system pressure. The details of DC discharge polymerization are described in Chapter 13.

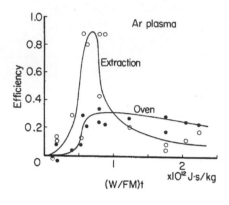

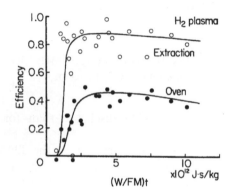

Figure 8.5 Efficiency of Ar and H_2 plasma treatment of PVC sheet as depicted by the parameter $(W/FM)t$.

3.2. Normalized Deposition Rate and Energy Input Parameter in DC Discharge

The implications of the correlation shown in Figure 8.7 are as follows: (1) The energy input parameter (based on the luminous gas phase) does not control the deposition of material onto the cathode surface. (2) The current density of a DC glow discharge is the primary operational parameter. (3) The flow rate of monomer does not influence the film thickness growth rate. (4) The film thickness growth rate is dependent on the mass concentration of monomer (cM) in the cathode region rather than the mass input rate (FM). (In these experiments, the system pressure was maintained at a constant value of 50 mtorr, and thus c was a constant.)

The cathodic deposition, in general cases, can be expressed by the following equation;

$$D.R./[M] = \alpha_c[I/S][c] \tag{8.5}$$

Since the variation of voltage is small in wattage-mode operation, $[I]$ could be replaced by $[W]$ for practical comparison:

$$D.R./[M] = \alpha_c[W/S][c] \tag{8.6}$$

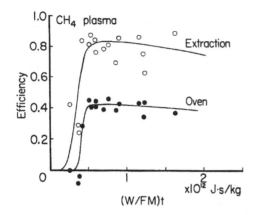

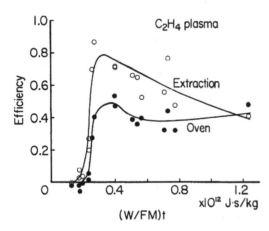

Figure 8.6 Efficiency of CH$_4$ and C$_2$H$_4$ LCVD on PVC sheet as depicted by the parameter $(W/FM)t$.

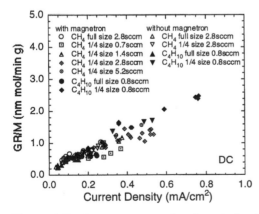

Figure 8.7 A master curve for the relationship between GR/M and the current density for DC cathodic polymerization; data obtained under various conditions for methane and *n*-butane, at a fixed system pressure of 50 mtorr.

The concentration of gas c is given by $c = p/RT$, and deposition rate can be given in a similar manner as Eq. (8.3):

$$\text{D.R.}/[M] = \alpha'[W/S][p] \tag{8.7}$$

The equation indicates that cathodic polymerization is controlled by the conditions of the local environment near the cathode. The normalized deposition rate in DC (deposition E) is D.R./[M], not D.R./[FM], and the normalized power input parameter is Wc/S, not W/FM. In DC discharge, the dissociation glow virtually adheres to the cathode surface. Therefore, the equation proves that the dissociation glow controls the deposition rate on the cathode surface.

3.3. Deposition on an Electrode in Alternating Current Discharges

An electrode in an AC discharge is the cathode for half of the deposition time and the anode for the other half of the time. Comparing Eq. (8.2) and Eq. (8.7), the contribution of the cathodic polymerization can be estimated by examining the system pressure dependence of the deposition rate (at a fixed flow rate). If plasma polymerization (deposition G) is the dominant factor, it is anticipated that the deposition rate would be independent of the system pressure. If cathodic polymerization (deposition E) is the dominant factor, the deposition rate onto an electrode is dependent on the system pressure, and the value of deposition rate is expected to be one-half of that for DC cathodic polymerization.

The system pressure dependence of the deposition rate onto the electrode surface in DC, 40-kHz, and 13.56-MHz discharges are shown in Figure 8.8. As anticipated from the deposition rate equation, the deposition rate of DC cathodic polymerization is linearly proportional to the system pressure. The deposition rate in 40-kHz discharge was found pressure dependent also, but that in 13.56 MHz was found to be independent of system pressure, actually slightly negative. The deposition rate in the 40-kHz discharge is one-half of that in the DC discharge, and the slope of pressure dependence is also roughly one-half of that obtained from DC discharge.

These findings indicate that cathodic polymerization takes place on the electrode in a 40-kHz discharge. This assessment agrees with the observation that

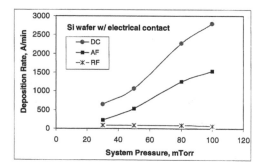

Figure 8.8 The system pressure dependence of deposition rate of TMS on Si wafer with electrical contact to the powered electrode in DC (cathode), 40-kHz, and 13.56-MHz plasma polymerization processes; discharge conditions are 1 sccm TMS, 5-W power input.

the dissociation glow is virtually adhering to the electrode in 40-kHz discharge. As the frequency increases to 13.56 MHz, however, the electrode does not act as the cathode such as in DC or 40-kHz discharges, and the *plasma polymerization* then governs the deposition onto the electrode, i.e., pressure independent. This assessment also agrees with the observation that the dissociation glow in radio frequency discharge is away from the electrode, which is in dark space, as described in Chapter 4. The details of effects of electrical contact on metal surface attached to the cathode surface are described in Chapter 13.

4. PARTITION BETWEEN DEPOSITION E AND DEPOSITION G

4.1. DC Discharge

The material formation in LCVD is caused by the dissociation glow (DG) and the ionization glow (IG). In DC discharge, the material formed in the cathode glow deposits nearly exclusively on the cathode surface due to the adherence of DG to the cathode, but some of them could deposit on surfaces in the reactor. The situation with the material formed in the negative glow is the same, i.e., it could deposit on the cathode, the anode, and surfaces placed in the reactor. Distribution of the deposition (to the cathode and the anode) is dependent on the distance between the cathode and the anode. Consequently, the total deposition on the cathode is also dependent on the distance.

In a DC discharge, the deposition rate on the substrate (silicon wafer) placed on the cathode surface without electrical contact differs significantly from that with electrical contact to the cathode (cathodic polymerization), and the pressure dependence is marginal as depicted in Figure 8.9. The deposition onto a floating substrate can be characterized as typical plasma polymerization that occurs in the luminous gas phase. A major difference in refractive index is seen in deposition *E* and deposition G. If one considers that the overall DC plasma polymerization is a mixture of cathodic polymerization (in the dissociation glow) and plasma polymerization (in negative glow), the deposition on the cathode is primarily cathodic polymerization. With an insulating layer between the substrate and the cathode surface, there is no cathode glow and, hence, no cathodic polymerization that deposits polymer on the substrate. The substrate on the cathode surface without electrical contact receives mainly the products of plasma polymerization, which occurs in the negative glow, with probably some stray deposition from the dissociation glow, which account for a slight pressure dependence.

The characteristics of the deposition on the cathode surface (deposition E) and the deposition on the electrically floating surface placed in gas phase (deposition G) in DC discharge LCVD are compared as follows.

DC discharge	Deposition on cathode surface (deposition E)	Deposition on substrate in gas phase (deposition G)
Operational parameter	$[W/S][p]$	$[W/FM]$
Normalized D.R.	$D.R./[M]$	$D.R./[FM]$
D.R. is dependent on	$[p]$	$[FM]$
D.R. is independent of	$[F]$	$[p]$

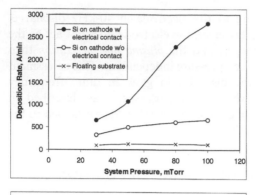

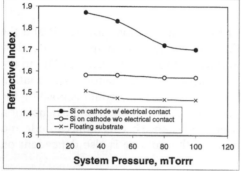

Figure 8.9 The system pressure dependence of deposition rate and refractive index of TMS in DC LCVD; 1 sccm TMS, 5W power input.

4.2. 40-kHz Discharge

In a 40-kHz discharge, the deposition onto the surface of the electrode, regardless of electrical conductivity or contact, is significantly different from deposition onto a floating substrate as depicted in Figure 8.10. The cathodic aspect of the electrode is less (one-half of DC discharge), but because of this the overall cathodic aspects of polymerization extend beyond the surface of the electrode yielding cathodic plasma polymer on an electrically insulated substrate placed on the electrode. Thus, the features of cathodic polymerization dominate in the vicinity of the electrode regardless of electrical contact.

The characteristics of the deposition on the cathode surface (deposition E) and the deposition on the electrically floating surface placed in gas phase (deposition G) in 40-kHz discharge LCVD are compared as follows:

40-kHz discharge	Deposition on electrode surface (deposition E)	Deposition on substrate in gas phase (deposition G)
Operational parameter	$[W/S][p]$	$[W/FM]$
Normalized D.R.	$D.R./[M]$	$D.R./[FM]$
D.R. is dependent on	$[p]$	$[FM]$
D.R. is independent of	$[F]$	$[p]$

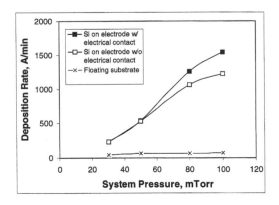

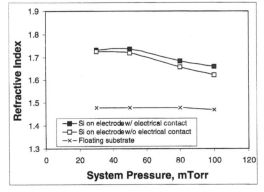

Figure 8.10   The system pressure dependence of deposition rate and refractive index of TMS on Si wafer with electrical contact and without electrical contact to powered electrode or floating substrate in 40 kHz LCVD; 1 sccm TMS, 5W power input.

4.3.   RF (13.5-MHz) Discharge

In 13.56-MHz discharge, the deposition onto the surface of an electrode, regardless of electrical conductivity or contact, is also appreciably different from that onto the floating substrate, although the magnitude of the difference is much smaller than those found in DC or 40-kHz discharges as shown in Figure 8.11. However, the deposition onto the electrode has no feature of cathodic polymerization because DG is not on the electrode in 13.56-MHz discharge. The higher deposition rate on the electrode could be explained by the effect of the dissociation glow being very close to the electrode surface. In radio frequency discharge the site of activation has shifted away from the electrode surface and the features of cathodic polymerization are diminished. However, the main mechanisms of LCVD due to the dissociation glow and the ionization glow are still operative in radio frequency discharge. The core of the onion layer structure of luminous gas phase shifted from the electrode surface to the gas phase but remains close to the electrode surface.

The characteristics of the deposition on the cathode surface (deposition E) and the deposition on the electrically floating surface placed in gas phase (deposition G)

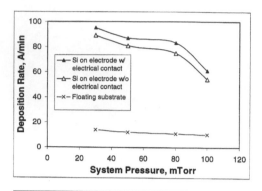

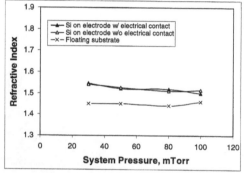

Figure 8.11 The system pressure dependence of deposition rate and refractive index of TMS on Si wafer with electrical contact and without electrical contact to powered electrode or floating substrate in 13.56-MHz LCVD; 1 sccm TMS, 5 W power input.

in radio frequency (13.5-MHz) discharge LCVD are compared as follows:

Radio frequency discharge	Deposition on electrode surface (deposition E)	Deposition on substrate in gas phase (deposition G)
Operational parameter	$[W/FM]$	$[W/FM]$
Normalized D.R.	D.R./$[FM]$	D.R./$[FM]$
D.R. is dependent on	$[FM]$	$[FM]$
D.R. is independent of	$[p]$, slightly negative	$[p]$

5. PROPERTIES OF PLASMA POLYMERS AND DEPOSITION KINETICS

5.1. Type A and Type B Plasma Polymers

Plasma polymerization is system dependent, and a monomer does not yield a well-defined polymer that can be identified by plasma polymerization. Plasma polymers formed at the high W/FM end of the power-deficient domain as well as in the monomer-deficient domain are tight three-dimensional amorphous networks, that do not contain discernible functional groups (type A plasma polymers).

Plasma polymers formed at the low W/FM end of the power-deficient domain could contain functional groups in the monomer, but the structure is much looser and often consists of oligomeric deposition (type B plasma polymer).

The coating that has tight network structure with functional groups cannot be obtained by a single-step plasma polymerization of a monomer. The retention of functional groups of monomer by some efforts, such as pulsed discharge, remote-plasma deposition, and so forth, can be achieved at the expense of the unique characteristics of the type A plasma polymers such as good barrier characteristics and good adhesion to the substrate.

5.2. Location of Deposition Within a Reactor

As discussed above, deposition kinetics is dependent on the mode of discharge. Accordingly, the location within a reactor where the major deposition occurs also changes depending on the mode of discharge. The relative position of a substrate with respect to the DG greatly influences both depositions.

In DC discharge, roughly 80% of polymer deposition is deposition E because the DG adheres to the cathode and appears as the cathode glow, as described in Chapter 3. The anode is a passive surface so far as the deposition is concerned, and by keeping the anode far away from the cathode or eliminating the anode plate by using the conducting wall of a reactor as the anode, the deposition E could be increased above 90%, implying that the deposition G could be reduced to less than 10% of the total deposition [5,6]. This is a great advantage if the substrate is electrically conducting and can be used as the cathode of DC discharge.

In alternating polarity discharge, each electrode acts as the cathode of DC discharge in one-half of the reaction time up to roughly 100 kHz. Consequently, the deposition rate onto an electrode surface drops to one-half of the deposition rate onto the cathode in DC discharge. However, the distribution between deposition E and deposition G does not change from that in DC discharge, i.e., roughly 80% deposition E and 20% deposition G because the DG still adheres to the electrodes on their cathodic cycle, and the total surface that can act as the cathode doubles. Unless some measures were taken to shift the location of the DG, the AC discharge suffers the relatively low production yield in luminous gas phase (deposition on products/total monomer) due to the low deposition G. The magnetic field superimposed on the electrode surface (magnetron discharge) seems to alter the shape and location of the dissociation glow and yield higher deposition G. Magnetron discharge is described in Chapter 14.

As the frequency increases to radio frequency range, an electron cannot complete the path from the cathode to the anode before the polarity changes, and the mode of electron movement changes from the linear motion to the oscillation according to the frequency of the power source. This change of the mode of electron movement shifts the location of the DG from the electrode surface to the gas phase as described in Chapter 4. Consequently, the relative contribution of deposition E and deposition G changes in favor of the latter. Due to the expansion of the glow volume particularly in low pressure, the whole deposition G could not be utilized in the practical processing. Due to the oscillating electrons, the magnetic field superimposed on the electrode cannot confine the volume of glow. Consequently, the benefit of magnetron is marginal in radio frequency discharge.

6. DEPOSITION KINETICS IN LOW-PRESSURE CASCADE ARC TORCH

In low-pressure cascade arc torch (LPCAT), the electrical power is applied in the cascade arc generator, in which only carrier gas, generally Ar, is activated to create luminous gas. The luminous gas created in the cascade arc generator is blown into the second expansion chamber, in which the monomer is introduced. Thus, the luminous gas of Ar neutrals primarily creates polymerizable species, and following these two steps should treat the deposition kinetics. Principles described in this chapter apply to each of the two steps. Details of deposition kinetics in LPCAT are described in Chapter 16.

The characteristics of the deposition on the cathode surface (deposition E) and the deposition on the electrically floating surface placed in gas phase (deposition G) in LPCAT LCVD are compared as follows:

LPCAT	Deposition on electrode surface (deposition E)	Deposition on substrate in gas phase (deposition G)
Operational parameter	N.A.	$[W(FM)_c/(FM)_m]$
Normalized D.R.	N.A.	$D.R./[FM]$
D.R. is dependent on	N.A.	$[FM]$
D.R. is independent of	N.A.	$[p]$

7. POWDER FORMATION IN GAS PHASE

The formation of powders in luminous gas phase has significant implications in the processing of LCVD. In an attempt to obtain a uniform nanofilm on a substrate, the powder formation in gas phase ruins the product. On the other hand, the analysis of powder formation as a function of operational parameters provides important information pertinent to the growth mechanism and the deposition mechanism of LCVD, by which growing species deposit on the surface.

The excessive formation of powders occurs only under limited conditions, although powder formation has been observed in reactors of different designs and types of discharge and with various monomers, particularly in a specific section of a reactor that is related to the flow pattern of gas. Therefore, powder formation provides an excellent opportunity for examining the basic principles of the polymer deposition mechanism.

In 1972, Liepins and Sakaoku [7] reported that polymeric powders were formed nearly exclusively in the radio frequency reactors shown in Figures 8.12 and 8.13, in which an organic vapor was introduced into the glow discharge of a carrier gas. The monomers that formed powders nearly exclusively and the yield of powder formation are summarized in Table 8.1. Monomers that did not form powders exclusively (i.e., formed plasma polymer in the form of a film or a film with powders) are shown in Table 8.2. The significant points about these experiments are as follows:

1. A minimal inert gas pressure of 0.6 torr is necessary to form powders exclusively. The upper limit of pressure is due to the reactor, which cannot be operated above 3.0 torr, at which pressure the plasma extinguished.

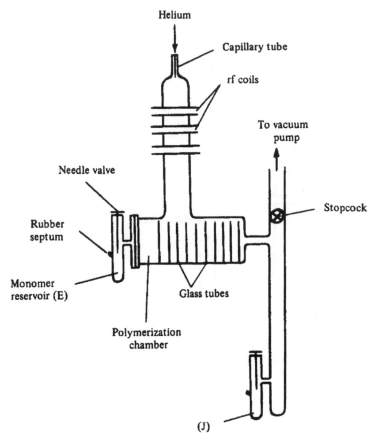

Figure 8.12 Schematic system arrangement. The chamber design consists of a modified low-temperature asher reaction chamber. Adapted from Ref. 7.

2. The yield of styrene powder formation is dependent on the carrier gas used. The yield decreases in the following order: He (18%), N_2 (16%), Ne (15%), Ar (7%), and air (3%).
3. Among the monomers investigated, the aromatic hydrocarbons (group I monomers) were most efficient in forming powders.
4. Most powders consisted of a large portion of the soluble polymer (as high as 90%) in tetrahydrofuran, which indicates that the kinetic pathlength in powder formation is very short.

On the other hand, the distribution of polymer depositions in an inductive radio frequency discharge reactor described in Chapter 20, in which monomer is introduced into luminous gas phase and does not go through an inductive coil zone just like in the cases of powder formation studies described above, indicates the following trends:

1. The distribution has a peak at the monomer inlet.
2. The peak becomes sharper as the monomer flow rate (consequently the system pressure) increases.

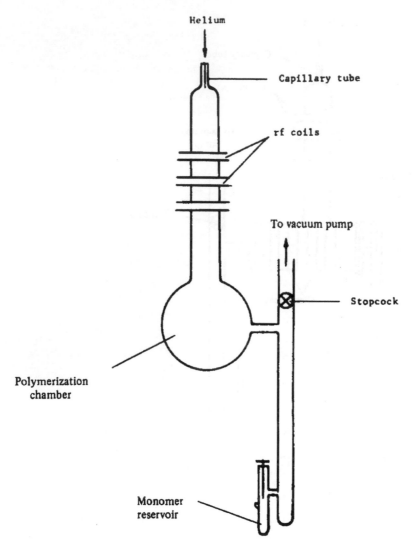

Figure 8.13 Schematic system arrangement: the chamber design consists of a long-necked 50-ml round-bottom flask. Adapted from Ref. 7.

3. The addition of an inert gas into the monomer flow, which increases the system pressure, produces a sharper polymer distribution peak.
4. The location of the peak approaches the monomer inlet as the reactivity to form plasma polymer increases. This trend follows the order: organic compounds with an aromatic ring and/or a triple bond (group I monomers) > compounds with a double bond and/or a cyclic structure (group II monomers) > compounds without any of these structures (group III monomers), i.e., group I > group II > group III.

The reactors used by Liepins and Sakaoku [7] can be characterized by the following conditions (in terms of the parameters dealt with in the

Table 8.1 Monomers Polymerized into Powder

Monomer	Amount of monomer used (g)	Amount of powder (g)	Conversion (%)	Polymerization time (min)	Inherent viscosity η_{inh} [a]	Solubility[b] (%)	Color
Styrene	2.10	0.38	18	13	0.06	90	Light tan
Toluene	2.60	0.39	15	16	0.05	80	Light tan
Benzene	2.41	0.43	18	28	0.03	90	Tan
p-Xylene	2.95	0.41	14	20	0.05	50	Light tan
Hexane	5.20	0.17	3	35	0.03	60	Light tan
Isoprene	2.30	0.31	13	25	—	Insoluble	Tan
Acetonitrile	3.05	0.40	13	20	0.04	80	Dark tan
Vinyl chloride	—	0.21	—	30	0.04	80	Dark brown
Tetrabutyltin	2.00	0.19	10	25	0.02	70	Tan
Styrene-divinylbenzene[c]	3.10	0.36	12	20	—	Insoluble	Light tan
Styrene-1,2-dibromoethane[c]	2.90	0.29	10	16	0.04	90	Brown

[a]Determined on solutions from 0.123 to 0.315 g/100 ml of tetrahydrofuran at 30.0°C.
[b]In tetrahydrofuran; the data represent the highest solubility observed of material collected during the first 5–10 min of polymerization.
[c]A 1:1 mixture (by weight).
Source: Ref. 7.

Table 8.2 Monomers Forming Predominantly Film[a]

Monomer	Type of product
Ethylene	Film
Acetylene	Film and powder
Propylene	Film
Butadiene	Film and powder
1,2-Dibromoethane	Film
1,2-Dichloroethane	Film
1,1,2,2-Tetrachloroethane	Film
Tetrafluoroethylene	Film and powder
Perfluoropropionitrile	Film and powder
4-Vinylpyridine	Film
Methyl methacrylate	Film
Trimethly borate	Film and powder
Borazine	Film and powder
Tris-β-diethylaminoborazine	Film and powder
Diphenyldiethoxysilane	Film and powder
Benzene/borazine[b]	Film and powder
Trimethylborate/ethylenediamine[b]	Film and powder

[a]Better than 50% by weight of the product is in the form of a coherent film.
[b]A 1:1 mixture (by weight).
Source: Ref. 7.

distribution study):

1. A large flow rate of monomer is used.
2. A carrier gas is introduced through the radio frequency coil, and a monomer is introduced in the downstream.
3. The system pressure is kept much higher (> 0.6 torr) than the system pressure range of < 0.1 torr employed in the distribution studies described above.

Combining these trends, we can postulate the mechanism of polymer powder formation as follows. When a relatively high concentration (pressure) of monomer vapor meets with the luminous gas phase of a carrier gas in a relatively small volume at sufficiently high pressure, the formation occurs quickly in a relatively small-volume element. Because a sufficient quantity of reactive species are created in the small-volume element, the polymer formation steps approach a critical level above which particles cannot stay in the gas phase without the reactive species diffusing out of the volume element. In other words, the kinetic pathlength is very short under such conditions, as proved by the fact, shown in Table 8.1, that most powders are soluble in solvent.

When these relatively small growing species coalesce in the gas phase without creating strong bonds, it creates gas phase–borne surface, and further deposition of growing species occurs onto the gas phase–borne surfaces. Conversely, the film formation can be visualized as the coalescence of small particles at the substrate surface. In fact, the presence of identifiable particles in a film and the existence of microspherical morphological structures have been reported by many investigators for various kinds of plasma polymer depositions. Such structures could be in the nanometer range, requiring X-rays for identification.

Because powder formation can be characterized as the rapid formation of polymeric species in a localized gas phase, the quantity of particles or powders mixed in a coherent film that forms at a substrate surface should also be related to the rate of film formation. Thompson and Smolinsky [8] found a direct correlation between the particle density on the surface of a plasma-polymerized film and the growth rate of the film as depicted in Figure 8.14.

The inclusion of particles in a film of plasma polymer was once considered by some investigators to be a characteristic problem due to the plasma polymerization mechanism, which hampers the practical use of plasma polymers in some applications. In contrast to this view, the formation of powder or the inclusion of particles in a film is related to the polymer *deposition* part of polymerization–deposition mechanisms. The inclusion or elimination of particles, therefore, could be accomplished by selection of the proper operational parameters and reactor design. The data of Liepins and Sakaoku [7] are a typical demonstration that powders can be formed nearly exclusively if all conditions are selected to favor powder formation. An important point is that the monomers used in their study were those commonly used by other investigators for the study of film formation by plasma polymerization; in other words, no special monomer is needed to form powders exclusively.

Because powder formation depends on the polymer deposition portion of the polymerization–deposition mechanisms of LCVD, its dependence on operational parameters such as the flow rate and system pressure is not necessarily the same in

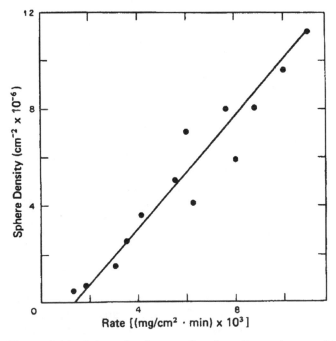

Figure 8.14 Sphere density as a function of growth rate of trimethylsilane film. Adapted from Ref. 8.

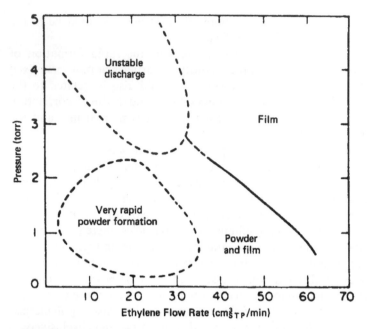

Figure 8.15 Powder and film regions at 100 W. Adapted from Ref. 9.

reactors of differing design. Kobayashi et al. [9] found that the powders formed in a capacitively coupled radio frequency discharge in a bell jar type of reactor, in which the conditions of powder formation in terms of operational parameters were completely different from those found by Liepins and Sakaoku [7].

Kobayashi et al. [9] used a capacitively coupled 13.56-MHz radio frequency discharge by 6-in.-diameter electrodes (5 cm apart). Powder formation was reported to be a strong function of the gas pressure and the flow rate. The region where powder formation occurred is shown in Figure 8.15. This region shifted as the discharge power was lowered, and at 50 W the diagram changed to that shown in Figure 8.16.

In such a capacitive discharge, powder formation was stated to occur in the region of low pressure and low flow rate, whereas in the inductively coupled reactor used by Liepins and Sakaoku [7] it was found to occur under conditions of high flow rate and high pressure. The critical parameter for powder formation, however, was reported by Kobayashi et al. [9] to be the energy input per mole of gas. They reported that for a pressure of 2 torr, the approximate values of the critical energy input of 50, 100, and 150 W correspond to dosages of 2.68×10^6, 2.9×10^6, and 2.08×10^6 J/mole, respectively.

An important point here is that the low-pressure region in the work of Kobayashi et al. [9] is well above the minimal pressure found in Liepins and Sakaoku's [7] work (i.e., 0.6 torr). Moreover, the maximal energy input at a given discharge power is obtained at a low flow rate. The critical dosage reported by Kobayashi et al. corresponds to about 10^8 J/kg in the W/FM parameter described previously. Because the flow of monomer in a bell jar type of reactor is not well defined, such an apparent value of W/FM cannot be directly compared with the

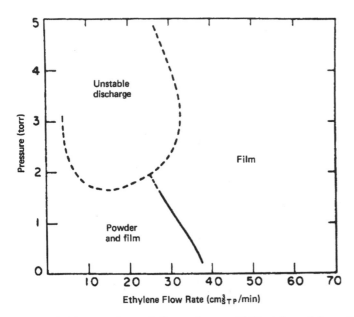

Figure 8.16 Powder and film regions at 50 W. Adapted from Ref. 9.

values for tubular reactors. Nevertheless, it is clear that the study of Kobayashi et al. covers a much lower W/FM region than that of Liepins and Sakaoku. A small-volume (1- to 5-liter) inductively coupled radio frequency reactor generally operates at about $2-5 \times 10^8$ J/kg for relatively small molecular weight monomers, whereas the higher end of the experiments of Kobayashi et al. seems to be about 10^8 J/kg.

Furthermore, it is important to recognize that the molecular dissociation glow is located very close to the electrode surface in the radio frequency discharge at 2 torr and that the material formation, including powder formation, occurs in this localized zone. It is difficult to estimate the local energy input from the energy input and flow rate of monomer into the overall reactor. In contrast to this situation, i.e., the small inductively coupled radio frequency reactor used by Liepins and Sakaoku [7], the energy input into the reactor corresponds closely to the local energy input where powder formation occurs. More importantly, the energy input applies to Ar and monomer does not go through radio frequency coil region.

Therefore, the apparent discrepancy is due to the use of inadequate parameters to describe the conditions of powder formation. Thus, powder formation can be described as the rapid formation of polymers at a localized space in luminous gas phase, which requires (1) sufficient mass and (2) sufficient energy density in the localized space. Kobayashi et al. [10] found later that powder formation is also dependent on the size of the interelectrode gap, as shown in Figure 8.17, which also shows the correlation between powder formation and rate of polymer deposition. These trends are also expected from the shift of the molecular DG from the electrode surface as a function of the interelectrode gap.

Concerning the design factors of reactors, Liepins and Sakaoku [7] stated that of the three polymerization chambers investigated (including a bell jar reactor in addition to two reactors shown in Figures 8.12 and 8.13), *the design depicted in*

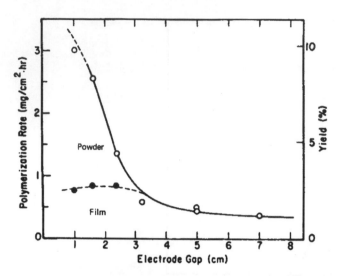

Figure 8.17 Dependence of the type of polymer formed on the size of the electrode gap. Power, 100 W; pressure, 2 torr; flow rate, 80 cm³/min. Adapted from Ref. 10.

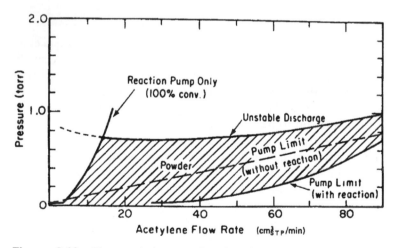

Figure 8.18 Characteristic map for the plasma Polymerization of acetylene at 50 W. Adapted from Ref. 10.

Figure 8.13 outperformed the other two in the rate of powder formation despite its much smaller total chamber volume. Indeed, the small total volume and the design that allows high-density glow to meet the high concentration of monomer in a localized space are the keys to enhancing powder formation. Conversely, avoiding these conditions is the key factor in eliminating the inclusion of powders in coherent films.

Powder formation is therefore highly system dependent. For example, in the bell jar reactor used by Kobayashi et al. [10], acetylene always forms powders as depicted in Figure 8.18, whereas coherent films are always obtained from acetylene

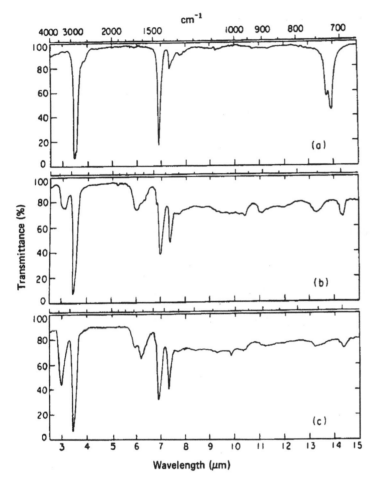

Figure 8.19 Infrared spectra of (a) low-density polyethylene film, (b) plasma-polymerized film, (c) plasma-polymerized powder. Adapted from Ref. 11.

in the lower-pressure plasma polymerization in an inductively coupled reactor as described in the deposition profiles in Chapter 20.

Thus, powder formation can be viewed as a polymer deposition process occurring by the same polymer formation mechanism that applies to the formation of films. In other words, no special monomer, reactive intermediate, or a different agent of polymer formation is needed for powder formation. The infrared spectra of films and powders are virtually identical [11], as shown in Figure 8.19 and Table 8.3.

Powder formation in an LCVD system is a reflection of the polymer deposition mechanism. The size and number of particles may be taken as a measure of the polymerization–deposition mechanism or the status of an LCVD system. At one extreme is exclusive powder formation, as reported by Liepins and Sakaoku [7]; at the opposite extreme is the formation of a continuous film in which no visible particles can be found. Even in the latter case, however, the work of Havens et al. [12] involving the use of small-angle X-ray scattering indicates that detectable domains

Table 8.3 Assignment of Infrared Absorption Bands

Low-density polyethylene film (cm^{-1})	Plasma-polymerized film (cm^{-1})	Plasma-polymerized powder (cm^{-1})	Assignment
—	3400	3400	OH stretch
2924	3000	3000	
2899	2960	2960	CH$_2$ stretch
2857	2900	2900	
2850	—	—	
—	1700	1700	Co stretch in $-CH_2-\overset{\overset{\displaystyle O}{\|}}{C}-CH_2-$
—	1680	—	CO Stretch in $-CH=CH-\overset{\overset{\displaystyle O}{\|}}{C}-$
—	1600	1600	C=C stretch
1473	—	—	
1463	1463	1463	CH$_2$ bend
1369	1369	1369	
1353	—	—	CH$_2$ wag
1303	—	—	CH$_2$ twist
—	960	960	CH out-of-plane deformation in $-CH=CH-$(trans)
—	900	890	CH out-of-plane deformation in CRR'=CH$_2$ or CHR=CHR' or CH$_3$ rock in a chain of three to four C atoms
—	750	750	CH$_2$ rock in $+CH_2+_2$
730	—	—	CH$_2$ rock in $+CH_2+_3$
720	—	—	CH$_2$ rock in $+CH_2+_n \geq 5$
—	700	700	CH out-of-plane deformation in $-CH=CH-$(cis)

Source: Ref. 11.

corresponding to particle sizes lower than the micrometer range are present in apparently structureless films. Therefore, it is consistent that polymer formation initiated in the gas phase is carried out up to continuous film formation on the substrate.

The presence of surface is included as the third body that dissipates the excess energy carried by chemically reactive species in the polymer formation mechanisms depicted in Figure 5.3, although the surface was not shown in the figure in order to avoid cluttering of diagrams. Under the conditions that favor the powder formation, the critical kinetic pathlength for deposition is reached in the gas phase, and the powder floating in the gas phase acts as the third body to enhance further growth of powder particles. Polymer deposition can be conceived as a phenomenon that occurs whenever a gaseous species (in a kinetic path of growth) fails to bounce back on collision with a surface, including the surface of particles already formed in the gas phase, which is the case of powder formation.

REFERENCES

1. Yasuda, H.; Hirotsu, T. J. Polym. Sci ., Polym. Chem. Ed. **1978**, *16*, 743.
2. Yasuda, H. *Plasma Polymerization*; Academic Press: Orlando, FL, 1985; 277–333.
3. Miyama, M.; Yasuda, H. J. Appl. Polym. Sci. **1998**, *70*, 237–245.
4. Iriyama, Y.; Yasuda, H. J. Appl. Polym. Sci.; Appl. Polym. Symp. **1988**, *42*, 97.
5. Yasuda, H.K.; Yu, Q.S. J. Vac. Sci. Tech. A Vacuum Surfaces Films **2001**, *19* (3), 773–781.
6. Yu, Q.S.; Yasuda, H.K. Plasmas Polym. **2002**, *7*, 41–55.
7. Liepins, R.; Sakaoku, K. J. App. Polym. Sci. **1972**, *16*, 2633.
8. Thompson, L.F.; Smolinsky, G. J. Appl. Polym. Sci. **1972**, *16*, 1179.
9. Kobayashi, H.; Bell, A.T.; Shen, M. J. Appl. Polym. Sci. **1973**, *17*, 885.
10. Kobayashi, H.; Shen, M.; Bell, A.T. J. Macromol. Sci. **1974**, *8*, 373.
11. Kobayashi, H.; Bell, A.T.; Shen, M. Macromolecules **1974**, 7, 277.
12. Havens, M.R.; Mayhan, K.S.; James, W.J.; Schmidt, P. J. Appl. Polym. Sci. **1978**, *22*, 2793.

9
Ablation by Luminous Gas (Low Pressure Plasma)

1. PHYSICAL SPUTTERING AND CHEMICAL ETCHING

Ablation of materials by the action of low-pressure luminous gas is an important factor that cannot be ignored in any process that utilizes luminous gas reactions. Ablation under consideration is not limited to the substrate material but also extends to materials used as an electrode and reactor wall. It is also important to recognize that the ablated materials could deposit in different locations, as described in Chapter 10.

Unexpected elements in a plasma polymer often are due to the redeposition of ablated materials. The presence of nitrogen found in a plasma polymer of a monomer that does not contain nitrogen can be traced to contamination of the reactor, which has been used for plasma polymerization of nitrogen-containing monomers [1]. The ablation of electrode material has been utilized to create a graded metal–polymer and polymer–metal interfaces to obtain an excellent adhesion [2,3]. Ablation, therefore, could be utilized in a beneficial way in the engineering of interfaces if we know the nature of ablation and how to control it.

Two major mechanisms are responsible for the ablation of material exposed to glow discharge: (1) physical sputtering of elements and (2) chemical etching by reactive species in luminous gas phase. The sputtering of metal by argon glow discharge is a typical example of physical sputtering, which is essentially a momentum–exchange process. The energy of the impinging Ar^+ is transferred to the colliding atom and dislodges the atom from the crystalline structure, transferring its energy to a neighboring atom. This energy transfer process continues until one of the atoms is knocked out into the vapor phase. Chemical etching involves the chemical reaction of the impinging species with the dislodging elements, such as oxidation by O_2 plasma and fluoride formation by CF_4 plasma.

Ablation under the influence of luminous gas has been used in the removal of materials and the cleaning of surfaces, but it has also been utilized in the vacuum deposition of sputtered materials. Vacuum deposition by sputtering of a target material is often called *sputter coating*. Metal deposition by sputtering is a widely used in vacuum deposition of metals and is a typical example of sputter coating. The sputtering of electrode material is an important factor to be considered particularly in direct current (DC) to roughly 100 kHz discharge employed in luminous chemical vapor deposition (LCVD).

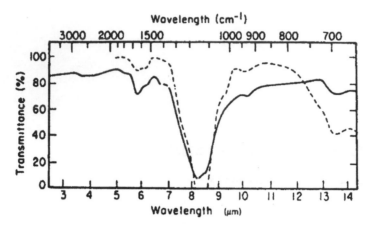

Figure 9.1 Infrared spectra of sputtered (dotted line) and plasma-polymerized polytetra-fluoroethylene (solid line). Adapted from Ref. 4.

Similar sputtering processes have been applied to sputter coating of polymeric materials by using a polymer as the target. The mechanism involved, however, is significantly different from sputtering of metals, and the term sputter coating applied to polymer materials is not an appropriate description of the actual processes that take place under the technically identical setup. The small Ar^+, regardless of how energetic it might be, cannot sputter a polymer molecule, of which the mass is greater than that of the impinging ion by the order of 10^3–10^6. A fraction of a polymer molecule can be split off by an impinging ion, but such a fragment cannot be deposited in the same way that a sputtered metal is deposited on a substrate. What actually takes place is the fragmentation of polymer molecules by impinging species and the subsequent LCVD of the gaseous ablation products. Tibbitt et al. showed that all analytical data for the plasma polymer of tetrafluoroethylene and sputter-coated polytetrafluoroethylene (PTFE) are virtually identical [4]. Furthermore, both of them are significantly different from PTFE. The infrared (IR) spectra and dynamic electrical properties of these polymers are shown in Figures. 9.1 and 9.2.

2. ABLATION OF POLYMERS

2.1. General Principle for Polymer Ablation

The ablation of polymer surface can occur by chemical reactions with reactive species in the luminous gas phase. In this case also, the degradation or fragmentation of polymer occurs before the ablation, and the ablated materials could be chemically different from the constituent segments of the target polymer depending on the nature and the extent of chemical reactions that occur in the overall ablation processes. Whether luminous gas can etch a surface is dependent on the sensitivities of elements involved in the solid surface with the luminous gas phase. Consequently, the ablation of a polymer depends on the nature of the polymer and the nature of the luminous gas phase, and the photolysis of polymers plays a significant role.

The effectiveness of chemical etching is largely dependent on the volatility of fragmented species. For instance, oxygen plasma is an effective etching gas for most

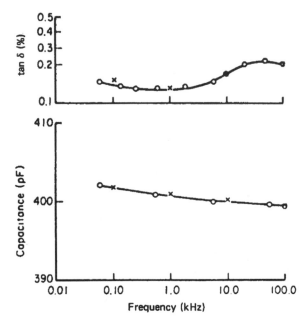

Figure 9.2 Dynamic electrical properties of polytetrafluoroethylene thin films at 20°C; Key: ○, sputtered polytetrafluoroethylene; ×, plasma-polymerized polytetrafluoroethylene. Adapted from Ref. 4.

organic polymers but not for silicon-containing polymers because oxygen plasma forms stable silicon oxides (solid). Plasmas of fluorine containing a compound such as CF_4 are effective etching gases for silicon and silicon-containing polymer because Si and F forms volatile compounds, but not for most organic polymers. The ablation of a polymer by luminous gas (glow discharge) can be generally observed by measuring the loss of weight as a function of the exposure time. The weight loss is generally linearly proportional to the exposure time as depicted in Figure 9.3, and the ablation rate could be calculated from the weight loss rate. However, in some case the weight loss is not linear with the exposure time as shown in Figure 9.4. With water vapor glow discharge, the initial rate of ablation is greater than that by O_2 glow discharge, but the weight loss levels off after a relatively short exposure time [5]. Even in such a case, the general trends due to the chemical structure of the polymer are observed. The rates of weight loss for various polymers on plasma exposure are shown in Table 9.1 [6]. It should be noted that CF_4, which is recognized as a good etching gas for Si-containing materials, is not a good etching gas for most organic polymers that do not contain Si in their backbone chain.

The total weight loss of polymers due to the exposure to glow discharge depends on the two major factors: (1) the susceptibility of elements involved in a polymer to a specific luminous gas phase to which the polymer is exposed and (2) the subsequent chemical reactions of the free radicals that were caused by the detachment of the elements from the polymer. If a glow discharge–susceptible element such as O is in the main chain of a polymer, these two factors could not be distinguished and a very high weight loss results, as in the case of polyoxymethylene (POM).

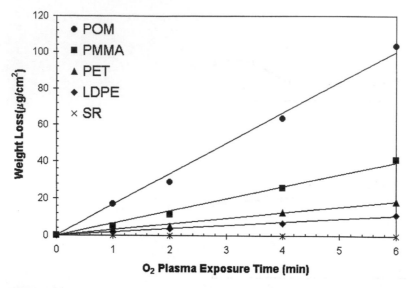

Figure 9.3 Weight loss as a function of exposure time with oxygen glow discharge; POM: polyoxymethylen, PMMA: poly(methyl methacrylate), PET: poly(ethylene terephthalate), LDPE: low density polyethylene, SR: silicone rubber (polydimethylsiloxane).

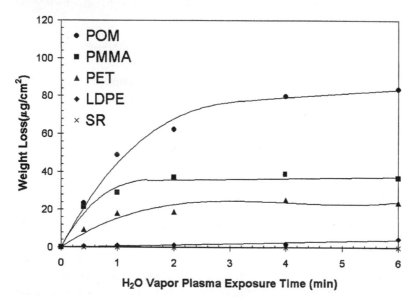

Figure 9.4 Weight loss as a function of exposure time with H_2O glow discharge; POM: polyoxymethylene, PMMA: poly(methyl methacrylate), PET: poly(ethylene terephthalate), LDPE: low density polyethylene, SR: silicone rubber (polydimethylsiloxane).

The main effect of exposure to a luminous gas phase is the detachment of small molecules such as H_2, CO, CO_2, and so forth, depending on the nature of gas used, which leaves free radical (unpaired electron) sites on the polymer chains (polymer free radicals described in Chapter 6). The weight loss rate greatly depends on how

Table 9.1 Plasma Susceptibility of Polymers

Type	Schema	Structural formula	Fiber	Air (50 W, 1.0 torr) Rate	Mean	He (5 W, 1.5 torr) Rate	Mean	CF$_4$ (50 W, 1.0 torr) Rate	Mean
				Weight loss rate, (mg/cm^2·min)×10^3					
1	—O—	—O—	POM	68.5	68.5	2.4	2.4	8.0	8.0
	—O— (structure)	sugar-ring structure	Triacetate	18.3		0.8		0.7	
			Cupra	15.7	16.2	2.4	1.4	2.0	1.3
			Acetate	14.7		1.1		1.2	
		aromatic diester	PET	9.3	9.3	0.5	0.5	0.8	0.8
	(C=O)	(C=O)	Vinylon	11.3	11.3	0.4	0.4	1.8	1.8
2	—C(=O)—N—	—C(=O)—N—	Nylon 6	8.9		1.0		1.1	
			Nylon 66	8.9	8.9	1.0	0.9	0.8	1.3
			Silk	8.8		0.8		2.0	
3	≡N	≡N	PAN	5.1	5.1	0.1	0.1	0.5	0.5
	—	—	PE	8.1		0.7		1.9	
			PP	3.4	5.5	0.3	0.4	0	0.9
			PVC	5.1		0.3		0.7	

the polymer free radicals react subsequently. The role of the two factors mentioned above could be seen clearly in the case of PTFE in O$_2$ glow discharge shown in Table 9.2 in which weight loss rates due to O$_2$ glow discharge for linear polymers are compared [7].

Due to the high electronegativity, F detaches relatively easily from the polymer. This aspect is discussed in Chapters 7 and 10. It is also evidenced in many glow discharge experiments in which Teflon parts are used as vacuum components that cause the deposition of F-containing material from monomers that do not contain fluorine atom. On the other hand, the weight loss rate observed is the lowest except silicone rubber. This low weight loss rate can be attributed to the stability of the polymer free radicals left in the chain of PTFE, which cannot abstract F from the neighboring segments for further degradation of the polymer. According to Kuzuya's interpretation [8,9], the degradation of hydrocarbon polymers exposed to Ar glow discharge starts with the abstraction of H in the neighboring environment by the polymer free radical. In the case of PTFE, the polymer free radical cannot abstract F and there is no H. Consequently, the Ar glow discharge–treated PTFE

Table 9.2 Weight Loss Rates of Polymers by O_2
Glow Discharge Under a Set of Conditions

Polymer	Weight loss rate $(\mu g/cm^2 \, min)$
POM	16.1
PMMA	6.01
PET	3.02
LDPE	1.80
PTFE	0.087
Silicone rubber	0.00

does not degrade, although sufficient numbers of F could be detached to make the surface substantially hydrophilic.

The extremely low weight loss rate observed with silicone rubber is due to the change of chemical structure of polymer by chemical reaction of the reactive oxygen. The oxidation of Si forms nonvolatile oxides, whereas the oxidation of C leads to volatile oxides. Consequently, Si in the organic polymer is converted to inorganic oxides, which are stable in the luminous gas phase in vacuum.

2.2. Influence of Chemical Structure of Polymers

The impingement of ions and other reactive species existing in plasma to a polymer surface is much more complex than the same species impinging on a metal surface because of the chemical nature and conformational shape of polymer molecules. It is far from a simplified model of a ball colliding with another ball or an assembly of balls. Some portion of a molecule or repeating unit could be more vulnerable than other portions. This aspect is extremely important if one considers that a significant portion of energy transfer occurs in the form of photons penetrating into the polymer exposed. Consequently, there exists a correlation between the chemical structure of repeating unit of polymers and plasma susceptibility (chemical etching tendency) of polymers as shown in Table 9.1 and 9.2.

2.3. Ablation and Degradation of Polymer

Ablation by luminous gas in a glow discharge is the loss of matters by the action of energetic and/or chemically reactive species that constitute the luminous gas phase. The weight loss observed after a polymer sample is exposed to a glow discharge is a typical measure of the extent of the total ablation. However, the total ablation could occur in several steps, and the phenomena are much more complicated than the sputtering of metals and inorganic materials. A good example to distinguish ablation and degradation of polymer is the case of Teflon described above. Under the influence of luminous gas, a Teflon sample ablates F atoms relatively easily due to the high electronegativity of F, but the remaining polymer molecules do not degrade and consequently shows very low weight loss, as shown in Table 9.2. In contrast to this situation, the weight loss of POM is due to the high degradation rate of the glow discharge–treated sample. Because of high electronegativity of O atoms in the backbone of POM, it is highly likely that C-O bonds will be broken as the direct and primary effect of glow discharge treatment, which leads to simultaneous degradation and ablation of gaseous entities. Without O in the backbone, most hydrocarbon polymers follow essentially the same steps seen with Teflon but with H instead of F, e.g., ablation of H and degradation of polymer chain by the further abstraction of H.

It was recently observed that the surface dynamic stability of glow discharge–treated polymer films follows the inverse order of the weight loss rate [5]. The weight loss value emphasizes the extent of ablation, which is completely a different issue from what is left on the ablated surface. The trend indicates that highly ablatable materials leave rather unaffected surface behind, but a polymer, which has lower weight loss rate, leaves affected moieties that could not be ablated but sufficiently degraded to influence the surface dynamic stability. The degraded material left on the surface greatly influences the adhesion of coating applied on the glow discharge–treated polymer surface. However, such an influence of glow discharge treatment cannot be judged by the ablation characteristic measured by the weight loss. The details of this aspect are described in Chapter 30.

2.4. Influence of Discharge Gas

Plasma etching has been developed mainly in the processing of microelectronic materials where etching of silicon and silicon derivatives is the major concern. Therefore, the concept of good etching gases is largely based on the etching of silicon. Plasma etching of polymers that do not contain silicon, particularly in the backbone, does not follow the etching tendency of silicon, and a "good etching gas" (for silicon) is not necessarily a good etching gas for most polymers. This situation is seen in Table 9.1.

The addition of an inert gas, e.g., Ar, tends to increase physical etching compared to that without inert gas (a carrier gas). Physical etching enhances the effectiveness of chemical ablation process in many cases. Adding Ar to H_2 plasma enhances the removal of oxides from a steel surface [10–13]. Although O_2 or $(O_2 + Ar)$ plasma cleans metal surfaces by chemically removing organic contaminants, the process cannot distinguish organic contaminants and organic polymers and consequently cannot be used for the cleaning of a polymer surface in a similar manner to the metal surface cleaning.

2.5. Dependence of Ablation on Operational Parameters of Glow Discharge

The creation of reactive species that cause ablation is essentially the same process as that occurs in LCVD, except that the final result is completely opposite, i.e., ablation vs. deposition. In this context, ablation by luminous gas could be described as luminous chemical vapor treatment (LCVT). Therefore, the dependence of ablation on operational parameters in LCVT is very similar to that of LCVD, which is discussed in more detail in Chapter 4. The chemical ablation of polymeric materials by O_2 plasma [6] is described here to demonstrate how oxidative ablation is influenced by the operational parameters of discharge.

Chemical etching by oxygen glow discharge obviously depends on the concentration of chemically reactive species, which is created by the interaction of

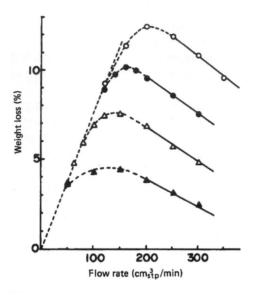

Figure 9.5 Effect of gas flow rate on weight loss of nylon 6 treated by air plasma for 5 min.; key: ○, plasma discharge at 100 W; ●, 70 W; △, 50 W; ▲, 30 W.

O_2 molecules with the electromagnetic energy input. Therefore, flow rate, which determines the mass of O_2 in a system, and discharge power, which determines the energy input, are major factors in determining the etching rate. In a practical system, however, one of these factors tends to dominate depending on the conditions of the discharge.

In Figure 9.5 the etching rates observed with nylon 6 at different discharge wattages are plotted against the flow rate of air. The etching rate increases with increasing flow rate of air, indicating that the etching rate is proportional to the rate of supply of reactive species of O_2 into the system; however, the etching rate reaches a maximal value and starts to decrease as the flow rate increases further. This happens at all discharge wattages, but the maximal etching rate as well as the flow rate required to reach the maximum are higher when a higher discharge wattage is employed.

After passing the maximum, the etching rate decreases in spite of the fact that more O_2 is being supplied. This phenomenon can be explained by the two limiting cases of the actual discharge process. At regions of lower flow rate, ample electric energy is supplied to create reactive oxygen species; in that region, the supplying rate of O_2 is rate determining. Thus, the increase in flow rate increases the etching rate linearly until the supply of O_2 reaches a point where the supply of electric energy becomes deficient. Therefore, the downhill side of the curves in Figure 9.5 (right side of the maximum) can be recognized as the energy-deficient region where the rate-determining factor is the energy input rate. A similar situation is discussed for LCVD in Chapter 8.

If we apply the same energy input parameter to the luminous gas phase used in LCVD (plasma polymerization), W/FM, where W is the discharge energy (watts), F the volume flow rate (cm^3_{STP}/min), and M the molecular weight of gas, this situation can be understood by the same principle. In Figure 9.6, the etching rates shown in Figure 9.5 are plotted against W/F, where F is the flow rate of air,

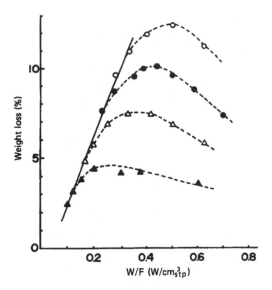

Figure 9.6 Relation between values of W/F and weight loss for nylon 6 treated by air plasma for 5 min; Key: ○, plasma discharge at 100 W; ●, 70 W; △, 50 W; ▲, 30 W.

which is proportional to the flow rate of O_2. Because O_2 is the only major concern, W/F is essentially proportional to $W/F(O_2)M(O_2)$. As seen in Figure 9.6, all points that are not in the linear portion of plots in Figure 9.5 fall in a straight line, indicating that in this region the energy input rate is the rate-determining factor.

Thus, the chemical etching rate is dependent on whether the process is performed in the flow rate–controlled region or in the discharge power–controlled region. In the former case the etching rate increases with flow rate of the etching gas, and in the latter case the etching rate increases with the discharge wattage.

The etching of organic polymers by nonreactive plasma can also occur due to the photolysis effect of luminous gas. As mentioned earlier, sputtering of organic polymers requires the fragmentation of macromolecules. Therefore, etching rates of polymers by discharge of nonreactive gas such as argon or helium are highly dependent on the chemical nature of the macromolecules as shown in Table 9.1.

The structure dependency of the etching rate is enhanced in chemically reactive luminous gas, but the nature of dependence remains the same. Plasma-sensitive structures such as –O– in the backbone of a polymer- and oxygen-containing pendant group play a dominant role. It is important to note that LCVT, chemical etching of polymers by non-polymer-forming gas plasma, can be well described by the same discharge power parameter for LCVD, which is W/FM.

3. SIMULTANEOUS SPUTTER COATING AND LCVD

3.1. RF Discharge with Metal Cathode

As mentioned earlier, the process that utilizes the deposition of sputtered material is often termed sputter coating. The material is sputtered from a target by plasma, and the sputtered material is deposited onto a substrate, which is usually placed out of the plasma region to avoid interaction of plasma with the substrate material. In such a process, a chemically nonreactive gas such as argon is generally used for glow discharge.

When a monomer of LCVD is used (either alone or mixed with argon), simultaneous sputter coating and LCVD occur. The resultant deposition is a mixture of the product of LCVD (plasma polymer) and the material of the sputter coating. Dilks and Kay [14] studied such a simultaneous sputter coating and plasma polymerization, and their experimental setup is shown in Figure 9.7. One of the parallel electrodes is grounded, and a radio frequency power of 13.56 MHz is applied to the other electrode. (The ungrounded electrode is termed the cathode because it attains an overall negative potential due to the greater mobility of the electrons than the ions in plasma.)

The positive ions arrive at the cathode with increased kinetic energy, and the target material (used as the cathode) is removed from its surface by cumulative physical and chemical etching, the magnitude of which is dependent on the nature of the plasma gas. The substrate is placed on the surface of the anode (electrically insulated). Perfluoropropane was used as the plasma gas, and silicon, germanium, molybdenum, tungsten, and copper were employed as the cathode (target materials to be sputtered).

Among the cathode materials used, silicon and germanium were not incorporated into the plasma polymer because these two materials form volatile fluorides and were pumped out of the system. In contrast, copper, which forms nonvolatile fluoride, was incorporated into the plasma polymer. Once metals were incorporated into the plasma polymers, they were uniformly distributed throughout the plasma polymer matrix.

The general trends found in this kind of simultaneous sputter coating of molybdenum and LCVD of perfluoropropane are as follows:

1. Although the amount of metal incorporated into the film varies within narrow limits (18–26% by weight), it is directly related to the ratio of the etching rate to the polymer deposition rate.

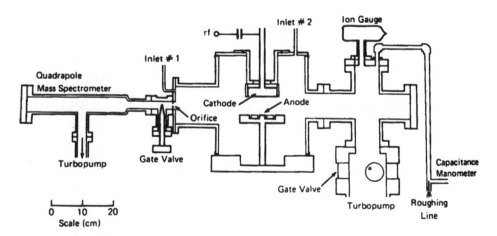

Figure 9.7 Schematic of the plasma reactor configuration. Adapted from Ref. 14.

2. The structure of the metal-containing entities is essentially unchanged in this range.
3. Although the polymer structure varies very little as a function of the plasma parameters, there is a distinct tendency for the polymers formed at the higher discharge power to be a tighter three-dimensional network (more highly cross-linked), as evidenced by a greater amount of CF and carbon that is not directly attached to fluorine (based on the XPS spectrum), as well as a decrease in the overall F/C ratio of the film.

The core level XPS spectra of the polymers synthesized in three separate experiments using germanium, molybdenum, and copper cathodes, respectively, and gold substrates (maintained at 16°C) electrically insulated from the anode are shown in Figure 9.8. The designations $Ge-C_3F_8$, $Mo-C_3F_8$, and $CU-C_3F_8$ refer to the cathode material and the monomer gas combinations. Empirical formulas for those materials based on XPS data are $(C_3F_{4.0}O_{0.6}Mo_{0.3})_n$, $(C_3F_{3.9}O_{0.3}CU_{0.3})_n$, and $(C_3F_{3.6}O_0Ge_0)_n$.

Although germanium was not incorporated into the plasma polymer in the $Ge-C_3F_8$ system, the cathode weight loss was found to be about 2.6 times greater than the corresponding figure for molybdenum. This means that the effective chemical etching (by forming highly volatile product) prevents the incorporation of the product into the deposition of LCVD. Therefore, the chemical etching rates of metal and the incorporation of the sputtered metals into the plasma polymer could be inversely related. These data provide excellent illustrations of simultaneous, opposing processes, i.e., the ablation of material at one location and the incorporation of the ablated material into plasma polymer at another location.

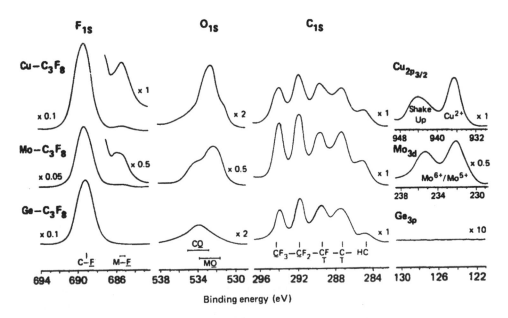

Figure 9.8 ESCA spectra of the polymers formed in three experiments involving germanium, molybdenum, and copper cathodes, respectively. Adapted from Ref. 14.

3.2. Magnetron Discharge

Another example of simultaneous sputter coating and plasma polymerization is the incorporation of aluminum, which is used for the electrodes of a magnetron glow discharge, into the plasma polymer of CH_4 under certain discharge conditions. (In order to avoid this phenomenon metals that have low sputtering yields are used for the magnetron electrode in LCVD.) In glow discharge of hydrocarbons, which are generally in the polymer-forming group under glow discharge conditions, polymer formation predominates under most conditions used in plasma polymerization, and the sputtering of electrode metal does not occur under normal operational conditions. However, in glow discharge of CH_4, which is the smallest of hydrocarbons, with Al electrode, which has high sputtering yield, the balance between plasma polymerization and ablation could shift toward sputtering of Al depending on conditions of discharge, particularly on the energy input level or the plasma energy density.

When LCVD of CH_4 is carried out using a magnetron discharge system with relatively low frequency (e.g., 10 kHz), a certain section of electrode surface facing the toroidal glow is exposed to an intensive glow ring created by the magnetic field. The ion bombardment on this section of electrode surface evidently becomes strong enough to sputter metal such as the aluminum used for the electrode. Because CH_4 does not create chemically reactive species that cause chemical etching, the sputtering of metal is considered to occur by physical sputtering. Once aluminum is sputtered from the electrode, the metal is incorporated into plasma polymer in just the same way that metals are incorporated into a perfluorocarbon discharge, as described earlier. Aluminum is distributed evenly throughout the entire thickness of deposition.

The rate of sputtering of aluminum from the electrode used in a magnetron plasma polymerization system is dependent on the plasma energy density, which can be stipulated by the external parameter V/p, which is the acceleration potential in the vicinity of the cathode, for Ar discharge, while the deposition of CH_4 is dependent on W/FM in joules per kilogram of CH_4 for LCVD as described in Chapter 8.

When argon is used as the plasma gas, the sputtering rate measured by the deposition rate of aluminum on a quartz crystal thickness monitor was found to be linearly proportional to W/p as depicted in Figure 9.9 [15]. In the domain of glow discharge, W increases by the increase of current at nearly constant discharge voltage, and W/p could replace V/p.

Because of the polymerization that predominates in the plasma of CH_4, the linear dependence of sputtering on W/p seen in Figure 9.9 was not observed in the case of CH_4 glow discharge. The Al ablation, in this case, can be examined by measuring the amount of aluminum in plasma polymers of CH_4 prepared at various values of W/FM as depicted in Figure 9.10, in which the amount of aluminum is given by XPS Al2s peak height in arbitrary units.

At a lower range of W/FM, no aluminum is found in the plasma polymer. It is incorporated only when W/FM exceeds a certain threshold value (about 5×10^9 J/kg), indicating that the sputtering of aluminum in CH_4 glow discharge also occurs only when the plasma energy density exceeds a certain threshold value.

The sputtering of metal from the electrode by Ar glow discharge depends primarily on the kinetic energy of the impinging ions. Therefore, it is easy to

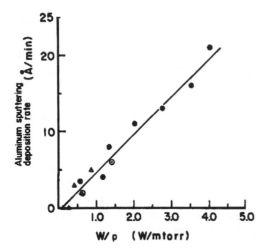

Figure 9.9 Sputtering on the discharge parameter W/p; the aluminum sputtering is measured by the deposition onto the thickness monitor sensor in the glow discharge of argon, flow rate (cm^3_{STP} /min): ●, 2.3; ⊙, 5.6; △, 12.4; ▲, 23.3. Adapted from Ref. 15.

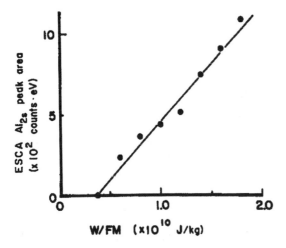

Figure 9.10 Discharge parameter of CH_4 glow discharge; the amount of aluminum deposited is determined by XPS.

understand its dependence on W/p. In the case of metal sputtering by CH_4 (polymer-forming) plasma, the basic principle should be the same; however, because of the polymer-forming nature of plasma, the kinetic energy of the impinging ions cannot be given by the same parameter W/p. Likewise, the energy that contributes to the sputtering of metal cannot be given by the overall energy input because the polymer-forming process consumes a large portion of the energy input. Further-more, as described in Chapter 3, what ion is responsible for the sputtering of aluminum in CH_4 glow discharge is not clear. Nevertheless, it is important to recognize that the sputtering of metal used as electrode could be controlled by the operational parameter of LCVD. Conversely, the LCVD process could result in sputtering of the electrode metal depending on the operational conditions and the nature of metal. Because of this aspect, metals that have low sputtering yield, such as tungsten or titanium, are used as electrodes of magnetron discharge for ordinary LCVD.

Although incorporation of metal might be considered to constitute contamination of an LCVD product, this phenomenon can be utilized advantageously. An example is the improvement of adhesion of plasma polymers to metal (substrate) surfaces. Adhesion of a polymer film or coating to a metal is an important factor in many practical uses of polymers. Poor adhesion of a coating is a serious problem when the coating is exposed to high humidity or liquid water. Without physical interpenetration of the coating layer into the substrate, as is the case obtained by roughing a metal surface by sand blasting, the adhesion of a polymer film to a metal surface is relatively weak.

When a very thin film (e.g., thicknesses of less than $1\,\mu m$) of a polymer is applied to a smooth surface of platinum, most polymers peel off within minutes upon immersion in liquid H_2O. This is also true for most plasma polymers applied to platinum surfaces. However, when an ultrathin film of CH_4 LCVD was deposited under the conditions that provide plasma energy density sufficiently high to sputter aluminum from the electrodes, tenacious adhesion that survived over $10\,h$ of boiling in saline solution was obtained, probably due to the incorporation of electrode metal at the interface.

4. ABLATION-INFLUENCED LCVD

The term *ablation,* with respect to organic materials, represents two types of reactions. One is the splitting of a portion of solid materials, and the other is the splitting of a portion of the molecules in luminous gas phase. The first process might be closely related to the etching of material, but the second process influences the deposition process and the nature of the deposits. The ablation process associated with monomer can be visualized as the splitting of a molecule or the fragmentation of the monomer. This aspect is very important in plasma polymerization, and it is not an exaggeration to state that no monomer polymerizes without fragmentation in glow discharge.

As soon as plasma is created, the gas phase is no longer the vapor of the original monomer but becomes a complex mixture of the original monomer, ionized species, excited species of the original monomer, excited species of fragments from the monomer, and gas products that do not participate in polymer formation such as H_2 and F_2. Perfluorocarbons represent perhaps the most extreme case of ablation competing with polymer formation. The principle found with perfluorocarbons, however, should be applicable to nearly all cases.

Although tetrafluoroethylene normally undergoes cycle 2 polymerization via olefinic double bond, the splitting of the C-F bond becomes the dominating factor at a high level of energy input (W/FM), and ablation prevails. This situation can be easily visualized by the dependence of deposition rate of tetrafluoroethylene (TFE) on LCVD conditions. Figure 9.11 depicts the wattage dependence of TFE deposition

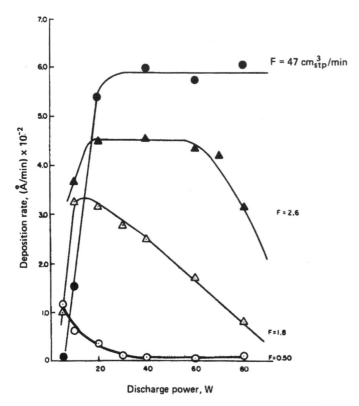

Figure 9.11 Dependence of polymer deposition rate on discharge power at various flow rates for tetrafluoroethylene.

rates at various flow rates [16]. At low flow rate (high W/FM), the deposition rate decreases with increasing discharge wattage.

Because solid materials must maintain the vacuum, the reactor wall always exists in an LCVD reactor. The plasma also interacts with wall materials as well as any other materials that exist in the plasma, such as substrate and support. Therefore, polymer-forming intermediates and gaseous by-products may also originate from solid materials with which plasma interacts by virtue of the ablation caused by the luminous gas. In this sense, any material that interacts with plasma becomes a source of monomer for plasma polymerization.

When plasma polymerization of TFE was carried out in a glass tube and an NaCl powder, which was spread on a glass plate, was used as the substrate to measure the IR spectrum by forming pellets of NaCl coated with plasma polymer, conspicuous IR peaks appeared at 1130, 1080, and 1035 cm^{-1}, as shown in Figure 9.12 [17]. When the same polymerization is carried out with a KRS-5 internal reflection prism as the substrate instead of NaCl powder, the IR spectrum showed no peak at this region [18]. The peaks are attributed to Na$_2$SiF$_6$ formed by the reaction of NaCl and a fluorosilicon compound, which is an ablation product of the glass wall formed by the reaction between glass and reactive species of fluorine in the plasma (atomic fluorine, F$^-$, etc.). The intensity of these peaks increases as the flow rate of TFE was lowered at a fixed wattage of 70 W (W/FM was increased).

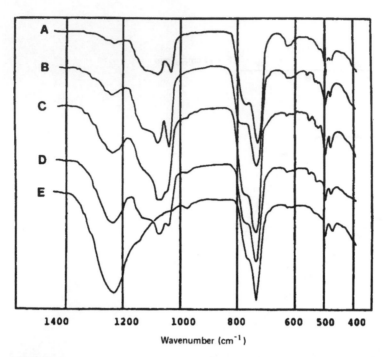

Figure 9.12 Infrared spectra of plasma-treated NaCl powder under various monomer flow rate conditions (70 W), $F(C_2F_4)$ (cm^3/min) for each spectrum is as follows: A, 0.1; B, 0.15; C, 0.27; D, 0.51; E, 0.86.

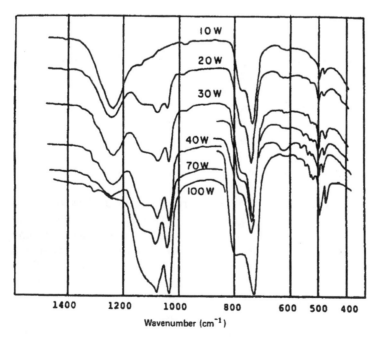

Figure 9.13 Infrared spectra of plasma-treated sodium chloride powder under various wattages at a fixed flow rate ($F = 0.155$ cm^3/min).

At the highest flow rate used (i.e., $0.86 \, cm^3_{STP} /min$), no peak attributable to Na_2SiF_6 was observed. Similar results obtained under conditions of fixed flow rate and varying discharge power are shown in Figure 9.13. Under such conditions, no peak for Na_2SiF_6 appears at low wattage, and the intensity of peaks increases with increasing wattage. From Figures 9.12 and 9.13 we can conclude that when the W/FM reaches a certain level the ablation of glass by plasma of TFE becomes evident and increases with the increasing W/FM.

Thus, it is clear that the balance between polymer formation and ablation is dependent not only on the chemical nature of the monomer, the wall material, the substrate, and so on, that come into contact with the luminous gas phase, but also on the discharge conditions, particularly the energy density given by W/FM.

REFERENCES

1. Engelman, R.A.; Yasuda, H.K. J. Appl. Polym. Sci., Appl. Polym. Symp. **1990**, *46*, 439.
2. Sun, B.K.; Cho, D.L.; O'Keefe, T.J.; Yasuda, H. Chemically graded metalization of nonconducting substrates by glow discharge plasma polymerization technique. In *Metallized Plastics 1*, Mittal, K.L., Susko, J.R., Eds.; Plenum Press: New York, 1989; 9–27.
3. Yasuda, H.; O'Keefe, T.J.; Cho, D.L.; Sun, B.K. J. Appl. Polym. Sci., Appl. Polym. Symp. **1990**, *46*, 243.
4. Tibbitt, J.M.; Shen, M.; Bell, A.T. Thin Solid Films **1975**, *29*, L43.
5. Weikart, C.M.; Yasuda, H.K. J. Polym. Scie. A: Polym. Chem. **2000**, *38*, 3028.
6. Yasuda, T.; Gazicki, M.; Yasuda, H. J. Appi. Polym. Sci., Appl. Polym. Symp. **1984**, *38*, 201.
7. Data to be published.
8. Kuzuya, M.; Niwa, J.; Ito, H. Macromolecules **1993**, *26*, 1990.
9. Kuzuya, M.; Kondo S.; Sugito, M. Macromolecules **1998**, *31*(10), 3230.
10. Wang, T.F.; Yasuda, H.; Lin, T.J.; Antonelli, J.A. Prog. Organic Coatings **1996**, *28*, 291.
11. Yasuda, H.; Wang, T.F.; Cho, D.L.; Lin, T.J.; Antonelli, J.A. Prog. Organic Coatings **1997**, *30*, 31.
12. Lin, T.J.; Antonelli, J.; Yang, D.J.; Yasuda, H.K.; Wang, F.T. Prog. Organic Coatings **1997**, *31*, 351–361.
13. Ohuchi, F.S.; Lin, T.J.; Yang, D.J.; Antonelli, J.A. Thin Solid Films **1994**, *10*, 245.
14. Dilks, A.; Kay, E. A: CS Symp. Ser. **1979**, *108*, 195.
15. Yasuda, H. *Plasma Polymerization*; Academic Press Orlando, FL, 1985; 193.
16. Yasuda, H.; Hsu, T. J. Polym. Sci., Polym. Chem. Ed. **1978**, *16*, 415.
17. Masuoka, T.; Yasuda, H.; J. Polym. Sci., Polym. Chem. Ed. **1981**, *19*, 2937.
18. Masuoka, T.; Yasuda, H. J. Polym. Sci., Polym. Chem. Ed. **1982**, *20*, 2633.

10
Competitive Ablation and Polymerization

1. CHEMICAL ETCHING AND MATERIAL FORMATION

As mentioned in Chapter 9, many reactions occur simultaneously in luminous chemical vapor deposition (LCVD) systems. In order to understand material formation in the luminous gas phase one should recognize the nature of other processes as well because it is nearly impossible to single out one particular process by eliminating others completely.

From the viewpoint of LCVD as a material production process, there are two opposing processes: polymer formation, which leads to deposition of material, and ablation, which leads to removal of material. For instance, saturated fluorine-containing compounds are widely used in the etching process of silicon and silicon-containing materials for the production of microelectronic circuitry. On the other hand, perfluorocarbons are used as monomers of plasma polymerization to deposit materials in the form of coatings. In the early stage of the development of plasma polymerization, those two processes were considered to be independent, and for some years it was thought that certain perfluorocarbons used for plasma etching could not be polymerized by glow discharge.

The interactions of these two opposing processes and their coexistence in luminous gas phase were recognized in the early 1970s. Kay demonstrated that the main effect of CF_4 plasma can be shifted from etching to polymer deposition by adding H_2 into gas and that, depending on the amount of H_2 added to the CF_4, the balance between material deposition and ablation can be controlled [1]. Yasuda reported at the same meeting a similar phenomenon attributable to ablation in the plasma polymerization of tetrafluoroethylene, which is described in Figure 9.11 [2].

These two early findings lead to the concept of competitive ablation and polymerization (CAP) which emphasizes the importance of a balance between polymer formation and ablation [3]. The first finding demonstrated the control of ablation due to extremely reactive fluorine-related species (atomic fluorine, F^-, etc.) by chemical reactions, and the second demonstrated the role of discharge conditions that control the production of highly ablative species.

CF_4 cannot polymerize without breaking C-F bonds because the creation of C-C bond constitutes the fundamental step of polymer formation. Once fluorine is split off from the original molecules, its ablative effect becomes dominant, which makes CF_4 a very effective etching agent in glow discharge. The addition of H_2 to such a system has a stabilizing effect. This situation can be easily visualized by comparing bond energies. The high reactivity of fluorine-based species can be seen from the fact that the F-F bond energy is only 154 kJ/mol. The bond energy for H-F is 565 kJ/mol, indicating that 411 kJ/mol of stabilization is gained by reacting with H_2. Thus, the role of H_2 is that of a scavenger of ablative species.

2. CLASSIFICATION OF GASES FOR GLOW DISCHARGE

As far as plasma polymerization and plasma treatment of materials, particularly organic polymers, are concerned, the luminous gas phase (low-pressure plasma) can be divided into three major groups based on the mode of consumption of the gas used to create the plasma: (1) chemically nonreactive plasma; (2) chemically reactive plasma; and (3) polymer-forming plasma. The terms *chemically reactive* and *chemically nonreactive are* based strictly on whether the gas used in glow discharge is consumed in chemical processes yielding products in the gas phase or being incorporated into the solid phase by chemical bonds.

Chemically nonreactive types of plasma are mainly those of monoatomic inert gases. Argon plasma can ionize other molecules or sputter materials but is not consumed in chemical reactions. Chemically reactive types of plasma are those of inorganic and organic molecular gases such as O_2, N_2, CO, NH_3, CF_4 and so forth, which are chemically reactive but do not form polymeric deposits in their pure gas plasma. Polymer-forming plasmas are obviously chemically reactive but form a polymeric solid deposit by themselves. Many organic and inorganic vapors are in the polymer-forming group. It is important to note that only the polymer-forming type of gases create the molecular dissociation glow that is described in Chapter 3. In other words, only gases that are capable of LCVD create the dissociation glow.

An important point is that the ablation of material can occur in all types of plasma, although the extent of the process and the dominating mechanisms vary with the type of gas and discharge conditions. Ablation by chemically nonreactive plasma occurs by the momentum-exchange process (physical sputtering). Ablation by chemically reactive plasma can occur by both physical and chemical etching processes depending on the stability and volatility of the products of the chemical reaction between the etching gas and the material. For instance, CF_4 plasma (chemically reactive plasma) can etch metals by either mechanism depending on the discharge conditions. Silicon, germanium, molybdenum, tungsten, and copper are chemically etched by C_3F_6 plasma, but the incorporation of the etched metal into the deposition depends on the volatility of the etched entities [4].

Because polymer formation and ablation are competitive and opposing processes, polymer-forming plasma has the least ablative effect; however, ablation in such plasmas cannot be completely ruled out. Sputtering of metals used as the internal electrodes for plasma polymerization has been recognized as a contamination of plasma polymers. Under certain conditions, the sputtering of the electrode materials becomes significant and plays an important role in the engineering of interface as described in Chapter 9.

The sputter coating of polymers (misrepresentation of the real phenomenon) is another important demonstration of how important the CAP mechanism is not only in plasma polymerization but also in other plasma processes. Thus, inadvertent plasma polymerization can take place in an attempted etching process; conversely, etching can occur in an effort to deposit a polymer by plasma polymerization. Therefore, a thorough understanding of the CAP principle seems to be important for the successful operation of any plasma process.

3. PLASMA SENSITIVITY OF ELEMENTS

What happens in a low-pressure plasma process cannot be determined in an a priori manner based only on the nature of the plasma gas or on the objective of the process. The plasma sensitivity series of elements involved, in both the luminous gas phase and the solids, that make contact with the luminous gas phase seems to determine the balance between ablation and polymerization by influencing the fragmentation pattern of molecules in the luminous gas environment.

Plasma sensitivity refers to the fragmentation tendency of gaseous materials and surfaces that come into contact with the luminous gas phase, which contains various energetic species including ions, electrons, excited species, metastables, photons, and chemically reactive species. Ionizing radiation with energetic species such as electron beams, ion beams, and X radiation causes much more severe (penetrating) fragmentation than the exposure to the luminous gas (low-pressure plasma). Plasma sensitivity refers to the latter case rather than sensitivity to ionizing radiation.

Plasma sensitivity series refers to the order of element sensitivity to plasma in a manner similar to the expression of ionization of metals in solution by the galvanic series. There is no clear-cut plasma sensitivity series established today. However, some trends seem to be closely related to the plasma sensitivity series. These are trends found in the order of weight loss rates when polymeric materials are exposed to plasmas [5]. The early recognition of this effect was expressed as the "iN-Out rule" of thumb, which explains that, in a plasma environment, oxygen has a high tendency to be removed from a molecule but nitrogen has a tendency to remain in the molecule [6]. This rule was originally outlined in plasma treatment of polymers; however, a similar rule seems to apply to the fragmentation of gas molecules in the luminous gas phase.

The ionization of an organic molecule does not follow the same process applicable to a simple monoatomic gas. The bond energies that link elements are much smaller than the ionization energies of monoatomic gases used as carrier gases in low-temperature plasma processes. Consequently, the creation of a plasma state of an organic molecule, by itself or with a carrier gas, causes fragmentation of the molecule. This dissociation or fragmentation of molecules by low-energy electrons is responsible for the formation of molecular dissociation glow, which appears as the cathode glow in some discharge, as described in Chapter 3. The dissociation of gas also seems to follow the "iN-Out" rule that is applicable to a solid organic material.

In order to elucidate the mechanisms by which a polymeric material deposits and also those by which surface modification of a polymeric material by plasma proceeds, comprehension of the CAP principle and the plasma sensitivity series seems to be vitally important.

4. CAP PRINCIPLE

The schematic diagram of CAP principle is shown in Figure 10.1. In this scheme, four major processes are necessary to complete the mass balance in the reactor: (1) monomer feed in; (2) ablation; (3) material deposition; and (4) escape from the system (pump out). The formation of reactive species is an ablation process because considerable fragmentation of the monomer (starting material) occurs in general cases.

Material deposition occurs via plasma formation of reactive species; however, it is not a simple step of forming a polymeric material from a set of reactive species. The reactive species do not necessarily originate from the monomer because the ablation process can and does contribute. Gaseous reactive species can originate from once-deposited material (plasma polymer) and also from the reactor wall or any other solid surfaces that are in contact with the luminous gas. How these complex reactive species lead to the material deposition is described in Chapter 5.

The complexity of polymer deposition, which is depicted in Figure 10.1, could be best understood by comparing corresponding schemes for other (hypothetical) mechanisms. Let us first consider a hypothetical case in which a monomer polymerizes in vacuum and deposits onto an exposed surface without any complication. In other words, if a polymer can be formed in vacuum by free radical polymerization, the polymerization mechanism can be depicted as shown in Figure 10.2. In this case, the chemical structure of a plasma polymer can be predicted from the structure of the monomer, and only unreacted monomer escapes

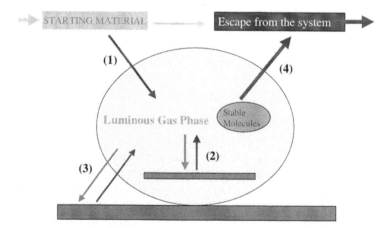

Figure 10.1 Schematic diagram of the Competitive Ablation and Polymerization (CAP) principle: (1) dissociation (ablation) of monomer to form reactive species, (2) deposition of plasma polymer and ablation of solid including plasma polymer deposition, (3) deposition to and ablation from nonsubstrate surfaces, and (4) removal of stable molecules from the system.

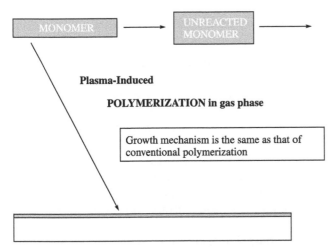

Figure 10.2 Schematic diagram of hypothetical free radical polymerization in vacuum.

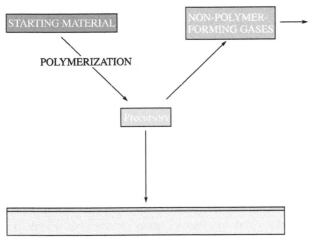

Figure 10.3 Plasma polymerization by the precursor concept: chemical structure of precursors determines structure of plasma polymers.

from the system. This kind of vacuum deposition of polymer by plasma does not occur in the experimental setups recognized as plasma polymerization reactors or LCVD reactors. It should be noted that a chain growth polymerization couldn't occur in the gas phase under vacuum as described in Chapter 5.

In an attempt to consider some extent of fragmentation of the monomer as well as to explain polymerization of simple organic molecules that are not considered monomers, plasma polymerization mechanisms are often explained by assuming plasma-induced precursors, which have polymerizable structures. The precursor concept is detailed in Figure 10.3. It is significantly different from the simple process described in Figure 10.2; however, it still depends on a simple deposition process from precursors to plasma polymer. This concept intuitively assumes that the structure of a plasma polymer can be predicted from the structures of precursors.

By comparing Figure 10.1 with Figures 10.2 and 10.3, the following important factors that constitute the CAP principle can be discerned.

1. *Ablation*, which is fragmentation or dissociation, is involved in every process, i.e., monomer to reactive species, plasma polymer to reactive species, wall surface to reactive species, and escape of fragmented species from the system. Fragmentation of molecules is the primary effect of plasma exposure to a material. The importance of ablation can be visualized in the well-established fragmentation patterns of many organic materials, which constitute the foundation of secondary ion mass spectroscopy (SIMS).
2. The *reactive species* (precursors in the precursor concept) are created not only by fragmentation of the monomer but also by fragmentation of the plasma polymer formed and of materials existing on the various surfaces that come into contact with the plasma.
3. The *escaping species* consist of "non-polymer-forming" stable species and some unreacted monomer depending on system conditions, such as power input level, flow rate, flow pattern, pumping rate, and shape and size of reactor.
4. The *species that do not contribute to polymer formation* are basically stable molecules, such as H_2, HF, and SiF_4. When these species are created in the luminous gas phase, the balance between deposition and ablation shifts. However, the manner in which the balance shifts is dependent on the specific system. The formation of stable escaping species is crucially important in determining the nature of the depositing materials. Some examples that illustrate the factors described above are given in the following sections.

4.1. Fragmentation of Monomer

The extent of fragmentation in the step from monomer to reactive species can be visualized by examining the pressure change in a closed-system plasma polymerization of trimethylsilane (TMS) depicted in Figure 4.2. In a closed system, an increase in pressure means an increase in the total number of gaseous species, which can be achieved only by fragmentation of the original molecules. In such a system, the composition of the gas phase changes with reaction time. Accordingly, the composition of deposited polymer also changes with reaction time. This change is evident in the XPS profile of polymer deposition from TMS in a closed system, the pressure change of which is shown in Figure 4.2. Figure 10.4 contains XPS profiles of plasma polymer deposited in a closed system and of that deposited in a flow system.

TMS, $HSi(CH_3)_3$, contains one Si, three C, and ten H in its original form, and the C/Si ratio is 3.0. In a flow system, the deposited plasma polymer has a very uniform C/Si ratio throughout the film (the lower line in Figure 10.4), indicating that the gas phase composition does not change with reaction time. This also shows that more Si than C is incorporated into the plasma polymer in a flow system plasma polymerization. In a closed system, the ratio changes with reaction time, reflecting the change in the composition of the gas phase (the upper curve in Figure 10.4). This also indicates that Si deposits in the early stages of closed-system polymerization, yielding an Si-poor gas phase. The depletion of Si in the luminous gas phase is reflected in the sharp increase of the C/Si ratio in the plasma

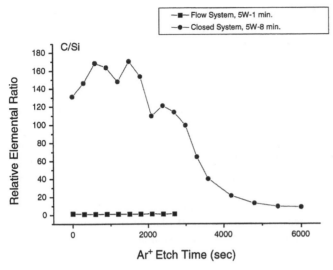

Figure 10.4 XPS cross-sectional profile of TMS plasma polymers deposited in a flow system and in a closed system.

polymer with increased reaction time. The closed system was run for a longer time than the flow system experiment in order to show the depletion of Si in a closed system.

Data presented in Figure 4.2 and Figure 10.4 show how much fragmentation occurs during plasma polymerization and how much influence the fragmentation has on the chemical composition of the resulting plasma polymer. It should be emphasized that the extent of fragmentation is dependent on the various system conditions of plasma polymerization. In the case of deposition onto an electrically floating, conducting substrate and also onto a nonconducting substrate, the energy input level manifested by W/FM is the predominant factor. In the case of deposition onto an electrode surface, [(the current density) $\times p$] is the predominantly important parameter as described in Chapter 8.

The system pressure of a flow system plasma polymerization reactor changes when plasma is initiated. This change depends on the extent of fragmentation of the monomer used, the nature of fragmented molecules, the polymer deposition rate, and the pumping rate of escaping gases. The decrease or increase of system pressure cannot be taken as the tendency or reluctance of the monomer to polymerize. TMS generally causes a large pressure increase; however, its polymerization or deposition rate is much greater than other hydrocarbon monomers that cause a large decrease in flow system pressure.

4.2. Formation of Stable Molecules in Luminous Gas Phase

When stable molecules are formed in the luminous gas phase, the balances considered in Figure 10.1 will shift in a significant manner. The first observation of polymerization in an etching plasma environment by Kay et al. is an extreme case showing dramatic change in plasma processing [1]. In this case, the formation of

HF completely shifts the balance from ablation to polymerization. In the absence of hydrogen, fluorine-containing gases act as effective etching gases for the etching of silicon because they form stable silicone fluorides that can be easily removed from the system. When this process is interrupted by the formation of a more stable and easily removable HF in the presence of hydrogen gas, CF_4 plasma begins to deposit plasma polymer.

When the crucial balance-dictating element comes from somewhere other than the gas phase, the adhesion of plasma polymer can be seriously damaged. A phenomenon that can be explained in this manner was found in a recent study investigating interface engineering of aluminum alloys [7,8]. Consecutive plasma polymerization of TMS and then hexafluoroethane (HFE) on an appropriately prepared aluminum alloy surface resulted in excellent adhesion of a primer, which was applied to the surface of the second plasma polymer. As the primer could not be peeled off by any chemical means, the adhesion of plasma polymer layers to the substrate metal was obviously excellent. When the plasma reactor was contaminated with fluorine-containing moieties due to a mistake in the operation procedure, extremely poor primer adhesion resulted in the same coating system, and the primer layer was easily peeled off from the substrate as a film. This incident provided an opportunity to investigate the interface between the first plasma polymer layer and the substrate metal, since the plasma polymers were strongly adhered to the peeled-off primer film.

In Figure 10.5, XPS cross-sectional profiles of two plasma polymers are compared: (1) that of the paint that peeled off and (2) that of a normal sample of well-adhered plasma polymer layers without primer. [The XPS data for (2) were

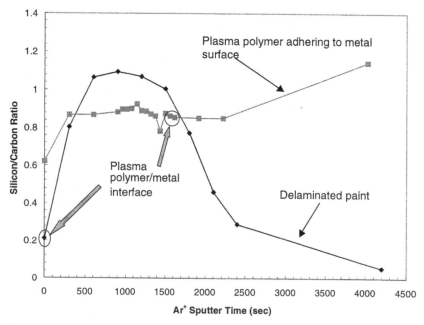

Figure 10.5 XPS cross-sectional profiles of plasma polymers deposited on an aluminum surface, and that adhering to primer film peeled off from the aluminum substrate.

available before the incident occurred.] The interface between the first plasma polymer and the substrate metal can be reached only by sputtering the plasma polymer (without primer), if the adhesion of the plasma polymer to the metal substrate is good. In the delaminated film, the plasma polymer layer was intact adhering well to the primer film, indicating that the delamination occurred as the interfacial failure. In this case, the top surface represented the interface between the plasma polymer layer and the metal substrate, which failed. The two arrows in Figure 10.5 indicate the location of the interface (TMS film/Al) in the cross-sectional profiles. The TMS plasma polymer/substrate interface of the failed sample shows a significantly lower Si/C ratio (0.2–0.3) than that of the normal sample (0.9).

This situation can be illustrated by the principle schematically depicted in Figures 10.6 and 10.7. Figure 10.6 depicts the normal mode of TMS plasma polymerization. Si-containing moieties and C-containing moieties deposit more or less independently. The normalized deposition rate (after the difference of mass is taken into account) of Si-containing moieties is roughly six to seven times greater than that of C-based moieties [9], implying that more Si deposits than C in a flow system plasma polymerization. Figure 10.7 depicts the plasma polymerization that caused the interfacial failure, i.e., the interference in the deposition of Si-containing moieties. Si-containing species are intercepted by the F-containing moieties emanating from the contaminated substrate. Once the stable moiety (represented by Si-F) is formed, it is pumped out of the system readily, and the Si-containing moieties, which otherwise deposit, are diverted to the pump. This situation leads to a decrease in Si content at the metal/plasma polymer interface, which is the exact case found in the XPS profile analysis shown in Figure 10.7. How the oxygen plasma pretreatment of the substrate, which was inadvertently omitted in the problematic case, prevents this situation needs further explanation in terms of the nature and cause of the wall contamination as described in the following sections.

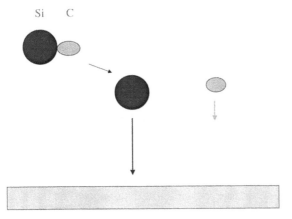

Figure 10.6 Schematic representation of LCVD of TMS: fragmentation of TMS depicted by a oval Si-containing moiety and circular C-containing moiety to its constituent moieties, and different deposition rates depicted by the size of arrow for Si moieties and C moieties.

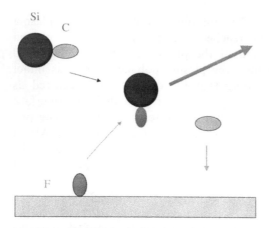

Figure 10.7 Schematic representation of the change in plasma polymerization of TMS caused by fluorine-containing moiety emanating from the substrate surface (depletion of Si moieties in plasma polymer).

4.3. Oligomerization and Polymerization

The way in which a plasma polymer is formed has been explained by the rapid step growth polymerization mechanism, which is depicted in Figure 5.3. The essential elementary reactions are stepwise recombination of reactive species (free radicals) and stepwise addition of or intrusion via hydrogen abstraction by impinging free radicals. It is important to recognize that these elementary reactions are essentially oligomerization reactions, which do not form polymers by themselves on each cycle. In order to form a polymeric deposition, a certain number of steps (cycle) must be repeated in gas phase and more importantly at the surface. The number of steps is collectively termed the kinetic pathlength.

If growing species stray from the active LCVD environment, they will deposit on the reactor walls, which is out of the luminous gas phase, as oligomers (not polymers). The presence of oligomer deposition on the reactor walls could cause serious interference with subsequent plasma polymerization, which is carried out in a contaminated reactor.

The important general aspects of plasma oligomers are listed below:

1. Some oligomers formed in LCVD process of F-containing gas, which adhere to some surfaces (metal oxides) by chemisorption, cannot be pumped out under vacuum, even under the high vacuum used for XPS. Consequently, XPS analysis depicts plasma oligomers as indistinguishable from plasma polymers unless detailed analysis of the interface is performed.
2. The adhesion of the plasma oligomers to the general wall surface is poor, and plasma oligomers can therefore migrate within a plasma reactor during the regular evacuation process. The vapor pressure is not high enough to be pumped out but not low enough to stay as a stable condensed phase material.
3. Most plasma oligomers are soluble in water or organic solvents.
4. Plasma-induced oligomers could result from plasma treatment of a polymer, as well as from plasma oligomerization of monomers intended for LCVD.

The presence of oligomers in plasma polymers, as well as in the surface state of plasma-treated polymers, has highly important implications in the interpretation of surface analysis data. XPS does not distinguish oligomers from polymers. Scanning electron microscopy results in inherent damage due to electron bombardment, which in turn has different effects on oligomers and the polymeric network. SIMS could distinguish the fragmentation patterns of oligomers and polymers, if quantitative data relating to oligomers were known. Unless these factors could be distinguished, an elaborate surface analysis could lead to confusion rather than clarification.

5. EFFECT OF CONSECUTIVE PLASMA POLYMERIZATIONS

C_2F_6 (HFE) is not a monomer of general plasma polymerization but an effective etching gas because it does not polymerize in the absence of hydrogen, as described in Chapter 7. However, when it is applied to a TMS plasma polymer surface, hydrogen is abstracted from the surface and forms a very thin layer of plasma polymer as explained above by the formation of a stable entity, H-F. It is essentially a self-terminating deposition process leading to an extremely thin layer of HFE plasma polymer. XPS analysis indicated that the thickness of the F-containing layer in the TMS/(HFE) system is less than few nanometers [7].

This film system was seen to outperform others not incorporating the adhesion-promoting HFE film when a primer is applied to the LCVD coated alloy surface. The alloy panels were always treated with O_2 plasma to remove any organic contaminants from the alloy surface prior to film deposition. The entire steps involved in the plasma coating process are:

1. Solvent cleaning of the alloy surface with acetone
2. Oxygen plasma treatment of the cleaned surface
3. Cathodic polymerization of TMS, and
4. Cathodic polymerization of HFE

Steps 2–4 were carried out in a vacuum reactor consecutively. Thus, the plasma reactor follows the cycle: (1) pump down from ambient environment; (2) O_2 plasma; (3) TMS plasma; (4) HFE plasma; (5) exposure to ambient air. The sequences of plasma processes are; O_2/TMS, TMS/HFE, and HFE/O_2, or can be expressed –(O2/TMS/HFE/)–.

After a large number of sample preparations by this set of plasma coatings, one set of panels inadvertently missed the O_2 plasma treatment but still had the plasma film system deposited on them. These panels were destined for testing with various primers in an effort to evaluate various primers in a pollution prevention program. After paint application, they were seen to experience miserable adhesion failures in scribed wet-tape testing. The entire paint layer delaminated where it was in contact with tape. The inadvertent omission of the O_2 plasma treatment changed the coating system from one with superadhesion to one with extremely poor or practically no adhesion. This dramatic change of adhesion is depicted in Figure 10.8. Figure 10.8a shows the properly treated sample surviving a very severe paint stripping test, while Figure 10.8b indicates absolute failure caused by the omission of the O_2 plasma treatment on a simple water soaking test.

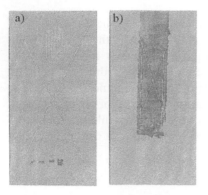

Figure 10.8 Scanned image of the surface of two alloy panels showing adhesion failure caused by the omission of O_2 plasma treatment of the substrate prior to plasma film deposition and application of the primer (Deft 44-GN-72 MIL-P-85582 Type I Waterbased Chromated Control Primer). a) Panel after Skydrol LD4® fluid resistance test, which had the O_2 plasma treatment prior to film deposition and primer application. b) Panel after scribed wet (24-h immersion in tap water) tape test, which had not been treated with the O_2 plasma treatment prior to film deposition and primer application.

5.1. Adhesion Failure Caused by Wall Contamination

It is important to note that plasma deposition occurs predominantly on the cathode surface in the cathodic polymerization, and a new cathode (substrate) was used in every plasma coating operation. In other words, the contamination of the reactor is considered to be minimal.

A particular point should be made regarding the application cycle of the plasma processes. In the general scheme of the progression of process steps, ignoring venting and substrate replacement, O_2 plasma treatment followed the HFE plasma treatment of the previous run, and HFE plasma treatment followed TMS plasma application in the regular operation; i.e., –(O2/TMS/HFE/)–. With the omission of the O_2 plasma treatment, the order of the sequential plasma processes has changed, i.e., the plasma polymerization of TMS followed the HFE plasma treatment, which was performed in the preceding run, i.e.,–(HFE/ TMS)–.

This reversal of the sequence from TMS/HFE to HFE/TMS is the most important issue. The second major issue is the effect of the O_2 plasma treatment on the fluorine-containing contaminants. The XPS analysis of an initial sample revealed virtually no silicon on the alloy surface beneath the lifted primer but did indicate a rather substantial fluorine presence. The appearance of a strong silicon signal on the interface side of the removed primer indicated that the entire plasma film had likely delaminated at the interface with the alloy. Analysis of additional samples confirmed that the entire film and primer system had delaminated from the alloy panels.

Figure 10.9 shows the core spectra taken from the exposed alloy beneath the removed primer from a panel. Using changes in take-off angle (angle between the sample surface and direction of emission/collection of electrons) and light sputtering, it is seen that the fluorine on the alloy is in two distinct chemical states—one on the outer surface (10-degree take-off angle) and one just below (90-degree take-off

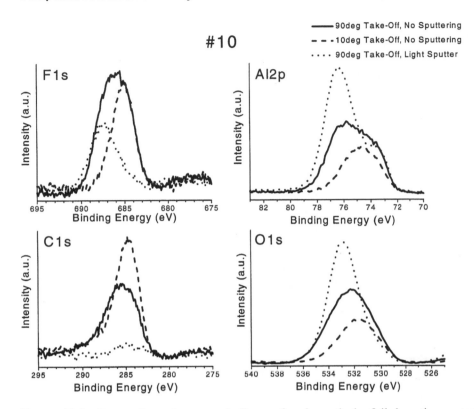

Figure 10.9 Spectra from the exposed alloy surface beneath the failed coating system on sample #10; the spectra shown were acquired at two different take-off angles prior to sputtering and then also at normal emission after a light sputtering.

angle and the remnant after sputtering). This same figure shows that the carbon level on the panel is associated with the outer fluorine state. It is also apparent that the presence of this outer contamination is associated with a modification of the aluminum oxide.

However, the carbon contamination associated with the fluorine on the surface of the alloy panels is not a fully formed plasma polymer like that formed on the TMS films. Figure 10.10 shows C 1s and F 1s spectra from an HFE plasma polymer on a TMS film taken with the monochromatic source at high resolution. There is some tail at higher binding energy in the C 1s spectra in Figure 10.9, but nothing like that from the HFE plasma polymer film. Also, the position of the fluorine signal in the failed samples is not consistent with the polymer formation. The higher binding energy component on the contaminated samples is associated with the underlying inorganic modified alumina, and not with the increased carbon level on the surface. The conclusion that must be reached based on the structure of the spectra is that the contaminating fluorine deposit is not a well-developed plasma polymer film.

The surface of the primers opposing the alloy panels was found to have the full TMS film still strongly adhered to the primer, with the HFE film within the TMS/primer interface. Figure 10.11 shows the elemental sputter depth profile from

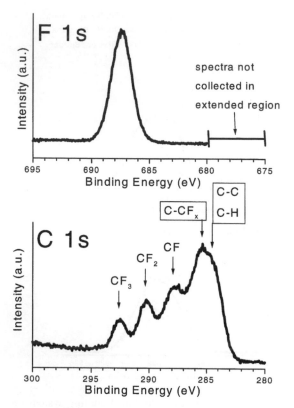

Figure 10.10 Monochromatic generated F 1s and C1s spectra from HFE film on a TMS plasma polymer.

the interface side of the removed primer of one of the failed panels. The region of strong silicon signal in the plot indicates that the tenacious adhesion between the plasma polymer and the primer is still intact and that the mode of failure was purely adhesive at the plasma polymer/aluminum alloy interface. The bottom panel in Figure 10.11 shows the summary Si/C elemental ratio information from the same interface combined with the information obtained from depth profiles of additional lifted coatings.

The increase in fluorine signal coupled with the decrease in silicon signal in the depth profile indicates the interface region between the TMS film and the primer, where the HFE film formed the strong adhesion. The slightly increased fluorine at the beginning of the depth profile (indicating the surface where the coating was originally in contact with the alloy) shows that fluorine was present when cathodic polymerization of TMS was applied to a new, freshly cleaned aluminum alloy panel and interfered with the normal deposition of TMS.

The effect on the TMS film formation is apparent in the bottom panel of Figure 10.11 where the Si/C ratio is much lower than the bulk level, persisting into the TMS film. The contribution from carbonaceous contamination cannot be responsible for the lowered Si/C ratio at the initial portion of the depth profile, since this lowered level persists through more material than the thickness of the

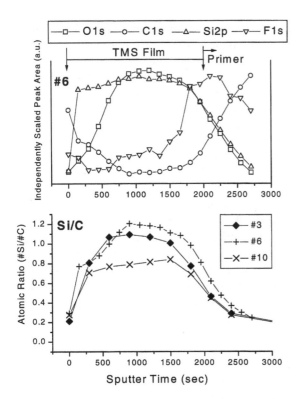

Figure 10.11 (Top Panel) Elemental area plot from XPS sputter depth profile of interface side of lifted primer on one failed sample; (Bottom Panel) Summary Si/C ratio information from depth profiles of the interface side of the primers lifted from three failed panels.

deposit on any test panel. This indicates that the contaminant interfered with the normal TMS deposition, causing preferential deposition of carbon-rich films at the outset.

An obvious question that arises is how did the fluorine come to the surface of the new substrate before the first plasma process (of TMS) was applied, and what was its origin? The answer is that fluorine must come from the fluorine-containing contaminants in the plasma reactor. This fact was confirmed by a series of experiments to identify the source [7].

Knowing the source of the contamination, the questions are narrowed to the following:

1. How are fluorine-containing contaminants brought to the surface that would become the interface between alloy and TMS plasma polymer?
2. How does the O_2 plasma treatment eliminate the effect of fluorine-containing contaminants and restore adhesion at the interface?

A study of the pretreatment application and the surface prior to deposition indicates that the aluminum alloy panels have a marked sensitivity to the buildup of a fluorocarbon background in the plasma reactor. This study also showed that the application of the O_2 plasma treatment modified the alloy surface, changing it from one composed of aluminum oxide to a surface composed of mixed oxide and aluminum fluoride, and in extreme cases, to a mostly mixed fluoride chemistry incorporating some oxygen.

Fluorine contamination has been reported in various environments and applications in the past. It has shown up in plasma processing [10–18], as cross-contamination from storage in contaminated containers or with contaminated samples [14,18], and modification of aluminum deposited on fluoropolymer substrates and other polymers having fluorine-based plasma treatments has also been observed [19–21]. Fluorocarbon lubricants have also been noted to modify the oxide structures on aluminum alloys [22,23], and the degradation of Al_2O_3 catalytic supports has been associated with fluoride conversion during reactions with fluorocarbons [24]. Alloy oxide modification has also been well noted in the presence of fluorine compounds not of the fluorocarbon family [25].

Most of these discussions regarding fluorine contamination of aluminum surfaces have focused on the conversion of aluminum oxide to fluoride or oxyfluoride. Evidence for similar conversions was included, and in extreme cases conversion to aluminum bonding quite similar to that in AlF_3 was found. However, the poor adhesion of the samples skipping the O_2 plasma treatment is related not to the fluorine contamination as such, but rather to the carbonaceous nature of the adsorbed materials, which is subjected to the plasma polymerization of TMS. Oxygen plasma cleaning removes this carbonaceous component, while the surface fluorine concentration is enhanced.

In order to get insight into the adsorption of F-containing contaminants, Al alloy panels with Si pieces attached to them were subjected to the evacuation process after previous HFE treatments and loading of a fresh sample and then the oxygen plasma treatment. The results of XPS analysis of the Si pieces and the alloy panels are shown in Figure 10.12. The top panels in the figure correspond to the samples being exposed to the evacuation process and the bottom panels correspond to the samples being oxygen plasma treated. In order to show the details of line shapes, the top and the bottom parts of the figures are shown with different vertical scale. The F 1s signal in the bottom figure is roughly five times greater than that in the top figures at the same vertical scale.

After the evacuation process, the alloy panel has both fluorine and the large, low binding energy carbon signal, whereas the Si wafer piece has no fluorine on it and only carbon associated with adventitious hydrocarbons. After oxygen plasma treatment (in the contaminated reactor), the fluorine level on the aluminum panel substantially increases. There is fluorine on the silicon wafer piece after oxygen plasma cleaning, but aluminum was also observed on it as a result of sputtering effects, which complicates any interpretation of the origin of the fluorine on the silicon. The oxygen plasma treatment then removed the large carbon concentration from the alloy panel, while converting the surface oxide into the mixed oxyfluoride structure, incorporating more fluorine than initially present.

The results shown above indicate that the following sequence of events occurs. The F-containing oligomers, residing on the reactor walls, are mobile enough to be transferred to the surface of new substrates when the reactor is evacuated, and bind with aluminum oxides so strongly that they are not disturbed by exposure to atmosphere and cannot be pumped out in vacuum, even in a high-vacuum environment for XPS analysis. Without a specific strong interaction on other

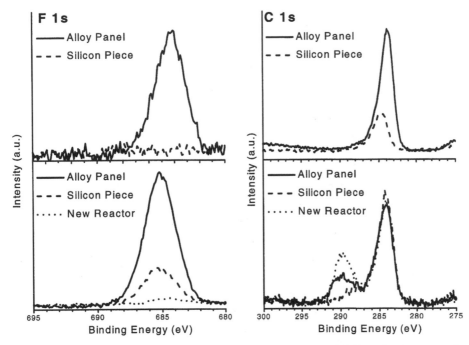

Figure 10.12 F 1s and C 1s spectra from both alloy panels and Si wafer pieces attached to them. Top figure panels are from samples exposed to the evacuation process and bottom panels are from samples exposed to the O$_2$ plasma cleaning process. The third trace in each of the bottom panels is from an alloy panel that was O$_2$ plasma cleaned in a new reactor with minimal fluorine contamination, indicating that the high binding energy C 1s peak is not associated with fluorocarbon bonding.

surfaces, including the sample silicon wafers attached to the panels, the adsorbed materials, if adsorption occurred, would be pumped out in high vacuum or dissociate upon exposure to atmosphere, and could not be detected by ex situ XPS analysis.

The role of this contaminant on the interface characteristics between the TMS film and the alloy is also quite evident in the sputter depth profiling results shown in Figure 10.13. These results were generated from two samples representative of the two process steps in question from the contaminated reactor; the first had the TMS and HFE films deposited on Alclad 7075-T6 after acetone wiping, whereas the second sample was O$_2$ plasma treated prior to film application. The Si/C ratio showed the effect of the fluorocarbon adsorption on the interface film properties. The O$_2$ plasma–treated sample maintained a consistent ratio through the interface, while the sample with no O$_2$ plasma cleaning exhibited a definite decrease in this ratio, indicating decreased silicon concentrations at the interface. This Si/C ratio was seen to be a good indicator of the integrity of plasma film/alloy interface, with extremely poor films having a markedly lower ratio.

The effect of the oxygen plasma treatment in the presence of the contaminant is again seen in the fluorine trace from the depth profile, showing the increase

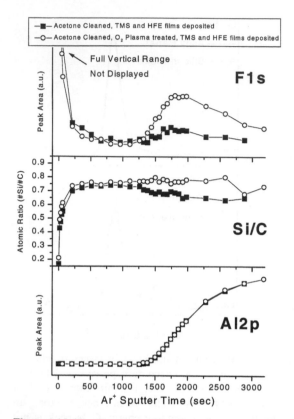

Figure 10.13 Depth profiles focusing on the increase in fluorine concentration at the interface on the O₂ plasma treated sample and the decrease in the Si/C ratio at the interface on the panel without the O₂ plasma treatment. The aluminum peak area serves as a reference to when the film is removed from each panel and shows that the film thickness and sputtering rates are quite similar. The both samples were prepared by using the contaminated reactor.

in fluorine concentration at the interface on the oxygen plasma–treated sample. The fluorine depth profile from the sample without plasma treatment also shows the more moderate fluorine level at the interface from the adsorption on the alloy oxide, associated with the carbon increase.

5.2. Migration of Oligomeric Wall Contaminants

Once a reactor is contaminated to a certain level, it is very difficult to get rid of the persistent influence of the contaminants. During an arbitrarily chosen 1-month period, a contaminated rector was used only for the following two kinds of samples. One sample designated as (Ace) in the legend of Figure 10.14 was prepared by placing an acetone-cleaned Alclad 7075 coupon in the reactor and pumped down to the base vacuum of approximately 1 mtorr and then removed from the reactor. Another sample designated as (Ace/O₂) was prepared by applying O₂ plasma treatment for 2 min. Figure 10.14 depicts the persistence of the contamination examined by following the XPS fluorine contents on the Al alloy surfaces as a

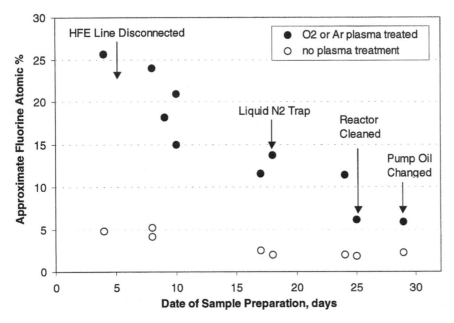

Figure 10.14 The decay of fluorine contamination with continued reactor use, involving multiple intermittent chamber evacuations and O_2 and Ar plasma-treatments of new substrates; other than this intermittent use and times when chamber-cleaning procedures were employed, the reactor is continually under vacuum, indicating that this contamination does gradually pump out.

function of time. The following conclusion could be drawn from the results shown in Figure 10.14.

1. O_2 plasma treatment incorporates more fluorine (approximately five times) on the Al alloy panel than the sample just loaded to the reactor and pumped down.
2. XPS studies on the panels prepared with the HFE line disconnected, the liquid N_2 trap, and the vacuum pump oil change confirmed that these are not possible sources of fluorine contamination. The F content gradually decreases with the evacuation time and intermittent O_2 plasma discharges.
3. From the gradual decrease of the level of contamination with evacuation time with intermittent O_2 or Ar processing, it can be speculated that the fluorine level in the regular operation never reached the level shown at the end of 30 days in Figure 10.14. This is because the HFE/O_2 sequence, which was maintained in the regular operation before the inadvertent omissions occurred, served as a cleaning process for the contaminant in the reactor.

It is important to emphasize that O_2 plasma treatment collects much more F-containing contaminants on the alloy surface than without O_2 plasma treatment, although this treatment virtually eliminated the interference of the contaminants to the subsequent TMS deposition as described in the previous section. In other words, O_2 plasma treatment does not reduce the amount of fluorine on aluminum alloy surface but reduces plasma-ablatable fluorine on aluminum alloy surface.

5.3. Role of Electronegativity of Element in Consecutive LCVD

The decrease of Si due to F-containing contaminants and the role of the oxygen plasma treatment can be explained by the principle of CAP. The key factor to explain the change of elementary composition at the interface is the plasma sensitivity of elements involved on the surface and in the plasma phase. The ablation of materials exposed to plasmas appears to follow the plasma sensitivity series of the elements involved, which is in the order of the electronegativity of the elements, i.e., elements with higher electronegativity in the condensed phase are more prone to ablate in plasma that contains elements with lower electronegativity [5].

When the thin layer of F-containing oligomers is exposed to the TMS plasma, some F atoms are removed from the layer and form Si-F moieties in the plasma. Fluorine atoms (high electronegativity) in the contaminants are easily detached from the surface by the interaction of plasma of low-electronegativity Si (TMS), forming stable species (with Si-F bonds) in the plasma phase, which will be pumped out of the system.

Oxygen plasma treatment of the Al alloy surface with F-containing oligomers is a similar situation but with different consequences because oxygen plasma does not form polymeric deposition. Oxygen (lower electronegativity) plasma ablates F-containing oligomers from the substrate surface. In plasma phase, F atoms are detached from the organic moieties and become F-containing plasma, which reacts with elements of lower electronegativity in the condensed phase such as O in metal oxides. Thus, plasma-sensitive F-containing oligomers are converted to more stable (in plasma environment) F-containing inorganic compounds such as aluminum fluoride-oxide, aluminum fluoride, and so on, although the details of species are not well known. Although F increases on the surface after the oxygen plasma treatment, the concentration of plasma-ablatable F at the surface is reduced, which virtually eliminates the interference of TMS polymerization by F-containing oligomers.

The following experiments were performed to confirm the major principles discussed above: (1) the interference of TMS plasma polymerization by fluorine ablating from the substrate surface, and (2) the elimination of plasma-ablatable F by O_2 plasma pretreatment of the substrate surface. First, a very thin layer (a very few nanometers) of $(HFE + H_2)$ plasma polymer was deposited on an aluminum sheet by 5 s of plasma polymerization. One sample was prepared by depositing TMS plasma polymer directly onto this surface. The second sample was prepared by treating the $(HFE + H_2)$ plasma–modified surface with oxygen plasma before depositing TMS plasma polymer. Without oxygen plasma treatment, the TMS plasma polymer did not adhere to the substrate surface, whereas with the oxygen plasma treatment, excellent adhesion was obtained. These results confirm the mechanisms by which the wall contamination caused catastrophic damage to the adhesion of film formed by the plasma CVD process. These summaries are schematically shown in Figures 10.15 and 10.16.

6. MULTISTEP GLOW DISCHARGE TREATMENT OF POLYMER SURFACE

An attempt was made to create a fabric with one hydrophilic side and the other hydrophobic by applying a different kind of plasma surface modification to each [5].

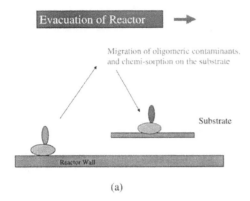

(a)

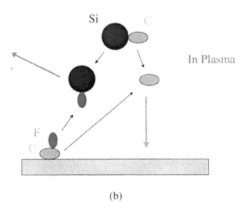

(b)

Figure 10.15 Schematic representation of the mechanism how F-containing contaminant interfere with plasma polymerization of trimethylsilane (TMS); (a) migration of F-containing oligomers, (b) the interference of TMS deposition by F-containing moieties.

Air plasma treatment was used to make one surface hydrophilic, and CF_4 plasma treatment was used to make the other hydrophobic. Such a fabric with a different set of surface characteristics on each side can be made; however, the success of this undertaking is contingent on which treatment is applied first. The sequence dependency of plasma treatments may be explained by the concept of plasma sensitivity of the elements involved in the two steps. Results are summarized in Tables 10.1 and 10.2.

Examining these results, the following factors involved in this experiment should be kept in mind:

1. Fabrics are porous. Consequently, the plasma treatment applied to one side of a fabric penetrates to the other side, even though the second side is not exposed to the plasma directly. This penetration effect of plasma treatment was previously known [10].
2. Likewise, a second plasma treatment will influence the effect of the first plasma treatment, even though the surface treated first is not directly exposed to the second plasma treatment.

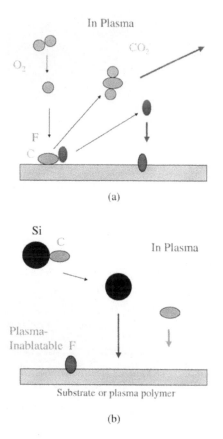

Figure 10.16 Schematic representation of mechanism how oxygen plasma treatment prevents the interference of plasma polymerization of TMS by F-containing oligomers; (a) Oxygen plasma treatment, (b) Plasma polymerization of TMS.

As is evident in Table 10.1, two characteristically different sides can be obtained only when the air plasma treatment is applied first. This can be explained by the plasma sensitivity of the elements involved. Fluorine is the most electronegative element involved in this experiment. Consequently, it can be removed relatively easily by exposure to the plasma, which consists of less plasma sensitive elements (elements of lower electronegativity).

When the two steps of plasma treatment are carried out successively in the same reactor without copious cleaning of the reactor between the two treatments, the influence of the wall surface on the second plasma treatment also becomes an important factor. Table 10.2 reveals the following important factors:

1. Fluorine contamination of the reactor wall is clearly evident. Air plasma treatment introduced a small but clearly identifiable amount of F on the treated surface.
2. The efficiency of F incorporation into the surface structure is higher on the surface that is not directly exposed to the CF_4 plasma. This trend is found consistently in the data presented in Table 10.2.

Table 10.1 Consecutive Treatment of Two Sides of a PET Fabric by Different Plasmas[a]

Side	1st Treatment	Contact angle of H_2O (degree)	2nd Treatment	Contact angle of H_2O (degree)
A	Air plasma	Water penetrates[b]		58.6[c]
B		Water penetrates	CF_4 plasma	111.1
A	CF_4 plasma	112.7		Water penetrates[b]
B		54.5	Air plasma	Water penetrates[b]

[a]Plasma treatment penetrates through a porous structure, i.e., it is not limited to the exposed surface.
[b]Contact angle cannot be measured because a water droplet penetrates quickly into the fabric.
[c]Contact angle can be measured but a water droplet penetrates slowly into the fabric.

Table 10.2 Consecutive Treatment of Two Sides of a PET Fabric by Different Plasmas[a]

Side	1st Treatment	XPS F counts	2nd Treatment	XPS F counts
A	Air plasma	4092		74375[b]
B		4542	CF_4 plasma	63555
A	CF_4 plasma	56376		53921[c]
B		68408	Air plasma	24802[c]

[a]Plasma treatment penetrates through a porous structure, i.e., it is not limited to the exposed surface.
[b]Contact angle can be measured but a water droplet penetrates slowly into the fabric.
[c]Contact angle cannot be measured because a water droplet penetrates quickly into the fabric.

These findings clearly show that the principle of CAP also applies to glow discharge treatment, which is not intended to deposit plasma polymers, in a similar manner with respect to the interaction of luminous gas with materials.

REFERENCES

1. Kay, E. Invited Paper, Int. Round Table Plasma Polym. Treat., IUPAC Symp., Plasma Chem. **1977**.
2. Yasuda, H.; Hsu, T. J. Polym. Sci., Polym. Chem. Ed. **1978**, *16*, 415.
3. Yasuda, H. ACS Symp. Series **1979**, *108*, 163.
4. Dilks, A.; Kay, E. A CS Symp. Ser. **1979**, *108*, 195.
5. Yasuda, H.; Yasuda, T. J. Polym. Sci., Polym. Chem. Ed. **2000**, *38*, 943.
6. Yasuda, T.; Gazicki, M.; Yasuda, H. J. Appl. Polym. Sci., Appl. Polym. Symp. **1984**, *38*, 201.
7. Moffitt, C.E.; Reddy, C.M.; Yu, Q.S.; Wieliczka, D.M.; Yasuda, H.K. Appl. Surf. Sci. **2000**, *161* (3–4), 481.
8. Yasuda, H.K.; Yu, Q.S.; Reddy, C.M.; Moffitt, C.E.; Wieliczka, D.M. J. Vac. Sci., Tech. **2001**, *A19* (5), 2074–2082.
9. Yu, Q.S.; Yasuda, H.K. J. Polym. Sci., Polym. Chem. Ed. **1999**, *37*, 967.
10. Karulkar, P.C.; Tran, N.C. J. Vac. Sci. Tech. **1985**, *B3*, 889.
11. Jimbo, S.; Shimomura, K.; Ohiwa, T.; Sekine, M.; Mori, H.; Horioka, K.; Okano, H. Jpn. J. Appl. Phys. **1993**, *32*, 3045.
12. Shacham-Diamond Y.; Brener, R. J. Electrochem. Soc. **1990**, *137*, 3183.

13. Bohling, D.A.; George, M.A.; Langan, J.G. in IEEE/SEMI Advanced Semiconductor Manufacturing Conference and Workshop: ASMC '92 Proceedings (IEEE, New York, 1992), p.111.
14. Ernst, K.-H.; Grman, D.; Hauert, R.; Holländer, E. Surface Interface Anal. **1994**, *21*, 691.
15. Thomas, III, J.H.; Bryson, III, C.E.; Pampalone, T.R. Surface and Interface Anal. **1989**, *14*, 39.
16. Grman, D.; Hauert, R.; Holländer, E.; Amstutz, M. Solid State Tech. **Feb. 1992**, *35* (2), 43.
17. Thomas, III, J.H.; Bryson, III, C.E.; Pampalone, T.R. J. Vac. Sci. Tech. **1988**, *B 6*, 1081.
18. Grman, D.; Ernst, K.-H.; Hauert, R.; Holländer, E. Microcontamination, **1994**, *12* (7), 57.
19. Seidel, C.; Gotsmann, B.; Kopf, H.; Reihs, K.; Fuchs, H. Surface and Interface Anal. **1998**, *26*, 306.
20. Du Y.; Gardella, J.A., Jr. J. Vac. Sci. Tech. **1995**, *A 13*, 1907.
21. Wu, P.K. In *Polymer/Inorganic Interfaces II*; MRS Proceedings; Drzal, L.T., Opila, R.L., Peppas, N.A., Schutte, C. Eds.; MRS: Pittsburgh, 1995; Vol. 385, 79.
22. John, P.J.; Liang, J. J. Vac. Sci. Tech. **1994**, *A 12*, 199.
23. Kasai, P.H.; Tang, W.T.; Wheeler, P. Appl. Surf. Sci. **1991**, *51*, 201.
24. Farris, M.M.; Klinghoffer, A.A.; Rossin, J.A.; Tevault, D.E. Catal. Today **1992**, *11*, 501.
25. Strohmeier, B.R. Appl. Surf. Sci. **1989**, *40*, 249.

11

Internal Stress in Material Formed by Luminous Chemical Vapor Deposition

1. CAUSE OF INTERNAL STRESS IN PLASMA POLYMER

Plasma polymers [materials formed by luminous chemical vapor deposition (LCVD)] have their unique advantage in ultrathin film applications (e.g., less than 100 nm), due to their chemical inertness, good adhesion to nearly any substrate, and excellent barrier properties. However, these advantageous characteristics are specific to type A plasma polymer, and when their thickness grows beyond a threshold value (e.g., several micrometers), plasma polymer films tend to crack or buckle from a rigid substrate due to the internal stress in the film, which develops during the deposition process. When either cracking or buckling occurs, plasma polymer films will lose their excellent protective properties. Therefore, the consideration of the internal stress is crucially important for achieving plasma polymer protective coatings because the thicker is not the better.

The internal stress in plasma polymer films is generally expansive, i.e., the force to expand the film is strained by external compressive stress. According to the concept presented by Yasuda et al. [1], the internal stress in a plasma polymer stems on the fundamental growth mechanisms of plasma polymer formation. A plasma polymer is formed by consecutive insertion of reactive species, which can be viewed as a wedging process. The internal stress is related to how frequently the insertion occurs as well as on the size of inserting species. The both factors are dependent on the operational factors of plasma polymerization.

When a thick layer (e.g., 1 µm) of plasma polymer of styrene is deposited on a rigid surface such as a glass plate, the layer of plasma polymer tends to buckle up and often cracks when the coating is subjected to ambient air. This phenomenon was originally thought to be associated with the absorption of moisture from the atmosphere and consequent swelling of the layer. However, it seems to be more closely related to the characteristic properties of the plasma polymer and the adhesion to substrate than to the swelling of the film. It was also observed that a composite film consisting of a thin layer of plasma polymer deposited on a flexible polymeric substrate such as a polyethylene film often shows a strong tendency to bend and curl. A close examination of the curling phenomenon revealed that (1) the curling always occurs in such a way that the coated film curls up, keeping the plasma polymer layer outside, and (2) the curling takes place during the process of plasma polymerization, in which no swelling due to the absorption of moisture can occur. The moisture-induced buckling mentioned earlier is due to the breaking of adhesion by the action of moisture causing the partial delamination of the coated layer, which expands. The internal stress causes the buckling when the adhesion is broken.

This curling can be attributed to an internal stress arising in the plasma polymer during polymer deposition. It is important to recognize that the internal stress in the plasma polymer (in the "as-polymerized" state) is an expansive stress and that this is in marked contrast to what would be expected if the adsorbed monomer were polymerized at the surface of the substrate, which would create, with very few exceptions, contractive stress due to the contraction of volume on the polymerization of a monomer.

In order to understand the internal stress in plasma polymers, it is necessary to recall the growth and deposition mechanisms of LCVD described in Chapter 5. The first important aspect is the rapid step growth mechanism depicted in Figure 5.3, which does not form a polymer by chain reactions. The second is the deposition of reactive species (rather than of polymer or of monomer), which is the key factor of the plasma polymer deposition mechanism. The deposition of a reactive species (e.g., a free radical) that is created in the gas phase is viewed as a loss of kinetic energy on collision with the wall of the reactor, due either to chemical reaction with the surface or to the increased molecular weight of the species, and the important point is that such a deposited species is still involved in the overall growth mechanism illustrated in Figure 5.3. In other words, plasma polymerization and the deposition of polymer might be illustrated in such a way that the building blocks are formed in the plasma phase but the actual material formation proceeds, in most cases, at the surface. Therefore, the formation of a coherent film by plasma polymerization can be visualized as a continuous wedging process. The repeated wedging processes cause the characteristic expansive stress.

2. CURLING OF BILAYER

The curling of a plasma polymer–coated film is schematically illustrated in Figure 11.1. The substrate polymer film has thickness D and Young's modulus E; the plasma-polymerized layer has the corresponding parameters given by d and e. If the substrate is constrained to its original shape, a "swelling stress" σ_s develops in the plasma-deposited layer. This stress exerts a bending moment, which is partly relieved when the composite film is allowed to bend. Bending creates reactive stresses in the substrate, and an equilibrium is reached when the moment of the stress in the thickness d of the plasma-polymerized layer is opposite to and equal in quantity to the moment of the stress in the thickness D of the substrate film, i.e., $M_d + M_D = 0$, where the moments M are those with respect to the neutral axis shown by the dashed line in Figure 11.1. A second equilibrium condition is that the stress integrated over the cross-section perpendicular to the neutral axis must be zero because no external force is applied. These two equilibrium conditions are sufficient to calculate the location of the neutral axis and to derive the relationship between σ_s and the radius curvature R of the composite film.

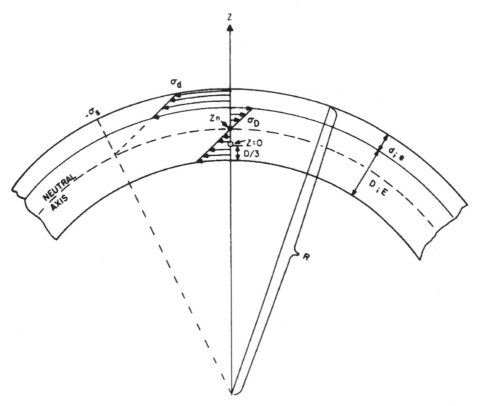

Figure 11.1 Model of a composite membrane, bending because of a stress σ_s in the thin layer deposited by plasma polymerization onto a flexible polymeric substrate; layer and substrate have thickness d and D and Young's module e and E, respectively.

For a simplified case in which $d \ll D$ and in which E and e are of the same order of magnitude, σ_s is given by

$$\sigma_s = ED\,2/6Rd \tag{11.1}$$

The force per unit width of film that causes curling of the composite film is given by the product of stress σ_s and layer thickness d, namely, $\sigma_s d$. This curling force increases with increasing layer thickness as seen in Figure 11.2. At small values of plasma polymer thickness, the curling force $\sigma_s d$ increases linearly with thickness d, indicating that σ_s is a constant. The curling force tends to deviate from this linear dependence on d at a greater thickness of plasma polymer; that is, stress σ_s decreases with increasing thickness in this region. This seems to be related to the cracking of the layer, which has been observed to occur in relatively thick layers and which certainly relieves a portion of the internal stress. Figure 11.2 indicates that a meaningful value of σ_s can be determined for Eq. (11.1) if the coating thickness is limited to the linear region at lower thickness (e.g., in those cases shown in Figure 11.2, thickness less than 4000 Å). The values of internal stress measured at a 4000-Å coating thickness for some plasma polymers prepared under an arbitrary set of conditions are listed in Table 11.1 [1]. It should be cautioned, however, that

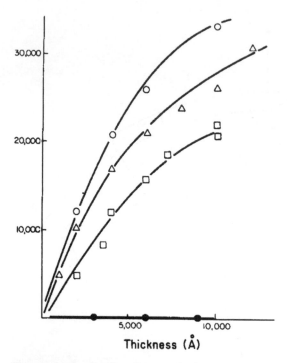

Figure 11.2 Curling force $\sigma_s d$, the product of the internal stress σ_s and thickness d of the plasma-deposited layer, versus d; the three curves correspond to layers obtained by plasma polymerization of the following monomers: ○, pyridine; △, acrylonitrile; □, thiophene; ●, tetramethyldisiloxane.

Table 11.1   Internal Stress of Plasma Polymers[a]

Monomer	Structure	σ_s (dynes/cm^2)
Thiophene		2.7×10^8
Pyridine		5.2×10^8
Acrylonitrile	$H_2C\!\!=\!\!CH\!-\!C\!\equiv\!N$	4.3×10^8
Furan		7.0×10^8
Styrene	$H_2C\!\!=\!\!CH$	4.3×10^8
Acetylene	$HC\!\equiv\!CH$	3.8×10^8
Methyloxazoline	H_2C	0
Tetramethyldisiloxane	$\overset{CH_3}{\underset{CH_3}{H\!-\!Si}}\!-\!O\!-\!\overset{CH_3}{\underset{CH_3}{Si}}\!-\!H$	0

[a]Coating thickness: 4000 Å. Reaction conditions: monomer pressure, 30 μm Hg; power, 80 W.

like any other properties of LCVD materials, the internal stress depends on how LCVD process is carried out, and it is not a sole function of the molecular structure of the monomer.

3. INTERNAL STRESS IN GLOW DISCHARGE POLYMERS

The internal stress of plasma polymers is dependent not only on the chemical nature of monomer but also on the conditions of plasma polymerization. In the plasma polymerizations of acetylene and acrylonitrile, apparent correlations are found between σ_s and the rate at which the plasma polymer is deposited on the substrate [2], as depicted in Figure 11.3. The effect of copolymerization of N_2 and water with acetylene on the internal stress is shown in Figures 11.4 and 11.5. The copolymerization with a non-polymer-forming gas decreases the deposition rate. These figures merely indicate that the internal stress in plasma polymers prepared by radio frequency discharge varies with many factors. The apparent correlation to the parameter plotted could be misleading because these parameters do not necessarily represent the key operational parameter.

The practical importance of the internal stress is that if too thick a layer is deposited by plasma polymerization, the overall force reaches the point where the internal stress exceeds the cohesive force of the plasma polymer or the internal stress at the interface exceeds the adhesive force of the plasma polymer to the substrate.

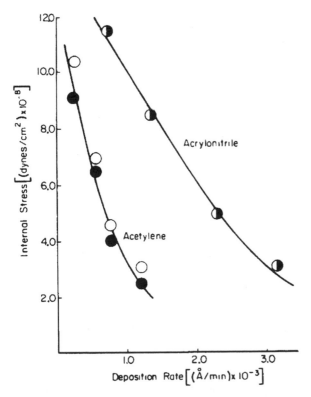

Figure 11.3 Dependence of internal stress on the deposition rate ($\bigcirc$, 100 W; $\bullet$, 45 W).

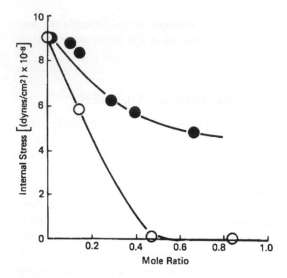

Figure 11.4 Changes in the internal stress of plasma polymers by copolymerization of acetylene with nitrogen (●) or water (○).

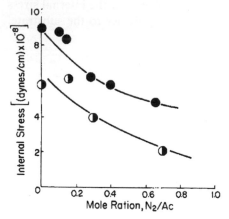

Figure 11.5 Changes in the internal stress of plasma copolymers of acetylene/N_2 (●) and of acetylene/H_2O/N_2 (◑) as a function of the mole ratio of nitrogen to acetylene.

If the adhesion of plasma polymer to the substrate is good, cracking of the plasma polymer results, as evidenced by many studies of transport characteristics of plasma polymers as a function of the thickness of plasma polymer (see Chapter 34). If the adhesion is poor, buckling of the plasma polymer deposited onto a rigid substrate such as a glass plate occurs.

Thus, recognition of the characteristic internal stress buildup in a plasma polymer is important for estimating the upper limit of thickness of a plasma polymer for a practical application. Poor results with respect to such parameters as adhesion and barrier characteristics are often due to the application of too thick a plasma polymer layer. The tighter the network of plasma polymer, the higher is the internal stress. Consequently, the tighter the structure, the thinner is the maximal thickness or threshold thickness, which can be applied without causing the self-destruction of the coating.

Plasma polymers of certain kinds of monomers have very little, if any, internal stress, and thickness is not a limiting factor of application. However, because of this very feature such polymers may not provide certain coating functions that are sought for the application of plasma polymerization. In other words, the internal stress is not a drawback of plasma polymer but an important characteristic of the materials formed by LCVD.

The expansive internal stress in a plasma polymer is a characteristic property that should be considered in general plasma polymers and is not found in most conventional polymers. It is important to recognize that the internal stress in a plasma polymer layer exists in as-deposited plasma polymer layer, i.e., the internal stress does not develop when the coated film is exposed to ambient conditions. Because of the vast differences in many characteristics (e.g., modulus and thermal expansion coefficient of two layers of materials), the coated composite materials behave like a bimetal. Of course, the extent of this behavior is largely dependent on the nature of the substrate, particularly its thickness and shape, and also on the thickness of the plasma polymer layer. This aspect may be a crucial factor in some applications of plasma polymers. It is anticipated that the same plasma coating applied on the concave surface has the lower threshold thickness than that applied on a convex surface, and its extent depends on the radius of curvature.

Many correlations were found between internal stress and factors involved in glow discharge polymerization. The results obtained by Morinaka and Asano [3] showed the association of high internal stress with high power of discharge. The internal stress of plasma polymers is also dependent on the chemical nature of monomers. Morinaka and Asano's study has shown that, although internal stresses were observed when hexamethyldisiloxane was plasma polymerized, the amount of the stresses is one-tenth that observed for hydrocarbons [3]. Letts et al. [4] also reported that an internal stress was observed under their plasma conditions for *trans*-2-butene but not for the perfluorinated analog. Chen et al. [5] showed that the polarity and value of bias have a considerable effect on the structure of plasma-polymerized styrene. However, these correlations to an arbitrarily selected parameter do not provide information needed to construct the general mechanism how the internal stress is built in the plasma polymer layer, and they could possibly lead to misleading conclusions if interpreted literally.

4. INTERNAL STRESS IN PLASMA POLYMERS PREPARED BY LPCAT

Since LPCAT utilizes argon metastables rather than ions and high-energy electrons, it was thought at first that the fragmentation of monomer molecule is less than in conventional plasma polymerization, and LPCAT could yield a thicker layer with lesser internal stress. However, it turned out that the trend toward higher levels of fragmentation at higher energy input would still hold for LPCAT polymerization, and LPCAT polymers, so far as the internal stress is concerned, are by and large the same as glow discharge polymers [6]. The ability of LPCAT to utilize higher molecular weight (high boiling and low vapor pressure) liquid allows investigation of a wider range of monomers with high boiling point, which are difficult to handle by the conventional glow discharge polymerization. Internal stress was investigated in a more systematic manner under various operating conditions; this was missing in data collected with glow discharge polymers appearing in the literature.

First, internal stress is plotted against arbitrarily selected single parameters just as the data in glow discharge polymers were presented. All films prepared by LPCAT showed the same kind of curling force, from which internal stress can be calculated. Figure 11.6 shows the influence of LPCAT operation parameters on

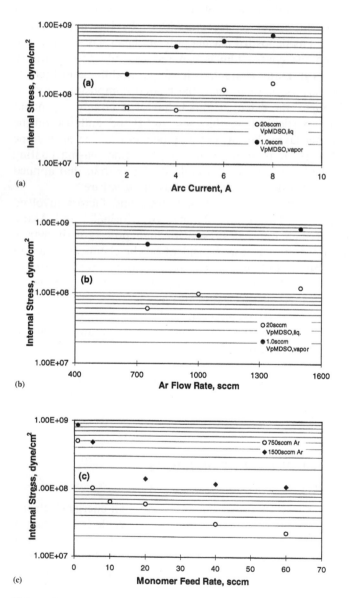

Figure 11.6   Dependence of internal stress in VpMDSO plasma polymers on (a) arc current, (b) Ar flow rate, and (c) monomer flow rate, 750 sccm, 4.0 A arc current.

internal stress in vinylpentamethyldisiloxane (VpMDSO) plasma polymer films. It can be seen from Figure 11.6a and b, that higher arc current and higher argon flow rate resulted in a higher internal stress in the plasma polymer films.

Figure 11.6c shows the monomer feed rate dependence of internal stress in VpMDSO plasma polymer films at different argon flow rates. The overall values of internal stress in plasma films obtained with argon flow rate at 1500 sccm are much higher than those obtained at 750 sccm.

From Figure 11.6c it can also be noted that the internal stress in CAT polymers deceased with increasing VpMDSO monomer feed rate. In plasma deposition process, when the other plasma parameters are kept the same, the increase of monomer feed rate indicates that the same amount of energy input is consumed by a larger number of monomer molecules. In other words, when the other plasma parameters are kept constant, the increase of monomer feed rate will actually reduce the relative energy input in plasma polymerization process.

Figure 11.7 compares the effect of monomer flow rate on the internal stress for different monomers. Figure 11.7a shows the internal stress dependence

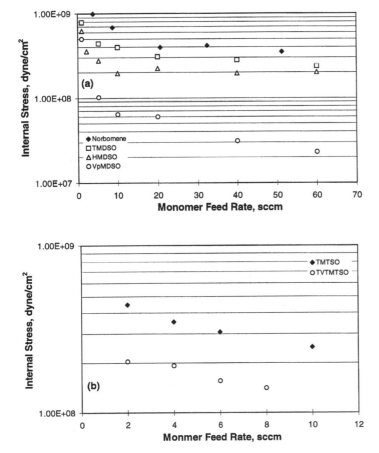

Figure 11.7 Monomer feed rate dependence of internal stress in plasma polymers; 750 sccm Ar, 4.0 A arc current.

of plasma polymer films on the monomer feed rates with four different monomers: 5-vinyl-2-nobornene (norbornene), tetramethyldisiloxane (TMDSO), hexamethyl-disiloxane (HMDSO), and VpMDSO. With increase of monomer feed rate, similar decreasing trends of internal stresses were observed for the films from all these monomers.

From Figure 11.7a, it was noted that the hydrocarbon-type monomer 5-vinyl-2-nobornene gave rise to plasma polymer films with higher internal stress, though it has six-carbon atom ring structure and a vinyl group. The other three siloxane type monomers—TMDSO, HMDSO, and VpMDSO—provided plasma polymer films with lower internal stress. As can be seen from Fourier transform infrared (FTIR) spectra shown in Figures 11.8–11.10, the Si-O-Si structure from siloxane type monomers has been largely preserved in the plasma polymers.

Figure 11.7b shows the internal stress in LPCAT films of cyclic siloxanes: 1,3,5,7-tetramethylcyclotetrasiloxane (TMTSO) and 2,4,6,8-tetravinyl-2,4,6,8-tetra-methylcyclotetrasiloxane (TVTMTSO). The large siloxane ring structure in these two monomers did not provide any decrease of internal stress in resultant plasma polymer films, compared with simple siloxane monomers, i.e., TMTSO, HMDSO, and VpMDO.

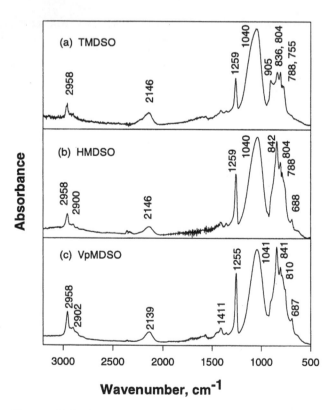

Figure 11.8 FTIR spectra of plasma polymers prepared by cascade arc torch from (a) tetramethydisiloxane (TMDSO), (b) hexamethydisiloxane (HMDSO), and (c) vinylpenta-methyldisiloxane (VpMDSO); 750 sccm argon, 4.0 A arc current, 5.0 sccm monomers.

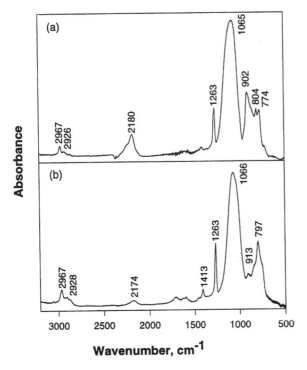

Figure 11.9 FTIR spectra of plasma polymers of (a) 1,3,5,7-tetramethylcyclotetrasiloxane (TMTSO) and (b) 2,4,6,8-tetravinyl-2,4,6,8-tetramethylcyclotetrasiloxane (TVTMTSO) prepared by cascade arc torch; 750 sccm argon, 4.0 A arc current.

5. INTERNAL STRESS AS A FUNCTION OF NORMALIZED ENERGY INPUT PARAMETER

In general plasma polymerization processes it has been established that the deposition rate and properties of a plasma polymer primarily depend on the value of the normalized energy input parameter W/FM, as described in Chapter 8. In LPCAT polymerization processes, as described in Chapter 16, the deposition rate of a plasma polymer primarily depends on the value of the normalized energy input parameter, which is given by $W(FM)_c/(FM)_m$. In this composite parameter, W is the power input applied to arc column, $(FM)_c$ is the mass flow rate of carrier gas (argon), and $(FM)_m$ is the mass flow rate of monomer that is injected into the cascade arc torch. The quantity of $W(FM)_c/(FM)_m$ can be considered as the energy, which is transported by carrier gas plasma, applied to per mass unit of monomers.

Figure 11.11 shows the dependence of internal stress on the value of the normalized energy input parameter, $W(FM)_c/(FM)_m$. It can be seen that the internal stress in all the plasma polymers showed an increasing dependence on the value of $W(FM)_c/(FM)_m$. Since $W(FM)_c/(FM)_m$ is an energy factor, a higher value of this composite parameter represents more energy being applied to the LPCAT polymerization process. Thus, a higher internal stress was developed in the plasma polymer film during deposition process. From Figure 11.11 it was also noted that,

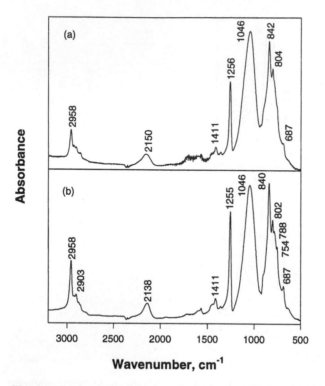

Figure 11.10 FTIR spectra of vinylpentamethyldisiloxane (VpMDSO) plasma polymers prepared by cascade arc torch at monomer flow rate of (a) 1 sccm and (b) 40 sccm; 750 sccm argon, 4.0 A arc current.

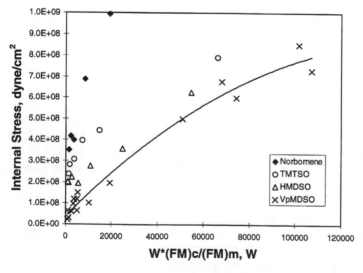

Figure 11.11 Dependence of internal stress on the energy input parameter, $W*(FM)_c/(FM)_m$, for cascade arc torch polymerization.

against energy factor $W(FM)_c/(FM)_m$, the internal stress in hydrocarbon plasma polymer film was built up much faster than siloxane counterparts. In this figure, the dependence of internal stress on molecular structure of monomer is clearly seen by the separated lines representing different monomers.

Since the deposition rate increases as a function of the same parameter (see Chapter 16), i.e., the normalized energy input parameter, the trend shown in Figure 11.3 might appear contradicting with the principle shown by Figure 11.11. However, the trend shown in Figure 11.3 is actually in accordance with the principle shown by Figure 11.11. In the operation for data shown in Figure 11.3, the deposition rate was increased by increasing monomer flow rate. The normalized energy input parameter for glow discharge polymerization (Figure 11.3) is W/FM, and the increase of F actually decreases the value of W/FM in the upper end of the transitional domain and also in the monomer-deficient domain of LCVD (see Chapter 8). Thus, the decrease of the internal stress with increasing deposition rate is correct correlation, but only within the boundary condition of the experiment for Figure 11.3. However, the correlation is not the general correlation because the parameter used is not the basic parameter of LCVD. This points out the importance of finding the basic governing principle of the process involved.

The dependence of internal stress on the value of the composite parameter $W(FM)_c/(FM)_m$ supports the postulated wedging plasma deposition process. The $W(FM)_c/(FM)_m$ is the normalized energy input parameter of LCVD by LPCAT. Accordingly, with higher value of this combined parameter, the total energy input to the luminous gas phase and to the depositing film increases. As a result, a higher internal stress was observed in the plasma polymer films that was formed at a higher energy input. Thus, the internal stress could be viewed as the accumulation of energy carried by the impinging reactive species, i.e., the higher energy must be put into the layer to build the higher internal stress.

6. CORRELATION BETWEEN INTERNAL STRESS AND REFRACTIVE INDEX

A larger refractive index usually represents a higher density or tighter network structure of the plasma polymer films. From a point of view that LCVD is a wedging process of reactive building block entities, it is anticipated that there exists a correlation between the refractive index and the internal stress. Figure 11.12 shows the refractive index change of plasma polymers with arc current, argon flow rate, and monomer flow rate. It can be seen that the refractive indices in the films increased with the increase of both arc current and argon flow rate.

Figure 11.13 shows the dependence of refractive index in plasma polymer film on energy factor $W(FM)_c/(FM)_m$. The refractive indices of all the plasma polymers showed an increasing dependence on the value of energy factor $W(FM)_c/(FM)_m$.

It was also noted that, under comparable values of energy factor $W(FM)_c/(FM)_m$ as shown in Figure 11.13, the hydrocarbon plasma polymer showed much higher refractive index than siloxane counterparts. This phenomenon perhaps resulted from their chemical structures in the monomers, specifically the difference in the elements that become the building blocks for LCVD. In the effort to interpret the difference, one must bear in mind that all molecules dissociate under the

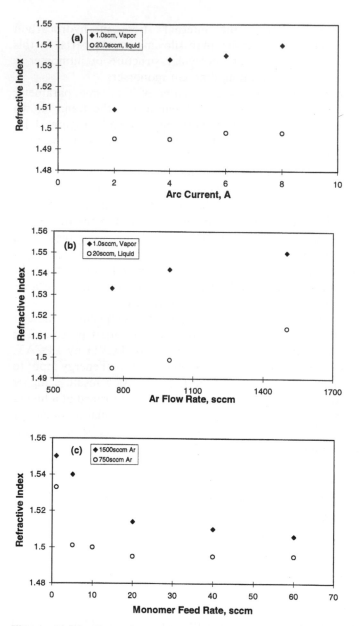

Figure 11.12 Dependence of refractive indices in VpMDSO plasma polymers on (a) arc current, (b) Ar flow rate, and (c) monomer flow rate; 750 sccm, 4.0 A arc current.

environment of luminous gas. With hydrocarbons, the building block element is C. With siloxane derivatives, the building block unit can be considered Si-O-Si, which can be detected by FTIR spectra. Because of strong affinity of Si to O, Si-O-Si structure is largely retained when the dissociation of a monomer occurs under the environment of LPCAT. It should be cautioned, however, it does not mean that the structure $-(\text{Si-O-Si})_n-$ exists in the film. Furthermore, this building block is

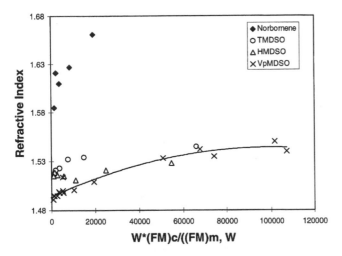

Figure 11.13 Dependence of refractive index on energy factor $W*(FM)_c/(FM)_m$ for cascade arc torch polymerization.

inorganic entity in nature and should not be misinterpreted as the flexible repeating unit of silicone polymers, which is $-(Si-O)_n-$ with two organic ligands on Si.

At the same plasma conditions, plasma polymer films from hydrocarbon monomers gave rise to larger refractive indices than those films from siloxane monomers. There seems to be a trend, within Si-containing monomer group, that the lower the atomic ratio of C to Si the higher is the internal stress, i.e., the highest internal stress is observed with TMTSO ($C/Si = 1.0$), the intermediate with HMDSO ($C/Si = 3.0$), and the lowest with VpMDSO ($C/Si = 3.5$). This trend appears, at first glance, as contradicting with the high refractive index for hydrocarbon LPCAT films. However, it actually supports the concept of the wedging process in the following manner.

The packing density obtainable in LCVD process is dependent on the size and its distribution of impinging entities. For the simplicity of discussion, let us assume that (1) CH_2 is the building block for hydrocarbon LCVD film and (2) Si-O-Si is the building block for siloxane LCVD film, but siloxanes also produce CH_2 by dissociation of the monomer. First, the main building block for siloxane film, Si-O-Si, is larger than CH_2, and consequently the packing density cannot be as high as that of smaller CH_2. This is the reason why hydrocarbons yield a higher refractive index than that for siloxanes. Second, the highest packing density could be obtained with single impinging entity, and the packing density would decrease if the second impinging entity is added regardless of the size of the second entity. The packing density would decrease with the ratio of the second entity up to the point where the second entity becomes the major component. The trend found within siloxanes can be explained by this principle.

If a comparison is made between Figures 11.11 and 11.13, a correlation of higher internal stress to larger refractive index in plasma polymer films was observed. In the plasma polymer films with larger refractive index, the internal stress that developed during the deposition process is more difficult to release afterward due to the tighter structure.

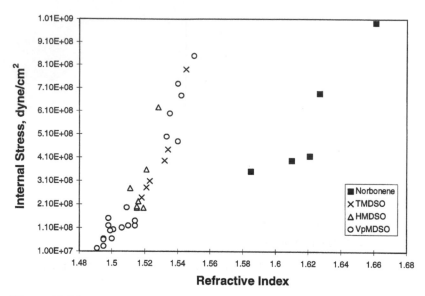

Figure 11.14 Qualitative correlation of internal stress with refractive index of cascade arc torch plasma polymer films.

The above observation can be seen more clearly from Figure 11.14, in which the internal stress was plotted against refractive index. In the plasma coatings with larger refractive index, the internal stress that developed during the deposition process is more difficult to be released due to corresponding tighter structure. Therefore, such a correlation of higher internal stress to larger refractive index in plasma polymer films is not only reasonable but also points to the origin of the internal stress in plasma polymers.

The refractive index is a good measure of the packing density of atoms or chemical moieties (not the mass density of material). The high internal stress built in the solid of LCVD materials is the reflection of the high packing density. Any operational and/or molecular parameters that increase the packing density of the deposition will increase the internal stress. It is important to note that such a high packing density is one of the important and unique characteristics of LCVD materials, and the effort to reduce the internal stress could be counterproductive. Once the internal stress is successfully reduced, the material might not have the unique characteristics of type A LCVD films.

The correlation between the refractive index and the internal stress is not unique, i.e., there are multiple lines. The quantitative correlation depends on the molecular nature of monomers and the modes of operation of LCVD. For example, TMDSiO, which showed no internal stress by radio frequency glow discharge deposition, showed significant level of internal stress when it was deposited in the cascade arc torch deposition. However, it can be considered that the tighter the network structure, the higher is the internal stress. It should be noted that the range of refractive index of Si-containing LCVD polymers by LPCAT is lower than that of LCVD polymer formed by other discharges, particularly glow discharge polymers by cathodic polymerization, which have refractive index higher than 2.0.

The high internal stress is an important factor in the tempering of the surface and constitutes a very important feature of LCVD. However, it requires special caution with respect to where it should be used and how it should be created.

REFERENCES

1. Yasuda, H.; Hirotsu, T.; Olf, H.G. J. Appl. Polym. Sci. **1977**, *21*, 3179.
2. Yasuda, H.; Hirotsu, T. J. Polym. Sci. Polym. Chem. Ed. **1977**, *15*, 2749.
3. Morinaka, A.; Asano, Y. J. Appl. Polym. Sci. **1982**, *27*, 2139.
4. Letts, S.A.; Myers, D.W.; Witt, L.A. J. Vac. Sci. Technol. **1981**, *19* (3), 739.
5. Chen, M.; Yang, T.C.; Ma, Z.G. J. Polym. Sci. Polym. Chem. Ed. **1998**, *36*, 1265.
6. Yu, Q.S.; Yasuda, H.K. J. Polym. Sci., A: Polym. Chem. **1999**, *37*, 1577–1587.

12
Modes of Operation and Scale-up Principle

1. MODES OF OVERALL PROCESSING

From the viewpoint of industrial scale application of LCVD and LCVT, probably the most important factor is the continuity of processing, namely, whether the processing can be operated in a continuous mode or should be done in batch mode. The continuous mode contains two modes of operation, i.e., operation of glow discharge and handling of substrate. This distinction of two modes of operation could be seen in an example of coating or treating a long web of substrate, such as film, cloth, or wire. The true continuous processing can be accomplished by feeding substrate from outside of a vacuum reactor and retrieving the proceed product outside of the reactor, i.e., "air-to-air" operation. This can be done by feeding substrate through a narrow opening and applying stepwise pumping through small-volume segments, and the reverse steps on the other end of the reactor. Another way is by placing a roll of film in a vacuum reactor, followed by the feeding and retrieving of substrate in vacuum, i.e., "vacuum-to-vacuum" operation. This is a batch process, but LCVD or LCVT could be run in a steady state of luminous gas phase.

From the viewpoint of LCVD or LCVT, whether a steady-state luminous gas phase can be used or not makes the most important difference in the feasibility of processing. In the example used above, the choice is only in the feeding mechanism, and the batch operation of LCVD or transient state LCVD could be a factor only in the beginning of vacuum-to-vacuum operation.

However, when noncontinuous substrates are employed, the process could be run either by continuous operation or by batch operation, probably equally well, and the choice heavily depends on factors other than LCVD. The requirement for a continuous operation is that the total numbers of products must be large enough to justify a continuous operation for a year, for instance, and there is no sense in building a continuous reactor for processing only a few items. If the number, size, and shape of substrate to be handled changes, e.g., parts and components of machine or aircraft, the butch operation is mandatory choice.

In batch processing of LCVD or LCVT, substrates are placed in a reactor, and LCVD or LCVT is carried out as a unit operation. Repeating the same operation treats a large number of substrates. Batch processing is the primary mode for nearly all laboratory scale operations. It can be done in a closed system or in a flow system. Because the number of molecules in a reactor under low pressure is small, it is often necessary to use a flow system in order to obtain a sufficient amount of coating or extent of treatment.

This is not always the case, however, depending on the size of reactor and the specificity of the deposition on the substrate. For instance, the deposition in DC discharge occurs mainly to the cathode surface and very efficient LCVD can be done in a closed-system reactor. The details of this aspect are covered in Chapter 13 dealing with DC discharge. Dealing with complex metallic substrates in various lengths and widths, a large reactor is needed just to receive the substrates. In such a case, the ratio for (reactor volume)/(surface area of substrate) is large, and the reactor contains enough gas molecules in a closed system to coat the substrate.

The closed system has a distinctive advantage that the total mass in the system is well defined and does not change by the electrical discharge while the number of gas molecules changes, which is an advantageous feature in studies of basic chemical reactions. The gas phase changes on the onset of glow discharge are described in Chapter 3. The fragmentation of a molecule increases the total number of gas molecules in a system, and the polymerization or deposition of material decreases the total number of gas molecules. The balance between these two opposing reactions shows in the change of pressure of a closed system. Hence, a very simple pressure measurement could provide very important information on how the process is progressing.

The continuous operation of noncontinuous substrates, e.g., contact lenses, video disks, microsensors, etc., is performed by placing a certain number of substrate in an evacuation/transfer chamber, in which the evacuation is carried out and samples are transferred to the adjacent sample holding chamber in vacuum. The evacuated sample holders are placed on a conveyer one by one and pass through glow discharge zones. The coated substrates follow the reverse process at the downstream end of a reactor to be taken out in the ambient environment. Thus, the substrate charge is done in butch mode, but the LCVD process is done continuously.

Operational parameters that control LCVD can be divided into two major categories: (1) characteristic parameters of a reactor, which can be altered but in most cases are not a variable of operation, and (2) parameters that require adjustment for each run and often during a run. Size of electrodes, distance between electrodes, and frequency of electric power are parameters of the first category. Monomer flow rate, system pressure, and discharge power are operational parameters of the second category. The parameters of the first category are important in the design of an LCVD reactor, but the parameters of the second category are critically important in the execution of an LCVD to produce a desired product.

Because a proper understanding of the parameters of the second category is essential for the selection of those of the first category, and the majority of research reactors are batch reactors, our discussion of operational parameters begins with the latter parameters in batch mode operation.

2. PRESSURE OF A STEADY-STATE FLOW SYSTEM (WITHOUT PRESSURE CONTROL)

The system pressure of a flow is determined by the feed-in rate of a gas and the pumping-out rate of a vacuum system. The monomer flow-in rate is determined by the opening of an orifice (e.g., a metering valve) and the differential pressure applied across the orifice. The pumping-out rate, however, is determined by the overall pumping-out capability of a pump system. The latter is determined by the capacity of the pump, the size and length of the vacuum line that connects the pump and reactor, and the type of cold trap (or absence of it) used in the system. As will be shown, the cold trap, particularly a liquid N_2 trap, acts as an excellent pump for many organic compounds. The system pressure of a flow can be controlled to a predetermined value by employing a throttle valve of which the opening is regulated by the pressure of the system within a certain range of flow rates. In order to establish a meaningful flow system LCVD operation, it is mandatory to have a pressure-controlling throttle valve in the pumping system.

Although a cold trap is an excellent pump and prevents organic vapors from entering the mechanical rotary pump, it also increases the concentration of potentially dangerous chemical compounds, which are created by the low-pressure electrical discharge. Therefore, special caution should be exercised in deciding whether a cold trap should be used in the first place, and if one is used, a careful procedure for disposing the trapped substances should be established according to the nature of the monomer. Careless use and handling of a cold trap could lead to an explosion of trapped chemicals. The advantages and potential hazards should be weighed for the type of monomer employed.

When the system pressure is measured at a fixed point in a reactor, without the pressure control by a throttle valve, the reading of the pressure is empirically related to the flow rate of a monomer by

$$F = ap^b \tag{12.1}$$

The value of exponent b lies in the range $1 < b < 2$, but often it is close to 2 [1]. The relationship given by Eq. (12.1) is well illustrated in Figure 12.1, where the flow rate of ethylene is plotted against the pressure of the system with and without a liquid N_2 trap. According to Figure 12.1, the liquid N_2 trap provides a pumping rate six to seven times higher than that of the mechanical pump used in the system.

3. SYSTEM PRESSURE UNDER A GLOW DISCHARGE (WITHOUT PRESSURE CONTROL)

When a glow discharge is initiated with a steady-state flow of monomer (without system pressure control), the system pressure changes to a new steady-state value as a steady-state luminous gas flow is established. This change of the system pressure from that of a pure monomer flow to that of a flow under LCVD conditions is caused by the following two major factors: (1) the gas phase changes because of the creation of the luminous gas phase, and (2) the pumping-out characteristics

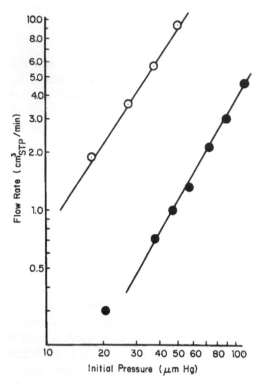

Figure 12.1 Flow rate versus pressure for ethylene obtained with (⊙) and without (●) a liquid-nitrogen trap.

of gases change accordingly, i.e., the capability of the pump to exhaust a certain amount of gases in a unit time changes when the composition of the gases is altered.

Under LCVD conditions, original organic monomer molecules undergo significant fragmentation. For instance, most hydrocarbons yield H_2 as a major gaseous product of LCVD. Therefore, the major portion of the original organic monomer molecules disappears from the gas phase under consideration by depositing a solid material and leaving non-polymer-forming H_2 in the gas phase. Thus, as soon as the LCVD reaction starts, the gas to be pumped out of the system changes from monomer to the gaseous products of the LCVD. In many cases, the gases to be pumped out are a mixture of unreacted monomer and various gaseous products formed by the LCVD.

This change in the composition of the gas phase causes change in the pumping-out capability of the system. For instance, a liquid N_2 trap that provides six to seven times greater pumping capability than a mechanical pump for ethylene does not work for H_2 because H_2 cannot be frozen at the liquid N_2 temperature. Some turbomolecular pumps, for instance, have a much lower pumping rate for H_2 than for larger gases.

Although the change in system pressure from that of monomer flow to that of flow under plasma conditions depends on many factors, the relationship expressed by Eq. (12.1) still holds for the LCVD system, but the flow rate to be used in the equation is no longer the flow rate of the original monomer. This situation can best be explained by data obtained in a closed system.

In a closed-system LCVD, the total number of gas (organic monomer) molecules, n, changes to γn when the monomer is subjected to a glow discharge. Because the pressure of a closed system is directly proportional to the number of gas molecules in the system, the value of γ can be obtained by the ratio p_g/p_0, where p_g is the pressure in the glow discharge and p_0 the pressure before the glow discharge is initiated.

Because the material formation in LCVD can be given by a simple formula, $nM \rightarrow M_n$, LCVD reduces the total number of molecules in the gas phase, and the LCVD of an organic monomer acts as a pump. This reduction in the number of molecules is counteracted by the production of nonpolymerizable gases. The value of γ is dependent on the balance of the polymerization and the gas production.

With a monomer, such as acetylene, that polymerizes quickly and produces little H_2 gas, the value of γ is close to zero, and the plasma polymerization of acetylene acts as an efficient vacuum pump. As a consequence of this pumping effect, the system pressure of acetylene plasma polymerization drops sharply, and if a relatively low flow rate is employed, the glow discharge is extinguished due to the drop in system pressure and is initiated again as the system pressure builds up after a certain time. Such a pulsating glow discharge of acetylene is often observed with reactors of relatively small volume at relatively low flow rates.

The change of pressure in a flow system due to LCVD is not determined solely by the characteristic parameter γ of a monomer. Although the value of γ obtained by a closed-system experiment indicates the fragmentation characteristics of a monomer, a change in system pressure of a flow system due to plasma conditions does not provide information relevant to the value of γ. This is because the initial system pressure (before the glow discharge is initiated) is not a unique function of the flow rate of monomer but is highly dependent on the pumping rate.

For instance, when the plasma polymerization of a hydrocarbon is carried out in a flow system that has a liquid N_2 trap, an increase in system pressure due to plasma polymerization is observed because the liquid N_2 trap cannot trap the H_2 produced and the pumping-out rate of the system drastically decreases for the product gas. However, when the same plasma polymerization is carried out without a liquid N_2 trap (in this case the initial pressure of the system is much higher than in the previous case), a decrease in system pressure due to plasma polymerization is observed because a mechanical pump generally pumps H_2 faster than a larger organic molecule.

One can manipulate the value of initial pressure p_0 (before glow discharge initiation) by controlling the pumping rate (e.g., by throttling the pumping side of a reactor), but the manipulation of p_g (pressure under a glow discharge) is not easy, and sometimes it is nearly impossible (e.g., the plasma polymerization of acetylene can proceed even though the pumping system is disconnected from the reactor).

Therefore, the control of p_0 (before glow discharge initiation) does not mean the control of p_g. Figure 12.2 shows the change of p_0 to p_g for cases in which p_0 is manipulated to a constant value of 60 mtorr at various flow rates. Thus, in such a case, depending on the flow rate, the system pressure may increase or decrease, or no change may be observed when the glow discharge is initiated.

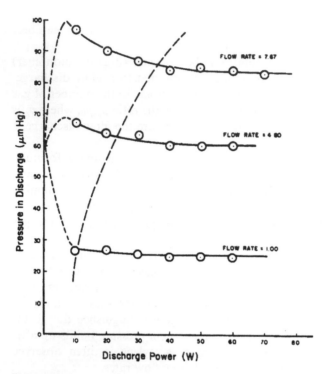

Figure 12.2   Pressure p_g in a glow discharge versus discharge power at a fixed flow rate and initial pressure p_0 for ethylene (different pumping rates are used to adjust p_0 to 60 μm Hg.

4. FACTORS THAT DETERMINE PRESSURE UNDER A GLOW DISCHARGE IN FLOW

Because the pressure before glow discharge initiation does not determine the pressure under a glow discharge, it is important to find the factors that determine this parameter. As far as the gas phase is concerned, the plasma polymerization in a closed system can be described as

$$n \rightarrow \gamma n \tag{12.2}$$

where n is the number of gas molecules. If we assume that the same reaction occurs in a flow system, Eq. (12.2) can be written as

$$dn/dt \rightarrow \gamma \, dn/dt \tag{12.3}$$

Thus, in a flow system, LCVD is visualized as a process that changes the flow rate of gas.

The flow rate F_1 of monomer coming into the reactor is given by the pressure p_1 of the flow system before glow discharge initiation, according to Eq. (12.1):

$$dn/dt = F_1 = a_1 p_1^{b1} \tag{12.4}$$

As soon as the reaction (plasma polymerization) occurs, dn/dt changes to $\gamma dn/dt$, and this flow of gases must be pumped out by the vacuum system for a steady-state flow to be maintained. As far as the system pressure is concerned, we can assume that $\gamma dn/dt$ or $F_2 = \gamma F_1$ is introduced into the system, and the same equation can be used [i.e., Eq. (12.1)] to find the new system pressure p_2:

$$F_2 = a_2 p_2^{b2} \tag{12.5}$$

Thus, a change in flow rate is manifested by a change in system pressure. Therefore,

$$a_2 p_2^{b2} = \gamma F_1 \tag{12.6}$$

Consequently, the pressure p_g in a glow discharge started from a steady-state flow F_0, which has the system pressure p_0, is given by

$$\log p_g = \log c + d \log p_0 \tag{12.7}$$

where

$$c = (\gamma a_1/a_2)^{1/b2} \tag{12.8}$$

and

$$d = b_1/b_2 \tag{12.9}$$

or

$$\log p_g = \log c' + d' \log F_0 \tag{12.10}$$

where

$$c' = (\gamma a_1/a_2)^{1/b2} \tag{12.11}$$

and

$$d' = 1/b_2, \tag{12.12}$$

Equation (12.7) relates p_g to p_0, but both constants c and d contain parameters for the monomer and the product gas. Therefore, it is anticipated that Eq. (12.7) would hold between p_g and p_0 but that the intercept and the slope of a straight line would depend on the pumping characteristics of the monomer. Figure 12.3 depicts the p_g–p_0 relationship for the plasma polymerization of ethylene with and without a liquid N_2 trap in the system. Although there exists a relationship given by Eq. (12.7), p_g cannot be uniquely related to p_0. In other words, the manipulation of p_0 by manipulation of the pumping rate does not control the value of p_g.

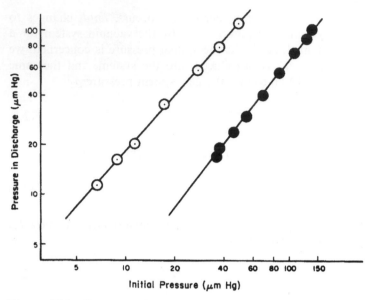

Figure 12.3 Pressure p_g in a glow discharge versus initial pressure p_0 for ethylene with (⊙) and without (●) a liquid-nitrogen trap.

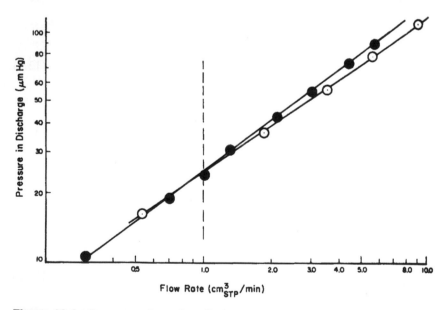

Figure 12.4 Pressure p_g in a glow discharge versus flow rate F_0 for ethylene with (⊙) and without (●) a liquid-nitrogen trap.

In contrast to this situation, Eq. (12.10) relates p_g to F_0 and the constants c' and d' contain parameters for the product gas only. Therefore, p_g can be uniquely related to F_0, as shown in Figure 12.4, in which the same data presented in Figure 12.3 are plotted against F_0.

An important point is that the change from p_0 to p_g, recognized as an increase or a decrease in the system pressure, is dependent on the value of p_0, which varies with the pumping rate of the monomer at a fixed flow rate. Therefore, in a flow system, the value of $\delta = p_g/p_0$ is not a parameter describing the characteristics of a monomer in a glow discharge; i.e., δ in a flow system is not identical to γ in a closed system, which is a characteristic parameter of a monomer.

The plots of $\log p_g$ versus $\log F_0$ for different monomers that produce the same product gas (i.e., H_2) yield parallel straight lines that intercept $F_0 = 1$ at different values depending on the value of γ for the monomer, according to Eqs. (12.10) to (12.12). Figure 12.5 contains plots of $\log p_g$ versus $\log F_0$ for ethylene and acetylene; both produce H_2 as the main product gas but the values of γ differ (0.10 for acetylene and 0.65 for ethylene). The intercept of the straight line at $F_0 = 1$ is related to the value of γ by

$$(\log p_g)F_0 = 1 = (\gamma/a_2)^{1/b2}$$

In order to obtain a γ value from a flow system plasma polymerization, it is necessary to have knowledge of b_2 and a_2, both of which are parameters of the pumping characteristics of the product gas given by Eq. (12.1).

The ideal situation just described is observed with a reactor that has a well-defined flow pattern without any bypassing of the monomer flow and under conditions that yield very high conversion (nearly 100%) of monomer to polymer. With a relatively large volume bell jar type of reactor with a relatively small plasma volume, such an ideal relationship may not be observed due to the high bypass ratio of monomer and consequent low conversion of monomer to polymer.

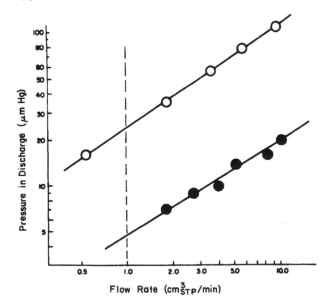

Figure 12.5 Pressure p_g in a glow discharge versus flow rate F_0 for acetylene (○) and ethylene (●).

Nevertheless, the fundamental relationships that govern the system pressure under a glow discharge should also apply to such a system.

The pressure of a LCVD system is a very important factor because it determines the concentration of gaseous chemically reactive species. However, strict control of the pressure under glow discharge conditions is difficult and sometimes nearly impossible. The monomer flow rate F_0 and the system pressure p_0 of a steady-state flow before plasma polymerization can be independently adjusted by manipulation of the inlet valve (for F_0) and outlet valve (for p_0) openings. The system pressure p_g in a glow discharge, however, is generally different from p_0 (depending on the value of γ for the monomer and also on F_0). The use of the throttle valve that is controlled by the pressure of system could take care of the change of system pressure caused by the onset of glow discharge, but it is not always possible to control p_g by manipulating the outlet valve opening without changing F_0. Particularly with a monomer that has a low γ value, the p_g of the system cannot be adjusted unless a very high flow rate is employed to increase the bypass ratio and/or to decrease the conversion ratio of monomer to polymer, in which the pressure change becomes small.

It is important to recognize that p_g is different from p_0, which requires the measurement of both pressures, and also that the initial pressure p_0 of the system is not as important a parameter of plasma polymerization as has been considered. The important parameter is the pressure under the conditions of plasma polymerization described as p_g.

The measurement of p_g requires a pressure transducer system that is not influenced by the electric power used for the plasma polymerization, particularly when a high-frequency radio frequeny power is employed. Some pressure transducers that give pressure readouts independent of the nature of a gas are ideally suited for plasma polymerization. Some electronic gauges the readout of which depends on the nature of the gas (e.g., thermal conductivity) do not provide accurate readings of p_g because in most cases the composition the gas mixture in the LCVD reactor is unknown and there is no way to calibrate the meter for an unknown gas mixture.

Because the velocity of a gas molecule and the dissociation of gas in a glow discharge are functions of $1/p$, the value of p_g (not p_0) is important in controlling the distribution of polymer deposition and the properties of plasma polymer; however, p_g cannot be considered as a completely manipulatable processing factor of LCVD. The value of p_g can be manipulated to a certain extent, but it is determined largely by the nature of the starting material (i.e., gas production characteristics) and the pumping capacity of the system. For instance, if a certain value of p_0 is attained with throttle valve fully open, the increase of p_g cannot be controlled by the throttle valve.

A change in the system pressure also changes the volume and the intensity of the luminous gas phase (plasma), which not only changes the relative position of the polymer-collecting surface in the plasma, but also changes the ratio of polymer collected on the surface to the total amount of polymer formed. Consequently, a change in pressure may cause a change in the apparent deposition rate of plasma polymer.

The value of p_g largely reflects the pressure of the non-polymer-forming gases in the plasma (by-product of the plasma polymerization) when the conversion ratio of monomer to polymer is high. Therefore, the control of p_g is important with respect to the ionization characteristics of the product gas, which is important for maintaining the glow discharge, and the movement of chemically reactive species, which influences the distribution of polymer deposition. It should be noted, however, that p_g in the same context of the pressure of simple gas discharge such as that of argon or helium cannot be applied to polymer-forming species. Therefore, p_g should be treated as a partially controllable operational parameter that depends largely on the fragmentation characteristics of the monomer used.

5. MONOMER FLOW RATE

The flow rate of a monomer is generally given by the volume of the gas at standard temperature and pressure (273 K and 1 atm) per unit time (e.g., cm^3_{STP}/min, which is often designated as standard cubic centimeters per minute, "sccm"). In the gas phase, the pressure and volume determine the total number of molecules of the gas under consideration. In other words, in order to define a system one must define pV; p or V alone cannot define the system. In contrast to this situation for a gas, with a noncompressible liquid the volume alone can be used to define the system. Thus, $1\,cm^3_{STP}$, means 1 cubic centimeter at 1 atm and 273 K, the dimension of which is not L^3 but L^3p. Therefore, the flow rate given by cubic centimeters (STP) per minute, sccm, is proportional to moles per minute. Thus, it is important to recognize that cubic centimeters (STP) does not represent the volume of a gas but gives a value proportional to the number of moles of a gas.

In LCVD, the simplest parameter that can be correlated with the flow rate of monomer is the polymer deposition rate, which is generally and most logically expressed by (mass)/(area)(time). As long as the dependence of polymer deposition rate on monomer flow rate is sought for a given monomer only, the monomer flow rate given by sccm can be used without difficulty; when such a correlation is extended to different monomers and the polymer deposition characteristics are compared, however, the flow rate based on cubic centimeters per minute cannot be used because the mass of a mole of gas depends on the molecular weight of the monomer. The polymer deposition rates of various monomers should be compared on the basis of the mass flow rate; otherwise, polymer deposition rates are not directly proportional to the polymerization rates.

If the polymer deposition rate is compared on the basis of monomer flow rate in cubic centimeters (STP) per minute (i.e., F rather than F_W), there is an obvious dependence on the molecular weight of monomer, as depicted in Figure 12.6. The use of normalized deposition rate, (Deposition Rate)/FM, takes care of this situation as described in Chapter 5.

The flow rate F (based on volume at standard temperature and pressure) can be easily determined by the measurement of $d(pV)/dt$. In a vacuum system, which has a constant volume, measuring $V(dp/dt)$ is the easiest and perhaps the most accurate way to determine the flow rate. This can be done by the following procedure.

First, a steady-state flow of monomer is established, and the system pressure p_1 is read. Then, at time zero, the valve that connects the reactor to the pumping system is closed and the increase in pressure dp/dt, which is given by an initial straight line of the pressure versus time plot, is read. The flow rate F is given by

$$F = (dp/dt)V(273/T)(1/ps) \tag{12.13}$$

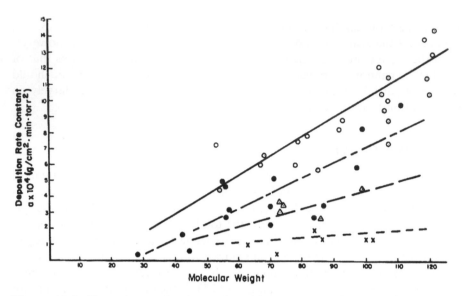

Figure 12.6 Dependence of polymer deposition rate on molecular weight of monomer. Group I (○): triple-bond-containing, aromatic, and heteroaromatic compounds; group II (●): double-bond-containing and cyclic compounds; group III (△): compounds without aforementioned structures: group IV (×): oxygen-containing compounds.

where V is the volume of the reactor (cc), T is the temperature of the measurement (Kelvin), and p_S is 1 atmospheric pressure in the same units used for the measurement of p, e.g., 760 if torr is used.

The volume of a reactor can be determined by several methods; however, a sufficiently accurate value (to three digits) can generally be obtained for a reactor of more than 1 liter volume by calculating the volume based on the size, shape, and length of the reactor, and connecting pipes placed between the inlet valve and the outlet valve.

Because most flow meters and flow control devices require a calibration factor for each gas, it is best to determine the flow rate by the procedure given above even when such a flow meter or flow controller is used in the system, and confirm the calibration factor. The use of flow controllers and throttle valves is the common practice in recent years.

When an easily condensed organic vapor is used, the flow controller and/or the flow meter is not as useful as it is for a gas monomer due to the condensation of the vapor in the meter or the controller, as well as to the differential pressure available to the device, which is too low. Consequently, using a simple metering valve and determining the flow rate by the method described could achieve the delivery of a high-boiling organic monomer. The flow controller system, in which a liquid reservoir, flow controller, and connecting pipes are contained in a temperature-controlled vessel to prevent condensation of the vapor, provides the best control of condensable vapors.

In the case of a mixture of gases, one can determine the flow rate of each component gas, following the principle just described, the step-by-step addition of each component gas. First, the flow rate of the first gas F_1 is determined. Then, the second gas is added, and the total flow rate $F_1 + F_2$ is determined. Next, F_2 is calculated from the value of $F_1 + F_2$. On such a step-by-step addition of component gases, the gas that has the highest available (delivery) pressure should be used as the first gas because the high pressure available will minimize the change of flow rate due to the increase in reactor pressure by the addition of subsequent gases. A good flow controller will eliminate this possible variation; however, the two gas flows should be thoroughly mixed before feeding into the reactor by passing through a tube with baffle plates or other mixing device.

6. MEANING OF FLOW RATE IN AN LCVD SYSTEM

In an incompressible system (such as a liquid), flow rate has two significant meanings: (1) the feed-in rate of mass and (2) the "sweeping rate" of the molecules in the system, from which an estimate of the residence time of a molecule in the system can be calculated. However, the second meaning of flow rate requires careful examination when applied to a gas flow, depending on the conditions used. In general, the "flow" of gas in a vacuum should not be conceived as being similar to the flow of a liquid. In a flow of gas under a vacuum, the absolute velocity of the gas molecules and the diffusional displacement velocity (brownian motion) are large. The gas flow rate does not represent the velocity of individual gas molecules but only the total flux. It therefore does not represent the sweeping rate applicable to an individual molecule. In a polymer-forming luminous gas phase (reactive system), the flow rate of monomer should not be taken beyond the meaning of the rate of feed-in into the reactor because the flow rate is determined in the nonreactive gas state and no information about the gas under luminous gas phase is generally available. The distribution of polymer deposition is indicative of this point.

Another important factor to be considered in conjunction with the flow rate of monomer is that LCVD occurs predominantly in the glow region. Therefore, the true reaction volume is close to the volume of glow, V_g. However, the volume of glow is not always the same as the volume of the reactor, V_r; and in certain cases $V_g \ll V_r$.

The flow rate is usually determined and presented as the rate of feed-in into V_r but not into V_g. For instance, in a bell jar type of reactor with parallel electrodes, V_g is only a small percentage of V_r. Therefore, the bypass ratio of monomer flow, which may be roughly proportional to V_r/V_g, should be taken into consideration when plasma polymerizations in different reactors are compared. The important point here is that the bypass ratio thus considered is itself a variable parameter depending on the conditions of plasma polymerization because an effective plasma polymerization (i.e., with a low value of γ) acts as a pump and draws the monomer into the reaction volume V_g from the surrounding volume. Conversely, in the case of a monomer with large value of γ, the pressure in V_g becomes greater than that in the volume $(V_r - V_g)$ and reduces the rate of monomer supply from the surrounding volume to the glow volume.

Although the sweeping rate of flow does not apply to a gas flow, as mentioned earlier, if such a parameter is compared (e.g., for a rough comparison of resident time of monomer in the reactor), the flow rate in the context of the flow velocity in a vacuum reactor can be given by F/p, where F is given in sccm's and p is the system pressure in atmospheres. Therefore, a flow rate 10 sccm at 0.1 torr has a gas velocity 100 times higher than 10 sccm at a pressure of 10 torr.

The linear velocity of gas flow in a vacuum is also dependent on the cross-sectional area of the reactor in which the flow takes place. Therefore, a parameter describing the linear velocity of a flow is given by $F/A*p$, where A is the cross-sectional area (square centimeters). The value of $F/A*p$ for a flow of 50 sccm in a bell jar of 50-cm diameter at 1 torr is 19.4 cm/min, whereas the value of $F/A*p$ for a flow of 5 sccm in a tube of 5-cm diameter at 0.1 torr is 1940 cm/min.

The parameter that is proportional to the resident time should also be calculated on the basis of $F/A*p$, not F. The parameter V/F or $L*A/F$, where V is the volume of a reactor and L the length of a tube, does not yield a number in units of length because F is equal to $d(pV)/dt$, not dV/dt for a gas flow.

Thus, a comparison of flow rates reveals a difference in monomer feed-in rates only. As is clear from these examples, 50 sccm in a bell jar system under the given conditions is actually a much slower flow (in the context of the sweeping rate) than 5 sccm in the tube system under the given conditions for the latter case, despite the value of F being 10 times greater.

The ratio F/V_g is an important factor that indicates how fast the monomer is fed into the reaction system, and the value of $F/A*p$ is an important parameter related to the resident time of a flow system (i.e., $V*p/F$ or $L*A*p/F$).

Another important factor of a LCVD system, in conjunction with the monomer flow rate F, is whether the plasma polymerization is carried out in (1) the diffusion-dominating case or (2) the flow-dominating case. If we have a mixture of two gases, 1 and 2, in a space, the transport of gas 1 and 2, respectively, occurs according to the gradient of concentration of each gas. This process is generally recognized as a diffusion process similar to diffusion in the liquid or solid phase. The diffusion constant D of a gas mixture is defined in exactly the same way as in the liquid or solid phase. In the gas phase, however, the medium in which the diffusive transport takes place and the diffusing entities are not as clearly defined as in the case of diffusion in the liquid or solid phase. Consequently, the diffusion coefficient is often referred to as the interdiffusion coefficient.

Interdiffusion coefficients of gases are much larger than those of liquids. The diffusion coefficients of most gases are of the order of a few tenths of a square centimeter per second, cm^2/s, at the standard temperature and pressure, whereas diffusion coefficients of liquids are less than 10^{-5} cm^2/s. The diffusion coefficient D at pressure p (torr) and temperature T (K) can be derived from the diffusion coefficient D_0 at the standard temperature and pressure ($p_0 = 760$ torr, $T_0 = 273$ K) by

$$D = D_0(T/T_0)^n(p_0/p) \tag{12.14}$$

where n lies between 1.75 and 2.0.

Einstein showed that the displacement x executed by a particle during time t in a medium with a diffusion coefficient D is

$$x^2 = 2Dt \tag{12.15}$$

At standard temperature and pressure x for most gases is about 1 cm/s. The number of collisions one particle exercises while traveling distance x is $(3/2)(x/\lambda)^2$, where λ is the mean free path. Because the diffusion coefficient is inversely proportional to the pressure of a gas, the diffusional displacement becomes significantly larger in a vacuum, far greater than what one might conceive by the term *diffusion*. This is one reason that the more uniform deposition can be obtained at the lower pressure.

Thus, the movement of reactive species in LCVD environment is highly dependent on the system pressure. Accordingly, the value of F/p, where F is the flow rate in sccm and p is in atmosphere (e.g., $p = 6.58 \times 10^{-5}$ atm for 50 mtorr), is an important parameter for LCVD. However, the value of F/p measured before glow discharge is initiated determines the initial rough domain of plasma polymerization, and as soon as glow discharge is initiated, the system pressure changes according to the fragmentation characteristics of the monomer. The use of a pressure-controlling throttle valve partially takes care of the pressure change due to the creation of luminous gas phase; however, it hides the pressure change, which is a characteristic feature of the monomer γ. It is important to know value of γ for the monomer to be used, which can be measured by using a closed system before using the monomer in a flow system.

Regardless of a closed system or a flow system, it is generally considered that the discharge time is the key factor in the LCVD processing. However, the actual processing depends on many other factors of operation, such as degassing time before the discharge is initiated (degassing time), the time after a monomer is introduced into the system until discharge is initiated (flow establishing time), and the adsorption of LCVD gas by the surfaces, which are not of the substrate. The actual processing of LCVD operation starts with the preparation of substrate for placing in LCVD reactor, which is especially important for hygroscopic or hydrophilic substrates. Some example of these factors influencing the overall processing of LCVD could be seen in the following sections.

Inagaki and Yasuda [3] investigated transient-stage polymer deposition by using mixed monomers, of which one component is N_2. N_2 is a non-polymer-forming reactive gas that does not form polymer by itself but copolymerizes with another monomer. In one type of experiment (method A), a steady-state flow of mixed monomer is established and maintained for 5 min without discharge, in which period the adsorption of organic monomer onto the substrate surface (quartz thickness monitor) and other surfaces occurs.

In another mode (method B), an N_2 gas flow is established and a glow discharge of N_2 alone is initiated; then the organic monomer is introduced into the N_2 plasma in such a manner that the same mixture of N_2/monomer can eventually be established. In this case, no adsorption of organic monomer onto the substrate surface and other surfaces can take place. Figure 12.7 depicts thickness growth as a function of discharge time for styrene/N_2 system, and Figure 12.8 depicts the same for acetylene/N_2 system. The slope of the thickness growth curve gives the deposition rate. In the initial stage, the deposition rate is 2.7 nm/min for method A and 0.53 nm/min for method B. The steady-state deposition rate is 0.69 nm/min for method A and 0.96 nm/min for method B.

There are three major factors involved in LCVD that can be detected by the data shown in these two figures: (1) adsorption of monomer on surface during the process of establishing a steady-state flow of monomer, which could be liberated

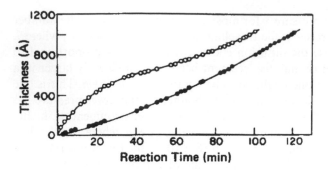

Figure 12.7   Polymer film thickness growth in a styrene/N_2 system as a function of discharge time for two different mixing method of gases; the upper curve for the method A, the lower curve for the method B.

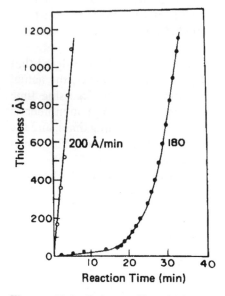

Figure 12.8   Polymer film thickness growth in a acetylene/N_2 system as a function of discharge time for two different mixing method of gases; the upper curve (left) for the method A, the lower (right) curve for the method B.

on the inception of glow discharge; (2) the time necessary to mix two gases in a flow system reactor; and (3) change of the gas phase from molecular mixture to the luminous gas phase that causes the change of the total number of gases in unit time in the reactor.

As shown in Figure 12.7, an initial deposition rate that is roughly four times greater than the steady-state deposition rate is observed with styrene, which is an easily condensable vapor (high adsorption onto the surface), by method A. The gaseous monomer acetylene, on the other hand, shows an initial deposition rate identical to the steady state is obtained immediately on the inception of discharge, as shown in Figure 12.8. Whether the adsorbed monomer polymerizes at the surface or generates gas phase monomer on the inception of glow discharge is difficult to ascertain. Because the higher deposition rate (the slope of the curves shown in these figures) persists for a considerable period (about 30 min), it is more likely that the adsorbed monomer (at all surfaces in the reactor) contributes to the increased concentration of monomer on the inception of glow discharge. The high deposition rate in the early stage, in the case of styrene/N_2 in method A, can be explained by the fact that desorption of the adsorbed styrene occurs under the environment of discharge causing higher deposition rate.

It is important to note that the true steady state deposition rate that is independent of the mixing modes (method A or B) was obtained only after more than 100 min operation in the case of styrene-N_2. Without monomer adsorption in the case of acetylene-N_2 system, the steady-state deposition rate is established much sooner than that in the styrene-N_2 system. However, even in this case, it took more than 30 min to reach the steady-state deposition rate under the conditions of method B.

These data clearly indicate that the same factor, which is not included as an operational parameter, has an important role at the beginning of plasma polymerization in a closed-system reactor and in a flow system. That is the change of gas phase on the inception of discharge or the creation of luminous gas phase described in Chapter 3. This factor is further influenced by the extent of adsorption or condensation of gas on the surface before discharge is initiated, i.e., the condensability of monomer and the gas mixing time have significant influence on the initial deposition rate.

It is important to point out that if a flow system LCVD process is carried out as a batch process that takes only a fraction of the time lag to establish the steady-state deposition rate, e.g., 5 min processing of styrene-N_2 system, of which the time lag is more than 100 min, the deposition rate observed does not represent the characteristic deposition rate of the monomer. Such an operation would be perfectly acceptable for obtaining an LCVD coating of a certain thickness, if it is reproducible; however, such data cannot be used to investigate the fundamental behavior of the monomer in an LCVD environment. It is worth noting that the temperature dependence of LCVD discussed in Chapter 5 was investigated with the steady-state deposition rates in a continuously operating glow discharge system (not a batch-operated flow system).

Data shown in Figure 12.7 and Figure 12.8 indicate that a steady-state glow discharge could not be attained in many batch operations within the duration of experiment, although the steady–state glow discharge is intuitively assumed. They also indicate the advantage of a continuous glow discharge operation, in which substrates pass through a steady-state glow discharge. Because of this reason, the continuous operation (of glow discharge) generally yields much more reliable products than the batch operation of the same process.

7. ADSORPTION OF MONOMER ON SUBSTRATE AND REACTOR WALL SURFACES

Establishing a constant system pressure in a closed system or in a flow system with substrate materials in the reactor is not as simple and straightforward as it might appear. A gas is generally fed into the system by a constant flow rate controlled by a mass flow controller. The system pressure, on the other hand, depends on how much of the gas fed into the system is adsorbed on the surface, i.e., the adsorbed gas is not registered in the gas pressure. Consequently, the total mass of gas fed into a system to establish a constant system pressure differs greatly depending on the sorption characteristics of the substrate placed in the reactor.

When a porous polymer film, which acts as a good "getter" for organic vapors, is used as the substrate, the sorption (physical adsorption and chemical sorption) effect of a monomer can be seen by the change of the system pressure of a closed-system discharge [4]. Figure 12.9 depicts the change of the system pressure of a closed system with the discharge time for LCVD of 4-vinylpyridine with a different substrate. The pressure change was examined for three substrates with different sorption characteristics: i.e., glass, which has the minimum sorption; Millipore film, which has a very high sorption characteristics; and a porous polysulfone film, which has a moderate sorption with the monomer. An arbitrarily selected system pressure of 200 mtorr was established by feeding the monomer at a constant flow rate.

The monomer 4-vinylpyridine has a very low hydrogen yield, and the system pressure decreases quickly on the inception of glow discharge, as depicted in the bottom line of Figure 12.9 for glass substrate. The pressure increase at the inception of glow discharge indicates that the sorbed monomer emanates from the substrate to the luminous gas phase. With Milllipore film, the system pressure increases more than twice that of the system pressure before it decreases with more or less the same

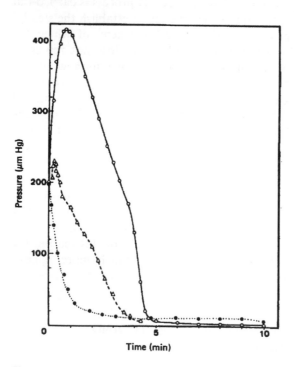

Figure 12.9   Pressure change observed in a chamber containing a substrate pseudo-saturated to an initial pressure of 200 mm Hg of 4-vinylpyridine during glow discharge; Key: ○, Millipore filter; ●, glass slide; △, porous polysulfone film.

rate observed with glass substrate, indicating that more than an equal amount of monomer in gas phase was sorbed in the substrate.

It is important to recognize that the most critical stage of LCVD is the very early stage in which the depositing material interacts with the substrate and the critical interface is created. In the later stage, the luminous gas phase interacts with the deposited LCVD material. The critical initial stage is influenced by the sorbed monomer regardless of whether LCVD is carried out in a closed or a batch-operated flow system. This implies that the truly dependable LCVD operation could be obtained only by means of a continuous in-line operation, in which a steady state of a luminous gas phase is established and well-preconditioned substrates move into the luminous gas phase and out by a linear motion.

It is important to recognize that all surfaces that contact with the luminous gas phase participate and influence LCVD operation. Therefore, in principle, in a batch operation, the first run with clean reactor wall could not be replicated in the second run with contaminated reactor wall. Thus, it is necessary to include the step for cleaning the reactor. If only hydrocarbons were used in an LCVD, the cleaning could be done by O_2 discharge prior to the normal LCVD operation. (The influence of wall contamination was described in Chapter 10.) In this respect, the effort to minimize the deposition on nonsubstrate surfaces is important even in batch operation of LCVD. Magnetron discharge is quite effective in this respect, as described in Chapter 14.

8. CONTINUOUS LCVD

In the continuous processing, a steady-state flow of luminous gas is established and maintained for the duration of operation, e.g., 1 month, without interruption. Due to the factors described above, it takes some time, e.g., 30 min, to establish a steady-state flow of luminous gas. Once a steady state is established, it can be maintained it for sufficient time to allow continuous processing. Substrates are fed into the steady-state flow of luminous gas in a cross-flow pattern. The rate of transport of substrate and the length of the path in the luminous gas phase determine the treatment time.

It is common that a substrate or a section of continuous substrate such as fibers and films passes multiple sets of luminous gas flow. Figure 12.10 depicts a schematic diagram of continuous operation of LCVD with two sets of luminous gas flow. Each chamber is pumped individually to avoid cross-contamination, and a buffering chamber separates the two sets of luminous gas flow. Thus, multiple of LCVD can be applied on a substrate according to this principle. A continuous substrate, such as fibers, tubes, or film, is fed vertically. Noncontinuous substrates should be handled differently with substrate holders. The horizontal feeding of substrates shown in Figure 12.10 requires multiple of substrates holding devices that travel through the reactor.

In a continuously operated flow system, those factors associated with a closed system or a batch-operated flow system mentioned above are virtually eliminated except at the very beginning of the operation. Therefore, the reproducibility of LCVD obtained by continuously operated LCVD is superior to that obtained by the batch operation. For instance, the reproducibility of reverse osmosis membranes

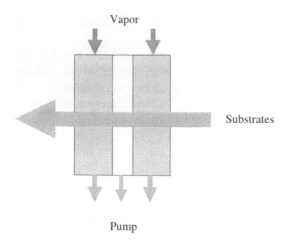

Vapor

Substrates

Pump

Figure 12.10   Schematic representation of continuous operation of LCVD.

prepared by the continuous mode processing, in which the initial stage sample is discarded, is significantly better than that of membranes prepared by batch-operated plasma polymerization [5].

In order to achieve a continuous operation of LCVD, it is necessary to consider the following factors: (1) deposition on electrodes, (2) deposition on reactor wall, and (3) flaking off of the deposition from these surfaces. Magnetrons for LCVD were originally developed for the continuous operation of LCVD for production of RCA video disks in early 1970s. Magnetron discharge confines the volume of luminous gas phase, which minimizes the deposition onto the reactor wall. It is a general practice to place removable liners on the walls of reactor, which will be removed and replaced by new liners on each maintenance stoppage of the operation.

Planar magnetron electrodes are placed in vertical position in order to avoid the flaked-off materials falling onto the substrate surfaces. It is also necessary to find the optimal condition of operation that depends on the combined effects of the nature of vapor or mixture of a vapor and a gas, the operating pressure, the discharge power, frequency of electrical power, magnetron configuration, and the strength of magnetic field. In an ideal situation when those parameters were optimized, a part of electrode surface, corresponding to the toroidal glow, remains uncoated and no sputtering of the metal occurs providing a sustained steady-state glow discharge. The details of magnetrons are described in Chapter 14. The efficiency of the magnetic confinement requires the linear motion of electrons within a cycle of an alternating polarity of the electrical power, and hence a power source in the range of 10–50 kHz is preferred.

The continuous operation of LCVD could be controlled, in principle, by the strict controls of the flow rate, the system pressure, and the discharge power. The addition of a residual gas analyzer in each pumping system is advantageous in detecting abnormality in gas flow, occurrence of leakage, etc. The addition of more sophisticated diagnostic devices, such as an optical emission spectrometer, in-line XPS, etc., is often counterproductive because the stability of performance and the maintenance requirement cycle of these devices do not match those for the continuous mode operation of LCVD. The information obtainable by these devices should not be included in the specification of the operation; otherwise the troubles associated with these devices interrupt the operation while the LCVD process is normal.

The control of the deposition thickness is crucially important in the LCVD operation, regardless of the modes of operation. However, it is very difficult to measure the actual thickness of ultrathin film deposited on polymeric substrates. If an LCVD film is deposited on polymeric substrates, such as films, fibers, and molded articles, it is nearly impossible to determine the thickness by a simple nondestructive method that can be done quickly enough to monitor the LCVD operation. In order to circumvent this problem, the use of special substrates added or attached to the normal substrate is found to be satisfactory. A small piece of Si wafer is a typical case of this approach. Ellipsometer can measure the deposition on the Si wafer easily and quickly, which provides thickness and refractive index values.

9. SCALE-UP PRINCIPLE

It is tempting to scale up a laboratory scale operation that showed promising results by increasing the size of reactor. However, such an attempt is bound to fail because the onion layer structure of the luminous gas phase changes completely if the volume of a reactor, the size of the electrode, and the interelectrode distance are increased. An even worse attempt is changing the mode of glow discharge in order to cope with the change of the reactor volume. It is the general rule of thumb that the curiosity-based laboratory scale experiments, however successful were the results, should not be scaled up simply by increasing the size of the reactor proportionally.

The large-scale operation should be conceptually built first, considering all requirements necessary to produce products in an industrial operation. The shape and nature of the substrate, e.g., continuous sheet of film or fibers, large or small disks, etc., dictate what kind of operation could be feasible in the industrial scale operation. From the conceptual operation, the key factors of LCVD process should be extracted, and then a laboratory scale reactor should be designed and constructed. In other words, a specific laboratory reactor should be built for a specific industrial scale operation. When this approach is followed, the scale-up of a successful laboratory operation is actually the scale-back to the original conceptual operation.

On the scale-back process, the key factor is to not change the luminous gas phase as much as possible, which could be done by multiplying the unit process in the laboratory scale reactor in a larger volume vessel, instead of increasing the size of the unit process, e.g., the size of electrodes, the distance between electrodes, and so forth.

Since the dissociation glow can be considered to be the major medium in which polymerizable species are created, the location of the dissociation glow, i.e., whether on the electrode surface or in the gas phase, has the most influence on where the most of the deposition by the plasma polymerization occurs. The deposition of plasma polymer could be divided into deposition E and deposition G as described in Chapter 8. The deposition kinetics is completely different for deposition E and deposition G.

Since the continuous operation relies on the deposition G, the choice is limited to radio frequency discharge or high-frequency magnetron discharge. This depends on the nature and the number of substrates to be handled. In general, however, the use of magnetrons in plasma polymerization coating has significant advantages in (1) confining the volume of glow, (2) reducing the breakdown voltage in low-pressure regime, which allows operation in lower pressure than RF discharge could operate, (3) lowering the overall impedance of the glow discharge, and (4) increasing deposition G. These advantages enable us to produce superior coatings by industrially feasible operations. The magnetron discharge increases deposition G and decreases deposition E with additional advantage of the capability to operate in the low-pressure regime without expanding the glow volume. The production of type A plasma polymers more or less require the low-pressure (e.g., <100 mtorr) operation.

The use of magnetrons in plasma polymerization is not in the mainstream of plasma polymerization research in laboratory scale; however, the use of magnetrons is a dominant factor in large-scale industrial plasma polymerization; starting with RCA video disk coating in early 1970s to the latest successful contact lens coating. In these industrial applications, substrates are continuously fed into a steady-state glow discharge in a cross-flow mode, and the steady-state glow discharge is maintained for a prolonged period, e.g., 1 month, without disruption, while the treatment time (the resident time of a substrate) is a few seconds to a few minutes.

Such a long sustainable operation is possible only if (1) discharge conditions do not change during the period and (2) the contamination of the reactor wall can be minimized so as not to interfere with the coating operation during the period. The magnetron discharge fulfills these two crucially important requirements. Although no explicit information is available, many seemingly successful applications without magnetron had to be abandoned within a couple of years due to the failure to fulfill the above two requirements. In any case, the careful planning before laboratory scale research is the key to a successful industrial scale LCVD operation.

REFERENCES

1. Yasuda, H.; Hirotsu, T. J. Appl. Polym. Sci. **1978**, *22*, 1195.
2. Yasuda, H. Macromol. Rev. **1981**, *16*, 199.
3. Inagaki, N.; Yasuda, H. J. Appl. Polym. Sci. **1981**, *26*, 3557.
4. Yasuda, H.; Lamaze, C.E. J. Appl. Polym. Sci. **1973**, *17*, 201.
5. Yasuda H.; Matsuzawa, Y. I&EC Product Research & Development **1984**, *23*, 163.

13

Direct Current and Alternating Current Discharge

1. CATHODIC POLYMERIZATION VS. PLASMA POLYMERIZATION IN NEGATIVE GLOW

Direct Current (DC) discharge can be used primarily for cathodic polymerization using the substrate metal as the cathode. Since deposition E (deposition on electrode) is around 80%, the cathodic polymerization is a great advantage, but it should be recognized that the cathodic polymerization is meant to be a short-term batch operation for nanofilm coating of the cathode because the thicker deposition of dielectric material on the cathode extinguishes the discharge. In the same vein, the attempt to utilize deposition G (deposition in luminous gas phase) is not warranted. Plasma polymerization coatings by high-frequency (HF), e.g., 40 kHz, and radio frequency (RF) discharge are primarily used for coating of dielectric substrate placed in luminous gas phase. The basic characteristics of these three types of glow discharges should be well understood in order to permit selection of the right discharge for a specific aim of plasma polymerization coating.

The DC cathodic polymerization, 40-kHz (HF) and 13.5-MHz (RF) plasma polymerization of trimethylsilane (TMS) were compared in a bell jar type of reactor [1–3]. The bell jar has the dimensions of 635 mm height and 378 mm diameter. A pair of stainless steel plates (17.8 × 17.8 × 0.16 cm) was placed inside the bell jar with spacing of 100 mm and used as parallel electrodes. The substrate used in the plasma deposition process was an aluminum alloy panel positioned in the midway between the two parallel electrodes.

Figure 13.1 depicts the two configurations of the electrode–substrate arrangement used for TMS deposition in different glow discharges. In configuration a, the Al substrate was used as the powered electrode, which is the cathode in DC process. In such a configuration, the two parallel stainless steel plates were used as grounded electrodes, which are on the same electrical potential. In configuration b, the Al panel was used as a floating substrate positioned between the two parallel electrodes. In this configuration, one of the two parallel electrode plates was used as the powered electrode (or cathode in DC process) and another was used as the grounded electrode (or the grounded anode in DC process).

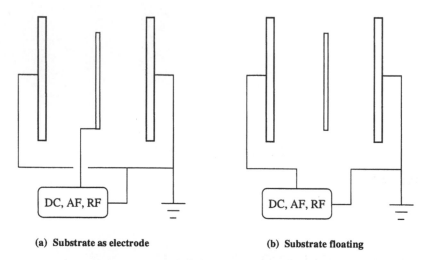

(a) **Substrate as electrode**   (b) **Substrate floating**

Figure 13.1 Schematic diagrams of the two configurations of the electrode setup used in the glow discharge polymerization process, (a): Al substrate was used as powered electrode and (b): Al substrate was floating in between the two parallel electrodes.

1.1. Trimethylsilane Deposition on Electrode

In DC cathodic polymerization, the activation of reactive species and deposition of polymers mainly occur in the cathode glow (molecular dissociation glow, which touches the cathode surface). In a glow discharge initiated by an alternating current (AC) power source, e.g., 40 kHz, the electrode can be visualized as an alternating cathode in half of the discharge time and anode in another half. In radio frequency, the oscillating electrons in the glow discharge are mainly responsible for the creation of polymer-forming species. As a result, in RF discharge the molecular dissociation glow no longer touches the electrode surface but remains very close to the electrode surface. Consequently, the role of electrode with respect to the deposition onto the electrode changes dramatically (see Chapters 3 and 4).

As described in previous chapters, the material formation in luminous chemical vapor deposition (LCVD) takes place in the molecular dissociation glow and in the ionization glow. In a diffused luminous gas phase such as a relatively large-volume RF discharge, the contributions of these two types of reactions are difficult to be identified. In DC discharge, the chemical reactions in the molecular dissociation glow are dominant and the majority of deposition occurs to the cathode surface. Accordingly, the analysis of deposition onto the electrode surface would provide important information for comprehending the difference due to the nature of the power source.

Comparing the deposition rate dependence on operational parameters for the deposition in the diffused luminous gas phase and for the cathodic deposition [Eqs. (8.2) and (8.7)], the contribution of the cathodic polymerization can be estimated by examining the system pressure dependence of the deposition rate (at a fixed flow rate). If the material formation in the diffused luminous gas phase is the dominant factor, it is anticipated that the deposition rate would be independent of the system pressure. If the material formation in the molecular dissociation glow is

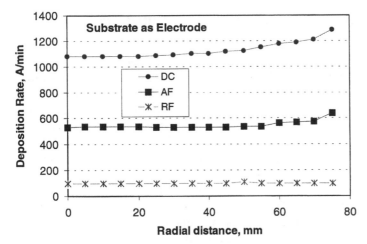

Figure 13.2 The deposition rate profiles of TMS in DC, 40-kHz, and 13.56-MHz LCVD with substrate as electrode; 1 sccm TMS, 50 mtorr system pressure, 5 W power input.

the dominant factor in a discharge created by an alternating polarity power source, the deposition rate onto an electrode is dependent on the system pressure, and the deposition rate is expected to be one-half of that for DC discharge.

TMS deposition rate profiles in DC, 40-kHz, and 13.56-MHz discharges are shown for electrode in Figure 13.2. It can be seen that, regardless of the frequency of electrical power source used, a uniform deposition of TMS polymers was observed in the three plasma processes, although an appreciable edge effect occurred in the DC and a less pronounced effect occurred in the 40-kHz discharge when the substrate was used as the cathode or powered electrode. The uniform distribution of deposition rates justifies the use of single measurement at the center of the electrode to represent the characteristic deposition rate of a system.

The system pressure dependencies of the deposition rate onto the electrode surface in DC, 40-kHz, and 13.56-MHz discharges are shown in Figure 8.8. As anticipated from the deposition rate equation, given in Eq. (8.7), the deposition rate of DC cathodic polymerization is linearly proportional to the system pressure. The deposition rate in 40-kHz discharge was found pressure dependent also, but that in 13.56-MHz discharge was found to be independent of system pressure (actually slightly negative). The deposition rate in the 40-kHz discharge is one-half of that in the DC discharge, and the slope of pressure dependence is also roughly one-half of that obtained from DC discharge. These findings indicate that the material formation in the molecular dissociation glow takes place on the electrode in a 40-kHz discharge. As the frequency increases to 13.56 MHz, the electrode does not act as the cathode such as in DC or 40-kHz discharges, and the material formation in the diffuse luminous gas phase then governs the deposition onto the electrode. Because the dissociation glow exists very close to an electrode, the deposition rate onto an electrode is higher than that on a floating substrate; however, the deposition on an electrode of RF discharge is that of glow discharge polymerization that occurs in the diffused luminous gas phase.

1.2. Trimethylsilane Deposition on Floating Substrate

TMS deposition rate profiles on floating substrate in DC, 40-kHz, and 13.56-MHz discharges are shown in Figure 13.3, in which the scale of deposition rate is 0–300, whereas that in Figure 13.2 is 0–1400. The deposition rate on floating substrate is smaller than that on electrode in all cases, but uniform deposition occurs on the substrate. In DC discharge, there is a significant difference in the deposition rate on the surface facing the cathode and that on the surface facing the anode. This difference is proof that the cathodic polymerization prevails in DC discharge, and the distance from the cathode is a factor that controls the deposition onto substrate placed in gas phase.

The system pressure dependence of the deposition rate onto the electrode surface in DC, 40-kHz, and 13.56-MHz discharges are shown in Figure 8.8. In order to see the influence of electrical contact, some silicon wafers were electrically insulated from the substrate plate used as the cathode by placing a thin slide cover glass between the silicon wafer and the substrate. The influence of the electrical contact on deposition rate onto the electrode and onto the floating substrate is shown in Figures 8.8–8.10 as a function of system pressure. In the lower part of the figures, the influence of the same factors on the refractive index is shown. The scale of the deposition rate axis is different for each case in order to show the system pressure dependence clearly in each case.

In a DC discharge (Fig. 8.9), without electrical contact, the deposition rate differs significantly from that for *cathodic polymerization* (with electrical contact), and the pressure dependence is marginal. If we consider the substrate without electrical contact on cathode surface as the floating substrate at 0 distance from the cathode, the deposition rate on the 0 distance floating substrate should be proportional to the deposition rate on the cathode, which explains the marginal pressure dependence. The deposition onto a floating substrate can be characterized as typical *plasma polymerization* (material formation in the diffused luminous gas phase). A major difference in refractive index is seen between the electrode and nonelectrode use of the substrates.

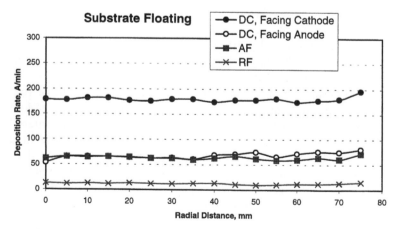

Figure 13.3 Deposition rate profiles of TMS in DC, 40-kHz, and 13.56-MHz LCVD on floating substrate; 1 sccm TMS, 50 mtorr system pressure, 5 W power input.

If one considers that the overall DC plasma polymerization is a mixture of cathodic polymerization (material formation in the dissociation glow that is adhering to the cathode surface) and plasma polymerization (material formation in the diffused luminous gas phase), the deposition on the cathode is primarily cathodic polymerization. With an insulating layer between the substrate and the cathode surface, there is no cathode glow, and hence no cathodic polymerization that deposits polymer on the substrate. The substrate on the cathode surface without electrical contact or any noncathode surface receives the products of plasma polymerization in negative glow.

In a 40-kHz discharge (Fig. 8.10), the deposition onto the surface of the electrode, regardless of electrical conductivity or contact, is significantly different from deposition onto a floating substrate. The cathodic aspect of the electrode is less (one-half of DC discharge), but because of this the overall cathodic aspects of polymerization extend beyond the surface of the electrode yielding cathodic plasma polymer on an electrically insulated substrate placed on the electrode. Thus, the features of cathodic polymerization dominate in the vicinity of the electrode regardless of electrical contact.

In 13.56-MHz discharge (Fig. 8.11), the deposition on the substrate placed on an electrode, regardless of electrical conductivity or contact, is also appreciably different from that on the floating substrate, although the magnitude of the difference is much smaller than that found in DC or 40-kHz discharges. However, the deposition onto the electrode has no feature of cathodic polymerization. In 13.56-MHz discharge, the deposition of materials is primarily by plasma polymerization, of which deposition rate is given by Eq. (8.3). The higher deposition rate on the electrode could be explained by the effect of the molecular dissociation glow being close to the electrode surface. In RF discharge the site of activation has shifted away from the electrode surface and the features of cathodic polymerization diminished, i.e., the pressure dependence of deposition rate on an electrode is slightly negative in RF discharge.

The material formation in LCVD is caused mainly by the dissociation glow, and the ionization glow consists mainly of non-polymer-forming species. In DC discharge, the material formed in the cathode glow deposits nearly exclusively on the cathode surface due to the adherence of dissociation glow to the cathode, but some of them could deposit on surfaces existing in the reactor. The situation with the material formed in the negative glow is the same, i.e., it could deposit on the cathode, the anode, and surfaces placed in the reactor. With electrically floating substrates, the deposition rates, as well as the refractive indices, are nearly the same for DC and 40-kHz glow discharges. Under the set of conditions employed, 13.5-MHz discharge yielded the lower deposition rate, but the refractive index was found nearly the same as those samples formed in DC and 40-kHz discharges. This implies that the material formation in the diffused luminous gas phase of DC and those in 40 kHz and 13.56 MHz are essentially the same.

The major difference between the cathodic polymerization and the glow discharge polymerization is the influence (or absence) of ion bombardment during the process of material deposition. With an organic compound as monomer, ions that bombard the cathode are mainly those of hydrogen and other non-polymer-forming elements.

2. ROLE OF ANODE IN DC CATHODIC POLYMERIZATION

In DC cathodic polymerization conducted in a bell jar reactor, the cathode (substrate) is positioned in the middle between the two anodes. In such electrode arrangement, the distance between the cathode and the anode is expected to have some effects on the deposition rate and deposition profile with respect to those without anode assembly. Figures 13.4 and 13.5 show the influence of the distance between two anodes (one-half of which is the cathode–anode distance) on TMS deposition rate on cathode (i.e. substrate) and anode, respectively.

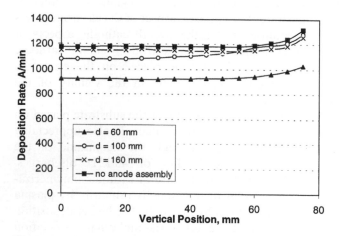

Figure 13.4 The influence of electrode distance on the deposition rate on Cathode in DC cathodic polymerization; 1 sccm TMS, 50 mtorr, DC 5 W, d the distance between two anodes, d/2 is the distance between the cathode and an anode.

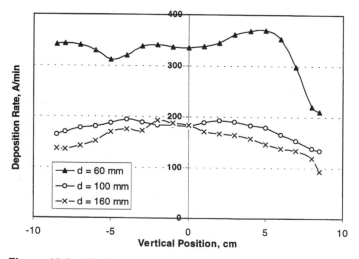

Figure 13.5 The influence of electrode distance on the deposition rate on Anode surface in DC cathodic polymerization; 1 sccm TMS, 50 mtorr, DC 5 W, d the distance between two anodes, d/2 is the distance between the cathode and an anode.

From Figure 13.4 it can be seen that, with the increase of anode spacing from 60 mm to 160 mm, the deposition rate on cathode (substrate) showed an increasing trend. The deposition on the cathode (substrate) surface seemed to reach the maximum when the anodes were removed from the plasma system, i.e., no anode assembly was present and the grounded reactor wall functioned as anode. In contrast, it is noted that, from Figure 13.5, the deposition on the anode surface decreased with the increase of anode spacing. These results clearly indicated that the too-close anode spacing not only reduced the preferred plasma polymer deposition on substrate (cathode) but also induced more undesired deposition on the anode surface. In other words, DC cathodic polymerization without anode assembly seems to be a more efficient and realistic approach in its practical applications.

As seen in Figure 13.4, DC cathodic polymerization of TMS without anode assembly gave rise to higher deposition than that with anode assembly. Therefore, the nature of anode and its role in DC cathodic polymerization should be further clarified. In DC discharge, as noted from Figures 13.4 and 13.5, the much smaller amount of deposition occurs onto the anode surface than that occurs onto the cathode. Figure 13.6 shows the effect of the floating panels positioned between anode and cathode on the deposition rate on anode and cathode surfaces in DC discharge. The floating panels did not affect the plasma deposition on the cathode. They acted just as a surface cover (2 cm away) of the anode and showed a similar deposition rate to that on anode.

It is interesting to note that a reduced deposition near the covered edges and no deposition were detected in the center of the anode due to the presence of the floating panels. These results indicate that the anode is a passive surface as far as the plasma polymerization is concerned, and the deposition does not differ from the floating substrate placed between cathode and anode. This also means that the anode, as a passive surface, collects polymerizing species created by the glow discharge polymerization and that the glow discharge polymerization deposits on the cathode surface as well. When the passive surfaces are eliminated, the majority of the deposition due to the (negative) glow discharge polymerization also occurs on the cathode surface. Thus, as noticed in Figure 13.4, deposition on the cathode increases with removal of the anode assembly.

3. TMS DEPOSITION ON MULTIPLE CATHODES WITHOUT ANODE ASSEMBLY

DC cathodic plasma polymerization of TMS was carried out with multiple cathodes in a relatively large (178 liter) barrel-type metal reactor [2]. In such a reactor, no anode assembly was arranged inside the chamber. During the plasma polymerization process, the DC power was directly applied to the aluminum panels, which were loaded inside the reactor and used as cathodes. The grounded reactor wall functioned as the anode during the DC discharge.

The arrangement of the panels (cathodes) is shown schematically in Figure 13.7. The deposition profiles on each panel (cathode) with three different panel spacing of 2, 4, and 6 cm are shown in Figure 13.8. It can be seen in Figure 13.8a that 2-cm panel spacing is so close as to affect the electrical field near the cathode surface and

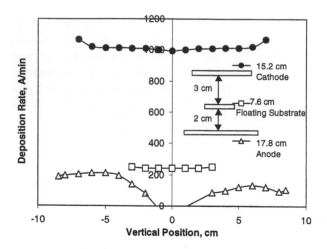

(a) 1/2 piece of Al panel in front of Anode

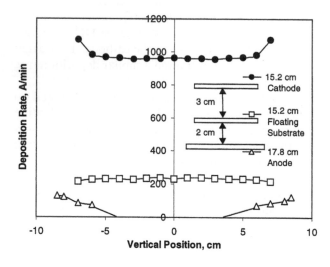

(b) 1 whole piece of Al panel in front of Anode

Figure 13.6 The effect of the floating panels positioned in front of the anode on the deposition rate on Anode surface and Cathode surface in DC cathodic polymerization; 1 sccm TMS, 50 mtorr, DC 5 W, anode spacing $d = 100$ mm, $d/2$ is the distance between the cathode and an anode.

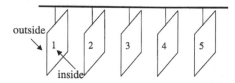

Figure 13.7 Schematic diagram of the arrangement of 5 panels used as the cathodes in DC cathodic polymerization of TMS without using anode assembly.

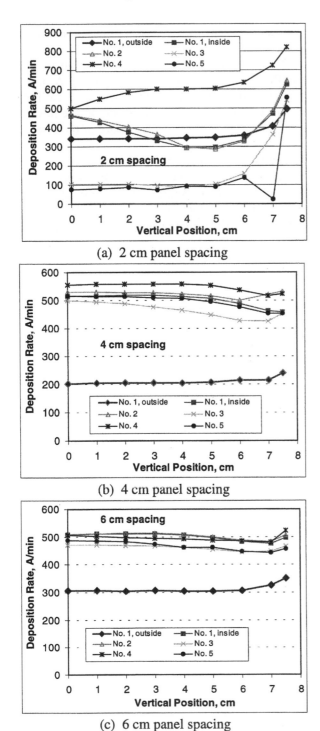

(a) 2 cm panel spacing

(b) 4 cm panel spacing

(c) 6 cm panel spacing

Figure 13.8 The effect of panel spacing on the deposition profile with 5 panels in a row as the cathodes of DC cathodic polymerization of TMS; TMS, 1 sccm, 50 mtorr, DC 5 W, outside deposition was measured only on No. 1 panel.

the monomer diffusion in the space between two panels. Thus, nonuniform deposition profile was found on all five panels (cathodes).

It can be seen from Figure 13.8 that very uniform TMS deposition profiles were achieved on all five panels in both cases of 4- and 6-cm panel spacing. As noted from Figure 13.8b and c, one interesting aspect is that, in DC cathodic polymerization with a row of cathodes (panels), the deposition rate on the internal panels is about 200–300 Å/min higher than that on the outside surface of the external panels.

Figure 13.9 shows the TMS deposition profile with five panels in a row at 100 mtorr under different DC power inputs. The increase of DC power input from 5 W to 25 W significantly increased the deposition rate of TMS on the cathodes. Two interesting facts were noted from Figure 13.9. The first is that the deposition difference disappeared between the two sides of the outer panels as noted in

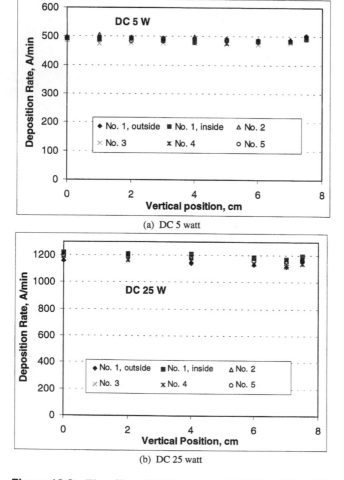

(a) DC 5 watt

(b) DC 25 watt

Figure 13.9 The effect of DC wattages, (a) 5 W and (b) 25 W, on the deposition profiles with 5 panels (cathodes) in a row with panel spacing of 6 cm; TMS, 1 sccm, 100 mtorr, 6 cm panel spacing, outside deposition was measured only on No. 1 panel.

Figure 13.8. The second is that the edge effect was depressed in comparison with that at 50 mtorr as observed in Figure 13.8. As a result, an even more uniform plasma deposition was achieved with DC cathodic polymerization that was carried out at higher system pressure (at a fixed flow rate).

Figure 13.10 shows the change of voltage, current, and the current density with DC power input in TMS discharge. It is noted from Figure 13.10a that the increase of DC power input had little effect on the voltage but significantly increased the DC current flowing through the discharge. Regardless the system pressure difference, as shown in Figure 13.10b, a linear dependence of DC current density on power input was observed in DC glow discharge of TMS.

Figure 13.11 shows the change of TMS deposition rate with DC power input at system pressures of 50 and 100 mtorr. As anticipated, a linear dependence of the deposition rate on DC power input was observed at both 50 and 100 mtorr system

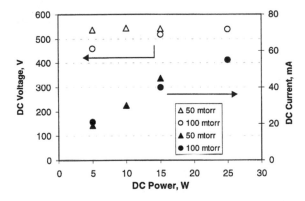

(a) Change of DC voltage and current

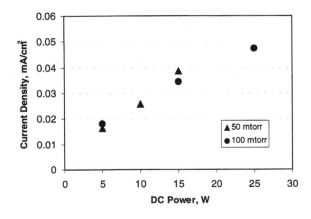

(b) Change of DC current density

Figure 13.10   The change of (a) voltage and current, and (b) the current density with DC power input in TMS glow discharge; 1 sccm TMS, 5 panels, 6-cm panel spacing.

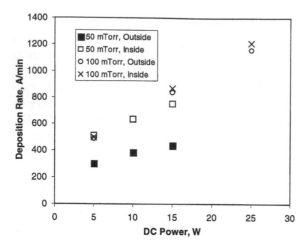

Figure 13.11 The dependence of TMS deposition rate on DC power input under different system pressures with 5 panels in a row; TMS flow rate was 1 sccm and the deposition rates were obtained on panel #1 as shown in Figure 13.7 with panel spacing of 6 cm.

pressures because current density had a linear dependence on DC power input. It should be noted that despite use of multiple cathodes (substrate panels), a high deposition rate (> 1200 Å/min) was also achieved in DC cathodic polymerization of TMS. These results indicated that in a large-scale plasma reactor the deposition rate of TMS on multiple panels could be significantly increased by proper adjustment of the DC discharge parameters.

The effects of the number of panels on the plasma parameters are shown in Figure 13.12, in which TMS DC cathodic polymerization on multiple panels was conducted with power operation mode under a fixed plasma condition of DC 5 W, 1 sccm TMS, and 50 mtorr. As seen from Figure 13.12a, with the same DC power input of 5 W, the increase of the panel numbers showed little effect on the DC voltage and current during the deposition. As the panel number increased from 5 to 30, the voltage decreased by about 50 V on a 500-V base and the current increased by about 2 mA on a 20-mA base. Figure 13.12b shows the panel number dependence of the current density and TMS deposition rate on the panel surfaces. With the increase of the panel numbers, a decreasing trend was observed for the deposition rate due to the decrease of DC current density. When the deposition rates on the panels (cathodes) were plotted against the current density, a linear dependence of deposition rate on current density is confirmed.

4. DC LCVD IN A CLOSED REACTOR SYSTEM

In flow system LCVD, the system pressure is continuously adjusted by controlling the opening of a throttle valve connected to the pumping system. Because of fragmentation of the original monomer in the plasma state, the composition of the gas phase changes on the inception of the plasma. The increase in the total number of gas molecules is compensated by the increased pumping rate in a flow system, and

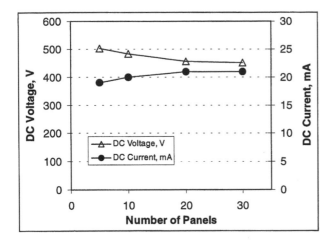

(a) DC voltage and current

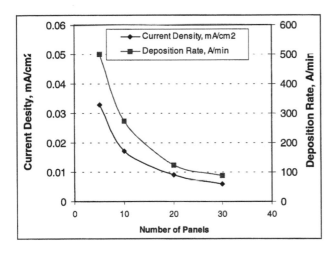

(b) current density and TMS deposition rate

Figure 13.12 The influence of panel numbers on (a) the plasma voltage and current, as well as (b) the current density and TMS deposition rate on the cathode (the panels); TMS, 1 sccm, 50 mtorr, DC 5 W, 6-cm panel spacing.

a steady-state flow of a consistent gas phase composition is established at a predetermined system pressure.

In closed-system LCVD, a fixed number of monomer molecules are contained in the reactor, and glow discharge is initiated. Figure 13.13 depicts the schematics of a closed-system reactor. The pressure in such a system (in a given volume) is proportional to the total number of gas phase molecules. The fragmentation of monomer molecules as well as the ablation of gaseous species from the deposited material will increase the system pressure, whereas deposition will decrease the

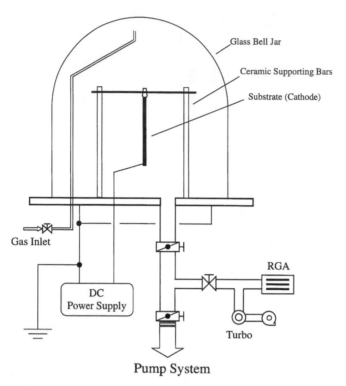

Glass Bell Jar

Ceramic Supporting Bars

Substrate (Cathode)

Gas Inlet

DC
Power Supply

RGA

Turbo

Pump System

Figure 13.13   Schematic presentation of the reactor used.

system pressure. Thus, the change in system pressure with plasma polymerization time indicates the change in the overall balance between the plasma fragmentation/ablation and the plasma film deposition.

In DC glow discharge of TMS in a closed reactor, the system pressure increases when the discharge is initiated. The change of system pressure is shown in Figure 4.2. The system pressure continuously increases while the glow discharge is on but remains at a constant value as soon as the glow discharge is turned off. This indicates that the total number of gas phase species increases with time in spite of the deposition of plasma polymer of TMS. A residual gas analyzer (RGA) characterized the gas phase composition of TMS plasma in the closed reactor system, of which results are shown in Figure 4.1. The most significant gas phase species identified in TMS plasma system are summarized in Figure 13.14. It can be seen that the main gas phase species from the fragmentation of TMS monomers are hydrogen molecules and carbon containing or silicon-containing molecular segments.

In the early stage of the discharge, i.e., the first 60 s, silicon-containing species were the dominant species in the gas phase system. Because of the fast deposition characteristics of silicon species, the silicon-containing species disappeared very quickly in the early stage of glow discharge. After 60 s of glow discharge, very few silicon-containing species remained in the gas phase in the plasma system. Similar trends were reported in a closed system polymerization of hexamethyldisiloxane by radio frequency glow discharge [4].

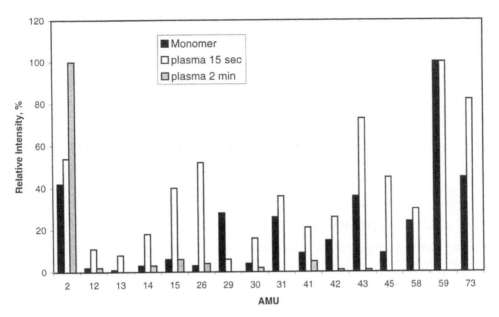

Figure 13.14 The most significant gas phase components observed in a closed system TMS plasmas; TMS 25 mT, 2 panels of Alclad 7075-T6, DC 1000 V. ("Monomer" is not exposed to plasma, and represents the cracking pattern of TMS in RGA).

In the early stage of discharge, the organic carbon species (no silicon content) had lower intensity than silicon-containing species. After 60 s of deposition, the carbon-containing species outnumber the silicon-containing species in the plasma polymerization system. At this stage, the deposition is dominated by carbon-containing species due to previous consumption of silicon moieties.

In the final stage, after 120 s, both the silicon-containing and carbon-containing species have been consumed by the LCVD deposition. In the gas phase, only hydrogen is left in the plasma system and no further deposition occurs. Therefore, it is anticipated that there will be no further thickness growth of the TMS plasma coatings after 120 s.

Figure 13.15 shows the time dependence of the thickness and refractive index of TMS plasma coatings on the plasma polymerization time in a closed reactor system. It can be seen that the coating thickness increases very fast in the first 90 s. After 90 s, the TMS coating stops growing with additional deposition time. However, the refractive index of the TMS coating continues to increase with the deposition time. This increase obviously results from the continuing bombardment by the active species. It should be noted that the times referred to in this discussion are confined to the specific conditions used in the specific reactor. The characteristic times (for TMS) vary depending on size of the reactor, system pressure, size of the cathode, and current density.

It was found that the characteristic plasma deposition rate of Si-containing organic compounds is nearly six times than that of hydrocarbons [5]. It is therefore anticipated that Si-containing moieties would deposit faster than C-based moieties leading to an Si-rich depositions from TMS, which contains one Si and three C in the

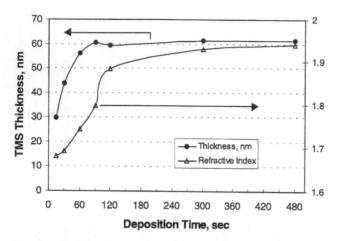

Figure 13.15 Changes in thickness and refractive index of TMS plasma coatings with discharge time in a closed reactor system; TMS 25 mT, 2 panels of Alclad 7075-T6, DC 1000 V.

original molecule. This difference in the characteristic deposition rates would be amplified if the plasma polymerization is carried out in a closed system because the gas phase composition with respect to Si and C changes rapidly and continuously, and the composition of the deposition changes accordingly. This is exactly what is found in closed-system cathodic polymerization of TMS. Therefore, it is anticipated that the closed-system deposition of TMS would lead to a graded composition film.

According to this scheme of closed-system plasma polymerization of TMS, it is anticipated that the atomic composition of the plasma polymer should continuously change with the plasma polymerization time. The comparison of XPS depth profile–generated C/Si ratios for plasma polymers deposited in both a flow system reactor and in a closed-system reactor is shown in Figure 4.10. The TMS plasma coating prepared in a closed system shows gradual composition changes from lower carbon content (C/Si ratio of about 1.7) at the interface with the substrate to carbon rich (C/Si ratio of about 4.7) at the surface as deposition proceeds. In contrast, the coating obtained in the flow system had a uniform composition throughout the film. These results clearly show that closed-system plasma polymerization of TMS produces a film with graded composition; i.e., with increasing carbon content from the film/metal interface to the surface of the nanofilm.

Considering the fact that the refractive index continues to increase after most of the polymerizable species are exhausted in the gas phase, DC LCVD of TMS in a closed system contains the aspect of LCVT of once-deposited plasma polymer coating by hydrogen luminous gas phase. In the later stage of closed-system LCVD, oligomeric moieties loosely attached to a three-dimensional network are converted to a more stable form, and significantly improved corrosion protection characteristics (compared to the counterpart in flow system polymerization of TMS) were found, details of which are presented in Part IV. Thus, the merit of closed-system cathodic polymerization is well established.

REFERENCES

1. Yasuda, H.K.; Yu, Q.S. J. Vac. Sci., and Tech., A-Vacuum Surfaces & Films **2001**, *19* (3), 773.
2. Yu, Q.S.; Yasuda, H.K. Plasmas & Polymers **2002**, 7, 41.
3. Yu, Qingsong.; Moffitt, C.E.; Wieliczka, D.M.; Yasuda, Hirotsugu. J. Vac. Sci., and Tech. **2001**, *A19* (5), 2163.
4. Poll, H.-U.; Meichsner, J.; Arzt, M.; Friedrich, M.; Rochotzki, R.; Kreyssig, E. Surface and Coatings Technology **1993**, *59*, 365.
5. Yu, Q.S.; Yasuda, H.K. J. Polym. Sci.: Part A: Polym. Chem. **1999**, *37*, 967.

14
Magnetron Discharge for Luminous Chemical Vapor Deposition

1. APPLICABILITY OF MAGNETRONS IN LCVD

An electrode system equipped with a superimposed magnetic field is generally referred to as a "magnetron" [1]. Magnetrons have been widely used in sputter deposition of metals and alloys; however, their use in the plasma deposition of organic and inorganic materials [plasma polymerization or luminous chemical vapor deposition (LCVD)] has been relatively uncommon [2]. The reason magnetrons have not been used widely in plasma polymerization might be associated with the opposing concepts of sputter deposition and plasma polymerization. Namely, the main use of magnetrons is to sputter the target material placed on the cathode for the physical deposition of the sputtered materials. In many applications of plasma polymerization, on the other hand, the inclusion of electrode material in the polymeric deposition has been viewed as contamination. Thus, magnetron glow discharge has been widely conceived as incompatible with plasma polymerization.

In certain applications of plasma polymerization, the incorporation of electrode material, particularly in a controlled and designed manner, is extremely useful and becomes a great asset in LCVD. For instance, a thin layer of plasma polymer of methane with a tailored gradient of copper has been shown to improve the adhesion of the thin layer to a copper substrate as well as the adhesion of metal to a polymer film [3,4]. In general applications of LCVD, in which the metal contamination should be avoided, it is important to select the electrode material that has low sputtering yield. Titanium has been used successfully in such cases.

The use of magnetron glow discharge for the formation of thin films by plasma polymerization has several obvious advantages over the use of glow discharges, which do not have magnetic enhancement. Perhaps the most obvious advantage is the confinement of glow discharge in the low-pressure regime of plasma polymerization, i.e., polymer deposition can be carried out effectively at a system pressure of less than 100 mtorr. Without a magnetic confinement of glow discharge, the glow expands as the system pressure decreases, often to the entire volume of a reactor. This expansion of glow discharge occurs even when small internal electrodes are used in a relatively large bell jar reactor. In this case, the deposition of material onto the target substrates becomes very inefficient, and the excessive deposition onto the reactor wall

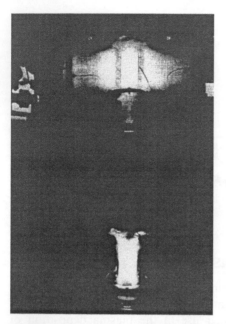

Figure 14.1 Confining effect of magnetron glow discharge. The upper and lower reactors have identical electrode systems, except that the upper electrodes have no magnetic enhancement and are operated at identical glow discharge conditions. Argon, $0.40\,\mathrm{cm^3_{STP}/min}$; 46 mtorr; 5 W; 10 kHz.

practically prohibits the use of such a reactor for the continuous coating of moving substrate in a large-scale operation.

The glow-confining effect of magnetron discharge is shown in Figure 14.1, which represents glow created in a tandem reactor having two identical reactor systems [5]. The magnets in the upper chamber were removed, and the identical discharge conditions (wattage, flow rate, and pressure) were applied for Ar discharge in each reactor. The difference in the two glow discharges is solely due to the presence or the absence of magnetic confinement.

Perhaps the most important, but not so obvious, aspect of the magnetron glow discharge for LCVD is that polymer deposition can be performed in a low-pressure regime in which plasma polymerization by a nonmagnetron glow discharge is difficult or impossible. The low-pressure operation is favored in obtaining better properties (uniform in thickness and without inclusion of particles) of polymers formed by plasma polymerization.

The chemical and physical properties of plasma polymers derived from a monomer are dependent on many factors of overall conditions of plasma polymerization. In other words, a monomer does not yield a well-defined polymer in plasma polymerization. The variation of properties is largely influenced by the energy input parameter of plasma polymerization, W/FM, as described in Chapter 8.

Use of a magnetron enables us to operate plasma polymerization at rather high energy input levels by virtue of low FM and the small glow volume, which often cannot be achieved by the ordinary glow discharge without magnetic confinement. Certain characteristics of plasma polymers can be obtained only under conditions of low pressure and high energy input. In such cases, the advantageous features of magnetron glow discharge in (1) its capability in low-pressure regime and (2) its relative ease of providing high energy input are highly appreciated.

Plasma polymerization in a low-pressure regime inevitably reduces the deposition rate of the process. However, the deposition rate obtainable by magnetron discharge is significantly higher than that by nonmagnetron discharge. Hence, the reasonable deposition rates could be obtained in low-pressure regime by employing magnetron discharge. Whereas publications dealing with magnetron glow discharge polymerization are few compared to those of general glow discharge polymerization without magnetic enhancement, nearly all relatively large-scale industrial applications (outside of microelectronic applications) of glow discharge polymerization are more or less limited to the use of magnetron glow discharge [6,7]. This trend undoubtedly reflects the importance of magnetron glow discharge in large-scale applications of plasma polymerization.

An important advantageous factor of magnetron discharge that is difficult to appreciate or recognize by laboratory scale experiments is the minimal deposition on surfaces other than that of the substrate. This is a crucially important factor in relatively large-scale industrial operations of LCVD that requires continuous operation for a long period e.g., 1 month. A particularly important aspect is that the operational conditions of magnetron discharge could be set so that a portion of an electrode surface under the intense toroidal glow could be kept open without deposition of plasma polymer. The deposition and sputtering of the deposited material reaches a dynamic equilibrium and the steady-state glow discharge could be maintained for a long time.

2. PROFILE OF MAGNETIC FIELD

2.1. Planer Magnetron

Magnetic field configuration of a simple planer magnetron is shown in Figure 14.2, in which a steel circular ring plate and a steel circular center plate are used for the bases of magnetic poles. These two paramagnetic plates, a circular center plate and a ring plate, are placed coaxially on the back of a nonmagnetic (titanium) electrode plate. A varying number of permanent magnets are placed bridging the base plates maintaining the same polarity on each base plate so that two steel plates become two opposing magnetic poles. The strength of the magnet and the number of magnets determine the magnetic flux of the magnetron.

In a cooled magnetron for LCVD, the assembly of magnets is placed on back of a thin aluminum block, which has a channel of cooling water, and a thin plate of an electrode material (e.g., titanium) is placed on the other side of the aluminum block by using insulating spacers (e.g., on four corners) to separate the electrode plate from the magnetron assembly. Thus, a cooled magnetron assembly is insulated from the electrode to prevent creation of glow on the backside of a magnetron. The main purpose of cooling a magnetron is to maintain a constant temperature of magnets for constant magnetic field, and also to prevent excessive heating of the electrode.

A typical profile of the magnetic field at the surface of the electrode is depicted in Figure 14.3 as changes of two components of the magnetic field as functions of

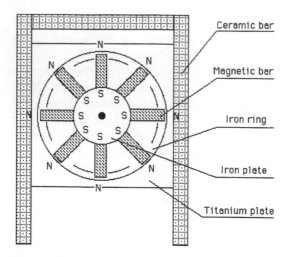

Figure 14.2 Structure of a magnetron electrode.

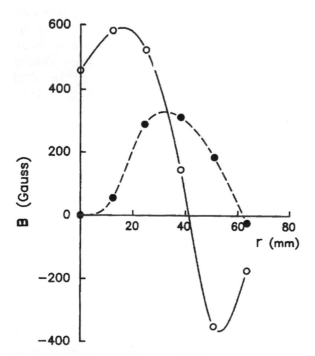

Figure 14.3 Profile of the magnetic field at the surface of electrode. Solid line: perpendicular (to the electrode surface) component. Dotted line: parallel component. r: distance from the center of electrode.

distance from the center of electrode, r. The maximum of the parallel (to the electrode surface) component (dotted line) is observed at the location of the inflection point of the perpendicular component. This location, about 30–40 mm from the center, coincides with the middle of the intense glow ring, which is a characteristic of a magnetron glow discharge. The intense glow ring (toroidal glow) does not touch the electrode surface; however, a corresponding ring appears on the electrode surface after a certain period of glow discharge polymerization due to the remarkably different polymer deposition characteristic near the toroidal glow. Depending on the conditions of glow discharge, the ring pattern, which develops on the surface of the electrode, varies from the heavy coating of plasma polymer to the bare metal surface without polymer deposition.

Figure 14.4 depicts the change in the perpendicular component of the magnetic field at the electrode surface as a function of number of magnets employed. Figure 14.5 (a collapsed three-dimensional profile of the magnetic field) shows the change in the magnetic profile (for the eight-magnet case) as the distance from the electrode surface increases. The results indicate that about 30 mm away from the electrode surface there is no measurable magnetic field in the interelectrode space. The profile of magnetic field in the interelectrode space of a magnetron glow discharge system is dependent on the strength of the magnetic field, manifested by the perpendicular component B_{max} of a magnetron and the distance between two electrodes. This situation is shown in Figure 14.6, which depicts the profiles of the perpendicular component of the magnetic field in the interelectrode space for three different reactors, in which slightly different magnetrons are used with different interelectrode distances. In this figure, the origin of the ordinate is shifted to the center of the interelectrode space, with respect to the abscissa of Figure 14.5, and the perpendicular component of the magnetic field is measured on the line connecting the centers of two electrodes, i.e., $r = 0$. Therefore, the value of ℓ extends up to the half of the interelectrode distance of each reactor.

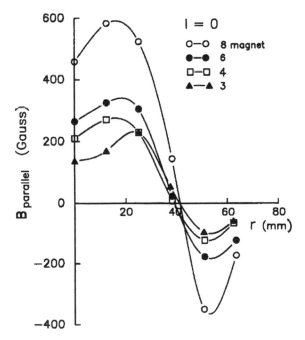

Figure 14.4 The effect of the number of magnets employed on the profile of magnetic field at the surface of electrode ($\ell = 0$) as a function of radius r.

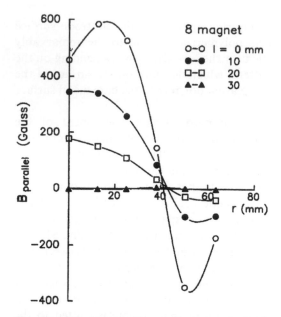

Figure 14.5 The profile of the magnetic field measured at the center of an electrode at various distance from the electrode.

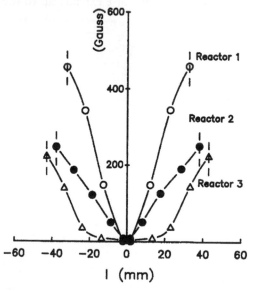

Figure 14.6 The profile of the magnetic field in the interelectrode space at the center of electrode.

2.2. Cylindrical Electromagnetron

While the use of permanent magnets provides relatively simple magnetrons, study of the influence of magnetic field, which is an important variable of plasma polymerization by a magnetron, requires the change of the electrode system for each level of magnetic field strength. The use of electromagnets, instead of permanent magnets, avoids this inherent shortcoming of permanent magnetrons and provides great flexibility in operation of plasma polymerization.

External variable magnetic fields have been used to enhance glow discharge deposition of a-Si:H layers [8–10] and other inorganic compounds [11], as well as some organic monomers [12]. However, in all of these experiments the variable magnetic field was applied from outside of the chamber and was not integrated into the discharge electrodes. The externally applied magnetic field may change the shape of the luminous gas phase; however, whether and how the magnetic field would change the location of dissociation glow is not known.

Yasuda and Olcaytug carried out fundamental studies on the integrated electromagnetron for plasma polymerization [13]. An enamel-insulated wire (diameter d_w) is wound on a core cylinder (length L) for the thickness of M depicted in Figure 14.7. The dimension of an electromagnetron with two magnets is shown in Figure 14.8.

The magnetic fields produced by such a solenoid are calculated as functions of parameters L, M, and d_w. Basic characteristics of electromagnets as functions of current and wattage are measured and compared with calculated values. Magnetic field profiles of a single magnet and magnetrons that consist of double magnets arranged in N-S/N-S (type A) and N-S/S-N (type B) sequences are measured as functions of current and the distance from the surface of a magnetron. Each magnetron is equipped with a water-cooling line in the center portion of the core in order to maintain a constant temperature of the electrode surface.

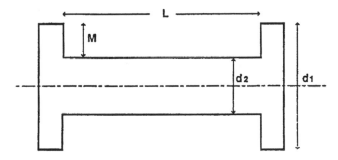

Figure 14.7   Diagram of the core tube for an electromagnetron.

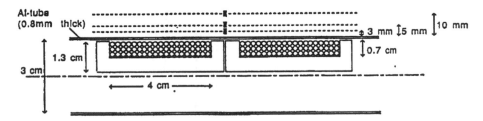

Figure 14.8   Dimension of an electromagnetron with two magnets encased in an aluminum tube: the magnet has a small hole in the center for the alignment.

2.2.1. Single Magnet

The parallel and perpendicular components (with respect to the surface of a cylindrical magnetron) at a current of 3 A are shown in Figure 14.9. The parallel components are measured at 3 and 6 mm from the surface, and the perpendicular components are measured at 0 and 3 mm. The profile of magnetic field is quite symmetrical. Since the magnetron has a cylindrical shape, it is anticipated that the profile is symmetrical along the circumference.

2.2.2. Double Magnets

The two electromagnets are placed in an aluminum tube as shown in Figure 14.8, and the influence of the adjacent magnet on the profile of the magnetic field was investigated. Depending on the alignment of magnetic dipoles, which can be changed by switching the individual coils according to the direction of the current, the two magnets can be arranged either in N-S/N-S or N-S/S-N.

   N-S/N-S Combination.   This should yield one long continuous magnet. The effect of increasing current on the magnetic field profile for the perpendicular component (at 0 mm from the surface) and the parallel component (at 3 mm from the surface) is shown in Figure 14.10a and b, respectively. The vertical walls of the cylinders separate the two solenoids. Because the highly permeable core material touches the surface at this part, the magnetic field near the surface is distorted.

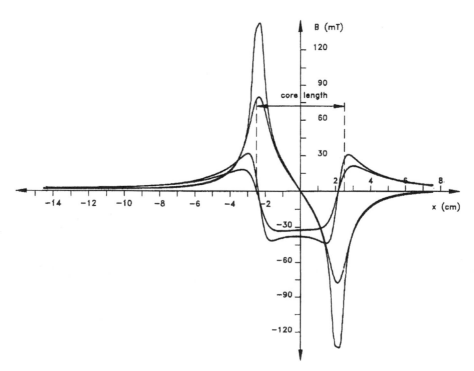

Figure 14.9   Magnetic field profile of a single electromagnet at 3 A: the perpendicular component at 0 and 3 mm, and the parallel component at 3 and 6 mm from the surface.

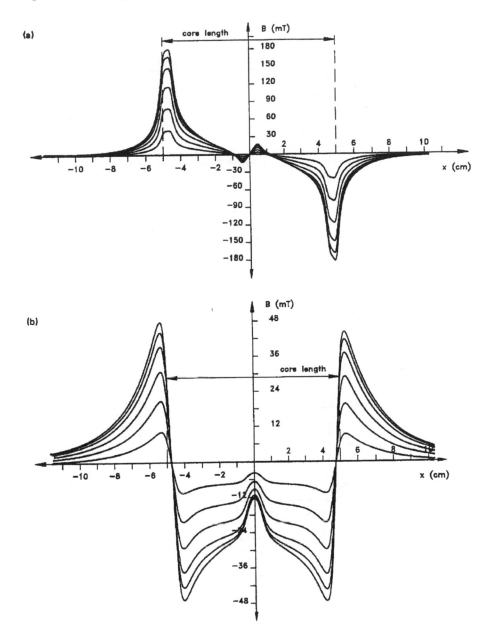

Figure 14.10 (a) The influence of current on magnetic field profile (the perpendicular component) of N-S/NS double magnets. Current: 1, 2, 3, 4, 5, 6 A at the surface (0 mm); (b) The influence of current on magnetic field profile (the parallel component) of N-S/N-S double magnets: Current: 1, 2, 3, 4, 5, 6 A at 3 mm from the surface.

The increase of current magnifies the similar profile except in the middle section of the magnetron, particularly in the parallel components. The influence of distance from the surface for the perpendicular and parallel components is shown in Figure 14.11a and b, respectively. The distorted magnetic field at the middle of the magnetron is limited to the region near the surface. Away from the surface the

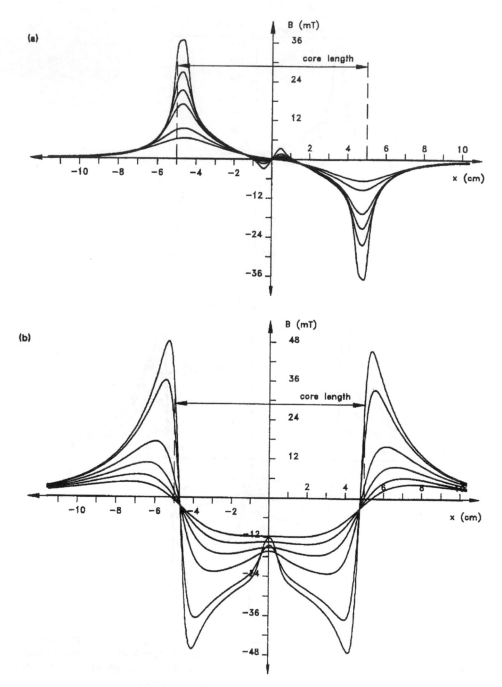

Figure 14.11 (a) The influence of the distance from the surface on magnetic field profile (perpendicular component) of N-S/NS double magnets, distance: 0, 2, 3, 5, 10 mm, at 6 A, (b) the influence of the distance from the surface on magnetic field profile (parallel component) of NS/N-S double magnets, distance: 3, 5, 10, 15, 20, 25 mm, at 6 A.

magnetic field profile becomes quite smooth and similar to that of the magnetron that is made of a single magnet.

N-S/S-N Combination. In this combination, the perpendicular components are superimposed at the end where the two poles of identical polarity meet. The influence of the current on the perpendicular component and on the parallel component is shown in Figure 14.12a. The influence of the distance from the surface on the same components is depicted in Figure 14.12b.

2.2.3. Electromagnetrons for LCVD

Two types of electromagnetrons (two for each type) are constructed for the deposition study based on data obtained by using elementary magnetrons described above. Each magnetron was prepared using a long single core, instead of putting multiple small electromagnets together to make a magnetron.

Type A: Segmented Multidipole Magnetron. The core is divided into seven segments, and a separate wire is wound in each segment. The direction of magnetic dipoles of each magnet can be determined by the direction of electric current. Selecting the appropriate flow direction of the current one obtains either S-N/S-N- or S-N/N-S- combinations. Magnets are connected in the latter combination. The profile of magnetic field for type A seven-dipole electromagnetron (S-N/N-S-) is shown in Figure 14.13.

Type B: Single-Dipole Magnetron. A wire is wound continuously for the entire length of a magnetron. The profile of magnetic field for type B magnetron is shown in Figure 14.14.

The toroidal glow, which is characteristic of a magnetron, appears at the position where the perpendicular component (with respect to the surface of electrode) of the magnetic field changes its sign. This position is seen where the perpendicular component crosses the baseline. At this location all magnetic field lines are parallel to the electrode surface and perpendicular to the electric field. The secondary electrons ejected from the cathode surface will accelerate toward the anode in the cathode fall region; however, those electrons are trapped by the existing magnetic field that is perpendicular to the direction of motion of electrons. The apparent width of the toroidal glow is proportional to the cosine of the angle where the perpendicular component crosses the baseline. The intensity of the toroidal glow is proportional to the magnetic field strength of the parallel component at this location.

In the case of a type A magnetron, the perpendicular component of the magnetic field crosses the baseline seven times at a sharp angle, and the parallel components at those locations are much stronger due to the much shorter separation of magnetic poles in each segment. Therefore, a type A magnetron yields seven narrow and strong toroidal glows (a ring surrounding the cylindrical electrode), each located at the middle section of an individual magnet.

In the case of a type B magnetron, the perpendicular component of the magnetic field crosses the baseline only once at a very small angle, and the parallel component at this location is weak due to the long separation of magnetic poles. Consequently, a type B magnetron yields a wide but weak toroidal glow at one-half of the length of the magnetron.

The direction of two magnetic dipoles seems to influence the shape of the toroidal ring glow when two type B magnetrons are used for discharge. The parallel

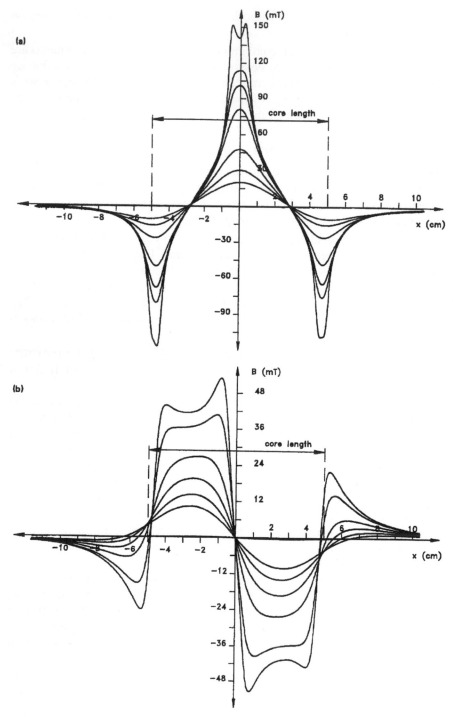

Figure 14.12 (a) The influence of current on magnetic field profile (the perpendicular component) of N-S/S-N double magnets, current: l, 2, 3, 4, 5, 6 A at the surface (0 mm); (b) The influence of current on magnetic field profile (the parallel component) of N-S/S-N double magnets, current: l, 2, 3, 4, 5, 6 A at 3 mm from the surface.

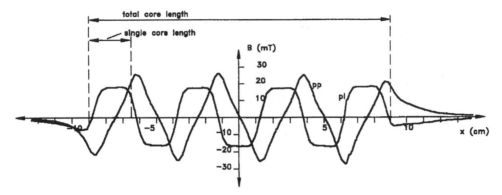

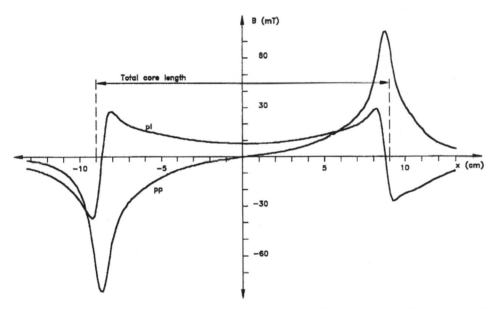

Figure 14.13 Parallel and perpendicular components of a long (one piece) seven-dipole (type A) electro-magnetron.

Figure 14.14 Parallel and perpendicular components of a long single-dipole (type B) electromagnetron.

field tends to enhance the glow in between two magnetrons, and the opposite field tends to push out the glow on the opposite sides of two magnetrons. This trend is obscured in a multidipole magnetron (type A).

3. EFFECT OF MAGNETIC FIELD ON THE BREAKDOWN VOLTAGE

3.1. Non-Polymer-Forming Gases

The influence of the magnetic field on the breakdown voltage of non-polymer-forming gases such as He, Ar, N_2, and O is rather complex and is not as pronounced as for polymer-forming gases [14]. The latter trend is due partly to the fact that the breakdown voltages of inert gases are much lower than those for organic monomers, and further decrease of the breakdown voltage is marginal.

The breakdown voltage V_b is the voltage that is necessary to create an electrical discharge of a gas phase. Once the breakdown occurs, the voltage drops to the voltage that is necessary to sustain the discharge V_d, i.e., the breakdown voltage is not the discharge voltage of a glow discharge but indicates the minimal voltage required to create discharge under a given set of conditions. The breakdown voltage is a function of the dielectric property of the gas and a function of pd, where p is system pressure and d is the distance between two electrodes.

The breakdown voltage of a gas in an LCVD reactor with a fixed value of d decreases as the system pressure decreases and reaches the minimum and increases rather sharply at the lower pressure (low-pressure regime). Because of this phenomenon it is difficult to operate LCVD in the low-pressure regime. The reduction of breakdown voltage by the magnetron primarily addresses prevention of the increase of V_b in the low-pressure regime in order to take advantage of LCVD in this regime. The increase of breakdown voltage, V_b, in the low-pressure regime is generally proportional to $1/pd$, and the plot of V_b against $1/p$ for a reactor with fixed d yields a straight line, of which slope can be used as a measure of the effectiveness of breakdown voltage reduction by a magnetron, i.e., the smaller the more effective.

The presence of a magnetic field generally decreases the breakdown voltage at a given pressure (at fixed d) and significantly lowers the minimal pressure that can produce glow discharge. Figure 14.15 depicts the breakdown voltage dependence on the system pressure of argon with and without magnetic field.

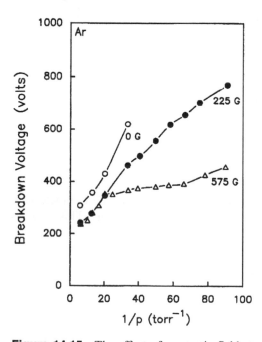

Figure 14.15 The effect of magnetic field strength on the breakdown voltage of Ar as a function of $1/p$, where p is the system pressure.

3.2. Polymer-Forming Gases

The effect of magnetic field on the breakdown voltages of gases that are considered as monomers of plasma polymerization is more pronounced, showing more systematic influence of the magnetic field strength. The general trends are shown in Figure 14.16 for C_2H_4 and in Figure 14.17 for C_2H_3F. The magnetic field reduces the slope of V_b versus $1/p$ plots. The stronger the magnetic field, the larger is the extent of reduction.

The slope observed in the pressure dependence of the breakdown voltage depends on the chemical nature of gas. The fluorine-containing monomers have the smaller slopes (without magnetron), and the effect of the magnetic field on the breakdown voltage of fluorocarbons is less pronounced than the effect on the corresponding hydrocarbons. Figure 14.18 depicts the influence of the number of fluorine atoms in a series of monomers that can be given by $C_2H_nF_{4-n}$, observed at $B = 575\,G$. Without a magnetic field, the completely opposite order is seen, as shown in Figure 14.18, for the same monomers. The hydrocarbon ($n = 4$) has the highest value of breakdown voltage and also has the steepest slope, but with the magnetron these values become the lowest.

4. EFFECT OF MAGNETIC FIELD ON VOLTAGE–CURRENT DIAGRAM

Due to the electron trapping under the influence of a magnetic field, electron loss by lateral diffusion decreases. As a result, ionization efficiency increased, as manifested

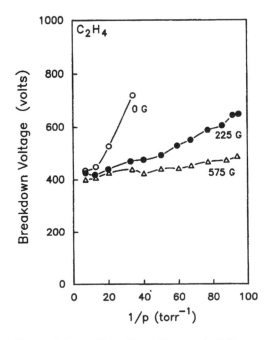

Figure 14.16 The effect of magnetic field strength on the breakdown voltage of C_2H_4 as a function of $1/p$, where p is the system pressure.

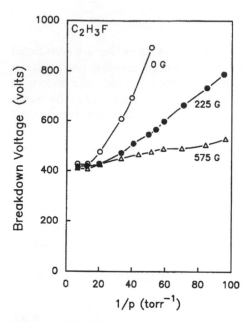

Figure 14.17 The effect of magnetic field strength on the breakdown voltage of C_2H_3F as a function of $1/p$, where p is the system pressure.

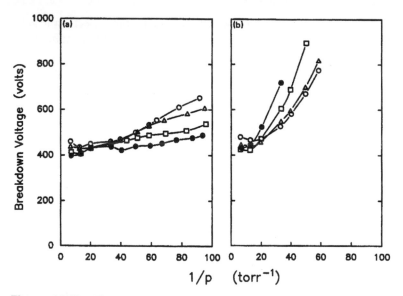

Figure 14.18 The change of the breakdown voltages of monomer series $C_2H_nF_{4-n}$, as a function of $1/p$ ($B_{max} = 575\,G$): (a) eight magnet and (b) no magnet, ($\bigcirc$) $n = 0$; ($\triangle$) $n = 1$; ($\square$) $n = 3$; ($\bullet$) $n = 4$.

by an increase in the current at the same applied voltage. Figures 14.19 and 14.20 shows the V–I curve for the argon plasma and the methane plasma, respectively. Both curves show trends similar to those observed with typical magnetron sputtering systems, which follow the relationship $I = kV^n$, where n increases as the trapping of

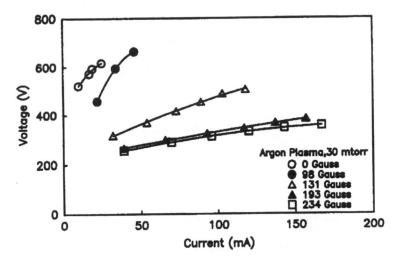

Figure 14.19 *V–I* curves for the argon magnetron glow discharge plasma at a pressure of 30 mtorr.

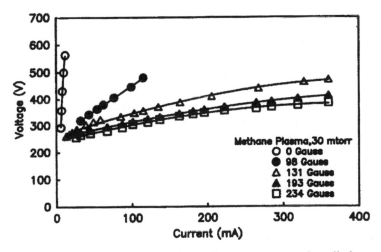

Figure 14.20 *V–I* curves for the methane magnetron glow discharge plasma at a pressure of 30 mtorr.

electrons becomes more efficient [15,16]. As is seen in the figures, the influence of magnetic field on the efficiency of electron trapping, as manifested by the decreasing steepness of the "curves" (or the increasing n value), is more pronounced in methane plasma than in argon plasma.

Figure 14.21 depicts the effect of the magnetic field on the discharge voltage. The voltage is plotted against magnetic field strength in an argon plasma at various system pressures for a given discharge power input. At high magnetic field strength the effect of pressure on the voltage drop became less significant. Furthermore, the voltage drop with increasing magnetic field strength was sharper at high discharge power and low pressure. As the mean free path of an electron increases with decreasing pressure, the loss of electrons by lateral diffusion increases. In such a case,

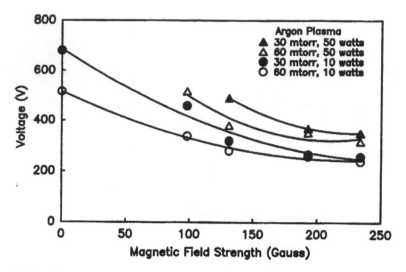

Figure 14.21 Drop of voltage as a function of magnetic field strength in argon magnetron glow discharge plasmas.

the effect of the magnetron for electron trapping is more pronounced. At constant magnetic field strength and discharge power, a reduction in voltage and thus an increase in current were observed at high argon pressure due to the presence of more collision targets.

Because the voltage required to increase the current is relatively high in glow discharge plasmas with no magnetic enhancement, it is very difficult to apply high power to such a nonmagnetron glow discharge plasma without arcing or creating local sparks. Thus, the voltage range that can be used to create a stable glow discharge is relatively low and narrow. In contrast, the usable range of voltage is much wider when a magnetron is used.

5. EFFECT OF MAGNETIC FIELD ON THE SPUTTERING OF ELECTRODE MATERIALS

5.1. Argon Plasma

Figures 14.22 and 14.23 show the sputter deposition rates of copper from the copper electrode surfaces in argon plasmas at pressures of 30 and 60 mtorr as a function of discharge power. The deposition rate increased with increasing magnetic field strength and discharge power. No appreciable deposition of copper was observed in the argon plasma at a pressure of 30 mtorr for a maximal parallel magnetic component of less than 100 G.

Similar trends were found in the sputter deposition of aluminum, as is shown in Figures 14.24 and 14.25. The nearly linear proportional relationship between power and deposition rate found in the sputter deposition of Cu was not seen in the case of aluminum sputtering. Furthermore, the sputter deposition rate of copper was higher than aluminum at the same discharge power input, as would be expected from the sputtering yield data reported by Laegreid and Wehner [17]. The result is also

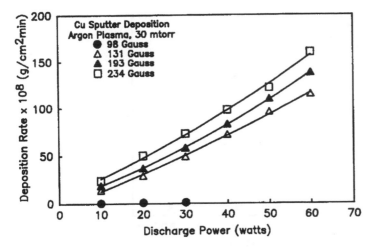

Figure 14.22 Rate of sputter deposition of copper as a function of discharge power in argon plasmas (30 mtorr) with various magnetic fields.

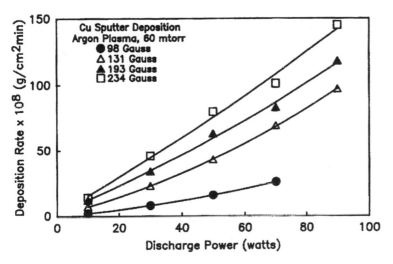

Figure 14.23 Rate of sputter deposition of copper as a function of discharge power in argon plasmas (60 mtorr) with various magnetic field strengths.

consistent with the deposition-efficient data for Cu and Al magnetron electrodes reported recently by Rossnagel [18].

Figure 14.26 shows the sputter deposition of copper and aluminum in argon plasmas at various system pressures. As is expected from the effect of magnetic field on the voltage drop described previously, the sputter deposition of metal was more pronounced at low pressure. In addition, the deposition rate was higher at a low sputtering pressure for a constant discharge power. This may be due to the effect of gas scattering of sputtered materials, which lowers the deposition efficiency at the substrate plane, as has been commonly observed in magnetron discharge systems. Furthermore, if a maximal parallel magnetic component of 234 G was employed,

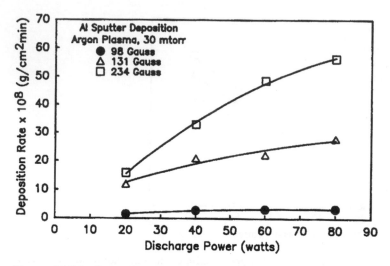

Figure 14.24 Rate of sputter deposition of aluminum as a function of discharge power in argon plasmas (30 mtorr) with various magnetic field strengths.

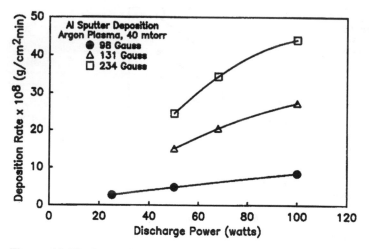

Figure 14.25 Rate of sputter deposition of aluminum as a function of discharge power in argon plasmas (40 mtorr) with various magnetic field strengths.

a ring-shaped eroded area having a size one-fifth of the entire electrode surface area was observed on each electrode surface.

5.2. Methane Plasma

In contrast to argon plasma, in which the sputtering of metal from the electrode is the primary process, the deposition of polymeric materials via plasma polymerization predominantly takes place in methane plasma. In such a polymer-forming plasma, the sputter deposition of electrode materials is considered as a secondary process, and the extent of the sputtering of metal depends on the plasma polymerization conditions, the nature of the electrode material, and the magnetic field strength.

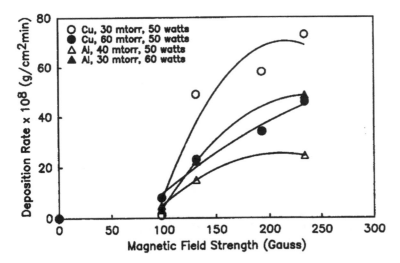

Figure 14.26   Rates of sputter deposition of copper and aluminum as a function of magnetic field strength in argon plasmas.

Furthermore, in this type of plasma the polymeric materials deposit not only on the substrate but also on the electrode surfaces. Thus, the sputter deposition of metal in polymer-forming plasmas is more complicated than that in non-polymer-forming plasmas.

The deposition pattern of polymeric materials seen on the electrode surface with or without magnetic enhancement in methane plasmas was similar to the sputtering pattern observed in argon plasmas. While the entire electrode surface was covered with polymeric deposits in the absence of magnetic field, a ring-shaped deposition area was observed in the presence of the magnetic field. Figures 14.27 and 14.28 show the width of the ring-shaped deposition area for copper and aluminum electrodes as a function of magnetic field strength and the energy input parameter W/FM. The width decreased with increasing W/FM and magnetic field strength, a trend that is similar to the variation of erosion area with discharge power delivered to the target and magnetic field strength in a DC planar magnetron discharge. The color of the deposited film in the ring became darker as the width decreased. At high levels of energy input, the ring-shaped polymer deposition area disappeared, and the deposition of materials took place at other surface areas on the electrodes.

The sputtering of electrode materials by methane plasmas was investigated by analyzing the metal content in the film formed on the quartz crystal surface by XPS. The results are presented in Figures 14.29 and 14.30 for the magnetron systems with copper and aluminum electrodes, respectively. Even though the data are quite scattered, the percentage of metal incorporated tends to be higher in films deposited at high W/FM and at high magnetic field strengths. It is also noted that a significant metal deposition was observed only when the energy input for the glow discharge manifested by W/FM exceeded a certain value, indicating that a threshold energy input exists for the sputtering of metal in polymer-forming plasmas. The threshold value of W/FM decreased with increasing magnetic field strength. The existence of this threshold energy, which was not observed in the argon

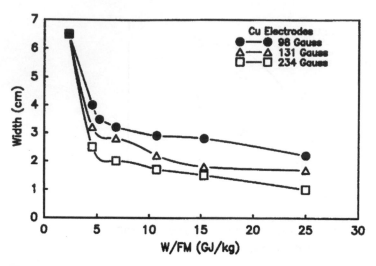

Figure 14.27 The effect of W/FM on the width of the ring-shaped polymer deposition area on copper electrodes observed in methane plasmas with various magnetic field strengths.

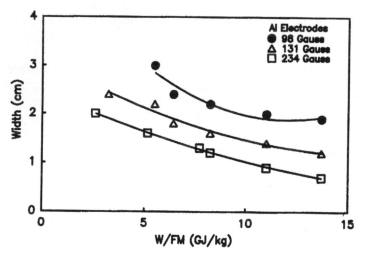

Figure 14.28 The effect of W/FM on the width of the ring-shaped polymer deposition area on aluminum electrodes observed in methane plasmas with various magnetic field strengths.

plasma, might be utilized in attempts to incorporate electrode materials in plasma polymers to create a component-graded nanofilm. It also indicates that the contamination due to electrode material could be minimized by the selection of metal and the plasma polymerization conditions. Figure 14.31 depicts the comparison of sputter deposition rates of copper and aluminum by argon and methane plasmas for a parallel magnetic component of 234 G. It is clear that the sputtering rates of copper and aluminum are much lower in methane plasma than in argon plasma.

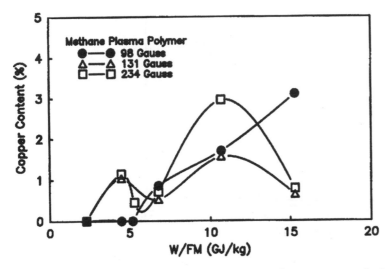

Figure 14.29 The effect of W/FM on the percent of copper content in the plasma polymer of methane obtained in methane magnetron glow discharge plasmas.

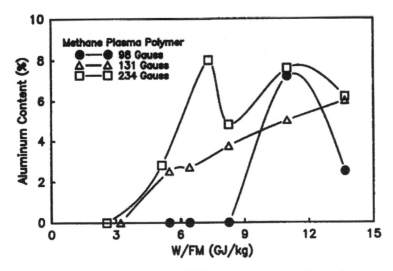

Figure 14.30 The effect of W/FM on the percent of aluminum content in the plasma polymer of methane obtained in methane magnetron glow discharge plasmas.

6. EFFECT OF MAGNETIC FIELD ON THE DEPOSITION RATE OF PLASMA POLYMER

The effect of magnetic field superimposed on an electrode on the efficiency of material formation in LCVD could be best illustrated by data obtained with electromagnetron, with which magnetic field strength can be altered even after a glow discharge is created. The effect of magnetic field on the deposition rate of methane was investigated using type A and type B electromagnetrons, of which details of operations are summarized in Table 14.1.

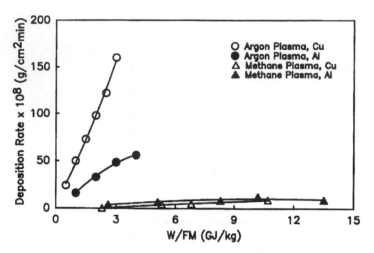

Figure 14.31 The effect of W/FM on the rates of sputter deposition of copper and aluminum in the argon and methane magnetron glow discharge plasmas with a maximum parallel magnetic component of 234 G.

At a given combination of flow rate and system pressure, the minimal voltage and current needed to create glow discharge without magnetic field, V_0 (discharge) and I_0 (discharge), were determined. When the magnetron voltage V (magnetron) is applied and increased while maintaining a constant V_0 (discharge), the typical magnetron glow discharge with toroidal glow rings appears at the critical magnetron voltage V_c (magnetron).

When the glow discharge is transformed into the magnetron glow discharge, the voltage of glow discharge V (discharge) drops appreciably from V_0 (discharge) to the magnetron discharge voltage V_m (discharge) and the discharge current increases from I_0 (discharge) to the magnetron discharge current I_m (discharge).

When V (magnetron) is increased above the critical voltage V_c (magnetron), the magnetron discharge voltage V_m (discharge) remains constant and the discharge current I_m (discharge) increases with increasing magnetron voltage V (magnetron). The increase of discharge current I_m (discharge) at a fixed discharge voltage V_m (discharge) is evidence of the better efficiency of a magnetron glow discharge at higher magnetic field strength.

Because of the decrease of glow discharge impedance due to the increasing magnetic field, the overall discharge power increases at a given constant discharge voltage as the magnetron current is increased. Figure 14.32 depicts the comparison of deposition rates for nonmagnetron and magnetron discharges. The points corresponding to V (magnetron) = 0 are shown by open circles. The increase in deposition rate for a given symbol is due to the increase in magnetron current at a given discharge voltage. While the discharge voltage is maintained at a constant value, the discharge current increases due to the reduction in the discharge impedance. An additional set of data represented by "nonmagnetron glow discharge" was obtained by using a type B electromagnetron without magnetron current under different glow discharge conditions.

The influence of magnetic field is seen in the significant increase in the deposition rate, 5- to 10-folds compared to the corresponding conditions without

Table 14.1 Deposition Rates by Electromagnetrons

Monomer: CH$_4$		Pressure = 55 mtorr			Flow Rate = 5.0 sccm	
V (mag) (V)	I (mag) (A)	V (dis) (V)	I (dis) (A)	P (dis) (W)	Pressure (mtorr)	Dep. rate Å/min
Type A Magnetron						
Low-power discharge						
0.00	0.0	316	38	12.0	48.1	6.0
2.65	1.4	316	42	13.3	46.3	10.0
3.44	2.0	315	43	13.5	46.5	16.0
4.70	3.0	315	47	14.8	45.6	18.6
7.25	5.0	315	51	16.1	48.9	21.6
8.16	6.0	316	57	18.0	43.0	33.0
High-power discharge						
0.00	0.0	414	118	48.9	45.2	20.0
2.67	0.4	414	136	56.3	42.2	28.5
3.55	2.0	416	167	69.5	42.6	61.0
5.00	3.0	417	243	101.3	43.7	80.0
6.25	4.0	414	178	73.7	55.6	65.0
7.45	5.0	414	149	61.7	67.8	45.0
8.17	6.0	414	257	106.4	54.0	75.0
Type B Magnetron						
Low-power discharge						
0.00	0.0	320	41	13.1	55.5	8.0
5.07	5.0	316	57	18.0	53.6	18.0
6.16	6.0	316	96	30.3	48.9	30.0
7.26		316	143	45.2	45.7	40.6
	7.0					
8.15	8.0	316	170	53.7	46.0	47.0
High-power discharge						
0.00	0.0	414	140	58.0	55.0	8.6
5.00	5.0	390	430	167.7	—	72.0
6.07	6.0	387	480	185.8	39.6	75.0
7.25	7.0	375	549	205.9	40.4	83.7
Nonmagnetron						
0.00	0.0	316	41	13.0	48.6	1.8
0.00	0.0	350	47	16.5	49.4	9.9
0.00	0.0	400	105	42.0	39.9	14.6
0.00	0.0	450	260	117.0	36.7	19.3

magnetic field. However, the way by which the deposition rate increases is different in low- and high-power discharges. The deposition rate increased sharply with the discharge power under the low discharge power conditions. Under the high discharge power conditions, the difference between the deposition rates in magnetron glow discharge and in nonmagnetron glow discharge becomes more pronounced, but the deposition rate increase due to the increase of magnetic field is relatively small.

The data shown in Figure 14.32 also indicate that much higher discharge power can be put into the system with magnetrons compared to the nonmagnetron with the

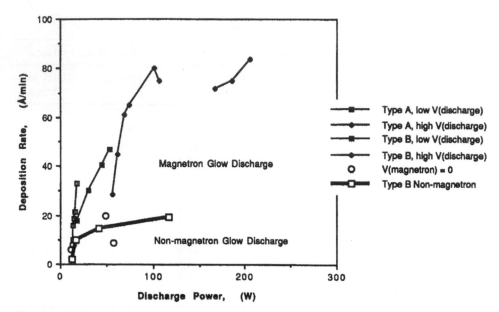

Figure 14.32 The enhancement of deposition rate by type A and type B electro-magnetrons.

identical configuration. Experiments of this nature, in which a glow discharge is initiated without magnetic field and then the varying magnetic field is applied, are a demonstration of a unique feature of electromagnetrons that cannot be performed with conventional magnetrons that utilize permanent magnets with a fixed magnetic strength.

Magnetrons can be operated at very low W/FM level, since glow discharge can be created at low voltages where glow discharge could not be created without magnetic enhancement, i.e., V_m (discharge) is much smaller than V_0 (discharge). The use of magnetrons in the very low W/FM domain is the unique characteristic of magnetron, which is not well recognized.

Due to the relatively high electric power consumption, e.g., 60 W per magnet for 1000 G, in the seven-segment version, the practical use of electromagnetrons might be limited to certain specialized cases in which the change of magnetic field is required during a deposition process. Use of core materials with high magnetic permeability could reduce this drawback.

The advantages of magnetron for plasma polymer deposition can be summarized as follows:

1. Magnetron glow discharge enables us to operate at much lower pressure where nonmagnetrons cannot create a glow discharge.
2. The deposition rate, observed in the interelectrode space, at a given discharge power is increased significantly by the presence of a magnetic field.
3. Glow discharge of a monomer at a given pressure can be operated at much higher wattage levels with magnetrons.
4. Glow discharge can be operated at high-energy input, W/FM, regime, which cannot be performed by nonmagnetron discharge.

5. The impedance of a magnetron discharge system is reduced according to the magnetic field strength of magnets. Accordingly, magnetron discharge can be operated at low W/FM regime, where glow discharge cannot be created without magnetron.

7. SIGNIFICANCE OF MAGNETRON DISCHARGE IN LCVD

When an electrical discharge of an organic compound (monomer) is created, roughly 98% of the monomer leaves the gas phase in a relatively short time, and the nonpolymerizable gases (mainly hydrogen) remain in the gas phase. The molecules left the gas phase deposit on surfaces existing in the reactor. In a nonmagnetron DC glow discharge polymerization, roughly 80% of polymer deposition occurs on the surface of the cathode and the remaining part on other surfaces including the anode. In high-frequency alternating current, e.g., 10 kHz, this balance does not change because both electrodes act as cathode in half of the discharge time, although the deposition on each electrode drops to one-half that on DC cathode. In these cases, the dissociation glow adheres to the electrode(s).

In radio frequency discharge, the dissociation glow does not adhere to the electrode surface, the deposition onto the electrode drops sharply (in comparison to DC and 10-kHz discharges), and the deposition onto substrates placed in the interelectrode space significantly increases. However, the deposition rate on the substrate does not increase as the shift of the mass balance would indicate because the glow volume expands in comparison with the cases of DC or radio frequency discharge and the fraction of material depositing on the target substrate decreases. In order to reduce the glow volume of radio frequency discharge, higher pressure could be used. However, it changes the domain of plasma polymerization entirely.

The presence of magnetic field near the surface of the electrode changes the mechanism for the excitation of gas in magnetron discharge. Electrons traveling from the cathode to the anode are trapped by the magnetic field that is perpendicular to the motion of electrons and circle around the magnetic field line. The collision of circling electrons and gases causes the luminous gas phase in magnetron discharge, whereas the collision of oscillating electrons and gases causes the luminous gas phase in radio frequency discharge.

The significance of the magnetron in plasma polymerization could be envisioned that the toroidal glow lifts up the dissociation glow from the electrode surface and shifts the main medium of plasma polymerization to the interelectrode space in somewhat similar manner to radio frequency discharge. In magnetron discharge, however, the volume of glow is confined more or less to the interelectrode space or smaller volume, which leads to significantly high deposition rates obtainable to the substrate surface placed in the space. It is important to recognize that characteristics of plasma polymers deposited by magnetron discharge at high W/FM, which is much higher than what can be obtained by nonmagnetron discharge, are more or less the same as those of ordinary plasma polymers [19], implying that the high W/FM mainly enhances the deposition rate without causing excessive fragmentation of monomer. In a magnetron glow discharge, the energy put into the electrical circuit, W, is largely transferred to the toroidal glow, enabling us to utilize the energy input more efficiently for LCVD.

Considering all advantageous features described above, an appropriate use of magnetron discharge seems to play the key role in plasma polymerization aimed at the interface engineering of dielectric materials. Dealing with electrically conducting materials, DC cathodic polymerization seems to be the best method as described in Chapter 13. In DC discharge, magnetron can be used in a different manner for different purpose, which is discussed in Chapter 15. The profile of electron temperature and electron density under the influence of magnetic field are collectively described for cathodic and anodic use of magnetrons in DC discharge.

REFERENCES

1. Waits, R.K., In *Thin Film Processes*; Vossen, J.L.; Kern, W. Eds.; Academic Press: New York, 1978; 131–170.
2. Sato, K.; Iriyama, Y.; Cho, D.L.; Yasuda, H. J. Vac. Sci. Technol. **1989**, *A7*, 195.
3. Sun, B.K.; Cho, D.L.; O'Keefe, T.J.; Yasuda, H. Chemically graded metalization of nonconducting substrates by glow discharge plasma polymerization technique. In *Metallized Plastics 1*; Mittal, K.L., Susko, J.R. Eds.; Plenum Press: New York, 1989; 9–27.
4. Yasuda, H.; O'Keefe, T.J.; Cho, D.L.; Sun, B.K. J. Appl. Polym. Sci.: Appl. Polym. Symp. **1990**, *46*, 243.
5. Yasuda, H. *Plasma Polymerization*; Academic Press: Orlando, FL, 1985; 319–333.
6. Ross, D.L. Coatings for video discs. RCA Rev. **March 1978**, *39*,136.
7. Leahy, M.F.; Kaganowicz, G. Solid State Technol. **1987**, *30* (4), 99.
8. McKenzie, D.R. J. Appl. Phys. **1984**, *56*, 2356.
9. Taniguchi, M.; Hirose, M.; Osaka, Y.; J. Non-Cryst. Solids **1980**, *35 & 36*, 189.
10. Hamasaki, T.; Kurata, H.; Hirose, M.; Osaka, Y.; Appl. Phys. Lett. **1980**, *37*, 1084.
11. Yuzuriha, T.H.; Mlynko, W.E.; Hess, D.W.; J. Vac. Sci. Technol. **1985**, *A3*, 2135.
12. Janca, J.; Necasuva, M.; Kucirkova, A. J. Phys. **1981**, *B31*, 1391.
13. Yasuda, H.; Olcaytug, F. J. Vac. Sci. Technol. **1991**, *A9* (4), 2342.
14. Cho, D.L.; Yeh, Y.-S.; Yasuda, H. J. Vac. Sci. Technol. **1989**, *A7*, 2960.
15. Wilson, R.W.; Terry, L.E. J. Vac. Sci. Technol. **1976**, *13*, 157.
16. Thornton, J.A.; Penfold, A.S.; *Thin Film Processes*; Vossens, J.L., Kern, W. Eds.; Academic Press: New York, 1978; 76.
17. Laegreid, N.; Wehner, G.K. J. Appl. Phys. **1961**, *32*, 365.
18. Rossnagel, S.M. J. Vac. Sci. Technol. **1988**, *A6*, 3049.
19. Sato, K.; Yeh, Y.S.; Yasuda, H. J. Vac. Sci. Tech. **1989**, *A7*, 3188.

15
Anode Magnetron Discharge

1. CATHODE AND ANODE MAGNETRONS

Important and interesting questions that arise from the magnetron discharge described in the previous chapter are as follows:

1. How does the magnetron electrode used in a high-frequency discharge behave when it is on the anodic cycle?
2. Can the anode magnetron be used in the precleaning process of the metallic substrate that is used as the cathode in the cathodic plasma polymerization? In a more broad sense, can the surface of metal (cathode) be cleaned by argon direct current (DC) discharge sputtering?

When magnetrons are used in alternating polarity discharge, the magnetron in the anodic cycle is the anode magnetron, and the discharge is between a cathode magnetron and an anode magnetron. Therefore, it is important to understand characteristics of the anode magnetron. In this chapter, the characteristic features of the anode magnetron and its applications are described by comparing characteristics of discharges with cathode magnetron, anode magnetron, and nonmagnetron electrodes.

As discussed in Chapter 14, the best domain of magnetron discharge for luminous chemical vapor deposition (LCVD) is plasma polymerization by high-frequency alternating current (AC) power source in low-pressure regime. The DC cathodic polymerization is a very powerful tool for interface engineering of metallic or electrically conducting material; however, magnetron cathode cannot be used because the substrate is the cathode. However, if a magnetron is used as the anode of DC discharge, some advantageous feature of magnetron discharge, such as the lowering of breakdown voltage, stable discharge in low-pressure regime, and so forth, can be utilized in cathodic polymerization using the substrate as the cathode. The characteristics of magnetron used as the cathode against a plain anode, and magnetron used as the anode against a plain cathode are investigated by means of electron temperature and electron density measurements by Tao et al. [1]. The following modes of DC glow discharges of argon were examined:

1. Conventional mode of DC glow discharge (C-A), where the cathode (C) and the anode (A) are planar electrodes without magnetic enhancement.
2. Cathodic magnetron discharge (CM-A), where the cathode magnetron (CM) is used against a planar electrode without magnetron (A).

3. Anode magnetron discharge (C-AM), where a planar cathode without magnetron (C) is used against a magnetron anode (AM).

Characteristic glows by DC discharge (C-A) are shown in Figure 3.2. The distribution profile of T_e and of N_e is shown in Figures 3.4 and 3.5, respectively. The glow characteristics of cathode magnetron (CM-A) discharge are depicted in Figure 15.1 At all pressures in the range from 10^{-3} to 1 torr a stable, doughnut-shaped toroidal glow ring was observed, and the general shape of the plasma is relatively independent of pressure. Figures 15.1a and b are schematic representations of the shape of the glow in the CM-A discharge. The cathode dark space is very thin, probably less than 1 mm, and is difficult to identify. The main plasma ring is very intense and extends several millimeters from the cathode surface. It was observed that the intense glow ring became brighter and moved closer to the cathode surface

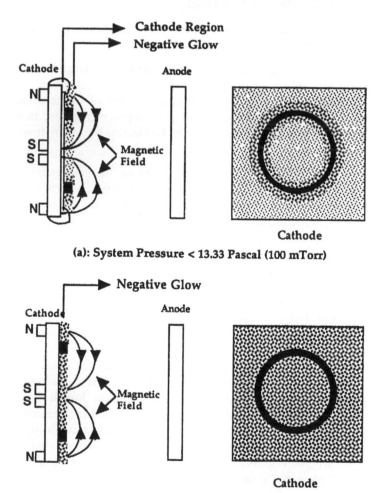

Figure 15.1   Schematic view of cathode-magnetron DC glow discharge of argon in the inter-electrode space; interelectrode distance = 102 mm, the luminous regions are shown shaded.

as the pressure increased, while a diffuse glow occupies the center and edge portions of the cathode surface, and a much more diffuse plasma extends from the outer edge of the ring to the anode. In higher pressure systems, the diffuse glow beyond the toroidal glow becomes brighter.

Figure 15.2 depicts the glow characteristics of anode magnetron (C-AM) discharge. Unlike the nonmagnetron discharge and the cathode magnetron discharge, in the low-pressure system region (less than 80 mtorr), the cathode dark space becomes kidney shaped, i.e., the center region thickness is thinner than the edge thickness as can be seen in Figure 15.2a. In other words, the negative glow is pushed closer to the cathode in the center portion of the cathode. The thickness of the dark space decreases as the discharge power, the pressure, and the anode magnetic field strengths increase. The negative glow forms a

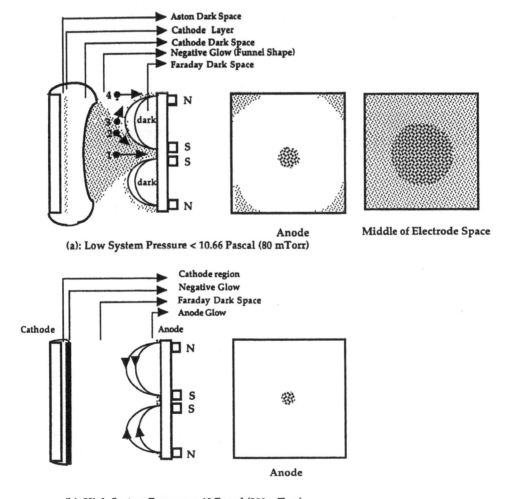

Figure 15.2 Schematic view of anode-magnetron DC glow discharge of argon in the interelectrode space, interelectrode distance: 102 mm; the luminous regions are shown shaded.

funnel-shaped glow extending from the center of the anode toward the cathode. The Faraday dark space forms two hemispheric dark spaces in the cross-sectional view. The toroidal intense glow in the CM-A mode discharge becomes the toroidal dark space in the C-AM mode discharge. The positive column is also "consumed," leaving only the negative glow and dark spaces adjacent to each electrode.

Figure 15.2a shows the three regions in which the movement of electrons is significantly different. Electron 1, which originated from the center region of the cathode, will move toward the anode with a minimum of influence by the magnetic field. Electrons 2 and 3, which are in the radial middle region of the electrodes' space, are confined by the magnetic field, and may travel along the $\mathbf{E} \times \mathbf{B}$ drift motion path and/or along with the arch-shaped magnetic field lines to the center and periphery of the anode electrode surface, so that no further electrons can pass through the magnetic field lines to extend glow, which causes the two hemispheric dark spaces adjacent to the anode electrode. Electron movement along the $\mathbf{E} \times \mathbf{B}$ drift motion path also increases the center region electron density and glow intensity, which explains the formation of the funnel-shaped glow. Electron 4, which is in the radial edge region of the electrodes' space, is accelerated by the electric field to ionize and/or to excite neutrals.

The effect of anode magnetron on glow can be seen by comparing glow discharge of oxygen with and without the anode magnetron shown in Figure 15.3. A center cathode plate is paired with two anode magnetrons. Without anode magnetron negative glow is diffused as shown in the upper picture. With anode magnetron, the negative glow is focused as funnel-shaped glows on both sides of the cathode as seen in the bottom picture.

2.   ELECTRON TEMPERATURE AND ELECTRON DENSITY

2.1.   Ar Discharge Without Magnetron (C-A)

The distribution of electron temperature (T_e) and electron density (N_e) in interelectrode space of DC discharge is shown in Figures 3.4 and 3.5, respectively. The electron temperature T_e increases from the cathode to the leading edge (20–25 mm from the cathode) of the negative glow, where the maximal value of T_e was observed. The energetic electrons undergo inelastic collisions (ionization and excitation) at the boundary of the negative glow, which make the negative glow the brightest part and the electrons lose their energy. T_e decreases from the leading edge of the negative glow to the center region of electrodes space (about 45 mm from the cathode) and then varies little to the anode. The electron temperature is higher in the edge of the glow (corresponding to the edge of the electrode), particularly near the cathode.

Figure 15.4 depicts electron temperature and electron density as a function of distance from the cathode surface. The profiles are taken at the center of electrode and also at the radial distance $R = 4.5$ cm, which is roughly the middle of the toroidal glow if the cathodic magnetron was used. Figure 15.5 depicts the distributions of T_e and N_e within a plane parallel to electrodes as a function of distance from the center of electrode. The distributions within a plane, which is parallel to the cathode surface, are uniform except for the edge effects appearing near the both ends of glow if the plane is at or near the cathode surface.

Without anode-magnetron

With anode-magnetrons

Figure 15.3 Effect of anode magnetron on glow, oxygen, flow system.

The number of electrons N_e is low near the cathode and does not increase while electrons are accelerated in the dark space and the electron temperature is increasing. When the electrons gain enough energy to ionize the Ar atoms, ionization occurs. At the onset of ionization, electrons lose their energy and the number of electrons increases. This situation is clearly seen in the distribution profiles of T_e and N_e. Where the decrease of T_e is observed, N_e increases. N_e continues to increase toward the anode and decreases sharply near the anode as the anode captures electrons. There is clearly a general trend that the values of T_e and N_e are inversely proportional ($T_e \times N_e =$ constant); the maximal value of one is found where the minimal value of the other is located.

The increase of N_e from cathode to anode may be due to the electron movement from cathode to anode and also to the occurrence of electron impact ionization in this region. In a 10-W and 50-mtorr argon DC glow discharge, T_e ranged from 17 to 48 eV and N_e ranged from 1×10^7 to $7.5 \times 10^8/cm^3$ in this region.

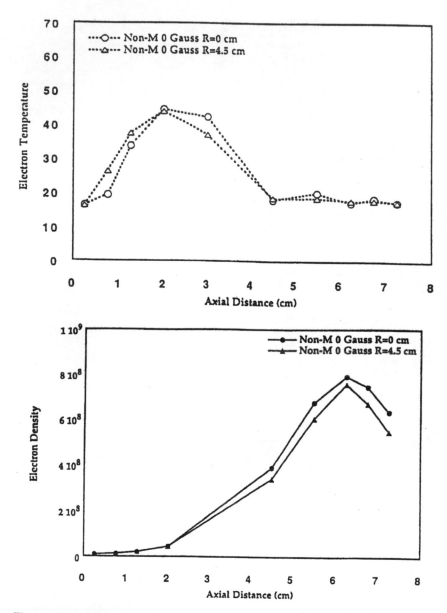

Figure 15.4   Electron temperature (eV) and electron density (#/cm^3) versus axial distance (from cathode to anode) for two radial positions (R) in Ar DC glow discharge; 10 W, 2 sccm, 50 mtorr, and 76 mm electrode distance.

2.2.   Ar Discharge With Cathode Magnetron (CM-A)

The same magnetron is used in both cathodic and anodic magnetron discharges against a plain counterelectrode. Figure 15.6 depicts the distributions of T_e, and Figure 15.7 shows N_e in cathode magnetron DC discharge of Ar. Because of the circular magnetic field configuration, electron temperature and electron density

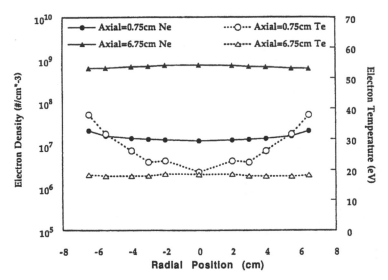

Figure 15.5 Electron temperature and electron density versus radial position fro two axial positions in Ar DC glow discharge; 10 W, 2 sccm, 50 mtorr, 76 mm electrode distance.

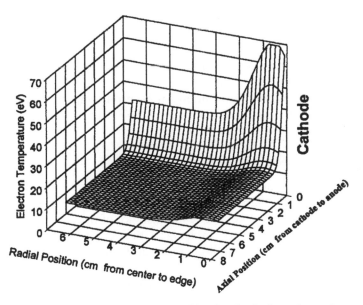

Figure 15.6 The distribution profile of T_e in the interelectrode space on the center plane in the CM-A mode (cathode magnetron) discharge; 2 sccm, 50 mtorr, 10 W, electrode separation distance 76 mm.

depends on the radial position. The distribution of electron temperature and electron density for two radial positions, $R = 0$ (center of electrode) and $R = 4.5$ cm, which is approximately at the maximal peak of parallel component of magnetic field (the middle of toroidal glow), is shown in Figure 15.8. It is important to note that the

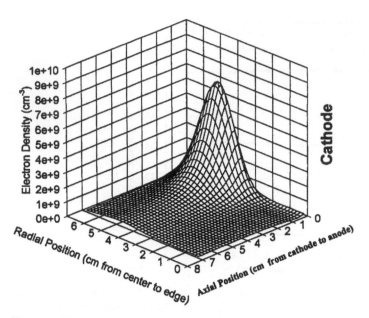

Figure 15.7 The distribution profile of N_e in the interelectrode space on the center plane in the CM-A mode (cathode magnetron) discharge; 2 sccm, 50 mtorr, 10 W, electrode separation distance 76 mm.

toroidal glow of the cathode magnetron collects large number of electrons with lower electron temperature; consequently, the electron density is highest at the toroidal glow. Away from the toroidal glow, the distributions of electron temperature and electron density smooth out as the distance from the cathode surface increases. Radial distributions at two axial positions are shown in Figure 15.9.

2.3. Ar Discharge with Anode Magnetron (C-AM)

The distributions of T_e and N_e are shown in Figures 15.10 and 15.11, respectively. The cathode dark space has a kidney shape, as shown in Figure 15.2, and the center thickness is smaller than the other region, i.e., the center distance of the leading edge of the negative glow is smaller. Thus, the energy exchange with electrons occurs more rapidly at the center than at other regions, resulting in lower T_e at the center radial position.

Since the magnetron is placed on the anode, the electron trapping by the magnetic field is completely different from that in the CM-A mode discharge. Electrons are not accelerated beyond the fringe of negative glow. Magnetic field does not trap these electrons but diverts the electron path to avoid magnetic field. The path traveled by the electrons from the cathode to the anode will be different depending on the part of the cathode in which the electron originated.

Electron temperature and electron density as a function of axial distance (from cathode) are depicted for two radial positions (same as for cathode magnetron depicted in Figure 15.8) in Figure 15.12. Radial distributions at two axial positions are shown in Figure 15.13. Negative glow is pushed toward the cathode and

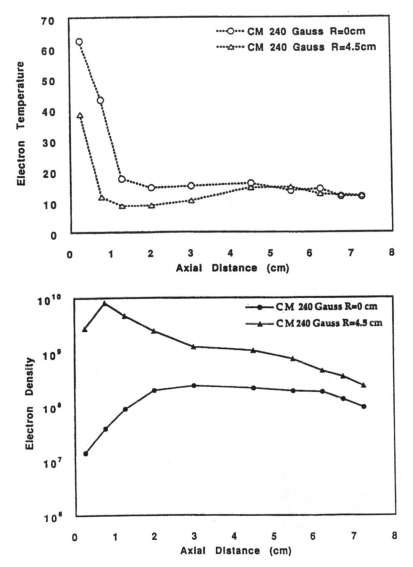

Figure 15.8 Electron temperature and electron density as a function of axial distance (from cathode to anode) for two radial positions, $R=0$ and $R=4.5\,\text{cm}$; 10 W, 2 sccm, 50 mtorr, 76 mm electrode distance.

electrons are pulled into the relatively narrow center part of the anode in anode magnetron DC glow discharge.

2.4. Comparison of Cathode Magnetron, Anode Magnetron, Nonmagnetron Discharges

Axial distributions of electron temperature and electron density for the three discharges at the center line of electrode are compared in Figure 15.14. At $R=0$

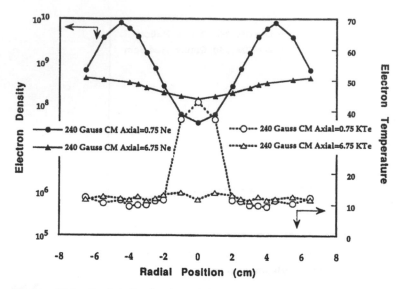

Figure 15.9   Radial distribution of electron temperature and electron density for two axial positions in cathode magnetron DC glow discharge of Ar; 10 W, 2 sccm, 50 mtorr, and 76 mm electrode distance.

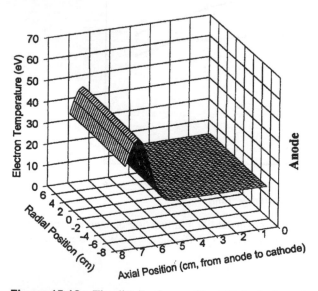

Figure 15.10   The distribution profile of T_e in the interelectrode space on the center plane in the C-AM mode (anode magnetron) discharge; 2 sccm, 50 mtorr, 10 W, electrode separation distance 76 mm.

position, both cathode magnetron and anode magnetron push the location of the maximal electron temperature much closer to the cathode surface, and the electron density by the anode magnetron is significantly higher than the rest of two discharges.

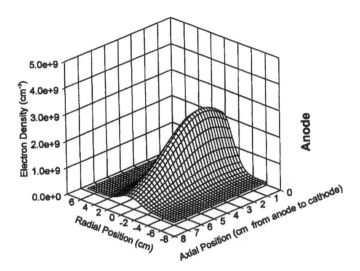

Figure 15.11 The distribution profile of N_e in the interelectrode space on the center plane in the C-AM mode (anode magnetron) discharge; 2 sccm, 50 mtorr, 10 W, electrode separation distance 76 mm.

Axial distributions of electron temperature and of electron density for nonmagnetron and anode magnetron are essentially the same with those at $R = 0$ shown in Figure 15.14. In cathode magnetron discharge, however, electron temperature drops sharply near the cathode and remains at more or less the same level. Electron density reaches the maximum at around the distance where electron temperature reaches the minimum and then gradually decreases toward the anode. The slight decrease of electron density approaching the anode is seen only in the negative glow with cathodic magnetron.

This situation changes completely when the distributions are compared at $R = 4.5$ cm, as depicted in Figure 15.15, because of the cathodic toroidal glow located at this position. In cathode magnetron discharge the electron temperature drops sharply near the cathode and then gradually increases after passing the minimal value. Electron density increases sharply as the electron temperature drops reaching the maximum at around the distance where electron temperature is at the minimum and then steadily decreases significantly toward the anode. The significant decrease of electron density approaching the anode is seen only in the negative glow with the cathode magnetron.

Radial distributions of electron temperature and electron density are compared at axial distance 2.5 mm and 7.5 mm, respectively, in Figures 15.16 and 15.17. In DC discharge of Ar without magnetron, the distributions of electron temperature and electron density near the electrode surface (2.5 mm from the cathode) are uniform, but both show the edge effect, more pronounced in electron temperature. At this position (in cathode dark region), there are small numbers of electrons that have low electron energy.

The radial distribution of electron temperature and electron density shown in Figures 15.16 and 15.17 clearly show the difference due to the presence of the cathode magnetron toroidal glow. In cathode magnetron discharge, there are two

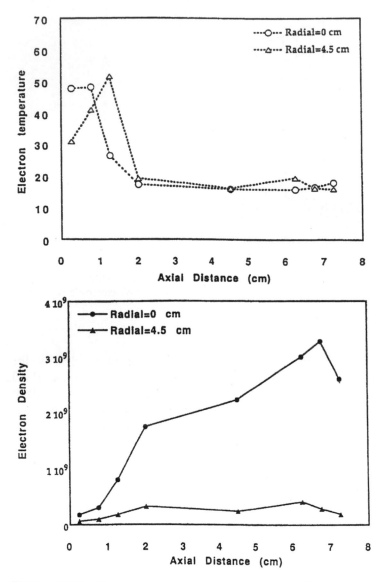

Figure 15.12 Electron temperature and electron density as a function of axial distance (from cathode surface) for two radial positions, $R=0$ and $R=4.5$ cm, in anode magnetron DC glow discharge of Ar; 10 W, 2 sccm, 50 mtorr, 76 mm electrode distance.

peaks corresponding to the cathode magnetron toroidal glow, in which electron density is very high but electron temperature is much lower than that in the center part. The electron temperature peaks corresponding to the toroidal glow virtually disappear and electron temperature drops significantly (nearly one-half) at the 7.5-mm position, but electron density in the peak increases over that in the 2.5-mm position. These observations indicate that the toroidal glow exists very close to the cathode surface, in the cathode dark region of corresponding nonmagnetron DC glow discharge, filled with large numbers of electrons with low energy (about 10 eV)

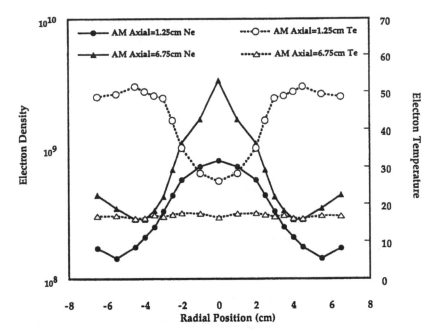

Figure 15.13 Radial distribution of electron temperature and electron density at two axial positions, 1.25 cm and 6.75 cm from the cathode surface, in anode magnetron DC glow discharge of Ar; 10 W, 2 sccm, 50 mtorr, 76 mm electrode distance, and 240 G.

compared to the maximal electron temperature (about 60 eV) observed in the nontoroidal glow region of the cathode magnetron discharge. This implies that the cathodic toroidal glow is more dissociation glow than ionization glow in LCVD, in which an organic compound is used instead of Ar.

2.5. Influence of Magnetic Field Strength of an Anode Magnetron

The influence of magnetic field strength of anode magnetrons is depicted in the axial distribution of electron temperature and electron density in anode magnetron DC glow discharge with references to those in nonmagnetron discharge in Figure 15.18. Influence of electrode distance is depicted in Figure 15.19, which shows the same effect of shifting the distribution profiles to left as the electrode distance becomes smaller, which is seen as the influence of magnetic field strength. Anode magnetron discharge pushes the distribution patterns of nonmagnetron DC glow discharge toward the cathode surface, i.e., negative glow is pushed closer to the cathode surface. The maximal electron temperature is not affected.

3. SPUTTER CLEANING OF CATHODE SURFACE

How effectively a DC discharge of argon could clean the cathode surface can be seen by observing the disappearance of color of the cathode, which is created by the deposition of TMS on a cold-rolled steel, by Ar discharge treatment [2]. The cathodic

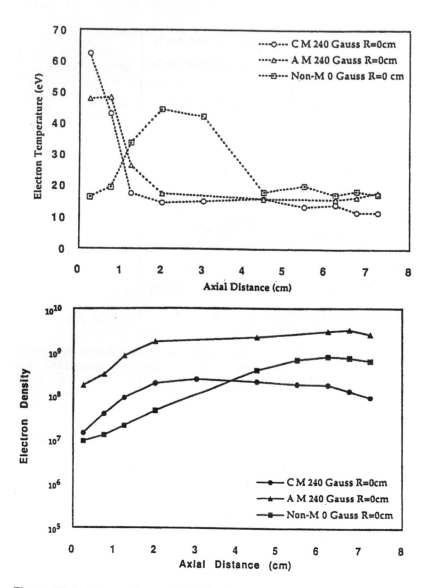

Figure 15.14 Comparison of axial distribution of electron temperature and electron density at center radial position in cathode magnetron, anode magnetron, and nonmagnetron DC glow discharge of Ar; 10 W, 2 sccm, 50 mtorr, 76 mm electrode distance, 240 G magnetron.

polymerization coating of TMS, thickness about 70 nm, provides a deep blue color when deposited on a CRS surface, which can be used as the color indicator for sputter cleaning efficiency. The color disappearance under various DC discharge conditions is a good indication of how effectively the cathode surface can be sputter cleaned by Ar DC discharge. When the surface is completely cleaned, the color of the cathode plate becomes that of bare metal.

Figure 15.20 depicts the color left after certain Ar glow discharge treatments, in which only the magnetic field strength of the anode magnetron is changed, including

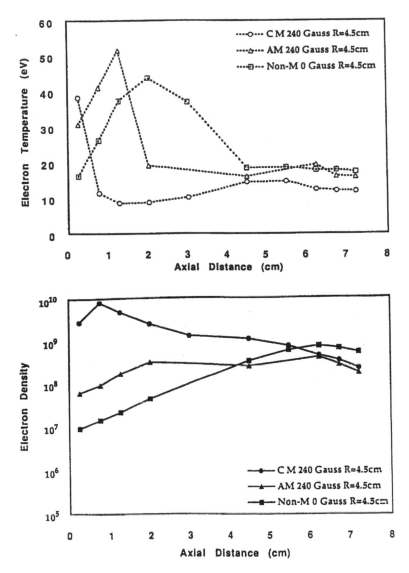

Figure 15.15 Comparison of axial distribution of electron temperature and electron density at radial position $R = 4.5$ cm in cathode magnetron, anode magnetron, and nonmagnetron DC glow discharge of Ar; 10 W, 2 sccm, 50 mtorr, 76 mm electrode distance, 240 G magnetron.

0 G for nonmagnetron discharge. Argon discharge without anode magnetron (0 G) for 10 min sputter cleaned only the edges of the cathode as depicted in Figure 15.20d, and further treatment for 5 h still left nearly untouched TMS coating in the middle portion of the cathode surface as seen in Figure 15.20e. This edge effect is due to the concentration of electrical field on the edges of the cathode, which makes the cleaning of the center part of cathode surface by argon discharge difficult under the set of conditions used.

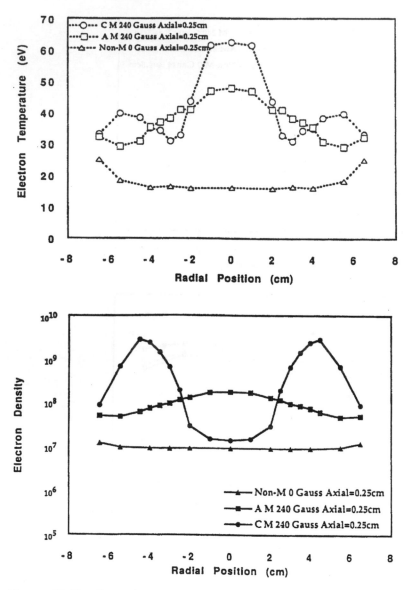

Figure 15.16 Comparison of radial distribution of electron temperature and electron density at 2.5 mm axial position in anode magnetron, cathode magnetron and nonmagnetron DC glow discharge of Ar; 10W, 2 sccm, 50 mtorr, and 240 G (anode and cathode magnetron).

The use of anode magnetron concentrates glow in the center portion of the cathode and makes complete cleaning possible without changing other parameters of glow discharge. The magnetic field strength of an anode magnetron is an important factor in this process as seen in Figure 15.20b and c. With 180-G anode–magnetron, the cathode surface was completely cleaned in 10 min, while 100-G anode–magnetron discharge still left an incompletely cleaned center portion in the same treatment time.

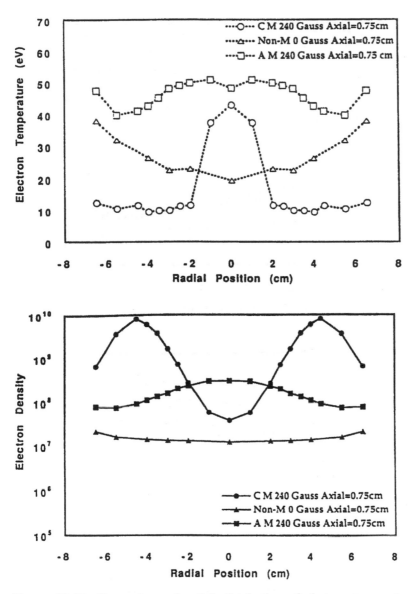

Figure 15.17 Comparison of radial distribution of electron temperature and electron density at 7.5 mm axial position in anode magnetron, cathode magnetron and nonmagnetron DC glow discharge of Ar; 10 W, 2 sccm, 50 mtorr, and 240 G (anode and cathode magnetron).

Figure 15.21 shows the influence of the interelectrode distance on the sputter cleaning of the cathode surface. Since the magnetic field strength decreases with the distance from the surface of a magnetron, the anode magnetron effect diminishes as the interelectrode distance increases. Since the energy of ions impinging on the cathode depends on E/p, where E is the electrical field and p is the system pressure, the effectiveness of the anode magnetron system depends on the system pressure as depicted in Figure 15.22.

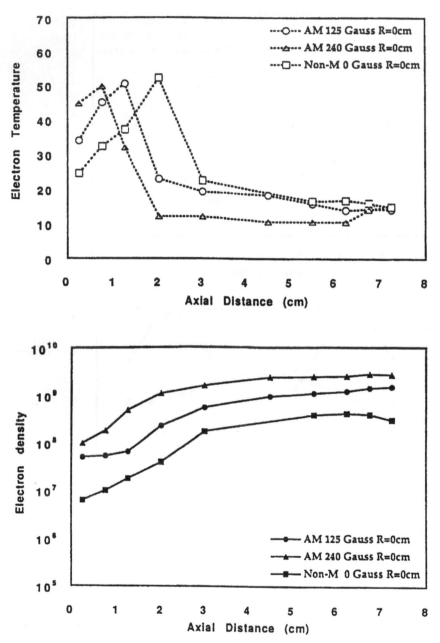

Figure 15.18 Influence of magnetic field strength on axial distribution of electron temperature and electron density, at center radial position, of an anode magnetron DC glow discharge of Ar; 800 V, 2 sccm, 50 mtorr.

The effects depicted in these figures show the effectiveness of the anode magnetron in sputter cleaning of the cathode surface but also indicate the limitation of the sputter cleaning by means of the anode magnetron. While the treatment of well-defined surface, such as a flat sheet or a moving foil, could be treated efficiently

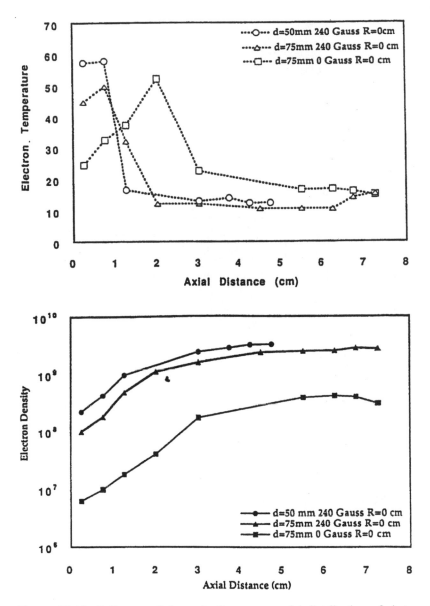

Figure 15.19 Influence of electrode distance on axial distribution of electron temperature and electron density; 800 V, 2 sccm, 50 mtorr.

by a fixed anode magnetron, the substrate (cathode) in a complex shape could not be sputter cleaned well because parameters described above change with irregularly shaped substrate, if the irregularity extends beyond the range of factors described here. This problem could be solved if an anode magnetron could be used in a scanning mode. Nevertheless, an anode-magnetron system does provide the capability of sputter-cleaning metallic substrate used as the cathode for DC cathodic polymerization, which is indispensable.

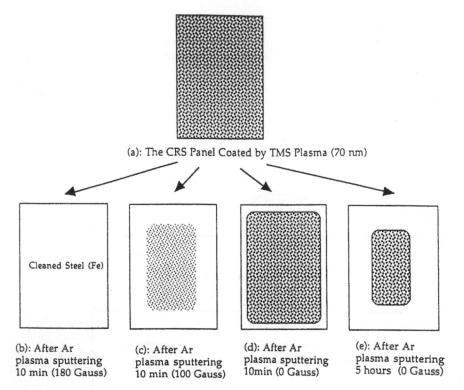

(a): The CRS Panel Coated by TMS Plasma (70 nm)

Cleaned Steel (Fe)

(b): After Ar
plasma sputtering
10 min (180 Gauss)

(c): After Ar
plasma sputtering
10 min (100 Gauss)

(d): After Ar
plasma sputtering
10min (0 Gauss)

(e): After Ar
plasma sputtering
5 hours (0 Gauss)

Figure 15.20 Effect of anode magnetron in DC discharge cleaning of the cathode surface by argon discharge; 2 sccm, 50 mtorr, and 76 mm interelectrode distance.

4. ANODE MAGNETRON FOR CATHODIC POLYMERIZATION

4.1. Effect of Magnetic Field Configuration

When two anode magnetrons are employed against a substrate used as the cathode, the combination of configurations of magnetic field in both magnetrons become a factor to be considered in the anode magnetron cathodic (AMC) polymerization [3]. The effects of the combination of anode magnetron configuration on AMC polymerization, with respect to the deposition rate and its distribution, are described in this section. This setup of magnetron is a model case for the magnetron glow discharge with alternating power source, in which alternating current power source is applied on both magnetron and the center plate becomes floating substrate. The effect of magnetic field configuration one can find by this particular set of experiments would very likely be applicable in high-frequency discharge in a typical range of 10–50 kHz, where magnetron glow discharge is most effective. As mentioned in Chapter 14 on electro-magnetron plasma polymerization, the polarity of the magnetic field employed in the magnetron pair has the subtle but important effect on the deposition rate of plasma polymer on the substrate placed in the interelectrode space.

The experimental setup is schematically depicted in Figure 15.23. Each anode was a 7×7 in. titanium plate. Eight permanent magnetic bars were placed on the

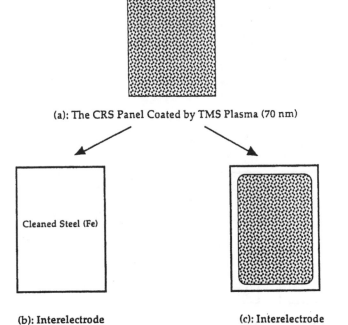

(a): The CRS Panel Coated by TMS Plasma (70 nm)

Cleaned Steel (Fe)

(b): Interelectrode
distance of 50 mm
or 75 mm

(c): Interelectrode
distance of 100 mm
to 175 mm

Figure 15.21 Effect of electrode distance in anode magnetron DC argon cleaning of the cathode CRS plate coated with plasma polymer of TMS, 6.66 Pa (50 mtorr), 180 G.

backside of each anode in a circular configuration bridging a center iron plate and an outer iron circular ring plate, with the same poles oriented toward the center (see Figure 14.2). The magnetic field strength of each anode was 110 G, which is expressed by the maximal Gauss meter reading of the perpendicular component with respect to the electrode surface. The distance between the two anode electrodes was adjustable. The 3×6 in. substrate panel (cathode) was placed midway between the two anode electrodes. Figure 15.24 depicts the combinations of polarity of magnetic field.

A parallel magnetron (PM) discharge system has a magnetic field configuration in which the south poles of the permanent magnet bars are oriented toward the center of the electrode on both electrodes. In PM configuration, the radial magnetic field fluxes from the two magnetrons in the midpoint of two anodes are in the same direction as shown in Figure 15.24a.

In an opposing magnetron discharge system (OM), the south poles of the permanent magnet bars are placed on the center in one anode while the north poles of the magnet bars are placed on the center of the other anode. In OM configuration the radial magnetic field fluxes from the two magnetrons in the midpoint of two anodes are in the opposite direction as shown in Figure 15.24b. A discharge system with No magnetron (NM) is shown in Figure 15.24c as the reference case.

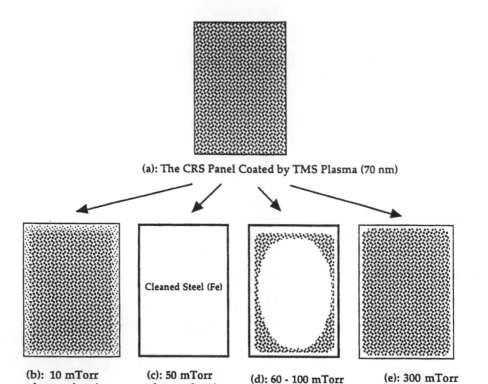

(a): The CRS Panel Coated by TMS Plasma (70 nm)

Cleaned Steel (Fe)

(b): 10 mTorr
plasma cleaning

(c): 50 mTorr
plasma cleaning

(d): 60 - 100 mTorr
plasma cleaning

(e): 300 mTorr
plasma cleaning

Figure 15.22 Effect of the system pressure in anode magnetron argon DC discharge with TMS coated cathode CRS plate, 76 mm interelectrode distance, and 180 G.

4.2. Comparison of Deposition Rates by PM, OM, and NM Configurations

The distribution of the deposition rate in an AMC plasma polymerization system with PM, OM, or NM anodes is shown in Figure 15.25. These results show that the anode magnetron shifts the deposition pattern to high in the center of the cathode. The edge effect diminishes but a small peak appears in the center. The distribution with OM configuration seems to be more uniform than that with PM configuration, but because of the uniformity the deposition rate at center portion is slightly lower.

In plasma polymerization, the character of the dissociation glow that occurs near the surface of the cathode is more important than ion bombardment in determining the deposition rate and its distribution described in previous chapters. The edge effect seems to be less pronounced in LCVD than in the sputtering of the cathode material. The anode magnetrons seem to overcompensate the edge effect in LCVD.

4.3. Effect of Electrode Distance

Figure 15.26 depicts the effect of electrode distance on the distribution pattern of deposition rate with PM and OM configurations. The distance from the counter-electrode influences the magnetic field strength near the substrate surface in a

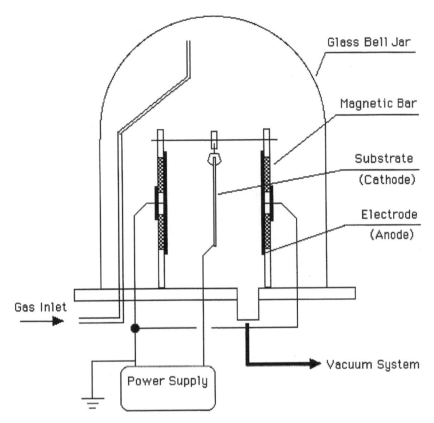

Figure 15.23 Schematic diagram of the bell jar reactor system.

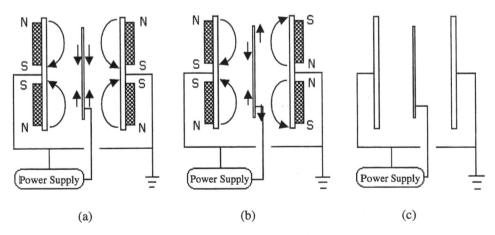

Figure 15.24 Schematic diagram of different magnetic field configurations on the backside of anode electrodes: (a) parallel magnetic field configuration (PM), (b) opposite magnetic field configuration (OM), and (c) no magnetron (NM).

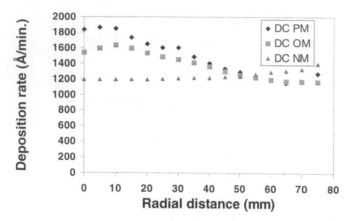

Figure 15.25 The dependence of the deposition rate distribution on magnetic field configuration, TMS, 50 mtorr, 1 sccm, 5 W, $d = 100$ mm.

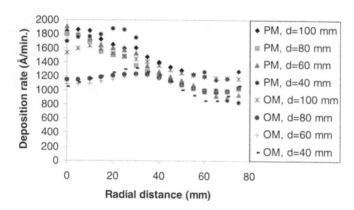

Figure 15.26 The influence of electrode distance on the deposition rate in AMC plasma polymerization; TMS, 50 mtorr, 1 sccm, 5 W.

magnetron system; however, the effect is small compared to the effect of the magnetron configuration.

4.4. Effect of Cathode Surface Area

The influence of the surface area of cathode on the plasma deposition rate is shown in Figure 15.27. It is clear that the plasma deposition rate decreases with the increase of cathode surface area. This is because plasma deposition rate is proportional to current density in cathodic polymerization as described in Chapter 8. The patterns of distribution due to magnetron configuration are similar, but the trends are magnified as the deposition rate increases with smaller cathode area.

4.5. Effect of System Pressure

The influence of system pressure on deposition rate is shown in Figure 15.28. The results indicate that the characteristic features of cathodic polymerization overwhelm

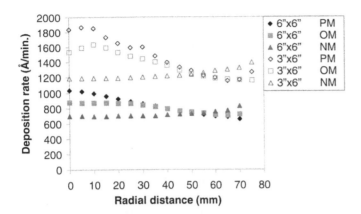

Figure 15.27 The influence of substrate area on the deposition rate; TMS, 50 mtorr, 1 sccm, 5 W.

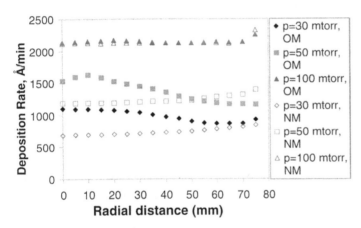

Figure 15.28 The dependence of deposition rate distribution on system pressure (p) in AMC plasma polymerization; TMS, 1 sccm, 5 W, $d = 100$ mm.

the influence of the magnetic field. The deposition rate in cathodic polymerization is proportional to the current density and the concentration of monomer in the cathode region that is proportional to system pressure, not the flow rate as mentioned previously.

The concentration of monomer in the cathode region increases proportionally to the system pressure. The deposition rates shown in Figure 15.28 are nearly proportional to the system pressure while the flow rate is maintained at a fixed value. The typical influences of magnetic field discussed in the previous sections diminish as pressure is increased, and the edge effect of deposition appears at 100 mtorr.

The main purpose of having a magnetically enhanced glow discharge for plasma polymerization (deposition G), e.g., 40-kHz magnetron discharge, is to make it possible to carry out plasma polymerization effectively in a low-pressure regime (e.g., less than 100 mtorr). Using a magnetron as anode against a cathode, which is a

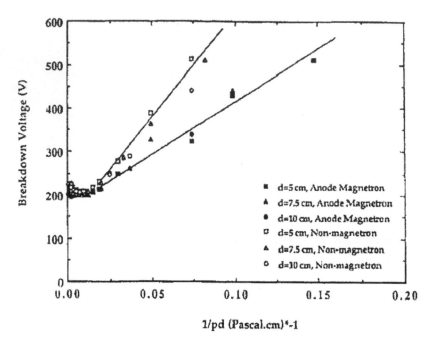

Figure 15.29   Effect of anode magnetron on the breakdown voltage of DC discharge of Ar in low-pressure regime.

metallic substrate, anode magnetron does decrease the breakdown voltage and make a DC discharge easier, as depicted in Figure 15.29. The results shown in Figure 15.29 are in accordance with this objective, and also confirm that there is no need for an anode magnetron for a confined discharge in higher pressure operation of plasma polymerization.

The effects of anode magnetron in sputter cleaning and in cathodic plasma polymerization could be summarized as follows.

1. Anode magnetron reduces the breakdown voltage of DC discharge, which provides more effective sputter cleaning and better uniformity of deposition profile of DC cathodic polymerization, in a low-pressure regime, e.g., less than 50 mtorr. (No anode magnetron is needed in high-pressure regime, e.g., more than 100 mtorr.)
2. The pairing of magnetic field configuration has notable effects on the deposition rate on the center cathode. PM configuration provides higher deposition rate than OM configuration but causes the deposition peak at the center. OM configuration seems to provide overall advantages of anode magnetron in DC cathodic polymerization.
3. The effect of magnetic field configuration could be important when two magnetrons are used in AC power discharge. However, the increase of deposition rate is generally attained at the expense of the uniformity of deposition rate profile, and the true advantage would depend on other factors involved in each application. If substrates move in the interelectrode space, PM configuration would be advantageous to obtain a higher deposition rate.

REFERENCES

1. Tao, W.H.; Prelas, M.A.; Yasuda, H.K. J. Vac. Sci. Tech., **1996**, *A14* (4), 2113.
2. Tao, W.H.; Yasuda, H.K. Plasma Chem. Plasma Proc. **2002**, *33* (2), 313.
3. Zhao, J.G.; Yasuda, H.K. J. Vacuum Sci. Tech. **2000**, *18* (5), 2062.

16
Low-Pressure Cascade Arc Torch

1. CASCADE ARC

Arc discharge is a kind of direct current (DC) discharge that is characterized by low voltage and high current, which is generated by a pointed small wire (cathode) and surrounding metal surface (anode) [1–7]. The cathode and the anode are contained in a small nozzle (arc generator). The luminous gas phase created in the gap between the cathode and the anode is blown out of the arc generator by the flow of the carrier gas, which forms a jet of the luminous gas. The pressure of the carrier gas is generally superatmospheric at the entrance of the arc generator. When the luminous gas is blown into atmospheric pressure, the temperature of the flame is very hot, and the high-pressure arc torch is used as a heat source for cutting metals and welding processes. Cascade arc is the arc that is created by a special mode of arc generation, in which the cathode and the anode are separated by a series of conducting walls that are separated each other by insulators.

2. CASCADE ARC GENERATOR AND LOW-PRESSURE CASCADE ARC TORCH REACTOR

Figure 16.1 depicts the structure of a cascade arc generator. A disk of copper or brass, e.g., diameter about 6 cm, and thickness about 1 cm, which has a center hole, e.g., about 3 mm diameter, is the major element of the cascade arc generator. Roughly 7–10 rings are assembled with insulating gasket between rings so that 7–10 electrically floating conducting walls are aligned in the 2- to 3-mm diameter nozzle. Circulating temperature-controlled water cools each ring. A metal wire with low sputtering yield, typically tungsten wire, is used as the cathode, and the last ring is typically used as the anode.

When electrical voltage is applied between the cathode and the anode, the breakdown of the gas in the interelectrode space, which is surrounded by numbers of electrically floating conducting surfaces, occurs. The first arc discharge occurs between the cathode and the first ring, which establishes its electrical potential that is higher than the next ring. The discharge propagates from the first ring to the second ring, and so on to the anode surface, in a cascading mode. Because of this mode of the propagation of arc discharge, this type of arc is termed as *cascade arc* or *wall-stabilized arc*. The advantage of cascade arc over other types of arc is in the ease of generating the arc and the stability of the arc generation process.

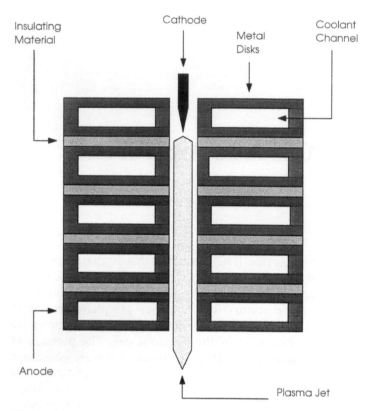

Figure 16.1   General schematic of the cascade arc generator.

A single monoatomic gas, e.g., argon or helium, is used as the carrier gas of the cascade arc discharge. When the luminous gas is injected into an expansion chamber under low pressure, e.g., 1 torr or less, the flame extends a significant length (e.g., 1 m), which depends on the flow rate, input power, diameter of the nozzle, and pressure of the expansion chamber. This mode of cascade arc torch is termed low-pressure cascade arc torch (LPCAT), which is useful in the surface modification by means of low-pressure cascade arc torch treatment and low-pressure cascade arc torch polymerization.

LPCAT can be used without employing the second gas, which is injected into the expansion chamber, or with addition of single or multiple gases or vapors in the expansion chamber. Because of a very high velocity of luminous gas jet stream, the location of the second gas injection is not a critical factor because the second gas is sucked into the jet stream. However, the second gas is generally introduced near the nozzle. Figure 16.2 depicts assembled cascade arc generator that can be attached to a vacuum system. In this configuration, the second gas injection port is integrated into the cascade arc generator. A pictorial view of an LPCAT reactor is shown in Figure 16.3, which displays a stream of luminous gas injecting from the nozzle into the expansion chamber in vacuum.

As seen in Figure 16.3, the LPCAT flame is relatively narrow, implying that a uniform diffused luminous gas phase is not created in the expansion chamber. Consequently, the treatment that can be achieved by an LPCAT is governed by the

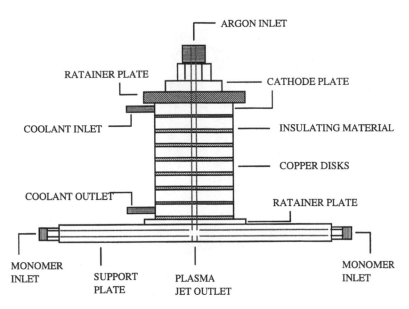

Figure 16.2 Assembled cascade arc generator.

Figure 16.3 Pictorial view of cascade arc torch reactor.

line of sight process, regardless of whether the substrate touches the luminous gas flame or not, and limited to a relatively small area that is exposed to the flame or is near the tip of flame. When a substrate is placed along the line of the jet stream, the well-identifiable flame is destroyed, and gaseous species scatter in the downstream of the substrate. The scattered species could cause surface treatment effects; however, their extents are much smaller than that by the jet because of the reduced density of reactive species.

Plurals of integrated cascade arc generators, which are strategically placed on an expansion chamber, can be used to achieve wider and more uniform treatment.

Figure 16.4 LTCAT model reactor for large-scale applications.

The relative motion of substrate with respect to the luminous gas jet is more or less mandatory for the uniform treatment. Figure 16.4 depicts a reactor equipped with three cascade arc generators, of which two are used to treat substrates placed on a rotating plate.

Another potential mode of LPCAT processing is that the integrated cascade arc generator, such as that shown in Figure 16.2, is placed in a vacuum chamber held by a robot arm. In this mode, LPCAT jet could scan over a complex shaped substrate by the robotic operation.

With low-pressure cascade arc, plasma formation (ionization/excitation of Ar) occurs in the cascade arc generator, and the luminous gas is blown into an expansion chamber in vacuum. The majority of electrons and ions are captured by the anode and the cathode, respectively, of the cascade arc generator, and there is no external electrical field in the expanding plasma jet. Consequently, the photon-emitting excited neutrals of Ar cause the majority of chemical reactions that occur in the plasma jet. The luminous gas coming out of the nozzle interacts with gases existing in the space into which it is injected or the surface that is placed to intercept the jet.

LPCAT can be utilized in the following three modes in the surface modification of materials:

1. Without addition of the second gas or vapor, i.e., jet of excited argon neutrals
2. With addition of the second gas that does not form the deposition of material (non-polymer-forming gas), i.e., jet of excited neutral species of the second gas, and
3. With addition of the second gas that causes the deposition of material (polymer-forming gas; monomer), i.e., plasma polymerization by means of excited neutral species of argon.

The second gas is introduced in the expansion chamber. Because of an extremely high velocity of gas injecting from a small nozzle (e.g., 3 mm in diameter), the second gas injected into the expansion chamber in vacuum cannot migrate into the cascade arc generator. Thus, the activation of Ar in the cascade arc generator and deactivation of the activated species of Ar in the expansion chamber, which activate the second gas introduced in the expansion chamber, are temporally and spatially separated. The LPCAT treatment and polymerization occur under such a totally (temporally, spatially, and kinetically) decoupled activation/deactivation system.

3. LUMINOUS GAS IN CASCADE ARC TORCH

In the LPCAT process, only an inert gas such as Ar exists in the cascade arc generator, and DC voltage is applied between the cathode and the anode. Therefore, it is a DC discharge of Ar, but it occurs under much higher pressure than in most low-pressure DC discharges, and the gas travels very fast in one direction in the generator. The basic process of ionization of Ar takes place in the cascade arc generator, which can be depicted as follows.

The ionization of Ar by high-energy electron, e^*,

$$Ar + e^* \rightarrow Ar^+ + 2e \tag{16.1}$$

The creation of excited neutrals by an electron having the same level of energy also occurs simultaneously:

$$Ar + e^* \rightarrow Ar^* + e \tag{16.2}$$

The cathode captures most of Ar^+, and the anode captures the majority of electrons. When the luminous gas phase created in the cascade arc generator is blown out of the nozzle, the majority of species in the luminous gas jet are excited Ar neutrals and some strayed electrons.

In an expanding cascade arc plasma jet, the main characteristic is that electrons have very low kinetic energy (low electron temperature). The electron temperature estimates range from 0.3 to 1.5 eV in argon plasma jets by Langmuir double-probe measurements [5,8]. Since the double probes sample only the higher energy electrons, and not the bulk of electron distribution, the average electron temperature could be significantly lower than the above reported values. Since ions cannot be created at such low electron temperatures in the expansion chamber, only neutral lines are emitted in a pure argon or helium luminous gas jet.

In the expansion chamber, no externally applied electrical field exists, and no acceleration of electrons occurs. In such a passive environment, the number of electrons follows typical first-order decay as a function of the distance from the nozzle [8], as shown in Figure 16.5. The decay is probably due mainly to the decrease of electron density by the expansion of the jet stream width. Excited Ar neutrals outnumber electrons and dominate subsequent dissociation/excitation phenomena. The cascade arc luminous gas jet could be viewed as a jet stream of excited neutrals of the carrier gas.

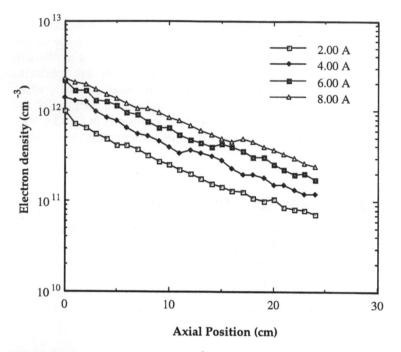

Figure 16.5 Electron density (cm^{-3}) as a function of axial position and arc current. Other conditions are 2000 sccm argon and 560 mtorr (75 Pa).

3.1. Low-Pressure Cascade Arc Torch (Without Addition of the Second Gas)

In the expansion chamber without addition of the second gas, excited species Ar* decay by emitting photons,

$$Ar^* \rightarrow Ar^{(*)} + h\nu_3 \tag{16.3}$$

where Ar* represents Ar atom in an excited state, of which the threshold energy (the excitation energy from the ground state) is given by E_B; $Ar^{(*)}$ represents the Ar atom in a lower electronic energy after losing energy by the photon emission ($h\nu_3$).

The typical emission spectra of argon and helium are shown in Figure 16.6 from 310 to 920 nm, with few significant emissions outside of this region observed over the 200- to 1050-nm range. The emissions from the argon and helium plasma jets correspond exclusively to neutral argon or helium excited species, with no argon or helium ion lines [9].

Some representative emission lines from the luminous gas jets are summarized in Table 1. The dominant features for argon plasma jet correspond to 4p → 4s transitions, with higher energy levels, such as $5p[2\,\tfrac{1}{2}] \rightarrow 4s[1\,\tfrac{1}{2}]^0$, being quite clearly observed as well. The excited atoms in a helium luminous gas jet have much higher energy levels than those in an argon luminous gas jet. With such a high energy over 23 eV, the excited helium neutrals can ionize nearly any molecular gases (excluding organic vapors, which mostly dissociate) injected into the expansion chamber.

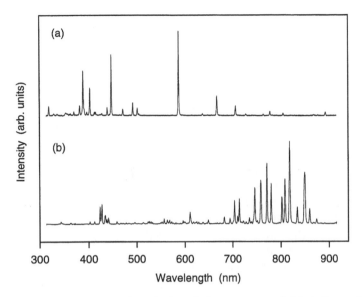

Figure 16.6 Typical optical emission spectra of (a) helium plasma jet and (b) argon plasma jet, the spectra was obtained at an axial position 2.7 cm from the jet inlet; conditions are: (a) 3000 sccm helium, 1.35 kW, and 89 Pa; (b) 2000 sccm argon, 0.64 kW, and 75 Pa.

Table 16.1 Most Intense Emission Lines Observed in Argon and Helium Plasma Jets

Species	Emission wavelength (nm)	Transition	Energy of emitting state above ground state, E_B (eV)
Ar	420.1	$5p[2\tfrac{1}{2}] \to 4s[1\tfrac{1}{2}]^0$	14.50
	696.5	$4p'[\tfrac{1}{2}] \to 4s[1\tfrac{1}{2}]^0$	13.33
	750.4	$4p'[\tfrac{1}{2}] \to 4s'[\tfrac{1}{2}]^0$	13.48
	763.5	$4p[1\tfrac{1}{2}] \to 4s[1\tfrac{1}{2}]^0$	13.17
	772.4	$4p'[\tfrac{1}{2}] \to 4s'[\tfrac{1}{2}]^0$	13.33
	794.8	$4p'[1\tfrac{1}{2}] \to 4s'[\tfrac{1}{2}]^0$	13.28
	811.5	$4p[2\tfrac{1}{2}] \to 4s[1\tfrac{1}{2}]^0$	13.08
	842.5	$4p[2\tfrac{1}{2}] \to 4s[1\tfrac{1}{2}]^0$	13.09
He	388.9	$3p\,^3P^0 \to 2s\,^3S$	23.01
	402.6	$5d\,^3D \to 2p\,^3P^0$	24.04
	447.2	$4d\,^3D \to 2p\,^3P^0$	23.73
	587.6	$3d\,^3D \to 2p\,^3P^0$	23.07
	667.8	$3d\,^1D \to 2p\,^1P^0$	23.07

The ratio of the concentration of excited species (energy carriers such as Ar) to that of the unexcited carrier gas is proportional to the power input given by W. This relationship can be expressed in Eqs. (16.4) and (16.5) [10].

$$\frac{[Energy\,Carrier]}{[FM]_c} = \frac{[FM]_c^*}{[FM]_c} = \alpha_1 W \tag{16.4}$$

where α_1 is proportionality constant, in unit of s/J.

$$[EnergyCarrier] = [FM]_c^* = \alpha_1 W[FM]_c \tag{16.5}$$

In Eqs. (16.4) and (16.5), $[EnergyCarrier]$ is the concentration (mass density) of excited species of the carrier gas in the jet stream blown out from the arc column into the expansion chamber (not the concentration in the chamber). The subscript "c" represents the carrier gas, and superscript "*" represents the excited carrier gas. F is molar flow rate, and M is the molecular weight of the gas. FM is the mass flow rate. W is the power input applied to the arc column.

Figure 16.7 shows the change of argon emission (763.5 nm) intensity against a composite experimental parameter $W(FM)_c$. Other argon emission lines, such as 706.7 nm and 604.3 nm, also showed the same dependency. The argon emission intensity was proportional to the arc current at a fixed argon flow rate and proportional to the argon flow rate at a fixed arc current. In other words, if the emission intensity was plotted against W or $(FM)_c$, plurals of lines corresponding to values of $(FM)_c$ or W, respectively, were obtained. These are the similar dependencies observed for the deposition in glow discharge on the discharge wattage and flow rate of monomer when deposition rates are plotted against F or W individually, instead of the composite parameter W/FM, which is the true operational parameter (see Chapter 8).

The argon emission intensity showed a linear dependence on the combined parameter, $W(FM)_c$, i.e., the total energy applied to the carrier gas in the LPCAT process, as described by Eq. (16.5), can be expressed by this combined experimental parameter, $W(FM)_c$.

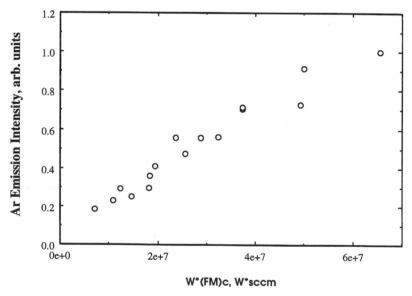

Figure 16.7 The argon emission (763.5 nm) intensity as a function of the combined experimental parameter; W is the power applied to the arc column.

3.2. Addition of Non-Polymer-Forming Gases

The introduction of the second gas into the expansion chamber causes two significant visible changes: (1) the shrinking of the luminous gas jet (quenching) and (2) the change of color of the luminous gas jet. These phenomena indicate that the excess energy carried by excited Ar neutrals are consumed in the creation of a new luminous gas phase, which consists of a mixture of the second gas and Ar (carrier gas), and the color depends on the amount and the nature of the second gas. Thus, the length of the mixed luminous gas jet (flame) depends on the flow rate of the second gas (at a fixed flow rate of Ar), i.e., the higher the flow rate, the shorter the length of flame. Tables 16.2–16.4 [9] depict characteristics of the luminous gas phase when a second gas is added to the expansion chamber.

The electron impact ionization does not occur in the cascade arc torch, and the energy transfer between excited neutrals of the carrier gas and the added gases becomes the dominant mechanism, i.e., the Penning-type reaction or resonance

Table 16.2 Identified Emission Lines and Bands

Species	λ (nm)	Transition	E_B (eV)	Remarks
N_2	337.1	$C^3\Pi_u \rightarrow B^3\Pi_g$	11.1	2nd pos.
	662.3	$B^3\Pi_g \rightarrow A^3\sum_a^+$	7.4	1st pos.
N_2^+	391.4	$B^2\Sigma_u^+ \rightarrow X^2\Sigma_g^+$	18.7	1st neg.
O	777.2	$3p^5P \rightarrow 3s^5S^0$	10.7	
	844.6	$3p^3P \rightarrow 3s^5S^0$	11.0	
O_2^+	525.1	$b^4\Sigma_g^- \rightarrow a^4\Pi_u$	18.2	1st neg.
H	656.2	$3d^2D \rightarrow 2p^2P^0$	12.09	H_α
	486.1	$4d^2D \rightarrow 2p^2P^0$	12.75	H_β
	434.0	$5d^2D \rightarrow 2p^2P^0$	13.06	H_γ
H_2		$a^3\Sigma_g^+ \rightarrow b^3\Sigma_u^+$	11.79	Continuum
NH	336.0	$A^3\Pi \rightarrow X^3\Sigma^-$	3.7	3360-Å system

Table 16.3 Spectra Emitted in Argon Plasma Jet with Addition of Reactive Gases

Reactive gas	Visual appearance	Observed spectra
None	Orange	Ar atom lines
N_2	Pink	N_2 end pos.
O_2	Orange	O atom lines
Air	Pink	N_2 end pos. O atom lines
H_2	Light blue	H_2 molecular continuum H_α, H_β lines
$N_2 + H_2$	Pink	N_2 1st & 2nd pos. NH band H_α line
NH_3	Light blue	NH band H_α line

Table 16.4 Spectra Emitted in Helium Plasma Jet with Addition
of Reactive Gases

Reactive gas	Visual appearance	Observed spectra
none	White	He atom lines
N_2	Green	N_2^+ 1st neg.
O_2	Light blue	O_2^+ 1st neg. O atom lines
Air	Green	N_2^+ 1st neg. O atom lines
H_2	White	NH band
		H_α, H_β lines

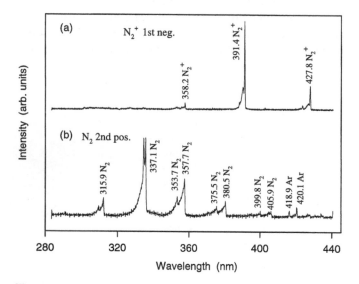

Figure 16.8 Typical emission spectra of nitrogen in low temperature cascade arc plasma
jets: (a) N_2 in helium plasma jet, (b) N_2 in argon plasma jet.

reaction between gas molecules and excited neutrals of the carrier gas is the principal
generation process for the reactive species.

Upon addition of the second gases to the plasma jets, the emission of argon or
helium plasma jets is generally (but not always) highly quenched. The dominant
features in the emission spectra are due to excited species corresponding to the added
gases, with only the strongest argon or helium lines remaining visible.

3.2.1. Nitrogen

When nitrogen is added to plasma jets, all argon or helium emission lines are highly
quenched, and a very strong pink flame for argon plasma jet and a green flame
for helium plasma jet are formed. The dominant features in the spectra are due to
nitrogen species. Figure 16.8 depicts OES signals of luminous gas jet of N_2 added to
(1) Ar plasma jet and (2) He plasma jet.

In radio frequency plasmas, which are characterized by low ionization degree
and high electron temperature, the emission is mainly due to electron excitation and
pooling reaction of the molecular metastable. Figure 16.9 depicts various emission

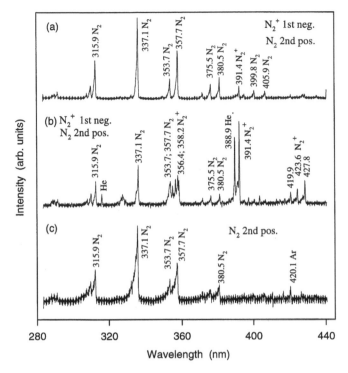

Figure 16.9 Typical emission spectra of RF plasmas. (a) pure nitrogen, 60 sccm N$_2$, 10 Pa, RF power 40 W; (b) mixtures of 60 sccm nitrogen and 2000 sccm helium, 75 Pa, RF power 100 W; (c) mixture of 60 sccm nitrogen and 2000 sccm argon, 75 Pa, RF power 250 W.

observed in radio frequency discharge of nitrogen. The main ion is N$_2^+$, which also can be excited by electrons to give the molecular ion emission. In expanding cascade arc helium plasma jet, only the spectrum of the first negative system of N$_2^+$ was observed as shown in Figure 16.10, which depicts comparison of nitrogen luminous gas in radio frequency and in LPCAT.

Argon-excited neutrals were consumed by the increasing nitrogen addition, and the emission intensity of the N$_2$ second positive, emission band keeps increasing with increasing nitrogen flow rate from 0–60 sccm, which shows that more activated nitrogen species were produced with increased N$_2$ flow rate at the expenses of Ar emissions, as shown in Figure 16.11.

The electronic energy level of excited argon or helium neutrals plays a dominant role in the creation of excited species of nitrogen in the plasma jets. As can be seen in Table 16.1, excited argon neutrals (E_B: ~13 eV) have more than enough energy to excite either state of N$_2$ first positive (E_B: ~7.4 eV) and N$_2$ second positive (E_B: ~11.1 eV). Excited helium neutrals (E_B: ~23 eV) have more than enough energy to excite any state of N$_2$ first positive, N$_2$ second positive, and N$_2^+$ first negative (E_B: ~18.7 eV). However, the creation of excited species does not follow a simple rule of energy. Some anticipated excited species could not be found by the emission spectra of the luminous gas created by adding N$_2$ into low-pressure cascade arc plasma jets of He, which indicates that a selective excitation of nitrogen species occurred. Only the activated species of N$_2$ second positive (E_B: ~11.1 eV) in argon plasma jets

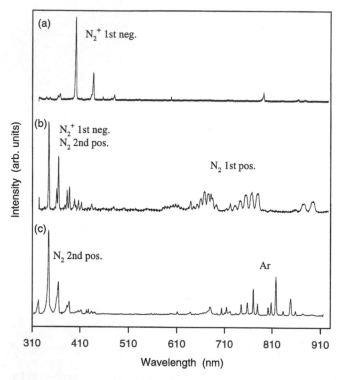

Figure 16.10 Typical emission spectra of 60 sccm N_2 (a) in helium plasma jet, (b) RF plasma of pure N_2, 10 Pa, RF power 30 W, (c) in argon plasma jets.

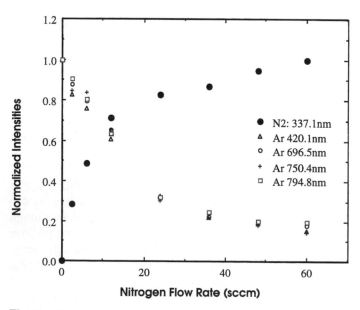

Figure 16.11 The dependence of emission argon lines and nitrogen 2nd pos. band (337.1 nm) emission intensities at an axial position 2.7 cm from jet inlet on the flow rate of nitrogen added to argon plasma jet. Conditions are 2000 sccm argon, 60 sccm nitrogen, 75 Pa, 0.64 kW.

and N_2^+ first negative (E_B: ~18.7 eV) in helium plasma jets were observed, which have the electronic energy levels close to those of excited argon atoms (E_B: ~13 eV) or helium atoms (E_B: ~23 eV), respectively. No excited species of nitrogen with higher or much lower energy levels than that of excited argon or helium neutrals was identified.

In creating reactive nitrogen species in the LPCAT, only species that have close energy match to the excited species of carrier gas occur, e.g., N_2 second positive and excited argon atoms or N_2^+ first negative and excited helium atoms. The selective excitation of nitrogen species was not observed in radio frequency plasmas from either pure nitrogen or its mixture with argon or helium, probably reflecting the contribution of the electron impact ionization in the radio frequency glow discharge of nitrogen.

3.2.2. Oxygen

When oxygen was added to an argon plasma jet at 60 sccm, neither obvious quenching of argon emission nor color change of the plasma jet was observed. The emission due to Ar atoms dominated the spectrum. However, the optical emissions due to excited oxygen atoms at 777.2 nm (E_B: ~10.7 eV) and 844.6 nm (E_B: ~11.0 eV) were clearly observed.

When oxygen was introduced into a helium plasma jet, rapid quenching of helium emission occurred and a very short light-blue flame was formed, and the emission due to O_2^+ first negative system (E_B: ~18.2 eV) appeared in addition to the oxygen atom emission. Due to the obvious quenching, the emissions due to He atoms became very weak. Figure 16.12 shows the presence of oxygen atom emission in both Ar and He plasma jets with addition of oxygen. Figure 16.13 shows

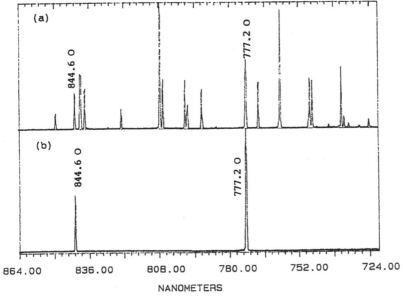

Figure 16.12 The emission from excited oxygen atom; (a) oxygen in Ar LPCAT and (b) oxygen in helium LPCAT.

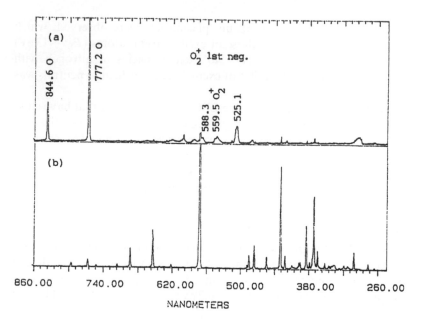

Figure 16.13 The presence of oxygen ion and the quenching of He LPCAT; (a) O_2 added to He LPCAT, (b) He LPCAT (without O_2).

(1) the presence of O_2^+ in He LPCAT and the quenching of He excited species with addition of oxygen, and (2) the He LPCAT without addition of oxygen.

This result can be clearly understood from the different energy levels between the excited argon and helium neutrals, i.e., the excited helium atoms have more than enough energy to ionize oxygen molecules and excite them from the ground state to an excited state, but excited argon atoms do not have enough energy. The reactions

$$He^* + O_2\left(X^3\Sigma_g^-\right) \rightarrow O_2^+\left(b^4\Sigma_g^-\right) + e + He$$
$$O_2^+(b^4\Sigma_g^-) \rightarrow O_2^+(a^4\Pi_u) + h\nu$$

are energetically possible for the production of O_2^+ first negative system species. The generation process for excited oxygen atoms in argon plasma jet can also be clearly described by the following excitation transfer reaction [11]:

$$Ar^* + O_2\left(X^3\Sigma_g^-\right) \rightarrow O^* + O + Ar$$

3.2.3. Air

Since the main components of air are nitrogen and oxygen, the main feature of the emission spectra of air in argon or helium plasma jets is also a typical mixture of nitrogen and oxygen emission spectra in the plasma jets. When air was introduced into the argon plasma jet, a strong pink flame was formed and the main emissions were due to N_2 second positive molecules and excited oxygen atoms. When it was added to a helium plasma jet, the dominant emissions were due to N_2^+ first negative molecular ions and excited oxygen atoms, but no O_2^+ first negative emission was observed.

3.2.4. Hydrogen

When hydrogen gas was introduced into argon plasma jet, all the argon emissions were strongly quenched and a very short and brilliant flame with a light-blue color was formed. The emissions from excited hydrogen atoms (E_B: $\sim$12 eV) and molecules (E_B: $\sim$11.8 eV) were observed. However, when hydrogen gas was added to helium plasma jet, neither the strong quenching effect nor color change was observed. Only very weak emissions from excited hydrogen atoms was detected, and no emission from excited hydrogen molecules was observed, since the energy level of excited helium atoms exceeds the ionization energy for both H atoms (E_i: $\sim$13.6 eV) and H_2 molecules (E_i: $\sim$15.4 eV).

This system forms highly ionized so-called Penning mixtures [12,13]. The higher excited states of H_2^+ are partly stable and partly unstable, depending on the quantum numbers of the electron present. The stable excited states have, however, only very shallow minima of the potential curves [14]. That is the reason why no spectrum of H_2^+ is observed for the helium plasma jet. The argon excited neutrals, on the other hand, cannot ionize hydrogen atoms or molecules, but could produce excited H_2 molecules, which can be detected by optical emission spectroscopy.

3.2.5. Ammonia

When ammonia gas was introduced into an argon plasma jet at a flow rate of 60 sccm, all argon emission lines disappeared, and a very short but brilliant light-blue flame was formed. A very strong NH emission band was observed. In the ammonia radio frequency plasma, some very weak N_2 emission bands due to N_2 second positive appeared, but in the ammonia flame formed in an argon plasma jet no emission related to N_2 species was observed.

3.2.6. Nitrogen and Hydrogen Mixture

Figures 16.14 and 16.15 show the emission spectra of an argon plasma jet with the addition of a nitrogen and hydrogen at different compositions. The N_2 emission due to N_2 second positive intensities decreased greatly, even with a small amount of hydrogen. At a higher hydrogen composition in the mixture, very weak emission bands due to N_2 second positive were observed. Quite strong emissions due to N_2 first positive appeared with different levels of hydrogen in the gas mixture. Adding hydrogen to the system can limit the selective excitation of nitrogen species in low-pressure cascade arc plasmas, though the mechanism is not clear.

3.3. Addition of Polymer-Forming (Material-Forming) Gas

The creation of chemically reactive species from polymer-forming gas (monomer) in plasma jet of LPCAT follows the same principle described for non-polymer-forming gases, but the major reaction is molecular dissociation of monomer by energy transfer mechanism. Upon addition of monomers to the argon luminous gas jet, the emissions of argon luminous gas are highly quenched. The dominant features

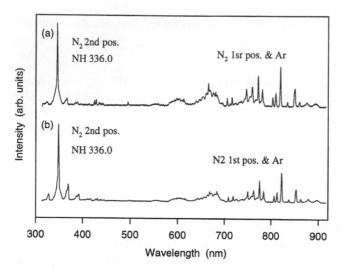

Figure 16.14 he optical emission spectra of argon plasma jets with addition of nitrogen and hydrogen mixture; (a) 10 sccm nitrogen and 10 sccm hydrogen, (b) 60 sccm nitrogen and 2.7 sccm hydrogen, 2000 sccm argon, 0.64 kW, and 75 Pa.

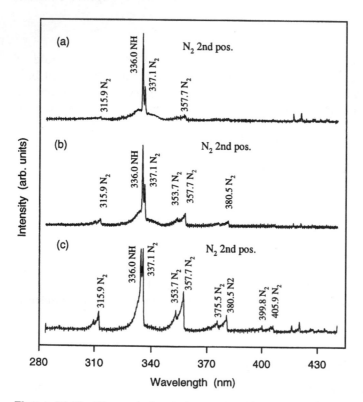

Figure 16.15 The optical emission spectra of argon plasma jets (a) with addition of 10 sccm nitrogen and 10 sccm hydrogen, (b) with addition of 60 sccm nitrogen and 2.7 sccm hydrogen, and (c) with pure nitrogen addition, 60 sccm nitrogen. The other conditions are 2000 sccm argon, 0.64 kW, and 75 Pa.

in the emission spectra are due to excited species corresponding to the relevant monomers, with the strong argon lines remaining visible. Photon emission aspects of some organic molecules are described below.

3.3.1. CH_4 or CH_3OH

The spectral features of OES spectra of low-temperature cascade arc plasmas of CH_4 and CH_3OH are identical for both of these systems. Emission bands of CH radicals and emission lines of H atoms constitute the dominant emission characteristics of OES spectra for both plasma systems. The CH emission ($A^2\Delta \rightarrow X^2\Pi$ at 431 nm) intensities are almost the same for both CH_4 and CH_3OH plasmas, and follow similar trends with the change of arc currents.

If one assumes that the densities of the electronically excited states, from which the observed emissions originate, are directly proportional to those of the ground state, the emission intensity profiles reflect approximately the species concentrations, especially at constant pressure. Therefore, the above results may indicate that, under the same operating conditions, almost the same quantity of CH radicals exist in CH_4 and CH_3OH cascade arc plasmas. From the OES spectra assignment, CH radicals are the only reactive species that could be attributed to the growth of plasma polymers in both CH_4 and CH_3OH cascade arc plasmas.

If one speculates only based on OES data, CH_4 and CH_3OH cascade arc plasmas should give the similar depositions at similar deposition rates. Contrary to speculation, the deposition rates of CH_3OH in LPCAT are less than one-tenth of that for CH_4, and the properties of depositions are appreciably different. Furthermore, the results of deposition rates indicated that there existed the inverse correlation between OES signals and the deposition rate. This implies that photon-emitting species observed are not necessarily the actual polymerizable species. OES cannot detect the poisoning effect of oxygen atom in methanol on the growth mechanisms of LCVD (see Chapter 5), which has a significant effect on the deposition rate, simply because those chemical reactions occur via species that do not emit photons.

3.3.2. CF_4 or C_2F_4

Clear-cut demonstration for the case in which polymerizable species that do not emit photons are created exclusively by photon-emitting species via molecular dissociation by energy transfer principle was found with addition of CF_4 or C_2F_4 to Ar LPCAT. The finding implies that the photon emission per se is not the essential indicator of chemical reaction pertinent to plasma polymerization, while the photon emission is the essence of luminous gas phase. The CF_4 and C_2F_4 plasmas are the well-investigated fluorinated carbon plasma systems by OES diagnostics according to investigations with conventional plasma sources. However, without the influence of electron impact ionization in LPCAT, these cases clearly show that the formation of chemically reactive species occurs primarily by the energy transfer mechanism that does not necessarily involve photon-emitting species.

The OES spectra of low-temperature cascade arc plasmas for carrier gas Ar only and with CF_4 and C_2F_4 addition were identical, although quenching of the flame occurred when the second gas was added, as shown in Figure 4.8. Low-temperature cascade arc plasmas of CF_4 and C_2F_4 did not show any additional peaks or continuums compared to the Ar plasmas, like CF_2^+ or the CF lines that were found in the OES spectra for radio frequency plasmas of $(Ar + CF_4)$ and $(Ar + C_2F_4)$, as seen in Figure 4.11. None of the species previously reported for CF_4 radio frequency plasma, i.e., CF_2 and CF bands, as well as fluorine atom emissions, exists in the OES spectra of CF_4 or C_2F_4 LPCAT [15]. The addition of CF_4 or C_2F_4 quenched the cascade arc torch flame, indicating that excess energy is transferred to the second gas, but did not create reactive species of the added gas that emits photons identifiable by OES.

XPS data of substrate polymer exposed to the quenched plasma jet, on the other hand, showed that the low-temperature cascade arc torch treatment yielded just as good, if not better, fluorination of PET fibers as radio frequency plasma treatment with $(Ar + CF_4)$ and $(Ar + C_2F_4)$ [15,16]. These examples clearly demonstrate that the main polymerizable species in plasma polymerization (free radicals) are not photon-emitting species in most cases. This is in accordance with the growth and deposition mechanism based on free radicals, which account for the presence of large number of dangling bonds in most plasma polymers. The radio frequency discharge of a mixture of Ar and CF_4 or C_2F_4 shows F-containing species, indicating that the route to create polymerizable species in LPCAT is different from that in radio frequency plasma due to the absence of electron or ion impact ionization of monomer. However, the deposition characteristics are nearly identical, indicating that the influence of electron impact ionization of organic molecules on LCVD is minimal. The formation and the dissipation of luminous gas phase of organic vapor by excited Ar neutrals in the cascade arc polymerization is described Chapter 4.

4. DEPOSITION IN CASCADE ARC TORCH

4.1. Activation of Monomer by Luminous Gas

In comparison with conventional electrical discharge processes, LPCAT is a very different process in that its activation of carrier gas and the creation of polymerizable species by the activated carrier gas are temporally and spatially separated. When discharge power is applied to the cascade arc generator, the plasma of carrier gas (usually argon) is produced in the cascade arc column and the luminous gas phase is blown into a vacuum chamber where monomers are introduced. The deactivation of the reactive species, some of which lead to the creation of polymerizable species in the luminous gas phase, occurs within the relatively narrow beam of an argon luminous gas jet. The higher the flow rate of Ar, the narrower is the beam and the longer the luminous gas flame.

In such a system, energy input into the cascade arc is responsible only for the formation of the luminous gas phase of argon, and the generation of the reactive species of monomers that is introduced into the luminous gas jet of Ar in the expansion chamber is caused by the argon luminous gas phase. Therefore, the power input into the cascade arc cannot be used as the energy input parameter for LCVD that takes place only in the second expansion chamber.

The amount of energy applied to the monomers in the second chamber depends on how much energy is carried out from the cascade arc column by argon luminous gas phase. This feature of CAT LCVD is very different from conventional electrical discharge processes in which the electrical energy input is applied directly to the monomers. Therefore, a different controlling factor specific to CAT LCVD should be identified.

4.2. Two Steps in Low-Pressure Cascade Arc Torch

Adapting the principles of deposition kinetics in LCVD described for conventional plasma polymerization in Chapter 8, the deposition kinetics of CAT-LCVD can be described in the following manner.

4.2.1. The First Step of Generating Luminous Gas Phase of a Carrier Gas

The power input applied to the cascade arc column only generates argon luminous gas phase (plasma) inside the arc column. Since the argon luminous gas jet is the direct energy source for the activation of monomers, it is appropriate to consider that the amount of energy brought out by argon luminous gas phase from the cascade arc column is the true energy applied for the CAT LCVD process in the deposition chamber. In this first step, the electrical power input is applied to the carrier gas (e.g., argon) inside the arc column to start a discharge and generate energy carriers (mainly excited neutrals).

The relationship between the mass density of excited species of carrier gas and energy input to the cascade generator is described in Eq. (16.5), and its validity is confirmed by Figure 16.7. The argon emission intensity showed a linear dependence on the combined parameter, $[W(FM)_c]$. The total energy applied to the monomers in the CAT-LCVD process can be expressed numerically with this combined experimental parameter, $[W(FM)_c]$.

4.2.2. The Second Step of Creating Reactive Species of Monomer

Since the carrier gas plasma is the direct energy source for the activation of monomers, the concentration of energy carrier $[EnergyCarrier]$ represents the amount of energy available to activate plasma polymerization in the cascade arc torch. Similar to Eq. (8.3), as defined in conventional glow discharge polymerization, LPCAT polymerization can be described by the following expressions by replacing W (for general glow discharge polymerization) by $[EnergyCarrier]$ in Eq. (8.3).

$$Normalized(DR) = \alpha_2 \frac{[EnergyCarrier]}{[FM]_m} \tag{16.6}$$

By virtue of Eq. (16.5),

$$Normalized(DR) = \alpha_2 \frac{\alpha_1 W[FM]_c}{[FM]_m} \tag{16.7}$$

where α_2 is unitless proportionality constant. Thus:

$$Normalized(DR) = k' \frac{W[FM]_c}{[FM]_m} \tag{16.8}$$

In equations, subscript "m" represents monomer, and the other experimental parameters are the same as described earlier.

4.3.   Normalized Energy Input Parameter of LPCAT LCVD

The term $W[FM]_c/[FM]_m$ is an energy input factor (J/s), the quantity of which can be considered as the energy, transported by carrier gas luminous gas phase, applied to per mass unit of monomers in unit time.

The plotting deposition rates of different monomers with different feed rates in *Normalized*(DR) versus $W\,(FM)_c/(FM)_m$ coordinates would provide the most objective comparison of their tendencies with regard to deposition in CAT LCVD. In LCVD, it is very important that the polymerization (material formation) is "atomic" rather than "molecular" processes, implying that the depositing entities are fragmented species of the original monomer molecule. Therefore, the deposition rate in LPCAT polymerization is determined largely by the type of atoms contained in the monomer structure rather than by molecular structures.

The dependence of normalized deposition rate of butane at two flow rates as a function of the energy input parameter $[W(FM)_c/(FM)_m]$ is shown in Figure 16.16. The general deposition characteristics of LPCAT polymerization are depicted in Figure 16.17, in which different monomers, different flow rates, and different electrical power inputs are all pooled together. These figures show that (1) the deposition rate in cascade arc torch can be well represented by Eq. (16.8), and (2) the deposition characteristics of silicon-containing monomers and hydrocarbon monomers are significantly different.

The silicon-containing monomers exhibited a much higher deposition yield than hydrocarbon monomers. The wider scattering among various Si-containing organic molecules is due to the fact that each molecule contains a different amount of C-based moieties. It can be estimated from this figure that Si-based moieties polymerizes nearly six times faster than C-based moieties. Since there is a significant difference in the deposition behaviors of Si-C compounds and hydrocarbons, these results also indicate that the type of atoms contained in the monomer structure plays an important role in plasma polymerization in LPCAT, just as in the case of LCVD in other modes of electrical glow discharge.

4.4.   Deposition Profile

Due to the relatively narrow beam of plasma jet, the deposition occurs in a relatively narrow area [18]. Deposition rate profiles as a function of axial position are shown in Figure 16.18. Experimental conditions are 8.00 A arc current, 2000 sccm argon, 10.0 sccm methane, and 560 mtorr (75 Pa). The centerline deposition rate decreases sharply and the deposition rate at the shoulders of the profile increases with increased axial position. The profile broadening can be attributed to the diffusion of polymerizable species from the jet axis (expansion of the cross-sectional area of plasma jet). The total mass deposition rate over a circle of 50 mm radius is given in Table 16.5, along with the conversion of methane to plasma polymer.

Although the distribution profile broadens as the axial distance increases and the center line deposition rate decreases, the total deposition rate on a

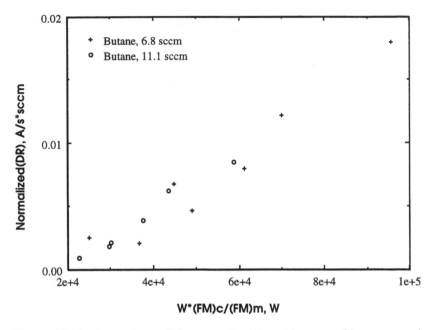

Figure 16.16 Dependence of the normalized deposition rate of butane cascade arc plasmas on the parameter $W^*(FM)_c/(FM)_m$. The deposition rates were obtained at an axial position of 27.5 cm from the luminous gas jet inlet.

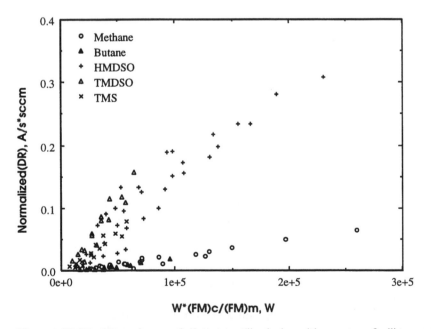

Figure 16.17 Dependence of the normalized deposition rate of silicones and hydrocarbon monomers on the parameter $W^*(FM)_c/(FM)_m$ in cascade arc torch polymerization. The deposition rates were obtained at an axial position of 27.5 cm from the luminous gas jet inlet.

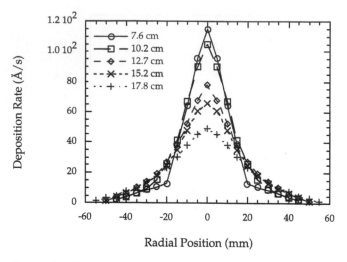

Figure 16.18 Variation of the radial deposition rate distribution profile with axial position, conditions: 8.0 A, 2000 sccm argon, 10.0 sccm methane, and 560 mtorr.

Table 16.5 Total Mass Deposition Rates and Percent Conversion of Methane to Plasma Polymer at Various Axial Positions

Axial position (cm)	Total rate ($\mu g/cm^2 s$)	Conversion (%)
7.6	9.9	10.2
10.2	11.2	11.5
12.7	11.9	12.3
15.2	11.3	11.7
17.8	11.3	11.7

[a]Conditions are 8.00 A, 2000 sccm argon, 10.0 sccm methane, and 560 mtorr.

plane perpendicular to the jet axis remains nearly constant within the distance examined. This implies that the distance from the nozzle, beyond a certain distance, is not a critical factor in practical deposition process because the velocity of gas is so high that the variation of the distance is insignificant in the practical deposition process. On the other hand, the narrow distribution with little broadening indicates that there is little interaction between chemically reactive species in the luminous gas jet stream. This means that the kinetic pathlength of growth mechanism is very short and the building block of depositing material is small.

LPCAT polymerization or coating could be considered more or less the same as the plasma polymerization or coating by other conventional plasma processes, except that the kinetic pathlength of growth is short. The ultrathin layers prepared by LPCAT polymerization have the general characteristics of plasma polymers, i.e., amorphous (noncrystalline), high concentration of the dangling bonds (free radicals trapped in immobile solid phase), and the high degree of the internal stress in the layer, which is described in Chapter 11.

5. CHARACTERISTIC FEATURES OF LPCAT PROCESSING

The major effects of LPCAT treatment as well as LPCAT polymerization coating are by and large the same as conventional glow discharge treatment and plasma polymerization coating. However, the processing of LPCAT treatment or coating is significantly different in that it is a spray processing that requires the relative motion of the substrate with respect to the torch flame for the substrate larger than the diameter of LPCAT jet. Although the local treatment or deposition rate is high, it does not translate directly to the high overall processing rate depending on size and shape of the substrate.

The process requires nearly three orders of magnitude higher flow rate of the carrier gas than typical flow rates of monomer or a mixture of monomer and a carrier gas in conventional glow discharge processes under low pressure. The high consumption of carrier gas might necessitate the inclusion of carrier gas recovery system in an industrial scale operation. Therefore, the process is not an alternative means to carry out conventional plasma processes, and adaptation of the process should be done with careful identification of the specific goal that cannot be attained by other conventional plasma processes. The major features of the process that could distinguish LPCAT processes from other conventional plasma processes are as follows:

1. The mode of creation of polymerizable or chemically reactive species, i.e., chemically reactive species, is the interaction with excited neutrals of the carrier gas (no bombardment of electrons and ions).
2. The kinetic pathlength is short due to the high one-directional transport velocity of polymerizable species.

Whether any of these characteristic features is an advantage or a disadvantage entirely depends on the objectives to be accomplished by the process. However, these two characteristic features indicate that LPCAT process is better suited for the surface treatment by excited neutrals of Ar than the surface coating.

5.1. LCVD by LPCAT

The feature that too high one-directional gas flow hinders interactions of gaseous species seems to be responsible for very poor adhesion and corrosion protection characteristics of LPCAT coatings of TMS applied on metal surfaces, which were significantly inferior (virtually no adhesion and corrosion protection) to the superior adhesion and corrosion protection attained by DC cathodic polymerization of TMS. The very high one-directional transport velocity hinders the interaction between polymerizable species in the luminous gas phase before they reach the target surface. The velocity of gaseous species leaving the nozzle of the cascade arc generator is estimated to be in the supersonic range, and the beam of excited species reaches the substrate surface without allowing the reactive species to interact each other.

The short kinetic pathlength means that smaller or oligomeric species deposit on the substrate, which is similar to the formation of powders in gas phase. Consequently, the barrier characteristics and the adhesion of the coating become poorer than could be obtained by conventional plasma polymerization with a higher kinetic pathlength. On the other hand, the same feature could provide a unique advantage when larger species should be avoided, e.g., in the case of deposition onto the surface of nanoparticles, which do not have large enough surfaces to accommodate larger species.

5.2. Surface Modification by Ar LPCAT

The first feature that chemically reactive species are created by the interaction of molecules (in gas phase or on the surface) with excited neutral species of Ar has a very significant influence in the surface modification of polymers. When a polymer surface is exposed to Ar or O_2 plasma, the energetic ions and electrons (at the level of ionization energy) bombard the surface. The influence on the surface is determined by the energy level. The chemical bonds involved in the molecules that constitute the surface are relatively low (3–4 eV) compared to the ionization energy of the gas used in plasma (over 10 eV).

The cession of a σ bond yields two free radicals. The free radical on the surface subsequently reacts with ambient oxygen when the treated substrate is exposed to air rendering the surface hydrophilic (in the case of Ar plasma). The high energy of impinging entities (electrons and/or ions) tends to yield excessive cession of bonds, which creates a weak boundary layer slightly below the top surface. This situation could be visualized by the trends that plasma treatment of hydrophobic surface makes the surface paintable, but the paint does not adhere well because it could delaminate through the weak boundary sublayer.

In Ar LPCAT treatments, the excited neutrals of Ar, instead of Ar^+, interact with polymer molecules at the surface. The energy level of those excited species are nearly as high as those for ions and electrons, i.e., E_B values of excited neutrals are close to the ionization of Ar, E_i; however, the interaction is through the energy transfer process in which the energy matching principle plays an important role, i.e., the whole E_B is not utilized. Consequently, the indiscriminate bombardment of highly energetic species does not take place in LPCAT treatment, which leads to a much less damaging surface treatment of polymers. Figure 16.19 schematically depicts the creation and absence of the weak boundary layer.

Ar LPCAT treatment of thermoplastic olefins (used as the fascia of automobile bumper) yields much more durable and stable adhesion of paint applied on the treated surface than conventional plasma treatment processes could render [19]. Ar LPCAT treatment of fibrous polypropylene reinforcing materials for concrete also showed the much improved energy absorbing characteristics of the fiber-reinforced concrete by the same principle [20]. TMS coating on zirconium oxides, which are used as the radiopaque pigments in polymethyl methacrylate bone cement recipe, create bonding between the pigments and the polymer matrix and improves the fatigue life of the bone cement [21]. These aspects are described in some details in Part IV.

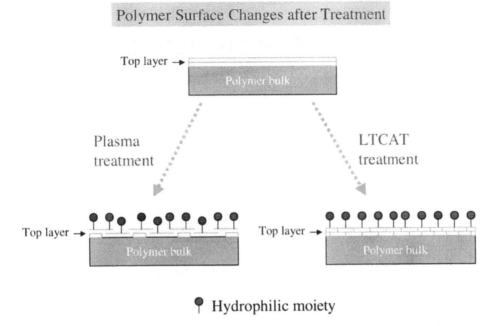

Figure 16.19 Comparison of the net effects obtainable by plasma treatment and LTCAT treatment.

5.3. Surface Modification of Particulate Matters

The very high one-directional transport velocity of gas in LPCAT becomes a very unique advantage in the surface treatment of submicrometer or nanoparticles. As the particle size decreases to submicrometer or nanometer range, particles tend to aggregate strongly and the primary particles are hardly observed. One of the major reasons for the surface modification is the prevention of aggregation of primary particles. If surface modification is attempted on the aggregated particles, the treatment or coating tends to stabilize the aggregates, which defeats the purpose of applying the surface modification.

The ideal surface modification of powders is the process that breaks down the aggregates and modifies the surface of the primary particles simultaneously. LPCAT treatment could be very close to this ideal situation. The supersonic velocity of reactive species breaks up the existing aggregates to a significant level, if not completely, allowing the chemically reactive species to interact with the primary particle surface.

In such a process, particles are dropped into the horizontal LPCAT jet a few centimeters away from the nozzle, and blown to the downstream of the expansion chamber and separated from the gas stream. Figure 16.20 depicts the principle of the LPCAT powder treatment process. Figure 16.21 depicts the schematics of the system, and Figure 16.22 shows a pictorial view of the system. Such a system could treat a large amount of powders in relatively short time, e.g., 10 kg of powders could be treated in less than 1 h, while conventional plasma treatment of powder generally

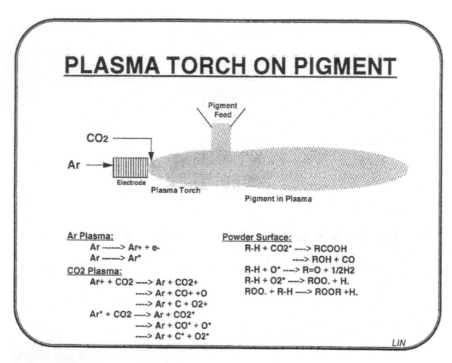

Figure 16.20   Schematic illustration of CO_2 LPCAT treatment of fine particles.

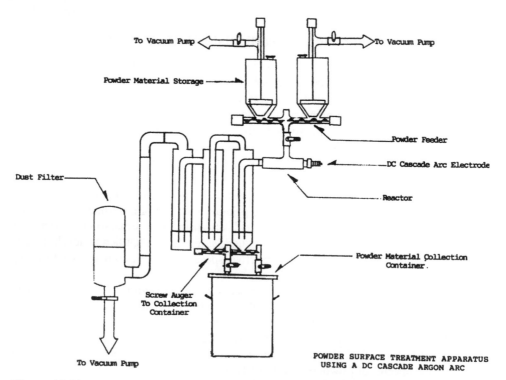

Figure 16.21   Schematic of a LPCAT powder treatment system.

Figure 16.22 A pictorial view of a LPCAT powder treatment system.

requires more than 1 day to treat the same quantity. The rate-determining step of LPCAT treatment of powders is the rate of powder feeding, whereas that in conventional plasma treatment is the mixing of powder to enhance the exposure of powder surface to the luminous gas phase, which is generally slow, ineffective, and uncertain. Thus, LPCAT treatment has a very unique advantage in powder surface modifications.

REFERENCES

1. Kroesen, G.M.W. *Ph.D. thesis*, Eindhoven University of Technology, Eindhoven, The Netherlands, 1988.
2. Beulens, J.J. *Ph.D. thesis*, Eindhoven University of Technology, Eindhoven, The Netherlands, 1992.
3. Beulens, J.J.; Kroesen, G.M.W.; Schram, D.C.; Timmermans, C.J.; Crouzen, P.C.N.; Vasmel, H.; Schuurmans, H.J.A.; Beijer, C.B.; Werner, J. J. Appl. Polym. Sci.: Appl. Polym. Symp. **1990**, *46*, 527.
4. Kroesen, G.M.W.; Schram, D.C.; van de Sande, M.J.F. Plasma Chem. Plasma Process. **1990**, *10*, 49.
5. De Graaf, M.J.; Dahiya, R.P.; Jauberteau, J.L.; De Hoog, F.J.; van de Sande, M.J.F.; Schram, D.C. Colloq. Phys. **1990**, *51*, C5-387–C5-392.
6. Buuron, A.J.M.; Otorbaev, D.K.; van de Sanden, M.C.M.; Schram, D.C. Phys. Rev. E **1994**, *50*, 1383.
7. Maecker, H. Z. Naturforsh **1956**, *11a*, 457.
8. Fusselman, S.P.; Yasuda, H. Plasma Chem. Plasma Proc. **1994**, *14*, 251.
9. Yu, Q.S.; Yasuda, H.K. Plasma Chem Plasma Proc **1998**, *18* (4), 461.

10. Yu, Q.S.; Yasuda, H.K. J. Polym. Sci., Polym. Cehm. Ed. **1999**, *37*, 967.
11. Ricard, A. In *Plasma-Surface Interactions and Processing of Material*; Auciello, O., Ed.; Kluwer Academic: The Netherlands, 1990, Chap. 1.
12. Acton, J.R.; Swift, J.D. *Cold Cathode Discharge Tubes*; Academic Press: New York, 1963; 224–227.
13. Krogh, O.; Wicker, T.; Chapman, B. J. Vac. Sci. Technol. **1986**, *44*(3), 1796.
14. Herzberg, G. *Molecular Spectra and Molecular Structure, I. Spectra of DiatomicMolecules*; Van Nostrand Reinhold: New York, 1950; 359–367.
15. Krentsel, E.; Fusselman, S.P.; Yasuda, H. J. Polym. Sci.: Polym. Chem. Ed. **1994**.
16. Yasuda, T.; Okuno, T.; Miyama, M.; Yasuda, H. J. Polym. Sci. A Polym. Chem. **1994**, *32*, 1829.
17. Fusselman, S.P.; Yasuda, H.K. Plasma Chem. Plasma Proc **1994**, *14* (3), 277.
18. Lin, Y.-S.; Yasuda, H.K. J. Appl. Polym. Sci. **1998**, *67*, 855.
19. Zhang, C.; Gopalaratnam, V.S.; Yasuda, H.K. J. Appl. Polym Sci., **2000**, *76*, 1985.
20. Kim, H.Y.; Yasuda, H.K. Appl Biomater **1999**, *48*, 135.

17

Anode Magnetron Torch

1. ANODE MAGNETRON TORCH

As described in Chapter 15, an anode magnetron (against a plain cathode) creates a funnel-shaped negative glow, and electrons near the surface of the anode are focused on the center part. In such an arrangement, Ar is fed into the reactor. In the cascade arc torch described in Chapter 16, the luminous gas created in the cascade arc generator is blown into the reaction chamber, in which electrons and ions do not have the major role; i.e., low-pressure cascade arc torch (LPCAT) creates a jet stream of excited Ar neutrals. If one feeds Ar only from a small hole created in the center of an anode magnetron, it might be possible to focus the negative glow in DC discharge and the ion bombardment on the cathode while maintaining the overall electric power input relatively low.

Probably the most important aspect that has not been examined so far is the capability of the process to scan a larger surface area without creating a specific reactor for substrates with specific size and shape. One potentially important aspect of anode magnetron torch (AMT) is the capability to scan a larger surface area without creating a large surface anode to treat differently sized substrates (used as the cathode of DC discharge).

An anode magnetron system, which generates a funnel-shaped negative glow, can yield a relatively uniform sputtering rate on the cathode surface. However, for a very large substrate surface treatment, if we use a similarly large electrode system, a high input power is required to keep the desired current density. Such a high level of input energy is limited in actual application. In a large system it would also be difficult to maintain the uniformity in the sputtering rate. Consequently, the experimental parameters found in a laboratory scale experiment cannot be directly applied to a large-scale reactor.

If AMT glow discharge could be effectively confined in a local area, the AMT plasma system could be considered as a method to treat a large substrate surface via a scanning process at a low level of input power without being hampered by the edge effect of Ar DC discharge described in Chapter 15. Furthermore, the operating parameters used in a laboratory scale experiment could be directly applied in a large substrate surface treatment or possibly coating, which has not been explored so far.

2. STRUCTURE OF ANODE MAGNETRON TORCH

The AMT plasma system is an asymmetrical electrode system (small anode and large cathode), which has the capability to overcome the problems mentioned above [1]. Figure 17.1 depicts the structure of an anode magnetron electrode, which is essentially a magnetron electrode with a gas inlet in the center. A schematic diagram of the anode magnetron in the AMT plasma system is shown in Figure 17.2. An electrode with magnetic field enhancement is placed snugly inside a glass tube. However, its position within the tube can be easily varied. The distance between the end of the glass tube to cathode surface (gap distance), a, and the distance between the anode surface to the cathode surface (electrode distance), b, can be varied independently. The kidney-shaped open space in front of the cathode represents the cathode dark space, which is represented by the cathode dark space thickness (CDST). CDST e1 represents the minimal thickness, and DCST e2 represents the thickness at the inner diameter of the glass tube. The dashed space represents the luminous gas phase created by the anode magnetron discharge of Ar. The carrier gas, Ar, is fed through the center hole of the electrode. The glass tube provides physical confinement of the gas flow, as well as support and insulation of the anode. This anode system is called the anode magnetron torch (AMT). If an anode system has no magnetic enhancement, it is simply called an anode torch (AT). A schematic diagram of the entire reactor system is shown in Figure 17.3.

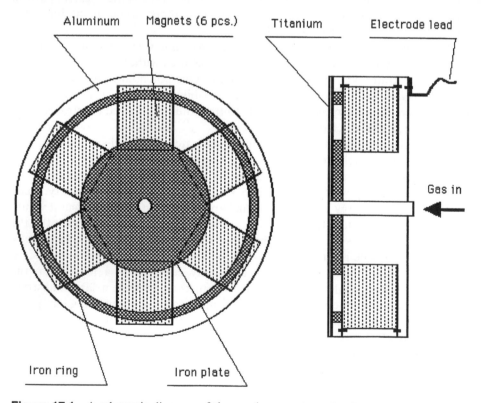

Figure 17.1 A schematic diagram of the anode magnetron structure.

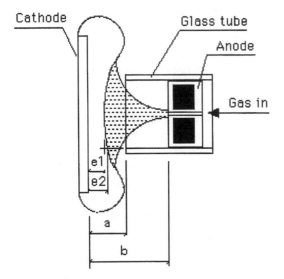

Figure 17.2 A schematic diagram of the anode magnetron torch (AMT).

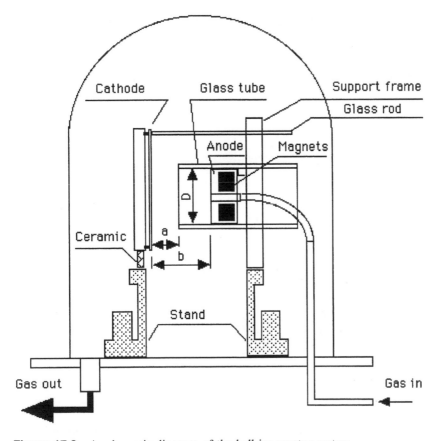

Figure 17.3 A schematic diagram of the bell-jar reactor system.

The glass tube is fixed on a stand by an aluminum support frame. A glass rod is used to fix the distance between the electrodes, as they may be pulled each other and attract together under the strong magnetic field, if a cold-rolled steel (CRS) panel is used as the cathode. The gas is fed through a small hole placed in the center of the anode. The cathode is made of a $7 \times 7\,in^2$ CRS plate. The ratio of cathode surface area to anode surface area is 4.5:1.

It is well known that the color of a thin film deposited on a light-reflecting surface corresponds to its thickness. If a CRS panel is coated with trimethylsilane (TMS) using a plasma technique, the significant color changes—light brown, brown, purple, dark blue, light blue, yellow—will appear on the surface respectively as the TMS coating thickness increases. Therefore, various TMS coating thicknesses can be differentiated by the color changes. In order to determine the influence of gap distance and electrode distance on AMT glow discharge sputtering characteristics, argon plasma was used to sputter CRS, which had been coated with TMS and showed a dark blue color. In such an experiment, if the color resulting from the argon plasma treatment is in the brown or purple ranges, it indicates that the sputtering process is predominant. On the other hand, if the resulting color is in the light blue or yellow ranges, the deposition process is predominant.

3. DISTRIBUTION OF ANODE MAGNETIC FIELD STRENGTH

The spatial distributions of anode magnetic field strength are shown in Figures 17.4–17.7. For strong and weak magnetic field strengths, the maximal values are about 1550 (parallel), 1800 (perpendicular) gauss and 280 (parallel),

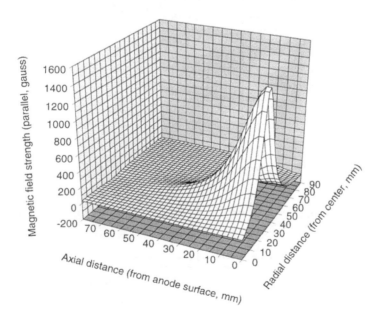

Figure 17.4 The spatial distribution of parallel magnetic field strength employing strong magnets ($B_{//stmax} = 1550\,G$) as measured with a Walker-MG 3D gauss meter.

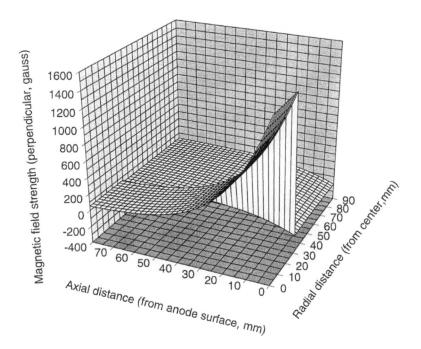

Figure 17.5 The spatial distribution of perpendicular magnetic field strength employing strong magnets ($B\perp_{stmax} = 1800\,G$) as measured with a Walker-MG 3D gauss meter.

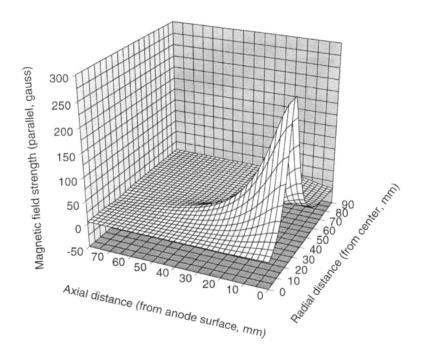

Figure 17.6 The spatial distribution of parallel magnetic field strength employing weak magnets ($B_{//wkmax} = 280\,G$) as measured with a Walker-MG 3D gauss meter.

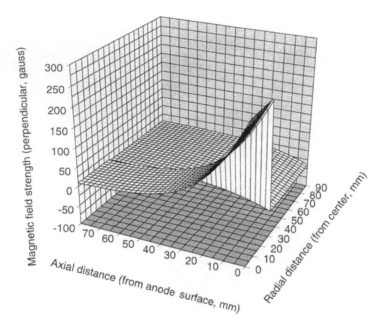

Figure 17.7 The spatial distribution of perpendicular magnetic field strength employing weak magnets ($B\perp_{wkmax} = 310\,G$) as measured with a Walker-MG 3D gauss meter.

310 (perpendicular) gauss, respectively. Using the same anode magnetron configuration, the shapes of the distributions of the magnetic field strength are almost identical, but the magnitudes differ. For those figures, r is defined as the radial distance from the center of the anode surface; z is the axial distance from the surface of the anode. For parallel magnetic field strength, the maximal value appears when r is approximately 3 cm and z is approximately 0 cm. For perpendicular magnetic field strength, the maximal value exists when r is about 2 cm and z is about 0 cm. The parallel magnetic field strength decreases as the axial direction distance increases. In the radial direction, following its increase to the maximum, the parallel magnetic field strength decreases as the radial direction increases. Finally, it approaches zero. The distribution of parallel magnetic field strength is almost uniform on the cathode surface when z is larger than 8.0 cm for the strong magnetic field ($B_{//stmax} = 1550\,G$), or larger than 4.0 cm for the weak magnetic field ($B_{//wkmax} = 280\,G$).

The distribution of magnetic field strength varies little when the cathode is made of materials such as aluminum or titanium. It varies somewhat more when the cathode is made of CRS; however, the degree to which this influence is observable depends on the distance between the anode and the CRS panel. (It was more or less necessary to use CRS because the deposition of TMS on other nonmagnetic metal, such as aluminum alloy, does not yield deep blue color necessary for thickness analysis by the colorimetric principle.)

The influence of the CRS substrate (cathode) on the distribution of magnetic field is shown in Figures 17.8–17.10. The presence of a CRS plate can create an inductive magnetic field, which partially offsets the original magnetic field effect in

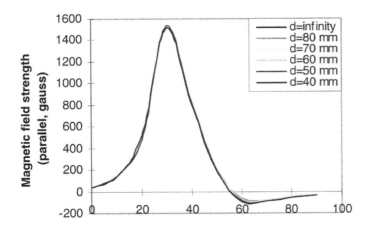

Figure 17.8 The dependence of parallel magnetic field strength distribution on electrode distance at $z = 0.0$ mm ($B_{//\text{stmax}} = 1550$ G, d is the distance between the anode surface and the cathode surface, z is the distance from the anode surface to the measurement location, cathode material is a 7×7 in^2. cold-rolled steel plate).

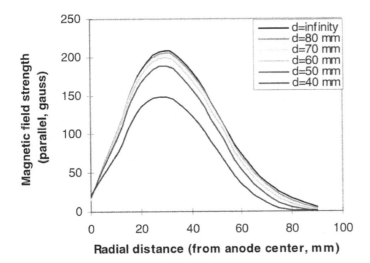

Figure 17.9 The dependence of parallel magnetic field strength distribution on electrode distance at $z = 30$ mm ($B_{//\text{stmax}} = 1550$ G, d is the distance between the anode surface and the cathode surface, z is the distance from the anode surface to the measurement location, cathode material is a 7×7 in^2. cold-rolled steel plate).

the vicinity of the CRS plate. The influence of the anode magnetron in the vicinity of the cathode attenuates as the distance between the two electrodes increases. The effect of cathode material can be neglected for large electrode distances (8.0 cm for $B_{//\text{stmax}} = 1550$ G and 4.0 cm for $B_{//\text{wkmax}} = 280$ G).

4. A COMPARISON OF AT AND AMT GLOW DISCHARGE

The anode magnetron depicted in Figure 17.1 can be operated without magnet. Such a mode of discharge is termed anode torch (AT). A schematic representation of AT glow discharge is shown in Figure 17.11. The leading edge of the cathode dark space is flat in AT glow discharge. If the system pressure is lower than 30 mtorr,

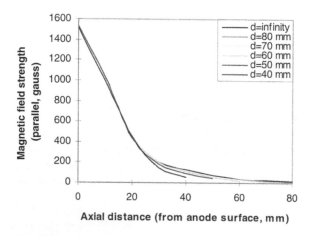

Figure 17.10   The dependence of parallel magnetic field strength distribution on electrode distance at $r = 30$ mm ($B_{//stmax} = 1550$ G, d is the distance between the anode surface and the cathode surface, r is the radial distance from the center axis of the anode to the measurement location, cathode material is a 7×7 in^2. cold-rolled steel plate).

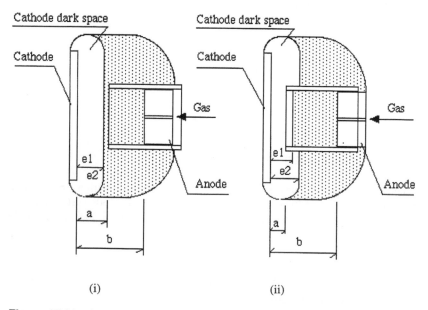

Figure 17.11   A schematic diagram of the shape of AT glow discharge; the shaded area is the glow zone, (i) large gap distance and (ii) small gap distance.

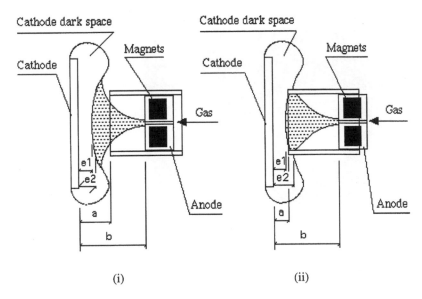

Figure 17.12 A schematic diagram of the shape of AMT glow discharge; the shaded area is the glow zone, (i) large gap distance and (ii) small gap distance.

it is difficult to establish a stable glow discharge. There is little change in the shape of the cathode dark space when the gap distance is varied. However, the leading edge of the cathode dark space in the focused area is pushed closer to the cathode surface with the decrease of the gap distance. Outside of the glass tube, the CDST varies little with changes in the gap distance.

A schematic diagram of AMT glow discharge is shown in Figure 17.12. In contrast to the case of AT glow discharge, the cathode dark space in AMT glow discharge is kidney shaped, which is typical of negative glow in an anode magnetron DC glow discharge. The CDST in the center region, e1, is less than that at the edge region, e2. In other words, the negative glow is pushed closer to the cathode surface at the center of the focused area. The negative glow is funnel shaped and extends from the center of the anode toward the cathode. The Faraday dark space forms two hemispherical dark spaces in the cross-sectional view. The shape of the glow discharge is similar to that of conventional anode magnetron glow discharge. The funnel-shaped negative glow still exists even when system pressure is very low ($p = 10\,\text{mtorr}$), at which discharge cannot be created by AT.

5. EFFECTS OF PHYSICAL PARAMETERS IN AMT GLOW DISCHARGE

Due to physical confinement, the relationship between the gap distance and the CDST becomes an important factor when large substrate surfaces are to be treated via the AMT technique. When Ar plasma was used to treat a TMS plasma polymer–coated CRS panel dark blue in color, the resulting multicolored

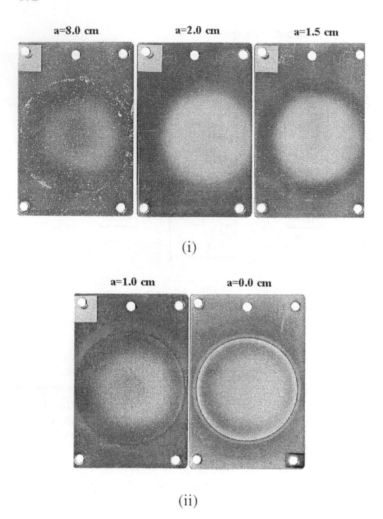

a=8.0 cm   a=2.0 cm   a=1.5 cm

(i)

a=1.0 cm   a=0.0 cm

(ii)

Figure 17.13 The dependence of Ar sputtering effects on gap distance; $B_{//\text{stmax}} = 1550\,\text{G}$, $b = 8.0\,\text{cm}$, $w = 10\,\text{watts}$, $p = 50\,\text{mtorr}$, sputtering time (i) 20 min, (ii) 10 min.

pattern on the CRS panel showed that the materials removed by sputtering may redeposit on the substrate surface outside of the edge of the glass tube as shown in Figure 17.13. When the value of a, defined as the gap distance between the cathode surface and the edge of the glass tube, is greater than the value of the e2, the multicolored pattern on the treated CRS panel does not show any light blue or yellow color. This indicates there has been no redeposition on the substrate surface, if $a > e2$. However, when the value of a is smaller than the value of the CDST e2, (i.e., $a < e2$) the light blue color will appear outside of the edge of the glass tube, indicating that redeposition has occurred.

Therefore, whether or not materials removed by sputtering redeposit on the substrate (cathode) surface depends on the relationship between the gap distance and the CDST e2. If the gap distance is equal to or greater than the CDST e2, materials removed by sputtering will not be redeposited on the substrate surface. Otherwise, the redeposition of such materials occurs outside of the edge of the glass tube. The value of e2 varies with experimental factors, as described below, but is roughly 15 mm. Therefore, if the gap distance is kept larger than 15 mm, no redeposition of sputtered materials has been observed.

Application of the AMT technique in dealing with large substrate surfaces could be accomplished using a scanning process at a low level of input power. If redeposition on the substrate surface occurs, the cleaned surface can be contaminated again during the AMT scanning process. The actual AMT scanning process must maintain a gap distance equal to or greater than the CDST e2. Therefore, the CDST plays an important role in the AMT process.

The distance between electrodes, b, is a vital parameter that can affect the luminosity distribution of the negative glow in the AMT glow discharge. In AT glow discharge, the luminosity distribution of the negative glow does not vary with changes in the distance between electrodes. However, in AMT glow discharge, the luminosity distribution of negative glow depends on b. A schematic diagram of the luminosity distribution of AMT glow discharge with a small electrode distance is shown in Figure 17.14. The tail width of negative glow expands with the decrease of b. The luminosity distribution of the funnel-shaped negative glow is relatively uniform when b is large (more than 7.0 cm under $B_{//max} = 1550$ G); it becomes nonuniform when b is small. The luminosity at the center of the funnel-shaped negative glow is less than that at the edge of the glow area. In other words, the luminosity at the center of the funnel-shaped negative glow diminishes as the electrode distance decreases.

When argon gas was used to sputter a colored CRS panel at different electrode distances using the AMT plasma technique, the multicolored profile shown

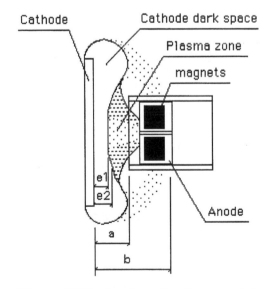

Figure 17.14 A schematic diagram of the shape of AMT glow discharge with a small electrode distance (< 6.0 cm for $B_{//stmax} = 1550$ G); the degree of shading in the glow zone indicates the level of brightness, i.e., the center of the glow zone is not as bright as the edges.

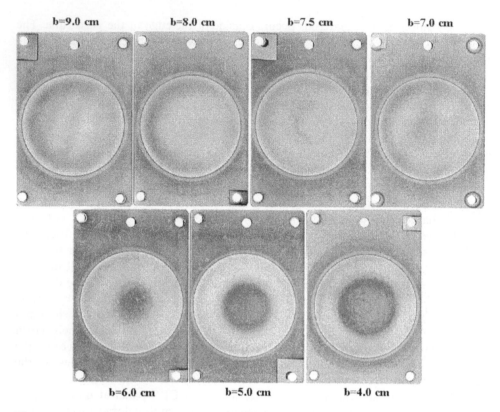

b=9.0 cm   b=8.0 cm   b=7.5 cm   b=7.0 cm

b=6.0 cm   b=5.0 cm   b=4.0 cm

Figure 17.15   The dependence of sputtering effects on electrode distance; $B_{//\text{stmax}} = 1550\,\text{G}$, $a = 0.0\,\text{cm}$, $w = 10\,\text{watts}$, $p = 50\,\text{mtorr}$, sputtering time $= 10\,\text{min}$.

in Figure 17.15 appeared on the CRS panel. When b is small, the big blue eye (unsputtered area) appears at the center of focused area, and the luminosity of AMT negative glow becomes nonuniform. The diameter of the blue eye expands as b is decreased, and the luminosity of the AMT negative glow becomes more nonuniform. Conversely, the blue eye disappears as b is increased, and the luminosity of AMT negative glow becomes uniform. The luminosity of the glow discharge is the result of the excitation and ionization collisions of electrons and Ar. In AMT plasma treatment, creation of a uniform AMT negative glow is essential to achieve a uniform sputtering rate distribution in the focused area.

6.   INFLUENCE OF VARIOUS PARAMETERS ON THE CDST IN AMT GLOW DISCHARGE

The influence of input power on the CDST is shown in Figure 17.16. The CDST decreases with the increase of input power. System pressure also influences the CDST in an AMT plasma system. The CDST expands with the decrease of system pressure. These two features are general characteristics of DC glow discharge, although the thickness of dark space has not been quantitatively examined.

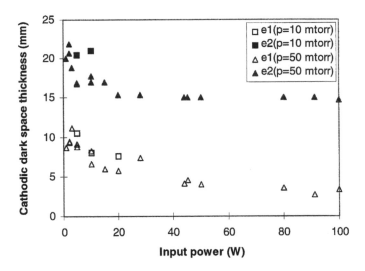

Figure 17.16 The influence of input power on the cathodic dark space thickness; $B_{//stmax} = 1550\,G$, $a = 1.5\,cm$, $b = 8.0\,cm$, Ar mass flow rate = 1 sccm.

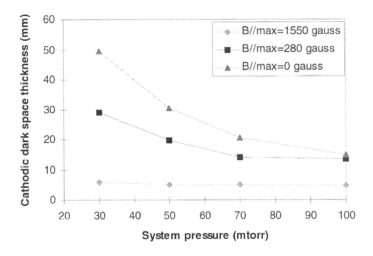

Figure 17.17 The influence of magnetic field strength on the cathodic dark space thickness; input power = 50 W, $a = 1.5\,cm$, $b = 8.0\,cm$, Ar mass flow rate = 1 sccm.

The magnetic field strength is a vital parameter in the AMT plasma system. The influence of magnetic field strength on the CDST in the AMT system is shown in Figure 17.17. The CDST e1 shrinks with the increase of magnetic field strength, and CDST is nearly independent of the system pressure with a 1550 G magnetron.

Figure 17.18 depicts the influence of gap distance on the cathodic dark space thickness. CDST e1 become slightly smaller while the CDST e2 expands with a decrease of the gap distance in the AMT glow discharge. Since a constant Ar is fed

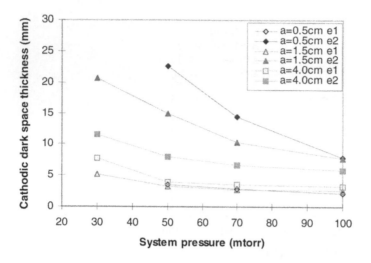

Figure 17.18 The influence of gap distance on the cathodic dark space thickness; $B_{//\text{stmax}} = 1550\,\text{G}$, input power $= 50\,\text{W}$, $b = 8.0\,\text{cm}$, Ar mass flow rate $= 1$ sccm.

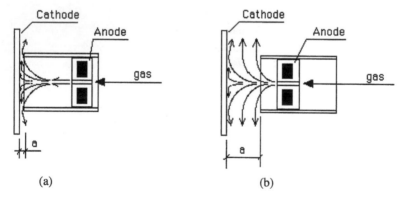

Figure 17.19 A schematic diagram of the influence of the gap distance on the gas stream; (a) small gap distance, (b) large gap distance.

into the AMT system, the change of gap distance changes the gas flow pattern as depicted in Figure 17.19. At a very small gap distance, even the pressure would become higher in the confined space. Consequently, the influence of gap distance has the same trends found with the change of system pressure.

In AMT glow discharge, the influence of the electrode distance, b, on the CDST is complicated because changes in electrode distance cause not only a change in the magnetic field strength near the cathode surface, but also a change in the distribution of the magnetic field strength. Such changes result in variance in the luminosity of AMT negative glow. Figure 17.20 depicts the effect of electrode distance on cathodic dark space thickness for nonmagnetron and two magnetrons. Figure 17.21 shows the influence of electrode distance on the dependence of cathodic dark space distance on system pressure.

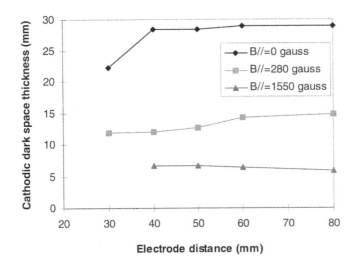

Figure 17.20 The influence of electrode distance on the cathodic dark space thickness; input power $= 10\,\mathrm{W}$, $a = 0.0\,\mathrm{cm}$, Ar mass flow rate $= 1\,\mathrm{sccm}$.

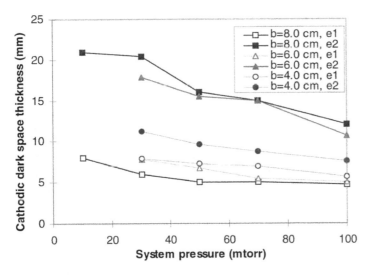

Figure 17.21 The influence of electrode distance on the cathodic dark space thickness; $B_{//\mathrm{stmax}} = 1550\,\mathrm{G}$, input power $= 10\,\mathrm{W}$, $a = 1.5\,\mathrm{cm}$, Ar mass flow rate $= 1\,\mathrm{sccm}$.

7. INFLUENCE OF OPERATING PARAMETERS ON SPUTTER CLEANING

7.1. Influence of Physical Parameters of AMT

In the AMT Ar sputtering process, gap distance is an important parameter. It influences the pathway of gas or sputtered particles (molecules or atoms). Sputtering rate will change when the gap distance is varied. Figure 17.22 shows the relationship

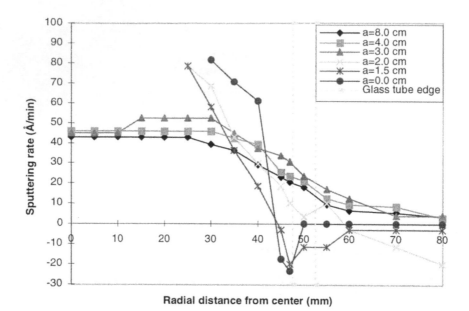

Figure 17.22 The dependence of the sputtering rate distribution on gap distance ($B_{//\mathrm{max}} = 1550\,\mathrm{G}$, $b = 8.0\,\mathrm{cm}$, Power $= 10\,\mathrm{w}$, $p = 10\,\mathrm{mtorr}$, Ar flow rate $= 1.0\,\mathrm{sccm}$).

between sputtering rate distribution and gap distance under a low system pressure ($p = 10\,\mathrm{mtorr}$). The distribution of sputtering rate according to radius is not uniform. When the gap distance decreases, the total sputtering rate increases in the focused area. Thus, gap distance plays an important role in achieving a high sputtering rate. At the experimental conditions shown in the figure, the sputtering rate distribution at the center is over $70\,\text{Å}/\mathrm{min}$ while a is between zero and $2.0\,\mathrm{cm}$. When a is greater than $2.0\,\mathrm{cm}$, the sputtering rate falls quickly. If a is greater than $4.0\,\mathrm{cm}$ (from $4.0\,\mathrm{cm}$ to $8.0\,\mathrm{cm}$), the sputtering rate hardly changes with variance of the gap distance.

In the absence of physical confinement ($a = b$), the sputtered particles move and diffuse without any physical barrier to the downstream because the electrode distance b is much higher than the mean free path λ of the sputtered particles. However, when the gap distance decreases, the influence of physical confinement ($b - a$) becomes significant because the path of sputtered particles becomes restricted. When a is smaller than e2, which is a function of mean free path λ, many sputtered particles cannot pass easily through the gap and deposit outside of the edge of the glass tube, where lower Ar ions bombard the surface (Figure 17.19). Therefore, redeposition occurs when a is smaller than e2.

Figure 17.23 shows the influence of gap distance on sputtering rate under higher system pressure ($p = 50\,\mathrm{mtorr}$). The results display almost the same trend as seen in Figure 17.22, aside from the magnitude of the Ar sputtering rate. A conspicuous sharp peak in sputtering rate near the wall of the glass tube appears with a small gap distance, which is observed under low pressure as a gradual shift of sputtering rate profile. The redeposition outside of the glass tube is also evident.

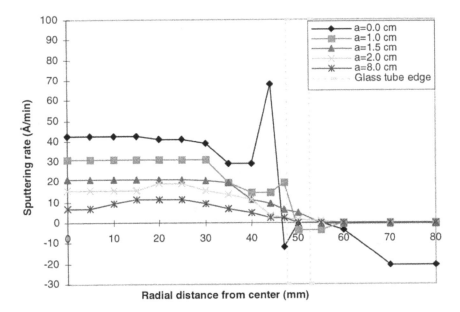

Figure 17.23 The dependence of the sputtering rate distribution on gap distance; $B_{//\text{max}} = 1550\,\text{G}$, $b = 8.0\,\text{cm}$, Power $= 10\,\text{w}$, $p = 50\,\text{mtorr}$, Ar flow rate $= 1.0\,\text{sccm}$.

When $a = 0$, the system is essentially a closed system, although the sealing is not good and gas leaks to the outside of the glass tube. Data made under such a condition still provide some fundamental aspects of sputtering, although the operation under such a condition has no relevance to practical use of AMT. Figures 17.24–17.26 depict the relationship between sputtering rate and the electrode distance under total physical confinement (gap distance $a = 0\,\text{cm}$). Figure 17.24 shows the case with a strong magnetic field ($B_{//\text{max}} = 1550\,\text{G}$). With the gap distance less than 6.0 cm, the materials sputtered near the edge of the physical confinement wall deposit in the center part. This trend decreases as the gap distance is increased. With the gap distance above 7 cm, the radial distribution of sputtering rate is uniform and nearly independent of the gap distance.

Figure 17.25 shows the relationship between electrode distance and sputtering rate distribution under AMT treatment with a weak anode magnetron ($B_{//\text{max}} = 280\,\text{G}$). The sputtering rate drops as compared to that in Figure 17.24. The Ar sputtering rate is uniform in the focused area when b is greater than 4.0 cm. This uniform sputtering rate is highest at $b = 4.0\,\text{cm}$ and then decreases with the increase of electrode distance. This differs from the sputtering rate distribution under AMT with a strong anode magnetron ($B_{//\text{max}} = 1550\,\text{G}$).

Figure 17.26 shows the influence of electrode distance on the sputtering rate without anode magnetron enhancement ($B_{//\text{max}} = 0\,\text{G}$). In this case, the sputtering rate changes little with variance of the electrode distance.

7.2. Influence of Magnetic Field Strength

The effect of magnetic field strength is shown in Figures 17.28 and 17.29 under input power 10 W and system pressure 50 mtorr and a and b as variable parameters.

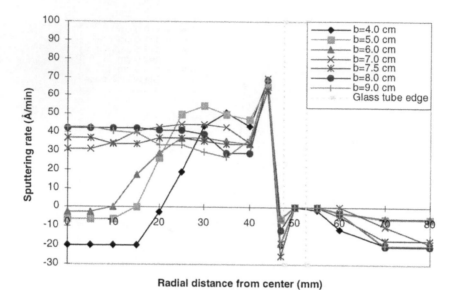

Figure 17.24 The dependence of the sputtering rate distribution on electrode distance; $B_{//max} = 1550\,G$, $a = 0.0\,cm$, Power $= 10\,w$, $p = 50\,mtorr$, Ar flow rate $= 1.0\,sccm$.

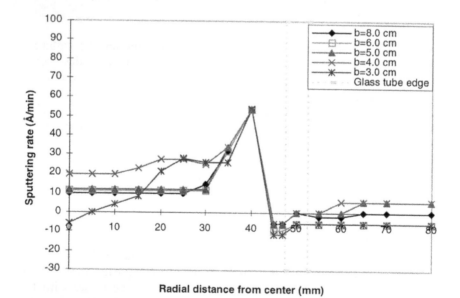

Figure 17.25 The dependence of the sputtering rate distribution on electrode distance; $B_{//max} = 280\,G$, $a = 0.0\,cm$, Power $= 10\,w$, $p = 50\,mtorr$, Ar flow rate $= 1.0\,sccm$.

Figure 17.27 shows that high magnetic field strength achieves a high and uniform sputtering rate if the electrode distance is large. A large electrode distance can maintain a uniform distribution of magnetic field strength near the substrate surface. However, if the electrode distance is small, such as $b = 4\,cm$ in Figures 17.28 and 17.29, the Ar sputtering rate distribution is no longer uniform.

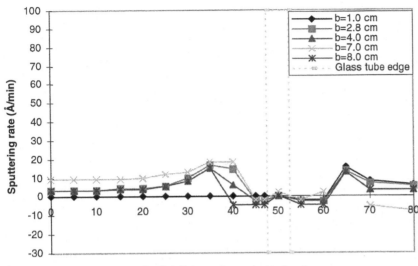

Figure 17.26 The dependence of the sputtering rate distribution on electrode distance; $B_{//\text{max}} = 0$ G, $a = 0.0$ cm, Power $= 10$ w, $p = 50$ mtorr, Ar flow rate $= 1.0$ sccm.

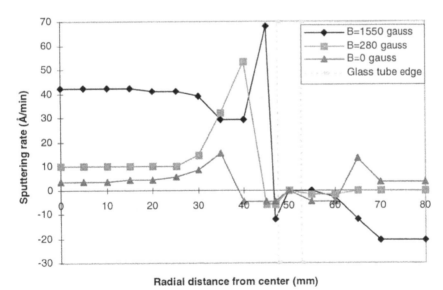

Figure 17.27 The dependence of the sputtering rate distribution on magnetic field strength; $a = 0.0$ cm, $b = 8.0$ cm, input power $= 10$ w, $p = 50$ mtorr.

This is because the distribution of magnetic field strength created by strong anode magnetron enhancement ($B_{//\text{max}} = 1550$ G) is not uniform near the cathode surface. For weak anode magnetron enhancement ($B_{//\text{max}} = 280$ G), the sputtering rate distribution still remains uniform in comparison to $b = 8$ cm due to a uniform magnetic field distribution near the substrate surface. Without anode magnetron

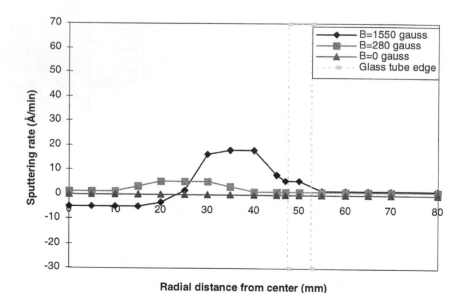

Figure 17.28 The dependence of the sputtering rate distribution on magnetic field strength; $a = 4.0\,cm$, $b = 4.0\,cm$, input power $= 10\,w$, $p = 50\,mtorr$.

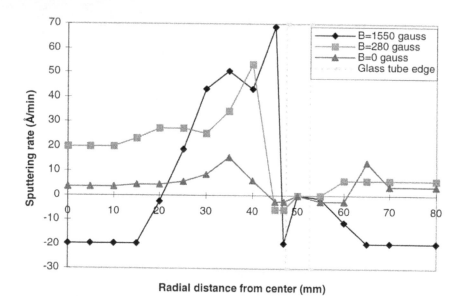

Figure 17.29 The dependence of the sputtering rate distribution on magnetic field strength; $a = 0.0\,cm$, $b = 4.0\,cm$, input power $= 10\,w$, $p = 50\,mtorr$.

enhancement $(B_{//\text{max}} = 0\,G)$, the sputtering rate is always very low no matter how other parameters change.

A strong magnetic field can focus or confine electrons more effectively, increasing the sputtering rate. Nevertheless, a uniform magnetic field strength near

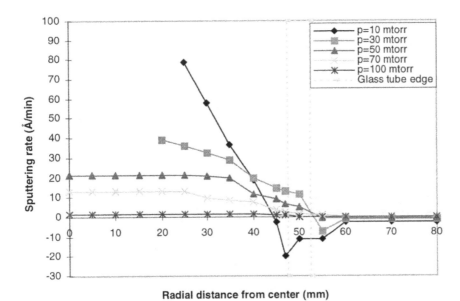

Figure 17.30 The dependence of the sputtering rate distribution on system pressure; $B_{//\max} = 1550\,\text{G}$, $a = 1.5\,\text{cm}$, $b = 8.0\,\text{cm}$, Power $= 10\,\text{w}$, Ar flow rate $= 1.0\,\text{sccm}$.

the substrate surface has an equally important role in the AMT plasma system. A uniform magnetic field strength can maintain a uniform distribution of electrons.

7.3. Influence of System Pressure

Figure 17.30 shows the relationship between Ar sputtering rate and system pressure with a strong anode magnetron. It shows that the Ar sputtering rate distribution is not uniform with respect to radial distance from the center. The Ar sputtering rate increases in the focused area with the decrease of system pressure. In the focused area, the sputtering rate at $p = 10\,\text{mtorr}$ is almost four times greater than that at $p = 50\,\text{mtorr}$. However, near the edge of the glass tube, when the system pressure is lowered, the Ar sputtering rate can decrease below zero, i.e., deposition occurs. This is because the CDST e2, which increases with the decrease of system pressure [1], becomes larger than the gap distance causing the redeposition of sputtered particles, as explained earlier.

Figure 17.31 shows the influence of system pressure under both a weak anode magnetron ($B_{//\max} = 280\,\text{G}$) and no anode magnetron ($B_{//\max} = 0\,\text{G}$). The redeposition of sputtered materials occurs at $p = 50\,\text{mtorr}$ because the gap distance, a, is smaller than the CDST e2.

7.4. Influence of Input Power

Figure 17.32 shows the influence of input power on the Ar sputtering rate distribution. The sputtering rate increases with the increase of input power, but the distribution of the Ar sputtering rate is not uniform with respect to radial distance. Sputtering rate is a function of the energy and the number of ions that bombard the

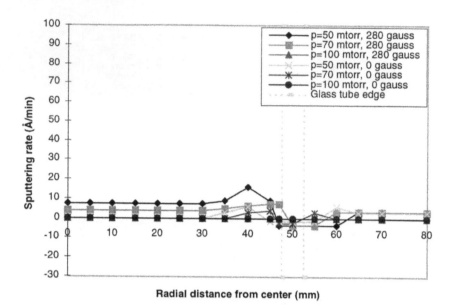

Figure 17.31 The dependence of the sputtering rate distribution on system pressure; $B_{//max} = 280\,G$ or 0 gauss, $a = 1.5\,cm$, $b = 8.0\,cm$, Power $= 10\,w$, Ar flow rate $= 1.0\,sccm$.

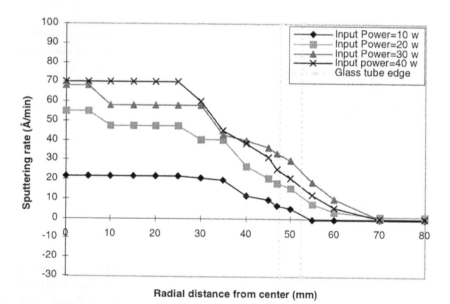

Figure 17.32 The dependence of the sputtering rate distribution on input power; $B_{//max} = 1550\,G$, $a = 1.5\,cm$, $b = 8.0\,cm$, $p = 50\,mtorr$, Ar flow rate $= 1.0\,sccm$.

cathodic surface. High input power increases the number of high-energy electrons, raising the probability of ionization collisions between electrons and gas atoms. Thus, high energy input increases the number of ions, but the energy of ions is determined by the ionization energy of Ar.

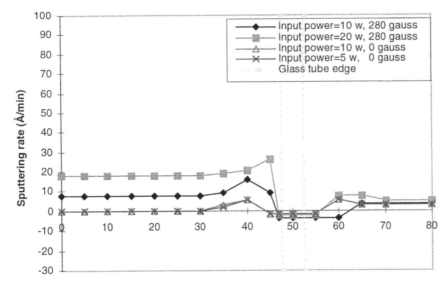

Figure 17.33 The dependence of the sputtering rate distribution on input power; $B_{//max} = 280\,\text{G}$ or 0 gauss, $a = 1.5\,\text{cm}$, $b = 8.0\,\text{cm}$, $p = 50\,\text{mtorr}$, Ar flow rate $= 1.0\,\text{sccm}$.

Figure 17.33 shows the influence of input power on the Ar sputtering rate under both the weak anode magnetron ($B_{//max} = 280\,\text{G}$) and no anode magnetron ($B_{//max} = 0\,\text{G}$). A trend similar to that evident in Figure 17.31 is shown. In both cases, the sputtered materials redeposit near the edge of the glass tube because the CDST e2 is larger than the gap distance ($a = 1.5\,\text{cm}$) under either of these conditions. The input power cannot be increased as high as that under a strong anode magnetron. Thus, one of the merits of the AMT system over plasma treatment systems without the influence of an anode magnetron is the ability to achieve a higher input power.

7.5. Influence of Gas Flow Rate

Figure 17.34 shows the influence of Ar gas flow rate (at fixed pressure) on the Ar sputtering rate distribution in the AMT plasma system. The sputtering rate distribution is almost uniform from the center to $r = 3.0\,\text{cm}$ at a low argon flow rate ($V = 1\,\text{sccm}$). With the increase of Ar gas flow rate, the sputtering rate increases at the center but hardly changes when r is greater than 2.5 cm. If r is larger than 2.5 cm, the sputtering rate distribution is basically unchanged.

7.6. Influence of Cathode Surface Area

The main purpose of developing the AMT system is to treat large substrate surfaces via a scanning process. In order to investigate the effectiveness of AMT plasma treatment of large substrate surfaces, large CRS panels of 17.8 cm × 17.8 cm and 25.4 cm × 25.4 cm were used as the substrate. The experimental results are shown in Figures 17.35 and 17.36.

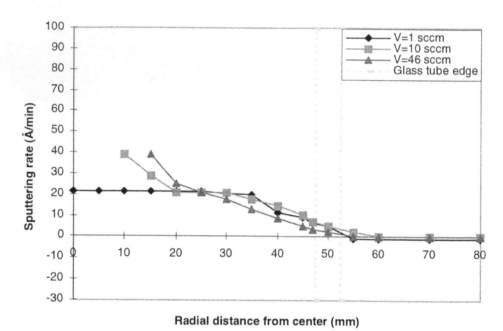

Figure 17.34 The dependence of the sputtering rate distribution on gas flow rate; $B_{//max} = 1550\,\text{G}$, $a = 1.5\,\text{cm}$, $b = 8.0\,\text{cm}$, Power $= 10\,\text{w}$, $p = 50\,\text{mtorr}$.

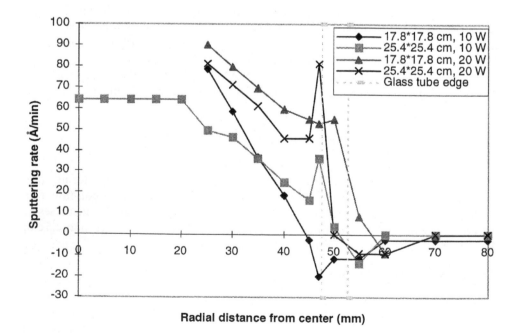

Figure 17.35 The dependence of the sputtering rate distribution on cathode surface area; $B_{//stmax} = 1550\,\text{G}$, $a = 1.5\,\text{cm}$, $b = 8.0\,\text{cm}$, $p = 10\,\text{mtorr}$.

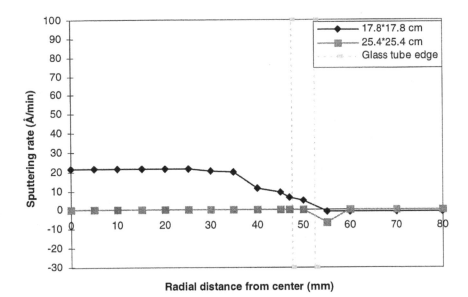

Figure 17.36 The dependence of the sputtering rate distribution on cathode surface area; $B_{//stmax} = 1550\,G$, $a = 1.5\,cm$, $b = 8.0\,cm$, $p = 50\,mtorr$.

The sputtering rate does not change much with the increase of cathode surface area at low system pressure ($p = 10\,mtorr$). The plasma treatment is localized and effectively confined. However, at a relatively high system pressure ($p = 50\,mtorr$), the sputtering rate is almost zero for the larger cathode. This case reiterates the importance of system pressure in AMT Ar sputtering treatment.

REFERENCES

1. Zhao, J.G.; Yasuda, H.J. Vac. Sci., Tech., **1999**, A *17* (6).
2. Zhao, J.G.; Yasuda, H.K. J. Vac. Sci., Tech., **2000**, A *18* (1).

18
Primary, Secondary and Pulsed Discharges

1. PRIMARY PLASMA AND SECONDARY PLASMA

The material formation in luminous chemical vapor deposition (LCVD) occurs mainly in the luminous gas phase (glow). However, the profile of deposition extends beyond the glow region depending on the chemical reactivity of species and operational conditions, although the deposition rate declines by orders of magnitude. In LCVT (treatment of surface by luminous gas without depositing material) such as O_2 glow discharge treatment of a polymer surface, on the other hand, the effect of chemical change of the surface (surface modification) extends far beyond the glow region depending on the lifetime of excited species and could penetrate porous substrate to a significant extent. In other words, the effect of LCVT is not limited to the top surface exposed to the luminous gas phase. The effect of LCVT could be attributed to the following major mechanisms: (1) photoirradiation and (2) chemical reactions of reactive or excited species.

The photoirradiation, particularly of ultraviolet (UV) to vacuum UV, could penetrate deep into the substrate exposed depending on the opacity of the material, and the net effect is often more damaging than beneficial. The reactive or excited species quench when they react with the surface, implying that the effect is limited to the top surface exposed. Thus, the most effective surface modification by the second mechanism occurs within a fraction of a second, but this very advantageous feature of the process makes the practical operation of LCVT more difficult. The prolonged treatment does not improve the surface modification by the chemically reactive species and the damaging effect of photoirradiation overwhelms the surface modification effect. Because of the undesirable irradiation effect, the surface modified by glow discharge treatment often lacks the durability needed for the surface modification.

In order to cope with this problem, many commercially available plasma surface modification equipments employ the mode of discharge often called *secondary plasma* or *remote plasma*. The term "secondary plasma" is misleading and incorrect in the strict sense of "plasma" in physics. The term *primary plasma* is only used in contrast to secondary plasma and refers to the space filled with luminous gas created by glow discharge. The term secondary plasma could be interpreted as the use of the secondary effect of plasma in the nonglow region of glow discharge. Likewise, the term "remote plasma" refers to the influence of plasma in the remote area where the luminous gas phase has quenched.

The glow region of a glow discharge, at least a part of it, could be considered as the nonequilibrium plasma state, but the nonglow region is not in the plasma state by any definition. The nonglow region of a reactor could contain many chemical species created in the glow region depending on the type of gas employed, which could be used to chemically modify the surface placed in the nonglow region of a reactor. In attempts to use chemical species created in the primary plasma (direct exposure to the luminous gas phase) without exposing the substrate to the primary plasma, a mode of surface modification process described by terms such as remote plasma, secondary plasma, and so forth is used.

The potential use of the secondary plasma is closely tied to the chemical reactivity of the gas used. Oxygen glow discharge, which does not form polymer, can be used in both the primary and the secondary plasmas. In this case, the use of the secondary plasma is well justified particularly when the substrate to be treated is sensitive to photoirradiation. On the other hand, in glow discharge of Ar, of which excited species do not exist beyond the glow region, the secondary plasma mode operation could not be used.

In plasma deposition, the secondary plasma does not make sense in most cases because no polymer deposits efficiently in the nonglow region, and the low yield of deposition process put it beyond reasonable consideration. Depending on the reactivity of species (too slow to complete the deposition in the glow region), some strayed deposition occurs in the nonglow region, but it is generally not practical to coat a substrate because of very low yield and the prolonged reaction time necessary. With very reactive species, such as acetylene in plasma, nothing happens in the nonglow region in a practical sense, and only deposition of strayed reactive species occurs with extremely low yield.

2. "SECONDARY PLASMA" REACTOR

It probably is necessary to describe the reactor, in particular the electrode arrangement, in some detail because it is a key factor to understanding primary plasma and secondary plasma. A typical secondary plasma reactor contains six sets of electrode assembly and five spaces for substrate trays (between electrode assemblies). The first two sets of electrodes and a space between them are shown schematically in Figure 18.1. The distance between the hot electrode 1 and the ground electrode 2 (the gap between electrode 1 and electrode 2, and also 3 and 4) is relatively small, i.e., approximately 3 cm. The distance between two sets of electrode assembly (distance between electrodes 2 and 3) is much greater, i.e., 25–30 cm.

In such a reactor, two kinds of electrode pairs exist. One is the pair represented by electrodes 1 and 2. Another is the pair represented by electrodes 2 and 3. Electrodes 1 and 3 are both "hot electrodes"; therefore, there is the possibility that glow discharge could be created between 1 and 2 or between 2 and 3. Glow discharge in such a situation depends on the breakdown voltage of two gas phases, which is dependent on a parameter given by the product of system pressure p and the distance d between two electrodes (see breakdown voltage in Chapters 14 and 15).

Electrodes and Shelves
in a "secondary plasma" reactor

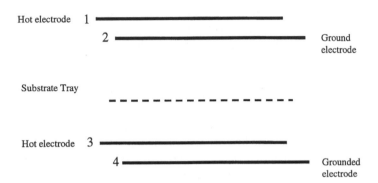

Figure 18.1 Schematic representation of electrodes and substrate tray arrangement.

The secondary plasma created in a "secondary
plasma" reactor at pressure > 200 mtorr

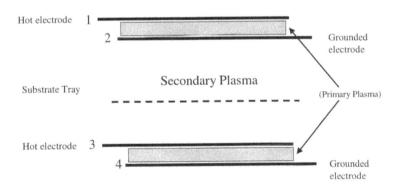

Figure 18.2 Schematic representation of the glow that develops in a secondary plasma reactor.

When a gas, such as oxygen, is introduced into the reactor at a relatively high flow rate to yield a relatively high pressure, e.g., more than 200 mtorr, and electrical power, typically radio frequency, is applied, glow discharge (the primary plasma) is created in the narrow space between two electrodes, i.e., 1 and 2, 3 and 4, etc., because the breakdown voltage is smaller with the smaller separation distance d, and a larger space between two sets of electrode assemblies (where a substrate tray is placed), i.e., 2 and 3, and 4 and 5, etc., remains dark, as depicted in Figure 18.2. The surface treatment in the dark space is expressed by the term "secondary plasma treatment" or "remote plasma treatment." The plasma reactor that has such a set of electrode assemblies and is capable to modify substrate surface in the dark space is often termed "secondary plasma reactor" or "remote plasma reactor."

In the case of oxygen plasma treatment, the chemically reactive species created in the primary plasma diffuse to the dark space and react with the surface of substrate. Since the substrate surface is not directly exposed to plasma, which contains various energetic species such as electrons, ions, excited neutrals, and photons in different energy levels, the damage, which might be caused by these energetic species, could be avoided. Furthermore, the line-of-sight treatment does not occur in the dark space, and the more uniform treatment of complex shape substrate could be obtained. This is the main reason why most plasma treatment equipment employs secondary plasma treatment or remote plasma treatment.

When the system pressure decreases (the mean free path increases) beyond a certain threshold value, which is determined by separation distance (the distance 1–2, or 3–4), discharge power, and surface area of electrodes, glow discharge cannot be created in a narrow gap, and electrons take a longer path, i.e., 1 to wall instead of 1 to 2, and 3 to 2 instead of 3 to 4. In this case, it is important to recognize that the substrate tray is now situated in the glow (the primary plasma). In other words, the mode of plasma process changed from the secondary plasma (for surface modification in the dark space) to the primary plasma exposing substrates directly to the primary plasma. This situation is shown schematically in Figure 18.3.

The primary glow discharge reactor in this configuration could be used for plasma coating in a low-pressure regime, in principle; however, in consideration of flow pattern in such a multiple shelves reactor and electrical field in multiple electrodes, it is difficult to achieve uniform coatings in such a mode of operation.

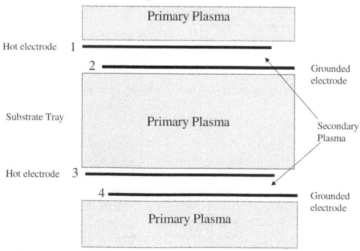

Figure 18.3 Schematic representation of glow develops at low pressure in a secondary plasma reactor.

It is important to recognize that gas used in such a secondary plasma treatment should be non-polymer-forming reactive gas such as H_2, O_2, NH_3, H_2O, CF_4, etc. If polymer-forming vapor or monomer is used, the deposition occurs mainly in the glow and the coating of electrodes occurs quickly. Thus, the secondary plasma reactor under the operational conditions selected for such a reactor cannot achieve an effective deposition of material. The species that escape from the primary glow discharge zone are oligomeric species that quench on contact with substrate surface, and the yield of material deposition is too low for serious consideration.

It is also important to note that the same glow discharge treatment could be achieved in the primary glow if the conditions are tailored for such a surface treatment by a very short exposure, e.g., a few seconds in contrast to a few hours of treatment time in the secondary plasma treatment. However, depending on the size and shape of substrate surface to be treated, a very short treatment is difficult to practice by a batch mode operation. This is where the merit of secondary plasma or remote plasma treatment is recognized.

3. COMPARISON OF IN-GLOW AND OFF-GLOW TREATMENTS

The results of surface modification of polytetrafluroethylene (PTFE) films with Ar radio frequency plasmas as a function of sample position (in-glow and nonglow) and plasma exposure time provide a comparison of the effectiveness of in-glow and nonglow (remote plasma) treatments [1]. Figure 18.4 shows a schematic representation of a tubular reactor indicating the sample positions. Figure 18.5 shows the contact angle change of radio frequency argon plasma–treated PTFE films with sample position and plasma exposure time. The sample position showed in Figure 18.5 is the distance of the sample away from the center of the copper coil

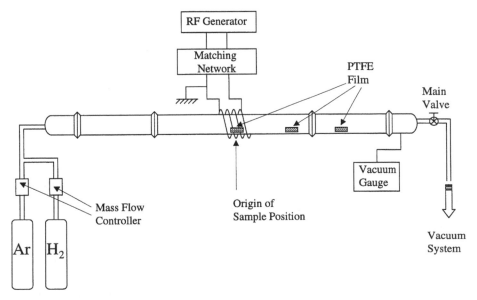

Figure 18.4 Schematic diagram of RF tubular reactor for surface modification of PTFE.

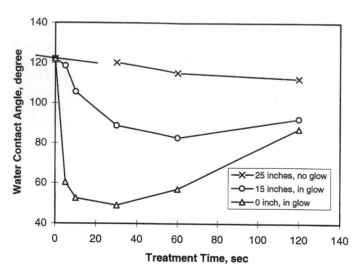

Figure 18.5 Surface contact angle changes of PTFE with exposure time in an argon RF plasma; 2 sccm argon, 100 mtorr, and 7 W RF.

electrode in the downstream of the glow. The sample position beyond the glow region (25 in. in Figure 18.4) gave little treatment effect in lowering water contact angles on the PTFE surface. The most efficient sample position is inside the copper coil, where an effective treatment was achieved on PTFE with a short plasma treatment of 5–10 s. However, a plasma treatment longer than 30 s gave rise to a higher water contact angle on the PTFE surface, which suggested that an overtreatment effect resulted at this position in 30 s.

The surface hydrophilicity achieved on PTFE films by argon radio frequency plasma treatment also showed a stable trend with aging time in air. As shown in Figure 18.6, after radio frequency argon plasma treatment, there is no significant recovery of the water contact angle observed on PTFE (hydrophobic recovery) for about 20 days. The surface modification that rendered the surface of PTFE more hydrophilic is an indirect effect of glow discharge treatment. The free radicals created by the bombardment of excited species of Ar react with ambient oxygen when the treated samples are exposed to air after the treatment. Because the free radicals formed on the PTFE backbone do not cause degradation of the polymer due to the lack of capability to abstract F atom from the neighboring segments (see Chapter 6), Ar glow discharge treatment is a very effective means to make the PTFE surface hydrophilic without damaging the bulk properties of the polymer. However, this process cannot be done in secondary plasma or remote plasma operation as mentioned earlier.

Yamada et al. [2] reported a very effective treatment by the similar remote hydrogen plasma. However, a comparative study of the same process under the same experimental setup reveled that hydrogen radio frequency plasma was not as effective as argon plasma in obtaining PTFE surface hydrophilicity. Figure 18.7 depicts the contact angle of water on hydrogen radio frequency plasma–treated PTFE. Hydrogen glow discharge in remote plasma mode didn't show any effect, and the effect by in-glow treatment is marginal. All experimental data obtained in the

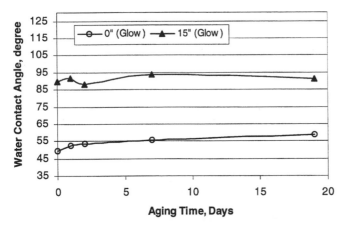

Figure 18.6 Water contact angle changes on argon RF plasma treated PTFE surface with aging time in air; 2 sccm argon, 100 mtorr, 7 W RF, 30 s treatment.

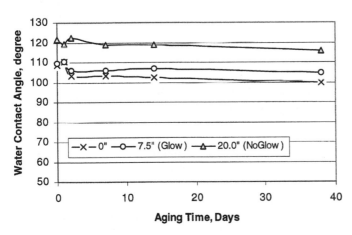

Figure 18.7 Surface contact angle changes of PTFE with exposure time in an hydrogen RF plasma; 10 sccm hydrogen, 100 mtorr, 15 W RF, and 2 min treatment.

study showed that the plasma treatment conducted in the glow region was much more effective in lowering the water contact angles on PTFE surface than that conducted beyond glow region as anticipated from the explanation of secondary plasma or remote plasma described earlier.

Figure 18.8 shows the water contact angle change on plasma-treated PTFE surface by radio frequency plasma of argon and hydrogen mixture. It was noted that, in improving the hydrophilicity of PTFE, the plasma from argon and hydrogen mixture was more effective than hydrogen plasma but less effective than argon plasma. These results also suggested that the excited species of argon have a significant impact on surface modification of PTFE, but not in remote or secondary plasma. Because of high electronegativity, F on Teflon surface can be removed by luminous gas of Ar much more easily than the bond energy of C-F might suggest.

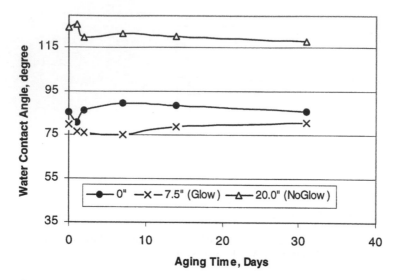

Figure 18.8 Surface contact angle changes of PTFE with exposure time in an RF plasma of argon and hydrogen mixture; 1 sccm argon, 1 sccm hydrogen, 100 mtorr, 7 W RF, and 2 min treatment.

4. IN-GLOW AND OFF-GLOW TREATMENT WITHOUT ION AND ELECTRON BOMBARDMENT

As described in Chapter 4, the luminous gas stream in the low-temperature cascade arc torch (LPCAT) is unique in that it consists of mainly excited neutral species of argon (Ar as the carrier gas), i.e., the influence of electron and ion bombardment is virtually neglegible. Thus, LPCAT argon treatment would provide a unique opportunity to examine the effect of Ar luminous gas on the surface modification of PTFE. The surface water contact angle changes on the surface modified by LPCAT are shown in Figure 18.9. It can be seen that in LPCAT a high argon flow rate over 1500 sccm is necessary to effectively reduce the surface contact angles of PTFE. In LPCAT, the flow rate of argon is proportional to the number of excited species as long as a sufficient power is supplied, as described in Chapter 16. Results shown in Figure 18.9 follow exactly this dependence.

The effect of sample position and plasma exposure time on LPCAT-treated PTFE surfaces are shown in Figure 18.10. When the LPCAT conditions were fixed at 1500 sccm argon and 6.0 A, the luminous torch length was about 12 in. from the torch inlet inside the reactor. In Figure 18.10, therefore, the 14-in. sample position was beyond the torch (off-glow region) and the 9-in. position is inside the torch (in-glow region). The LPCAT treatment of PTFE conducted beyond the torch had very little effect on surface hydrophilicity improvement of PTFE films. This clearly indicates that excited species of Ar do not exist in off-glow region. On the other hand, LPCAT treatment of PTFE conducted inside the torch (9-in. sample position) was very effective in improving surface hydrophilicity of PTFE films. As can be seen from Figure 18.10, a short treatment of 5 s significantly lowered the water contact angle of PTFE from 122 degrees to less than 60 degrees.

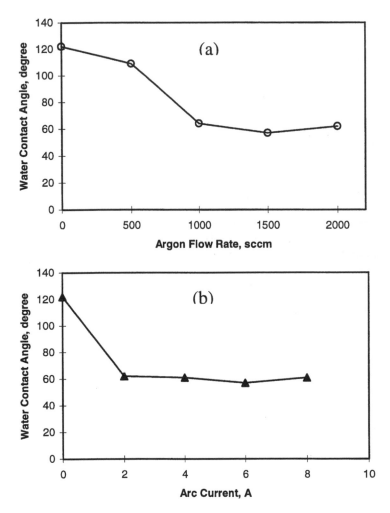

Figure 18.9 Surface contact angle changes of LTCAT Ar treated PTFE with varying (a) argon flow rate at 6.0 A arc current and (b) arc current at 1500 sccm argon for 10 s exposure to a low-temperature cascade arc torch.

Figure 18.11 depicts the hydrophobic recovery characteristics of LPCAT-treated PTFE films with aging time in air. A very stable hydrophilicity was obtained on the PTFE surface after LPCAT argon plasma treatment. As shown in Figure 18.11, little change in water contact angle was observed on LPCAT-treated PTFE surfaces for over 2 months aging in air. This high stability of argon LPCAT–treated PTFE surfaces can be ascribed to the same principle described for radio frequency Ar discharge–treated PTFE samples. Without influence of other energetic species such as ions and electrons, LPCAT Ar treatment provided the best surface modification of PTFE.

In LPCAT operation, the injection of a reactive gas into argon plasma torch can produce new reactive species through the energy transfer from argon neutrals as described in Chapter 16. Figure 18.12 shows the water contact angle changes

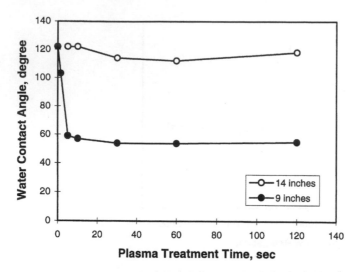

Figure 18.10 Surface contact angle changes of PTFE with exposure time in a low-temperature cascade arc torch at sample positions of 9 in. (in glow) and 14 in. (out of glow) from the torch inlet; 1500 sccm argon, 6.0 A arc current.

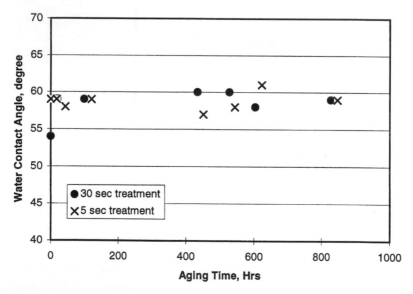

Figure 18.11 Surface contact angle changes of LTCAT treated PTFE film with aging time; 1500 sccm argon, 6.0 A arc current.

of PTFE with hydrogen feed rate and exposure time. It can be seen that the introduction of hydrogen decreased the treatment effectiveness of the argon torch on the PTFE surface. With the addition of hydrogen into the argon torch, as seen from Figure 18.12a, the decreasing step of water contact angle on LPCAT-treated PTFE became smaller. Figure 18.13 depicts the change of OES spectra due to the addition

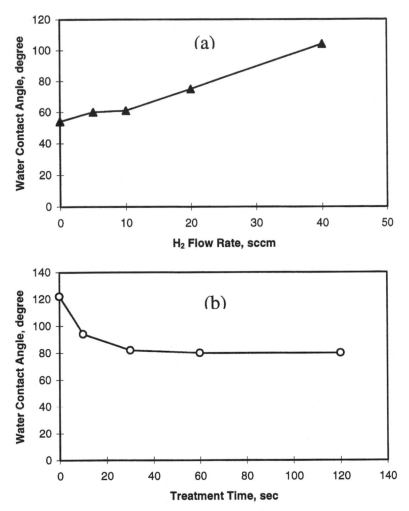

Figure 18.12 Surface contact angle changes of LTCAT treated PTFE; (a) dependence on hydrogen feed rate at 6.0 A arc current, 1500 sccm argon flow rate, and 1.0 min treatment, and (b) dependence on exposure time at 6.0 A arc current, 1500 sccm argon flow rate, 20 sccm hydrogen, 1.0 min treatment.

of H_2, and Figure 18.14 depicts the change of excited species in LPCAT as a function of H_2 added to the Ar plasma jet.

These results indicated that hydrogen consumed a part of excited argon neutrals, and the hydrogen-related species thus created was not as effective as excited argon neutrals themselves in modifying the PTFE surface. When the hydrogen feed rate was fixed at 20 sccm, the water contact angle on LPCAT-treated PTFE showed a decreasing trend with plasma exposure time, but not as efficient as argon torch treatment. From these results it is clear that unless long-lived chemically reactive species can be created by glow discharge, the secondary plasma or remote plasma mode operation cannot yield effective surface modification.

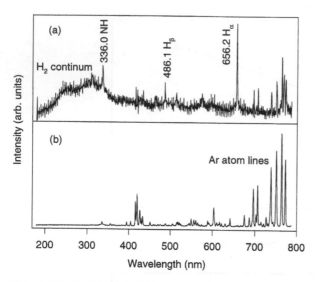

Figure 18.13 Typical OES spectra of LTCAT (a) argon torch with hydrogen addition and (b) argon torch only measured at an axial position of 2.7 cm from jet inlet; 2000 sccm argon, 8.0 A arc current, 50 sccm hydrogen.

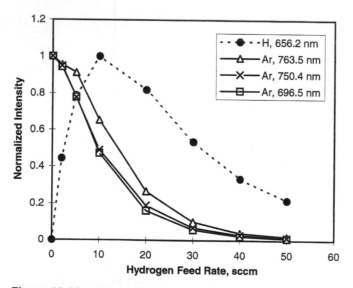

Figure 18.14 The dependence of excited argon and hydrogen atom emission intensities at an axial position of 2.7 cm from jet inlet on the feed rate of hydrogen added to argon plasma jet; 2000 sccm argon, 8.0 A arc current.

5. SURFACE MORPHOLOGY OF TREATED PTFE

Inagaki et al. reported that 30-s treatment by direct hydrogen plasma made a significant morphology change on PTFE surface observed by scanning electron microscopy (SEM) [3]. They attributed this surface morphology change to the heavy etching reactions induced by electrons and ions inside hydrogen plasma.

Figure 18.15 shows the SEM picture of original PTFE and plasma-treated PTFE by LPCAT argon plasma torch. At the same magnification as the data reported by Inagaki et al. [3], there is no surface morphology change detected by SEM on LPCAT-treated PTFE surfaces with less than 1 min plasma exposure. Although a little rough surface was observed with long plasma treatment, no significant morphology change was observed. These results indicated that LPCAT plasma treatment did not introduce much etching reaction or damage on the PTFE surface.

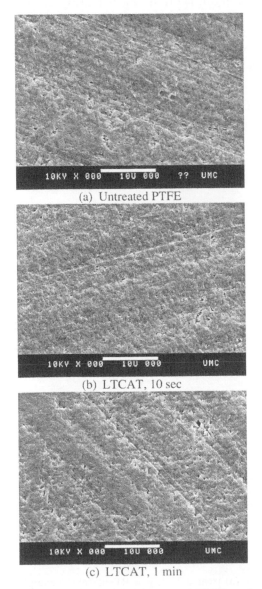

(a) Untreated PTFE

(b) LTCAT, 10 sec

(c) LTCAT, 1 min

Figure 18.15 SEM pictures of PTFE surfaces treated by a low-temperature cascade arc torch. (a) Untreated PTFE; (b) 10 s treatment; (c) 1 min treatment. Treatment conditions are 1500 sccm argon, 9 in. sample position, 6.0 A arc current.

The SEM picture of PTFE films treated by argon radio frequency plasma is shown in Figure 18.16. In comparison with LPCAT-treated samples, a comparable surface was observed on PTFE film treated for 2 min in the glow region. Since no dramatic surface morphology change was observed, it could be stated that there was no heavy etching reaction occurred in direct argon radio frequency plasma treatment under the experimental conditions (7 W vs. 100 W in the referenced work). It indicates that the excessive morphological change was due to the excessively high-energy input.

In LPCAT argon plasma jet, 4p level argon neutrals (E_B: ~ 13.0 eV) are the dominant species because of very low electron temperature and many fewer argon ions [4,5]. The energy level of 4p level argon neutrals is not sufficient to ionize hydrogen atoms (E_i: ~ 13.6 eV) and molecules (E_i: ~ 15.4 eV) [6]. The major hydrogen-related species in hydrogen-injected LPCAT are excited hydrogen atoms

(a) RF, 2 min

(b) LTCAT, 2 min

Figure 18.16 SEM pictures of PTFE surfaces; (a) treated by argon RF plasma at sample positions of 10 in. (inside glow) from copper coil electrode. Treatment conditions are: 2 sccm argon, 100 mtorr, 7.0 W RF power input, 2 min treatment, (b) treated by argon LTCAT for 2 min. Treatment conditions are 1500 sccm argon, 9 in. sample position, 6.0 A arc current.

and molecules. The effect of hydrogen ions on the treatment can be eliminated from the consideration. Therefore, the results of less effective hydrophilic modification of PTFE by hydrogen injected into an argon LPCAT suggested that excited argon neutrals are more efficient in improving the surface hydrophilicity of PTFE films than hydrogen-related species. This view is also supported by the more effective treatment effect of argon radio frequency plasma than hydrogen as observed from Figures 18.6 and 18.7, which were conducted under the same experimental conditions. As identified by SEM results, the application of this argon neutral beam by LPCAT can also avoid the surface etching effect induced by high-energy electrons and ions that usually occurs in conventional plasma processes, such as radio frequency plasmas [3].

In radio frequency argon plasma, however, besides excited argon neutrals, there exist many other energetic species, such as ions and high-energy electrons, that are more energetic than excited neutrals. It is usually considered that these ions and high-energy electrons can induce damage on polymer surfaces through the incision in polymer chains and etching of surface molecules. Surface morphology study by SEM also showed that a radio frequency power input as low as 0.12 GJ/kg could induce a certain surface changes on plasma-treated PTFE films. The composition difference in reactive species between LPCAT argon plasma and radio frequency argon might be the direct reason for the over treatment effect that was not found in LPCAT treatment of PTFE but was observed in radio frequency argon plasma treatment of PTFE film.

The treatment by secondary plasma reactor utilizes chemically reactive species created in glow discharge without influences of electron and ion bombardments and luminous gas phase. In-glow LPCAT treatment, on the other hand, utilizes luminous gas phase without the influence of ion and electron bombardment, and chemically reactive species are created on PTFE by energy transfer from the luminous gas phase. Thus, surface treatment by secondary plasma works only with gases that produce relatively long-lived chemically reactive species. Most secondary plasma treatments appear to be surface modifications by air or oxygen.

6. PULSED GLOW DISCHARGE

The effect of pulsed discharge on plasma polymerization may be viewed as the analogue of the rotating sector in photoinitiated free-radical chain growth polymerization. The ratio r of "off" time t_2 to "on" time t_1, $r = (t_2/t_1)$, is expected to influence the polymerization rate depending on the relative time scale of t_2 to the lifetime of free radicals in free-radical addition polymerization of a monomer. This method was used to estimate the lifetime of free radicals in conventional photon-initiated free radical polymerization.

The rate of vinyl polymerization decreases to a rate in the range $1/(1+r)$, (if $t_2 \gg$ average lifetime) to $1/(1+r)^{1/2}$ (when t_2 is close to the average lifetime) [7]. Contrary to this expectation, pulsed glow discharge exhibits somewhat abnormal effects on the pulse, indicating that the material formation in LCVD environment is not an ordinary free-radical chain growth polymerization although free radicals play key roles. Some monomers showed the increase, not decrease, of deposition rate, and some monomers showed marked increase of dangling bonds in the deposition [8].

The use of pulsed discharge provided key clues to elucidate the mechanisms of material formation and deposition as described in Chapter 5.

Some monomer show a more or less anticipated decrease in polymer deposition rates based on the concept that a pulsed discharge decreases the initiation rate, but some monomers show dramatically increased deposition rates. The most significant effect of pulsed discharge, however, can be seen in the concentration of free radicals trapped in plasma polymers (dangling bonds), which reflects the unique mechanisms of polymer formation in plasmas.

Two important aspects of the growth mechanisms shown in Figure 5.3 are that (1) cycle II leads to free radical "living polymer" and (2) the cross-cycle reaction shown by equation (3) in the figure is the terminator of living polymer during the on period of a pulsed discharge. In other words, during the on time some cycle II reactions are diverted to cycle I. Because of the balance of the contribution of cycle I and of cycle II, the free radical concentration in a plasma polymer can be correlated to the monomer types shown in Table 7.1. The greater the contribution of cycle II in the overall polymerization mechanisms, the more free radicals are found in the plasma polymers.

From the viewpoint of operation of LCVD, pulsed discharge provides effects somewhat similar to those that can be obtained by the remote or secondary plasma operation, namely, (1) reduction of the photoirradiation effect and (2) slow-down of the chemical reaction (surface modification or deposition of material), and allows uniform treatment or deposition. Pulsed discharge always decreases the substrate dangling bonds, which is the measure of the UV irradiation effect, as the value of r increases. Thus, the irradiation damage can be reduced dramatically. On the other hand, the concentration of chemically reactive species also decreases as the value of r increases.

The effect of pulsed discharge on material deposition is more complicated and requires careful examination of the overall merit that can be obtained by employing a pulsed discharge operation. One obvious advantage of pulsed discharge operation is that it requires only the addition of a pulsing unit to the radio frequency generator, and the value of r as well as the time scale t_1 can be changed easily. The downside is the reduction of monomer to polymer conversion yield and the prolonged reaction time, in most cases due to the decreased conversion yield. Pulsed discharge might improve or deprive the quality of the deposition depending on the aim of the LCVD operation.

From the viewpoint of growth and deposition mechanisms, pulsed discharge operation enhances the contribution of the molecular dissociation glow. This is because at the onset of glow discharge of a monomer the dissociation glow develops first and the ionization glow develops later as the composition of the gas phase changes due to the chemical reactions occurring in the dissociation glow. By pulsing the discharge, particularly at a short t_1 and a large r, the development of the ionization glow could be effectively deprived. This principle agrees with the findings that the photoirradiation effect decreases with pulsed discharge because the ionization glow contains higher energy photons. Results of recent studies indicate that the use of pulsed discharge reduces the excessive fragmentation of monomer molecule caused by continuous wave discharge [9–12].

Pulsed discharge, on the other hand, shortens the kinetic pathlength of material formation in gas phase, which results in deposition of oligomeric species. The lack of polymeric feature was reported when short t_1 and large r were employed [11]. This fact is crucial in evaluating the quality of the deposition. The short kinetic pathlength means that the material deposits without developing a three-dimensional network. On the other hand, the short kinetic pathlength preserves the original chemical structure of the monomer. The merit of pulsed discharge is most evident with monomers of conventional free radical polymerization, which can proceed the growth reaction during the off period, i.e., monomers with a double bond or triple bond.

The merit or demerit of employing pulsed discharge entirely depends on the objective of LCVD application. For example, if an LCVD is used to decorate the surface of material with some specific functional group, pulsed discharge allows placement of functional groups that can be predicted from the chemical structure of the monomer. On the other hand, an LCVD is used to lay down protective or barrier coating with superadhesion (type A plasma polymer), the deposition of oligomeric structure would not provide a good adhesion or a good protective layer, and hence a great advantage could not be anticipated. Pulsed discharge tends to yield type B plasma polymers.

Most surface analytical methods, such as XPS and IR, do not provide information pertinent to the functional aspects of the surface, such as solubility, durability, and mechanical integrity. In contrast to plasma polymer formed by continuous discharge, materials formed by pulsed discharge provide more resolved analytical data somewhat similar or closer to those for corresponding conventional polymers or monomer because the fragmentation of the original structure is minimal. This aspect may lead to the erroneous conception that pulsed discharges enable us to form more or less conventional polymers by glow discharge, whereas the information could come largely from oligomeric materials (type B plasma polymers).

REFERENCES

1. Yu, Q.S.; Reddy, C.M.; Meives, M.F.; Yasuda, H.K. J. Polym Sci. A Polym Chem **1999**, *37*, 4432.
2. Yamada, Y.; Yamada, T.; Tasaka, S.; Inagaki, N. Macromolecules **1996**, *29*, 4331.
3. Inagaki, N.; Tasaka, S.; Umehara, T. Surf. Coat Technol. *in press*.
4. Fusselman, S.P.; Yasuda, H.K. Plasma Chem. Plasma Process **1994**, *14*, 251.
5. Yu, Q.S.; Yasuda, H.K. Plasma Chem. Plasma Process **1998**, *18*, 461.
6. Lide, D.R. Editor-in-chief, *Handbook of Chemistry and Physics*; 73rd Ed.; CRC Press: Boca Raton, FL, 1992–1993; 10-211–10-217.
7. Burnett, G.M.; Melville, H.W. Proc. R. Soc. London **1947**, *A189*, 456.
8. Yasuda, H.; Hsu, T. J. Polym Sci., Polym. Chem. Ed. **1977**, *15*, 81.
9. Savage, C.R.; Timmons, R.B.; Lin, J.W. Chem Mater. **1991**, *3*, 575.
10. Han, L.M.; Rajeshwar, K.; Timmons, R.B. Langmuir **1997**, *13*, 5941.
11. Han, L.M.; Timmons, R.B.; Lee, W.W.; Chen Y.; Hu, Z. J. Appl. Phys. **1998**, *84*, 439.
12. Han, L.M.; Timmons, R.B.; Lee, W.W. J. Vac. Sci. Technol. **2000**, *B* (2), 799.

19
Reactor Size

1. TUBULAR REACTORS WITH DIFFERENT DIAMETERS

As described in Chapters 3, and 4, chemically reactive species in luminous chemical vapor deposition (LCVD) are created mainly by molecular dissociation. However, in the diffused luminous gas phase that constitutes the major volume of the luminous gas phase in most practical configurations of LCVD, the actual creation of chemically reactive species occurs at the tip of glow that contacts with the freshly fed monomers. Thus, the size and shape of glow within a reactor is critically important; however, these factors depend on the size and shape of the reactor.

While there are large numbers of previously published articles on plasma polymerization (LCVD), little information is applicable to other systems having different configuration and reactor size. This is due to the highly system-dependent characteristics of plasma processes. The reactor geometry influences the location of the core of the luminous gas phase as well as the size of the total diffused luminous gas phase as described in Chapter 3.

Without appreciating this aspect, therefore, interpretation of experimental data should not extend beyond the boundary of experimental conditions employed in the particular experiment, and the generalization of findings may not be justified and is often misleading. Thus, the influence of reactor size often becomes the major obstacle in an attempt to scale up reactor size for a larger scale operation of a process investigated in a laboratory scale equipment. In a tubular reactor system, in which the flow pattern is simple and straightforward, the diameter of a reactor tube might be the most relevant system factor. Effects of the size of a tubular reactor on the plasma polymerization of perfluoropropene might shed light on this problematic issue [1].

The tubular-type plasma reactor system used in the study consists of a reactor chamber, power supply, monomer feed, and pumping-out units, as depicted in Figure 19.1. One side of the glass tube is connected to a monomer inlet and the other side to a vacuum pump with O-ring joints. A radio frequency power generator of 13.56 MHz is coupled to two capacitive copper electrodes, which are 1 cm wide and 6 cm apart. The radio frequency power was controlled by an L-C matching network and monitored by power meter.

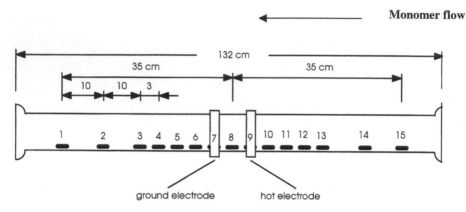

Figure 19.1 Arrangements of crystal sensors in the reactor, numbers on the crystal sensors represent positions; 15 sensors (Position No.1 to No.15) were used for large reactor, 13 sensors (Position No.2 to No.14) for medium reactor, 11 sensors (Position No.3 to No.13) for small reactor, 3 sensors (Position No.3, 8, and 13) for XPS analysis.

Three different size Pyrex glass tubes were used as reactor chambers, and their dimensions were as follows:

Reactor	Length (cm)	ID (cm)	Vol. (cm^3)	C.S. Area (cm^2)	Vol. between electrodes (cm^3)
S	132	1.8	336	2.54	15.3
M	132	3.3	1128	8.55	51.3
L	132	4.7	2289	17.3	104

The monomer gas, perfluoropropene, was fed to one end of the reactor tube. The system pressure was monitored by Baratron absolute pressure gauge, which was placed between tube end and vacuum pump. The system pressure was recorded by a recorder but not controlled by a throttle valve. In this system, the pressure and the flow rate are coupled, and the same system pressure does not correspond to a unique flow rate. The flow rate to yield a system pressure increases slightly with the size of the tube.

The plasma operating conditions performed in the study are summarized in Table 19.1. The monomer flow rates were determined by measuring the system pressure increase over the given time interval from the isolation of vacuum pump and then converted to flow rate (sccm). Each reactor volume was measured by gas expansion method: small size $= 557\,cm^3$, medium size $= 1278\,cm^3$, and large size $= 2357\,cm^3$. (The measured reactor volume includes the volume of connecting tubes beyond the center part glass tube.)

Crystal sensors for a thickness monitor were used as substrates. The placement of the substrates (crystal sensors) in the reactor is illustrated in Figure 19.1. Each sensor was masked with aluminum foil except for the center part ($r = 0.4\,cm$: the same diameter as the opening of a thickness monitor). The number of substrates

Table 19.1 Glow Discharge Operating Conditions

Experimental symbol	Reactor size	Input power (W)	Monomer pressure (mtorr)	Flow rate (sccm)
S15-30			30	0.07
S15-50		15	50	0.20
S15-75			75	0.48
S40-30	Small		30	0.07
S40-50	(ID = 1.8 cm)	40	50	0.20
S40-75			75	0.48
S80-30			30	0.07
S80-50		80	50	0.20
S80-75			75	0.48
M15-30			30	0.09
M15-50		15	50	0.21
M15-75			75	0.53
M40-30	Medium		30	0.09
M40-50	(ID = 3.3 cm)	40	50	0.21
M40-75			75	0.53
M80-30			30	0.09
M80-50		80	50	0.21
M80-75			75	0.53
L15-30			30	0.10
L15-50		15	50	0.26
L15-75			75	0.54
L40-30	Large		30	0.10
L40-50	(ID = 4.7 cm)	40	50	0.26
L40-75			75	0.54
L80-30			30	0.10
L80-50		80	50	0.26
L80-75			75	0.54

placed depended on the extension length of plasma. In a large reactor, 15 sensors (No.1 to No.15 in the figure) were placed because plasma could be extended to both ends of the reactor. In a medium reactor, 13 sensors (No. 2 to No.14) were placed. In a small reactor, 11 sensors (No. 3 to No. 13) were placed, since plasma was always confined around electrodes for all experimental conditions. The coated crystal sensors were used for thickness determination and XPS analysis of the depositions.

2. DEPOSITION CHARACTERISTICS

The effect of reactor size on the deposition characteristics was investigated by comparing deposition rate profiles of plasma polymerized perfluoropropene films in three reactors of different size. The local deposition rates were measured at various operating conditions (combinations of three monomer pressures and three discharge powers), which are listed in Table 19.1. Deposition rate profiles along the reactor tube in three reactors are shown in Figures 19.2, 19.3, and 19.4 for the corresponding

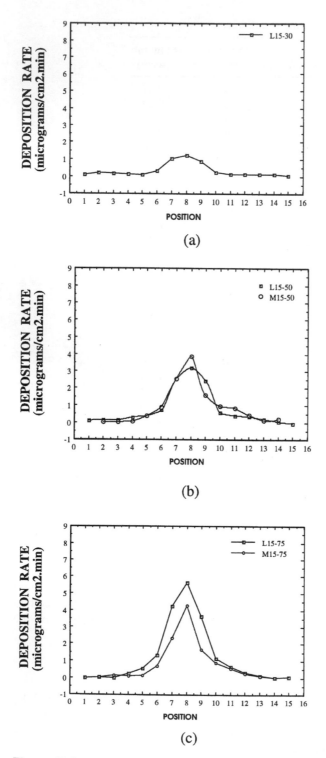

Figure 19.2 Deposition rate profiles of perfluoropropene plasma polymer at 15 W input power, (a) 30 mtorr (b) 50 mtorr (c) 75 mtorr (square: large reactor, circle: medium reactor).

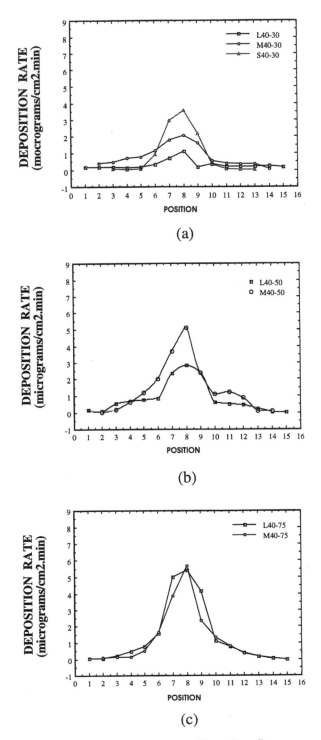

Figure 19.3 Deposition rate profiles of perfluoropropene plasma polymer at 40 W input power, (a) 30 mtorr (b) 50 mtorr (c) 75 mtorr (square: large reactor, circle: medium reactor, triangle: small reactor).

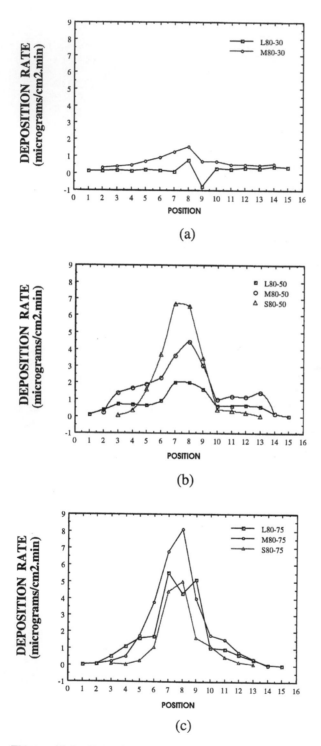

Figure 19.4 Deposition rate profiles of perfluoropropene plasma polymer at 80 W input power, (a) 30 mtorr (b) 50 mtorr (c) 75 mtorr (square: large reactor, circle: medium reactor, triangle: small reactor).

operating conditions indicated. Some deposition rate profiles are missing in the figures since plasma could not be sustained at certain conditions, especially in a small reactor. It is shown that most deposition occurs between two electrodes. The maximal peak of the local deposition rate coincides with the location where the strongest glow was observed.

The deposition rate profiles show that the influence of the operating conditions on the local deposition rate is dependent on the reactor size. As the monomer pressure increases (flow rate increases), the local deposition rates increase at any discharge power in a large reactor and also in a medium reactor. The difference between a large reactor and a medium reactor is the increasing rate with the monomer pressure. They increase at almost the same rate at any discharge power in a large reactor, whereas they increase more sharply at higher discharge power in a medium reactor.

As the discharge power increases, the local deposition rates decrease at any monomer pressure in a large reactor, whereas they decrease at a low monomer pressure but increase at a high monomer pressure in a medium reactor. In a small reactor, the same factor that increases the deposition rate in a large or medium reactor decreases the deposition rate (cf. Figures 19.4b and c). In a small reactor, high discharge powers are favored for the higher local deposition rates.

In plasma polymerization, the dependence of the deposition rate on the operating condition varies based on the domain of the plasma polymerization as described in Chapter 8. There are three domains: energy-deficient domain, transitional domain, and monomer-deficient domain. They are classified based on the dependence of the normalized deposition rate, D/FM, on the normalized energy input parameter, W/FM, where D is the deposition rate.

In order to clearly identify the change of domain due to the size of reactor, the D/FM vs. W/FM for each reactor size is plotted on one graph in Figure 19.5. In the large reactor, the deposition rate reaches its maximal value at low W/FM and decreases slightly as the input energy increases due to the fluorine etching. In the small reactor, the plasma polymerization starts at relatively high W/FM and the conversion increases continuously in the range of power input employed. The conversion in the medium reactor shows an intermediate behavior between the large and the small reactors, although the highest local deposition rate at a given W/FM is obtained with the medium reactor. The curve D/FM vs. W/FM shifts to the higher W/FM and lower D/FM side as the reactor size decreases.

The change of domain under the same operating condition but different reactor volume could be explained by the dissipation mechanism of energetic species in the luminous gas phase. The value of W/FM represents the energy input to the plasma volume in a reactor. As described in Chapter 8, the value of W/FM depends on design factors of a reactor and the exact value cannot be taken in a generic sense. Plasma loses its energy to the wall surface, and the minimal energy input to sustain plasma must be greater than the energy loss to the wall. Thus, $(W/FM)_c$ is a function of surface/volume ratio of a reactor.

As the size (diameter) of a tubular reactor decreases, the relative contribution of the dissipation of energy to the reactor wall increases since the surface-to-volume ratio of a tubular reactor increases proportionally to 1/(diameter of tube), in a tubular reactor. As a consequence, more input energy is needed to compensate the loss of energy to the reactor wall in a smaller reactor.

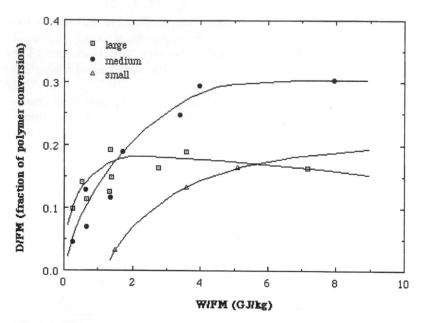

Figure 19.5 Specific conversion of monomer to plasma polymer as a function of W/FM for various size reactors (square: large reactor, circle: medium reactor, triangle: small reactor).

Therefore, the LCVD domain changes from the monomer-deficient to the energy-deficient domain as the reactor size decreases at the same operating conditions (based on external operating parameters) employed. The fact that the glow discharge could not be sustained under certain conditions, especially in a small reactor at low W/FM values, is also due to the same effect, i.e., the energy dissipation to the reactor wall is too large to sustain a glow discharge. Even when the glow could be sustained at high W/FM values in a small reactor, the glow was always confined around the electrodes.

Ihara and Yasuda investigated the deposition behavior of methane in the medium-sized tubular reactor with 13.56 MHz radio frequency discharge [2]. They observed that the critical W/FM value, $(W/FM)_c$, for methane was 8 GJ/kg, and nearly 100% of monomers were converted to the plasma polymer beyond this critical W/FM value. As shown in Figure 19.5, the critical W/FM value of perfluoropropene in the small reactor is around 6 GJ/kg and the D/FM is 15%, and the corresponding value in the medium is around 4 GJ/kg, and the maximal conversion is around 30%. In the large reactor, $(W/FM)_c$ is about 1 GJ/kg and the maximal D/FM is 20%. The lower value of the critical W/FM for C_3F_6 than that for CH_4 is explained in Chapter 7.

3. POLYMER COMPOSITION

The chemical composition and the surface structure of perfluoropropene plasma polymers were investigated using XPS. The XPS spectra were obtained only for the medium-sized and the large reactors. In each reactor, three crystal sensors

Table 19.2 Atomic Percent of Perfluoropropene Plasma Polymer and F/C Ratio

Symbol[a]	Fluorine	Carbon	Nitrogen	Oxygen	F/C
M15-75-C	56.5	42.3	N/A	1.1	1.33
M15-50-C	49.8	47.2	—	2.9	1.04
M15-30-C	45.2	50.1	—	4.5	0.89
M40-30-C	55.6	41.3	3.1	N/A	1.33
M80-30-C	57.1	37	5.9	—	1.56
L15-75-C	60.8	38.3	0.2	0.6	1.56
L15-50-C	60.6	38.4	0.4	0.7	1.56
L15-30-C	61	37.3	0.6	1.1	1.63
L40-30-C	61.6	37.1	0.5	0.8	1.63
L80-30-C	60.7	37.8	0.9	0.7	1.63
M15-75-D	50.3	46.3	N/A	3.2	1.08
M15-50-D	57.9	39.9	—	2.0	1.44
M15-30-D	40.5	54.2	—	5.2	0.75
M40-30-D	55.9	40.1	4.0	N/A	1.38
M80-30-D	56.5	38.2	5.3	—	1.5
L15-75-D	52.7	42.4	1.2	3.7	1.22
L15-50-D	57.7	40.2	1.0	1.1	1.44
L15-30-D	51.6	44.8	2.4	1.2	1.17
L40-30-D	64.1	34.3	0.7	0.8	1.86
L80-30-D	62.5	35.7	0.9	0.9	1.78
M15-75-U	47.5	49.3	N/A	3.1	0.96
M15-50-U	46.1	46.6	—	7.2	1.00
M15-30-U	56.7	41.7	—	1.5	1.38
M40-30-U	49.6	47.2	3.2	N/A	1.04
M80-30-U	54.4	41.3	4.3	—	1.33
L15-75-U	56.4	39.8	1.6	2.2	1.44
L15-50-U	56.9	40.5	1.1	1.6	1.38
L15-30-U	59.8	37.7	0.7	1.8	1.56
L40-30-U	60.7	37.5	0.7	1.1	1.63
L80-30-U	62.4	35.6	0.5	1.5	1.78

[a]M, medium-size reactor; L, large reactor; C, center position; D, downstream position; U, upstream position (e.g. M15-75-C: medium size reactor, 15 W input power, 75 mtorr monomer pressure, center position).

(substrates) were placed at the center between electrodes, 15 cm downstream and 15 cm upstream from the center, respectively (see Figure 19.1). Five operating conditions were chosen from Table 19.1 and shown in Table 19.2, including the lowest W/FM (15 W, 75 mtorr) and the highest W/FM (80 W, 30 mtorr). The order of arrangement in Table 19.2 (descending) corresponds to the increase in W/FM value at the same substrate position in each reactor.

The atomic ratios of fluorine to carbon at various operating conditions in two reactors are shown in Table 19.2. The F/C ratios at the center position in each reactor are compared in Figure 19.6 as a function of W/FM. In the large reactor, the F/C ratios are almost the same, approximately 1.6, regardless of the increase in W/FM. In the medium reactor, the F/C ratios changed at low W/FM level, but

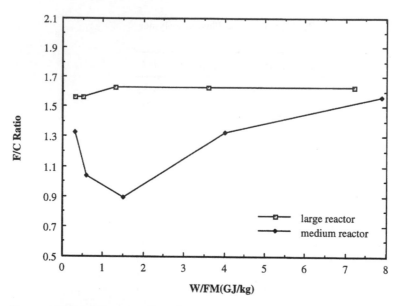

Figure 19.6 F/C ratio of perfluoropropene plasma polymer prepared at center position between elctrodes as a function of W/FM (square: large reactor, circle: medium reactor).

they are converged to the similar value of the large reactor as the W/FM increases. This change reflects the change of domain of LCVD.

As discussed in the previous section, plasma polymerization in the large reactor belonged to the monomer-deficient domain for all operating conditions so far as the deposition kinetics is concerned. The plasma polymerization in the medium reactor belonged to the energy-deficient domain at a low W/FM level and changed to the monomer-deficient domain as the W/FM increased. On the basis of these observations, the perfluoropropene plasma polymers prepared in the monomer-deficient domain may have similar composition and structure regardless of the reactor size and the operating conditions as shown in Figures 19.6 and 19.7.

The carbon 1s spectra were deconvoluted into six peaks using a nonlinear least-squares curve-fitting program. Peaks centered at 293.3, 291.2, 289.5, 288.3, 286.6, and 285.0 eV are due to CF_3, CF_2, $CF\text{-}CF_n$, CF, $C\text{-}CF_n$, and C, respectively. Large numbers of CF_3 and CF_2 groups and small numbers of C group are present in the polymer prepared in the monomer-deficient (energy-saturated) domain, and the F/C ratios are approximately 1.6.

As the W/FM value decreases to the energy-deficient domain, the composition and the structure are changed according to the operating conditions. This could be seen in the F/C ratios and C 1s spectra in the medium reactor at a low W/FM level in Figures 19.6 and 19.7b. In addition, the composition and structures of the polymer prepared far from the electrodes also show this pattern. The F/C ratios and C 1s spectra of the polymers prepared at 15 cm downstream are shown in Figures 19.8, and 19.9, and 15 cm upstream positions are illustrated in Figures 19.10, and 19.11. These data clearly demonstrate that the size of reactor affects the deposition kinetics and the polymer characteristics under otherwise identical operational conditions.

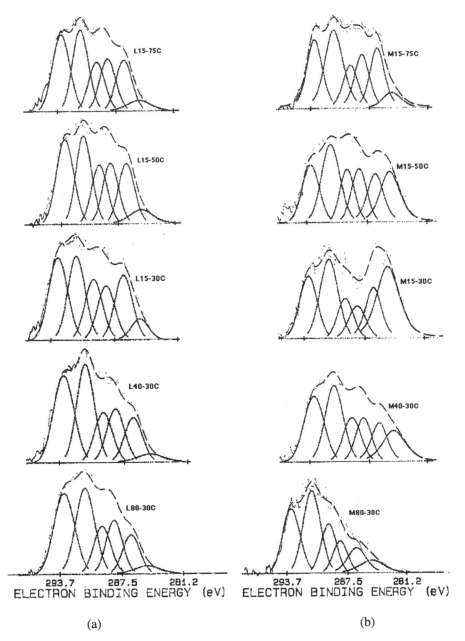

Figure 19.7 Carbon 1s spectra of perfluoropropene plasma polymer prepared at center position between electrodes (a): large reactor, (b): medium reactor (*W/FM* increases from up to down direction).

4. CHANGE OF DOMAIN BY SIZE OF REACTOR

Since the energy density would be different according to the relative distance from the electrodes, the different domains of plasma polymerization could exist along the

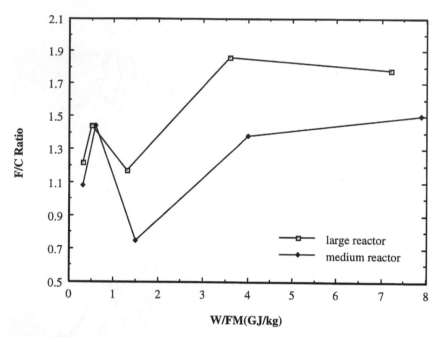

Figure 19.8 F/C ratio of perfluoropropene plasma polymer prepared at 15 cm downstream from the middle of two electrodes as a function of W/FM (square: large reactor, circle: medium reactor).

reactor axis under fixed operating conditions. Although the absolute determination of the domain along the reactor axis is difficult, the local deposition rates at same position as a function of W/FM reveal the changing domains. The following rough assignment of the domains could be made:

1. L-C (large reactor, center position): monomer-deficient domain for all operating conditions,
2. M-C (medium reactor, center position): transitional domain for low W/FM, and monomer-deficient domain for medium to high W/FM conditions,
3. L-D (large reactor, 15 cm downstream position): energy-deficient domain for low W/FM, transitional domain for medium W/FM, and monomer-deficient domain for high W/FM conditions,
4. M-D (medium reactor, 15 cm downstream position): energy-deficient domain for low W/FM, and transitional domain for medium to high W/FM conditions,
5. L-U (large reactor, 15 cm upstream position): energy-deficient domain for low to medium W/FM, and transitional domain for high W/FM conditions,
6. M-U (medium reactor 15 cm upstream position): energy-deficient domain for low to medium W/FM, and transitional domain for high W/FM conditions.

Based on the assignment of domains at each position as a function of W/FM, the F/C ratios plotted in Figures 19.6, 19.8, and 19.10, could be explained in a similar manner. In the energy-deficient domain, the fluorine contents in the polymer generally increase as the W/FM increases. When the W/FM value reaches the transitional domain, the fluorine contents decrease as the W/FM increases. However,

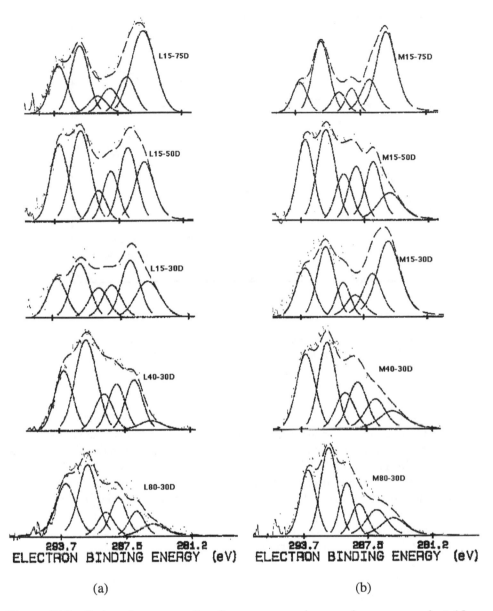

Figure 19.9 Carbon 1s spectra of perfluoropropene plasma polymer prepared at 15 cm downstream from the middle of two electrodes; (a): large reactor, (b): medium reactor, (W/FM increases from up to down direction).

the fluorine contents increase again as the W/FM continues to increase in the transitional domain up to near the monomer-deficient domain. In the monomer-deficient domain the composition of polymer is not changed by the W/FM as mentioned previously.

The polymer structures shown in Figures 19.7, 19.9, and 19.11, also follow the trend explained by the assignment of domains. In the energy-deficient domain

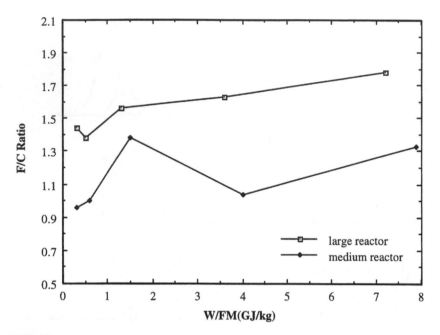

Figure 19.10 F/C ratio of perfluoropropene plasma polymer prepared at 15 cm upstream from the middle of two electrodes as a function of W/FM (square: large reactor, circle: medium reactor).

the C and C-CF$_n$ groups appear strongly. As the W/FM increase in this domain, these groups decrease whereas CF$_3$ and CF$_2$ groups increase until the W/FM reaches the transitional domain. In the transitional domain, C and C-CF$_n$ increase again as the W/FM increases. These groups as well as the F/C ratios decrease until the W/FM reaches the monomer-deficient region. In the monomer-deficient domain CF$_3$ and CF$_2$ groups are predominant species for all W/FM conditions.

However, these findings are in contradiction to the generally observed trend that CF$_3$ and CF$_2$ are found in the deposition that is formed with short kinetic pass length, which often occurs in the energy-deficient domain. In the monomer-deficient domain, C-CF$_n$- and C-rich structures with low F/C values are obtained. These discrepancies may be explained by consideration of the following factors.

1. The monomer used contains a double bond and a CF$_3$ group, with an F/C ratio of 2.0, and is a monomer that favors polymerization among perfluorocarbons, which are reluctant to polymerize. The general trend described above might not apply to C$_3$F$_6$. The results strongly suggest that this is the case.
2. The substrate is placed on the surface of tube.
3. The relative position of a substrate to the core of glow and also the distance from the tip of glow change depending on W, F, and the diameter of tube.
4. The deposition rates as well as XPS data are analyzed based on the externally controllable parameters whereas those factors are actually controlled by the local conditions.

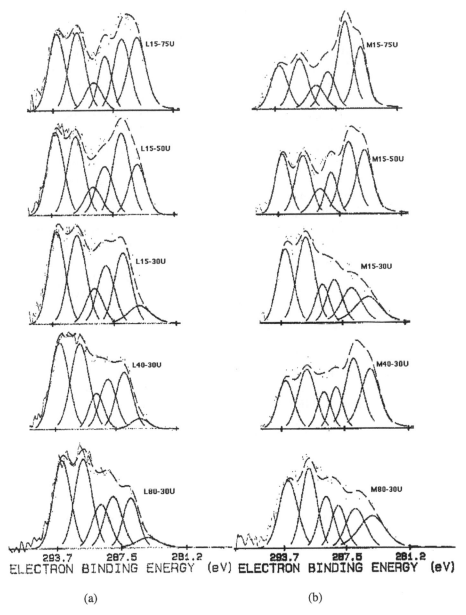

Figure 19.11 Carbon 1s spectra of perfluoropropene plasma polymer prepared at 15 cm upstream from the middle of two electrodes; (a): large reactor, (b): medium reactor, (W/FM increases from up to down direction).

5. Straight tube reactors have well-defined flow pattern, i.e., all monomer gas coming into a reactor pass through the reactor without any bypassing, and the sweeping rate can be given by flow rate F divided by the cross-sectional area A of the tube, F/A. The value of F/A for the medium tube M is twice of that for large tube L.

In such a straight tube reactor, it is clear that the activation of monomer, or the creation of polymerizable species, occurs at the tip of glow where the incoming monomer molecules interact with the luminous gas phase. Thus, the opening of double bond and detachment of F occur at this point, and the further reaction of the free radicals, the species created by the detachment of F, and the detached F's with gas phase species occurs while all gaseous species are moving toward the downstream side of the reactor. Under such a one directional flow conditions, particularly with monomer that contain $-CF_3$, the analysis based on the formation of $-CF_3$ from monomers that do not contain $-CF_3$ might become the focal point of discrepancy.

When the normalized energy input is increased, the volume of glow expands, and the tip of glow in upstream side moves away from the center thus increasing the distance from the sample in upstream side placed at the fixed position to the tip of glow increases. Based on the deposition profiles (Figures 19.2–19.4), the tip of glow seems to advance more in a larger diameter tube. This will cause the change of polymer composition because the substrate position is now further away from the tip of glow where the activation starts.

Insofar as an attempt to find the correlation between the reactor size and LCVD characteristics, C_3F_6 might not be the best monomer due to the complexity of reactions involved with perfluorocarbons. On the other hand, in an attempt to demonstrate the influence of the reactor size, the monomer provided sufficient data because of the ease of the analysis of F-containing species by XPS.

These data clearly show that the size of reactor is an important factor to be considered in dealing with data obtained by a reactor. The change of reactor size influences the overall performance of LCVD of a monomer, and consequently the description of operating conditions such as flow rate, system pressure, and discharge wattage cannot be used in a generic sense, unless the size factor of reactor, domain of plasma polymerization, and the relative position of the substrate with respect to the core of luminous gas phase and/or to the tip of glow could be identified. These data show the complicated system-dependent nature of LCVD, particularly that a monomer does not produce a polymer and that the externally operative parameters, such as W, p, and F, are not the actual parameters that control LCVD.

The apparent discrepancy described above is due in part to the fact that sample collection is done at different layer of an onion layer structure because substrates are placed on the fixed locations on the reactor wall. These factors associated with the size of a reactor are important in an attempt to increase the size of reactor in order to scale up the processing capability. Since the onion layer structure of the luminous gas phase changes with the size of the reactor, the straightforward increase of reactor size to cope with a larger substrate or larger numbers of substrates to increase the production rate often has disastrous consequences.

REFERENCES

1. Kim, H.Y.; Yasuda, H.K. J. Vac. Sci. Tech. **1997**, *A15*, 1837.
2. Ihara, T.; Yasuda, H. J. Appl. Polym. Sci. : Appl. Polym. Symp. **1990**, *46*, 511.

20
Flow Pattern

1. FLOW AND LUMINOUS GAS VOLUME

The deposition in a luminous chemical vapor deposition (LCVD) system occurs on surfaces that are either in contact with luminous gas phase (glow) or in the vicinity of glow. The amount of polymer deposition is influenced by three important geometrical factors of LCVD reactor. These are the relative position of polymer deposition with respect to (1) the location of electric energy input, (2) the monomer flow, and (3) the position within a reactor. The system pressure, which determines the mean free path of gaseous species, has a great influence on the distribution of polymer deposition. In general, the lower the system pressure, the wider or more even is the distribution [1,2].

Another factor that determines the pattern of polymer deposition is the reactivity of the active species involved. Some monomers produce more reactive species or a greater number of reactive species than others and appear to be more reactive in the overall polymer deposition process. These "reactive monomers" tend to deposit near the site of the electric energy input, if such a site (e.g., pair of electrodes) exists in relatively large gas phase (e.g., in a bell jar reactor) where the energy density of the luminous gas phase is the highest, or near the inlet where monomers contact with the luminous gas phase, if monomer cannot reach the energy input site without interacting luminous gas phase. In both cases, the distributions of polymer deposition within a reactor are uneven with the maximal deposition at the site, and the distribution depends on the flow pattern of monomer. The latter is the case for a simple straight tube reactor with external radio frequency coil of electrodes described in Chapter 19.

2. DEPOSITION OF FAST-POLYMERIZING MONOMER IN TUBULAR REACTOR

In order to examine the effect of flow pattern in a reactor, which is a crucially important design factor of an LCVD reactor, it is necessary to examine the profile of deposition in a simple reactor first. A tubular reactor with an external radio frequency power coupling is ideally suited to the study of the distribution of polymer deposition. In such a reactor, 100% of the monomer passes through the luminous gas phase in the reactor, and the situation is very close to the case in which no bypass of monomer occurs. The experimental setup used for

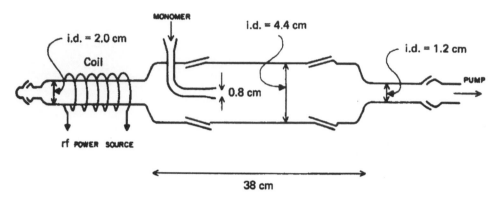

Figure 20.1 Schematic representation of the standard reactor used in the study of the distribution of polymer deposition under discussion.

the following examples of the distribution of polymer deposition is shown in Figure 20.1. It should be noted that monomer does not flow through radio frequency coil zone. Monomer interacts with the luminous gas phase at the monomer inlet.

The distribution of polymer deposition observed in the plasma polymerization of acetylene at different flow rates (and different system pressures under plasma conditions) is shown in Figure 20.2. It should be noted that acetylene is the fastest polymerizing hydrocarbon and the system pressure decreases on the inception of glow discharge. In this particular configuration of reactor, the monomer does not pass the radio frequency coil, and presents a typical case in which the creation of chemically reactive species occurs at the boundary where the monomer meets the luminous gas phase, i.e., activation by luminous gas, not by ionization.

The results indicate that a rather uniform distribution of polymer is obtained at a low flow rate (and low system pressure). Slightly higher deposition rates are found at the downstream side of the monomer inlet. At higher flow rates (and higher system pressures), the entire distribution curve is shifted up and a conspicuous maximum develops. The location of maximal deposition tends to shift slightly more to the downstream side of the monomer inlet as the flow rate increases.

The addition of a carrier gas (mixed into a constant flow of acetylene before the monomer inlet) changes the pattern of polymer deposition more drastically as depicted in Figure 20.3. It is important to note that the addition of H_2 or argon increases the peak height of the maximum without changing the average deposition rate, i.e., the distribution becomes narrower on the addition of these gases. Therefore, a considerable decrease in polymer deposition was observed at the locations on the upstream side (and on the downstream side) of the monomer inlet. The effect of argon, which does not participate chemically in the polymer formation, can be seen as a sharpening of the distribution curve. The possible reason for this is discussed later in this chapter.

The effect of N_2 may be due to two factors. The first is the same as that in the case of argon. The second factor is the chemical participation of N_2. Nitrogen

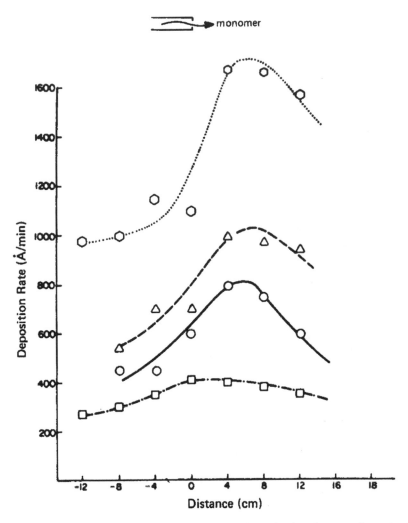

Figure 20.2 Distribution of polymer deposition in the plasma polymerization of acetylene at various flow rates; key: ($\bigcirc$) $F = 9.0$, $p_g = 35$; ($\triangle$) $F = 7.7$, $p_g = 28$; ($\bigcirc$) $F = 5.2$, $p_g = 20$; ($\square$) $F = 2.4$, $p_g = 10$; F denotes flow rate (sccm), and p_g mtorr.

in the plasma state is chemically reactive and should be considered as a comonomer of the plasma polymerization, although it does not polymerize alone in glow discharge. This effect is seen in the obvious increase in total polymer deposition. Perhaps the same is true with H_2 but because of the small contribution of the mass of H_2, this factor is not reflected in the deposition, which is measured by the mass of deposition.

Thus, the increase in polymer deposition near the monomer inlet is due to both a physical factor and a chemical factor (copolymerization of gas). Copolymerization obviously increases the total mass due to the incorporation of the second gas. This increase is clearly seen in the plasma polymerization of acetylene on the addition of Freon 12 (CCl_2F_2) as depicted in Figure 20.4.

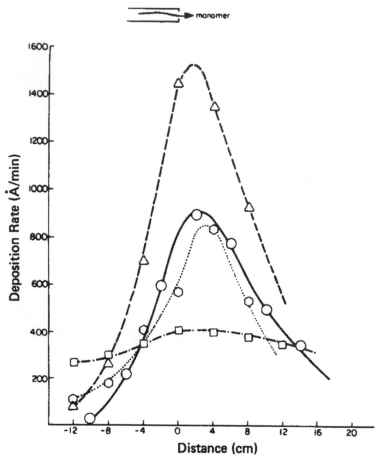

Figure 20.3 Distribution of polymer deposition in the plasma polymerization of acetylene with the addition of gas; key: (□) acetylene ($F = 2.5$ sccm), $p_g = 8$; (○) $H_2/Ac = 1.3$, $p_g = 70$; (◯) $Ar/Ac = 0.8$, $p_g = 8.8$; (△) $N_2/Ac = 1.0$, $p_g = 45$; N_2/Ac, H_2/Ac, and Ar/Ac denote the mole ratios of gas to acetylene, p_g in mtorr.

3. DEPOSITION OF SLOWLY-POLYMERIZING MONOMER IN TUBULAR REACTOR

With monomers that are less reactive than acetylene, such as ethylene, the distribution of polymer deposition is much flatter as depicted in Figure 20.5. It is interesting to note that (1) the distribution of polymer deposition is much more uniform than that of acetylene (Figure 20.2) and no obvious maximum is observed at low flow rates, and (2) the minimum (rather than the maximum) deposition is observed in the vicinity of the monomer inlet at high flow rates (and high system pressures in glow). Moreover, the glow in the vicinity of the monomer inlet is noticeably weaker than that in the rest of the reactor at the high flow rate that showed the minimal deposition in the region.

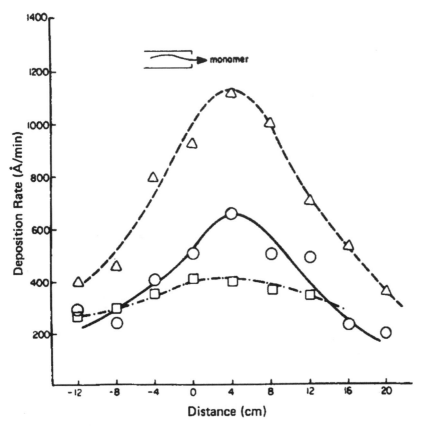

Figure 20.4 Effect of adding Freon 12 (CCl_2F_2) to acetylene at flow rate $F = 2.5$ sccm on the distribution of polymer deposition; key: ($\square$) acetylene, $p_g = 8$; ($\bigcirc$) Freon $12/Ac = 0.13$, $p_g = 11$; ($\triangle$) Freon $12/Ac = 0.75$, $p_g = 24$, Freon $12/Ac$ denotes the mole ratio of CCl_2F_2 to acetylene p_g in mtorr.

The effects of carrier gases added to ethylene are shown in Figure 20.6. The addition of N_2 or argon causes a slight narrowing of the distribution, but the effects are much smaller than those found for acetylene. A maximum with broad shoulders appears farther downstream (8 cm) than the maximum observed with acetylene, which appears near the inlet (0 cm). Thus, the enhancement of polymer deposition by the addition of an inert gas into the monomer feed flow is highly dependent on the nature of the monomer.

The effect of the addition of Freon 12 (CCl_2F_2) to ethylene at a flow rate of 2.5 sccm is shown in Figure 20.7. The appearance of a maximum and the increase in its peak height with the partial pressure of the Freon is similar to the case of acetylene, but the half-widths of the peaks are much wider with ethylene. At higher flow rates of ethylene, however, the distribution pattern is quite different as depicted in Figure 20.8. The minimum observed with ethylene at a flow rate of 9.6 sccm shown in Figure 20.5 is an indication of the existence of two peaks: one at the upstream side toward the radio frequency coil and another at the downstream

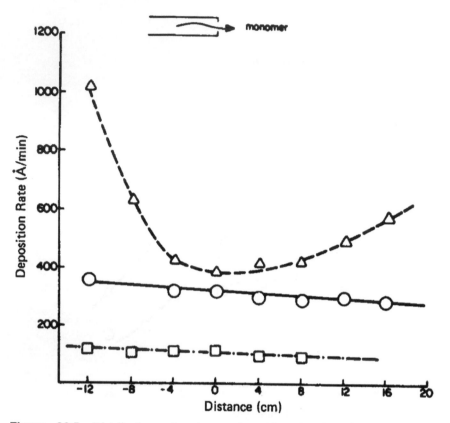

Figure 20.5 Distribution of polymer deposition in the plasma polymerization of ethylene; key: ($\triangle$) $F=9.6$, $p_g=117$; ($\bigcirc$) $F=2.5$, $p_g=44$; ($\square$) $F=1.1$, $p_g=24$; F denotes flow rate (cm^3_{STP}/min), and p_g in mtorr.

side with respect to the monomer inlet. The addition of Freon tends to shift these two peaks closer to the monomer inlet as the peak heights increase with the partial pressure of Freon. Freon 12 is a monomer of plasma polymerization, and the effect found is not the same as the addition of inert gases.

4. EFFECT OF FLOW PATTERN

The effect of flow pattern in a reactor was investigated with ethylene as the monomer by changing the position of monomer inlet and pumping exit of the system. A relatively high flow rate of about 9.8 sccm, which showed a conspicuous minimum at the location near the monomer inlet, was used throughout this series of experiments.

Various patterns of the flow in relation to the radio frequency coil and to the outlet (the pump) of the reactor are shown in Figure 20.9. In case I, the position of the monomer inlet is changed by the insertion of a tube of different length to the standard inlet by means of standard tapered glass joints. In case II, the position of

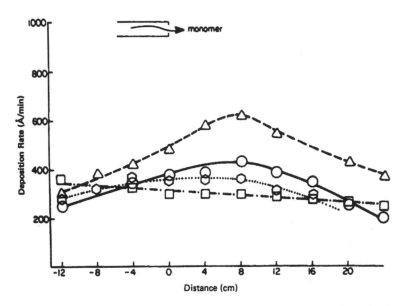

Figure 20.6 Distribution of polymer deposition in plasma polymerization of ethylene with the addition of gas; key: ($\triangle$) $N_2/Et = 0.9$, $p_g = 96$; ($\bigcirc$) $Ar/Et = 0.8$, $p_g = 110$; ($\bigcirc$) $H_2/Et = 1.1$, $p_g = 87$; ($\square$) Et ($F = 2.5$ sccm), $p_g = 44$; N_2/Et, H_2/Et, and Ar/Et denote the mole ratios of gas to ethylene, p_g in mtorr.

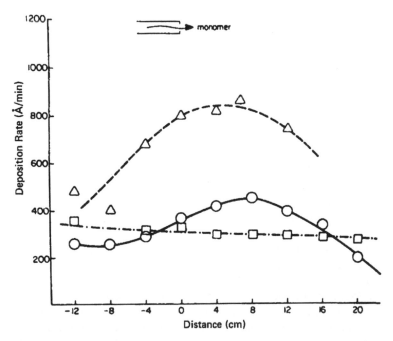

Figure 20.7 Distribution of polymer deposition in the plasma polymerization of ethylene with the addition of Freon 12 (CCl_2F_2); key: ($\square$) ethylene ($F = 2.5$ sccm), $p_g = 44$; ($\bigcirc$) Freon 12/Et = 0.10, $p_g = 43$; ($\triangle$) Freon 12/Et = 0.76, $p_g = 45$, (Freon 12/Et denotes the mole ratio of CCl_2F_2 to ethylene), p_g in mtorr.

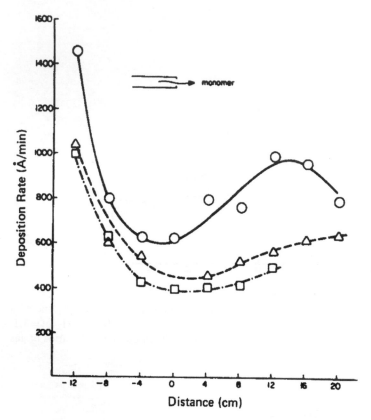

Figure 20.8   Effect of adding Freon 12 (CCl_2F_2) to ethylene at a high flow rate (9.6 sccm) on the distribution of polymer deposition; key: ($\square$) ethylene, p = 117; ($\triangle$) Ar/Et = 0.14, p_g = 143; ($\bigcirc$) Freon 12/Et = 0.15, p_g = 126, (Freon 12/Et denotes the mole ratio of CCl_2F_2 to ethylene), p_g in mtorr.

the monomer inlet is changed by means of a special A tube (without an inner sealed tube) or a special B tube (with three side-arm joints). In case III, a special A tube is used, and the connections to the monomer feed-in system and to the pump system are reversed. In cases IV and V, a special A tube is used, and the monomer inlet and outlet are reversed. The changes in polymer distribution due to flow patterns are compared in Figures 20.10–20.16, in which the origin (0) of the abscissas was chosen as the same point used in Figures 20.2–20.8.

   Figure 20.10 indicates that the direction of monomer flow (i.e., whether parallel or perpendicular to the pumping direction) does not cause a significant difference if the inlet location is the same. This means that the location of the inlet in a reactor is a more important factor than the way in which the inlet is inserted into the reactor, which is indeed found to be the case in Figure 20.11. The minimum in the distribution curve shifts with the location of the inlet. This trend was also found in an experiment employing a long extended inner tube as depicted in Figure 20.12.

   Whether or not the monomer is fed through the radio frequency coil also has an important influence on the distribution of polymer deposition. When

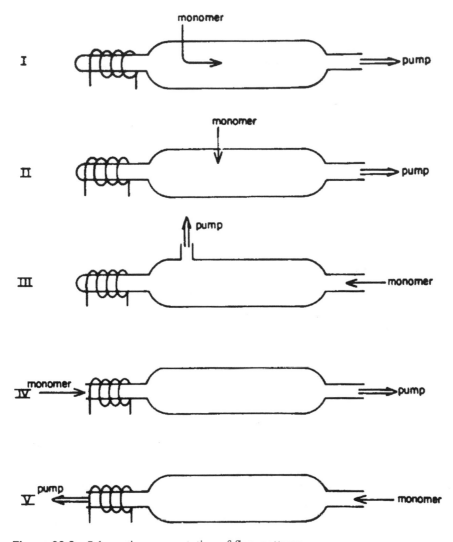

Figure 20.9 Schematic representation of flow patterns.

the monomer flows through the radio frequency coil, a large amount of polymer is deposited in the radio frequency coil zone, which is beyond the range of the measurements, and there is a considerable decrease in the polymer deposition in the tail-flame portion of the reactor as depicted in Figure 20.13.

The reversal of the flow direction causes a change in the polymer deposition pattern, which can be predicted by the location of the monomer inlet and the effect of flow direction as depicted in Figure 20.14. When the monomer flow direction is reversed (i.e., the major reactor is on the upstream side of the radio frequency coil, or the coil is located in the downstream end of a reactor), whether or not the monomer flows through the radio frequency coil has very little effect on the polymer deposition pattern as depicted in Figure 20.15. Regardless of whether the radio frequency coil is located at the upstream end or the downstream end of

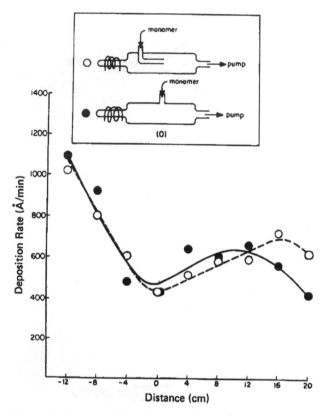

Figure 20.10   Effect of the direction of monomer injection on the distribution of polymer deposition.

a reactor, a reactor design that employs a flow pattern in which the monomer flows through the coil zone yields the most uneven and least efficient polymer deposition in the main reactor portion used in this type of tube reactor as depicted in Figure 20.16.

These experimental data indicate that three major factors influence the polymer deposition in a plasma reactor: the locations of (1) the energy input (radio frequency coil in these cases), (2) the monomer flow inlet, and (3) the monomer flow outlet. The local deposition rate d_i observed at a location is expressed as a function of the location of polymer deposition. These experimental data indicate that the major parameter that determines d_i is the distance from the energy input. Other factors (i.e., monomer inlet and outlet) determine the direction of flow, which can be either along or against the direction of the energy input to the point of polymer deposition. The direction of monomer flow has less influence on the polymer deposition than the distance from the energy input.

An LCVD system is somewhat similar to a gas flame in which the combustion rate and the gas flow rate establish a steady-state flame. In an LCVD system, the monomer flow rate and the polymer formation rate establish a steady-state polymer-forming luminous gas phase. This situation is expressed schematically in Figure 20.17, where (a) indicates the diffusional transport of the energy-carrying

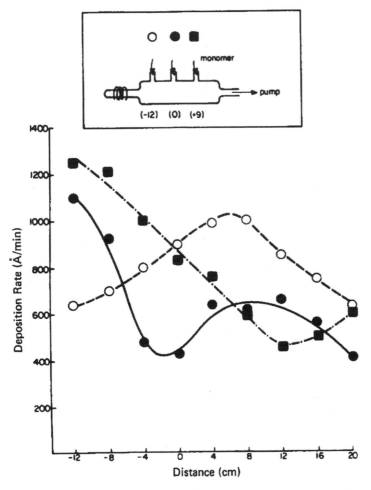

Figure 20.11 Effect of the location of the monomer inlet on the distribution of polymer deposition.

species in the luminous gas phase shown in Figure 5.3, (b) indicates the flow of monomer and product gases, and (c) indicates the diffusional transport of polymer-forming species.

Polymer-forming species are energy-consuming species as far as the energy transfer associated with the luminous gas phase is concerned. In contrast, metastable species such as those of Ar are clearly energy carrying. The effect of the addition of argon discussed earlier can be explained by the effect of adding energy-carrying species, which increases the effective energy transfer to create chemically reactive species at the location of the monomer inlet.

The density of the energy that is carried by energy-carrying species decreases with the distance from the energy input zone. Therefore, where the monomer is introduced with respect to the energy input is a very important factor. Upon the contact of monomer with the luminous gas phase, polymer-forming species, described in Figure 5.3, are created and diffuse in both upstream and downstream

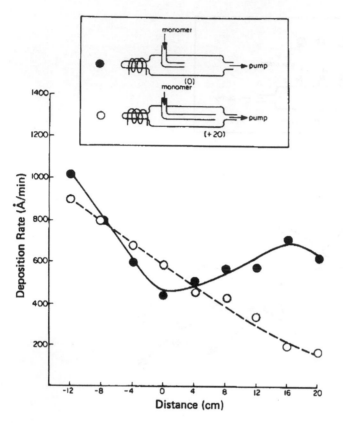

Figure 20.12   Effect of the length of the monomer inlet on the distribution of polymer deposition.

directions. The flow of monomer or the pumping may be in the direction of diffusional transport or against it. If the flow is against the diffusional transport, then quenching of plasma can occur, very little polymer deposition is observed in the nonglow (quenched) zone, and the deposition distribution shifts toward the energy input zone as seen in Figures 20.15 and 20.16.

5. COMBINED EFFECT OF FLOW PATTERN AND ENERGY INPUT

The importance of the glow zone (luminous gas phase) for polymer formation can be seen in the polymer distribution pattern obtained by a small straight-tube reactor (o.d. 11 mm) with a radio frequency coil placed in the middle of the tube [3]. In Figure 20.18 deposition rates of plasma polymers of tetrafluoroethylene, given by values of deposition rate divided by monomer flow rate, are plotted against the location. The location is measured from the center of the radio frequency coil in the direction of gas flow. The shaded band shows the extent of glow. The solid line represents the polymer distribution curve for $F = 5.6$ sccm, $W = 8$ W ($W/FM = 1.9 \times 10^7$ J/kg), and the dashed line represents that for $F = 0.56$ sccm,

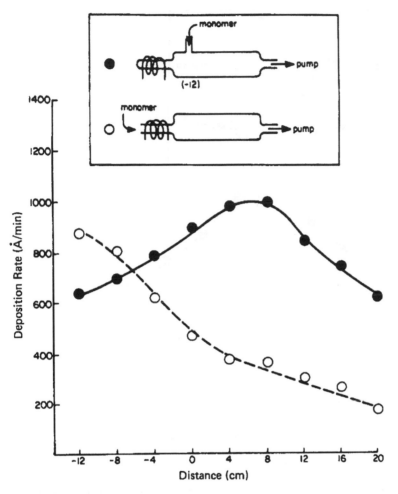

Figure 20.13 Effect of flow passing through the radio frequency coil on the distribution of polymer deposition.

$W = 32\,\text{W}$ ($W/FM = 7.7 \times 10^8\,\text{J/kg}$). Thus, the deposition occurs mainly in the glow zone, and any factor that diminishes the glow zone would cause a sharp decrease in the polymer deposition rate.

As described in previous sections, an increase in discharge power at a fixed flow rate means an increased rate of excitation or an increased rate of growth reactions. This increase in initiation takes place at the energy input zone (in an inductively coupled discharge) or near the electrode surface (in a capacitively coupled AC and radio frequency discharge), and its influence on polymer deposition rates or on the properties of polymers differs depending on many operational factors such as the type of discharge and the reactor design.

The effects of the discharge power on the distribution of polymer deposition in a tubular reactor (Fig. 20.1) are shown in Figures 20.19–20.22. Figure 20.19 depicts the change in polymer deposition pattern due to the discharge power observed in the plasma polymerization of styrene at a fixed flow rate of 5.6 sccm,

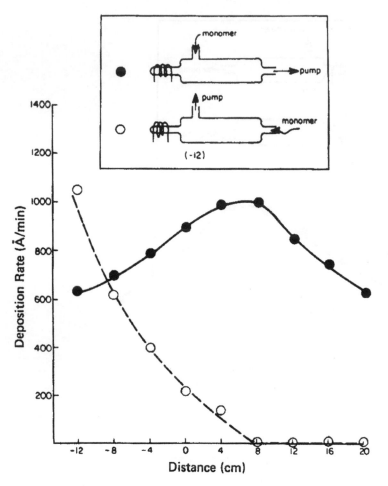

Figure 20.14 Effect of the direction of flow on the distribution of polymer deposition for cases in which there is no flow through the radio frequency coil.

and Figure 20.20 at a lower flow rate of 1.9 sccm. Similar results obtained for the plasma polymerization of acetylene are shown in Figure 20.21 for 5.6 sccm and in Figure 20.22 for 1.9 sccm.

At low flow rates, due to the lower system pressure (at a given pumping rate) the distribution curve becomes flatter. Because of the tendency for plasma polymerization to become flow rate controlled (rather than discharge power controlled) within the experimentally feasible range of discharge power (see the domains of plasma polymerization discussed in Chapter 8), the effect of discharge power on the polymer distribution becomes relatively small. The trend in the second case (i.e., in the discharge power–controlled domain) can also be seen in higher flow rates, i.e., the most drastic change in the polymer deposition pattern is observed when the discharge power becomes insufficient to sustain the full glow and in situations very similar to this case. With sufficient power, the polymer deposition pattern is not greatly influenced by the discharge power.

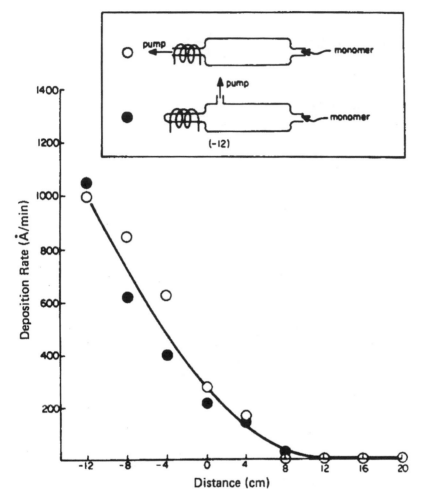

Figure 20.15 Effect of flow passing through the radio frequency coil in the reverse direction on the distribution of polymer deposition.

6. DEPOSITION ON ELECTRODE IN A BELL JAR TYPE OF REACTOR

A bell jar type of reactor used for plasma polymerization usually employs a set of parallel electrodes, and the glow is more or less confined to the space between the electrodes. In such a system, the total volume of the reactor is considerably larger (e.g., more than a factor of 10) than the plasma volume. A reactor that consists of a large tube and a pair of electrodes located inside can also be considered a bell jar type of reactor. In other words, whether a bell jar or a vessel of another shape is used is not a major issue.

In such a reactor, which has a large reservoir of monomer surrounding the plasma zone, the direct effect of the monomer flow is much less pronounced than in other reactors and the diffusional transport seems to predominate. Plasma polymer

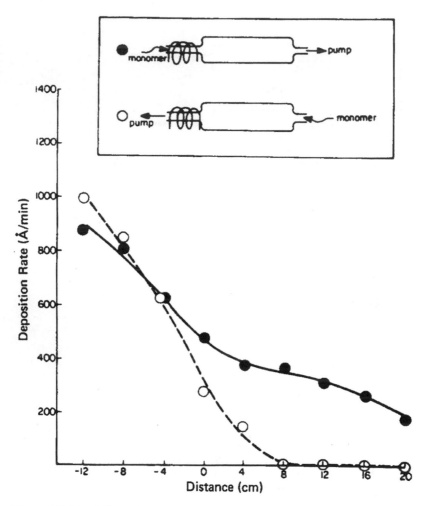

Figure 20.16 Effect of the direction of flow on the distribution of polymer deposition for cases in which flow passes through the radio frequency coil.

is collected either on the surface of the electrode (a substrate placed on the electrode) or on the surface of the substrate, which is placed between the electrodes. Obviously, the polymer deposition pattern is quite different depending on the mode of polymer collection.

The electrode surface is considered the energy input plane. In radio frequency discharge, the molecular dissociation glow, in which the major creation of chemically reactive species occurs, does not adhere to the electrode surface but is very close to the electrode surface. Therefore, in this case the major factor that determines the distribution of polymer deposition is the diffusional transport of monomer from the periphery of the electrode to the body of luminous gas phase that occupies the interelectrode space.

Therefore, the gaseous species that strike the polymer-depositing surface and form polymer (solid) differ depending on the distance from the tip of glow at

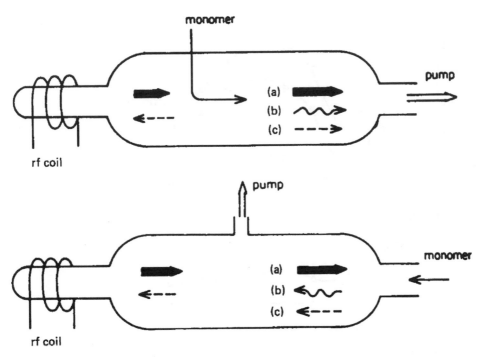

Figure 20.17 Schematic representation of the contribution of (a) diffusion transport of active species created by the radio frequency coil, (b) flow of monomer and/or product gas, and (c) diffusional transport of polymer-forming species.

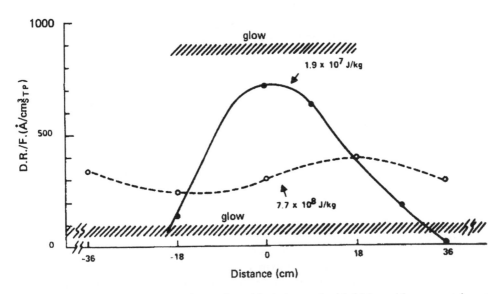

Figure 20.18 Distribution of polymer deposition observed with high and low energy input. (D.R: deposition rate).

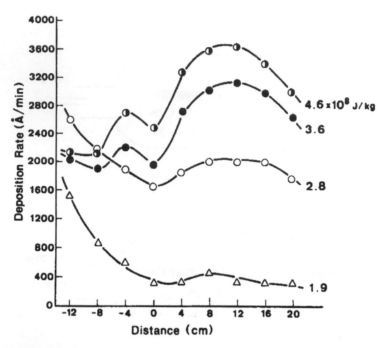

Figure 20.19 Effect of discharge power on the distribution of polymer deposition from styrene at a flow rate of $5.6 \, \text{cm}^3_{\text{STP}}/\text{min}$; key: ◑, 200 W; ●, 160 W; ○, 120 W; △, 80 W.

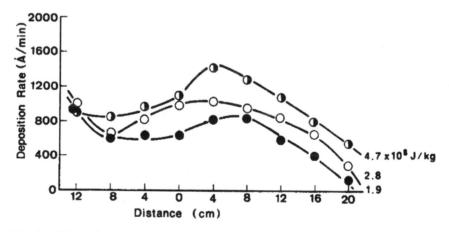

Figure 20.20 Effect of discharge power on the distribution of polymer deposition from styrene at a flow rate of 1.9 sccm; key: ◑, 70 W; ○, 40 W; ●, 27 W.

the periphery of the electrode. The mass of striking gaseous species in plasma increases with the distance from the periphery, which means that the deposition rate determined by the weight of polymer is highest at the center of the electrode. If the monomer is injected at the center of the electrode, then this trend should be reversed and the minimal deposition rate should be observed at the center of the electrode.

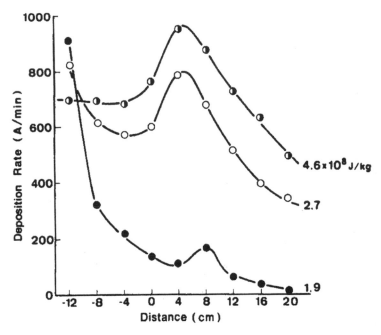

Figure 20.21 Effect of discharge power on the distribution of polymer deposition from acetylene at a flow rate of 5.6 cm$^3_{STP}$/min; key: ◑, 50 W; ○, 30 W; ●, 20 W.

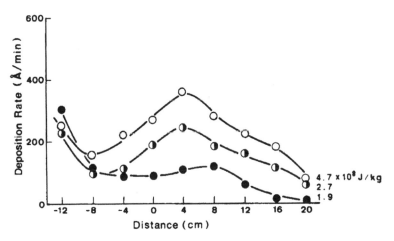

Figure 20.22 Effect of discharge power on the distribution of polymer deposition from acetylene at a flow rate of 1.9 cm$^3_{STP}$/min; key: ◑, 17.5 W; ○, 10 W; ●, 7 W.

This situation can be seen clearly in the distribution pattern observed by Kobayashi et al. [4] shown in Figure 20.23 (the arrangement of monomer inlet) and Figure 20.24 (the distribution of polymer deposition for corresponding cases). The slight asymmetry of the polymer deposition pattern can be attributed to the overall flow pattern existing in the entire reactor system. The principle of the polymer deposition is identical to that for the tubular reactor shown in Figure 20.17.

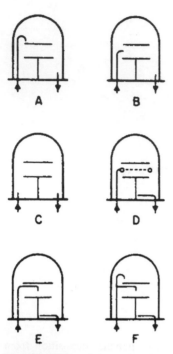

Figure 20.23 Schematic of inlet-outlet configuration of the bell jar type of reactor. Adapted from Ref. 4.

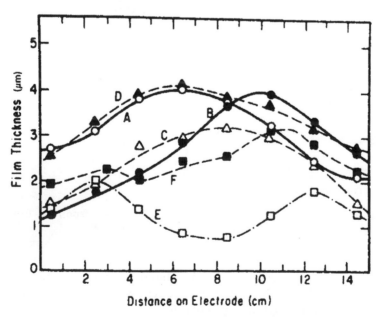

Figure 20.24 Effect of reactor configurations (refer to Figure 20.23 for configurations) on the distribution of film thickness on the electrode; bell jar type of reactor; in the flow direction; 2 torr; 100 W; ethylene flow rate 80 sccm; 60 min. Adapted from Ref. 4.

An important implication of the data obtained with both a tubular reactor and a bell jar reactor is that the polymer deposition onto a stationary substrate cannot be uniform due to the diffusional transport of polymer-forming species and the path-dependent growth mechanism. The variation of polymer deposition rates at various locations becomes smaller as the system pressure decreases because the diffusional displacement distance of gaseous species increases at lower pressure. It is important to recognize that a certain degree of thickness variation always exists when the plasma polymer is deposited onto a stationary substrate regardless of the type of reactor and the location of the substrate in the reactor.

7. PINCHING OF LUMINOUS GAS PHASE

In the growth mechanism explained in Figure 5.3, the steps or cycles are repeated while gaseous reactive species collide in the gas phase. Therefore, how large the species become and how quickly their size reaches the critical mass above which they cannot stay in the luminous gas phase are dependent on the density of the gaseous species and their flow pattern, which is determined by the size and shape of the reactor.

Because a polymer-forming luminous gas phase such as the tail-flame portion of an inductively coupled radio frequency glow discharge behaves as a fluid, the deposition mechanism can be investigated by examining the influence of the fluid mechanical aspects of luminous gas phase on the deposition rate of polymer.

When a constriction is introduced in a flow of polymer-forming luminous gas as depicted in Figure 20.25, an intensified glow is observed in the constriction [5]. In a steady flow of polymer-forming luminous gas, the total number of gaseous species in a plane perpendicular to the direction of flow must be identical within a unit time. Therefore, more gases (in a unit time) pass through a unit area in the constricted portion than in the rest of the reactor. Consequently, the density of chemically reactive species in the constriction increases according to the ratio of constriction. This increased glow density is introduced as a consequence of the change in the flow characteristics of the plasma, not because of increased electric energy input.

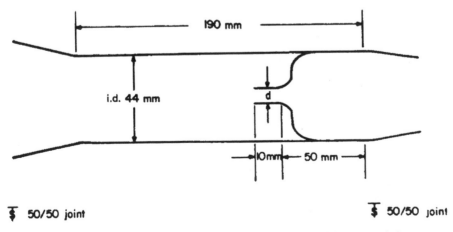

Figure 20.25 Schematic representation of a reaction tube with a constriction.

For simplicity of discussion, let us consider a simple monomer system and treat the monomer as an ideal gas. From the kinetic theory of gases, the ratio of gas–wall collisions to gas–gas collisions in the total system is proportional to the surface to volume ratio S/V for a given monomer at fixed pressure and temperature.

It is interesting to examine whether the polymer deposition rate is dependent on the surface-to-volume ratio S/V. Under the conditions employed for the experiment, S/V is given by $2/r$, where r is the radius of the tube in the constriction. Therefore, the dependence of polymer deposition rate on S/V can be obtained by a plot of deposition rates versus $1/d$, where d is the inner diameter of the constriction. Such plots are shown in Figure 20.26 for the plasma polymerization of tetramethyldisiloxane.

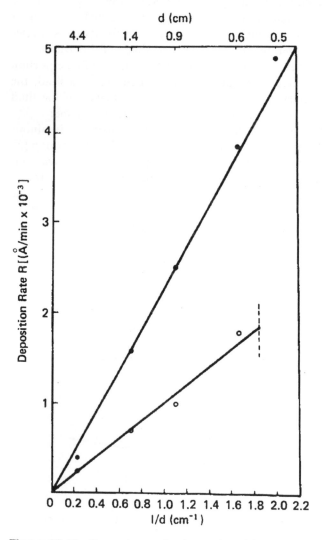

Figure 20.26   Dependence of polymer deposition rate on the diameter d of the constriction.

The solid circles represent the deposition rate obtained at a vapor pressure of 60 mtorr (the initial pressure) and the open circles at a vapor pressure of 40 mtorr. In these experiments, the system pressure was not controlled independent of monomer flow rate. The lower pressure was obtained by a lower flow rate. At the latter vapor pressure, glow does not penetrate into the 5-mm constriction and no polymer deposition occurs within the constriction. This is due to the increased gas–wall collisions that quench the luminous reactive species.

At a fixed flow rate of the equipment used, the slopes of two straight lines are in good agreement with the dependence of the deposition rate being approximately proportional to the square of the initial pressure, i.e., the deposition rate being proportional to the flow rate. The results show that the polymer deposition rate is indeed directly proportional to S/V. This means that neither the gas phase reaction nor the surface reaction is the predominating factor in LCVD unless the S/V ratio is extremely high.

The fluid mechanical aspect of luminous gas is also evident with non-polymer-forming plasma. It has been observed that the degradation of polymer exposed to N_2 plasma is very severe in the constricted portion: a polyethylene film inserted in the constricted portion suffered permanent deformation due to partial melting, whereas polyethylene films placed in wider portions of the tube (before and after the constriction) did not show any visible difference after they were exposed to N_2 glow discharge simultaneously as depicted in Figure 20.27.

Polyethylene (LD) samples at location 1 and 3 did not show any visible difference after they were exposed to N_2 glow discharge, but the film at location 2 (i.d. 0.9 cm) was partially melted due to the increased thermal effect of the intensified luminous gas that passes through the constricted portion of the reactor. Weight changes observed with different polymers are shown in Table 20.1, which also shows the intensified degradation of all polymers in location 2 (in constriction). The apparent intensity of glow also follows the same trend, i.e., the plasma is much more intense in the pinched portion, and the smaller the diameter of constriction, the more intense is the apparent intensity of glow.

It has been generally observed that polymer deposition occurs mainly on a surface exposed to glow. More precisely, the deposition rate of polymer onto a surface that does not make contact with glow is several orders of magnitude smaller than that onto a surface that contacts glow. The results outlined here clearly demonstrate that the rate of deposition of a polymer onto a surface that contacts glow is dependent on the S/pV of the glow. This factor seems to have important implications in the application of plasma polymerization, which may involve substrates of various sizes and shapes.

From the viewpoint of the polymer deposition mechanism, the significance of this effect is as follows. Each step involved in the growth mechanism (Fig. 5.3) requires the collision of reactive species (including gas–surface collision). With lower frequency of collisions, the kinetic diffusional pathlength becomes large, and the polymer deposition spreads widely from the location of inception of the monomer glow discharge. With the introduction of a constriction, the collision frequency increases within the constriction and the deposition rate of polymer increases according to the ratio of gas–gas and gas–surface collisions, which indicates that the gas phase reaction and gas surface reaction are inseparable mechanisms of polymer deposition.

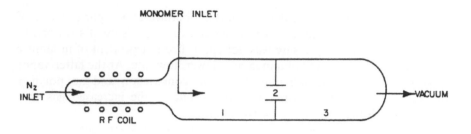

Figure 20.27 Picture of polyethylene (LD) films placed at location 1,2, and 3 (shown in the upper figure) and exposed to N_2 glow discharge.

Table 20.1 Effect of Glow Discharge Pinching on Degradation of Polymer Exposed to N_2 Glow Discharge[a]

Substrate polymer	Location 1	Location 2	Location 3
Polyoxymethylene	−14.6	−45.2	−0.492
PET	−1.98	−32.7	+0.252
Al-coated PET	−0.261	+0.526	+4.12
Polyethylene (LD)	0	−2.22	−0.728

[a]Weight Change in 20 min $(mg/mm^2) \times 10^4$.

REFERENCES

1. Yasuda, H.; Hirotsu, T. J. Polym. Sci., Polym. Chem. Ed., **1978**, *16*, 229.
2. Yasuda, H.; Hirotsu, T. J. Polym. Sci., Polym. Chem. Ed., **1978**, *16*, 313.
3. Yasuda, H.; Morosoff, N.; Brandt, E. S.; Reilley, C. N. J. Appl. Polym. Sci. **1979**, *23*, 1003.
4. Kobayashi, H.; Bell, A. T.; Shen, M. J. Macromol. Sci. Chem. **1976**, *10*, 491.
5. Yasuda, H.; Hsu, T. J. Appl. Polym. Sci. **1976**, *20*, 1769.

21
Composition-Graded Transition Phase

1. METAL–POLYMER INTERFACE

A luminous chemical vapor deposition (LCVD) layer could be used as transition phase between metal and polymer (coatings or films). The deposition of LCVD layer on metals can be done by the cathodic LCVD, in which the metal to be incorporated is used as the cathode of DC discharge. This method provides an excellent adhesion of LCVD nanofilm to the metal as well as the adhesion of coatings that will be applied on the surface of the LCVD layer [1,2]. This method is described in some detail in Part IV.

On the other hand, the deposition of metal layer on polymer surface is more problematic because the adhesion of metal deposition on a polymer surface is not so simple depending on how the metal layer is deposited, and also on the structure of the deposited metal layer. Metallized insulators such as polymeric and ceramic materials are widely used in the appliance, automotive, and electronics industries. Metallization of nonconducting substrates is technically difficult because of the structural incompatibility between the substrate and metallizing material, in terms of both chemical bonding and properties. The abrupt mismatch at the interface between them has been blamed for the major portion of failures of metallized parts under operating conditions.

Lamination of a metal foil to a nonconducting substrate with an adhesive has been widely used because of its simplicity. One of the disadvantages of this process is that it has limits on the shape of the substrate and film thickness that can be used. Such limits were excluded when autocatalytic chemical metallization was developed. However, this process has certain limitations as well, even though it is being widely used in the field. Several presteps, each requiring precise handling and conditioning of the materials involved, are necessary and effluent treatment should be considered, which could become the major drawback.

The continually growing demand for improved metallized nonconducting materials has created interest in developing new, efficient, and dependable processes. A metallization technique using glow discharge plasma polymerization has been developed in an attempt to generate a film capable of minimizing some of the difficulties cited above. Usually, a lower degree of adherence between the metal and the insulator is obtained if little or no primary chemical bonding is present. Except for some special cases, the substrate–metal adhesion is believed to be obtained by physical adsorption bonds, although activated chemisorption, diffusion, chemical reaction, electrostatic attraction, mechanical interlocking, or some combination of these processes may occur, depending on the process conditions [3,4]. Therefore, substrate surface cleaning, etching, or appropriate surface treatment to achieve favorable surface roughness for mechanical interlocking, and surface modification by oxygen plasma for forming chemical bonds such as organic–oxygen–metal bonds, etc., to improve adhesion between metal and nonconducting substrates, have been tried in the vacuum-metallizing field, and great improvements in adhesion of metal to insulators have been reported [5–7].

2. TRANSITIONAL BUFFERING FILM

The establishment of chemical bonds between a nonconducting substrate and metallizing material, based on research carried out in the field to date, seems to be one of the most crucial factors to improve the performances of metallized insulators under operating conditions. To accomplish this, the application of a transitional buffering film for use as a preplate (material ready for metal plating) for metallizing nonconducting materials can be used. The composition of the transitional buffering film, which changes from 100% organic to nearly pure metallic over the thickness of the film, eliminates sudden changes in physical and chemical structures between two dissimilar materials and assists in establishing a stronger, more durable interface. This increases the chance of obtaining improved chemical bonding between the polymer substrate surface and the plasma polymer layer at the bottom of the transitional buffering film, as well as the metallic bonding between the nearly pure metallic layer at the top of the film and the subsequent metallizing material. In this case, the process scheme consists of (1) layering a transitional buffering film by sputter deposition of a suitable electrode material with simultaneous plasma polymerization of methane, followed by (2) deposition of the desired metallizing material by electroless deposition or electroplating.

The first process is to impart electrical surface conductivity to the polymer so that the second process of laying down a thicker metal layer can proceed. The plasma reactor can do the deposition of metal alone also by using Ar alone as plasma gas; however, such a deposition of metal was avoided due to the following reasons. Vapor-deposited metal, e.g., by evaporation or sputter coating, develops a columnar structure in nearly all cases and consequently is quite vulnerable to corrosion. A better layer can be obtained by electrolytic or electroless deposition of copper in a more efficient way. Therefore, sputter coating is used only to create a composition-graded buffering layer, except in the case of ungraded plasma polymer with a metallic surface.

A capacitively coupled bell jar type of reactor operating at a frequency of 10 kHz was used for plasma polymerization. Either a single or double (Figure 21.1) diode magnetron electrode system was used to create glow discharge. Two copper or silver plates (18 × 18 cm) were used as the electrodes for a single-electrode system, and two titanium plates were added as the second set of electrodes for a double-electrode system. A sample-mounting disk was located in the middle of the diode and rotated at 50 rpm for uniform coating of a plasma polymer film. Methane was used

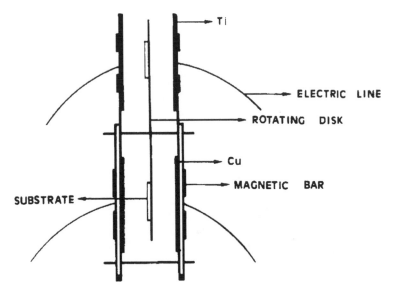

Figure 21.1 Double magnetron system for double-graded film deposition.

as a monomer for the plasma polymerization. In the double-electrode system, the metal that is to be sputter coated covers one set of electrodes, and another set is covered with metal with low sputtering yield for the deposition of pure methane plasma polymer. Each electrode system is used as the sole electrode system in one operation, which is followed by the second operation using another electrode system.

Substrates used included fiber-reinforced epoxy base polymer [FRP], nylon 66, polytetrafluoroethylene [Teflon], poly(ethylene terephthalate) [PET], phenolic resin, and thermoplastic polyimide [ULTEM, GE]. FRPs were the primary substrates used. Initially, they were cleaned with detergent in an ultrasonic bath followed by rinsing with deionized water and alcohol. For further cleaning, they were treated with oxygen plasma (1.33 sccm, 60 W, 5 min) followed by a hydrogen plasma treatment (3 sccm, 60 W, 5 min).

Copper was selected as the metallizing material because of its wide industrial application. Copper was then plated onto the preplate layer either by electroless or electroplating. A commercial solution was used for the electroless copper plating to obtain a 1.0-μm thick copper deposit. A 20 μm-thick layer of copper was deposited electrolytically under the following conditions.

For semiquantitative adhesion data, a tape test was carried out with a tape of peel strength 10 N/cm according to ASTM D3359-78. Tensile lap shear strength tests [8,9] were performed with an Instron floor model testing machine to obtain quantitative adhesion data. Circular copper films (1 cm diameter) were plated onto plasma-processed 1.5×9.0 cm coupons. This film was then glued onto another freshly cleaned unplated coupon surface of the same dimensions with epoxy glue, completely overlap bonded, and cured for 48 h at room temperature. The bonds were sheared at a cross-head rate of 0.025 cm/min. The structure of the lap–shear test and dimensions of the sample are depicted in Figure 21.2.

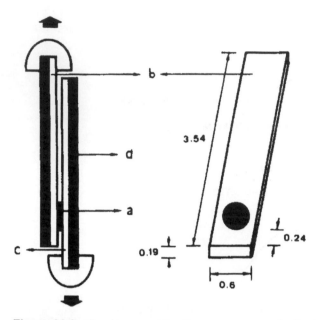

Figure 21.2 Specimen used for the measurement of adhesion; a, lap shear bond consisting of metallizing coating and epoxy glue; b, metallized substrate; c, unplated fresh substrate; d, mechanical support (unit, inch).

3. DEPOSITION OF METAL-CONTAINING POLYMER FILMS

Metal-containing polymer films were prepared by simultaneous sputter deposition of electrode material (Cu or Ag) with plasma polymerization of methane by magnetron discharge. The maximal flux density, perpendicular component to the electric field, on the electrode surface was 208 G. Methane was chosen as a monomer because its low molecular weight allows a higher energy input, which is favored for sputtering of the electrode surface. In addition, a methane plasma polymer provides unique water-insensitive adhesion and excellent lap–shear strength [8,10,11].

In magnetron discharges of methane with copper or silver electrodes, copper or silver particles were incorporated into a polymer film at higher W/FM values, above 2×10^9 J/kg and 1.5×10^9 J/kg, respectively, as shown in Figure 21.3. The volumetric fraction of metal in the polymer film is almost linearly dependent on W/FM. The metal content was estimated by XPS analysis assuming that the metal particles were uniformly distributed throughout the polymer film, which is indeed the case.

4. ACTIVATION OF NONCONDUCTING SURFACES FOR ELECTROLESS PLATING

The activating effect of metal-containing polymer films for electroless plating of metal was investigated by varying the copper and silver content in a polymer film. Electroless plating of copper was attempted after depositing a 20-nm-thick copper- or silver-containing methane plasma polymer film on FRP (CH_4 0.5 sccm, 80 W). Visual plating of copper was observed in the case of copper-containing films,

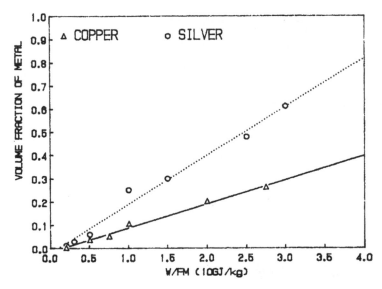

Figure 21.3 Effect of normalized energy input, W/FM, of the methane glow discharge on the sputter deposited metal content in the plasma polymer of methane film.

indicating the activation capability of sputtered copper particles in a polymer film. For coatings of 0.15 volume fraction Cu, almost spontaneous electroless plating with uniform distribution of copper particles covering the entire film was observed. As the copper content decreased, the uniformity and coverage of electroless copper also decreased. No plating of copper was observed on silver-containing films, possibly resulting from the nonautocatalytic character of silver for the electroless plating. The surface morphologies of an initial substrate, plasma-activated and electroless copper-plated surfaces, are shown in Figure 21.4. The figure shows uniform coverage of the surface with electroless copper.

Adhesion of electroless plated copper films was tape tested with an adhesive tape of 10 N/cm peel strength. The samples produced without plasma preconditioning, such as oxygen or hydrogen treatment, passed the test only about 70% of the time. Surface contamination was suspected, and further cleaning with oxygen plasma (1.3 sccm, 60 W, 5 min) was done before plasma polymerization. Even poorer results were obtained. All of the samples failed the test. One interesting phenomenon observed in this case was that copper particles seemed to partially dissolve into the plating solution before the surface was covered entirely by a plated copper film. This fact became more evident when electroless plating was carried out onto sputter-coated copper films in argon or helium plasma either with or without oxygen plasma treatment. Partially uncovered areas were observed, and even the sputter-coated copper disappeared in those areas. Since copper electrodes can be oxidized in air or by an oxygen plasma treatment, the particles sputtered from the electrode surface can consist of copper oxides instead of copper during initial plasma deposition. This could account for the dissolution of copper particles during electroless plating in highly basic solution (pH 12–13).

To eliminate the possible incorporation of copper oxides into a polymer film, a hydrogen plasma treatment was carried out after the oxygen plasma treatment

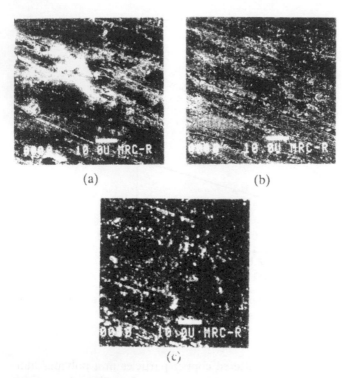

Figure 21.4 SEM micrographs of (a) as-received FRP substrate, (b) glow discharge activated (0.15 volume fraction copper), and (c) electroless copper plated surface ($\times$1000).

and before plasma polymerization. It was found in earlier work [12] that hydrogen plasma could remove the oxide layer on the electrode surface. Adhesion of electroless plated copper films improved considerably when the combined sequence (oxygen plasma treatment/hydrogen plasma treatment/plasma polymerization/ electroless plating) was used. No failures were observed during the tape test when this treatment was used. The same process was applied to nylon 66, phenolic resin, and ULTEM. In all cases, good adhesion was obtained.

The activation of nonconducting materials by the deposition of metal-containing polymer films could improve the autocatalytic metallization process by eliminating the aqueous etching and sensitizing steps. In addition, substrates that are hard to etch and activate by the aqueous process could be plated by this technique.

5. DIRECT ELECTROPLATING ONTO COMPOSITION-GRADED COMPOSITE FILMS

If the film, layered by the plasma polymerization technique, is sufficiently conductive and adheres to the substrate, direct electroplating can be performed. This simplifies the conventional process even further by removing the necessity of several presteps such as etching, neutralization, catalyzing, acceleration, and electroless plating. Five different process schemes were tried to further investigate the effectiveness of composition of transitional buffering films on the metallization of insulators in terms of adhesion and serviceability.

Composition-graded transitional buffering films were layered by simultaneous sputter deposition of electrode material with plasma polymerization using a methane (CH_4) monomer and argon carrier. The composition gradient through the film was accomplished by controlling the discharge power and the gas flow rates of methane and argon.

5.1. Substrate/Evaporation Deposited (EVD) Copper or Silver/Electrocopper (EVD Process)

EVD-copper or silver was deposited at a pressure of 2.1×10^{-3} Pa as a preplate for electroplating. A commercial unit utilizing a tungsten basket to contain and heat the metal was used. A 150-nm-thick copper or silver layer was obtained and followed by electrodeposition of copper.

5.2. Sputtered Copper or Silver/Electrocopper (Sputtering Process)

The following AC plasma sputtering conditions were used for depositing a copper or silver layer as a 1500 A preplate; Ar, 0.5 sccm, 140 W, pressure 1.4–1.9 Pa. The copper was then electroplated onto the sputtered copper or silver layer, as described previously.

5.3. Composition-Ungraded Film/ElectroCopper (Ungraded Process)

For these runs, flow rates of 1.5 sccm and 1.0 sccm, methane and argon, respectively, for the composition-ungraded step and 0.5 sccm as the flow rate of argon for the final sputtering step were used. The deposition of a normal layer of CH_4 was carried out first at 10 W for 10 min; then the methane flow was stopped and power was raised to 140 W for 18 min.

In the first step, there was simultaneous sputter deposition of electrode material and plasma polymerization, depositing a constant composition (ungraded) composite layer of copper or silver and methane plasma polymer. In the second step, the concentration of methane in the reactor gradually decreased. According to the pressure change in the reactor, the methane content in the reactor was nearly negligible 4 min after stopping the methane flow. Therefore, the deposition process in this stage can be considered to be just sputter deposition of copper or silver (depending on the electrode material). After an 18-min second-stage operation, the reactor was evacuated and sputter coating was carried out (140 W, 0.5 sccm Ar), untill the desired thickness (150 nm) was attained.

By the process described above, a plasma film could be obtained that had high enough electrical conductivity to allow direct electrodeposition of copper. The bulk resistivity of film measured by a four-point probe was 2.6×10^{-4} ohm-cm for the copper-containing polymer film when deposition was stopped after 18 min at 140 W. This value is critical if a uniform electrolytic deposit is to be obtained. For safety, deposition was carried out until a total film thickness of 150 nm was obtained, giving a nearly pure metallic layer thick enough to allow subsequent electroplating.

The final bulk resistivity was 7.8×10^{-4} ohm-cm and 6.750×10^{-4} ohm-cm for copper-containing and silver-containing films, respectively, taking the total thickness into account for calculation. The copper was then electroplated onto the composition-ungraded plasma film.

6. COMPOSITION-GRADED TRANSITIONAL BUFFERING FILM/ELECTROCOPPER

6.1. Single-Electrode Process (Single-Graded Process)

The combined gradual change in the methane-to-argon ratio and increase in discharge power resulted in the formation of a composition-graded film. During this cycle, the deposition rate of the methane plasma polymer decreased whereas that of copper or silver increased.

The composition depth profiles obtained from XPS data of the composition-ungraded and graded films in the case of a copper electrode used are given in Figure 21.5.

6.2. DOUBLE-ELECTRODE PROCESS (DOUBLE-GRADED PROCESS)

Titanium plates, which have lower sputtering tendency, were used as the other set of electrodes during these experiments and were arranged as shown in Figure 21.1. A discharge power of 80 W was supplied to the titanium plates in an attempt to obtain a higher degree of chemical bonding between the plasma polymer film and the polymer substrate. This reduced the incorporation of metal in the very initial film

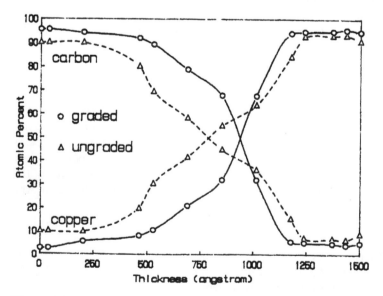

Figure 21.5 The depth profiles of the composition ungraded and graded (layered by single electrode process) films prepared by using copper electrode.

and kept the lower electrode set as clean as possible by avoiding deposition of the strongly adhering methane plasma polymer layer. Then the power was switched to the copper or silver electrode plates from the titanium electrode plates for sputter deposition.

7. ADHESION

7.1. Effects of Oxygen-Hydrogen Plasma Pretreatment

The results of tensile lap–shear tests of coatings produced on FRP substrates for different process schemes with and without plasma pretreatment are shown in Table 21.1. Examination of failed surfaces showed that separation occurred at the interface between the substrate and the bottom of the plasma film. No failure occurred between the plasma film and electrocopper layer.

The substrate condition is a crucial factor for attaining good film adhesion. Although detergent or solvent cleaning may provide a microscopically clean surface by removing certain contaminants such as scale, dust, or heavy grease, it is not satisfactory if a truly clean surface is needed. Also, the surface can be recontaminated in air during handling prior to the film deposition. These problems can be minimized or eliminated by conditioning the substrate with an oxygen plasma. Residual solvents or other contaminants from the air, which are normally hydrocarbons, can be easily removed by oxygen plasma through an ashing process.

Recontamination after the oxygen plasma treatment can be eliminated using in situ deposition of a plasma polymer film. However, the oxygen plasma treatment may give undesirable results. Weak zones on the substrate surface, through bond scission, and metallic oxides on the electrode surface can be formed. In particular, low molecular weight fragments formed by oxidation of the substrate surface have polar characteristics and may dissolve in aqueous solution, resulting in reduction of adhesion strength at the interface. The oxides, if any, on the electrode surface could be incorporated into plasma film and be attacked by acid in the plating solution penetrating into the initial interface probably through existing defects in the plasma film during electrodeposition of copper. However, a hydrogen plasma

Table 21.1   Effect of Pretreatment on Tensile Lap–Shear Strength

Preplate process	Lap–shear strength (MPa)	
	Copper electrode	Silver electrode
Sputtering, wo/c	3.9 ± 0.3[a]	3.7 ± 0.5
Sputtering, w/c	4.0 ± 0.4	3.7 ± 0.5
Single-graded, wo/c	5.3 ± 0.2	4.3 ± 0.2
Single-graded, w/c	6.7 ± 0.2	5.7 ± 0.5
Double-graded, wo/c	6.1 ± 0.6	6.0 ± 0.7
Double-graded, w/c	8.0 ± 0.7	8.2 ± 0.6

[a]95% confidence limits for five samples are shown.
wo/c, without oxygen and hydrogen pretreatment; w/c, with pretreatment.
If not specified, the flow ratio of methane to argon for the graded process scheme is 1.5/1.0. FRP was the substrate used.

treatment can minimize the undesirable effects of the oxygen plasma treatment by removing the oxygen of the substrate surface and reducing the surface oxides on the electrode surface. If the hydrogen plasma removes unstable oxidized carbon and polar groups formed by oxidation, the substrate surface will be more stable in aqueous solution. Therefore, the combination of oxygen and hydrogen plasma treatments can provide clean and favorable surface conditions for better adhesion.

Sputter cleaning with noble gas plasma such as argon might also be a possibility. However, in this case metal particles can be deposited during the cleaning process, especially when magnetron discharge is employed. In addition, the noble gas treatment may cause bond scission of the substrate surface, which will reduce the bond strength by ion bombardment [7].

No significant change in the surface morphologies of as-received, oxygen plasma–treated, and oxygen-hydrogen plasma–treated FRP substrates was detected. It indicates that there should not be appreciable improvement in bonding due to surface roughening.

No difference in bond strength for samples produced by the sputtering process can be attributed to the combination of etching, cleaning, and bond scission effects of the high-energy argon plasma. The high-energy argon plasma will modify the substrate surfaces and produce similar surface appearance, regardless of the previous pretreatments.

7.2. Composition Gradient Effects

The results of tensile lap–shear and chemical composition tests for different process schemes are shown in Tables 21.2–21.4. Metal analyses of the initial plasma coating layer and those of the failed surfaces after tensile lap–shear tests confirm that the failure occurred at the interface between the substrate surface and the bottom of the plasma polymer film. These results show that the intermediate layer provided by grading of the metal content throughout the plasma polymer film can improve the strength between the polymer and metal films. The graded metal-containing plasma polymer film can join a polymer and a metal with strong adhesion, and also reduce

Table 21.2 Effects of Plasma Process on Tensile Lap–Shear Strength

	Lap–shear strength (MPa)	
Preplate process	Copper electrode	Silver electrode
EVD	0.7 ± 0.0[a]	0.6 ± 0.0
Sputtering	4.0 ± 0.4	3.7 ± 0.5
Ungraded	5.8 ± 0.5	4.2 ± 0.4
Single graded	6.7 ± 0.2	5.7 ± 0.5
Single graded (1.0/1.0)	5.9 ± 0.4	4.8 ± 0.4
Double graded	8.0 ± 0.7	8.2 ± 0.6

[a]95% confidence limits for five samples are shown.
With oxygen and hydrogen pretreatment except for EVD. If not specified, the flow ratio of methane to argon for the graded process scheme is 1.5/1.0.

Table 21.3 Chemical Composition of the Initial Plasma-Coated Layer and the Failure Surface after Tensile Lap–Shear Testing (Copper Electrode)

Process	Initial Layer (At%)			Failure Surface (At%)		
	Cu	C	Others	Cu	C	Others
Ungraded (1.5/1.0)[a]	9.90	88.70	1.40	0.70	96.70	2.40
Single graded (1.5/1.0)	2.70	95.60	1.70	1.90	90.70	7.20
Single graded (1.0/1.0)	6.00	93.10	0.90	1.40	90.90	7.60
Double graded (1.5/1.0)	Ti	C	Others	Ti	C	Others
	0.10	98.50	1.40	0.05	94.10	5.85

[a]Represents the flow ratio of methane to argon.
The analysis was done on a fresh layer after sputtering off the very thin air-contaminated layer. All samples were pretreated with oxygen and hydrogen plasma.

Table 21.4 Chemical Composition of the Initial Plasma-Coated Layer and the Failure Surface. After Tensile Lap–Shear Testing (Silver Electrode)

Process	Initial Layer (At%)			Failure Surface (At%)		
	Ag	C	Others	Ag	C	Others
Ungraded (1.5/1.0)[a]	14.94	82.28	2.78	0.90	96.00	3.10
Single graded (1.5/1.0)	7.00	89.88	3.12	0.30	92.50	7.20
Single graded (1.0/1.0)	11.10	86.90	2.00	1.60	92.70	5.70
Double graded (1.5/1.0)	Ti	C	Others	Ti	C	Others
	0.11	96.30	3.59	0.02	93.17	6.81

[a]Represents the flow ratio of methane to argon.
The analysis was done on a fresh layer after sputtering off the very thin air-contaminated layer. All samples were pretreated with oxygen and hydrogen plasma.

the stress buildup at the interface caused by structural incompatibility between the polymer and metal by causing a more gradual structural change from "organic" to "metallic."

If the film is deposited by EVD or sputtering only, direct contact and less chemical bonding between the polymer and metal film result. Metallic bonding occurs between the sputter-deposited film and an electroplated film. However, the strength of the bonding layer between the substrate and EVD or sputtered metal is mainly dependent on mechanical interlocking or weaker physisorption and chemisorption interactions, which result in low tensile lap–shear strength. The data indicate that sputtering can bring about higher adhesion strength than evaporation deposition. This is probably due to the high energetic interaction of the sputtered metal with substrate surface, which will increase mechanical interlocking and atomic interfacial mixing during sputtering. If a composition-ungraded film is deposited as a preplate for electroplating, partial organic to organic bonding between the substrate and the plasma film is achieved and structural change is more or less stepwise. Thus, a slightly improved adhesion, compared with that for the sputtering process, is obtained.

Graded films have higher adhesion strengths because very firm organic-to-organic bonding at the interface between the substrate and initial plasma coating layer exists, and a gradual change of composition and structure over the thickness of film is obtained. If the single-graded process is used, copper or silver can be deposited at the very beginning of the film deposition due to the high sputter yield of copper or silver, and this may inhibit organic-to-organic bonding. This becomes clearer if adhesion values of the film at different argon-to-methane ratios are compared. A low methane-to-argon ratio gives high sputtering of copper or silver, thus decreasing the adhesive strength. The copper or silver content in the initial plasma polymer layer, as indicated in Tables 21.3 and 21.4, respectively, shows the expected trend.

With the single-electrode process, a metal-free substrate–plasma polymer interface can be obtained. However, this results in poorer adhesion of the plasma polymer because low-energy input must be supplied to prevent sputtering of electrode metal at the initial stage of the plasma operation. The adhesion strength of the plasma polymer itself increases with energy level [11]. When the double-graded process is used, the initial metal content can be kept quite low. Also, a high energy level (80 W) can be used to establish very firm organic-to-organic bonding between the substrate and the initial plasma polymer film using a titanium electrode because it has a low sputter yield.

7.3. Effects of Substrate Type

The results of the tensile lap–shear tests made on samples produced by the double-graded process on different substrates are shown in Table 21.5. SEM micrographs of the surface of the different polymer substrates are shown in Figure 21.6. Generally, phenolic resin and FRP seem to have the surface morphological features best suited for improved adhesion provided by mechanical interlocking. However, test results do not correlate with the prediction based on the morphology, indicating that the adhesion characteristic or chemical affinity of the methane plasma polymer for a substrate is more important and predominant in determining adhesion strength than surface roughness.

Table 21.5 Effect of Substrate Type on Tensile Lap–Shear Strength

	Lap–shear strength (MPa)	
Substrate type	Copper electrode	Silver electrode
Nylon 66	3.0 ± 0.4^a	3.0 ± 0.2
Teflon	3.2 ± 0.2	3.8 ± 0.0
Phenol	4.4 ± 0.3	6.0 ± 0.7
ULTEM	8.3 ± 0.4	9.0 ± 0.8
FRP	8.0 ± 0.7	8.2 ± 0.6

[a]95% confidence limits for five samples are shown.
Elecrocopper was plated onto the composition graded film layered by the double-graded process with oxygen and hydrogen treatment.

7.4. Effects of Electrolyte Additives

The results of the tensile lap–shear test obtained on samples made using different additives in the electroplating bath are shown in Table 21.6. Shear strength data show that an additive-free solution produces a more adhesive coating while the additives reduce the shear strength of the coating. It may be that the internal stress level is increased in the electrocopper deposit resulting from possible incorporation of additives into the copper lattice, bringing about a reduction of shear strength. Employment of rigorous solution purification seems to be a crucial

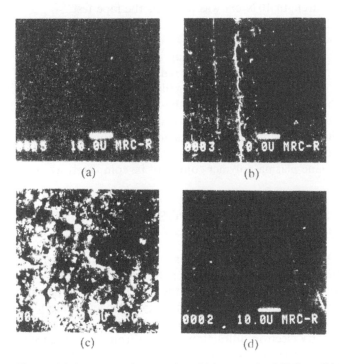

Figure 21.6 SEM micrographs of (a) as-received Teflon, (b) Nylon 66, (c) Phenolic Resin, and (d) ULTEM surfaces (×1000).

Table 21.6 Effects of Electrolyte Additives on Tensile Lap–Shear Strength

	Lap–shear strength (MPa)	
Additive	Copper electrode	Silver electrode
Blank	8.0 ± 0.7[a]	8.2 ± 0.6
Cl$^-$ 55 ppm	6.4 ± 0.5	7.9 ± 0.3
Brightener 0.5%	6.1 ± 0.7	7.3 ± 0.9
Brightener 0.5% and Cl$^-$ 55 ppm	6.3 ± 0.4	7.2 ± 0.4

[a] 95% confidence limits for five samples are shown.
Electrocopper was plated onto the compostion-graded film layered by the double-graded process with oxygen and hydrogen pretreatment. A commercial brighter was used. FRP was the substrate used.

factor for obtaining adherent electrocopper deposits, even though additive-free solution can produce a relatively sound and well adhering deposit, as shown in Table 21.6.

8. DURABILITY

Thermal cycling was performed on the coatings according to ASTM G5342-1972, followed by a semiquantitative tape test to check the adhesion strength. The temperature was cycled five times from 85°C to −40°C, unless failure occurred earlier. Adhesive scotch tape of peel strength 10 N/cm was used for the tape test.

Only the sample, which was produced by using the EVD-copper or silver preplate without pretreatment of oxygen and hydrogen, failed the durability test. Even though no visible defects appeared during the thermal cycling test, failure occurred by the tape test. No failure occurred with samples produced by employing composition-graded transitional buffering film. Even sputter-deposited films passed both tests. The good performance of the sputter-deposited sample during both tests can be attributed to inadvertent codeposition of resputtered residual plasma polymer, which had deposited in the reactor. Even though the inside of the reactor was thoroughly cleaned before sputtering, this kind of inadvertent codeposition of residual methane plasma polymer could not be completely avoided.

9. STRUCTURE AND STABILITY OF COMPOSITION-GRADED TRANSITIONAL BUFFERING PHASE

X-ray diffraction analysis, XPS analysis, and SEM and TEM analyses were performed to study the structure of composition-graded preplate films. In the X-ray diffraction pattern of composition graded films only the (111) orientation was in evidence. The TEM micrographs (Figure 21.7) show the very compact and random distribution of fine copper or silver crystalline particles as a major component with some residual carbon. The spotty electron diffraction rings indicate the existence of several microcrystals in the film. Two distinct particle sizes result from the process procedures. The compactly distributed fine particles come from the composition-graded step and the larger particles come from the final sputtering step during which the energy level of the plasma is higher.

The analysis of composition-graded films was obtained mainly by XPS examination. A composition-graded transitional buffering film deposited by the double-graded process using copper electrodes onto an FRP substrate was used as a sample for XPS measurements. This film was 150 nm thick and had an electrical resistivity of 3.9×10^{-5} ohm-cm. The XPS spectra and chemical composition of the composition-graded film as deposited (curve a) and after 1-min and 5-min sputtering (curves b and c) are shown in Figure 21.8 and Table 21.7, respectively. The XPS spectra were compared with the data in the literature [13] according to the binding energies. The corresponding binding energies are 932.4 eV (copper), 932.2 eV (Cu_2O), and 933.5 and 943.7 eV (CuO). The figure indicates that there is only a trace amount of copper oxide (CuO) at the surface. This probably resulted by reaction with the air due to the atmospheric exposure of the sample. No copper oxides are detected in the bulk of the film.

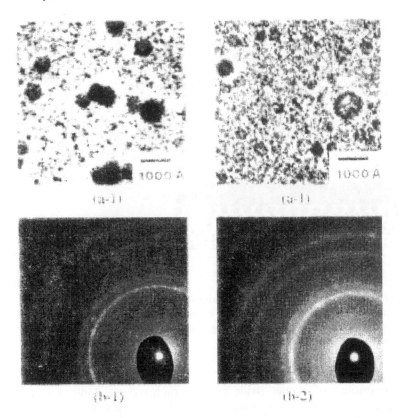

Figure 21.7 Transmission electron micrograph (a) (×88000) and electron diffraction pattern (b) of the composition graded transitional buffering film layered by double-graded process (1: copper electrode used; 2: silver electrode used).

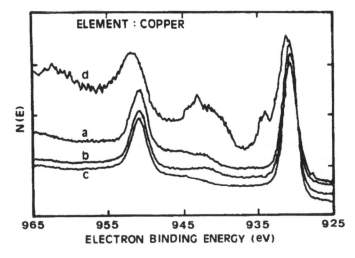

Figure 21.8 XPS spectra of the surface of composition graded transitional buffering film layered by double-graded process: curve a, surface curve b, 1 min sputter; curve c, 5 min sputter; curve d, oxidized copper surface.

Table 21.7 Surface Chemical Analysis of Composition-Graded
Film Layered by the Double-Graded Process [copper electrode
used] after Extended Time (48 h) Exposure to Atmosphere

Position	Copper	Carbon	Oxygen
0 min sputter	43.10	35.60	21.20
1 min sputter	93.10	4.10	2.60
5 min sputter	93.50	5.70	0.70

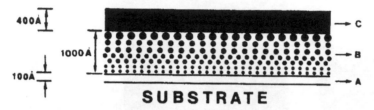

Figure 21.9 The proposed sectional model of a composition-graded transitional buffering
film layered by double-graded process: A, pure methane plasma polymer layer; B,
composition-graded layer of methane plasma polymer and metal; C, sputtered metal layer
with carbon contamination.

The negligible amount of copper oxide and the high oxygen content at the
surface suggest that most of the oxygen absorbed from the air exists in other forms
such as carbon monoxide or carbon dioxide. The surface of the composition-graded
film deposited by the double-graded process using silver electrodes was also tested by
XPS. No other significant chemical species besides silver in elemental form, carbon,
and oxygen was detected. The proposed sectional model of the composition-graded
film layered by the double-graded process is shown in Figure 21.9.

The stability of composition-graded films seems to be good because the
adhesion value and chemical analysis data obtained by XPS do not show any
significant changes after aging 2 months in a desiccator.

Composition-graded films, ranging chemically from organic to metallic by
their thickness, generated by a plasma polymerization technique can be used as an
intermediate layer for metallizing nonconducting substrates. Strong and durable
organic-to-organic chemical bonds are established between the substrate and the
initial layer while allowing a metallic bond between the top of the film and desired
metallizing material. The gradient is established through gradual changes in
structure and chemical composition, which is obtained by changing the deposition
parameters during film growth. The results may be summarized as follows:

1. Electroless plating on polymeric substrates activated by the sputter deposition of
   electrode material with simultaneous plasma polymerization is possible.
2. Composition-graded composite films of plasma polymer and metal can be
   controlled to give surfaces capable of being electroplated.
3. Surface conditioning by plasma pretreatment with oxygen and hydrogen is a
   crucial factor in obtaining very strong bonding at the interface between the
   substrate and composite plasma polymer film as a preplate for metallization.

4. The strongest and most durable film can be layered by initially establishing a high degree of organic-to-organic bonding by using the double-graded process with plasma pretreatment.
5. The shear strength of coatings metallized by employing composition-graded films is dependent on the extent and strength of chemical bonding between the substrate surface and the plasma polymer film.
6. The composition-graded film as a preplate for metallization is stable in the environment without moisture.

REFERENCES

1. Sun, B.K.; Cho, D.L.; O'Keefe, T.J.; Yasuda, H. In *Metallized Plastics 1*; Mittal, K.L., Susko, J.R., Eds.; Plenum Press: New York, 1989; 9–27.
2. Yasuda, H.; O'Keefe, T.J.; Cho, D.L.; Sun, B.K. J. Appl. Polym. Sci.: Appl. Polym. Symp. **1990**, *46*, 243.
3. Berry, R.W.; Hall, Peter M.; Harris, Murray T. *Thin Film Technology*; D. Van Nostrand: Princeton, 1968; Chap. 3.
4. Mittal, K.L.; J. Vac. Sci. Technol. **1976**, *13*, 19.
5. Benjamin, P.; Weaver, C. Proc. R. Soc. **1961**, *261*, 516–531.
6. Lindsay, J.H.; Lasala, Joseph. Plating Surf. Finishing **July 1985**, *72*, 50.
7. Celerier, A.; Machet, J. Thin Solid Films **1987**, *148*, 323.
8. Inagaki, N.; Yasuda, H. J. Adhesion **1982**, *13*, 201.
9. Dynes, P.J.; Kaelble, D.H. J. Macromol. Sci. Chem. **1976**, *A10*(3), 535–557.
10. Cho, D.L.; Yeh, Y.-S.; Yasuda, H. J. Vac. Sci. Technol., **1989**, *A7*, 2960.
11. Yasuda, H.K.; Sharma, A.K.; Hale, E.B.; James, W.J. J. Adhesion **1982**, *13*, 269.
12. Cho, D.L. Ph.D. Dissertation, University of Missouri–Rolla, 1986.
13. Osada, Y.; Yamada, K. Thin Solid Films **1987**, *151*, 71-86.

22
Tumbler Reactor

1. NEEDS FOR TUMBLER REACTOR

Developing a reactor for a luminous chemical vapor deposition (LCVD) operation to deal with large numbers of materials to be coated, probably the most difficult problem encountered is how to hold the substrates, place them in appropriate position, and move them in the luminous gas phase. In order to coat large number of substrate uniformly, the movement of substrate within the luminous gas phase is a mandatory requirement, with the exception of direct current (DC) cathodic LCVD. The difficulty progressively increases as the size of the substrate decreases, and it becomes virtually impossible to hold as the size reaches millimeters or less. For instance, the surface of small particulate matters cannot be treated or coated by the conventional modes of LCVD in which substrates are held by some kind of holder.

In handling unholdable, or difficult-to-hold, small substrates for luminous chemical vapor treatment (LCVT) or LCVD, substrates could be divided into two major groups. The first group is for the substrate that is small enough to be considered unholdable or difficult to hold, but is not considered as powders in the conventional sense. The size of such a substrate could be arbitrarily given by $1\,cm >$ largest dimension $> 1\,mm$. This group includes small-diameter rod, hollow tube, mesh, etc., and collectively termed *small-size substrates*. The second group is for powders, of which size is in the range $1\,mm >$ largest dimension $> 10\,nm$.

2. POWDER SURFACE TREATMENT OR COATING

Dealing with powders, it is necessary to either (1) drop or blow substrates through the luminous gas phase or (2) contain powders in the bottom of reactor and tumble the reactor wall in a manner similar to that of wet clothes in a drier.

The first method is a sure way to expose the surface of powder uniformly if one pass is sufficient to achieve the surface modification, but it is not easy to recycle the substrate in the luminous gas phase in vacuum. Therefore, the main issue in this approach is how to repeat the interaction of surface with the luminous gas phase efficiently, which entirely depends on the flow dynamics of powders. Multiple-step operation requires multiple discharge systems or repeated operation. The generation of discharge is more or less the same as the conventional modes used in LCVD reactors. External radio frequency electrodes or coil with glass tube is the most appropriate approach in this case. How to maximize the powder surface–luminous gas interaction and minimize the deposition on glass wall are the key factors of designing a reactor, which depends on flow characteristics of the powders to be handled. An example of powder coating by low-temperature cascade arc torch (LPCAT) is described in Chapter 16.

The second method is the tumbler treatment or coating of powders. So far as the handling of substrate is concerned, the tumbler coating of powder and the tumbler coating of small-size substrates are the same. However, many detailed factors involved in these reactors are not the same, and it is better to deal with these two processes separately.

In the tumbler reactor for powders, the surface of tumbler that holds powders must be a solid. A glass bottle that rotates between two radio frequency external electrodes or a coil, which is shown in Figure 22.1, is a simple tumbler reactor. Gas or monomer is fed by the tube located in the center of the bottle and pumped out through the outer tube of a double feed-through. While a steady-state flow of gas and a steady-state glow discharge is maintained in the bottle, an external motor rotates the bottle. The bottle has four baffle plates (indented glass wall), which carry powders to a higher angle and drop into the lower section creating mixing of powders. Increasing the rotating speed of the bottle can enhance the mixing. Factors that influence the efficiency of such a tumbler reactor are (1) density of powder, (2) sticking characteristics of powder (powder–powder and powder–wall), (3) total amount charged in the bottle, (4) discharge parameter (W/FM), and (5) reaction time.

Powders generally pack rather densely, particularly in comparison to small-size substrate, and the interaction of luminous gas (for deposition) with powder surface is limited to more or less the top powder layer. Consequently, the mixing by tumbling action is important to treat powder uniformly. The mixing along the axis of tube is also an important factor. If different color powders are placed in different

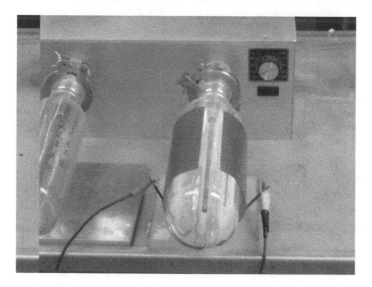

Figure 22.1   Glass bottle tumbler reactor with RF external electrodes.

locations in the tube, it is possible to estimate how long it takes to mix powders well, i.e., mixing time. The mixing time, of course, is a function of rotation rate, and depends on the design of baffle plates, which is placed on the surface of the bottle.

In principle, the rotating speed could be raised so that a considerable portion of powders is in gas phase; however, it causes difficulty in feeding monomer and establishing a uniform distribution of deposition rate along the axis of tube length. In other words, maintaining powders in the gas phase is not the prime purpose of the tumbler reactor.

The necessary reaction time is dictated by the mixing time. Therefore, the deposition rate should be adjusted to match the mixing time. Consequently, powder coating by a tumbler reactor is a time-consuming process by principle. If a condition that yields high deposition rate is employed without consideration of the mixing time, it merely creates unevenly treated powders. The required treatment time increases with the amount of powder charged in the reactor. Therefore, a larger reactor, which can be charged with a large amount of powders in one operation, often does not increase the production rate and reduces the uniformity.

Aggregating tendency of powders increases with decreasing powder size. Many submicrometer powders have strong aggregating tendency and exist as aggregated powders. One of the main reason for powder coating is often to reduce the aggregation of powders. Cautions should be taken not to coat aggregated powders, which will defeat the purpose of coating powders. This aspect is described in Chapter 16.

3. REACTOR WITH BASKETLIKE TUMBLER

Use of a basketlike tumbler, which has many holes on the substrate-retaining surface, can encompass the principle of tumbler reactor described above for coating of small-size substrates. Of course, the size of hole should be smaller than the minimal dimension of the substrate in order not to lose substrate during the tumbling operation. This approach has a great utility in handling large number of small-size substrates at relatively inexpensive reactor and operating costs, if certain requirements with respect to the nature of substrate were met.

The packing density of substrates is much less with small-size substrate, compared to that for powders, and luminous gas penetrates the space between substrates much more than in the case of powders. Consequently, rigorous tumbling is often not necessary. However, the efficiency of mixing described above still hold in handling of small-size substrate. The mass of a substrate is much greater than that of a powder, and the influence of the collision of substrate–substrate and substrate–wall cannot be ignored, particularly if the density of material that makes the substrate and the bulk density of a substrate are high.

If the material has a high density and soft bulk property, the external surface, on which LCVD and the collision with other surface occur, could be damaged considerably, including the damage to the once-deposited LCVD layer. If the internal surface, such as the inner surface of a tube-like substrate, is the main concern, the damage to the external surface is not a serious problem. On the other hand, if the external surface of small-size substrate is the main concern that necessitates the application of coating, some extra measure should be taken to avoid the effects of collisions. Thus, the shape, the bulk density of the substrate, and the main targeted surface are important factors that determine the feasibility of tumbler reactors.

If the coating of the inner surface of a tubelike substrate or the interior surface, which is protected from the surface damage, is the main goal of LCVD coating, the use of basketlike tumbler reactor is a straightforward process. In other words, if the damages on the exterior surface are not a concern, the use of basketlike tumbler reactor is the most effective approach. Depending on the size and shape of substrate, the main surface could be protected by attaching a substrate to a shield that protects the surface from colliding with other surfaces. Such a protective cap or basket should have the minimal surface area because the total surface area placed in the basketlike tumbler influence the deposition rate on the substrate.

If a substrate is placed in a metallic shield (cage), the substrate is placed in a Faraday cage. However, since glow discharge is created in the gas phase, which is outside of the Faraday cage, the luminous gas penetrates into the cage and the deposition occurs on the substrate surface, unless the total surface area of the cage is large enough to quench the luminous gas phase. The major reactive species in luminous gas phase in LCVD are reactive neutrals, not electrons and ions, which could penetrate a Faraday cage. This is a different situation from the case, in which a large metallic basket is placed in a glass tube, and external electrodes or a coil is used. In this case, the Faraday cage prevents the creation of glow inside the basket.

The dependence of deposition rate on number of substrates placed in a reactor is termed the *loading factor* of LCVD. The loading factor is not limited to a tumbler reactor but is applicable to all kinds of LCVD reactors. The loading factor is the sharing factor of deposition per substrate. In many cases, the total substrate surface placed in a reactor is relatively small in terms of the total surface area that collects deposition, and the loading factor is not appreciable or is ignored. However, dealing with small substrate to be coated by a tumbler coater, the loading factor is a key factor in designing a reactor and in determining how to operate such a reactor. Figure 22.2 depicts the loading factor calculated for a set of experiments at a fixed deposition rate, in which only the number of substrates is changed.

Since a relatively large volume reactor is employed, e.g., 200–500 liters for basketlike tumbler reactors, it is nearly mandatory to charge sufficient number of substrate, which makes experiments with small number of substrates difficult and inefficient. On the other hand, the total number of substrates that can be charged in the basketlike tumbler is restricted by other factors of a tumbler reactor. A smaller reactor by this principle is difficult, as will be made clear by the same reason described below.

4. ELECTRODES FOR BASKETLIKE TUMBLER REACTOR

The basketlike tumbler rotates in a vacuum reactor, into which the monomer is fed. Therefore, there are three basic volumes involved: (1) the volume of the vacuum reactor, V_1; (2) the volume of basketlike tumbler, V_2; and the volume of the internal electrode, V_3, as depicted in Figure 22.3. Since V_3 is a dead volume, the effective

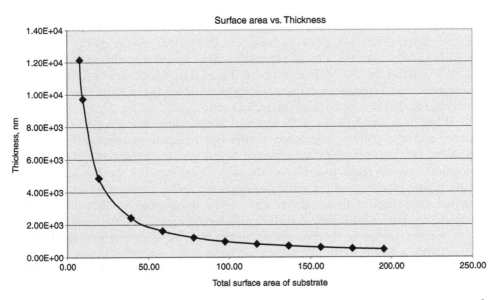

Figure 22.2 The loading factor of LCVD in a basket like tumbler reactor; surface area in m².

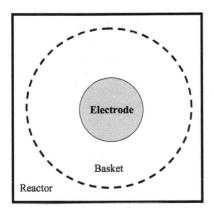

Figure 22.3 Relative position of basket and electrode in a vacuum reactor.

volume of reactor is $(V_1 - V_3)$, and the volume in the basket is $(V_2 - V_3)$. The volume available for substrates is roughly one-fourth of $(V_2 - V_3)$. The volume outside of the basket $(V_1 - V_2)$ holds a significant amount of monomer, which does not directly contribute to the deposition on substrates regardless of in which volume glow discharge is created.

In order to create glow discharge in the basket, it is necessary to insert the hot electrode coaxially placed in the center of the basket and ground the reactor and the tumbler basket as the counterelectrodes. Without the hot center electrode, glow discharge will develop between the reactor wall and the metal basket, i.e., in the volume $(V_1 - V_2)$ outside of the basket, and develop the same situation for the *secondary plasma* discussed in Chapter 18, which could be good for surface treating by the principle of the secondary plasma, but not for coating. If we place the hot electrode inside the basket, glow discharge can be created in volume $(V_2 - V_3)$ inside the basket. However, the size of electrode, more specifically the surface area of electrode relative to the surface area of the basket, which acts as the primary counterelectrode, becomes an extremely important factor for creating glow discharge in the basket that can be utilized for LCVD coating.

When two metal plates with identical surface area are placed in a bell jar reactor, and one plate is connected to the hot terminal of the radio frequency generator, and another plate is connected to the grounded terminal, and measure deposition rate on these two plates, the deposition rate on the hot electrode is always higher than that on the grounded electrode unless a special measure is taken to avoid this effect. This is mainly due to the unbalanced surface area of two electrodes in spite of the identical surface area for the two plates. All metallic surface in the reactor, including the bottom plate, the metallic skirt, the metallic bell jar, and so forth, act as the grounded electrode because the reactor is always grounded for the safety reason. Thus, the total surface area of the grounded electrode is much greater than the surface area of the hot electrode. The higher deposition to the hot electrode is due to this *unbalanced electrode surface*.

This effect of unbalanced electrode surface area is often neglected under the shadow of the self-biasing effect. It is considered that the unbalanced electrode area is the dominant factor that influences the deposition in luminous gas phase, of which the majority of reactive species are neutral. If a small-diameter rod is inserted to the center of the basketlike tumbler as the hot electrode, an extremely high deposition of black carbonaceous film occurs to the rod, and virtually no deposition occurs to the small-size substrates placed in the basket due to the unbalanced electrode surface effect. The extent of the unbalanced electrode effect is dependent to some extent on the nature of monomer. The monomers that deposit easily by LCVD show a pronounced effect and hence cannot be deposited on the substrate.

In order to create a luminous gas phase that can be utilized in a basketlike tumbler reactor for various monomers, it is necessary to increase the diameter of the center hot electrode to reduce the unbalance in electrode surfaces. The surface area of the center hot electrode and the surface area of the basket are the main issues in this situation because the basket is the primary counterelectrode and any other grounded surface does not play a role in this specific case. The increase of diameter of the center electrode reduces the extent of the unbalance but at the same time reduces the volume in which small-size substrates could be placed. Thus, there is a trade-off point between the electrode diameter and the utilizable basket volume. Roughly 40% of the basket diameter could be a reasonable diameter for the center electrode.

Thus, a tumbler reactor with basket sample holder could be a very efficient and inexpensive means to apply plasma polymer coating to suitable small substrate.

23
Surface Configuration

1. MOLECULAR CONFIGURATION VS. SURFACE CONFIGURATION

It is generally recognized that the surface of material is different from the bulk of the same material. However, there are few parameters that could be used to describe differences between bulk and surface properties of a material. With many surface phenomena, such as adsorption of proteins on polymer surfaces, a polymer surface is often treated as if the surface is a rigid and imperturbable plane on which proteins are adsorbed. The configuration of a protein that is adsorbed is one of the most important issues in dealing with the biocompatibility of polymers. However, little attention has been paid to the responsive movement of polymer segments to accommodate a protein in a specific configuration, such as a hydrophilic or hydrophobic moiety facing the surface. The responsive movement of polymer segments can be treated as perturbation of the surface by the contacting phase, which contains a certain kind of protein in the case of the protein adsorption.

The perturbability of a polymer surface is a function of chemical moieties on the polymer chain and the mobility of the chain. The importance of chemical moieties is obvious, but the role of the mobility of the polymer chain to allow a certain surface configuration is not well understood or is often ignored. There is no argument that poly(vinyl alcohol) (PVA) has many OH groups on a polymer chain, but it cannot be intuitively assumed that all or the majority of OH groups are located on the top surface facing outward. As a matter of fact, the majority of OH groups are not on the top surface of a dry PVA film.

In recent years it has been gradually recognized that the surface of polymers is highly perturbable and that the actual arrangement of chemical moieties of polymer at the surface changes when the surface is brought into contact with a new surrounding medium, such as when a polymer surface is immersed in water [1–14]. The change of surface characteristics has been a focal point of the general phenomena recognized by such terms as *surface dynamics, surface reconstruction, hydrophobic recovery,* and so forth, which are reflection of the change of chemical and morphological properties of polymer surface due to the change of the contacting medium.

Langmuir in 1938 made the following observations, which pointed out the important aspect of the surface [15]. (1) The wettability of a surface involves only short-range forces and depends primarily on the nature and arrangement of the atoms that form the actual surface and not on the arrangement of the underlying molecular layer. (2) Under special circumstances a layer at the surface may undergo an almost instantaneous reversal of orientation. (3) Anchoring the reactive groups by forming complexes can prevent such a change.

These observations point out the following important concepts:

1. It is necessary to distinguish the configuration of molecule and the surface configuration of the same molecule, which is a specific arrangement of atoms or ligands at the surface under a specific condition.
2. The surface configuration is a function of the contacting phase.
3. A surface configuration could be fixed to yield an imperturbable or less perturbable surface.

Whether or not the surface of a polymer is hydrophilic or hydrophobic is not determined by whether hydrophilic or hydrophobic moieties exist in a polymer molecule but by what kind of moieties actually occupy the top surface of the polymer. Namely, the surface properties are controlled by the surface configuration but not by the configuration of polymer molecules. The configuration of PVA can be represented by $-[CH_2-CH(OH)]_n-$, which indicates that the polymer is highly hydrophilic, and PVA is water soluble. However, the surface characteristics of a film of PVA depend entirely on how many OH groups exist on the top surface or buried in the bulk phase of the polymer that is described by the surface configuration.

2. SURFACE CHARACTERISTICS OF GELATIN GEL AND OF AGAR-AGAR GEL

Probably the most dramatic demonstration of the importance of surface configuration and its distinction from the molecular configuration is the surface characteristics of a hydrogel of gelatin, which contains more than 90% of water [13]. It is a general expectation that the surface of such a hydrogel of water-soluble polymer is highly hydrophilic and easily wettable by water. Contrary to this expectation, the surface of gelatin hydrogel (gel/air interface) is very hydrophobic as depicted by the contact angle of water in Figure 23.1. The high advancing sessile droplet contact angles indicate that the gel/air interface is hydrophobic, but the low receding contact angles indicate that the gel/water interface is hydrophilic. The figure also shows that the advancing contact angle of sessile water droplet is dependent on the humidity of ambient air, i.e., the lower the humidity, the higher is the sessile droplet contact angle of water.

In contrast to gelatin gel, hydrogel of agar-agar, which has approximately the same water content, shows very hydrophilic surface characteristics as depicted by the similar plots in Figure 23.2. Both the advancing and receding sessile droplet contact angles are low, which indicate that gel/air interface and gel/water interface are hydrophilic. This means that hydrophilic moieties are present at the interface of gel and air and of gel and water. It further implies that the presence of hydrophilic moieties is not influenced by the nature of the contacting medium (air or water). In the case of gelatin hydrogel, the presence of hydrophilic moieties is greatly

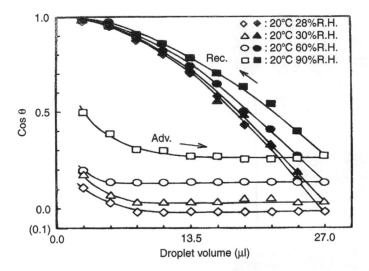

Figure 23.1 Effect of droplet volume on advancing and receding contact angles of water on a gelatin gel.

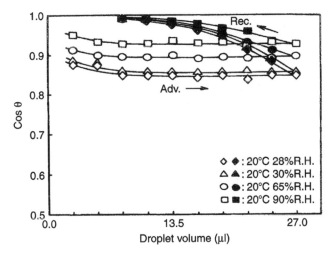

Figure 23.2 Effect of droplet volume on advancing and receding contact angles of water on an agar gel.

influenced by the nature of the contacting medium. In hydrogels that contain more than 90% of solvent (water), the short-range mobility of macromolecules is high, and the difference between gelatin and agar-agar hydrogels cannot be explained by the segmental mobility. The difference between the behavior of gelatin hydrogel and agar-agar hydrogen can be explained by the characteristic difference in molecular configuration of macromolecules that constitute hydrogels.

The surface configuration and the possibility of changing the surface configuration could be visualized by examining the configuration of several repeating units of a polymer. Figure 23.3 depicts a stretched conformation of the repeating

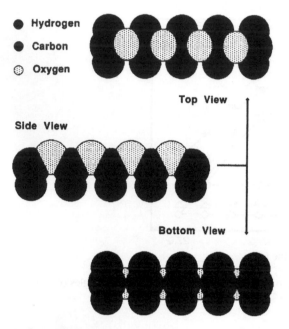

Figure 23.3 Top view, side view, and bottom view of a stretched segment of polyoxymethylene.

unit of polyoxymethylene (POM) in top, side, and bottom views. Although such a stretched conformation is unlikely to be found in a POM film, the figure illustrates the meaning and importance of surface configuration. Viewing this particular conformation of the polymer from the top, three surface configurations can be identified. If the surface configuration is as shown in the top view, in which all oxygen atoms face upward, the configuration would cause a very hydrophilic surface. If the surface configuration is as shown in the bottom view, in which no oxygen atom is on the top, the surface configuration would cause a very hydrophobic surface. If the surface configuration is as shown in the side view, in which oxygen atoms are partially exposed to the top, the surface configuration could cause a partially hydrophilic or partially hydrophobic surface.

The surface of POM is hydrophobic, indicating that oxygen atoms are mostly not exposed to the top surface and the surface can be more or less represented by the bottom view shown in the figure. It is important to recognize that the molecular configuration, i.e., $-(CH_2-O)_n-$, is fixed, and the three different surface configurations under consideration are the result of different orientations of the conformation of POM segment shown in the figure. The actual surface configurations that can be taken by POM molecules are much more complex because the surface configuration can be changed by changing the conformation of macromolecules (the conformation of the segment is fixed in the figure). Figure 23.3 clearly shows that POM could take numbers of surface configurations, which indicate that the surface of POM could be highly perturbable.

The number of possible surface configurations could be correlated to the stability or imperturbability of the surface. Figure 23.4 compares the possibility

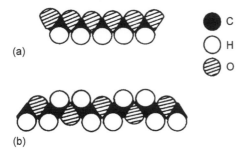

(a)

● C

○ H

◐ O

(b)

Figure 23.4 Comparison of possible surface configurations for polyoxymethylene and polyoxyethylene.

of surface configurations for POM and polyoxyethylene (POE). A striking difference between POM and POE is in the possibility of taking various surface configurations. As can be easily visualized from the figure, the rotation along the axis of the chain of POE yields very little change in the surface configuration, whereas the same action on POM yields numerous surface configurations. Thus, the POE surface is imperturbable, or at least less perturbable.

Why two water-soluble polymers (gelatin and agar-agar) in the hydrogel with the same water content show such a dramatic difference could be explained by the difference of configuration of macromolecules that allows or hinders the surface configuration change. The molecule of gelatin is the denatured collagen (single-strand protein), which takes random conformation in solution and in the hydrogel. The molecule of agar-agar is a polysaccharide, which is shown in Figure 23.5. The molecule of agar-agar also takes the random conformation in the hydrogel; however, the number of surface configurations that can be taken is very small because of the molecular configuration of the macromolecule. First, the repeating unit is planar in which the relative positions of atoms are fixed. Second, the hydrophilic moieties (OH groups) are located on both sides of the plane. The rotation along the chain axis does not yield the consequence explained by POM above. Namely, if an OH group is moved from the surface by the rotation, another OH group on the other side of the plane comes to the surface.

The difference in the surface characteristics of gelatin gel and agar-agar gel can be explained by the orientation of hydrophilic moieties at the surface as shown in Figure 23.6. In the case of gelatin gel, the hydrophilic moieties prefer to stay in the aqueous phase leaving the hydrophobic sections of a macromolecule on the top surface. The orientation changes immediately when the gel is placed in water or taken out of water because of the high mobility of segments in the low concentration (highly hydrated gel). Consequently, if the contact angle is measured at the gel/water interface rather than the gel/air interface, the sessile babble contact angle is low indicating that the same gel shows very hydrophilic characteristics.

In the case of agar-agar gel, in which the double-sided model represents the hydrophilic moiety, the number of the hydrophilic moieties at the air/gel interface and at the water/gel interface is nearly the same. Consequently, the sessile droplet contact angle of water is low and nearly identical to the sessile bubble contact angle of water.

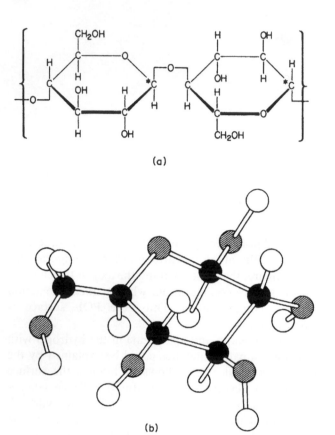

(a)

(b)

Figure 23.5 The molecular configuration of the repeating unit of agar-agar. (a) molecular configuration of the repeating unit; (b) three dimensional structure of a sugar unit.

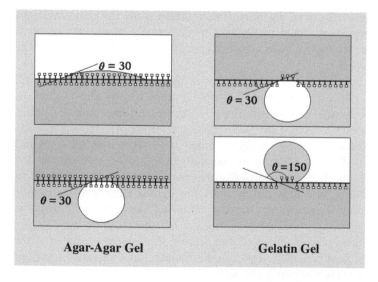

Figure 23.6 The effects of the orientation of hydrophilic moieties on the sessile droplet contact angle and sessile bubble contact angle.

3. OXYGEN ATOMS NEAR THE TOP SURFACE OF ETHYLENE–VINYL ALCOHOL COPOLYMER

Some evidence for the change of surface configuration due to the change of the contacting medium could be seen also with other less hydrophilic and hydrophobic polymers. A copolymer of ethylene–vinyl alcohol (EVA) contains a certain number of OH groups according to the copolymer composition. However, the OH groups are not uniformly distributed in the top surface region. On the surface equilibrated in air, the more O atoms exist away from the top surface, which can be observed by the angular dependence of XPS O 1s peak. When the film is immersed in water, the OH groups buried in the bulk phase migrate to the surface, which is now in contact with liquid water. The term "migrate" is used to describe the motion in a certain direction; however, it does not mean the long-range migration such as diffusive migration. This migration can be followed by XPS as a function of water immersion time by adopting the freeze-drying technique to eliminate water without allowing remigration or reorientation of OH groups near the surface. Figure 23.7 depicts a plot of the O 1s to C 1s ratio against $\sin \alpha$, where α is the take-off angle of XPS measurement, representing 90 degrees to 15 degrees as a function of water immersion time. The figure shows the change of the concentration of O atoms near the surface. The concentration profile completely changes its shape as a function of water immersion time. After 60 min immersion in water, the highest concentration is observed at the top surface [16].

In a dry film, the concentration of O is lower at the top layer and more O exists in the deeper region of the surface state. This implies that –OH groups tend to be buried in the inner part of the top surface region rather than being exposed at the top surface. The concentration of O at the top surface (represented by data points at $\alpha = 15$ degrees, the smallest value of $\sin \alpha$) increases as a function of immersion time that yields a funnel-shaped XPS angular dependence profile change shown in the figure. This figure clearly indicates that O atoms in a dry film exist away from the top

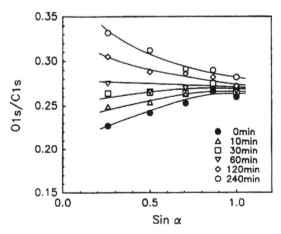

Figure 23.7 Change of the concentration of O atom near the top surface of ethylene-vinyl alcohol copolymer film as a function of water immersion time.

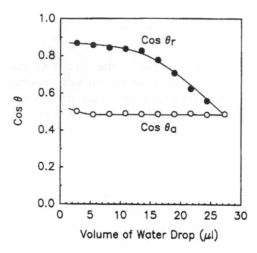

Figure 23.8 Advancing and receding sessile droplet contact angle of water on a dry ethylene/vinyl alcohol film.

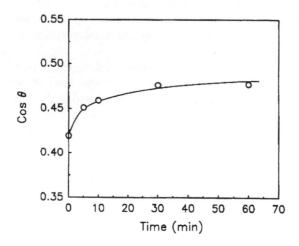

Figure 23.9 Change of sessile droplet contact angle with water immersion time at 40°C.

surface. When a film is immersed in water, O atoms are brought closer to the top surface and its extent increases with water immersion time.

The static advancing and static receding contact angles of water on a dry film of EVA showed a significant hysteresis effect as shown in Figure 23.8. The large hysteresis is due to the fact that the surface configuration under the water droplet is changed significantly and can be viewed as the evidence for surface configuration change due to the water contact. Sample immersed in 40°C water showed significantly lower advancing sessile droplet contact angle, and the value decreases with the immersion time. Figure 23.9 depicts the increase of cos θ as a function of immersion time in 40°C water.

While contact angle data show the influence of water immersion clearly, XPS analysis of atomic ratio for O/C measured at the normally employed take-off angle of 90 degrees shows only slight increases, as shown in Figure 23.10. While contact angle measurement reflects the change at the top layer very sensitively, XPS analysis reflects change in a much thicker layer of surface based on the electron escaping depth in order of 2.5–5 nm. However, the angular dependence of XPS data (Fig. 23.7) clearly shows the change of the location of –OH groups due to the water immersion.

An appreciable increase of the surface oxygen concentration, represented by XPS measurement at 15-degree take-off angle, with immersion time is observed with samples immersed in water at a higher temperature as shown in Figure 23.11. The increase of surface oxygen observed in this case can be viewed as a consequence of complex phenomena that involve transport of water from the surface to the

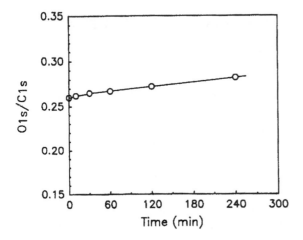

Figure 23.10 Change of XPS O 1s/C 1s ratio with water immersion time.

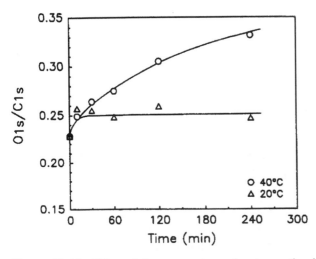

Figure 23.11 Effect of the temperature of water on the change of XPS O 1s/C 1s ratio as a function of water immersion time.

bulk phase of polymer, swelling of polymer, and change of surface configuration achieved by the rotational motion of the chain and the segmental motion, etc., which are functions of temperature. The influence of polymer surface property changes is discussed in Chapter 25.

According to the concept of the "equilibration of surface state," which is described in detail in Chapter 24, the surface configuration of such a copolymer is always changing according to the conditions of contacting medium, such as temperature and relative humidity. Therefore, the surface configuration of a copolymer depends on the history of a sample, and consequently it is very difficult to establish a reference state for a copolymer in a generic sense. For example, the values of advancing contact angle vary from sample to sample depending on the history of samples.

4. O ATOMS DETECTABLE BY XPS IN TEFLON PFA FILM

With hydrophobic polymers, the influence of water immersion is much simpler because water does not influence polymers as a solvent. Because of this, on the other hand, no significant change could be seen as a consequence of water immersion in general cases, except the surface charge buildup observed with Teflon. However, a very significant change due to water immersion was observed with Teflon PFA, which contains a small amount ($\sim$3%) of perfluoro(propylvinylether) [14]. This small amount of comonomer is sufficient to prevent the crystallization of the polymer and make the copolymer injection-moldable. This level of oxygen atoms in the copolymer are hardly detectable by XPS because the atomic percentage of oxygen in the copolymer is small and the oxygen atoms are not likely to be found in the outer regions of a dry surface. When the film is immersed in liquid water, however, a conspicuous O 1s peak appears on the XPS spectrum and its intensity increases with the immersion time.

Figure 23.12 depicts the change of XPS peaks as a consequence of water immersion. When similar experiments were carried out with a mixture of water/ethyl alcohol (50:50 volume ratio), the conspicuous appearance and increase with time were not observed as shown in Figure 23.13. The increase of O 1s peak as function of water immersion time is depicted in Figure 23.14, which includes immersion in a dilute solution of a surfactant and 50:50 mixture of water and ethanol. The reason why the increase of O 1s is influenced by the water mixtures is related to the driving force responsible for such changes. This aspect is discussed in Chapter 25.

5. LOCALIZED SURFACE CONFIGURATION CHANGE UNDER A SESSILE DROPLET OF WATER

Data shown above clearly show that the surface configuration of polymer changes when the contact medium is changed from air to liquid water. The same phenomenon occurs when a sessile droplet of water is placed on the surface of a polymer. In this case, however, the surface configuration change occurs only on the surface, which is under the sessile droplet. When the surface configuration change occurs, it creates the interaction force between the surface and the sessile droplet and holds

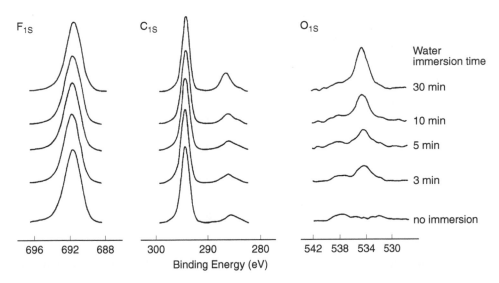

Figure 23.12 The change of XPS spectrum of Teflon PFA as a function of water immersion time.

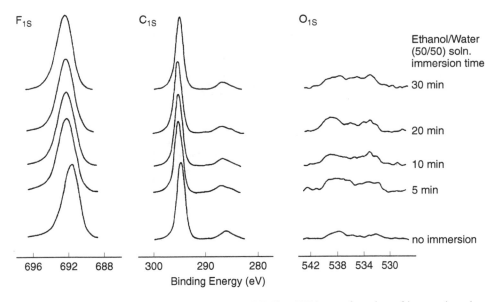

Figure 23.13 The change of XPS spectrum of Teflon PFA as a function of immersion time in 50/50 ethanol/water mixture.

the droplet on the surface. When the water droplet volume is reduced for the measurement of the receding sessile droplet contact angle, the droplet volume could decrease without changing the contact area of the sessile droplet, thus changing the shape to a flatter droplet. This is the major reason why receding sessile droplet contact angle decreases on the receding contact angle measurement. Figure 23.15

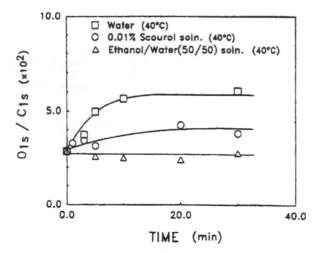

Figure 23.14 Effect of surface tension of water on the change of XPS O 1s/C 1s ratio as a function of water immersion time.

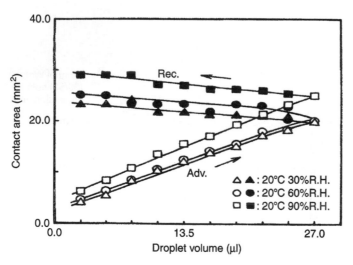

Figure 23.15 Effect of droplet volume on the contact area of water sessile droplet on a gelatin gel.

depicts the change of the contact area on the advancing and receding contact angle measurements observed with a gelatin gel [13]. In the receding process, the contact area increases slightly with time in spite of the decreasing volume of the droplet, indicating that that water sessile droplet is advancing by itself due to the strong interaction of water and gelatin molecules.

The more dramatic demonstration of the creation of interaction force between the surface and a sessile droplet of water can be seen in the measurement of the sessile droplet rolling-off angle, of which principle is depicted in Figure 23.16. If the sessile droplet contact angle is high and no surface configuration change occurs, such

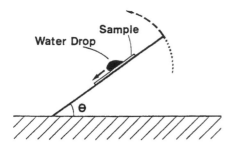

Figure 23.16 Schematic presentation of sessile droplet sliding angle measurement.

Figure 23.17 Pictorial view of a sessile droplet of water with high contact angle remaining on the vertical surface.

as in the case of Teflon surface in which no hydrophilic moiety is involved, the sessile droplet rolls off from the position where the sessile droplet was placed and rolls on the surface even with a slight incline of the surface, i.e., the rolling-off angle is very small.

If the high contact angle is due to the specific surface configuration of the surface in the air/surface interface, such as the case of gelatin gel, the sessile droplet does not roll off at the angle that is anticipated from the high contact angle. Figure 23.17 depicts a pictorial view of a sessile droplet of water with a high contact angle adhering to the vertically placed surface [17].

The data presented in this chapter clearly show the importance of the concept of surface configuration and its change due to a change of contacting medium. It is important to realize that the "surface" generally referred to is the interface with the contacting medium (generally air), and the "surface" is in equilibrium or in the process of attaining equilibrium with the contacting medium. Every surface is an interface, which cannot be fully described without specifying the contacting medium. The change of surface configuration due to the change of contacting medium is a very important subject, which is described in detail in Chapters 24 and 25. The details of contact angles are described in Chapter 26.

REFERENCES

1. Holly, F. J.; Refojo, M. F. J. Biomed. Mater. Res. **1975**, *9*, 315.
2. Baskin, A.; Nishino, M.; TerMinassian-Sarage, L. J. Collid Interface Sci. **1976**, *54*, 317.
3. Ratner, B.; Weatherby, P.; Hoffman, A.; Kelly, M.; Scharpen, L. J. Appl. Polym. Sci. **1978**, *22*, 643.
4. Thomas, R.; Trifilet, R. Macromolecules **1979**, *13*, 45.
5. Sung, C.; Hu, C. J. Biomed. Mater. Res. **1979**, *13*, 45.
6. Yasuda, H.; Sharma, A. K.; Yasuda, T. J. Polym. Sci., Polym. Phys. Ed. **1981**, *19*, 1285.
7. Hanazawa, E.; IshimotoNippon Sechaku Kyokaishi, R. **1982**, *18*, 247.
8. Hanazawa, E.; Ishimoto, R.; Nippon Sechaku Kyokaishi **1983**, *19*, 95.
9. Harttig, H.; Huttinger, K. J. Colloid Interface Sci. **1983**, *48*, 520.
10. Yasuda, T.; Okuno, T.; Yoshida, H.; Yasuda, H. J. Polym. Sci., Part B: Polym. Phys. **1988**, *26*, 1781.
11. Yasuda, T.; Yoshida, H.; Okuno, T.; Yasuda, H. J. Polym. Sci., Part B: Polym. Phys. **1988**, *26*, 2061.
12. Yasuda, H.; Charlson, E. J.; Charlson, E. M.; Yasuda, T.; Miyama, M.; Okuno, T. Langmuir **1991**, *7*, 2394.
13. Yasuda, T.; Okuno, T.; Yasuda, H. Langmuir **1994**, *10*, 2435.
14. Hirotsugu Yasuda, Tsumuko Okuno, Yuko Sawa, and Takeshi Yasuda, Languir **1995**, *11*, 3255.
15. Langmuir, I. Science **1938**, *87*, 493.
16. Yasuda, T.; Miyama, M.; Yasuda, H.; Langmuir **1994**, *10*, 583.
17. Yu Iriyama, Takeshi Yasuda, Cho, D.L.; Yasuda, H. J. Appl. Polymer Sci. **1990**, *39*, 249.

24
Surface State

1. SURFACE, INTERFACE AND SURFACE STATE

The surface we conceive is an interface with air or vacuum. In this sense, the term "surface" is a mispresentation of the interface of a material with air or in some cases with vacuum; i.e., in the cases of surface analysis in vacuum such X-ray photoemission spectroscopy (XPS), secondary ion mass spectroscopy (SIM), and so forth. Only when the contacting medium is other than air or vacuum does the term *interface* come into our conception. Thus, if the contacting medium is a gas phase, the term interface is not generally used, and the interface is more or less limited to the boundary between two matters in condensed phase, i.e., liquid/liquid, liquid/solid and solid/solid. However, according to the broader meaning of a boundary between two different matters, every surface we conceive is an interface with a contacting medium. Therefore, "surface" cannot be fully described without identifying the contacting medium. The change of surface properties, often dealt as surface dynamics, could stem, at least in part, to the mishandling of interfaces.

In dealing with surface-related phenomena it is important to recognize the surface region of a material as a state of matter, i.e., *surface state* [1]. The concept of the surface state is based on the recognition that the surface of a material is significantly different from the bulk phase, and the difference could not be just in the top atomic monolayer. In the bulk phase the interaction of an atom, a molecule, or a segment of a macromolecule with the neighboring entities is more or less the same in any direction. At the surface such a balance is broken by the absence of neighboring entities beyond the plane of the surface, resulting in anisotropic interactions.

The influence of this imbalance of interactions extends some distance into the material from the surface. The real surface of a material is not an absolutely flat and smooth array of atoms like that found on the surface of a single crystal, and a surface might contain many imperfections, voids, and boundary domains between different phases. The materials in this region, whose properties differ from those of the bulk phase, constitute the surface state. In this context, a "surface" is a two-dimensional plane and a "surface state" is a three-dimensional phase. Interfacial phenomena should be interpreted by examining the interaction of two surface states that contact at an interface.

The term *state* is used to describe a state of matter. Crook described plasma as the fourth state of matter (according to the order of the characteristic kinetic

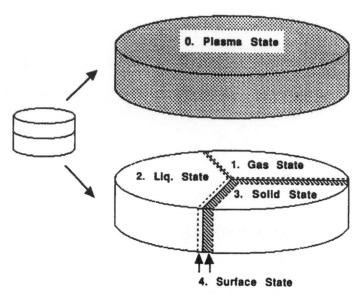

Figure 24.1   Surface states and equilibration of surface states in contact.

energy of entities in the state). The term state is used in *surface state* in precisely the same context. It is shown, however, that plasma is considered to be the zero state of matter (by reversing the order according to the cohesiveness of the state), and the surface state could be recognized as the substate of matter, which is adhering to its own phase. When two different states contact, the bulk phase of a state does not contact with the bulk phase of another state, but the corresponding surface states contact. The interface could be recognized as two surface states, each representing its own phase, in equilibrium. This situation is schematically depicted in Figure 24.1. At an interface between two states, e.g., liquid and solid, there exist two surface states, one in the solid and another in the liquid. These two surface states are in equilibrium or in the process of establishing equilibrium. In the latter case, the transitional process is recognized as surface dynamic change or surface reconstruction. Plasma treatment of a solid surface could be conceived, according to this concept, as the interaction of the surface state of plasma (in the vicinity of the substrate) and the surface state of the solid.

The influence of the imbalance in interactions at the surface may penetrate more deeply into a material for some characteristics than for others. Consequently, the depth or thickness of the surface state, which we can perceive, varies depending on what kind of properties or phenomena are being considered. For instance, the triboelectric property of plasma-deposited polymers, which is described later in this chapter, becomes independent of the substrate materials when the thickness of the plasma polymer reaches roughly 20–30 nm. Thus, the thickness of surface state corresponding to the contact electrification (surface electron transfer) is roughly in the order of 20 nm. However, the depth of surface state corresponding to the change of surface configuration of polymers, which results from the changes of the contacting medium, is probably significantly more than 20 nm.

The changes in the surface state of a polymer that occur when the surrounding medium is abruptly changed might be described as the process that leads to equilibration of surface states at the interface with the new contacting medium. When contact is established between two different materials, the surface states of the contacting materials begin to equilibrate. In the case of polymer immersed in water, the surface state of the polymer and the surface state of the water, which is recognized as the vicinal water in biological science, start to establish the equilibrium as soon as contact is made.

2. EQUILIBRATION OF SURFACE STATE ELECTRONS ON CONTACT

The concept of equilibration of surface states at an interface may be illustrated by the case in which the two contacting phases are solids. In such a case, the energy levels of the surface state electron can be used to explain the surface state equilibration that occurs on contact. When two dissimilar surfaces contact each other, the transfer of surface state electrons occurs to equilibrate the energy levels of surface state electrons at the newly created interface. When two surfaces are separated, each surface retains the equilibrium electron level, which has been just attained on the contact, leading to the creation of the static charge, if a material is, or both materials are, nonconducting. In such a case, the two surfaces stick together by the coulombic attraction and it is necessary to apply force to separate them.

The energy levels of the surface state of a solid can be illustrated by using the model for a semiconductor. Although most polymers are nonconducting, there exist free electrons on the surface of polymer, which are responsible for static charge buildup on the surface. By considering a polymer as a semiconductor with a large band gap and assigning quantized energy levels within the band gap to "surface state electrons," equilibration of two surface states on contact can be represented as a process involving electron transfer between two solid surfaces.

Figure 24.2 depicts the surface state electrons of two surfaces and the transfer of electrons when a contact is established between the two surfaces. When two surfaces are separated, each surface retains the newly established equilibrium surface electrons. Consequently, the surface that donated electrons becomes positively charged, and the surface that received electrons becomes negatively charged. This is the principle by which static charges are built up on surfaces due to contact and subsequent separation. The number of empty box in the figure can be viewed proportional to the electronegativity of a surface, i.e., surface B has higher electronegativity (capability to accept electrons) than surface A, and a surface with higher electronegativity charges negatively on the contact/separation process.

Completing a circuit and measuring the current that flows on contact or separation can quantify the transfer of electrons [2]. A typical result on contact electrification is shown in Figure 24.3. The time constants associated with the current peaks can be adjusted by inserting a series resistance in the measurement circuit. In real time, the equilibration of surface state electrons occurs instantaneously. In these experiments, the plasma polymer of tetrafluoroethylene (TFE) was deposited on two different substrate, nylon film and polished brass, and the contact and separation currents were measured with a (uncoated) brass probe

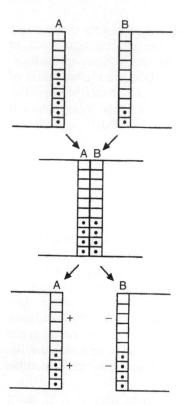

Figure 24.2 Schematic representation of the surface states equilibration illustrated by the transfer of surface state electrons.

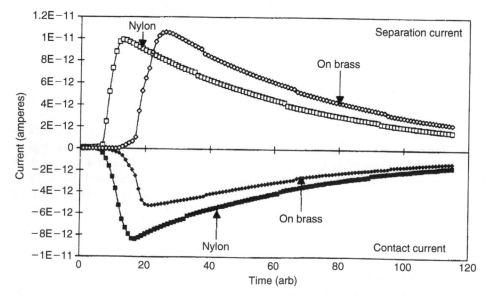

Figure 24.3 Measured contact- and separation-currents as a function of time.

as the reference surface. The time scale of the plots in the figure has been shifted so that the contact and separation currents are paired for each sample and the two different substrates can be distinguished. The thickness of the plasma polymer deposition is roughly 140 nm. The data shown in Figure 24.3 clearly show that the surface state equilibration illustrated in Figure 24.2 occurs on metal–polymer contact.

The peak current and the total area under the curve for the contact current and for the separation current are nearly identical, differing only in the sign. From the contact and separation current value a triboelectric series, which shows not only the position but also quantitative separation between two positions, is constructed as shown in Table 24.1. Triboelectric series based on the average current on contact and separation is given in Table 24.2, in which the zero point is set at nylon.

The sign of the static charge and rough extent of charge that will be built on a polymer surface can be predicted from the triboelectric series. The polymer that has a larger negative value will charge negatively and the polymer with a smaller negative value will charge positively. For instance, poly(ethylene terephthalate) (PET) will charge positively when it contacts with Teflon but will charge negatively when it contacts with polypropylene (PP). If a different reference is used, the value and the sign of charge change based on the reference surface used, but the relative position and the separation of points in the series do not change. It is important to recognize how far from the rest of group Teflon is located on the triboelectric series.

The electronegativities of elements involved in a surface state seem to dominate the overall electronegativity of the surface, i.e., the high electronegativity of F and the absence of other pending group in Teflon are responsible for the most

Table 24.1 Results of Contact Separation Current Measurements[a]

Polymer	PTFE	PET	KAPTON	PP	PS	PMMA	Nylon
Thickness (μm)	100	50	127	75	2000	1000	75
Contact charge (nC)							
Mean	−116.1	−13.1	−8.9	−3.1	−0.7	1.3	7.3
S D	4.6	0.5	0.5	0.0	0.1	0.2	0.4
Separation Cha. (nC)							
Mean	121.7	15.3	9.4	4.3	0.7	−1.8	−8.1
SD	3.7	0.4	0.3	0.2	0.2	0.1	0.4

[a]Reference surface: brass. Measurement was done in N_2 at room temperature. Samples were kept in N_2 for 4 days before the measurement. The contact and separation charge in nC per $6.12 \, mm^2$ sample (contact) area.

Table 24.2 Triboelectric Series Based on Contact Electrification/Separation Current

Polymer	PTFE	PET	KAPTON	PP	PS	PMMA	Nylon
Average value	−127	−22	−17	−11	−8.4	−7.5	0

electronegative polymer surface. Because of the highest electronegativity of Teflon, any surface attains a strong positive charge when the surface contacts with Teflon. Teflon films are often used to reduce the friction in handling films and fibers; however, this often causes a serious static charge problem when the humidity of ambient air is low.

Since most polymers are nonconducting, the electrons gained or lost in contact with a metal, such as a roll to control the movement of a film, remains on the surface whereas the countercharge built on a metal dissipates quickly because of its electrical conductivity and grounding of a machine. The charge built on the polymer surface is cumulative if the contact with the same metal is repeated. If a folded film touches metal rolls repeatedly and the number of contact is different for one side of film and for another side, the energy levels shown in Figure 24.2 for one side and for another side become different. In such a case, two films of the same polymer folded create static charges and cling to each other. If such surfaces are two surfaces of a plastic bag, the bag cannot be opened until the static charges are dissipated.

The problem associated with static charge of polymers is a good example for why the "polymer surface" should be treated as a "polymer/air interface." The humidity of air changes the surface state of a polymer due to the adsorption of water molecules on the surface state of the polymer. The reduction of static charge by an increase in the humidity is not due to the dissipation of charge to the gas phase, i.e., humid air, but due to the change of surface state in equilibration with humid air.

3. DEPTH OF SURFACE STATE AND INFLUENCE OF BULK PHASE

The data shown in Figure 24.3 indicate that the surface of plasma polymer of TFE shows similar behavior regardless of the substrate material on which it is deposited. The plasma polymer of TFE-coated brass and the plasma polymer of TFE-coated nylon 66 behave as if they are Teflon surface, i.e., the plasma polymer coating provided a new surface state on both substrate materials. This implies that the thickness of the surface state responsible for the surface electron exchange in this case is less than the thickness of the plasma polymer coating. If, on the other hand, the thickness of plasma polymer coating is less than the thickness of the surface state that participates the surface state electron exchange, the triboelectric characteristics would be influenced by the substrate material, and a plasma polymer of TFE coated on brass might behave differently for a plasma polymer of TFE coated on nylon 66. The threshold thickness of a plasma polymer coating above which the triboelectric property becomes independent of the substrate material would provide the critical thickness of surface state responsible for surface electron exchange.

Figure 24.4 depicts the change of surface electron energy level as a function of the thickness of a plasma polymer. In this case, plasma polymer of acetylene/N_2 was deposited on brass and the contact current was measured against nylon 66. The result indicates the following two important aspects of the surface state: First, the surface state electron energy level at a thin-coating thickness is influenced by that of the substrate material but becomes independent of the thickness above a threshold

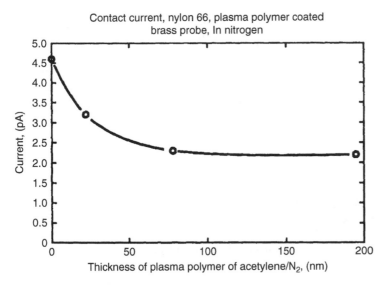

Figure 24.4   Effect of thickness of plasma polymer on contact current of nylon 66.

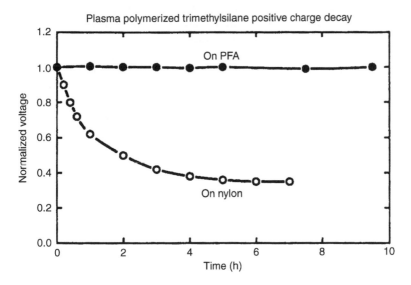

Figure 24.5   Dependence of charge decay on substrate polymer.

thickness. Second, the depth of the surface state, which is responsible for the contact electrification, seems to be around 20–30 nm. In order to create a new surface state that is independent of the substrate material by the deposition of a plasma polymer, it is necessary to deposit approximately 20-nm-thick layers.

Although the energy level of surface electrons becomes independent of the material beyond roughly 20 nm, the surface state is not totally independent of the bulk phase. This situation can be demonstrated by the influence of the bulk phase on the decay rate of static charge. Figure 24.5 depicts the influence of the bulk phase on the static charge decay. While the plasma polymer of trimethylsilane deposited on nylon shows dacay with time, the same polymer deposited on Teflon PFA does not show decay.

It is quite clear that the decay of static charge, both positive and negative, is highly dependent on the bulk properties of the substrate polymer, which confirms that the surface state is in equilibrium with the bulk phase. The decay of the positive charge residing on the hydrophobic plasma polymer seems to confirm that the dissipation of static charge occurs through the bulk phase of material.

4. INFLUENCE OF THE SURFACE STATE OF CONTACTING PHASE

The loss or gain of electrons at the surface is an important change of the surface state. The transfer of electron occurs even in the case of contacting liquid water, although its magnitude depends on the separation in the triboelectric series or the difference of work functions. The creation of surface charge by immersion and emersion of a polymer surface in water has been indeed observed with Teflon film. Therefore, polymeric surface, which is immersed in water and pulled out, has a different surface state from that before the immersion even though no discernible surface configuration change could occur. For most polymers, water immersion removes the residual charge, and it is a convenient mean to create a reference surface for the contact electrification experiment. The situation for Teflon is in strong contrast to most polymers with respect to water immersion. Due to the high electronegativity of fluorine atoms and large number of F on Teflon surface, nearly everything that contacts with Teflon causes negative charge on the surface. Liquid water is no exception in this situation.

The surface state is in equilibrium with the contacting medium, which means that the surface state changes when the contacting medium is changed. The surface configuration described in Chapter 23 is an important feature of the surface state of polymeric materials. The change of surface configuration means that the surface state of a polymeric material is changed.

The characteristics of water near a solid are significantly different from those in the bulk phase of water. This water is recognized as "vicinal water" in biological science. The vicinal water is the surface state of water in contact with a solid surface. The vicinal water has several transition temperatures that influence the properties of the vicinal water, among which the major transition temperature is 15°C.

The rate of surface configuration change, which occurs when the contacting medium is changed from air to liquid water, and also the change that occurs when the sample was kept in air, were investigated as a function of temperature by means of the surface tagging with CF_4 plasma treatment [1]. The rate of surface configuration change is determined from the initial rate of disappearance of XPS F 1s signal with time of water immersion or air exposure time. The resulting Arrhenius plot showed a conspicuous inflection point, indicating the presence of a transition temperature for the surface configuration change. When the same samples were exposed to dry air at different temperature, the rate of change was found to be two to three orders of

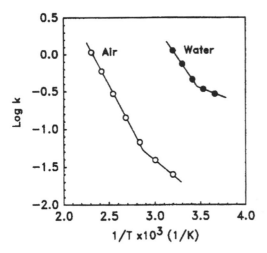

Figure 24.6 Temperature dependence of surface configuration change rate of PET in air and in water.

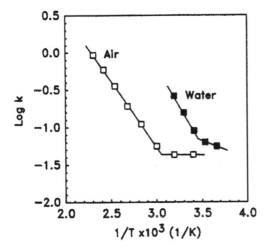

Figure 24.7 Temperature dependence of surface configuration change rate of nylon 6 in air and in water.

magnitude slower but again showed a transition temperature as shown in Figures 24.6, 24.7 and 24.8. Transition temperatures as well as activation energies above and below the transition temperature for the process in dry air and for the process in water are summarized in Table 24.3. The details of surface configuration changes are described in Chapter 25.

The transition temperatures observed in the dry experiment for various polymers coincided with the glass transition temperatures of polymers. This finding is in accordance with the process in which the segmental mobility plays the major role. However, the transition temperature found in water always occurred at around 15°C and was independent of the glass transition temperature of the polymer as

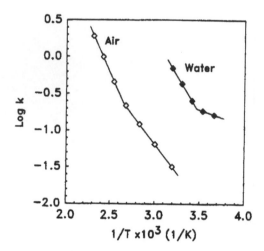

Figure 24.8 Temperature dependence of surface configuration change rate of PMMA in air and in water.

Table 24.3 Transition Temperatures and Activation Energies

			Temperature dependence of surface configuration change				
			In air			In water	
	T_g	T_s	E_a (kcal/mol)[a]		T_s	E_a (kcal/mol)[a]	
Polymer	(°C)	(°C)	$> T_s$	$< T_s$	(°C)	$> T_s$	$< T_s$
PET	60–85	74.5	10.6	4.6	15.2	8.2	2.2
Nylon 6	40–52	53.8	8.1	~0	15.4	9.7	2.0
PMMA	105	104.4	12.2	7.3	15.2	9.3	2.0

[a]Activation energy.

shown in Table 24.3. The temperature dependence showed a conspicuous inflection point at the same temperature for all polymers investigated. This temperature coincides with the major transition temperature of "vicinal water" known as Drost-Hansen temperature, implying that change of characteristics of vicinal water causes the change in the driving force for the process. The change corresponding to T_g should, in principle, exist even in the case of liquid water; however, due to the limited practical temperature range of water experiments, T_g was not observed in liquid water experiments with polymers examined. Figure 24.9 compares the temperature dependence of surface configuration change in water for PET, PMMA, PE, and nylon. In spite of vast difference in physical properties of polymers, the temperature dependence of surface configuration change in water is nearly the same both above and below the transition temperature.

If one considers that the surface state of polymer interacts with the surface state of water, not the bulk phase water, in the process of establishing the equilibrium, it is not surprising to find the same transition temperature for four

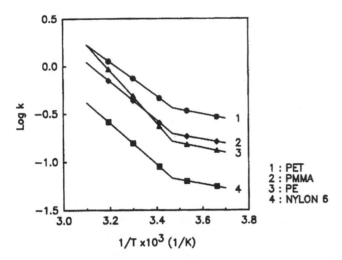

Figure 24.9   Temperature dependence of surface configuration change in water.

different polymers. Namely, what causes the change in the rate of surface configuration change is the activity of the surface state of water (vicinal water). If the activity of vicinal water changes at the transition temperature, it is logical to see that the rate of surface configuration change above and below the transition temperature of the vicinal water is nearly the same irrespective of substrate polymers.

5. DURABILITY AND BREAKDOWN OF SURFACE STATE

The surface state of a semicrystalline polymer, according to the three-phase model of semicrystalline polymers, consists of (1) a crystalline phase, (2) an amorphous (noncrystalline) phase, and (3) the transition phase between crystalline and amorphous phases. The macromolecules in the crystalline phase are considered immobile. The macromolecules in the amorphous phase are mobile, depending on $(T-T_g)$. The macromolecules in the transitional phase have restricted mobility due to the tie molecules to the crystalline phase. The transitional phase does not extend far from the surfaces of crystals, but the volume fraction of the transitional phase, though small, is proportional to the volume fraction of the crystalline phase. Thus, the higher the crystallinity, the higher is the effect of the transitional phase. The most widely used two-phase model ignores the presence of the transitional phase. However, for the integrity of the surface state, the role of the transitional phase seems to be very important because molecules cannot move as freely as in the amorphous phase (highly restricted mobility) but do not have well-assembled matrix as in the crystalline phase. Accordingly, the transitional phase acts as the weak boundary in the surface state.

Water permeates a flawless polymer film through the amorphous phase via "solution–diffusion" mechanisms. Therefore, water permeability is inversely proportional to the volume fraction of the crystalline phase (crystallinity). Water molecules are first dissolved into the polymer matrix at the interface; the dissolved water molecules diffuse through the polymer according to the chemical potential gradient across the film. Salt ions are hydrated with numbers of water molecules and are tightly associated with the counterions, which are also hydrated.

Accordingly, the transport of salt requires larger elementary free volume than does the transport of water molecules. Hydrated ions are much larger than water, and hydrated cation and hydrated anion must move together because of coulumbic attractive force between them. Consequently, salt ions cannot permeate an amphoteric hydrophobic/hydrophilic polymer, of which the hydration value is low, i.e., less than few volume percent, by the solution–diffusion principle. Therefore, salt permeation through a hydrophobic polymer film such as low-density polyethylene (LDPE) and parylene C film should not occur.

In reality, however, salts dissolved in water find or create paths into a hydrophobic polymer matrix and cause the breakdown of an insulating layer or the corrosion of the substrate metal. In contrast to the diffusion process described above, the process of salt going into polymer matrix could be termed *salt intrusion* because the breakdown of the surface state does not occur with water that does not contain salts. The salt intrusion starts with the breakdown of the surface state. In a study of the electric insulation characteristics of LDPE film, it was found that salt ions intrude into the polymer matrix by different mechanisms [3,4]. The exact mechanisms for salt intrusion are not known, but the phenomenological salt intrusion found can be summarized as follows, in an effort to explain the nature of salt intrusion that causes the breakdown of the surface state.

1. The alternating current (AC) resistivity of LDPE film does not change with water immersion even under electrical stress of $10\,kV/mm$ as shown in Figure 24.10, where the relative resistivity is plotted as a function of immersion time. The figures indicate that liquid water does not cause the breakdown of the surface state consisting of the three phases.

2. When an LDPE film is immersed in a salt solution (0.9% NaCl), the AC resistivity decreases as a function of the immersion time, as shown in Figure 24.11. These figures include the effect of a nanofilm of plasma polymer deposited on the surface of LDPE. With hydrophobic plasma polymer (HFE $+\,H_2$), the decrease of AC resistivity was not observed. These figures indicate that the surface state breakdown occurs when the salt intrusion takes place. The salt intrusion can be prevented by the application of a plasma polymer, which is an amorphous network (one phase and no weak boundary). The extent of protection seems to be dependent on the hydrophobicity of the network.

3. The insulation breakdown under electrical stress occurs in a fatigue mode, i.e., not a gradual deterioration but an abrupt failure, and is correlated to the salt intrusion characteristics of the film, as shown in Figures 24.12 and 24.13. The hydrophobic surface state created by plasma polymerization of (HFE $+\,H_2$) significantly prolongs the time when the breakdown of surface state by salt intrusion occurs.

Based on these observations, the distinction between water penetration and salt intrusion may be schematically represented as shown in Figures 24.14 and 24.15. In these figures, two interfaces—film/water and film/metal—are involved, and the effect of lack of water-insensitive adhesion at the film/metal interface is also depicted. In the absence of water-insensitive adhesion, water molecules that reach the polymer/metal interface cluster together by breaking weak polymer–metal interaction. If one

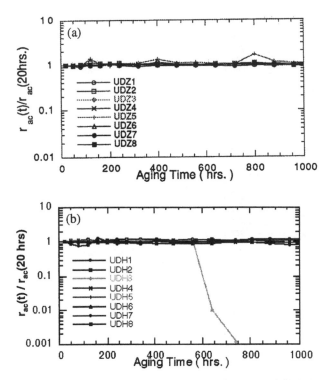

Figure 24.10 AC resistivity ratio versus aging time for untreated LDPE films in a deionized water environment: (a) unstressed, and (b) 10 kV/mm stressed.

assumes the diffusion constant of water through a polymer to be $10^{-8}\,cm^2/s$, the time lag of water diffusion through 15-μm-thick LDPE film is estimated to be 38 s. Any indication of the salt intrusion effect appears in a much longer time, i.e., days and months. Therefore, it is appropriate to assume that the intrusion occurs in a water-saturated polymer matrix. It is important to recognize that the surface state breakdown is the initial step of the salt intrusion. The effective prevention of salt intrusion by placing totally amorphous plasma polymer indicates that the transitional phase (between crystalline and amorphous phases) is responsible for the surface state breakdown.

The similar breakdown of the surface state was also observed with parylene C film. Parylene C is a semicrystalline polymer and one of the most effective barriers for gases and vapors according to permeability values. It was thought that such an excellent barrier film would provide superior corrosion protection of a metal when it was deposited on the metal surface. Contrary to the expectation, parylene C film didn't provide good corrosion protection due to the surface state breakdown described above. This conclusion was ascertained by studies using electrochemical impedance spectroscopy (EIS), which is described in Chapter 28.

The mechanisms involved in the deterioration of insulation characteristics of LDPE are analogous to the principle of water vapor permeation through it. Two distinctively different paths can yield the water vapor permeability at the same value, i.e., many water molecules moving rather slowly or few molecules moving

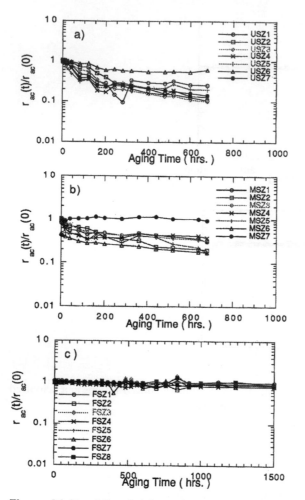

Figure 24.11 AC resistivity ratio versus aging time for unstressed LDPE films in a 0.9% saline environment: (a) untreated, (b) CH_4 plasma treated, and (c) $C_2F_6 + H_2$ (1 : 1) plasma treated.

rather rapidly, but in both cases the products of amount and rate could lead to a nearly identical value of water vapor permeability. (Polyethylene belongs to the latter case, i.e., low solubility and high diffusivity.)

The deterioration of insulating characteristic visualized by the tree growth phenomenon can be viewed in an analogous manner as the water vapor transport. Namely, there are two factors involved: initiation and rate of growth. The rate of tree growth has been the major research target in the past, but the numbers of trees or the seeding of trees (without artificially created nucleation site) has not been intensively investigated. It is interesting to see if plasma polymerization coating could prevent or reduce initiation of the fatigue process.

The tree growth rate is (bulk) characteristic of polymers, whereas the frequency of initiation of a tree is characteristic of a sample. While the bulk properties of a

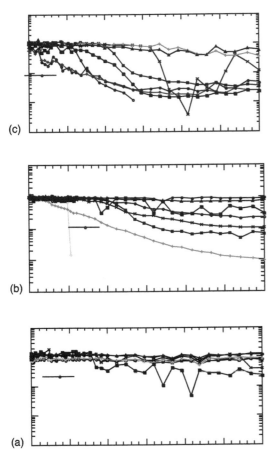

Figure 24.12 AC resistivity ratio vs. aging time for 5 kV/mm stressed samples in a 0.9% saline solution: (a) untreated, (b) CH$_4$ treated, (c) C$_2$F$_6$ + H$_2$ (1 : 1) treated.

polymer largely influence the surface characteristics of a sample, the latter cannot be uniquely related to the former. The sample (surface) characteristics largely depend on the factors involved in the processing of sample.

We should also distinguish the deterioration of the bulk properties and the breakdown due to the fatigue. The deterioration of bulk properties occurs as a consequence of the changes in the characteristics of polymer molecules such as degradation, oxidation, excessive crystallization, and cross-linking. The ultimate properties of the entire sample as a whole change gradually in such a deterioration mode. In the fatigue mode, the failure of a material property, such as tensile strength, electrical resistivity, and so forth, occurs after a sample is exposed to the stress level far below the ultimate breaking stress. In many cases, the localized failure is initiated at the surface and propagates through a specific path while the remaining portion of a sample stays unchanged. The state of surface of a polymer is a crucially important factor in the fatigue. The pattern of dielectric breakdown, shown in Figure 24.13, clearly indicates that we are not dealing with the change of bulk properties but rather with a breakdown in the fatigue mode.

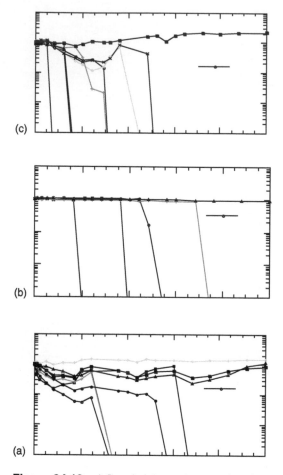

Figure 24.13 AC resistivity ratio vs. aging time for 10 kV/mm stressed samples in a 0.9% saline solution: (a) untreated, (b) CH_4 treated, (c) $C_2F_6 + H_2$ (1 : 1) treated.

The reduction of water vapor permeability (WVP) by such a nanolayer of plasma polymer is marginal at best. It is a general principle that water transport cannot be totally stopped by application of an organic polymer layer. Although characteristic WVP of most plasma polymers investigated in the study is two to three orders of magnitude smaller than that for the substrate LDPE, the time lag of diffusion increased only from roughly 170 min to 200–300 min due to the thinness of the plasma polymer layer. This means that within 5 h water molecules reach the opposite side of a film. The contribution of the coated layer becomes even smaller when a thicker substrate film (LDPE) is employed. In the case of a thick layer (on order of centimeters used in the cable), the contribution would be negligibly small. The contribution of a small resistance added in series to a larger resistance becomes smaller when the latter becomes larger.

On the other hand, it is important to recognize that the presence of a nanofilm, which caused a few hours delay time, is influencing the experimental results of electrical properties measured in time scale of 1000 h. This means that the major

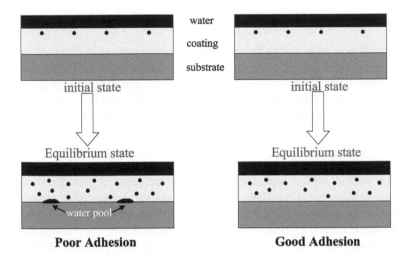

Figure 24.14 Schematic representation of water diffusion through a coating with and without water-insensitive adhesion of the coating to the substrate.

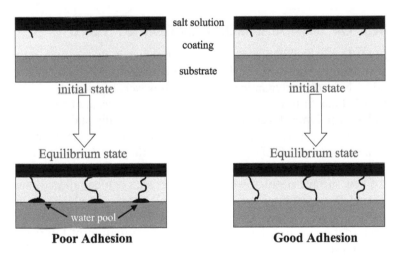

Figure 24.15 Schematic representation of salt intrusion through a coating with and without water-insensitive adhesion of the coating to the substrate.

issue is not the value of WVP or stopping of water transport but is the role of the presence of plasma polymer on the surface of LDPE. WVP and the surface dynamic stability of plasma polymers were found to be useful tools to investigate this important issue.

The application of a nanolayer of an optimal plasma polymer prevents the deterioration nearly completely under no electrical stress (see Fig. 24.11) and reduces significantly under the presence of electrical stress (see Fig. 24.12). This observation suggests that the plasma polymer coating shields existing surface flaws, which could act as potential initiation sites for the salt intrusion that triggers the fatigue mode failure.

When plasma polymers of hydrophilic monomers were deposited or the surface of LDPE grafted with hydrophilic poly(acrylic acid), the dielectric breakdown occurred in a very short time. These observations indicate that hydrophilic sites act as the gate for salt intrusion. Since the bulk properties of LDPE are believed to be unchanged with these surface modifications, the quick failure can be interpreted as the consequence of an interfacial phenomenon. The population of the potential salt intrusion sites is a very important factor that ultimately determines the breakdown of the insulation occurring in the presence of salt and electrical stress.

The stress level of $5\,kV/mm$ is sufficient to observe the onset of deterioration within the duration of experiment of 3000 h. At this level of stress, the complete breakdown did not occur (except one case), but the decay could be followed as a function of time. In saline solution, 5 out of 8 (62.5%) untreated samples showed the initiation of the deterioration within 1000 h. This value for plasma polymer–coated samples dropped to 2/8 (25.0%) for plasma polymer of methane and 1/8 (12.5%) for plasma polymer of $(C_2F_6 + H_2)$. The same frequency at 3000 h is 100% for untreated samples, 75% for plasma polymer of methane, and 25% for plasma polymer of $(C_2F_6 + H_2)$.

6. SIGNIFICANCE OF SURFACE MODIFICATION BY LCVD

The significance of LCVD is in the unique aspect of creating a new surface state that is bonded to the substrate material particularly polymeric material. The new surface state can be tailored to be surface dynamically stable. However, caution should be made that not all LCVD films fit in this category. Appropriately executed LCVD to lay down a type A plasma polymer layer creates surface dynamically stable surface state. In the domain, in which surface dynamic instability is a serious concern in the use of materials, a nanofilm by LCVD is quite effective in providing a surface dynamic stability, and other methods do not fare well in comparison to LCVD.

The creation of new surface state, of which characteristics can be tailored, has a significant implication in controlling surface defects of materials, which act as the initiation sites of the fatigue process. In view of the fact that most materials fail due to fatigue, the retardation or possible prevention of the initiation of fatigue is a very important and unique advantage of LCVD.

REFERENCES

1. Yasuda, H.; Charlson, E. J.; Charlson, E. M.; Yasuda, T.; Miyama, M.; Okuno, T. Langmuir **1991**, *7*, 2394.
2. Charlson, E. M.; Charlson, E. J.; Burkett, S.; Yasuda, H. K. IEEE Trans. Electrical Insul. **1992**, *27* (6), 1136.
3. Yasuda, H.; Charlson, E. J.; Cahrlson, E. M. EPRI Report, Effect of plasma polymer coating on the insulation breakdown, 1995.
4. Lee, S. Y. Dissertation, University of Missouri–Columbia, Effects of plasma polymer on the multi-stress aging of organic insulation and proposed degradation mechanisms, 1995.

25
Surface Dynamics

1. REFERENCE STATE FOR SURFACE DYNAMICS

Surface dynamics deals with the change of surface characteristics with time, which is often referred to as surface reconstruction. The observation of the surface reconstruction depends on the reference state from which the change occurs, and if one cannot define the reference state then the surface reconstruction cannot be dealt quantitatively or in a generic sense. This problem was found with moderately hydrophilic copolymer of ethylene–vinyl alcohol, which is described in Chapter 23. The reference state depends on the history of a sample, and the change cannot be reproduced without precise knowledge of the history of a sample [1]. In many cases dealing with surface reconstruction, the reference state is the original state of surface that was created by a film preparation process. In the case of hydrophobic recovery, the reference state is the surface immediately after making the surface hydrophilic.

According to the view that a polymeric surface is an ever-changing entity depending on the contacting medium [2], the restructured surface is not necessarily and most likely not the final one to stay, i.e., restructuring of once restructured surface or multiple repeated restructuring could occur with highly perturbable polymeric surfaces. The change of surface can be generally explained by the change of surface configuration.

In dealing with the dynamic change of characteristics of polymeric surfaces due to the change of contacting medium from air to liquid water, the appearance or disappearance of a specific atom such as fluorine or oxygen can be used to investigate the change of surface configuration in a semiquantitative manner. In the case of copolymers of ethylene and vinyl alcohol, the change of oxygen atom concentration profile in the surface region was followed to see the change in the surface state of film immersed in water as a function of (immersion) time. The same procedure was used to detect the dynamic surface configuration change of Teflon PFA as described in Chapter 23. In these cases, there is no need to introduce tagging species on the surface.

In absence of atom that can be used as the tagging atom by XPS or other analytical tools in the macromolecules, fluorine-containing moieties, or oxygen-containing moieties was introduced on the surface of polymers by plasma surface treatment, and the amount of the tagging atoms in the surface state was followed by XPS [2–7]. This method provided probably the most effective tool in the study of the surface dynamics of polymers because the tagging atoms are implanted on the top surface with the minimal alteration of the rest of the surface state, and the process establishes the reference state for the surface dynamics studies. The rates of disappearance and reappearance can be measured using the surface state on which the implantation took place as the reference state.

The general trends with respect to the change of surface configuration caused by the change of contacting medium is reviewed based on the state in which the sample is kept after the change of contact medium. Explaining the surface dynamic changes, the terms *migrate* and *move* are used; however, these motions are within the context of surface configuration described in Chapter 23. The term *migration* refers to the directional movement of moieties perpendicular to the surface of film in the context of surface configuration change, which is achieved by conformational change of macromolecules and should be distinguished from long-range migration such as that by diffusional transport.

2. INCORPORATION OF TAGGING ATOMS BY PLASMA TREATMENT

The low pressure plasma tagging of the top surface could be explained by the surface implantation of F-containing moieties. It is intended to implant F-containing moieties without forming a plasma polymer so that the movement of fluorine atoms could represent the conformational change of macromolecules on which F is attached. The perfluorocarbons that do not form polymer readily, such as CF_4 and C_2F_6, are used as plasma gas, and a relatively short treatment time (e.g., contact time less than 2 min) at a low W/FM level (e.g., 15 MJ/kg) is applied. Figure 25.1 depicts the principle of the method schematically.

Figure 25.2 depicts the sessile droplet contact angle of water on CF_4 plasma–treated PET with varying degree of crystallinity, and the influence of water immersion and the subsequent drying. Water-immersed samples were freeze dried to

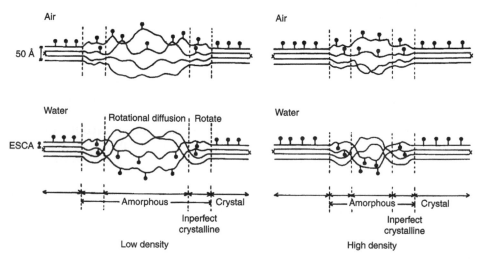

Figure 25.1 Schematic presentation of plasma surface implantation of F-containing moieties on a semi-crystalline polymer surface and the change occurring in water immersion.

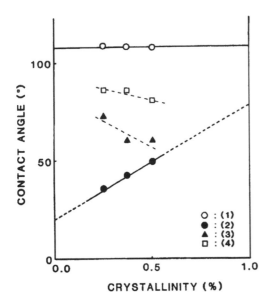

Figure 25.2 Sessile droplet contact angle of water on CF_4 plasma treated PET films subsequently treated differently; (1) samples kept in air (no water immersion), (2) samples immersed in water for 120 min, (3) water immersed samples heat treated at 100°C for 10 min, (4) water immersed samples heat treated at 180°C for 10 min.

remove water in order to preserve the new surface state attained by the water immersion. The contact angle measurement was carried out with freeze-dried sample, except samples kept in air. The heat treatment was applied to the freeze-dried samples. The contact angles on the dry surface represented by treatment 1 in Figure 25.2 indicate that practically identical treatment is applied on polymer surface independent of polymer crystallinity. It was also confirmed that CF_4 plasma treatment didn't change the crystallinity of PET films by X-ray analysis.

When samples are immersed in water a significant decrease of the contact angle of water is observed. The change from the solid line with open circles to the dotted line with closed circles indicates the extent of contact angle change. The extent of decrease was inversely proportional to the crystallinity of the sample, which indicates that the surface configuration change occurs mainly in the amorphous phase in the surface state, i.e., F atoms attached to the crystalline surface are immobile.

The change from a closed circle on the dotted line to corresponding open square or closed triangle represents the recovery on drying. The extent of recovery from a water-immersed sample is dependent on how the sample was treated. The heat treatment at higher temperature caused the greater extent of the recovery. The extent of recovery is inversely proportional to the crystallinity in both heat treatments, confirming the trend found on the immersion process.

The change of XPS C 1s spectrum corresponding to the sequential treatments (from the bottom figure to the top figure) is depicted in Figure 25.3. The signals for F-containing C decreases and the signal for C without F increases on water immersion. On the heat treatment of water-immersed sample, the reversed process

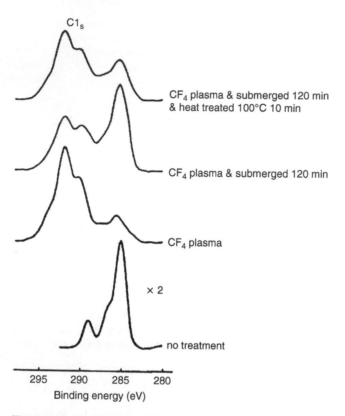

C1$_s$

CF$_4$ plasma & submerged 120 min
& heat treated 100°C 10 min

CF$_4$ plasma & submerged 120 min

CF$_4$ plasma

× 2

no treatment

295   290   285   280
Binding energy (eV)

Figure 25.3 Changes of XPS C 1s signals of PET film (crystallinity 25.4%) upon the sequential treatments.

is observed. The coupled changes of F-containing C and C without F is corresponding to the mode of surface configuration change depicted in Figure 25.1.

The decay of hydrophobicity manifested by the sessile droplet contact angles of water as well as of XPS F 1s peak intensity is observed in nylon 6 and PET films treated similarly. There is a direct linear correlation between the contact angles of water and the intensity of XPS F 1s peaks for each substrate polymer as shown in Figure 25.4. This correlation confirms that the increase and the decrease of the contact angle of water on the surface of plasma-treated films are due mainly to the change in the surface concentration of fluorine-containing moieties, which enable us to deal with the decay and the recovery phenomena by either measurement.

3. POLYMER SURFACES IN DRY CONDITION

3.1. Without Incorporation of Tagging Atoms

When a polymer is kept dry, all polar groups contained in the polymer molecule tend to move away from the interface; consequently, the interface is populated with relatively hydrophobic moieties of the polymer molecule. The surface state of a

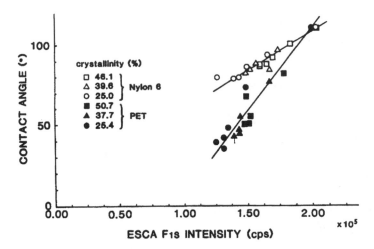

Figure 25.4 Correlation between sessile droplet contact angle of water and XPS F 1s intensity of CF$_4$ plasma treated Nylon 6 and PET films.

polymer is rich with (relatively) hydrophobic moieties or portions of the polymer molecule.

3.2. Surface Modification is Applied to Make the Surface Hydrophilic

The introduced hydrophilic moieties tend to migrate into the deeper section of the surface state while the surface is kept dry. Such a migration occurs in the direction to minimize the interfacial tension of polymer/air interface. This process has been recognized as the decay of the surface modification effect.

3.3. Surface Modification is Applied to Make the Surface Hydrophobic

This is the case of waterproofing treatments. The change is very slow, unless temperature is raised. In many cases, no significant change in the surface configuration, in short time scale, is observed as long as the surface is kept dry. The surface tension of the polymer surface is lowered by the surface modification, and consequently the interfacial tension of polymer/air interface, which is the driving force for surface dynamic change, is already minimized.

4. IMMERSION OF A DRY POLYMER IN LIQUID WATER

4.1. Without Incorporation of Tagging Atoms

When a dry polymer film is immersed into liquid water, an opposite process to what described above takes place. Hydrophilic groups migrate to the interface, and hydrophobic moieties move to the inner part of the surface state of a polymer. The main driving force for these movements is the interfacial tension. The greater the interfacial tension, the greater is the surface configuration change.

In the presence of both hydrophilic and hydrophobic moieties, the migrations of moieties occur in the direction to minimize the interfacial tension. Therefore, a greater extent of the surface configuration change is observed with amphoteric hydrophobic/hydrophilic polymers. In general, water immersion will make a polymer surface more hydrophilic than the dry film.

4.2. Surface Modification is Applied to Make the Surface Hydrophilic

The change due to water immersion is minimal, and the decay of surface modification is generally not observed.

4.3. Surface Modification is Applied to Make the Surface Hydrophobic

A significant extent of the decay of hydrophobicity is observed because the surface modification lowered the surface tension, which created a large interfacial tension with water, i.e., higher driving force is created.

5. EMERSION OR DRYING OF A WET SURFACE

When a polymer surface that has been kept in liquid water is emerged from liquid water and kept in dry condition, the reverse process to what is described in the water immersion occurs. This phenomenon is identical to what is described in the dry air situation above (Sec. 3) except that the starting point or the reference state is different. However, when this process is viewed as the recovery process of the decay of hydrophobicity, which took place when the sample was immersed in water the first time, there are significant differences in both the rate of change and the extent of recovery. This discrepancy is due to the fact that these processes are different processes, i.e., the driving force for the surface configuration change in the immersion and that in the emersion are completely different, and these processes are not a part of a reversible process.

The quantitative comparison of the rates and the extent of changes is further complicated by additional phenomena such as the imbibitions of liquid water, the diffusion of water molecules, subsequent swelling or plasticization of the polymer by absorbed water, and so forth. For instance, the swelling would greatly enhance the ease of migration of moieties due to the attained high mobility of polymer molecules. All these complications change the surface tension of polymer and shift the starting point for the subsequent drying (emersion) process.

5.1. Without Incorporation of Tagging Atoms

The influence of water immersion has little effect on very hydrophobic polymers; consequently, the effect of emersion is nearly identical to the case without water immersion. The influence of water immersion becomes progressively greater with the increase of hydrophilicity. With highly hydrophilic polymers, the surface of water-immersed polymer is completely different from that of dry polymer, and consequently the emersion becomes drying process of wet polymers.

5.2. Surface Modification is Applied to Make the Surface Hydrophilic

The surface modification is applied on hydrophobic polymers in this case, and the complication due to change of bulk phase due to water immersion is small. The decay of hydrophilic characteristics or hydrophobic recovery is similar to that occurring in air.

5.3. Surface Modification is Applied to Make the Surface Hydrophobic

The surface modification is applied on hydrophilic polymers in this case. Depending on the length of water immersion, significant levels of complications due to the change of bulk phase occur. Consequently, the decay of hydrophobicity or hydrophilic recovery is significantly different from that occurring in air.

6. FACTORS INVOLVED IN AN INTERFACE

The surface dynamic changes, irrespective of starting points, occur in the direction to minimize the interfacial tension between polymer and the contacting medium. The overall change in the surface configuration can be viewed as the product of two major parameters, i.e., (1) polymer chain mobility and (2) driving force:

Rate of surface dynamic change

$$= F(\text{polymer chain mobility}) \times G(\text{interfacial tension}) \qquad (25.1)$$

This is an analogous situation to the diffusive transport of small molecules in a polymer matrix. The flux J can be given by $J = D\, dc/dx$, where D is the diffusion constant (related to the mobility of polymer segments) and dc/dx (concentration gradient) is the driving force:

Diffusional flux $= F(\text{polymer chain mobility}) \times G(\text{concentration gradient})$

Consequently, any factor that changes the polymer chain mobility and any factor that changes interfacial tension cause the change in the rate of surface configuration change.

If the presence or absence of liquid water contacting with a polymer surface changes the mobility of polymer segments, the influence of the driving force term could be overwhelmed by the change of surface state such as swelling of the top surface. On the other hand, if one could find a system in which water does not influence the mobility of polymer segments, it might be possible to see clearly the role of the interfacial tension for the surface configuration change for such a polymer. In dealing with surface configuration change of a polymer surface, it is important to reexamine definitions and the validity of intuitive assumptions involved.

The following viewpoints are taken into consideration throughout the discussion of the subject.

1. A surface can be recognized only as an interface. In this strict sense, only interfacial properties of a solid surface can be considered, and generic surface properties of a solid based on atomic or molecular structure, particularly of a polymer, are misleading concepts.
2. Interfacial properties cannot be described without identifying the contacting medium. Interfacial properties of a polymer solid are dependent on the conditions under which the surface is equilibrated. The surface configuration of a polymer is a function of the contacting phase of polymer/contacting phase interface. In this context, the conventional sense of surface property (interface with air) is dependent on the history of the surface and the humidity of air.
3. The surface dynamic change occurs when the interfacial equilibrium is broken and is driven by the interfacial tension in the new environment.

7. MOLECULAR INTERACTION VS. INTERFACIAL INTERACTION

Liquid water as contacting medium (dealt as a solvent) interacts with polymer molecules in different degrees; some dissolve polymers, some swell polymers, and some are merely adsorbed at the polymer/water interface with no significant alteration of physical properties of polymers. The enthalpy term of the interaction could be represented by Flory-Huggins's interaction parameter χ. In order to distinguish the molecular interaction from the interfacial interaction, the molecular interaction could be termed χ interaction and the interfacial interaction γ interaction. When a solid polymer is in contact with water (solvent), the extent of χ interaction depends on the hydrophilicity or hydrophobicity of the polymer. However, the interfacial phenomena generally cannot be interpreted by molecular level parameters that describe the bulk phase of a polymer such as hydrophilicity or hydrophobicity.

The wetting of a polymer surface is characterized by the interfacial tension between the liquid and the surface. The contact angle at the three-phase line is a good measure of the wettability of the surface. Figure 25.5 depicts the force balance

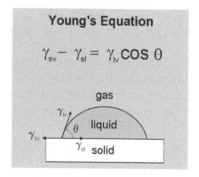

Figure 25.5   Force balance at three-phase line.

at the three-phase line. However, a high contact angle of water on a polymer surface does not uniquely relate to the hydrophobicity of the polymer molecules, which constitute the surface. A high sessile droplet contact angle of water merely indicates that the surface configuration of the surface attained in the environment just before a droplet was placed was in such that the contact angle is high, but does not mean that the polymer molecules, which constitute the surface, are hydrophobic in nature.

An excellent example of this situation can be seen in the phenomenon that exhibits a very high (sessile droplet) contact angle on a highly hydrated hydrogel, e.g., of gelatin, of a polymer that is known to be highly hydrophilic, as shown in Chapter 23. The very high sessile droplet contact angle of water on a gelatin hydrogel does not mean that gelatin molecules are hydrophobic. A highly hydrated gel cannot be formed with hydrophobic polymers to start with. A very high contact angle water on a gelatin hydrogel is due to the fact that most hydrophilic moieties are located or facing toward the bulk phase of a gel that contains abundant water, and the top surface is populated with the hydrophobic portion of gelatin molecules facing air.

Whether a polymer is hydrophilic or hydrophobic can be explained by the χ interaction. However, why the air/hydrogel interface of a gelatin hydrogel is hydrophobic cannot be explained by χ interaction because it is a strictly interfacial phenomenon that is governed by the interfacial interaction. The strictly interfacial phenomena could be explained by γ interaction. The first term in Eq. (25.1) is χ interaction and the second term is γ interaction. The overall rate of surface dynamic change is, therefore, dependent on both γ interaction and χ interaction [7]. Thus,

$$\text{Surface dynamic change} = F(\chi \text{ interaction}) \times G(\gamma \text{ interaction}) \tag{25.2}$$

8. INTERFACIAL TENSION BETWEEN POLYMER SURFACE AND LIQUID WATER

In the classical treatment of the interfacial tension, the force balance at the three-phase contact line (liquid–surface–air) is considered, and the interfacial tension, γ_{SL}, is given by the surface tension of the solid, γ_S, and the surface tension of the liquid, γ_L, as

$$\gamma_{SL} = \gamma_S - \gamma_L \cos \theta \tag{25.3}$$

where θ is the contact angle of the liquid at the three-phase contact line. This equation is known as Young's equation for contact angle. (The details of sessile droplet contact angles are discussed in Chapter 26.)

In the classical treatment of surface tensions, it is intuitively assumed that the surface tension of a solid, γ_S, can be assigned as if it is a material constant. In a practical sense, Eq. (25.3) is valid if the surface tension of the solid does not change after the contact with the liquid (sessile droplet) is made. While Young's equation describes the force balance at the three-phase line, it does not give information relevant to the true interfacial tension at the interface that is beneath the droplet, which is the major concern of surface dynamics. In general cases, γ_S and γ_{SL} are unknown parameters. The change of γ_{SL} under the sessile droplet is shown in Chapter 23.

An empirical method to estimate the surface tension of a solid is Zisman's plot (cos θ as a function of γ_L), which obtains the critical surface tension of wetting. In the absence of specific interaction between the surface and the liquids used for the measurement of contact angles, the critical contact angle of wetting can be accurately estimated and its value used as the surface tension of the surface. However, if a surface interacts with liquids used as the sessile droplet for the contact angle measurement, to the extent that the surface tension is altered, Zisman's plots deviate from the ideal linear relationship. In a strict sense, the plot is applicable only to imperturbable surfaces with which liquid contact does not alter surface configuration, i.e., no surface dynamics applies.

Surface dynamics is concerned with the kinetic aspect of this specific question of whether or not the surface tension of polymer changes by contacting with liquid water. If γ_S changes by the interaction of water with polymer molecules, γ_{SL} also changes accordingly.

It has been observed that a water droplet placed on a hydrophobic surface did not roll down when the surface is tilted to the vertical position while the contact angle remains over 90 degrees as shown in Figure 23.17. The large discrepancy between advancing and receding contact angles of a sessile droplet on a gelatin gel shown in Figure 23.1 is caused by the fact that the contact area of water droplet does not decrease as the volume of water is decreased, as depicted in Figure 23.15. The water droplet becomes flatter as water is withdrawn due to the increased interaction between water and polymer beneath the droplet, which resists the reduction of contact area [8].

These observations indicate that a strong attractive force is created between water and the surface under the droplet, and also that the decreased interfacial tension does not influence the force balance at the three-phase line. This implies that Young's equation given by Eq. (25.3) only applies at the three-phase line, and γ_{SL} in the equation does not represent the interfacial tension between the surface and liquid water that exists beneath the sessile droplet.

In many cases, the advancing rate of the three-phase contact line is much faster than the rate of the surface configuration change; consequently, a reasonably accurate estimate of the surface energy of polymer surface based on the value of contact angle can be made from the advancing contact angle. The fact that a constant advancing angle (independent of droplet volume, in the case of sessile droplet method) can be obtained indicates that such a requirement (i.e., the time scale of surface configuration change is much greater than the time scale of advancing the contact line) can be met in many cases. It should be noted, however, that this is not a generally acceptable assumption as shown by the case of sessile water droplet on a gelatin hydrogel surface.

In dealing with different phases, we have an inherent problem of not having the identical reference state for the thermodynamic parameters for different phases. In order to simplify the discussion, the following definitions and notations are used [7].

The standard solid surface tension may be defined as the surface tension of a solid in contact with its own vapor only, and the standard solid surface tensions designated by γ_S^0 can be assigned. The noncoherent phase, i.e., gas and vapor, can be considered to have zero surface tension.

Likewise the standard liquid surface tension may be defined as the surface tension of a liquid in contact with its own vapor and designated by γ_L^0 (which is often noted as γ_{LV} in literature). In order to further simplify the discussion, let us limit our consideration to a fixed temperature (e.g., 25°C) only.

Then the ideal interfacial tension between a solid and a liquid can be defined as the difference of surface tension of a solid and the liquid,

$$\gamma_{SL}^0 = \gamma_L^0 - \gamma_S^0 \tag{25.4}$$

The ideal interfacial tension for a polymer surface contacting with liquid water, γ_{SL}^0, is given by this equation, if water does not change the surface tension of the polymer, γ_S^0. As we know, however, the surface tension of a polymer changes as a consequence of contacting liquid water in many cases. The water-immersed polymer samples often show significantly different advancing contact angles from that for the dry sample as shown in the previous chapters and discussed in some detail in Chapter 26.

While Eq. (25.3) describes the force balance at the three-phase boundary line, our major concern is the interface between a polymer and liquid water without an air phase, which may be described by Eq. (25.4).

The actual surface tension of a polymer surface could differ significantly from γ_S^0 depending on the conditions of surrounding medium with which the surface is equilibrated. The surface tension of a polymer surface can be generally given by

$$\gamma_s = \gamma_s^0 + \gamma_s^x \tag{25.5}$$

where the superscript x refers to a specific condition of equilibration. In the case of a relatively hydrophobic (not very hydrophilic) polymer kept in air or in vacuum, the surface tries to minimize the value of γ_S (i.e., $\gamma_s^x \Rightarrow 0$), and after sufficient time γ_S becomes γ_S^0. The minimization of γ_S can be achieved by the change of surface configuration, without considering physical changes such as swelling or deswelling, which requires greater changes of surface state. The liquid surface tension can be dealt with in a more straightforward manner.

The interfacial tension may be dealt by the differential surface tension, $\Delta(\gamma_S)$, (the difference between surface tension of polymer and that of contacting phase):

$$\Delta(\gamma_s) = (\gamma)_{\text{final-contact-phase}} - (\gamma)_{\text{initial-contact-phase}} \tag{25.6}$$

where $\Delta(\gamma_s)$ indicates the direction of surface configuration change.

The interfacial tension in general cases can be given by the following equations: For the case of water immersion:

$$\Delta(\gamma_s)_{SW} = \gamma_L^0 - \left(\gamma_s^0 + \gamma_s^x\right) \tag{25.7}$$

For the case polymer is kept in air:

$$\Delta(\gamma_S)_{SA} = -\gamma_S$$
$$= -\left(\gamma_S^0 + \gamma_S^x\right) \tag{25.8}$$

The driving force for surface dynamic change is the interfacial tension, which is given by Eq. (25.7) or Eq. (25.8), depending on the contacting phase. It can be seen from these equations that the driving force in the case of water immersion is different from the case of wet surface drying as described before.

9. SURFACE CONFIGURATION CHANGE WHEN POLYMER IS IMMERSED IN WATER

The driving force and surface configuration change are depicted in Figure 25.6. The driving force for the surface configuration change is the interfacial tension, which depends on the interaction (attractive or repulsive) force between water and a specific moiety under consideration. In the case of Teflon PFA (described in Chapter 23), the attractive force between oxygen atoms in the ether linkages and water, and the repulsive force between water and fluorine atoms can be considered. Because of the dominance of fluorine atoms in a Teflon PFA type of polymer, the former does not contribute to the overall surface configuration change.

In the case of moderately hydrophobic (moderately hydrophilic) polymers, the cumulative effects of attractive force and the repulsive force determine the overall surface configuration change. The maximal extent of such surface configuration changes can be correlated to the magnitude of the interfacial tension because surface configuration changes occur in order to minimize the interfacial tension. The larger the interfacial tension, the greater is the maximal surface configuration change, which can be seen when equilibrium with the new environment is reached.

The magnitude of surface configuration change can be seen by the interfacial tension given by Eq. (25.7), and the value of γ_S^x varies depending on the equilibration conditions of the surface before the new interface was created. Considering that the surface tension of the pure liquid water is close enough to γ_L^0 and does not change, the reduction of $\Delta(\gamma_s)_{SW}$ can be achieved mainly by increasing γ_S^x.

In the case of perfluorocarbon polymers, γ_S^0 is of the order of 19 dyne/cm and $\gamma_s^x \approx 0$ because the surface configuration is not influenced by the contacting medium (air or vacuum). The value of γ_L^0 for water is of the order of 70 dyne/cm.

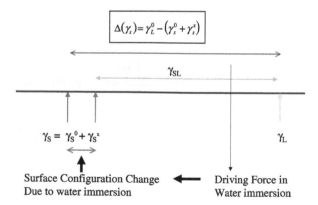

Figure 25.6 Driving force and surface configuration change when a polymer is immersed in water.

Table 25.1 Surface Tension of Liquid (40°C) Used for Immersion

Liquid	Surface tension (dyne/cm)
Water	70.3
0.01% scourol solution	42.8
50 Vol% ethanol solution	28.0

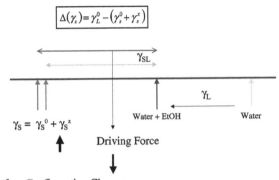

$$\Delta(\gamma_s) = \gamma_L^0 - (\gamma_s^0 + \gamma_s^x)$$

γ_{SL}

γ_L

Water + EtOH   Water

$\gamma_S = \gamma_S^0 + \gamma_S^x$

Driving Force

Surface Configuration Change

Figure 25.7 Driving force and surface configuration change with addition of ethanol.

The interfacial tension, therefore, is of the order of 51 dyne/cm. The increase of γ_S can be achieved either by moving polar moieties toward the top surface, in the case Teflon PFA in water, or moving hydrophobic moieties into the deeper section of the surface state, in the case of CF_4 plasma–treated polymers.

In the wetting process, if we decrease γ_L of water by adding small amount of surfactant or alcohol (50 vol%), the value of $\Delta(\gamma)_{SL}$ can be reduced significantly. The values of surface tension of the liquids used in the study are listed in Table 25.1. The influence of the reduction of water surface tension on the surface configuration change of Teflon PFA was given in Figure 23.14.

The interfacial tension, $\Delta(\gamma)_{SL}$ (assuming γ_S^0 of the sample to be 19.0 dyne/cm), decreases from 51.3 dyne/cm for pure water to 23.8 dyne/cm for the surfactant solution and 9 dyne/cm for the ethanol solution. This is a reasonable explanation of why the increase of O 1s peak is less for the surfactant solution and practically none for the ethanol solution. The surface dynamic change is the reflection of the driving force (change of interfacial tension) in these cases as depicted in Figure 25.7.

10. SURFACE CONFIGURATION CHANGE IN (DRY) AIR

The extent of surface configuration change is dependent on the differential surface tension given by Eq. (25.8), and the value of γ_S^x varies depending on the equilibration conditions of the surface before the new polymer/air interface was created. Because the value of γ_S cannot be smaller than γ_S^0, the extent of surface configuration change

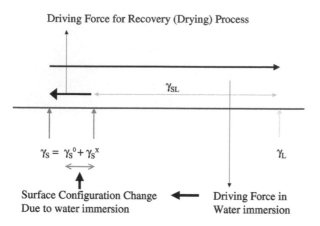

Figure 25.8 Comparison of driving forces in water immersion and drying.

is determined by γ_S^x. The difference of driving force in water immersion and drying of water-immersed surface is depicted in Figure 25.8.

Thus, the driving force for the reverse process is much less than for the wetting process. Consequently a simple drying of the wetted sample does not cause a significant hydrophobic recovery of the surface. From the viewpoint of interaction forces, the repulsive force between fluorine atoms and water molecules is not replaced by the attractive force in the similar magnitude, but the repulsive force is merely removed. The reverse process, therefore, may be more appropriately considered as a redistribution process of certain chemical moieties, which is driven by the local differential chemical potential of each species.

In the case of a polymer surface treated to make it hydrophilic, the value of γ_S is significantly large, and its reduction by virtue of surface configuration change is generally observed. A plasma-modified hydrophilic surface often loses its gained hydrophilicity in a week to 2 months depending on the nature of substrate polymer and of the treatment.

11. SURFACE DYNAMICS OF HYDROPHILIC POLYMERS

In contrast to very hydrophobic polymers, the interaction of water with hydrophilic polymers influences the surface state of polymer strongly, and the surface state cannot be defined without designating the activity of water in the contacting medium. Therefore, when liquid water contacts with the dry surface of a hydrophilic polymer, the change of surface state in terms of χ interaction, such as the degree of swelling, may play an overwhelming role in the surface configuration change. In such a case, water vapor changes the surface state, as recognized by the water sorption or the moisture regain, while the surface contacts with zero surface energy phase.

The interaction of water molecules with polymer molecules at the surface does not necessarily require contact with the condensed (liquid) phase. If the mobility of polymer segments increases with the sorbed water, a significant extent of surface configuration change could take place even with the change of relative humidity in the contacting air phase. It should be recognized, however, that such a change of surface configuration is not an interfacial phenomenon in a strict sense.

The role of χ interaction in surface dynamic change is examined using two model samples, cellulose (highly hydrophilic but highly crystalline) and cellulose acetate (moderately hydrophilic but amorphous), by extending the same experimental procedure for F-tagged surface previously employed [7]. The surface configuration change may be considered to be minimal for polymer molecules in the crystalline phase. It is, therefore, interesting to examine whether the crystallinity or the extent of water/polymer interaction is the dominant factor in the change of surface configuration when the contacting phase is changed.

A highly hydrophobic polymer such as Teflon PFA has weak χ interaction with water. The γ interaction is strong in liquid water but very weak in contact with water vapor. When the hydrophilicity of polymer increases the contribution of the χ interaction increases, and the relative contribution of the γ interaction changes according to the value of $\Delta(\gamma_S)$.

In the case of fluorinated surfaces of hydrophilic polymers, the difference between liquid water immersion and water vapor exposure can be described as follows. In liquid water immersion, $\Delta(\gamma_S)$ is large (because of the surface fluorination) and the γ interaction forces F-containing moieties to move away from the top surface; however, in water vapor exposure, $\Delta(\gamma_S)$ is negative (no driving force for the surface configuration change due to the γ interaction), and F-containing moieties are favored to remain on the surface. It is therefore considered that the movement of F-containing moieties represents the change of conformation due to the χ interaction.

The χ interaction increases the swelling ratio of these polymers and facilitates the easier movement of polymer segments, in both water immersion and water vapor exposure cases, and facilitates the surface configuration change in a passive manner. Because the swelling is faster in liquid water, the (passive) effect in the surface configuration change is anticipated to be greater in the water immersion case.

12. HYDROPHILIC POLYMERS IMMERSED IN LIQUID WATER

The CF_4 plasma treatment yields F/C peak area ratio of approximately 1.5 for both cellulose and cellulose acetate films. When those surface-tagged films were immersed in liquid water, the fluorine atoms, which are detectable by XPS, decreases as a function of the immersion time. Figure 25.9 depicts change of F 1s/C 1s ratio as a function of immersion time in water, which shows that the disappearance rate of F 1s peak is much higher for the semicrystalline cellulose film than that for the amorphous cellulose acetate film. These data actually compare the extent of χ interaction rather than the degree of crystallinity, indicating that χ interaction overwhelms the degree of crystallinity.

The recovery of fluorine in the surface region (within the escaping depth of electrons) of cellulose film was investigated by XPS, results of which are summarized in Figure 25.10. The values represent the average of five repeated experiments, with reproducibility of roughly $\pm 6\%$. The CF_4 plasma–treated cellulose films (dry) were immersed in water for 5 min (wet) and then the freeze-dried samples were kept under various conditions to allow observation of the posttreatment change. One group of

Effect of Liquid Water on F1s Signal

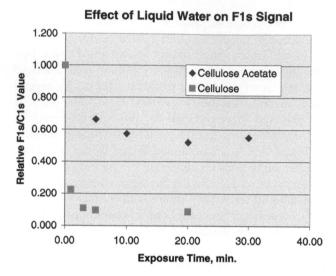

Figure 25.9   Decay of F 1s signal as a function of water immersion time.

Loss & Recovery of F/C

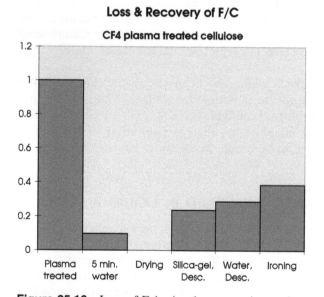

Figure 25.10   Loss of F 1s signal on water immersion, and recovery on drying.

samples was kept in a chamber with silica gels for 7 days (silica gel). One group of samples was kept in contact with saturated water vapor for 4 days (water vapor). One group of (freeze dried) samples was ironed at approximately 180°C for 20 s (ironing).

Immersion in water for 5 min reduced the F 1s/C 1s ratio to less than 10% of the value for the plasma-treated dry film. The postimmersion treatments shown on the right side of the figure brought back the ratio to 24–38% of the dry film (10–28% recovery). The slight increase is considered to be important because it negates the hypothetical argument that the loss of plasma-introduced moieties might be due to the washing effect of water immersion. It is interesting to note that (1) the maximal recovery was obtained by ironing (heat treatment); (2) exposure to water vapor provided the higher recovery than the dry air with silica gel. The latter factor seems to be particularly important with respect to the change of the overall surface state (not surface configuration) of hydrophilic polymers.

The reverse process of the surface configuration change when liquid water is removed is an entirely different process driven by the different driving force as described before. Therefore, the recovery is generally much slower than the disappearance in water immersion, and the complete or large extent of recovery is not anticipated. In dry air, dehydration of the top surface seems to occur creating a less hydrated skin, which slows down the dehydration of the remaining film and reduces the rate of recovery observable by F/C ratio. The wet environment (in water vapor) allows rehydration of the freeze-dried sample from the top surface or reduces the dehydration at the top surface to form skin and makes rearrangement of polymer segments easier. The ironing seems to raise the temperature of the top surface quickly without removing existing water in the surface state and facilitate the segmental motion of polymer near the top surface. All of these data indicate the importance of χ interaction in the surface dynamic change of hydrophilic polymers.

13. SAMPLES EXPOSED TO HIGH HUMIDITY (INSTEAD OF WATER IMMERSION)

The fact that the very fast disappearance of surface fluorine atoms on water immersion was observed for the highly crystalline cellulose rather than the amorphous cellulose acetate suggests that the hydrophilicity or swelling capability (by water) of a polymer is more important than the degree of crystallinity.

Results obtained with water vapor are summarized in Figure 25.11, which depicts the percent change (based on the dry film) as a function of exposure time.

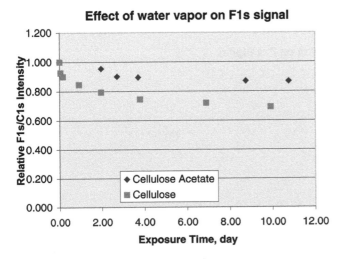

Figure 25.11 Decay of F 1s signal on exposure to water vapor.

The changes observed with water vapor–exposed films are much slower than those for liquid water–immersed films. The time scale is in order of days (minutes for water–immersed films), and the change levels off at 70% (cellulose) and 85% (cellulose acetate).

These experiments with water vapor exposure also confirm that the decrease of fluorine atoms observed by XPS is indeed mainly due to the migration of fluorine-containing segments away from the top surface region because no washing can occur in these experiments.

It is important to emphasize that the rate of change in water vapor (20°C) is roughly four orders of magnitude (1/10,000) smaller than that in liquid water (40°C). Within the time scale of a liquid water immersion experiment, no change occurred in the case of water vapor. This is in accordance with the difference of liquid water and water vapor in the γ interaction discussed earlier.

F 1s/C 1s values plotted against the water sorption values (by weight) of the cellulose film in Figure 25.12 shows the correlation between the surface configuration change and the swelling of film. The moisture uptake of the cellulose film (untreated) as a function of water vapor exposure time was first established. Values of F 1s/C 1s ratio at a given exposure time is plotted against the corresponding water uptake value at the same exposure time. The result shown in the figure seems to indicate that the decrease of the surface fluorine content in water vapor is directly proportional to the swelling of sample, implying that the decrease of fluorine is reflecting mainly the χ interaction, i.e., the change of mobility, because the interfacial tension (driving force) is small in a gas/solid interface.

The change of surface configuration when the contacting medium is changed is an observation of the change of surface state of a polymer on such an environmental change. The parameters necessary to define the surface state of a polymer include variety of physicochemical parameters. The degree of swelling, degree of crystallinity, chemical configuration of polymer, conformational state of macromolecules, surface configuration, and energy level of surface state electrons (related to the transfer of electrons on contact with another surface) are all important parameters to specify the surface state of a polymer under a specific environmental condition [2].

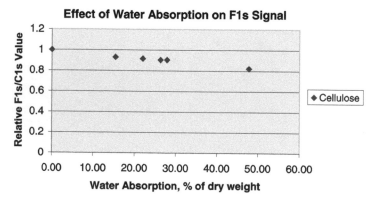

Figure 25.12 F 1s signal decay as a function of water vapor absorption on exposure to water vapor.

For a highly hydrophobic polymer surface, which has minimal χ interaction with water, the surface configuration change in the γ interaction may be the major factor for the overall surface state change in the case of liquid water immersion. For a hydrophilic polymer surface, the dominant factor in the overall surface state change is the change of the mobility of polymers in the surface state because the χ interaction of water with polymer is high (if the change involves the transformation of a dry polymer to a water-swollen polymer). If no change in the swelling (χ interaction) is involved, such as the case when the surface of a swollen hydrogel in water is exposed to air, the surface configuration change due to the γ interaction is the main event in the overall change of the surface state. In this case, the surface configuration change occurs in a nearly reversible manner because the chain mobility in highly swollen hydrogel is very high, and the surface configuration change can proceed nearly instantaneously. Because of this situation, the sessile droplet contact angle of water on a gelatin gel is always high, but the same surface under water shows very low contact angle (by bubble method) immediately after the surface was immersed in water.

REFERENCES

1. Yasuda, T.; Miyama, M.; Yasuda, H. Langmuir **1994**, *10*, 583.
2. Yasuda, H.; Charlson, E. J.; Charlson, E. M.; Yasuda, T.; Miyama, M.; Okuno, T. Langmuir **1991**, *7*, 2394.
3. Yasuda, H.; Sharma, A. K.; Yasuda, T. J. Polym. Sci., Polym. Phys. Ed. **1981**, *19*, 1285.
4. Yasuda, T.; Okuno, T.; Yoshida, K.; Yasuda, H. J. Polym. Sci., Phys. Ed. **1988**, *26*, 1781.
5. Yasuda, T.; Okuno, T.; Yoshida, K.; Yasuda, H. J. Polym. Sci., Phys. Ed. **1988**, *26*, 2061.
6. Yasuda, T.; Okuno, T.; Miyama, M.; Yasuda, H. Langmuir **1992**, *8*, 1425.
7. Yasuda, T.; Okuno, T.; Tsuji, K.; Yasuda, H. Langmuir **1996**, *12*, 1391.
8. Yasuda, T.; Okuno, T.; Yasuda, H. Langmuir **1994**, *10*, 2435.

26
Contact Angle and Wettability

1. ENERGY BALANCE AT LIQUID/SOLID INTERFACE

Contact angle is the angle at the three-phase (gas/liquid/surface) line of a sessile droplet or a sessile bubble. Contact angle is related to but not a direct measure of wettability of the surface by the liquid as seen in the case of water sessile droplet contact angle of a gelatin hydrogel (Fig. 23.1). Wettability is the ease with which an air/solid interface can be converted to a liquid/solid interface. The interfacial tension of a liquid/solid system is a direct measure of wettability; however, it is not a readily measurable quantity. The interfacial energy or interfacial tension of a solid with ambient air is generally referred to as the surface energy of the solid, and the smaller the difference between the surface energy of a solid and that of a liquid, the higher is the wettability of the surface by the liquid. In a strict sense, a "surface" is a hypothetical concept. The surface of a solid exists only as an interface as described in the previous chapters. When the contacting medium is ambient air or vacuum, such an interface is generally termed a surface. The difference of surface energies, therefore, does not represent the interfacial energy as described in Chapter 25.

The measurement of contact angle (see Fig. 25.5) is relatively simple, and the value of contact angle provides useful information in characterizing the interfacial properties from which some characteristic features of the surface could be withdrawn. For example, the contact angle of water is very valuable to judge the hydrophilicity, hydrophobicity, or the surface energy of a surface. However, the interpretation of the contact angle is not as simple as it might appear because the contact angle measurement could involve some factors that are not considered in the ideal situation expressed by Young's equation. Most abnormalities are caused by the change of surface itself as a consequence of creating the liquid/solid contact. Therefore, careful analysis of the abnormality could provide information pertinent to the surface dynamics of a surface. The sessile droplet (static) contact angle method is used to directly measure the contact angle of a sessile droplet; however, the dynamic contact angle measurement by Wilhelmy balance indicates the force, and the contact angle is calculated from the interfacial force contribution, i.e., an indirect method for measuring contact angle.

2. STATIC CONTACT ANGLE

The sessile droplet or sessile bubble contact angle is measured in the static mode, i.e., the droplet or bubble remains on the surface stationary. The advancing contact angle is measured by advancing the three-phase line by increasing the volume of a sessile droplet or bubble, but the measurement is done with a stationary sessile droplet. Depending on whether the contact angle is measured in the process of increasing or decreasing the volume of droplet, the contact angle is referred to as advancing or receding, respectively. The contact angle measured by "advancing" contact angle or "receding" contact angle is not the dynamic contact angle that is observed with a moving three-phase line.

3. TIME DEPENDENCE OF SESSILE DROPLET CONTACT ANGLE

Traditional goniometry and even more advanced techniques require visual and mechanical adjustments while the contact angle is measured, during which the droplet is resting on the substrate material and equilibrating with the surface and surroundings. The time between two consecutive additions of water, during which a sessile droplet rests on the surface and contact angle measurement is carried out, was reported between 5 and 3 min while many neglected to report this at all. The minimal time interval depends on how quickly the measurement could be completed. The modern equipment that utilizes computer image capturing technique enabled us to shorten the time interval to 2–3 s, and to minimize the change of contact angle while the measurement is carried out [1].

Figure 26.1 depicts the change of water droplet volume with time when the sessile droplet contact angle was measured by placing different size droplets on the

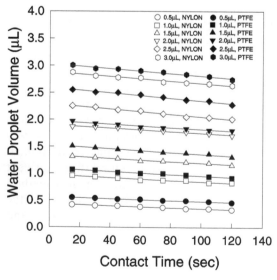

Figure 26.1 Effect of contact time on the water droplet volume of many different size droplets on Teflon and nylon surfaces.

surfaces of two polymers (nylon and Teflon). Both polymers were kept in dry air for about 6 months before contact angle measurement. The water droplets were allowed to equilibrate with the surface and ambient air for almost 2 min while contact angle measurements were taken every 15 s. Water droplet volume was calculated from the captured image of a droplet on a surface.

Water droplet volume was found to decrease at about the same rate on both surfaces. Since both Teflon and nylon will not adsorb a significant amount of water within the time frame of the experiment, the main reason for the decrease of water droplet volume can be attributed to the evaporation of water. The change in water droplet volume occurs a little more rapidly for larger water droplets since there is a larger gas/liquid interface for evaporation to take place.

Regardless of the initial droplet size, static contact angles were observed to increase with increasing water droplet volume but to decrease with contact time on both polymers. This is seen in Figure 26.2 as the linear decrease of the cosine of static contact angles, cos θ_S, with increasing droplet volume. The measured static contact angles were shown to change more quickly with time for smaller water droplet volumes. This effect was more pronounced on the more hydrophilic nylon surface. The slope of the time-dependent change in contact angles becomes smaller with droplet volume. The smaller water droplet has greater surface-to-volume ratio, and the losses of water by evaporation and absorption of water by the surface reflect more sensitively on the contact angle.

It is difficult to visually observe if the three-phase contact line is receding or advancing while the droplets are allowed to stand on the surface. By calculating the contact area at the water droplet/polymer interface, it is seen that the contact area decreases in the case of Teflon and increases in the case of nylon as depicted in Figure 26.3. This implies that the droplet spreads and

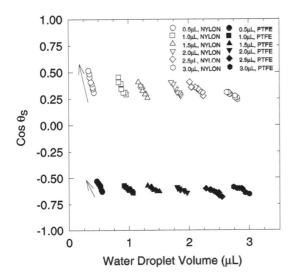

Figure 26.2 The effect of contact time on the cosine of static contact angles of many different size droplets equilibrating on Teflon and nylon polymer surfaces; the arrows indicate the increase of time; the cosine of the static contact angle decreases with droplet volume.

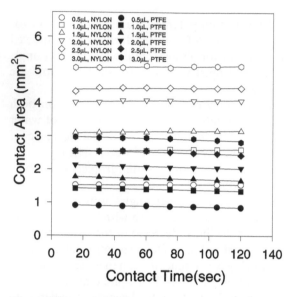

Figure 26.3   The effect of contact time on the contact area on nylon and Teflon; contact area is calculated from droplet width in contact with the surface.

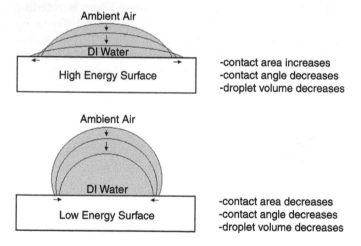

Figure 26.4   The effect of allowing a DI water droplet to equilibrate with a low energy (more hydrophobic) and a high energy (less hydrophobic) surface and ambient air on contact area, static contact angle, and droplet volume.

flattens with the resting time as the droplet volume decreases on nylon (higher surface energy). Conversely, the droplet retracts and flattens over the wet surface of Teflon (lower surface energy). The time-dependent changes of sessile droplet are schematically depicted in Figure 26.4. In general, the lower contact angle of water sessile droplet is obtained if the droplet is allowed to stand on the surface for a longer time.

4. ADVANCING AND RECEDING SESSILE DROPLET CONTACT ANGLES

Regardless of the term "advancing" or "receding," static contact angle measurement is performed while the three-phase contact line water/air/surface is stationary. By simply adding more water to the existing droplet a series of advancing contact angles are measured. The increase of droplet volume forces the three-phase contact line to always advance outward in the direction of the dry surface. The contact angle that is independent of the droplet size is taken as the advancing contact angle, which represents the surface property of a surface. The receding contact angle is measured after the advancing contact angles were measured in several steps by withdrawing a certain volume (the same volume aliquot used in the advancing contact angle measurements) from the droplet in several steps. Advancing and receding contact angles for various polymers are shown in Figures 26.5. Advancing and receding contact angles for the same polymers coated with plasma polymer of trimethylsilane (TMS) and of (TMS + O$_2$) are shown in Figure 26.6. The following 10 different conventional polymer samples were used: polytetrafluoroethylene (PTFE), ultrahigh molecular weight polyethylene (UHMWPE), polypropylene (PP), low-density polyethylene (LDPE), high-density polyethylene (HDPE), polycarbonate (PC), polyethyleneterephthalate (PET), polymethyl methacrylate (PMMA), polyvinylidene fluoride (PVDF), and nylon 6 (nylon). Each polymer plate was cut approximately 20×25 mm from a larger sheet with a thickness of 1 mm, except for PVDF which was 0.8 mm thick. Only untreated polymers including those prior to plasma treatment were cleaned in an ultrasonic soap water bath, thoroughly rinsed with deionized (DI) water, dried completely in air, and finally placed in a desiccator with calcium sulfate (CaSO$_4$).

As described in Chapters 23, 24, and 25, the receding contact angle is influenced by the surface configuration change and consequently dependent on the time of contact. While the value of advancing contact angle could represent the surface property of a surface, the value of receding contact angle cannot be used to represent the surface property. The discrepancy of the values of both contact angles, however, could be used to express the extent of the surface dynamic instability or stability in a qualitative manner. The smaller the difference, the more dynamically stable is the surface.

Both advancing and receding contact areas of untreated, TMS plasma–treated, and (TMS + O$_2$) plasma–treated PP film and PMMA sheet are shown in Figure 26.7. The change in advancing and receding contact area was by and large the same for all the untreated and TMS-treated polymers. The slope of the change in advancing contact area with water droplet volume is greatest for the (TMS + O$_2$)–treated samples (most hydrophilic). This implies that the water droplet spreads with greater ease on the more hydrophilic films. Conversely, the water droplet spreads to a much lesser extent on the more hydrophobic films.

Similar to (TMS + O$_2$) plasma–treated PP, many of the (TMS + O$_2$) plasma–treated polymers showed that the three-phase contact line hardly recedes on the wet surface. Specifically, (TMS + O$_2$) plasma–treated PTFE, UHMWPE, PP, HDPE, PVDF, and nylon showed hardly any change in contact area during the receding process. This is due to the strong specific attractive forces formed at the surface underneath the bulk droplet, i.e., the interfacial tension (underneath the droplet) changed due to the interaction of water with the surface.

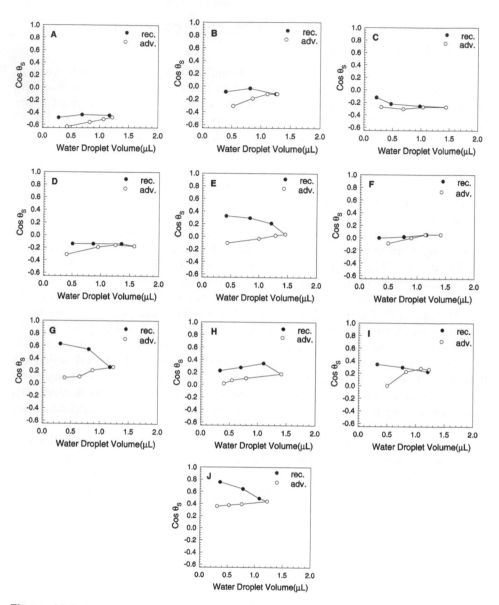

Figure 26.5 Static advancing and receding contact angles on untreated conventional polymers; A) PTFE, B) UHMWPE, C) PP, D) LDPE, E) HDPE, F) PC, G) PET, H) PMMA, I) PVDF, J) nylon.

Figure 26.8 depicts the effects of plasma polymer coatings on various polymers, which are arranged in order of hydrophilicity. TMS plasma polymer surface modification of 10 low-energy conventional polymers yielded similar surface, indicated by the average cosine of the static "advancing" contact angle cos θ_S = −0.0645 (θ_S = 94 ± 2.2). (TMS + O_2) plasma polymer modification greatly reduced the contact angle of 10 conventional polymers with an average cosine of the static

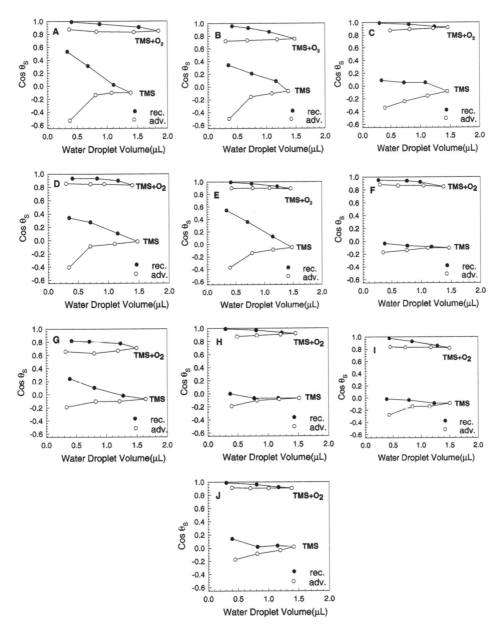

Figure 26.6 Static advancing and receding contact angles on the following TMS and TMS + O$_2$ plasma treated conventional polymers; A) PTFE, B) UHMWPE, C) PP, D) LDPE, E) HDPE, F) PC, G) PET, H) PMMA, I) PVDF, J) nylon.

"advancing" contact angle cos $\theta_S = 0.845$ ($\theta_S = 32 \pm 6.9$). Both plasma coatings are useful in modifying a variety of different polymers with a specific wettability of plasma polymer virtually independent of the underlying conventional polymer material. These data support the concept that plasma polymerization coating is an effective tool to create a new surface state implanted on various substrates.

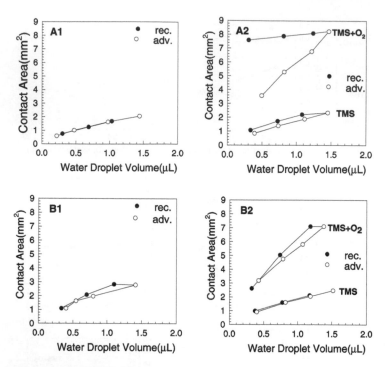

Figure 26.7 Advancing and receding contact areas at the DI water/polymer interface where rows (A) PP, (B) PMMA and columns (1) untreated, (2) TMS and TMS + O_2 treated.

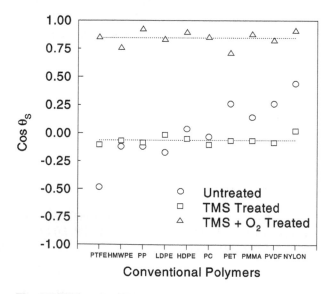

Figure 26.8 The effects of plasma polymerization coatings on the wettability of conventional polymers given by the cosine of the static "advancing" contact angle, θ_S; each value of the static contact angle was taken for the largest droplet size attained during the advancing process whereby the droplet size dependence is small, dotted lines indicate the mean cos θ_S of the TMS and (TMS + O_2) treated polymers.

5. DYNAMIC CONTACT ANGLE

In contrast to the sessile droplet method, Wilhelmy balance force measurements are obtained while the three-phase contact line is moving with respect to the polymer surface. This dynamic method requires the immersion and withdrawal of a sample of specific geometry through liquid water. Both advancing and receding contact angles are calculated from the force balance data; therefore, values of contact angle hysteresis can be obtained even for very fast changes in surface configuration. The Wilhelmy balance method is a more versatile and sensitive tool for monitoring the *surface dynamics* of polymeric surfaces than the sessile droplet method. By the same token, it should be recognized that contact angle is not a simple parameter to describe the surface property of a sample.

The Wilhelmy balance apparatus is composed of an electronic balance interfaced with a PC computer. The tensiometer measures the force exerted by a partially immersed thin-sample plate in water. A beaker containing water is moved up (increasing the immersion) and down (emerging the sample) by a constant speed, and the corresponding force change is recorded.

The total force exerted to the microbalance may be given by the following force balance equation:

$$F = Mg - \rho gtHd + L\gamma_L \cos\theta \qquad (26.1)$$

where F is the total exerted force on the balance, M is the mass of the plate, g is the gravitational acceleration, ρ is the liquid water density, t is the thickness of the plate, H is the width of the plate, d is the immersion/emersion depth, L is the plate perimeter [$L = 2 \times$ (thickness + width)], γ_L is the liquid water surface tension, and θ is the contact angle at the liquid/solid/air contact line. The first term is the gravitational force, the second term is buoyancy, and the third term is the interfacial tension. Since the balance automatically sets the starting point when the sample plate touches the water's surface, the gravitational force of the sample (the first term) may be neglected. Therefore, the actual measured force by the balance is given by:

$$F = L\gamma_L \cos\theta - \rho gtHd \qquad (26.2)$$

The measured force, F, in Eq. (26.2) is the difference in the interfacial force between the water and sample plate (wetting force) and the buoyancy force from the immersed portion of the plate. The total force, F, is divided by the sample plate perimeter, L, to give force per unit length, F/L (expressed in mN/m).

$$F/L = \gamma_L \cos\theta - (\rho gtH/L)\, d \qquad (26.3)$$

The surface tension of liquid, γ_L, is measured by the microbalance using a platinum wire ring. Thus, from the intercept of the extrapolated line of the second term, $\cos\theta$ is obtained.

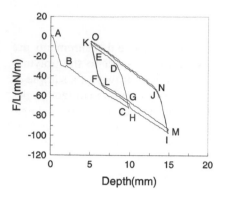

Figure 26.9 A typical Wilhelmy force loop of a relatively hydrophobic surface; the force loop is composed of three cycles each consisting of one immersion (advancing) and one emersion (receding) process. The sample plate is polycarbonate (PC) modified with TMS plasma.

Lowering and raising a platform, where a beaker with water rests, performs partial immersion and partial emersion of a plate in water. The force is measured during a series of typically three consecutive wetting cycles called a force loop. A typical force loop is shown in Figure 26.9 of a TMS plasma–modified PC plate. The first cycle starts when the bottom edge of the plate touches the water's surface and then immersed in the water to a depth of 10 mm (A → B → C) then emerged (withdrawn) to 5 mm (C → D → E). In the second cycle, the plate is immersed to a depth of 15 mm (E → F → G → H → I) then emerged to 5 mm (I → J → K). Lastly, in the third cycle, the plate is immersed again to 15 mm (K → L → M) and finally emerged to 5 mm depth (M → N → O). The immersion/emersion velocity was set to 5 mm/min for every wetting cycle. F/L and depth data were collected at a rate of 5 Hz.

Dynamic advancing and receding contact angles were calculated by extrapolating corresponding immersion and emersion F/L lines to zero immersion depth. It is assumed that the contact angle, θ, in Eq. (26.3) is unchanged during immersion and emersion where the meniscus shape is fully developed. For instance, in Figure 26.9, the first cycle dynamic advancing contact angle, $\theta_{D,a,1}$, is calculated from Eq. (26.3) by extrapolating the most linear portion of the F/L line (B → C) to 0 mm depth.

Each polymer plate is positioned perpendicular to the water's surface before every Wilhelmy experiment so that each point on the bottom edge of the plate contacts the water's surface at the same time. However, this is not a mandatory requirement, and adding a small weight that is attached to the bottom of the sample allows use of a nonrectangular sample. If two edges are not parallel and irregular, no linear line for buoyancy is obtained, and the simple extrapolation method cannot be used to calculate the interfacial tension. The slope of immersion or emersion line is determined by the geometrical factor of sample as seen in Eq. (26.3).

The Wilhelmy apparatus is suitable for both static and dynamic contact angle measurements. However, it is particularly suited for the measurement of dynamic wetting properties of polymer surfaces, and most Wilhelmy balance is used in the dynamic mode.

6. CHANGE OF MENISCUS SHAPE DURING IMMERSION AND EMERSION PROCESSES

The shape of the meniscus of water at the plate may either be depressed away or advanced toward the dry surface. In general, a hydrophobic surface would result in a depressed meniscus ($\theta_{D,a,1} > 90$ degrees, $\cos \theta_{D,a,1} < 0$) whereas a hydrophilic surface would result in an advanced meniscus ($\theta_{D,a,1} < 90$ degrees, $\cos \theta_{D,a,1} > 0$). The meniscus can be clearly seen visually for extremely hydrophilic and hydrophobic surfaces. Surfaces with moderate hydrophilicity/hydrophobicity exhibit flat meniscus, which seems to be perpendicular to the surface ($\theta_{D,a,1} = 90$ degrees, $\cos \theta_{D,a,1} = 0$).

Figures 26.10 and 26.11 show how the shape of the meniscus changes during one cycle of a force loop for hydrophobic (TMS-treated polymers) and hydrophilic [(TMS + O_2)–treated polymers] surfaces, respectively. After the initial contact of a hydrophobic surface with water, the three-phase contact line (dry surface/water/air) does not move, but the meniscus continues to become more depressed [(a → b → c) in Figure 26.10] with immersion depth. This "transition" stage corresponds to the line (A → B) in the force loop of Figure 26.9. During the transition stage a force is generated between the wet surface and the water in the direction of the dry surface [1–3]. This force continues to increase until it exceeds the adhesive force, which is coined as the adhesion tension of immersion, τ_i, by Guastalla, at the polymer surface/water interface [4]. For hydrophobic surfaces, the adhesive tension is weaker than the repulsive tension, both of which are both due to the interaction forces between the low-energy surface and the high-energy water molecules. Because of the strong repulsive tension, further immersion of the plate is required to generate enough force to exceed the small adhesive tension. Once the adhesive tension is broken, the three-phase contact line is forced up the vertical wall of the plate, and a dynamic steady-state meniscus (c → d) is developed with increasing depth seen as (B → C) in Figure 26.9.

There is no significant transition stage on initial immersion of a hydrophilic surface because the dynamic meniscus is immediately developed on contact [(a) in

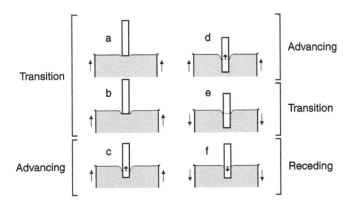

Figure 26.10 A schematic representation of the meniscus shape and position of the three-phase contact line (solid/liquid/air) during immersion and emersion of a hydrophobic surface (e.g., TMS treated polymers); the dual arrows indicate which direction the beaker is moving, the small arrow on the plate indicates the direction the three-phase contact line is moving.

Figure 26.11]. This is a result of the strong adhesive forces due to the strong interactions of the high-energy surface and water molecules. The three-phase contact line starts to advance toward the dry surface (b → c) immediately after contact is made with the water. In this case, the two-stage line (A → B → C) in Figure 26.9 appears as a straight line as in the force loops of (TMS + O$_2$)–treated polymers depicted in Figure 26.12.

Both hydrophobic and hydrophilic surfaces experience a transition stage at the start of the withdrawal process, seen as (C → D) in Figure 26.9, and (B → C) in Figure 26.12, which shows the case of hydrophilic surface. The meniscus again changes shape while the three-phase contact line remains stationary [(e → f) in

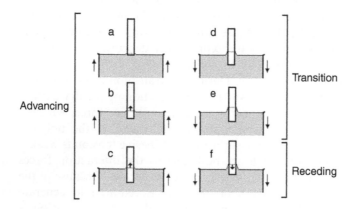

Figure 26.11 A schematic representation of the meniscus shape and position of the three-phase contact line (solid/liquid/air) during immersion and emersion of a hydrophilic surface (e.g. TMS + O$_2$ treated polymers); the dual arrows indicate which direction the beaker is moving, the small arrow on the plate indicates the direction the three-phase contact line is moving.

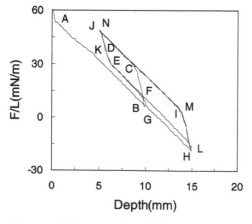

Figure 26.12 A typical Wilhelmy force loop of a relatively hydrophilic surface; the force loop is composed of three cycles each consisting of one immersion (advancing) and one emersion (receding) process. The sample plate is polycarbonate (PC) modified with TMS + O$_2$ plasma.

Figure 26.10, and (d → e) in Figure 26.11] until the force in the direction of the wet surface exceeds the adhesive tension of emersion, τ_e, at the polymer surface/water interface. After exceeding the adhesive tension, the three-phase contact line starts to move toward the prewetted surface, seen as (D → E) in Figure 26.9 and (C → D) in Figure 26.12. The adhesive tension in the receding process is usually less than that in the advancing process unless the surface was completely wetted by the first immersion (i.e., zero contact angle). When the complete wetting occurs on the first immersion, the first emersion line retraces the first immersion line, such is the case with O_2 plasma–cleaned glass.

It should be noted that on the receding cycle the wet plate surface has previously interacted with water molecules for a different period of time depending on the immersion depth of the plate. Therefore, the bottom deeper immersed portions of the plate interact with the water molecules for a longer period than the shallow immersed portions closer to the top of the plate. This causes small but continuous changes in the meniscus shape even after the three-phase contact line starts to move in the advancing and receding processes.

Depending on the nature of moieties in the surface state, functional mobility, and the previous environment with which they were equilibrated, the surface will move toward a new equilibrium state with water. Figure 26.9 clearly shows that successive F/L immersion lines (B → C), (F → G), and (L → M) are shifted slightly above the previous, respectively, indicating that each immersion cycle makes the surface more hydrophilic. The difference between advancing F/L lines is due to the surface configuration change due to wetting the surface. The driving force for such a change is the interaction force between the water molecules and the functional moieties of the polymer surface. More specifically, the repulsive interaction of hydrophobic moieties and water molecules causes rotational migration of the hydrophobic moieties toward the interior bulk material. Similarly, the attractive interaction of hydrophilic moieties and water molecules causes rotational migration of hydrophilic moieties toward the top surface that contacts with water. In both cases, functional group migrations result in minimization of interfacial tensions.

7. WILHELMY DYNAMIC CONTACT ANGLES AND WETTABILITY

All of the conventional polymers investigated possess relatively low surface energies, which are considered relatively nonwettable and moderately hydrophobic. As described in Chapter 27, *hydrophilic surface* could be defined as the surface of which contact angle of water is less than 45 degrees, and *hydrophobic surface* could be defined as the surface of which contact angle of water is greater than 90 degrees. The surfaces having a contact angle of water between 45 degrees and 90 degrees could be defined as the amphoteric hydrophobic/hydrophilic surface. Thus, the conventional polymers investigated have amphoteric surface characteristics.

Similar to static contact angles from the sessile droplet method, Wilhelmy dynamic contact angles are an excellent indication of the change in surface characteristics due to surface modification techniques such as plasma polymerization coating. The cosine of dynamic advancing contact angles from the first immersion, $\cos \theta_{D,a,1}$ of untreated, TMS-treated, and (TMS + O_2)–treated conventional

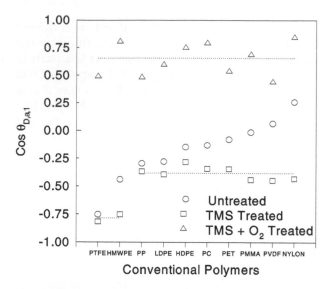

Figure 26.13   The effects of plasma treatment on wettability, which is given by the cosine of the dynamic "advancing" contact angles, cos θ_D calculated from the first immersion F/L lines of each force loop; dotted lines indicate the mean cos $\theta_{D,a,1}$ of TMS and TMS+O$_2$ treated polymers.

polymers, can be seen in Figure 26.13 arranged from the highest to the lowest contact angle according to the values of cos $\theta_{D,a,1}$ of untreated conventional polymers. (The same kind of plot for sessile droplet static contact angles is shown in Figure 26.8.) Figure 26.13 depicts the effects of plasma polymerization coatings of TMS and of (TMS + O$_2$) applied on the 10 conventional films described above. All the (TMS + O$_2$) plasma–treated polymers are clearly more hydrophilic than the TMS plasma–treated polymers because the values of cos $\theta_{D,a,1}$ are closer to cos $\theta = 1$ (completely wettable surface).

Eight of 10 TMS-treated polymers exhibited contact angles within an average cos $\theta_{D,a,1} = -0.381$ ($\theta_{D,a,1} = 112 \pm 3.6$), and all of them showed the negative cos θ. This indicates that the surface of plasma polymer of TMS-deposited films is clearly hydrophobic and independent of the substrate material for these polymers. TMS-treated PTFE and UHMWPE were slightly more hydrophobic with average cos $\theta_{D,a,1} = -0.785$ ($\theta_{D,a,1} = 141 \pm 4.2$).

The (TMS + O$_2$) plasma–modified polymers were made considerably more hydrophilic with average cos $\theta_{D,a,1} = 0.654$ ($\theta_{D,a,1} = 49.2 \pm 11.7$) but remain in the domain of amphoteric surface, under the conditions of plasma polymerization used. (TMS + O$_2$) plasma–deposited films were slightly more dependent on the nature of the conventional polymer substrates. This is probably due to the fact that substrate polymers have different oxygen plasma susceptibilities.

The force loop for untreated samples are shown in Figure 26.14. The force loops for TMS plasma–treated and (TMS + O$_2$) plasma–treated surfaces are shown in Figure 26.15. Any sign of deviation from the parallelogram force loop is an indication of surface dynamic instability. Plasma polymerization coating of (TMS + O$_2$) seems to cause some degree of surface dynamical instability depending on the nature of substrate polymer, e.g., PTFE, UHMWPE, HDPE, and PMMA.

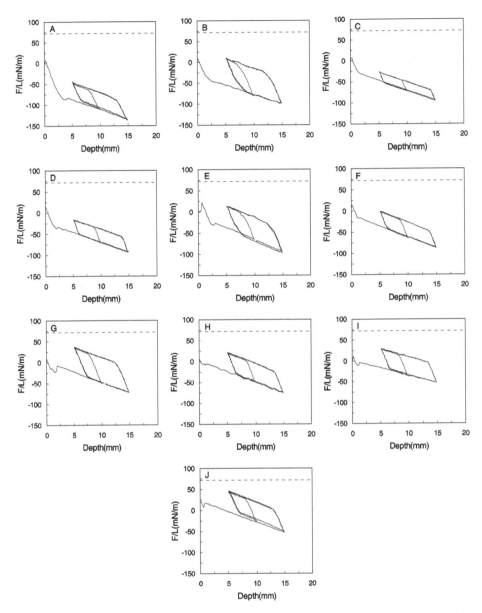

Figure 26.14 Force loops of the following untreated conventional polymer plates: A) PTFE, B) UHMWPE, C) PP, D) LDPE, E) HDPE, F) PC, G) PET, H) PMMA, I) PVDF, J) nylon; dotted lines indicate the surface tension of water, force loops were generated by a constant immersion/emersion rate of 5 mm/min.

8. CONTACT ANGLE HYSTERESIS

Probably the most apparent and common phenomenon observed in measurements of contact angle of water on a polymer surface is the discrepancy found between advancing and receding contact angles. This discrepancy, found in both sessile

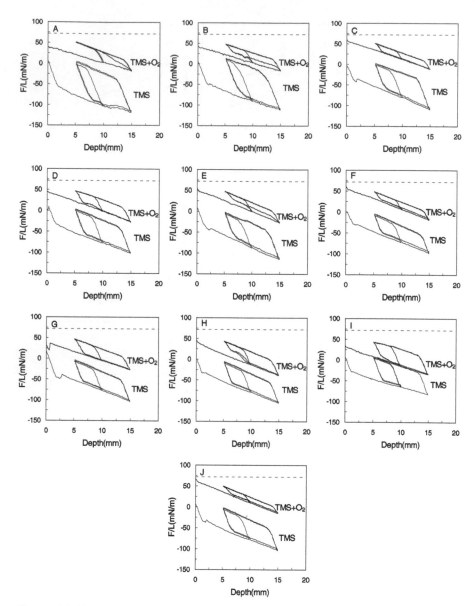

Figure 26.15 Force loops of the following plasma treated conventional polymer plates: A) PTFE, B) UHMWPE, C) PP, D) LDPE, E) HDPE, F) PC, G) PET, H) PMMA, I) PVDF, J) nylon; plasma deposition conditions listed in order of flow rate, pressure, input power and exposure time, respectively, are given by the following: TMS (1 sccm, 50 mtorr, 40 W, 1.6 min), TMS + O$_2$ (1 + 4 sccm, 50 mtorr, 115 W, 2 min). Dotted lines indicate the surface tension of water. Force loops were generated at a constant immersion/emersion speed of 5 mm/min.

droplet and Wilhelmy contact angle measurements, is referred to as contact angle hysteresis [1–3, 5–10]. Many factors are known to influence contact angle hysteresis, such as surface configuration change, swelling, crystallinity, surface roughness, adsorption/desorption, and the energy level of surface electrons. Contact angle is, in principle, a complex function of any one or more of these factors. For low surface energy polymers, contact angle hysteresis is mostly due to surface configuration change.

Contact angle hysteresis from either sessile droplet or Wilhelmy method is represented numerically as the difference between the receding and advancing contact angles, $\Delta\theta = \theta_r - \theta_a$. Contact angle hysteresis, in the context of dynamic wetting by the Wilhelmy balance method, is due to a combination of three main factors: (1) changing meniscus shape, (2) sample geometry, and (3) surface state change due to surface configuration change [11]. These factors influence two distinguishable types of contact angle hysteresis that could be termed dynamic and intrinsic hysteresis. *Dynamic hysteresis* is the vertical width of a force loop and could be expressed quantitatively as the difference in F/L lines on immersion and emersion at a given depth, e.g., 7.5 mm; $(\Delta F/L)_D = (F/L)_{D,r,1} - (F/L)_{D,a,1}$. *Intrinsic hysteresis* is the unrecoverable hysteresis resulting from consecutive wetting of the surface and can be expressed as the difference in the first and second immersion lines at a given depth, e.g., 7.5 mm, $(\Delta F/L)_I = (F/L)_{D,a,2} - (F/L)_{D,a,1}$.

Dynamic hysteresis is caused largely by the change of meniscus shape during a transition stage. Therefore, a hydrophobic surface shows the larger separation of the immersion line and the emersion line than a hydrophilic surface as seen in Figure 26.16, which depicts dynamic hysteresis for untreated, TMS-treated, and $(TMS + O_2)$–treated polymer films. However, dynamic hysteresis is probably not maximal with TMS treatment.

Intrinsic hysteresis is a direct result of surface configuration change, which occurs as a result of wetting the surface with water. The plates were purposely immersed to a deeper immersion depth in the second cycle to observe the extent of intrinsic hysteresis. Significant surface configuration change affects the calculated contact angles on immersion and emersion, which violates the assumption of

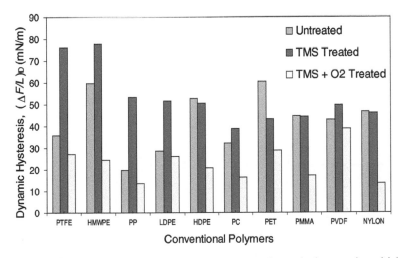

Figure 26.16 Effect of plasma treatment on dynamic hysteresis, which is the difference in the dynamic first cycle advancing line and the receding line, $(\Delta F/L)_D$ (mN/m) = $(F/L)_{D,a,1} - (F/L)_{D,r,1}$ is a direct result of the changing shape of the meniscus during a wetting cycle.

Eq. (26.3) that contact angles do not change during immersion and emersion. When the contribution of surface configuration is large, such as the case with hydrophilic (TMS + O_2)–treated polymers in Figure 26.15, deformations in the ideal parallelogram shape of a force loop occur. Intrinsic hysteresis of TMS-treated polymers in Figure 26.17 was independent of the underlying polymer within the time scale of dynamic wetting while (TMS + O_2)–treated polymers varied widely. This further implies that substrate-dependent etching of oxygen plasma species may be responsible.

Both dynamic and intrinsic hystereses are affected by sample geometry. Since the measured force on a plate is dependent on the sample geometry by the second term of Eq. (26.3), then the effects of the sample plate buoyancy become significant with plates of larger volume. The slopes of all the F/L lines during the advancing and receding processes where the dynamic steady-state meniscus is developed (B → C, D → E, F → G, H → I, J → K, L → M, N → O, in Figure 26.9), are determined by the sample plate geometry. During these regions of a force loop, it is assumed that the contact angle does not change, thereby making the interfacial term in Eq. (26.3) a constant. The immersion depth, d, is the only experimental parameter, which is designed to be linear with time. The slopes of the immersion and emersion lines are determined by the plate thickness, t, and width, H.

If a completely wettable plate ($\theta_{D,a} = 0$) is immersed to a depth of, say, 7.5 mm, a limiting value of F/L can be calculated from Eq. (26.3). This limiting value of F/L is totally dependent on the sample geometry and cannot be exceeded by the advancing or receding measured forces $(F/L)_a$ and $(F/L)_r$. For any plate geometry a limiting F/L line can be generated that is referred to as the ceiling buoyancy line. This limiting line is shown for TMS-treated (hydrophobic), (TMS + O_2)–treated (hydrophilic), and O_2-treated (completely wettable) glass plates of different size in Figure 26.18. Ceiling

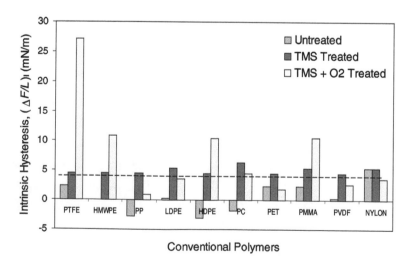

Figure 26.17   Effect of plasma treatment on the intrinsic hysteresis, which is the difference in second cycle immersion line and the first cycle immersion line, $(\Delta F/L)_I = (F/L)_{D,a,2} - (F/L)_{D,a,1}$; intrinsic hysteresis is directly proportional to the extent of surface configuration change of the surface state.

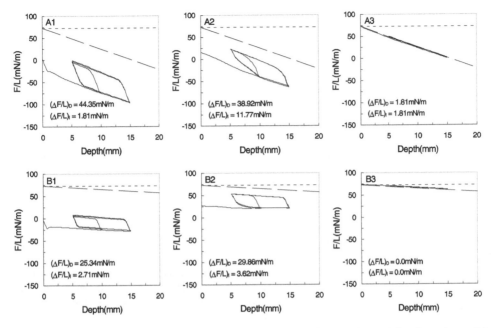

Figure 26.18 Rows A and B correspond to the force loops of the following glass slide dimensions (A) H = 25.4 mm, L = 1 mm, (B) H = 22 mm, L = 0.153 mm. Each column from 1 to 3 correspond to the following plasma treatments: (1) TMS treated, (2) (TMS + O_2) treatment (O_2/TMS = 4), (3) O_2 treatment. The dotted lines correspond to the surface tension of water, and the dashed lines correspond to the ceiling buoyancy of the plate immersed to 7.5 mm depth.

buoyancy threshold at an immersion/emersion depth of 7.5 mm for the larger (A) and the smaller (B) plates are 67.21 mN/m and 37.41 mN/m, respectively.

In general, dynamic hysteresis is smaller with smaller cross-sectional area because the effect of buoyancy is less. Thus, the dynamic hysteresis is not a measure of surface characteristics. Intrinsic hysteresis is a measure of the surface dynamic stability of the surface.

9. CORRELATION BETWEEN WILHELMY (DYNAMIC) AND SESSILE DROPLET (STATIC) CONTACT ANGLES

Static contact angles by the sessile droplet method are influenced by droplet size and contact time, whereas Wilhelmy dynamic contact angles are affected by the dynamic influence of a fluid moving on the plate surface. (The sessile droplet contact angle in this discussion means the advancing sessile droplet contact angle.) When the plate motion is slowed the advancing F/L lines increase and the receding F/L lines decrease yielding a narrower parallelogram (with smaller contact angle hysteresis) [7]. As the plate motion comes to a stop the dynamic advancing and receding contact angles approach the same value of static contact angle. This means that the advancing contact angle by the Wilhelmy method is always greater than the sessile droplet contact angle, and the Wilhelmy receding contact angle is always smaller than the

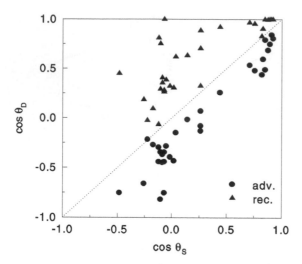

Figure 26.19 Relationship between the Wilhelmy "advancing" and "receding" contact angles, θ_D, and static sessile droplet "advancing" contact angles θ_S for all the untreated TMS treated and TMS + O$_2$ treated conventional polymers; the Wilhelmy contact angles are from the first cycle immersion and the static "advancing" contact angles are with a droplet size of 1.4 µl.

sessile droplet contact angle, and the difference is greater when a faster immersion and emersion rate is used. This difference is caused by the process of establishing dynamic meniscus at the beginning of immersion and also at the beginning of emersion.

Cosines of dynamic "advancing" and "receding" contact angles, $\cos \theta_{D,a}$ and $\cos \theta_{D,r}$, were plotted as a function of the cosine of static contact angles $\cos \theta_S$ on all untreated and plasma–treated polymers in Figure 26.19. As expected, $\cos \theta_{D,a}$ was smaller and $\cos \theta_{D,r}$ was larger than $\cos \theta_S$. Since dynamic contact angles are dependent on the velocity of immersion/emersion, a constant speed of 5 mm/min was selected to emphasize the major aspect of wettability.

The general relationship $\cos \theta_{D,a} < \cos \theta_S < \cos \theta_{D,r}$ was extended to a direct correlation proposed by Uyama et al. [12]. The correlation states that the mean of $\cos \theta_{D,a}$ and $\cos \theta_{D,r}$ equals $\cos \theta_S$. Application of this correlation to dynamic and static contact angle data of untreated and plasma–treated polymers was found to fit well as seen in Figure 26.20.

The sessile droplet (advancing) contact angle is a simple representation of the force balance at the three-phase line, which can be considered as the interfacial characteristics of a surface/liquid/air system and hence the surface characteristics of a sample. The advancing dynamic contact angle by the Wilhelmy balance method depends on the sample geometry, which can be represented by $tH/2(t + H)$, where t is the thickness and H is the width of a plate, and the immersion/emersion rate, and hence cannot be considered as a surface characteristics of a sample. If data obtained by Wilhelmy balance is intended to describe the contact angle in the context of surface characteristics, the average value of advancing and receding contact angles should be used.

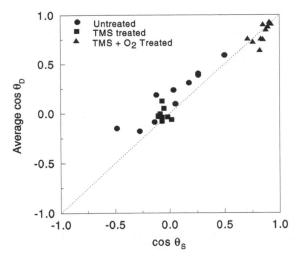

Figure 26.20   A correlation between the average of dynamic "advancing" and "receding" contact angles from the Wilhelmy method and static "advancing" contact angles from the sessile droplet method for untreated, TMS and TMS + O₂ treated polymers follows the relationship given by $\cos \theta_S = (\cos \theta_{D,r,l} + \cos \theta_{D,a,l})/2$.

The sessile droplet contact angle measurement is a simple and accurate method to obtain information pertinent to the surface energy of a sample. The Wilhelmy balance method, on the other hand, is a very useful method to investigate the surface dynamic aspect of a sample, which will be described in the following sections. The instability of some of plasma–treated polymer surface observed by the Wilhelmy balance method is also described in Chapter 30.

10.   LIQUID FILM ON SURFACE ON WILHELMY PLATE

It has been observed the formation of continuous liquids on high-energy solid surfaces, such as mica, quartz, and silica, when those surfaces are subjected to immersion and emersion in water [13,14]. A continuous film of water on a hydrophilic (i.e., high energy) plasma polymer–coated glass slide moments after it was immersed to a depth of 3 cm in a beaker of DDI water is shown in Figure 26.21A. The water film remained continuous as it receded to the bottom of the plate. After about 2 min the water film front receded to approximately 1 cm from the bottom of the plate, as shown in Figure 26.21B. The presence and stability of a continuous water film can be detected and quantified by the Wilhelmy method [15].

Most conventional polymers are moderately hydrophobic (i.e., they possess low surface energy); thus, spontaneous wetting by water does not occur on such polymers. The improvement of the wettability of polymeric solids usually requires some type of surface modification. Spontaneous spreading of a liquid on a solid surface occurs when the surface tension of the liquid is less than the surface tension of the solid, i.e., $\Delta(\gamma_s)_{SW}$ given by Eq. (25.6), is negative, in which case the contact angle is zero. Thus, "contact angle = 0" is the postulated necessary condition for the formation of stable and continuous water films, but some moderately to poorly

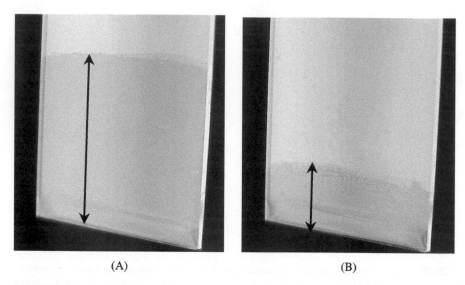

(A)   (B)

Figure 26.21   Pictures of a continuous film of water on a plasma polymer coated glass slide (A) just after wetting and (B) after 2 min of exposure to air. Arrows indicate the region of the continuous water film.

wettable surfaces are also capable of holding a continuous film of pure liquid water once the contact was made, by virtue of surface configuration change. The surface of Teflon is hydrophobic and the contact angle of water is above 90 degrees, and water cannot spread spontaneously on the surface. However, it does not mean that water cannot wet the surface or a water/Teflon interface cannot be created. Such a contact exists at the bottom of a sessile droplet of water residing on a Teflon surface. Holding of liquid water film on a surface is mainly concerned with the stability of the water film on a surface.

Aqueous film stability is dependent on the adhesive force or negative interfacial tension at the two-phase (i.e., solid/liquid) boundary. The force balance at the two-phase boundary may change independently from the three-phase force balance due to surface configuration change of interfacing surface state moieties, which occurs in order to minimize interfacial tension with water as described in previous chapters.

11.   WILHELMY FORCE LOOPS AND FLUID HOLDING TIME

The extent of intrinsic hysteresis can be determined by exposing a Wilhelmy plate to two consecutive wetting cycles, each consisting of one immersion step followed by an emersion step, and observing the extent of deviation between the first and second immersion lines. Most low-energy polymeric surfaces exhibit the common, ideal parallelogram-shaped force loops, since the intrinsic hysteresis is usually relatively small or even zero. The deviation between the first and second immersion lines is a result of surface configuration change during the time scale of Wilhelmy force measurement.

It has been observed that improving the wettability of moderately hydrophobic conventional polymers according to the ratio of TMS to oxygen gas mixture during plasma polymer deposition causes significant deviations from the ideal parallelogram-shaped force loops. Furthermore, the second immersion line starts to approach the first emersion line as the wettability increases. At some limiting wettability, a portion of the second immersion line retraces the first emersion line, as shown in Figure 26.22. This retracing is a special case that has now been determined to be due to the presence of a continuous water film adhering to the polymer plate much like the one depicted in Figure 26.21.

An extremely stable water film exists when the film is present over the entire wetted length of a Wilhelmy plate. This case is depicted in Figure 26.22 as the hook-shaped force loop formed during two wetting cycles. The continuous water film did not recede at all during the time scale of wetting, since the second immersion line retraced the first emersion line from a depth of 5 to 15 mm. The two contrasting force loops in Figure 26.22 are extreme cases of water film stability or the capability of polymeric surfaces to hold water. These and any intermediate cases in which the water film recedes during the time scale of wetting may be quantitatively characterized by a parameter referred to a fluid holding time (FHT):

$$\text{FHT} = d/v \tag{26.3}$$

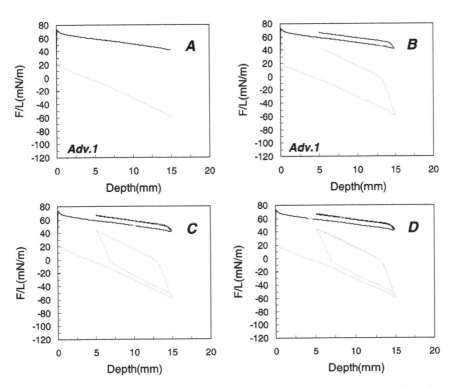

Figure 26.22 Wetting stages of Wilhelmy force loops on a polymeric plate; the gray and black lines depict the absence and presence of a continuous water film, respectively; the consecutive wetting stages of a Wilhelmy force loop are as follows: (A) first immersion (Adv.1), B) first emersion (Rec.1), (C) second immersion (Adv.2), and D) second emersion (Rec.2).

where d is the length of the continuous film up the side of the Wilhelmy plate and v is the velocity of the plate during the second immersion. The immersion/emersion velocities may be arbitrarily fixed, but the length of the continuous film is determined from the length of the region where the second immersion line retraces the first emersion line on the force loop.

Three different cases of aqueous film stability indicated by FHT are depicted in Figure 26.23. Corresponding force loops and diagrams of the continuous water film

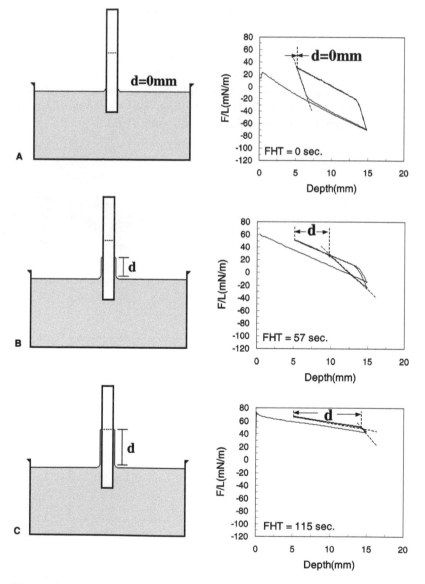

Figure 26.23 Interpretation of fluid holding phenomena on vertical plates immersed in solution to a depth of 15 mm (indicated by the dotted line on each plate) and then raised to a depth of 5 mm; corresponding force loop plots with fluid holding times of (A) 0 s, (B) 57 s, and (C) 115 s were calculated from a constant immersion speed of 300 mm/s.

position up the side of each Wilhelmy plate after immersion to a depth of 15 mm is presented. Figure 26.23A and C depicts two limiting cases of FHT, and (Figure 26.23B) is just one of many possible intermediate cases. The FHT in all three cases was determined with a fixed second immersion velocity of 5 mm/min.

12. EFFECT OF WILHELMY BALANCE PARAMETERS ON FLUID HOLDING TIME

Wilhelmy plate wetting parameters, such as plate velocity during immersion or emersion and halting the motion of the plate after immersion or emersion, have been shown to affect the intrinsic hysteresis, which consequently affects the overall shape of the force loop [3]. These wetting parameters were found to affect FHT differently depending on whether pure water or an aqueous artificial tear solution was employed as the wetting medium.

Like many other substances with hydrophilic and hydrophobic groups, proteins tend to spread across the interface between air and water forming a thin film. This results in a notably lower surface tension and consequently a lower force loop position than is obtainable with pure water, as shown in Figure 26.24, in which the left column represents water and the right column represents the artificial tear fluid, and the withdrawing speed increases from the top row to the bottom row. Withdrawing a Wilhelmy plate from protein solution results in deposition of a continuous aqueous protein film onto the surface by physical adsorption similar to Langmuir-Blodgett film deposition. The adsorbed film masked and altered the original surface properties of the plasma polymer–modified glass slides. This is indicated in Figure 26.24 by the large deviation between the first and second immersion lines; however, unlike the case of wetting with water, the second immersion line does not retrace the first emersion line. This suggests that a condensed tear film does not form within the time scale of wetting, but it is more than likely that a hydrated protein film forms on the wetted portion of the plate.

The effect of the second immersion velocity on the force loop shapes of water and artificial tear fluid on plasma polymer–coated glass slides can be seen in Figure 26.24 by comparing force loops vertically. Increasing the second immersion velocity had virtually no effect on the overall shape of the force loops when tear fluid was used but increased the retracing region of the force loops when water was used. This means the adsorbed film from the artificial tear solution did not recede but the continuous water films receded down the side of the coated slides. The FHTs of the tear fluid were calculated from the inflexion point of the second immersion line (limiting value) because the film did not recede, and compared with that of water as shown in Figure 26.25. Thus, the tear fluid FHT decreased with increasing second immersion velocity according to Eq. (26.3). The FHT for water is lower than the limiting value at 2 mm/min, since the water film recedes; however, it approaches the limiting FHT as the second immersion velocity is increased, since the water film does not have sufficient time to recede.

The effect of first emersion velocity on force loop shape is shown in Figure 26.26. The force loops for tear fluid behave independently from the first emersion velocity, since the second immersion velocity is fixed in this case. Therefore, the tear fluid FHT is constant, as depicted in Figure 26.27.

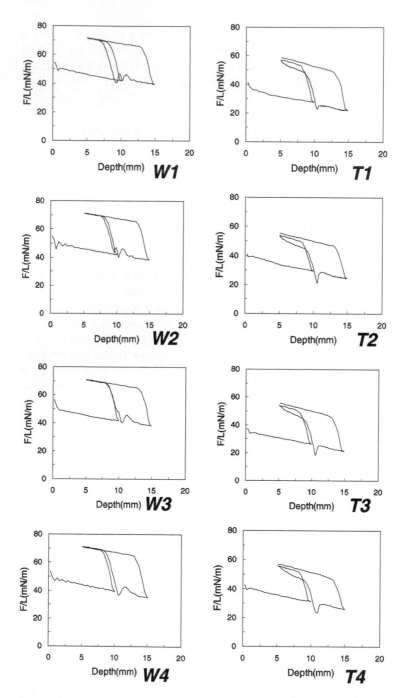

Figure 26.24 Wilhelmy force loops of (CH_4 + air) plasma treated glass plates in (W) DDI water and (T) artificial tear fluid at varying second immersion velocities: (1) 2 mm/min, (2) 5 mm/min, (3) 10 mm/min, and (4) 20 mm/min. Plasma discharge conditions were 38 W, 50 mTorr, 1 sccm CH_4, 2 sccm air, 20 min; first and second emersion and first immersion velocities were fixed at 20 mm/min, the use of water and tear fluid yielded advancing contact angles, $\theta_{D,a,1}$, means, and standard deviations of $47° \pm 1$ and $48° \pm 3$, respectively.

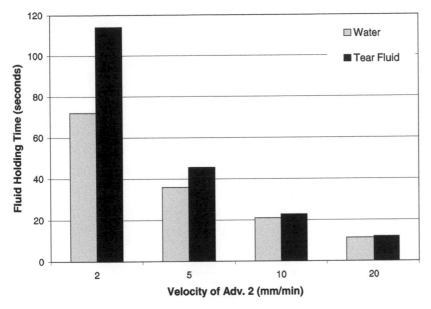

Figure 26.25 The effect of Wilhelmy plate second immersion velocity (Adv. 2) on the fluid holding time of water and artificial tear fluid on $CH_4 + air$ plasma treated glass slides.

As shown in Figure 26.26, the force loop shape for water is clearly dependent on the first emersion velocity, which controls the amount of time the plate and liquid are in contact before the next wetting cycle. Figure 26.28 depicts the effect of halting on the force loop shape, and Figure 26.29 compares the values of FHT. Halting the motion of the coated plates during immersion in the two liquids for 20 min before proceeding with the next wetting cycle had no significant effect on force loop shape or FHT. Halting the motion of the plates exposed to air after the first emersion was employed to force the continuous films to recede. The water film obviously receded whereas the hydrated protein film did not. However, the protein film dried somewhat, which lowered the position of the second immersion line slightly in Figure 26.28 (T3).

Most conventional polymers wetted with water exhibit typical parallelogram-shaped force loops with no indication of the presence of a continuous aqueous film [8]. However, when wetted with artificial tear fluid, a few conventional polymers of varying wettability displayed some second immersion line retracing of first emersion lines, as shown in Figure 26.30. This indicates that a continuous aqueous film of artificial tear fluid was present for a short time on the polymers even on highly hydrophobic Teflon. Halting the motion of polymer plates after the first emersion and exposing the wetted surface to air for extended periods caused the film to dry leaving behind adsorbed proteins. This rendered it impossible to characterize film stability, since the original surface properties were masked regardless of surface wettability.

Figure 26.31 depicts the same measuring procedure applied to a contact lens material sheet: (A) untreated, (B): $(CH_4 + air)$ plasma treated, and (C) $(CH_4 + air)$ plasma treated and then O_2 plasma treated. Characterizing aqueous film stability on untreated and plasma–modified contact lens materials using artificial tear fluid by

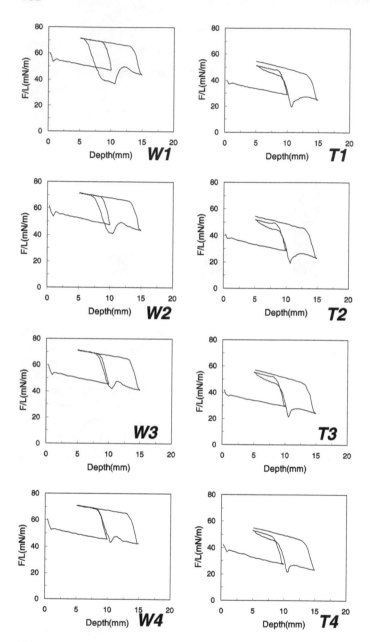

Figure 26.26   Wilhelmy force loops of $CH_4 + air$ plasma treated glass plates in (W) DDI water and (T) artificial tear fluid at varying first emersion velocities: (1) 2 mm/min, (2) 5 mm/ min, (3) 10 mm/min, and (4) 20 mm/min; first and second immersion velocities were fixed at 20 and 10 mm/min, respectively, and second emersion at 20 mm/min.

the Wilhelmy method was impossible due to the complications caused by the adsorption of proteins. On the other hand, the figures indicate that plasma coating decreases the contact angle of untreated material and reduces the adsorption of proteins from the artificial tear fluid, which is in accordance with the very low

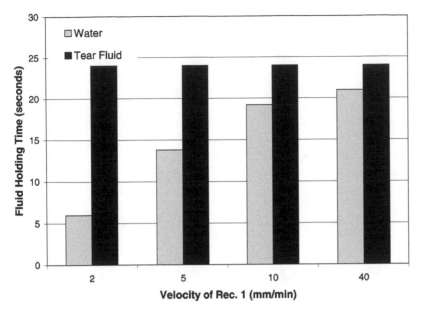

Figure 26.27 The effect of Wilhelmy plate first emersion velocity (Rec. 1) on the fluid holding time of water and artificial tear fluid on $CH_4 + air$ plasma treated glass slides.

protein adsorption by the coated contact lens observed by a separate protein adsorption measurement.

The FHT measured by the Wilhelmy balance method can be effectively used to compare the liquid holding capabilities of different surfaces. The value of FHT depends on the experimental parameters and cannot be used in an absolute sense. In general, the aqueous film stability is obtained when spontaneous wetting occurs on imperturbable surfaces. However, moderately hydrophilic and possibly even some hydrophobic surfaces that are perturbable by water were found to be capable of holding continuous films of water. Multicomponent fluid, such as a dilute solution of protein used as a simulated tear fluid, may yield misleading liquid holding characteristics of surfaces due to preferential adsorption of components on a surface.

13. MEANING OF CONTACT ANGLE

The terms "surface" and "surface energy" are misleading based on the intuitive assumption that such an ideal surface exists or on the lack of knowledge about the highly perturbable interfacial characteristics of realistic materials. Dealing with metals, ceramics, glasses, and other inorganic materials, of which surface is relatively imperturbable, the terms surface and surface energy might serve to describe the interfacial characteristics of the materials. Dealing with highly perturbable surface of polymers and other organic materials, however, the concept of surface is invalid and the so-called surface characteristics should be considered interfacial characteristics. We could use the term surface to describe the interface of air and solid in order to simplify the discussions. Contact angle is a measure of interfacial characteristics but not a measure of surface characteristics. This key point could be seen clearly by the

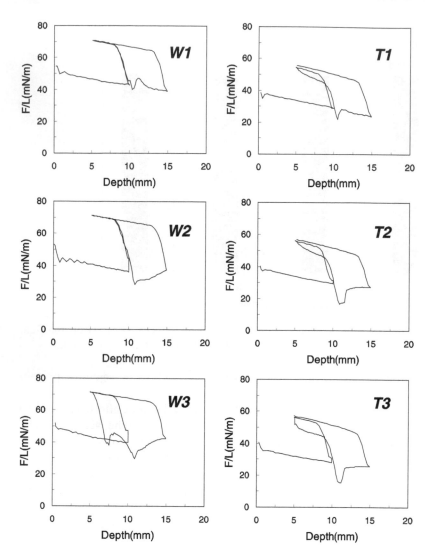

Figure 26.28 Wilhelmy force loops of $(CH_4 + air)$ plasma treated glass plates in (W) DDI water and (T) artificial tear fluid with (1) no halting, (2) halting after first immersion for 20 min while immersed in liquid after first immersion to a depth of 10 mm, and (3) halting after first immersion for 20 min while immersed in liquid at a depth of 10 mm followed by halting after first emersion for 10 min in air at a depth of 5 mm.

contact angle of water on a gelatin hydrogel, which shows a very high contact angle of sessile droplet contact angle of water but very low sessile bubble contact angle of water (Figure 23.1).

Sessile droplet contact angle (on polymer) indicates the force balance at the three-phase line on a surface just before the sessile droplet was placed. In the case of gelatin hydrogel, the surface is hydrophobic due to very fast change of the surface configuration. Even after cutting a gel in water to create a new fresh surface and placing a sessile droplet as quickly as possible, the surface shows a very high contact

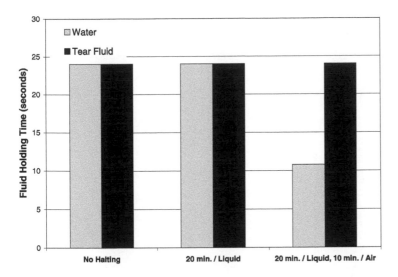

Figure 26.29 The effect of halting the motion of Wilhelmy plates of $CH_4 + air$ plasma treated glass slides in water and artificial tear fluid.

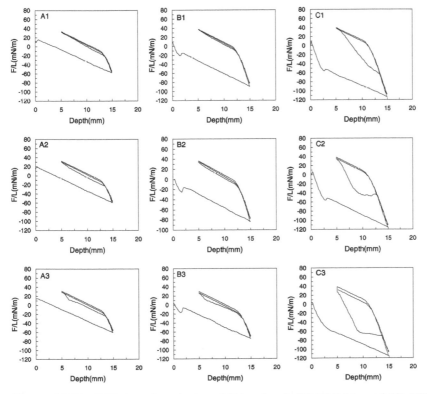

Figure 26.30 Wilhelmy force loops of (A) nylon-6, (B) PMMA, and (C) PTFE plates in artificial tear solution. At the end of the first emersion, the plates were held out of the solution for (1) 0 min, (2) 5 min, and (3) 40 min at a depth of 5 mm; the use of tear fluid on nylon-6, PMMA, and PTFE, yielded advancing contact angles, $\theta_{D,a,1}$, means, and standard deviations of $68° \pm 3$, $91° \pm 3$, and $130° \pm 1$, respectively.

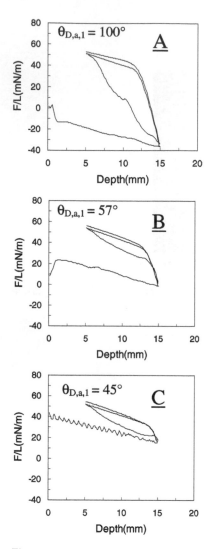

Figure 26.31 Wilhelmy force loops of (A) untreated, (B) (CH_4 + air) plasma treated, and (C) (CH_4 + air) plasma then O_2 plasma treated contact lens material using artificial tear fluid; the emersion/immersion velocities were all fixed at 5 mm/min.

angle of water. The sessile droplet contact angle of water on the gelatin hydrogel shows the interfacial characteristics of air/hydrogel. The sessile bubble measurement on the surface of the same gelatin hydrogel shows very low contact angle of water because the bubble contact angle shows the interfacial characteristics of water/(gelatin hydrogel). Both contact angles are correct representing the respective interfacial characteristics, and the discrepancy of numbers shows the difference of the interfacial characteristics of air/(gelatin hydrogel) and of water/(gelatin hydrogel). The case of contact angle of water on a gelatin hydrogel also points out that the contact angle of water does not always indicate the wettability of a surface with water.

In sessile droplet contact angle measurement, the advancing contact angle could be used to describe the surface characteristics, but the receding contact angle per se is a meaningless number unless its change with respect to the contact time can be utilized to analyze the surface dynamic nature of the surface.

The dynamic contact angles calculated from the Wilhelmy force balance cycle are the measures of characteristic interfacial characteristics, particularly with respect to the surface dynamics. The deviation from the normal parallelogram of force cycle provides valuable information pertinent to the change of surface configuration as well as of surface component, if the shape of sample does not deviate from a thin rectangular sample. In the interpretation of data, it is necessary to bear in mind that the Wilhelmy balance measures the force, not the contact angle, of liquid to solid. If it is intended to obtain the measure similar to that obtained by the sessile droplet advancing contact angle, i.e., the parameter that indicates surface energy or wettability, the mean value of advancing and receding contact angles should be used. The reason for this is that the Wilhelmy advancing contact angle gives a misleadingly high contact angle value because of the contributions of the geometrical factors and the mode of operation of measuring process.

REFERENCES

1. Weikart, C.M.; Miyama, M.; Yasuda, H.K. J. Colloid Interface Sci. **1999**, *211*, 18.
2. Chen, Y.L.; Helm, C.A.; Isrealachvili, J.N. J. Physical Chem. **1991**, *95*, 10736.
3. Miyama, M.; Yang, Y.; Yasuda, T.; Okuno, T.; Yasuda, H.K. Langmuir **1997**, *13*, 5494.
4. Guastalla, J. J. Colloid Interface Sci. **1956**, *11*, 623.
5. Holly, F.J.; Refojo, M.F. J. Biomed. Mater. Res. **1975**, *9*, 315.
6. Gagnon, D.R.; McCarthy, T.J. J. Appl. Polym. Sci. **1984**, *29*, 4335.
7. Hayes, R.A.; Ralston, J. J. Colloid Interface Sci. **1993**, *159*, 429.
8. Wang, J.H.; Claesson, P.M.; Parker, J.L.; Yasuda, H. Langmuir **1994**, *10*, 3887.
9. Yasuda, T.; Okuno, T.; Tsuji, K.; Yasuda, H.K. Langmuir **1996**, *12*, 1391.
10. Johnson, R.E., Jr.; Dettre, R.G.; Brandreth, D.A. J. Colloid Interface Sci. **1977**, *62*, 205.
11. Weikart M. Christopher; Miyama, Masayo; Yasuda K. Hirotsugu. J. Colloid Interface Sci. **1999**, *211*, 28.
12. Uyama, Y.; Inoue, H.; Ito, K.; Kishida, A.; Ikada, Y. J. Colloid Interface Sci. **1990**, *141*, 275.
13. Langmuir, I. Science **1938**, *87*, 493.
14. van Damme, H.S.; Hogt, A.H.; Feijen, J. J. Colloid Interface Sci. **1986**, *114*, 167.
15. Weikart M. Christopher; Miyama Masayo; Yasuda K. Hirotsugu. Langmuir **2000**, *16*, 5169.

27
Sessile Bubble Formation and Detachment

1. SESSILE DROPLET CONTACT ANGLE VS. SESSILE BUBBLE CONTACT ANGLE

The contact angle is defined as the angle of liquid at the three-phase line. Figure 27.1 depicts contact angles involved in sessile droplet and sessile bulb. Figure 27.2 depicts force balance at the three-phase line for a sessile droplet and for a sessile bubble. For a sessile bubble, Young's equation still holds. However, the location and the direction of γ_{sl} and γ_{sv} are different from the case for a sessile droplet. Figure 27.2b depicts a cross-sectional view of an air bubble on a solid surface immersed in liquid water (sessile bubble). The angle Θ, which is generally recognized as contact angle in the case of a sessile droplet, is the supplementary angle of the contact angle θ in the case of a sessile bubble. Thus, a surface that gives a large contact angle of water droplet gives a small Θ when a bubble develops on the surface immersed in water, i.e., an air bubble spreads on a hydrophobic surface.

The bubble injection method of contact angle measurement utilizes a sessile bubble below the surface that is immersed in water. It is nearly impossible to measure the sessile bubble contact angle on top surface of a sample immersed in water, of which contact angle is to be measured, because the buoyancy works in the direction to lift the bubble. Although the two methods should yield the identical contact angle, the values obtained by the two methods could deviate significantly depending on the solubility of air in water and the perturbability of the surface by water as described in the previous chapters.

When a small bubble develops on an immersed surface in water, by feeding air through a small hole, the three-phase line changes continuously as the size of bubble increases. When the bubble size reaches the critical volume, above which the buoyancy of the bubble becomes greater than the bubble attaching force, the bubble detaches from the surface and travels through liquid water as water-borne bubble. The analysis of this process provides important information pertinent to the forces involved in the interface that cannot be grasped easily without undertaking such an effort. Namely, it is possible to observe how the three-phase line advances on the surface as a function of the interfacial energy until the buoyancy takes over the interfacial force that keeps an expanding bulb on the surface. In the bubble trapping method, the buoyancy forces a bubble to attach to the surface above and the advance

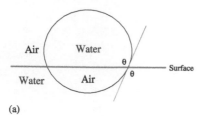

(a)

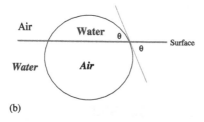

(b)

Figure 27.1   Correlation between the sessile droplet contact angle and contact angle of air bubble on the surface immersed in water; (a) hydrophobic surface, (b) hydrophilic surface.

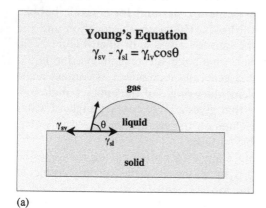

(a)

(b)

Figure 27.2   Force balance at the three phase line; (a) a sessile water droplet and (b) a sessile air bubble.

of the three-phase line cannot be clearly observed. In order to investigate this situation, the following experiments were carried out by taking advantage of plasma polymerization coating that enables us to create a desired surface energy on a plate on which a bubble attaches and spreads.

A well-defined hole was placed on a stainless steel sheet by a laser, and the surface of the stainless steel sheet was coated with a plasma polymer of $(TMS + O_2)$ by the cathodic polymerization using the stainless steel sheet as the cathode [1]. Changing the ratio of TMS to O_2 yields surfaces with varying contact angles of water. Figure 27.3 depicts the correlation between cosine of water sessile droplet contact angle and the oxygen mole fraction, which indicates that a desired surface tension can be obtained by changing the mole ratio of O_2 in a mixture of O_2 and TMS [2]. Two size of holes, 0.25 mm and 0.5 mm, were used. The bubble formation was recorded by a video camera for a predetermined period, and all bubbles were collected in a graduated cylinder. From the number of bubbles and the total volume, the average bubble diameter of a spherical bulb was calculated.

2. FORCE BALANCE IN A DEVELOPING SESSILE BUBBLE

The factors that influence the bubble expansion and the detachment from the surface may be seen in Figure 27.4, which depicts components of forces acting at the edge of an attached bubble. The surface tension of liquid that acts at the edge of the bubble contact base has two components that are dictated by the contact angle θ. The component parallel to the orifice surface is given by $\gamma \cos \theta$ and the perpendicular component by $\gamma \sin \theta$, where γ is the surface tension of the liquid.

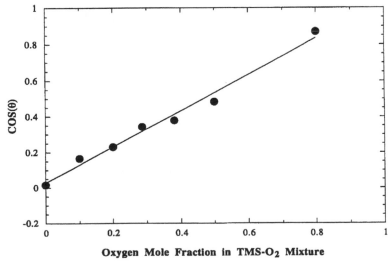

Figure 27.3 The correlation between $\cos \theta$ of water and the mole fraction of O_2 in TMS/O_2 mixture.

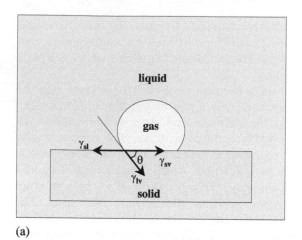

(a)

(b)

Figure 27.4 Force balance at the three phase line; (a) interfacial tension, (b) gas pressure.

With the consideration of extended base of contact, the attachment force for a bubble, f_1, can be given by

$$f_1 = \pi\varphi(\gamma \sin\theta + f_p \cos\theta)$$

where φ is the diameter of the contact base, f_p is the force due to the gas pressure acting at the contact point, which is numerically identical to the pressure p.

The force which resists against the expansion of bubble attachment base or the force for bubble closing, f_2, can be given by

$$f_2 = \pi\varphi\gamma \cos\theta$$

The expansion force of a bubble is the component of the internal pressure of air in the bubble acting at this point, and therefore at the edge of the bubble attaching to the surface, the component of the force parallel to the surface that expands the contact base of a bubble can be given by

$$f_3 = \pi \phi \gamma f_p \sin \theta$$

When the value of θ exceeds 90 degrees, f_2 and f_3 exchange their roles and f_2 becomes the bubble expansion force and f_3 the resisting force. The internal force of pressure, which acts perpendicular to the surface of the bubble, is balanced by the hydrostatic pressure of water acting on the opposite side of the bubble surface for each point, except at the three-phase contact point where θ is observed.

At $\theta = 90$ degrees, the internal force due to gas pressure is balanced against the hydrostatic pressure of water because the interfacial tension has no component in the direction of the solid surface. In this case, the interfacial tension is directly balanced against the buoyancy of a bubble. Such a direct balance between the interfacial tension and the buoyancy can be justified for only this special case of $\theta = 90$ degrees.

The detaching force, f_4, is the buoyancy of the bubble, which acts vertically upward (perpendicular to the surface under the conditions in consideration).

$$f_4 = B$$
$$B = V(\rho_1 - \rho_2)g$$

where ρ_1 is the density of the liquid and ρ_2 the density of air, g the acceleration of gravity, and V the volume of the bubble.

The critical conditions for contact base expansion and bubble detachment can be seen as

$f_3 > f_2$ contact base of a bubble expands

$f_3 = f_2$ maintains a constant contact base

$f_4 > f_1$ bubble detachment occurs

As seen in the sequence of still pictures presented, a spherical bubble with a finite contact area on the surface develops as soon as the boundary surface between air and water passes the orifice. The spreading of air phase on the membrane surface and the development of the bulk of a bubble occur simultaneously. However, a constant base is established in the very early stage, and the major portion of bubble development occurs using the fixed contact base. The base area differs depending on the interfacial tension between water and surface.

The critical factor is whether the size of base for a sessile bubble is larger or smaller than the size of orifice. If the size of orifice is larger than the base of a sessile bubble, a sessile bubble does not form. Since no one would anticipate creation of small bubbles from a large orifice, this domain is clearly beyond the scope of this consideration. For small orifices, whether an air bubble emerges from the orifice or emerges from surface is the critically important factor, which is determined by the balance of $\cos \theta$ and $\sin \theta$.

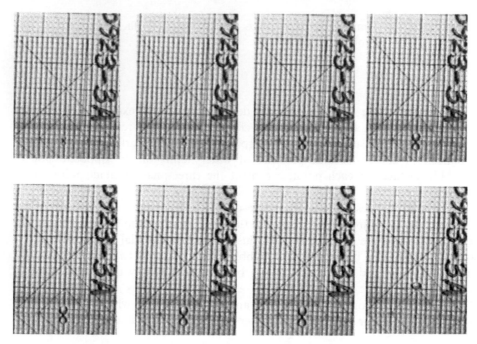

Figure 27.5 Bubble formation without expanding the contact base of a bubble beyond the orifice: contact angle of water 5.0 degree, orifice diameter 0.25 mm, airflow rate 0.26 ml/m, and bubble volume 0.0050 ml/bubble.

If the sessile bubble contact angle $\theta < \pi/2$ (hydrophilic surface), $\sin\theta < \cos\theta$. Accordingly, in this domain an emerging bubble cannot expand its contact base beyond the area of the orifice. In this domain, a bubble develops by using the area of the orifice as the contact base, maintaining the spherical shape of bubble, and detaches from the orifice. The ultimate size of the bubble depends on the volume, which creates enough buoyancy to detach the bubble from the edge of the hole. In this domain, the bubble volume is determined by the size of the orifice, and is independent of the contact angle of the surface (below $\pi/2$) that surrounds the orifice.

The formation of bubbles in this domain can be seen by sequentially captured images of bubbles shown in Figure 27.5. The picture of a bubble emerging from a hole appears as if two bubbles are attached because the bottom half is the reflection of the bubble (mirror image) on the surface. The separation of two images is the indication of the bubble detachment from the surface. (The double images appear in all pictures shown in subsequent figures. A 1-mm grid is placed as a reference of size.) A bubble maintains the spherical shape up to the point where it detaches from the orifice.

In the range $\pi > \theta > \pi/2$, $\sin\theta > \cos\theta$, and an air bubble emerging from an orifice expands its contact base beyond the orifice, if the size of hole is smaller than the maximal contact base that can be taken by a bubble. When the contact angle of water θ is not too far from $\pi/2$, the formation of an air bubble follows the same mechanism as the case of nonexpansion shown in Figure 27.5, except the base of the

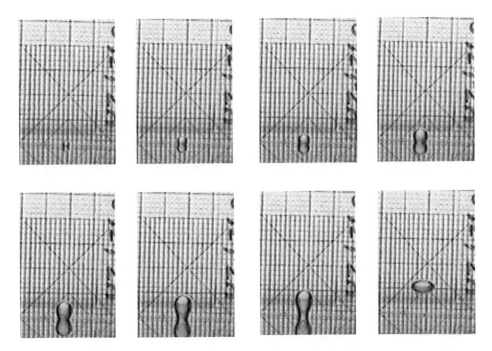

Figure 27.6 Bubble formation with expanded base of a bubble beyond the orifice; contact angle of water 70.0, orifice diameter 0.55, airflow rate 0.56 ml/m, bubble volume 0.037.

bubble is larger than the orifice. Figures 27.6 shows sequentially captured images of bubbles in this domain, i.e., $\pi > \theta > \pi/2$. The developing bubble maintains the shape of spherical bubble attached to the surface until just before the detachment occurs. The contact area of a bubble retracts to the size of the orifice just before the detachment occurs, and the shape changes to that of a hot-air balloon. The exact moment of detachment was not captured due to the limitation of shatter speed. The size of the next bubble following detachment suggests that detachment took place at the edge of the orifice.

The bubble formation in the domain $\theta > \pi$ can be seen in Figure 27.7, which are sequentially captured images of a bubble emerging from an orifice (diameter 0.25 mm) placed on a very hydrophobic surface. In this domain the shape of a developing bubble becomes a half sphere when the maximal contact area is established in the early stage of bubble development. After this point, a cylindrical bubble with half-sphere cap develops because the bulging of bubble beyond the contact base cannot occur due to the contact angle. The bubble, from this point on, expands its volume mainly by increasing the height of the cylindrical part. As buoyancy increases, the size of the attachment area starts to decrease, and the shape of the bubble changes to that of hot-air balloon also in this case.

It should be noted that the base of the bubble is nearly 10 times greater than the diameter of the orifice. (If the grid plate is not perfectly parallel to the surface, two unparallel lines show up as seen in this figure.) It should be reiterated here that the bubbles shown in Figures 27.5–26.7 emerge from the identical orifice with a diameter of 0.25 mm. The only difference is the surface energy of the plate, more

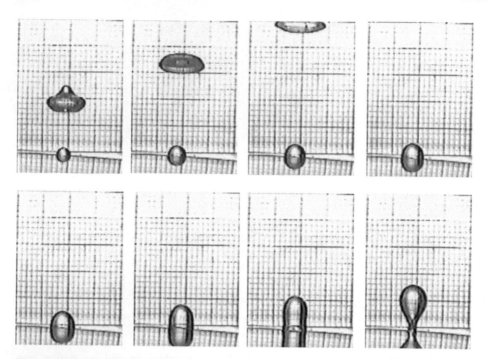

Figure 27.7 Bubble formation with cylindrical expansion: contact angle of water 99.7, orifice diameter 0.25 mm, airflow rate 0.82, and bubble volume 0.092 ml/bubble.

precisely the interfacial tension at the water/solid interface, which changed the characteristic sessile bubble contact angle.

These figures are convincing demonstrations of the principle that the size of a bubble is determined by the interfacial tension between air and surface. Figure 27.8 depicts the relationship between the size of bubble and contact angle (degree) of water on the surface, which surrounds the orifice. In order to show the difference due to the size of the orifice clearly, the surface area-to-volume ratio A/V, which is an important parameter in gas–liquid reactions, rather than V or bubble diameter, is used in this plot. The figure clearly shows the two distinctively different domains— (A) the orifice size controlled domain and (B) the interfacial tension controlled domain—and also that the demarcation line is contact angle of water 45 degrees. In the latter domain (B), larger bubbles emerge from a small orifice and the bubble size is independent of the orifice size.

The effect of size of orifice can be seen only in domain A. Without reducing the contact angle of water to less than 45 degrees, the small orifices do not produce small bubbles. From these observations of the advancing air phase on surface, the definition of hydrophilic and hydrophobic surface could be made, i.e., hydrophilic surface by $\theta < 45$ degrees, and hydrophobic surface by $\theta > 90$ degrees. Domain A mentioned above is the hydrophilic surface domain. The surface of which the contact angle of water falls in the domain, 90 degrees $\geq \theta \geq 45$ degrees could be considered as partly hydrophobic and partly hydrophilic (amphoteric). The most of commercially available polymers have the surfaces that belong to this amphoteric hydrophobic/hydrophilic domain.

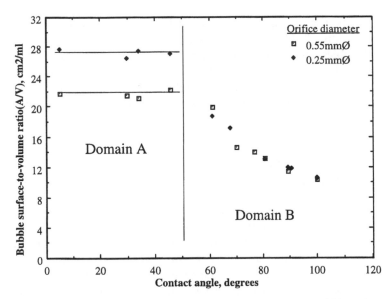

Figure 27.8 Domains of bubble formation: Domain A; bubble volume is controlled by the size of orifice, Domain B; bubble volume is controlled by the interfacial tension.

3. BUBBLE FORMATION FROM AN INCLINED SURFACE

In a tilted surface, the buoyancy of the developing bubble creates the drifting force parallel to the surface, changing the detachment mechanism and consequently the size of emerging bubbles [3]. If a bubble drift out of the orifice by the surface component of buoyancy, it means that buoyancy did not reach the critical level to detach the bubble from the surface. Therefore, in such a case, the drifted bubble does not detach from the surface by buoyancy, and the detachment occurs, *if* it occurs, by another force that didn't encounter the horizontal surface. In the absence of other forces, the drifted bubbles often do not detach from an inclined surface; instead they keep marching on the inclined surface until the end of the surface where tow bubbles collide and form a larger bubble, whose buoyancy is large enough to detach the merged bubble.

In the bubble formation from a horizontal surface, the bubble development and the bubble detachment are coupled. When the buoyancy of a developing bubble overcomes the bubble attachment force due to the interfacial tension, the bubble detaches from the surface and completes the process of the bubble formation. A higher flow rate of air in the low flow rate regime (e.g., 0.2–30 sccm) simply increases the frequency of the bubble formation but does not change the volume of bubble [1].

In the bubble formation from an inclined surface, however, the bubble development and the bubble detachment processes are decoupled because a developing bubble could drift out of the orifice due to the component of the buoyancy parallel to the inclined surface. Once a sessile bubble drift out of the orifice, the bubble development ceases because no air is fed into a sliding bubble. Since the bubble development and detachment are decoupled, the flow rate of air becomes an important factor, which controls the frequency of sliding bubble formation. In other words, the drifting velocity of a drifted bubble and the surface velocity of the next developing bubble become important factors. A higher flow rate creates more sliding bubbles on an inclined surface. The drifting velocity, on the other hand, is controlled by the interfacial tension and the drifting component of buoyancy. Consequently, the behavior of drifting bubbles provides an interesting demonstration of the importance of the interfacial tension that could be controlled by plasma polymerization coating.

4. BUBBLE DETACHMENT FROM THE HYDROPHILIC SURFACE

If the contact angle of water on the orifice surface is less than $\pi/2$, a developing bubble does not expand beyond the orifice, as discussed previously. In this case, the tilting of the orifice surface does not change the basic bubble formation, and a bubble develops and detaches from the orifice. Figure 27.9 depicts sequentially captured bubble formation from an inclined surface with a tilt angle of 30 degrees. The effective cross-sectional area decreases with the tilt angle because a bubble develops along the line of the buoyancy, and the drifting force causes premature detachment of a developing bubble. Consequently, an inclined surface, in this domain, creates the bubble, which is smaller than the bubble emerging from a horizontal surface under otherwise identical conditions. The tilting angle has little

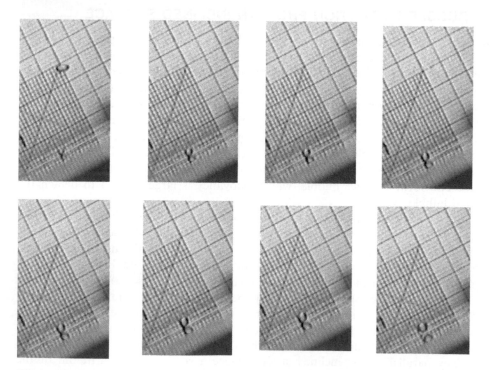

Figure 27.9 Bubble formation from a tilted orifice plate: tilt angle 30 degree, contact angle of water 41.5 degree, orifice diameter 0.55 mm, air flow rate 1.98 ml/m, bubble volume 0.011 ml/bubble.

influence in this domain (hydrophilic surface). (The effect of tilting angle is not shown.)

5.   DETACHMENT OF DRIFTED BUBBLES FROM THE AMPHOTERIC SURFACE

When a bubble drifts out of the orifice before its buoyancy increases to the critical level to detach the bubble from the surface, the bubble slides on the surface. The detachment of the bubble can occur after the bubble slides for a short distance on the surface by the inertia of the drifting, as shown in Figure 27.10. If a bubble cannot detach from the surface within a short distance, the bubble keeps sliding on the surface until the bubble reaches the end of the surface and merges with the preceding or following bubble as shown in Figure 27.11. The sliding velocity of a bubble on the surface and the frequency of creating sliding bubbles, which is dependent on the flow rate of air, determine the distance between two bubbles. The sliding of bubbles occurs regardless of the tilt angle of plate as shown in Figure 27.12 (tilting angle 45 degrees) and Figure 27.13 (tilting angle 60 degrees).

While the interfacial tension and the buoyancy of the bubble determine the sliding speed of an attached bubble, the flow rate of air determines the frequency of creating a sliding bubble. Figures 27.12 and 27.14 show the influence of the flow rate of air on the detachment of sliding bubbles on the surface. In the case shown in

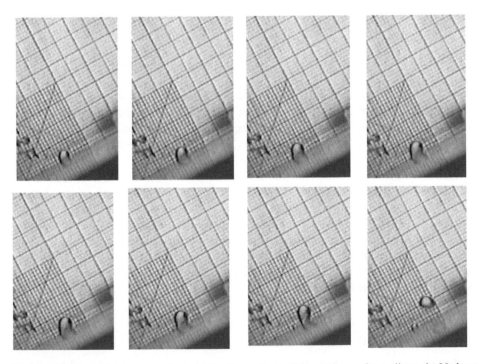

Figure 27.10   Detachment of a bubble after a short slide on the surface: tilt angle 30 degree, contact angle of water 82.7 degree, orifice diameter 0.55 mm, airflow rate 4.36 ml/m, and bubble volume 0.047 ml/bubble.

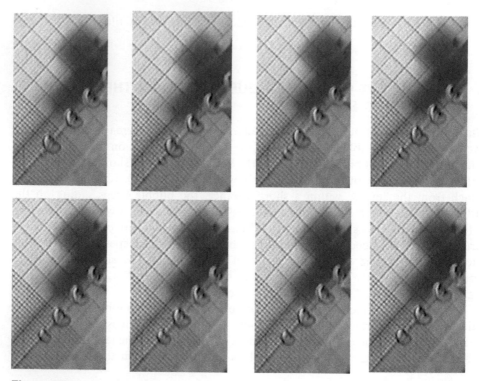

Figure 27.11 No detachment occurs and bubbles slide on the surface: tilt angle 45 degree, contact angle of water 82.7 degree, orifice diameter 0.55 mm, air flow rate 4.36 ml/m, bubble volume 0.047 ml/bubble.

Figure 27.12, numbers of sessile bubbles keep marching on maintaining a constant distance between them due to the slow feeding of air. When the flow rate of air is increased, the second drifting bubble develops before the first bubble establishes the sliding on the tilted surface and bumps into the first bubble as shown in Figure 27.14. When two bubbles merge, the buoyancy of the merged bubble becomes sufficient to detach the bubble from the surface. The marching bubbles on an inclined surface are an interesting demonstration of interfacial tension as a function of the surface tension of materials, which couldn't be done without the combination of the capability to use the identical orifice and the variable surface energies attained by the cathodic plasma polymerization. This is an interesting demonstration that the unique capability of luminous chemical vapor deposition to modify the surface state of materials creates the possibility of investigating the fundamental phenomena that could not be otherwise examined.

6. CRITERIA FOR WETTABILITY

The significance of sessile bubble development shown above could be rephrased as follows: When a surface is immersed in water, the surface is wetted regardless of the contact angle of water or the "wettability" of the surface. When air is injected

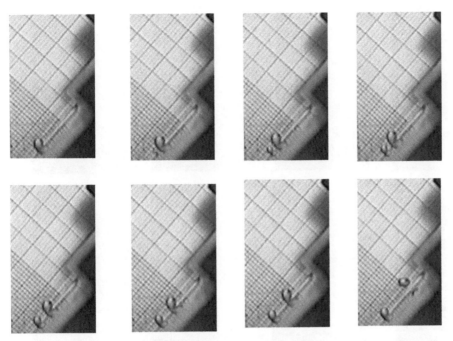

Figure 27.12 Effect of the flow rate on two drifted bubbles on the surface: tilt angle 45 degree, contact angle of water 62.3 degree, orifice diameter 0.55 mm, air flow rate 1.80 ml/m, bubble volume 0.019 ml/bubble.

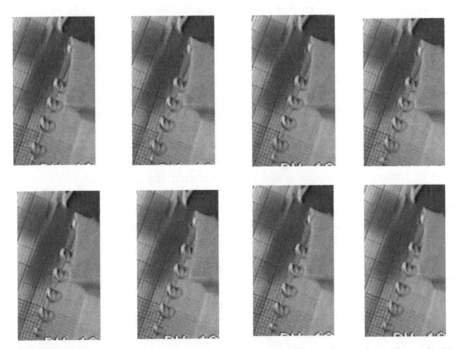

Figure 27.13 No detachment occurs and bubbles slide on the surface: tilt angle 60 degree, contact angle of water 82.7 degree, orifice diameter 0.55 mm, air flow rate 1.73 ml/m, bubble volume 0.047 ml/bubble.

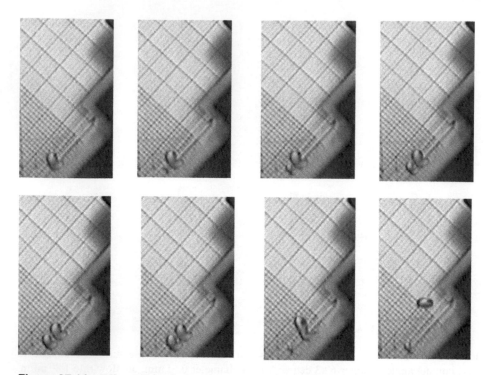

Figure 27.14 Effect of the flow rate on two drifted bubbles on the surface; detachment of a bubble by emerging two sliding bubbles: tilt angle 45 degree, contact angle of water 62.3 degree, orifice diameter 0.55 mm, air flow rate 4.76 ml/m (bubble volume 0.024 ml/bubble).

through a small hole to create an isolated air phase, the air bubble being developing spreads on the surface depending on the wettability of the surface with water. If the air bubble spreads on the surface beyond the periphery of the nozzle, the surface in contact with air is "dewetted" by air. Thus, the bubble spreading on the surface is the reverse process of the wetting. Accordingly, examination of the dewetting process can shed light on the wettability of a surface.

From examination of the dewetting process, the requirement for the wettable (by water) or hydrophilic surface could be defined that the contact angle of water should be below $\pi/2$. By the same principle, a contact angle above π could be set as the requirement for the nonwettable or hydrophobic surface. The surface whose contact angle is between π and $\pi/2$ could be viewed as the surface having amphoteric wettability or amphoteric hydrophilicity/hydrophobicity.

REFERENCES

1. Lin, J.N.; Banerji, S.K.; Yasuda, H. Langmuir **1994**, *10*, 936.
2. Wang, T.-H.F.; Yasuda, H.K. J. Appl. Polym. Sci. **1995**, *55*, 903.
3. Lin, J.N.; Banerji, S.K.; Yasuda, H. Langmuir **1994**, *10*, 945.

28
System Approach Interface Engineering

1. SYSTEM APPROACH INTERFACE ENGINEERING FOR CORROSION PROTECTION

System approach interface engineering (SAIE) is tailoring of interfaces to fit the requirements for the multilayer system to accomplish the objectives. A coated metal is an example of the system that is designed to protect the metal from the corrosion. The objective of the system is protection of the metal, and the coated metal is the total system. Although it could be conceived as a simple metal coated with a paint, i.e., two phases of metal and a coating, such a simple system actually involves many important interfaces if one looks at such a simple system from the view point of how the surface state of the two bulk phase changes at the transitional zones of each material. SAIE could be explained by examples of corrosion protection of aluminum alloys by applying a protective coating [1–4]. Figure 28.1 depicts the interface between a coating and a metal substrate, in which the modification of oxides, application of an ultrathin layer of plasma polymerization coating, and application of a 10- to 30-μm thick primer layer are involved. With consideration of the boundary of oxides and the metal bulk phase, which is not indicated in the figure, there exist 10 interfaces and boundaries to be considered.

The surface of pure aluminum sheet or film is covered by aluminum oxides, which are stable and provide excellent corrosion protection of the pure aluminum under normal environmental conditions. Aluminum corrodes in high (greater than 11.5) and low (less than 2.0) pH solutions with specific exceptions like strong acid solutions of high redox potential, e.g., concentrated nitric acid in which aluminum passivates. However, pure aluminum is too soft and the mechanical strength is not sufficient for many practical uses. Alloying provides improved mechanical strength and properties, but at the expense of the corrosion resistance because oxides are mixed oxides according to the content of alloying elements, which are not as stable as aluminum oxides and are vulnerable to corrosion. The surface oxides on Al alloys offer relatively good corrosion protection in mildly corrosive environments, especially in comparison with the case of steel; however, further protection of surface oxides is necessary.

A corrosion protection system should provide protection of the oxides and, in addition, should provide a good adhesive base for subsequent coating. The conventional corrosion protection system consists of alkaline cleaning and deoxidization of the surface followed by the application of a chromate conversion coating.

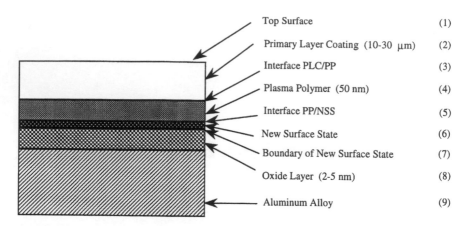

Top Surface	(1)
Primary Layer Coating (10-30 μm)	(2)
Interface PLC/PP	(3)
Plasma Polymer (50 nm)	(4)
Interface PP/NSS	(5)
New Surface State	(6)
Boundary of New Surface State	(7)
Oxide Layer (2-5 nm)	(8)
Aluminum Alloy	(9)

Figure 28.1 Schematic diagram of SAIE system.

The purpose of pretreatments is to remove the surface contaminants, and thus create a clean surface on which chromium oxide can be grown, which then acts as the corrosion protective layer and the adhesive base. The conventional corrosion protection system with chromate conversion coating can be visualized by replacing the plasma polymer layer in Figure 28.1 with chromate conversion coating. The SAIE approach in corrosion protection has an important advantage of being able to eliminate chromate conversion coating, which is an environmental and health hazard, and achieve the same or better corrosion protection.

According to the principle of SAIE, the combination of the best elementary factors does not necessarily yield the optimal performance of the assembled system. The change of one component, e.g., primer coating, may necessitate the change of pretreatment procedure. The best outcome of the change of primer could be obtained after all other factors are optimized to the new primer. This aspect could be visualized by the following examples.

The electrolytic deposition of a coating that is known as "E-coat" provides an excellent corrosion protection as evidenced by automotive coating. Today nearly all automobiles are corrosion protected by applying the cathodic E-coat, in which the steel body of a car is used as the cathode of the electrolytic deposition of a primer coat, on the surface of zinc phosphated steel. It is quite logical to consider that if an E-coat is applied to a chromate conversion–coated aluminum alloy surface, a significant improvement of the corrosion protection of aluminum alloys could be realized because such an attempt represents the combination of the two best components, i.e., chromate conversion coating and E-coat. We could find the best example that demonstrates the need of SAIE in such attempts.

2. CORROSION TESTS AND QUANTITATIVE EVALUATION OF RESULTS

The alloy panels with various low-temperature plasma interface–engineered, primer-coated surfaces were evaluated by accelerated corrosion tests. Two kinds of accelerated corrosion tests were conducted on all the samples, including the two types of control panels: SO_2 salt spray test performed per the American Standards for Testing Methods (ASTM) G85-94-A4, and Prohesion salt spray test performed per the American Standards for Testing Methods (ASTM) G85-94-A5, respectively, using two types of controls, i.e., chromate conversion–coated (Iridite 14-2) and then BASF E-coated or Deft primer (44-GN-36)–coated (CC Deft) Alclad 2024-T3 panels.

After completing corrosion testing exposure, the panels were rinsed with distilled water and visual observations were made. The panels were then subjected to Turco 5469 paint stripper solution to strip off the E-coat or spray primers (including the controls) from the scribed surface, so that the effect of corrosion beneath the coatings and away from the scribes could be viewed. These panels were then used to estimate the average corrosion creep widths, in order to compare the corrosion performance of the different sample systems [5].

The corrosion was quantified using image analysis software and a flat-bed scanner. A portion of each panel with the intersection of the scribe at the center, approximately 4 cm of scribe length on each side of center, was imaged. The scanned area of the panels was fixed at about $27 \, cm^2$ using the imaging software to fix the scanned area and the images were scanned at 450×900-pixel resolution for accurate image analysis. The corroded area is calculated from the numbers of pixels with the use of software. The corrosion width was determined by dividing the area by the total length of the scribe within the scanned area.

The image of scanned surface of corrosion test specimens as well as calculated corrosion area and corrosion width can be used to see the influence of the interface on the corrosion resistance of the overall system. Methods of pretreatment of substrate sheet, plasma conditions, and the sample identification codes are shown in Table 28.1.

Figure 28.2 shows the typical images of three Alclad panels (an aluminum alloy clad with nearly pure aluminum) tested with SO_2 salt spray. The panels are (1) chromate conversion coated followed by priming with E-coat; (2) chromate conversion coated followed by priming with Deft (chromated spray paint); and (3) plasma coated followed by priming with E-coat. The corroded area and corrosion width are also given in the figure.

Although the panel with the plasma deposited film followed by priming with E-coat is visually better, the use of the corrosion width provides a method for quantifying the improvement in the corrosion performance. Also the factor of about 2 difference in corrosion width between the two chromate conversion–coated panels is difficult to obtain from the qualitative difference observed from the scanned images. It can be seen from this comparison of three panels that the use of the measured corrosion width makes the differentiation of corrosion performance much easier. This method of evaluating corrosion test results is used to determine if the combination of the two bests could indeed yield the better corrosion protection of aluminum alloys.

Figures 28.3 and 28.4 depict corrosion tests results obtained for 2024-T3 and 7075-T-6 respectively. The first picture (from left), CC Deft, represents the control sample prepared by chromate conversion coating and chromated spray primer (Deft). The second sample, CC E-coat, is prepared by the same procedure for the CC Deft except that E-coat replaced Deft primer. The third sample is prepared by applying the same E-coat directly (without chromate conversion coating) on the

Table 28.1 Sample Identification Codes

Identification code	Meaning and conditions
[2A]	Alclad 2024-T3 aluminum alloy
(Ace)[a]	Acetone wiping with Kimwipes tissue
(Alk)	Alkaline cleaning (65°C, 25 min)
(Dox)	Deoxidization (room temperature, 10 min, always preceeded by alkaline cleaning)
(O)	O_2 plasma treatment (2 sccm O_2, 100 mtorr, 40 W, 10 min)[b]
(N)	N_2 plasma treatment (2 sccm N_2, 50 mtorr, 80 W, 10 min)
(AH)	Ar + H_2 plasma treatment (1 sccm Ar + 2 sccm H_2, 50 mtorr, 80 W, 10 min)
T	TMS plasma polymerization (1 sccm TMS, 50 mtorr, 5 W, 1 min)
TH	TMS + H_2 plasma polymerization (1 sccm TMS + 2 sccm H_2, 50 mtorr, 5 W, 1 min)
TO	TMS + O_2 plasma polymerization (1 sccm TMS + 1 sccm O_2, 50 mtorr, 5 W, 1 min)
TN	TMS + N_2 plasma polymerization (1 sccm TMS + 1 sccm N_2, 50 mtorr, 5 W, 1 min)
F	HFE plasma polymerization (1 sccm HFE, 50 mtorr, 5 W, 1 min)
CC	Chromate conversion coating (Iridite 14-2)
E	Cathodic E-coat (nonchromated)
A	Deft primer 44-GN-36 (chromated)
/	Process separation mark

[a]Code used in parentheses is indicates the surface cleaning process; code used without parentheses is indicates coating process, [b]Plasma duration in minute was used as noted in table unless otherwise specified with a superscript on the code.

respective aluminum alloy treated with Deox process. The fourth sample represents a low-pressure plasma interface–engineered sample represented by the codes described by Table 28.1.

For both alloys, the combination of chromate conversion coating and E-coat not only didn't yield the best result but yielded the worst result among the four samples compared. However, when E-coat is applied directly on the pretreated surface of alloy, the result was better than control (chromate conversion coating and chromated primer) in the case of 2024 T-3, and slightly less but comparable with the corresponding control in the case of 7075 T-6. These results clearly demonstrate the importance of the interface tailoring in order to take advantage of a supposed-to-be-better primer, i.e., the combination of the two bests does not necessarily yield the better result. These results also point out an important fact that no pitting corrosion occurred when a nonchromated E-coat is applied without chromate conversion coating.

3. PITTING CORROSION BENEATH UNDAMAGED COATING

Pitting corrosion away from the scribe was observed on almost all panels of both controls, which have chromate conversion coatings, after the corrosion tests. Pitting

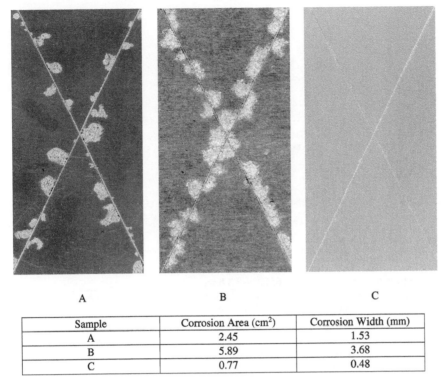

Sample	Corrosion Area (cm^2)	Corrosion Width (mm)
A	2.45	1.53
B	5.89	3.68
C	0.77	0.48

Figure 28.2   Alclad 2024 SO$_2$ Salt Spray Tested Panels (4 Weeks): A—chromate conversion coat with E-coat primer, B—chromate conversion coat with Deft primer, C—plasma polymer coat with E-coat primer; total scanned area is 27 cm^2.

corrosion away from the scribe occurs when the polymer paint film (in this case E-coat or Deft primer) has a high permeability to corrosive species like H$_2$O, O$_3$, and Cl$^-$. Further study of pitting corrosion was made using a penetrant dye inspection/ photography technique [6]. Penetrant inspection is routinely performed on structural aircraft parts to check for material defects such as cracks or porosity. Figure 28.5 shows the typical results of such a surface inspection of control panels and plasma polymer–enhanced [2B] panels. These images were made by exposing the panel to fluorescent dye, washing off the dye, applying a developer to the panel that pulls trapped dye to the surface, and finally photographing the panel under black light. The pitting corrosion on the control panels is quite obvious in the pictures shown in Figure 28.5.

Adhesion of the cathodic E-coat to the plasma polymer surfaces is an important parameter in the corrosion protection of Al alloys. In general, the adhesion performance of E-coat applied to plasma polymers was found to be far superior to that of the control panels. *N*-Methylpyrrolidinone (NMP) paint delamination was not observed after 120 min for E-coat on plasma polymer surfaces as compared to a maximal time for complete delamination of 5 min for E-coat on chromate conversion coating; [2B] CC/E panels. The adhesion performance of cathodic E-coat on the plasma polymer surfaces could not be differentiated by the conventional tape test (ASTM D3359-93B), since E-coat on all of the combinations

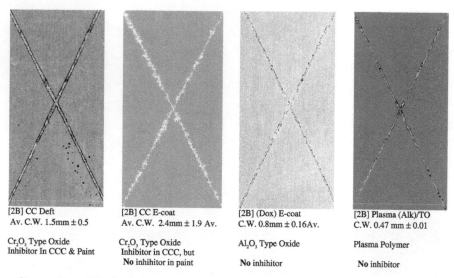

[2B] CC Deft
Av. C.W. 1.5mm ± 0.5

Cr₂O₃ Type Oxide
Inhibitor In CCC & Paint

[2B] CC E-coat
Av. C.W. 2.4mm ± 1.9 Av.

Cr₂O₃ Type Oxide
Inhibitor in CCC, but
No inhibitor in paint

[2B] (Dox) E-coat
C.W. 0.8mm ± 0.16Av.

Al₂O₃ Type Oxide

No inhihitor

[2B] Plasma (Alk)/TO
C.W. 0.47 mm ± 0.01

Plasma Polymer

No inhibitor

Comparison of Prohesion tested panels of 2024-T3 bare.

Treatment Type	Average Corrosion Area (cm²)	Average Corrosion Width (mm)	Average % Area of Pits
CC Deft	2.4	1.5 ± 0.5	1.8 ± 0.7
CC E-coat	3.9	2.4 ± 1.9	1.5 ± 0.3
(Dox) E-coat	1.28	0.8 ± 0.16	-0-
(Alk)/TO	0.75	0.47± 0.01	-0-

Figure 28.3 Effects of interface and primer on the corrosion protection characteristics of the overall system of 2024-T3 aluminum alloys.

of plasma polymers achieved the maximal rating of the tape test method. The major factors influencing the corrosion of a damaged (scribed) surface protection system and those for an undamaged system are depicted in Figure 28.6 for E-coated system and Figure 28.7 for chromate conversion and chromated primer system.

The corrosion resistance of a coated metal sheet could be considered to be dependent on at least five factors [1–4]. These factors are as follows:

1. Salt intrusion resistance of the top surface of a coating
2. Barrier characteristics with respect to water and salt and other corrosive chemicals
3. Function of passivating agents, if any
4. Level of adhesion of the coating to the substrate
5. Surface state of oxides on which the coating is applied

All of these factors are important in considering the corrosion resistance of a coating, which remains intact, i.e., undamaged coating. When the surface of a coating is scribed (damaged coating), the main barrier characteristics of a coating described in items 1 and 2 are bypassed. Then the major factors are reduced to items 3–5. Under such a condition, the exposed interface between metal and coating becomes the major factor. The role of passivating agents or corrosion inhibitors is focused on the exposed metal surface and the new metal surface that will be exposed by the corrosion-induced delamination of the coating near the scribed line.

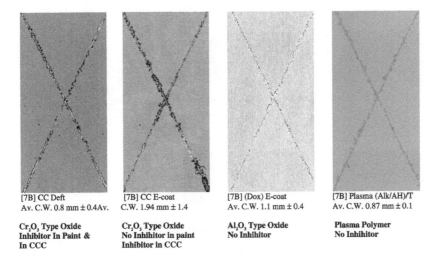

[7B] CC Deft Av. C.W. 0.8 mm ± 0.4Av.	[7B] CC E-coat C.W. 1.94 mm ± 1.4	[7B] (Dox) E-coat Av. C.W. 1.1 mm ± 0.4	[7B] Plasma (Alk/AH)/T Av. C.W. 0.87 mm ± 0.1
Cr₂O₃ Type Oxide Inhibitor In Paint & In CCC	**Cr₂O₃ Type Oxide No Inhibitor in paint Inhibitor in CCC**	**Al₂O₃ Type Oxide No Inhibitor**	**Plasma Polymer No Inhibitor**

Comparison of Prohesion tested panels of 7075-T6 bare.

Treatment Type	Average Corrosion Area (cm²)	Average Corrosion Width (mm)	Average % Area of Pits
CC Deft	1.28	0.8 ± 0.4	1.7 ± 0.7
CC E-coat	3.1	1.94 ± 1.4	1.5 ± 0.3
(Dox) E-coat	1.76	1.1 ± 0.4	-0-
(Alk/AH)/T	1.39	0.87 ± 0.1	-0-

Figure 28.4 Effects of interface and primer on the corrosion protection characteristics of the overall system of 7075-T6 aluminum alloys.

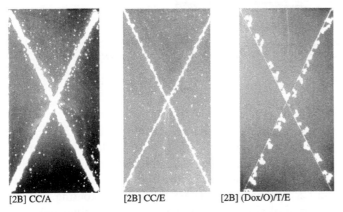

[2B] CC/A	[2B] CC/E	[2B] (Dox/O)/T/E

Figure 28.5 Penetrant inspection/photographs of chromate conversion coated controls and plasma polymer coated AA 2024-T3 ([2B]).

In a damaged (scribed) corrosion protection system, the wet adhesion of a coating becomes the most important factor because both liquid water and corrosive species attack the interface. A water-delaminated coating layer does not provide any corrosion protection. Thus, water-insensitive adhesion of a coating to a substrate is a mandatory requirement for the prevention of corrosion-induced delamination.

Scribe (0.3 mm)

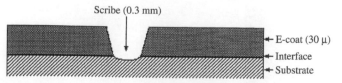

← E-coat (30 μ)
← Interface
← Substrate

Scribed surface of E-coated panel with different surface pretreatments for corrosion testing

Attack of Solution

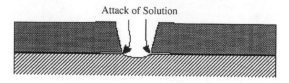

E-coat delamination from the interface at the scribe in corrosion testing solution.

Attack of Solution

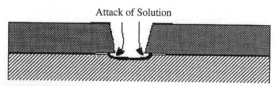

Corrosion takes place near the scribe for good corrosion protecting E-coated Al alloy surface

Attack of Solution

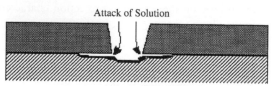

Corrosion creeping from the scribe to the interface of the E-coat and the substrate surface for the surfaces with poor adhesion. The corrosion depth is similar to that of away from the scribe

Figure 28.6 Schematic of corrosion process on E-coated Al alloys surfaces.

The plasma polymer–coated systems showed no pitting corrosion away from the damaged surface area (scribe), suggesting that such systems have good undamaged surface corrosion resistance characteristics. In an undamaged corrosion protection system, all chemical species involved in the corrosion of the substrate metal, such as H_2O, O_2, and Cl^-, primarily permeate through the coating. On the other hand, although the presence of chromate conversion coating might influence the barrier properties of E-coat, it is rather unlikely that the E-coat deposited on chromate conversion coating has significantly low barrier properties. Therefore, nearly the same amount of water permeates through an E-coat regardless of whether it is deposited on chromate conversion coating or on plasma polymerization coating. The difference, however, is due to the action of water at the interface.

When water molecules permeate through the coating and reach the interface, the water sensitivity of a coating adhesion becomes a crucial factor in an undamaged corrosion protection system. Water permeates through a flawless polymer layer

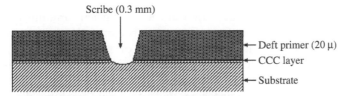

Scribed surface of chromate conversion coated and Deft primer coated Al alloy

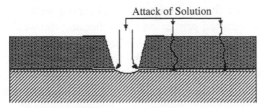

Deft primer lifting from the interface in the testing solution and permeation of the solution to the interface due to high permeability.

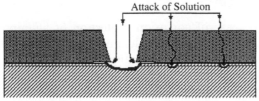

Corrosion initiation of the alloy surface and Deft primer interface from the scribe and at the solution permeated sites.

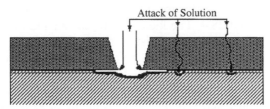

Corrosion creeps away from the scribe as well as corrosion takes place away from the scribe due to permeation of test solution.

Figure 28.7 Schematic of corrosion process on chromated primer coated on chromate conversion coated Al alloys surfaces.

by the diffusion of dissolved individual molecules but not as bulk (liquid) water. In general, water molecules that reach the interface have a stronger association with conventional polymers and with the metal surface than the adhesive interaction between the conventional polymer coating and the metal, leading to localized delamination of the coating. Once a void is formed at the interface, water tends to cluster together to form microscopic pools of liquid phase water. Water-insensitive adhesion is crucial in the prevention of this phenomenon. It is impossible to prevent transport of water within an organic polymer layer, although the water transport resistance of the coating determines its extent.

In an undamaged corrosion protection system, the corrosion characteristics of the surface of the substrate (i.e., pure aluminum oxide, mixed oxides, chromium oxide, or plasma polymer) become important as a sufficient amount of water and corrosive species reaches the interface. Corrosion becomes the driving force for the delamination of a coating and the propagation of surface corrosion only after the transport of corrosive species to the coating–metal interface and the creation of microvoids occurred. The extremely strong adhesion at coating–metal interface achieved by plasma interface engineering evidently could prohibit the creation of microvoids at the interface and hence protect the metal from corrosion.

Pitting corrosion away from the scribe lines was observed with both controls, i.e., E-coat on CC and Deft-chromated primer on CC, reflecting the weaker adhesion compared to plasma interface–engineered samples. The more pitting corrosions were found with the spray primer, which turned out to be the best performer (the smallest corrosion width) in a Prohesion salt spray test. In this case, the corrosion inhibitor is indeed working but cannot stop pitting corrosions under the undamaged coating.

4. STEPS INVOLVED IN SAIE

The first step of SAIE in the case of corrosion protection of aluminum alloys is the preparation of oxides. The top layer of an aluminum alloy is generally covered with hydrated mixed oxides. Either alkaline cleaning or a combination of alkaline cleaning and deoxidization removes major organic contaminants and this potentially unstable oxide layer. A thin layer of plasma polymer is deposited on the stabilized oxide layer thus created.

SAIE is based on the concept that the corrosion protection of a metal depends on the overall corrosion protective behavior of an entire system as a whole, as demonstrated in Figures 28.3 and 28.4. The factors to be considered in a corrosion protection system include the bulk characteristics of the coating(s), interfacial factors, and the surface state of the substrate on which a protective coating system is applied. The best surface state on which a plasma polymer or a primer coating will be applied directly depends on the preceding process step, i.e., alkaline cleaning, deoxidization, etc., which in turn depends on the subsequent process, i.e., plasma polymerization coating, E-coating, and so forth.

The concept of surface state, described in Chapter 24, is important for recognizing the nature of treatment that is applied to a surface. The essence of interface engineering lies in the tailoring of interfaces to facilitate the equilibration of surface states of different materials. Low-pressure plasma processes, such as gas plasma treatment and plasma polymerization coating, have unique advantages in that active or depositing species strongly interact with the surface state of the substrate and modify the surface state. An ultrathin layer of plasma polymer, e.g., thickness less than 50 nm, can be viewed as a new surface state because such a thin layer does not develop a characteristic bulk phase. Contact electrification measurements indicate that surface electrons are still influenced by those from the substrate, up to a film thickness of roughly 20 nm. Thus, plasma polymer modification or polymerization coating could be considered as a means to create an entirely new surface state grafted on a substrate.

Although low-pressure plasma treatment and plasma polymerization are the main tool of the interface engineering described in this book, SAIE does not necessarily require low-pressure plasma processes. Some excellent corrosion protection systems that do not include plasma processes have been found as a fringe benefit of the plasma interface engineering studies, as seen in the application of E-coat on various aluminum alloys. However, it is important to note that the deposition process in such systems (electrolytic coating) also strongly interacts with the substrate and is not a passive coating process. Regardless of whether or not luminous gas is used or electrical potential is applied, the coating process should be interactive at the interface in some ways in order to achieve the interface engineering.

The cathodic plasma polymerization employed in SAIE for corrosion protection of aluminum alloys is a 1-min deposition process, during which the temperature of the substrate alloy rises approximately 2–3°C. Minimization of this rise in temperature was found to be an important factor in yielding a good corrosion protection system in the case of aluminum alloys because the temperature rise causes the change of alloy components near the top surface yielding poor corrosion protection. The effectiveness of plasma polymerization coating depends on the preparation (interface engineering) of the surface onto which a plasma polymer deposits. In the interface engineering, gas plasma such as plasmas of oxygen, argon, and so forth are also used in the pretreatment (cleaning) of the chemically prepared surface prior to the deposition of a plasma polymer. In the case of corrosion protection of steel, plasma treatment to remove oxides was found to be very effective as described in Chapter 33. The optimal combination of these chemical and/or low-pressure plasma processes constitutes the system approach to low-pressure plasma interface engineering

Using the aluminum sheet substrate as the cathode of a direct current (DC) glow discharge, cathodic plasma polymerization is carried out. Dealing with metal surfaces, cathodic plasma polymerization is the most practical means to provide the best corrosion protection (see Chapter 13). A primer is applied on the surface of the plasma polymer. The thickness of the plasma polymer is roughly 50 nm on average and that of the primer layer is about 30,000 nm (30 μm). Primers used included E-coat (electrolytic deposition of paint) and spray primers, but no top coat was applied in the study of corrosion protection.

5. NONELECTROCHEMICAL PRINCIPLE OF CORROSION PROTECTION

Examining over 1000 corrosion test specimens [1–4] that were subjected to SO_2 salt spray (4 weeks) and Prohesion salt spray (12 weeks) tests, it became quite clear that the corrosion damage, which can be expressed by the widths of corrosion along the scribed line or by the percentage of corroded area, is the consequence of corrosion-induced delamination of the coating. In other words, the salt spray tests show how badly or how little the corrosion-induced delamination occurred during the period of salt spray exposure. Delamination of the coating seems to precede the occurrence of corrosion under the coating. It was also evident that the exposed surface (by scribing) didn't corrode much if the delamination of coating didn't occur.

Pitting corrosion, on the other hand, occurs without gross delamination of the coating during the normal duration of salt spray tests (e.g., 1–3 months), and only removal of the coating after the salt spray tests can reveal the presence of pits. Pitting corrosion occurs when the barrier characteristics of the coatings system are poor, and also when the adhesion of the coatings is poor under the influence of water permeation. In this case, localized microdelamination of the coating occurs at the site of pitting corrosion.

The combination of chromate conversion coating and chromated primer prohibits or minimizes the occurrence of corrosion at the damaged sites. Thus it diminishes, or minimizes, the driving force for the corrosion-induced delamination of the coating. On the other hand, the inclusion of corrosion inhibitors (chromates) in the coating damages the barrier characteristics of the bulk phase of the coating and also makes the adhesion water sensitive. Consequently, the chromated coating could effectively prevent damaged surface corrosion, but it does not prevent the occurrence of pitting corrosion and may even promote pitting corrosion.

There are obvious fundamental requirements that must be fulfilled to have an effective coating system with electrochemical inhibitors, or passivation agents, incorporated in a coating layer. First, the inhibitor must be able to migrate within a coating layer. Second, a sufficient amount should be incorporated in the coating to ensure continued passivation. The third, not so obvious, requirement is that the coating should have relatively high water permeability in order to provide enough water necessary to allow the function of aqueous electrochemical corrosion inhibition at the site of corrosion, particularly for undamaged surface corrosion. In other words, a superbarrier to water cannot and should not be used with corrosion inhibitors. These requirements make the barrier characteristics of the coating poorer and the adhesion more water sensitive.

In order to compare the relative extent of the damaged surface corrosion and of the pitting corrosion, the percentage of the corroded area in the tested surface is used, and summarized in Table 28.2. It is important to note that pitting corrosion was found only on chromate conversion–coated controls (one with chromated spray primer and another with E-coat). It is also important to note that E-coat directly

Table 28.2   Average Percent Corrosion Area of SO_2 (3 Weeks) and Prohesion Salt Spray (12 Weeks) Tested Aluminum Panels

Substrates	Coating systems	Pitting corrosion (%)		Along the scribe (%)	
		SO_2	Prohesion	SO_2	Prohesion
[2A]	CC/A	0.33 ± 0.06	-0-	20.4	10.1
	CC/E1	-0-	-0-	8.95	44.1
	(Ace)/E1	-0-	-0-	5.16	4.03
	(Ace/O)/TO/E1	-0-	-0-	4.39	3.79
[2B]	CC/A	1.05 ± 0.17	1.8 ± 0.7	4.56	8.89
	CC/E1	0.68 ± 0.04	1.5 ± 0.3	4.44	14.2
	(Dox)/E1	-0-	-0-	7.41	4.74
	(Alk)/TO/E1	-0-	-0-	2.07	2.79
[7B]	CC/A	0.41 ± 0.16	1.7 ± 0.7	3.08	4.74
	CC/E1	1.9 ± 0.4	1.5 ± 0.3	2.96	11.5
	(Dox)/E1	-0-	-0-	4.21	6.52
	(Alk/AH)/T/E1	-0-	-0-	19.5	5.16

applied on alloy surfaces, without chromate conversion coating, performed as well as or even better than chromate conversion–coated controls in salt spray tests, and no pitting corrosion was found with those systems.

The first control is a panel with chromate conversion coating and a chromated primer designated as A (see Table 28.1 for sample designation). The second control has chromate conversion coating and E-coat (E1). After the salt spray test, the primer was removed, and the now-exposed surface of each panel was placed on a scanner and the corresponding digital image was created. The control, CC/A, is a typical sample representing the corrosion protection afforded by means of electrochemical corrosion protection. The second control, CC/E1, is the sample representing the hybrid of electrochemical corrosion protection and barrier protection. The third sample, (Ace)/E1, represents the corrosion protection by barrier/water-insensitive adhesion principle without plasma coating. The fourth sample represents the corrosion protection by the "barrier adhesion" principle by means of plasma interface engineering. With a good corrosion-resistant surface, such as Alclad, the corrosion protection by barrier/water-insensitive adhesion seems to work much better than the electrochemical corrosion protection. These results also indicate that these two basic approaches are incompatible. Placing chromate conversion coating on the substrate ruins the good interactive coating aspect of E-coat.

A good barrier cannot be used in the electrochemical corrosion protection scheme as mentioned earlier. When corrosion inhibitors are incorporated in a primer, it is necessary to reduce the barrier characteristics of a coating in order to facilitate the migration of inhibitors within the coating. When the barrier characteristic is lowered, it also reduces the water-insensitive adhesion. Comparing two E-coats, of which one was intended to achieve what was just described, one can see the effect of these two interrelated factors on adhesion and corrosion protection. Two prototype E-coats, E1 and E2, are compared. E2 is designed to incorporate a proprietary (nonchromate) corrosion inhibitor (sacrificing barrier characteristic of the primer). Table 28.3 summarizes the adhesion characteristics of these two E-coats. While the dry tape test cannot distinguish the level of adhesion, NMP immersion time and the wet tape test clearly indicate that the adhesion of E2 is more water sensitive.

Table 28.3 Comparison of Adhesive Characteristics of Two E-Coat Systems[a]

| | | Tape test rating | | | | NMP time (min) | |
| | | E1 | | E2 | | | |
Substrate	Pretreatment	Dry	Water boiled for 1, 4, 8 h	Dry	Water boiled for 1, 4, 8 h	E1	E2
[2A]	(Ace/O)/TO	5	3, 3, 2	5	0, —	>120	5
[2B]	(Alk)/TO	5	3, 3, 2	5	4, 0, —	>120	75
[7B]	(Dox)/T	5	3, 3, 2	5	0, —	15	2

[a]Scale 0–5 indicates poor (0) to excellent (5) performance.

The coatings by the "barrier water–insensitive adhesion" principle protect aluminum alloys in a completely different manner. There is no corrosion inhibitor involved in the system, and no inhibition of corrosion occurs at the damaged site. However, because of the tenacious water-insensitive adhesion between the coating and the metal surface, delamination of the coating from the damaged sites cannot occur. Without opening a new surface, the corrosion at the damaged sites either cannot propagate or proceeds very slowly. In many cases, the exposed surface passivates itself, in all practical senses, and the corrosion width does not increase much beyond the original width of the scribed line. Because of water-insensitive adhesion and the good barrier characteristics of the coating, no pitting corrosion was found with coatings formed with the barrier water–insensitive adhesion principle. Pitting corrosion was found only on controls, which are chromate conversion coated and coated with a chromated primer. Thus, this approach was found to be very effective in minimizing both damaged surface corrosion and pitting corrosion.

Figure 28.8 shows the scanned images of panels after the Prohesion salt spray test. On both samples, E-coat was applied directly on alkaline-cleaned 2024-T3 aluminum alloys. Sample E2 showed numerous pitting corrosion sites, reflecting the lower barrier characteristic of this coating. When these E-coats were applied on chromate conversion–coated substrate, however, CC/E2 performed better than CC/E1 in a Prohesion salt spray test, which indicates that E2 achieved what it was designed to accomplish with the inhibiting agents. Without chromate conversion coating, the performance of E2 was found to be inferior to that of E1. These comparisons confirm the basic principle that the corrosion protection afforded by a particular scheme depends on the system as a whole. The modification of one component, e.g., primer composition, without consideration of matching with other factors usually does not yield improved corrosion protection by the system.

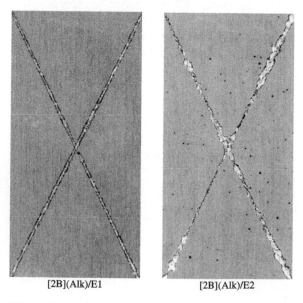

[2B](Alk)/E1   [2B](Alk)/E2

Figure 28.8 Comparison of the two E-coat coated aluminum panels after Prohesion salt spray tests; E-coats were removed after the test.

To this point, the results shown have been for E-coat, confirming that E-coat (E1) is a good primer for the creation of water-insensitive adhesion to an excellent barrier by means of an interactive coating application. Good corrosion protection by barrier water–insensitive adhesion can be also obtained with the application of spray primers. Water-borne spray primers, including both chromated and nonchromated primers, were applied on aluminum alloys with the appropriate surface preparation and plasma deposition of an ultrathin plasma polymer. Similar trends to the E-coated Al panels were observed on spray primer–coated Al alloys. As shown in Figure 28.9, after a Prohesion salt spray test, numerous pits due to pitting corrosion were observed on chromated primer–coated [7B] surfaces ([7B](Dox)/T/F/A) but not on nonchromated primer–coated [7B] surfaces ([7B](Dox)/Tfs/(Ar)/X).

A nonchromated, water-borne primer applied to [2B] alloy samples, with the appropriate surface preparation and plasma deposition of an ultrathin plasma polymer, was also compared to controls prepared by depositing a chromated primer on chromate conversion–coated Al substrate. The same comparison was also performed for IVD Al–coated 2024-T6 substrates (pure aluminum is deposited by ion vapor deposition process on aluminum alloy 2024-T6). In the latter case, the primer could not be removed from the IVD Al–coated panels that were treated with the plasma polymer prior to spray primer application. It is interpreted that the water-borne spray paint penetrates into the column structure of the top surface of the IVD Al–coated substrates when the surface energy was modified by the application of a plasma polymer. This effect could be viewed as interactive coating with a porous surface.

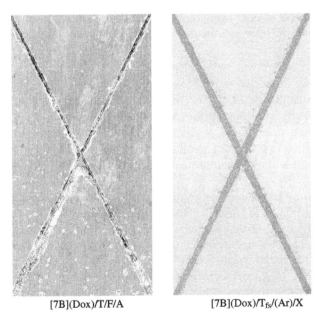

[7B](Dox)/T/F/A   [7B](Dox)/T$_{fs}$/(Ar)/X

Figure 28.9 Comparison of chromated primer coated Al panel ([7B](Dox)/T/F/A) with nonchromated primer coated Al panel ([7B]y(Dox)/T$_{fs}$/(Ar)/X) after Prohesion salt spray test; primers were removed after the test.

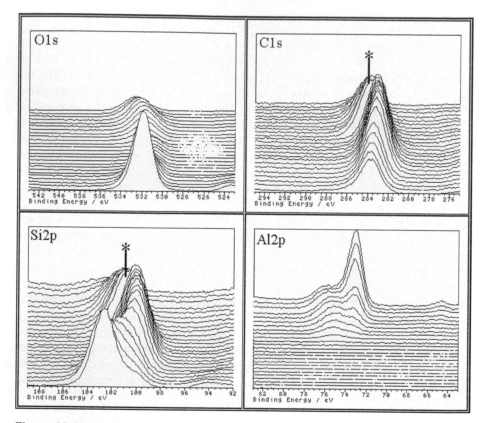

Figure 28.10 Combined spectra from depth profile of closed system TMS film treated with O_2 plasma after deposition; the asterisk marked line indicates the interface region: shifts in C 1s, O 1s, and Si 2p observed at the interface indicate the change of chemical bonds at the interface.

Cathodic plasma polymerization or LCVD of trimethylsilane (TMS) applied to an appropriately prepared aluminum alloy surface yields a roughly 50-nm-thick layer of amorphous $Si_xC_yH_z$ network, which is covalently bonded to aluminum oxide at the interface [7]. The XPS cross-sectional profiles given in Figure 28.10 show the conspicuous shifts in O 1s and Si 2p at the interface that indicate the changes of chemical bonds.

A primer coating applied on the surface of the nanofilm is covalently bonded via the reaction of peroxides, which are formed by the reaction of oxygen with Si dangling bonds (free radicals) when the coated surface is exposed to ambient oxygen [8]. Thus, the TMS nanofilm works as an interlayer to form a covalently bonded network between primer and the metal surface. An unparalleled level of water-insensitive adhesion between the two interfaces and the cohesive integrity and strength of the nanofilm distinguish it from other interfacial modifications such as the use of silane coupling agents.

When a test panel is subjected to an environment that is less corrosive, such as the Filiform test, the water sensitivity of the adhesion of a primer shows more clearly. In the Filiform test, scribed coated panels are exposed to HCl vapor for a

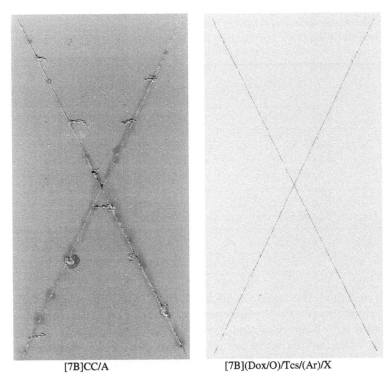

[7B]CC/A	[7B](Dox/O)/Tcs/(Ar)/X

Figure 28.11 Scanned images of Filiform tested 7075-T6 aluminum panels; paints were not removed in Filiform tests.

fixed period and then kept in a high-humidity/high-temperature chamber. Since the coating is scribed and the interface between coating and metal is exposed to HCl vapor and water vapor but not exposed to salt or chloride ions, the method seems to reflect the water sensitivity of adhesion and the corrosion-induced delamination does not play a role. The evaluation is made without removing the primer. The changes observable along the scribed lines are compared in this test. Figure 28.11 shows the scanned surface of Filiform-tested panels, which depicts the lateral advance of the delamination of a chromated spray primer (organic solvent) applied on a chromate conversion–coated panel (7075-T6), and also for a nonchromated water-borne spray primer with plasma interface engineering. The plasma interface–engineered samples coated with nonchromated, water-borne primer show no sign of damage to the scribed lines.

A general approach to increase adhesion is for polar groups to be introduced to the surface in order to improve the adhesion of water-borne paints. While the polar groups increase the adhesion characteristics, they also increase the surface energy, which makes the adhesion water sensitive and often increases the water permeability.

The same principle applies to plasma surface modification. It is generally observed that O_2 plasma treatment of a polymer surface dramatically increases the adhesion of paint applied on the treated surface. However, the adhesion thus created

Table 28.4 Surface Treatment Effect of TMS by Succeeding Plasmas on Adhesion of Spray Paint Primer [Spraylat EWDY048 (primer D)] to TMS Plasma–Coated [7A](Ace/O) Aluminum Panels[a]

		Tape test rating	
Plasma systems	Coating systems	Dry	Water boiled for 1, 4, 8 h
Flow system	T_{fs}/D	0	—
	$T_{fs}/(O)/D$	5	0, —
	$T_{fs}/(Ar)/D$	5	3, 3, 2
Closed system	T_{cs}/D	5	5, 5, 5
	$T_{cs}/(O)/D$	5	0, —
	$T_{cs}/(Ar)/D$	5	5, 5, 5

[a]Scale 0–5 indicates poor (0) to excellent (5) performance.

is sensitive to water, and the wet adhesion is poor [9]. This aspect can be seen in the difference of the wet adhesion and the dry adhesion depicted in Table 28.4. A plasma polymer of TMS deposited on an aluminum sheet in a flow system reactor is very hydrophobic, and the adhesion of a water-borne primer to this surface is poor (tape test result is 0). Oxygen plasma treatment of the plasma polymer increases the dry adhesion test dramatically (tape test result increases to 5); however, it does not survive boiling water for 1 h. The water-insensitive adhesion between the water-borne primers and the surface of the nanofilm depends on the interfacial tension and also on the cohesive integrity of the nanofilm.

The same plasma polymer deposited in a closed-system reactor has a graded elemental composition with a carbon-rich top surface, and the oligomer content is much lower [10], both of which increase the level of adhesion. The adhesion of the same water-borne primer is excellent and survives 8 h immersion in boiling water. When this surface is treated with O_2 plasma, the adhesion does not survive 1 h of boiling, while the dry tape test still remains at the level of 5. The water-sensitivity of adhesion depends on the chemical nature of the top surface as depicted by XPS data shown in Figure 28.12. Water-insensitive tenacious adhesion, coupled with good transport barrier characteristics, provides excellent corrosion protection, as supported by experimental data [1–4], and constitutes the basic principle for the "barrier-adhesion" approach.

6. ROLE OF ADHESION IN ELECTROCHEMICAL IMPEDANCE SPECTROSCOPY

Electrochemical impedance spectroscopy (EIS) is a valuable method with which to study the barrier property and corrosion protection performance of polymer-coated metals; it has been widely used in this field in recent years [11–15]. Many examples can be found in the literature, which illustrate the performance deterioration of different coatings on metals as well as pretreatment effects on the properties of

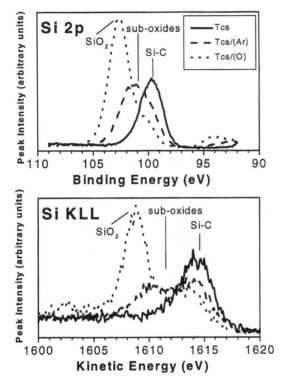

Figure 28.12 Si 2p photoelectron and Si KLL x-ray induced Auger spectra from the top surface of TMS plasma coatings (produced in a close reactor system) with and without second surface treatment by O_2 or Ar plasma corresponding to the adhesion data shown in the bottom half of Table 28.4.

a coating system [13–15]. EIS can also be applied to a freestanding film; however, few studies with this use of EIS have appeared in the literature [16–18].

The overall corrosion protection of a metal, as well as general protective coatings irrespective of their function of coating, depends on the performance of a system as a whole, including its many interfaces and coating layers. These factors are not only mutually dependent as a combination but also in the order of application (permutation). Any single factor cannot be treated as a dominant one. There is very little work in the literature that focuses on the role of interfacial factors in the corrosion protection performance of coated systems.

An ideal model system was selected to study the interfacial factors with EIS [19]. The model system was Parylene C–coated Alclad (aluminum-clad aluminum alloy). In this system, the surface state of the top surface (salt solution/coating interface) and the adhesion of the coating (coating/metal interface) were modified to study the influence of these factors on the corrosion protection performance of the system.

Parylene C film does not adhere well to any smooth surface due to its unique polymerization mechanisms as described in Chapter 5. A freestanding film can be easily peeled off of the substrate surface, although a film does not peel off by itself in many cases, even if immersed in water. This feature enables us to investigate a system

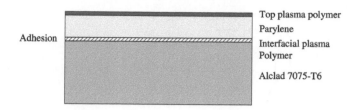

Adhesion

Top plasma polymer

Parylene

Interfacial plasma
Polymer

Alclad 7075-T6

Combination of Interfacial Factors for Each Sample Studied

Sample No. & Top Plasma Polymer		Adhesion of Parylene C to Metal	Liquid-contacting Surface	Polymer contacting Metal
(1)	None	No	PC	PC
(2)	None	Yes	PC	PP
(3)	Hydrophilic	No	Plasma Polymer of TMS / O₂	PC
(4)	Hydrophilic	Yes	Plasma Polymer of TMS / O₂	PP
(5)	Hydrophobic	No	Plasma Polymer of TMS	PC
(6)	Hydrophobic	Yes	Plasma Polymer of TMS	PP

PC: Parylene C,   PP: Plasma Polymer

Figure 28.13 Factors involved in Parylene C coated aluminum sheet.

with poor adhesion and also to investigate a freestanding film with EIS. Adhesion can be improved to an excellent level by means of plasma polymerization of an ultrathin layer to which a Parylene C film will bond covalently. Examination of these three cases—a freestanding film, a system with poor adhesion, and a system with good adhesion—made it possible to see how EIS data would change when a coating starts to delaminate from the substrate in the context of EIS.

The factors involved in the investigation are depicted in Figure 28.13. EIS data obtained for these samples are shown in Figures 28.14–28.20. These figures, with explanation of interfacial factors for each sample, provide step-by-step illustrations of factors involved in the interface engineering processes. The data revealed the following important aspects that have significant implications in the corrosion protection provided by a coating.

1. Parylene C, which has excellent barrier characteristics, did not perform well in corrosion protection. The salt intrusion resistance is poor, probably due to its semicrystalline nature of the bulk phase. The boundary phase between crystalline and amorphous phase seems to be vulnerable for the salt intrusion.
2. The change in Bode plot as a function of immersion time for a Parylene C–coated metal sheet is greatly influenced by the nature of the top surface (solution/coating interface).
3. A hydrophilic top surface accelerates time-dependent change, whereas a hydrophobic top surface decreases the extent of the change.
4. The change in Bode plot as a function of immersion time for a Parylene C–coated metal sheet is greatly influenced by the degree of adhesion of Parylene C to the metal substrate surface.

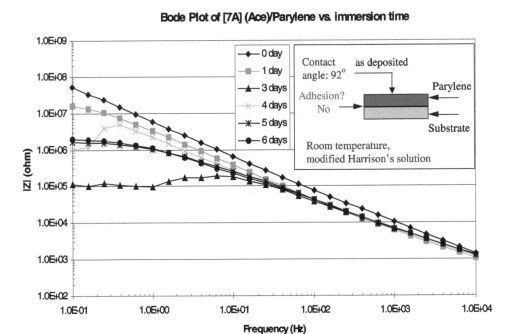

Figure 28.14 EIS Bode plots for Parylene C coated sample without pretreatment and surface treatment.

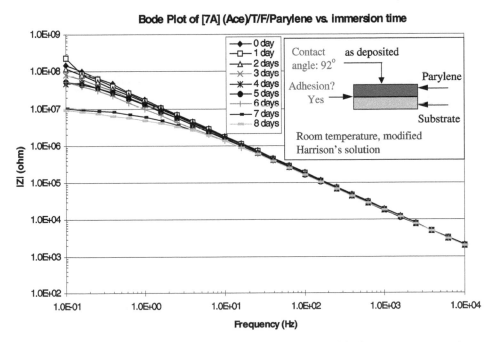

Figure 28.15 EIS Bode plots for Parylene C coated sample with the pretreatment to impart adhesion but no surface modification.

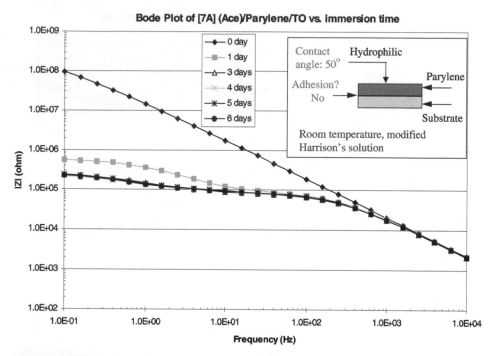

Figure 28.16 EIS Bode plots for Parylene C coated sample without pretreatment but with surface modification to make the surface hydrophilic.

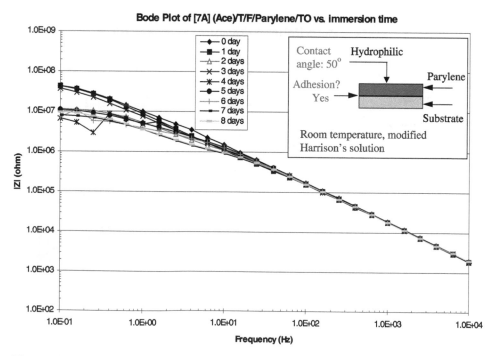

Figure 28.17 EIS Bode plots for Parylene C coated sample with pretreatment and the surface modification to make the surface hydrophilic.

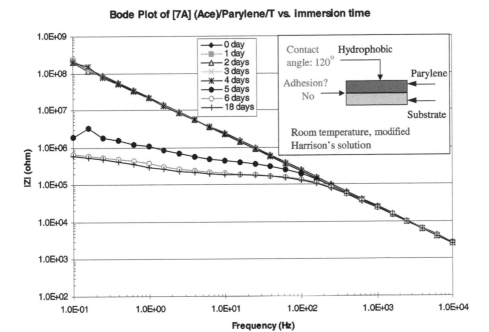

Figure 28.18 EIS Bode plots for Parylene C coated sample without pretreatment but with the surface modification to make surface hydrophobic.

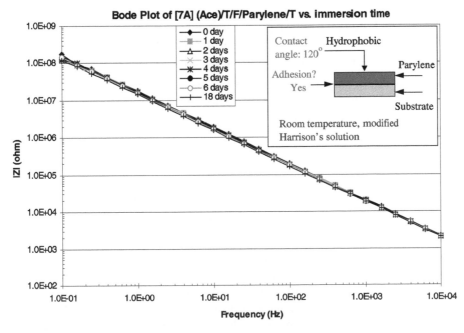

Figure 28.19 EIS Bode plots for Parylene C coated sample with pretreatment and the surface modification to make the surface hydrophobic.

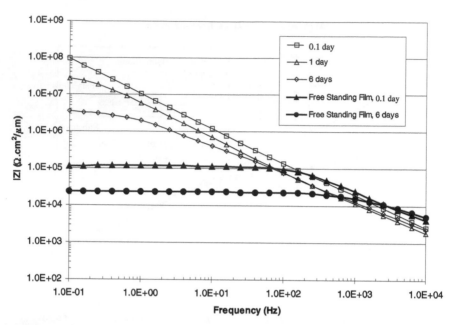

Figure 28.20 Changes of Bode plots as a function of immersion time for freestanding Parylene C film and Parylene C coated Alclad 7075-T6 aluminum sheets in 0.9% NaCl solution; 0.1 day indicates the initial run after 2 h immersion of the samples.

5. Without good adhesion, time-dependent change converges to the Bode plot for a freestanding film, which is strong evidence for the interfacial failure.
6. With a hydrophobic plasma polymer on top and good adhesion to the substrate, the Bode plot does not change with immersion time. Only with the combination of these factors was excellent corrosion protection by Parylene C obtained.

Thus, without SAIE, Parylene C film, which has excellent barrier and physical properties, cannot be utilized in corrosion protection of a metal. Conversely, SAIE is the key to yield an excellent corrosion protection systems. It is also important to recognize how a nanofilm of hydrophobic amorphous network of plasma coating can prevent the initiation of the salt intrusion process.

The incorporation of electrochemical corrosion inhibitors is the current mainstream approach for corrosion protection. These methods have been advanced nearly to the limit of this type of approach and often incorporate chemicals with undesirable environmental implications. In long-term corrosion protection, the loss of the inhibitors due to their leaching out of the system is a serious drawback. The difficulty of creating water-insensitive adhesion, while still maintaining the inhibitor mobility and water molecules in the coating layer, is another limitation. The barrier water–insensitive adhesion principle coatings are free of these problems and limitations. Furthermore, they are environmentally benign and free from the health hazards associated with conventional schemes. The natures of the top surface, the bulk phase, and the interface to the substrate are the key factors of SAIE, which are also applicable to any functional coating to protect the substrate.

7. EFFECTS OF INTERFACIAL DAMAGE ON CORROSION PROTECTION

It is a well-established practice to test corrosion resistance of a coated panel by exposing a scribed coating layer to a corrosive environment such as salt spray for a prolonged period. Corrosion resistance of the coating is qualitatively evaluated by examining the corrosion that took place near the scribed line. Such a method certainly provides an estimate of the level of corrosion resistance of the coating; however, this method does not yield information concerning the mechanisms of corrosion protection.

The scribing process is not only exposing the substrate metal but also exposing the interface between coating and metal. Once the coating/metal interface is exposed to a salt solution, the nature and extent of the adhesion of the coating to the substrate metal becomes the most important factor that dictates the occurrence of corrosion according to the data described above. EIS could detect the microscopic delamination of coating without artificially introduced defects. EIS study was extended to investigate the nature of scribing, specifically the damage to the coating/metal interface [20]. For this purpose, a severe scribing method was included to cause interfacial damage. Questions that arise are: Does a wider scribing width cause more corrosion? Is a deeper scribing depth more damaging? With these fundamental questions in mind, 32 sheets of AA2024-T3 were coated with an E-coat under the identical conditions of coating and curing, and were divided into four groups of different scribing modes. *The coating with moderate adhesion was used intentionally to magnify the interfacial aspect.* Four groups of scribed samples, in which the only difference is the mode of scribing, were exposed to a Prohesion salt spray test.

7.1. Salt Spray Test With Different Scribes

Panels were scribed by using a computerized engraver. It was noted that the cutter used had the tip geometry as shown in schematic diagram in Figure 28.21. Two depths (0.02 in. and 0.04 in.) of scribes were made using a cutter in the stationary or

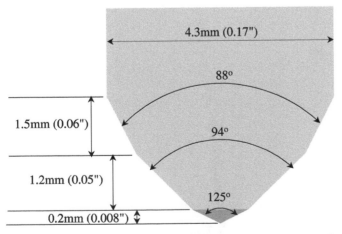

Figure 28.21 Schematic of the scribing cutter 42-037-000 Diamond Graver (New Hermes Inc., Duluth, GA) tip at 50× magnification.

spinning mode. Due to the cutting head geometry, a deeper cut yields a wider scribe line. To distinguish the difference of scribe mode, the scribe made in stationary mode, horizontal dragging of the tip across the panel surface, was designated as V shape, and the scribe made with a spinning cutter tip was defined as U shape. The V- or U-shape definition is used for identification of different scribes but not for the real scribe shapes. Adjusting the depth increment of each scribe and the numbers of scribes controlled the scribe depth.

Figure 28.22 shows the optical microscopic pictures of the four different scribes at 50× magnification. Flat scribes were produced by stationary mode (designated as V shape) with horizontal dragging of the cutter tip across the panel surface.

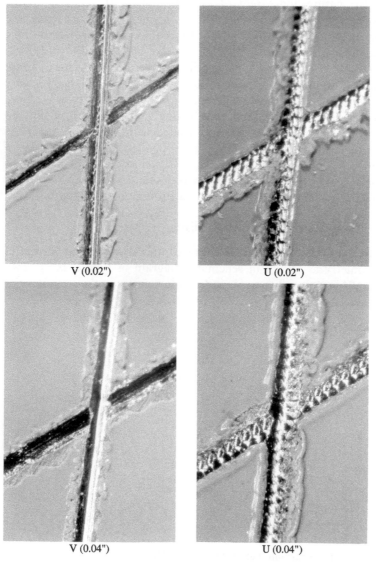

V (0.02")   U (0.02")

V (0.04")   U (0.04")

Figure 28.22   Optical microscopic pictures of the four different scribes at 50× magnification.

In contrast, the spinning tracks were clearly observed in the scribes made in spinning mode (designated as U shape). Spinning left burrs in the scribes and also wider scribes. Since they provided more exposed surface area, the left burrs in the scribe may have a greater chance to initiate the corrosion.

After completing test cycles in Prohesion salt spray, the panels were rinsed with distilled water and visual observations were made. Then the panels are subjected to a commercial paint stripper solution (Turco 5469) to strip off the E-coat on the scribed surface to see the corrosion underneath the E-coat film and away from the scribe.

Figure 28.23 shows the typical pictures of Prohesion tested sheets with different scribes and depths. From the visual examination of the tested panels, a general conclusion can be made that the U-type scribe resulted in much more corrosion through the corrosion test, but the scribe depth had very little effect on the corrosion test results.

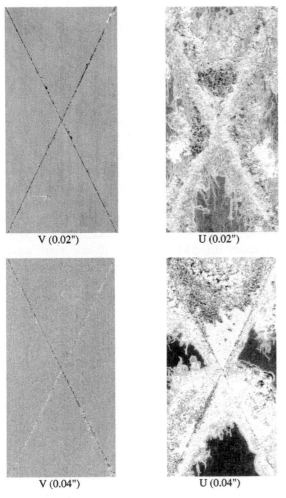

V (0.02")   U (0.02")

V (0.04")   U (0.04")

Figure 28.23 Scanned picture of Prohesion tested samples with different scribe types and depths; E-coat removed after test.

Due to the cutting head geometry, the increase of scribe depth also increased the scribed width. Consequently, a general conclusion can be extended to state that the scribing width and scribing depth has little effect on the corrosion test results. The most striking difference in the corrosion results was caused by the different mode of scribing. In a simplified view, the stationary mode (V shaped) scribe gave the least corrosion and the spinning mode (U shaped) scribe caused severe corrosion.

It is important to recognize that the major difference in two types of scribing lies on the extent of damage caused on the coating/metal interface as described in more detail in the following section. The scribing with spinning cutter head could cause an extensive damage to the interface between coating and metal, and the striking difference found between V-shaped scribe and U-shaped scribe could be attributed to the extent of damage to the interface. It is important to note that one of four samples showed opposite corrosion test results as if the sample were mislabeled. This means that the damage to the interface is not a sole function of the scribing mode. It simply means that the spinning head has a greater probability of inflicting severe damage. In any case, these figures clearly demonstrate the importance of the interface, in particular the adhesion aspect of interface.

7.2. Concurrent EIS Measurement with Prohesion Salt Spray

Some samples were also concurrently examined with EIS measurement during the regular Prohesion exposure period. During the 2000 h (83.3 days $\approx$ 12 weeks) of Prohesion exposure, the tested panels were taken out the chamber once a week and the EIS measurements were performed on the unscribed part of the scribed panels. The time to perform the electrochemical measurements was about 1 h, and during this time the scribed area was exposed to lab ambient conditions. EIS measurements were performed in dilute Harrison solution [0.05% NaCl and 0.35 $(NH_4)_2SO_4$ aqueous solution] using a Gamry potentiostat controlled by Gamry CMS100 software. Measurements were taken from 0.1–5 kHz with a 10-mV sign wave potential. Ten points were collected per decade. The reference and counter electrodes were a saturated calomel electrode (SCE) and a platinum electrode, respectively. Figure 28.24 shows the schematic of the sampling for EIS measurement of scribed panels. The circle between the scribe lines is the sampling area for the EIS measurements.

Figure 28.25 shows the dependence of impedance on different modes of scribing. It was noted that the sharp drop of impedance modulus at low frequency $(|Z|_{0.1Hz})$ for U-scribed samples of both U (0.02 in.) and U (0.04 in.) started after 14 days of exposure in the Prohesion chamber. The sharp drop of impedance modulus at low frequency $(|Z|_{0.1Hz})$ indicated the coating system failure, which is also evidenced by the onset of a second time constant suggesting partial delamination of the coating. In contrast, there is very little gradual drop of the $(|Z|_{0.1Hz})$ for V-scribed samples of both V (0.02 in.) and V (0.04 in.). The V-scribed samples did not start a large impedance modulus drop for a long exposure time until 70 days in the Prohesion test chamber.

In samples with varying scribe depth, the V (0.04 in.) samples with deeper scribe depth showed slightly better corrosion resistance than V (0.02 in.) samples at

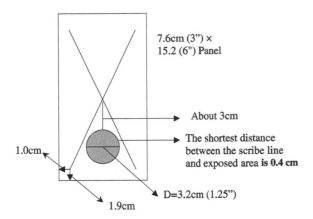

Figure 28.24 Schematic of the sampling for EIS measurement of scribed panels.

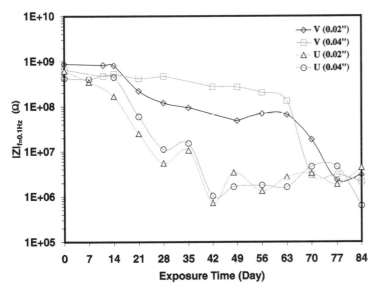

Figure 28.25 The impedance modulus at low frequency ($|Z|_{0.1Hz}$) dependence on the scribe types and depths and the exposure time in the Prohesion corrosion test chamber.

the early stage of the exposure. With prolonged exposure, there is little difference on their performance. The U (0.02 in.) and U (0.04 in.) samples showed no significant difference in corrosion resistance during all the Prohesion test period.

The results of the concurrent EIS measurement are consistent with what were found with Prohesion test results shown in Figure 28.23. The earlier failure of the coating system due to partial delamination at the interface resulted in severe corrosion on the U-shaped scribed panels. The concurrent EIS measurement reveals the importance of the lateral diffusion of salts initiating from the damaged interface. This situation could be explained by Figure 28.26, which schematically depicts the pathways of electrolyte to the sampling site of EIS measurement.

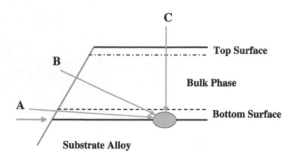

Figure 28.26   Pathways of corrosive chemicals to the site of EIS measurement.

Since the same primer with controllable coating thickness and characteristics was used, the different scribe modes could not have any effect on the pathways B and C in Figure 28.26. If pathway B or C is to be the dominant one, the difference of scribing modes should not influence the EIS impedance values as a function of the immersion time. In the case of V-shaped scribes (with minimal interface damage), it is likely that C might be the dominant pathway, and the difference in the scribing depth and width has little effect.

The most significant factor that can be seen in Figure 28.25 is the difference of the critical time at which the impedance modulus value starts to drop sharply, i.e., 14 days for U-shaped scribe and 70 days for V-shaped scribe. This large difference can be explained by the rapid transport by means of pathway A in the case of U-shaped scribes due to the extensive interface damage, which resulted in the earlier failure of the coating systems with partial delamination of the coating. Even in this case, the scribing depth and width has little effect. This means that the severe damage to the coating/substrate interface caused by the scribing process could cause the largest influence on the outcome of Prohesion salt spray test results. The interfacial damage is considered to be a function of adhesive strength of coating to substrate alloy. Good adhesion might minimize the interfacial damage. Without tenacious, water-insensitive adhesion of a coating layer to the substrate metal, the propagation of the interfacial damage could not be inhibited. (These samples were prepared without interface engineering.)

The influence of scribing itself on the outcome of corrosion test results is of important concern in establishing a test method. Obtaining the test results of narrow distribution is probably the main target of standardizing a test method. On the other hand, from the viewpoint of predicting the best performance of a set of coated samples, it would be better to apply the most severe conditions even though it might result in large scattering. The results shown clearly indicate that the interfacial damage is the most crucial factor in scribed surface corrosion tests.

In this chapter, corrosion protection of metal is used to illustrate the importance of interfacial factors and the role of system approach interface engineering. When a coating is applied in order to protect a substrate, the factors, particularly interfacial factors demonstrated by examples of corrosion protection, play important roles regardless of the nature of protection to be accomplished. An excellent coating cannot function well with poor interfaces. The creation of stronger interface is the core of SAIE.

REFERENCES

1. Reddy, C.M.; Yu, Q.S.; Moffitt, C.E.; Wieliczka, D.M.; Johnson, R.; Deffeyes, J.E.; Yasuda, H.K. Corrosion **2000**, *56*, 819.
2. Yu, Q.S.; Reddy, C.M.; Moffitt, C.E.; Wieliczka, D.M.; Johnson, R.; Deffeyes, J.E.; Yasuda, H.K. Corrosion **2000**, *56*, 887.
3. Moffitt, C.E.; Reddy, C.M.; Yu, Q.S.; Wieliczka, D.M.; Johnson, R.; Deffeyes, J.E.; Yasuda, H.K. Corrosion **2000**, *56*, 1032.
4. Yu, Q.S.; Reddy, C.M.; Moffitt, C.E.; Wieliczka, D.M.; Johnson, R.; Deffeyes, J.E.; Yasuda, H.K. Corrosion **2001**, *57* (9), 802.
5. Chandra M. Reddy; Yasuda, H.K.; Wieliczka, D.M.; Deffeyes, J. Plating and Surf. Finishing **1999**, *86* (10), 77–79.
6. Yasuda, H.K.; Yu, Q.S.; Reddy, C.M.; Moffitt, C.E.; Wieliczka, D.M.; Deffeyes, J.E. Corrosion **2001**, *57* (8), 670.
7. Yasuda, H.K.; Yu, Q.S.; Reddy, C.M.; Moffitt, C.E.; Wieliczka, D.M. J. Appl. Polym. Sci. **2002**, *85* (7), 1387.
8. Yasuda, H.K.; Yu, Q.S.; Reddy, C.M.; Moffitt, C.E.; Wieliczka, D.M. J. Appl. Polym. Sci. **2002**, *85* (7), 1443.
9. Yu, Q.S.; Lin, Y-S.; Yasuda, H.K. Prog. in Organic Coatings **2001**, *42*, 236.
10. Qingsong Yu; Moffitt, C.E.; Wieliczka, D.M.; Hirotsugu Yasuda. J. Vac. Sci. Tech. **2001**, *A19* (5), 2163.
11. van Westing, E.P.M.; Ferrari, G.M.; De Wit, J.H.W. The determination of coating performance with impedance measurements–IV. Protective mechanism of anticorrosion pigments, Corrosion Sci. **1994**, *36* (8), 1323.
12. Amirudin, A.; Thierry, D. Br. Corrosion J. **1995**, *30* (2), 128.
13. Cohen, S.M. Electrochemical impedance spectroscopy evaluation of various aluminum pretreatments painted with epoxy primer. J. Coatings Techn. **1996**, *68* (859), 73.
14. Tang, N.; van Ooij, W.J.; Górecki, G. Progress in Organic Coatings **1997**, *30*, 255.
15. Mertens, S.F.; Xhoffer, C.; De Cooman, B.C.; Temmerman, E. Corrosion **1997**, *53* (5), 381.
16. Barreau, C.; Massinon, D.; Thierry, D. SAE Transactions, **1991**, *100*, 1281.
17. Sussex, G.A.M.; Scantleburry, J.D. J. Oil Colour Chemists' Assoc. **1983**, *66*, 142.
18. Walter, G.W. Corr. Sci. **1991**, *32*, 1085.
19. Yasuda, H.; Yu, Q.S.; Chen, M. Prog. Organic Coatings **2001**, *41*, 273.
20. Yasuda, H.K.; Reddy, C.M.; Yu, Q.S.; Deffeyes, J.F.; Bierwagen, G.P.; He, L. Corrosion **2001**, *57* (1), 29.

29

Creation of an Imperturbable Surface

1. ORIGIN OF SURFACE DYNAMICAL INSTABILITY

The influence of the change of surface configuration due to the change of contacting medium is discussed in Chapter 25. However, when a coating is applied the surface dynamic stability or instability of the surface also depends on other factors, which depend on the coating process. The coating process generally involves the interfacial interaction of the coating with the surface. The smoothness, the uniformity with respect to the density, and the surface energy of the surface all influence the surface characteristics of the coating.

The surface of a semicrystalline polymer consists of crystalline phases and amorphous phases. If the interaction with the coating during the process of application is different in these different phases of the substrate, the surface dynamic stability of the coating could be different from the same coating applied on a completely amorphous surface made of the same polymer. If these are intentionally created phases having significantly different surface energies, the coverage of the surface greatly depends on the coating method and could cause surface dynamic instability due to the poor coverage of the surface by the coating. Even in such a case, the surface configuration change involved in the coating layer is important; however, the additional factor due to the surface morphology of the substrate should not be ignored. A typical case of this situation could be seen in the coating of phase-scparated silicone hydrogel polymers.

The oxygen permeability of a contact lens is crucially important for the extended-wear contact lens, which can be worn day and night without removal from the eye for a prolonged period, e.g., a month. The oxygen permeability of a hydrogel is determined by the volume fraction of water in the hydrogel. Since the oxygen permeability of water is not high enough to provide a hydrogel that can be used as an extended-wear contact lens, the phase-separated silicone hydrogels, which consists of a block copolymer of siloxane and hydrophilic vinyl monomer, have been developed [1]. Because of the characteristically high free volume of silicone polymers, such a phase-separated hydrogel, in which oxygen permeates mainly through the silicone polymer phase, could provide sufficiently high oxygen permeability necessary for an extended-wear contact lens.

While the silicone polymer phase provides a high enough oxygen permeability, it also dominates the surface characteristics, i.e., the surface of silicone hydrogel is too hydrophobic to be used as a contact lens. The silicone phase also readily absorbs lipids and lipid-soluble materials due to its high free volume, which is not desirable in the extended wear application.

It is also important to recognize that the surface of such a phase-separated silicone hydrogel is partly hydrophobic and partly hydrophilic, which could be expressed as amphoteric hydrophilicity or hydrophobicity based on the macroscopic phase separation. Although the term "amphoteric" has been used nearly exclusively in electrochemistry to describe the partly cationic and partly anionic nature of molecules and their reactions, the general meaning is "partly one and partly the other." The Greek word *amphoteros* means "both." Amphoteric hydrophilicity/hydrophobicity of homogeneous surface is described in Chapters 26 and 27. In the case of homogeneous film, amphoteric hydrophilicity/hydrophobicity stems on the surface configuration change of polymer molecules. In the case of phase-separated polymer, amphoteric hydrophilicity/hydrophobicity is dependent on hydrophilicity and hydrophobicity of the constituent phases. The surface dynamic instability is largely due to thus defined amphoteric hydrophilicity in both scales.

Based on these considerations, it is more or less necessary to apply the coating that has good surface dynamic stability, and the barrier characteristics to lipids and lipid-soluble substances in order to successfully utilize such a phase-separated silicone hydrogel for long-term biomaterial application. When a plasma polymerization coating is applied on the surface of phase-separated polymer, a coating layer covers the surface as depicted in Figure 29.1 because the depositing species in gas phase do not distinguish the two phases.

When the coating is applied by application of aqueous solution of polymer, on the other hand, it is a serious question whether such a structure could be attained because the surface energy of two phases is drastically different. It is unlikely that an aqueous coating solution could spread on the silicone phase in the same way it would spread on the hydrogel phase. This question could be answered by examining the coatings, one of which is coated with the layer-by-layer alternating application of very dilute solution of cationic polymer and anionic polymer (LBL coating), and another one is coated by luminous chemical vapor deposition (LCVD).

Because the hydrophilic polymer with negative charge in aqueous solution would not adhere to the silicone phase on the first dipping operation, it is likely that hydrophobic islands remain as the coating thickness increases with a repeated,

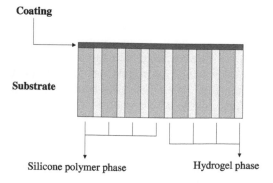

Figure 29.1 A simplified schematic view of a plasma polymerization coating applied on a phase-separated silicone-hydrogel polymer.

Wet (blotted) Surface   Dried Surface

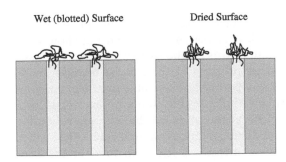

Figure 29.2 Schematic representation of an aqueous LBL coating layer on the phase separated silicone hydrogels surface in wet and dry conditions.

Table 29.1 Comparison of Oil-Soluble Dye Penetration Tests

Sample	Dry	Blotted	Wet
Uncoated	+	+	+
LBL coated	+	+	−
LCVD coated	−	−	−

alternating-polarity, dipping operation. Furthermore, it is likely that some LBL polymers might penetrate into the hydrogel domains and leave the relatively free poly-poly salt segments out of the surface, which can be visualized like seaweed at the bottom of the sea.

Figure 29.2 depicts a simplified schematic model of aqueous coating represented by LBL coating. The conformation of hydrophilic polymers in the coating depends on whether the surface is wet or dry as depicted in the figure, which determines the exposure of uncovered silicone phase. XPS data support the morphological structure of LBL coating applied on a silicone hydrogel polymer. The XPS Si/C ratio for the uncoated lens is 0.24. This ratio drops only to 0.16 when the LBL coating is applied (more C is added on the surface), indicating that a significant amount of Si exists on the surface. In contrast to this situation, the Si/C ratio for plasma polymerization coating of a hydrocarbon is 0.04, indicating uniform coverage of the surface by the coating depicted in Figure 29.1. The thickness of the coating is approximately 20 nm in both cases.

The coverage of coating and/or the barrier property of the coating can be examined by exposing the surface to an oil-soluble dye solution. In the oil-soluble dye test, oil and dye molecules permeate through the coating, if the coating has a poor barrier characteristics, and are absorbed in the silicone polymer phase, showing uniform stain if the surface is uniformly coated. If the surface is not uniformly coated, the dye molecules directly penetrate into the silicone phase showing spotty staining [2]. The results of the dyne penetration test are summarized in Table 29.1, which depicts the absence (−) or the presence (+) of stain when coated lenses were immersed in a vitamin E solution of Sudan black. The lenses in three different conditions—wet, blotted, and dried—were examined because lenses are kept in water before use.

The results indicate that Sudan black penetrated into the bulk phase of lens. The dye permeation test of LBL-coated lens in wet conditions did not show stain. However, when a thin layer of liquid water exists on the surface, the dye test losses validity in terms of evaluating the barrier property of the coating because the oil spreads on the water layer, which acts as a barrier to oil-soluble dye. Wet uncoated lens did show the dye stain, which probably indicates that water on uncoated lens is not as strongly held by the surface as LBL-coated surface. The dye permeation test of LBL-coated lenses under blotted and dried conditions showed stain. The stain in blotted sample was spotty and the color intensity is less than that for the dried surface, which is more intense than that for uncoated lens, i.e., dried LBL-coated lens has the strongest dye absorption.

The LCVD-coated lenses in wet, blotted, and dried conditions did not show stain at all. The dye test shows that plasma polymerization coating provided uniform coverage of silicone and hydrogel phases, and the coating is impermeable to Sudan black dissolved in vitamin E oil. The LCVD coating occurs as the interaction of reactive species in gas phase with the dry solid surface in vacuum. The gaseous reactive species strike both domains equally, and whether one phase consists of hydrophobic or hydrophilic polymer does not influence the basic mechanism of coating, which leads to the formation of a uniform coating layer depicted in Figure 29.1. The surface dynamic stability of plasma polymerization coating demonstrated by the dye test with Sudan black can be explained by the nature of the coating. One of the unique advantages of LCVD coating is that the influence of the nature of the substrate surface on which the coating is applied is minimal in comparison with other methods of coating.

2. CORRELATION BETWEEN SURFACE DYNAMIC STABILITY AND BARRIER CHARACTERISTICS

Angular dependence of XPS analysis of F-containing plasma polymer deposited on a low-density polyethylene (LDPE) film indicates that the amount of fluorine-containing and oxygen-containing moieties is slightly different toward the top surface. Data obtained with wet samples, which were immersed in water followed by freeze drying, show changes in both F/C and O/C ratios and their angular dependence. Wet surfaces have lower F/C ratios and higher O/C ratios than dry ones because when a surface is immersed in water, hydrophobic fluorine-coating moieties (such as CF_3, CF_2, CF) located near the surface migrate toward the inner bulk phase and hydrophilic oxygen-containing moieties migrate to the top surface region. The more surface configuration change occurs near the top surface. The term migrate is used in the context of the short-range movement by the conformational change of plasma polymer matrix.

The ease with which moieties rearrange themselves toward the inner bulk phase (F containing) or the outer top layer (O containing) when such a surface is immersed in water can be expressed by the ratio of (F/C ratio of water immersed sample)/(F/C ratio of dry sample) and (O/C ratio of water immersed sample)/(O/C ratio of dry sample). If no rearrangement occurs, the ratio is unity, and if all moieties migrate beyond the detection depth of XPS, the ratio is zero.

Since the F/C ratio decreases and the O/C ratio increases after water immersion, the ratio of (F/C ratio of water immersed sample)/(F/C ratio of dry sample) is less than 1, but the ratio of (O/C ratio of water immersed sample)/(O/C ratio of dry sample) is higher than 1. The lower change in the F/C ratio at inner layers (take-off angle > 15 degrees) is due to the net effect of two opposite directions of migration: fluorine-containing moieties located at one layer migrate into deeper layers and fluorine-containing moieties located at the outer layer migrate to this layer. The significance of the ratios (F/C ratio of water immersed sample)/(F/C ratio of dry sample) or (O/C ratio of water immersed sample)/(O/C ratio of dry sample) being near 1.0 is that such a surface of plasma polymers is surface dynamically stable, so that few changes of surface configuration occur after water immersion.

It should be emphasized that in order to be a good water barrier, a plasma polymer coating should fulfill the requirements of being surface dynamically stable and hydrophobic. A high contact angle observed with a polymeric material surface in the sessile droplet contact angle measurement does not necessarily mean the surface is hydrophobic, but rather that the surface configuration in equilibrium with a set of environmental conditions just before the measurement is such that the contact angle of water is high. There is no correlation between the contact angle of water and the effectiveness of water barrier characteristics.

A surface that shows less change due to water immersion means that the surface is tight or highly cross-linked so that the migration of fluorine-containing moieties into inner bulk phase is efficiently prevented. This also means that the surface molecules have low mobility. The mobility of surface molecules is related to the diffusivity of permeant molecules (water) through the surface layer. Consequently, it is logical that the surface dynamic stability is related to the water vapor permeation resistance of the surface. The role of surface dynamics of plasma polymers is more evident when the water vapor barrier property is compared for a particular perfluorocarbon plasma polymer over a range of energy input. In Figure 29.3, the ratio of (P uncoated/P coated) is plotted against the ratio of (F/C ratio of the water immersed sample)/(F/C ratio of dry sample) and (O/C ratio of the water immersed sample)/(O/C ratio of dry sample), respectively. The values of F/C and O/C ratios are measured at a take-off angle of 45 degrees.

Figure 29.3 clearly shows that the high water vapor permeation resistance is correlated to the minimal surface configuration change (wet/dry ratio being closer to 1.0) [3]. In other words, the optimal W/FM value, which yields the highest water vapor permeation resistance is also the one, which gives the minimal change in F/C and O/C ratios due to water immersion. This was the first confirmation of the correlation between the surface dynamic stability and water vapor permeation resistance of plasma polymer coatings.

3. SEGMENTAL MOBILITY OF THE SUBSTRATE POLYMER

When a nanofilm of plasma polymerization coating is applied on a polymer surface, the surface dynamic changes were still observed in spite of "tight" network structure of plasma polymers, and the rates of surface dynamic change differ depending on the nature of the polymers. The higher level of surface dynamic changes was often observed with polymers coated with plasma polymer of perfluorocarbons because

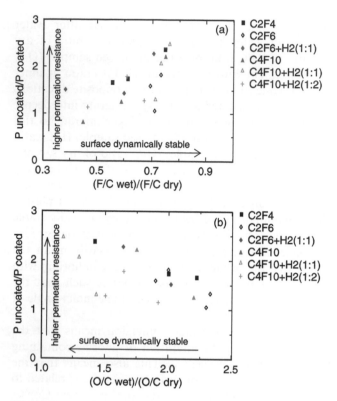

Figure 29.3 Correlation between water vapor permeability reduction rate with: (a) (F/C ratio of the water immersed sample)/(F/C ratio of dry sample), and (b) (O/C ratio of the water immersed sample)/(O/C ratio of dry sample) measured at take-off angle of 45 degrees.

plasma polymers generally introduce the amphoteric hydrophilicity due to the introduction of O-containing moieties by the reaction of dangling bonds with ambient oxygen when the sample is removed from the vacuum system. When the plasma surface labeling technique described in Chapter 25 is applied to plasma polymerization coatings, the influence of operational parameters on the surface dynamic stability of the coating can be examined [6].

The main effect of CF_4 plasma treatment may best be described as the surface implantation of fluorine-containing moieties (surface labeling) rather than the deposition of plasma polymer of CF_4. It is known that CF_4 does not form polymer in plasma unless hydrogen atoms are present in the plasma system as described in Chapter 7. In this case, the presence of H atoms in the plasma polymer of methane is an important factor. The effect of plasma polymerization coating is examined by comparing the surface dynamic change of the control, which is the substrate polymer without plasma polymerization coating of methane but the plasma surface labeling is applied directly.

Contact angles of water and of XPS F 1s peaks for the subsequently treated samples showed the effect of the operational parameters of plasma polymerization coating of methane on the perturbability of the final surface with CF_4 plasma treatment. The conditions of the plasma polymerization coating are manifested

Table 29.2 Conditions of Plasma Polymerization of CH_4 Applied on Nylon 6 Films

Sample	Wattage	Flow rate (sccm)	PO (mtorr)	P_g (mtorr)	Thickness (nm)	W/FM (GJ/kg)
Nylon 6/ M-1	75	6.30	56.1	65.1	60	1.0
Nylon 6/ M-2	75	2.52	31.0	34.7	60	2.5
Nylon 6/ M-3	75	1.58	24.0	29.6	60	4.0
Nylon 6/ M-4	75	1.15	18.5	23.5	60	5.5
Nylon 6/ M-5	75	0.90	15.4	20.2	60	7.0

Table 29.3 Conditions of Plasma Polymerization of CH_4 Applied on PET Films

Sample	Wattage	Flow rate (sccm)	PO (mtorr)	P_g (mtorr)	Thickness (nm)	W/FM (GJ/kg)
PET/M-1	85.17	4.36	43.6	46.3	120	1.6
PET/M-2	85.74	1.47	20.4	18.7	120	4.9
PET/M-3	84.15	0.48	9.22	7.2	120	14.7
PET/M-4	85.16	4.43	43.3	45.8	60	1.6
PET/M-5	85.78	1.42	20.2	19.0	60	5.1
PET/M-6	84.15	0.48	9.63	7.5	60	14.7

Table 29.4 Effects of CH_4 Plasma Polymerization on Contact Angle of Water on CF_4 Plasma–Treated Nylon 6 Films Immersed in Water, Contact Angle (Degree)

Immersion time (min)	No pretreatment	M-1	M-2	M-3	M-4	M-5
0	115.0	115.0	114.9	113.0	112.0	111.5
5	102.0	100.5	102.5	104.5	107.0	115.5
10	—	—	—	—	—	118.3
15	—	—	—	—	—	116.5
30	95.0	97.9	98.5	100.1	102.5	114.3
60	92.0	93.1	94.7	95.5	100.3	111.3
90	91.5	92.6	95.1	94.4	96.0	109.3
120	89.6	90.8	93.6	93.0	94.0	111.3
No CF_4 treatment	63.5	72.8	71.5	71.4	69.6	68.0

by the normalized plasma polymerization power input parameter W/FM. The conditions of plasma polymerization of methane are shown in Table 29.2 for nylon 6 films and Table 29.3 for PET films.

3.1. Nylon 6

Contact angles of water on various samples after they are immersed in water for various periods of time are shown in Table 29.4. The decay of hydrophobicity due to water immersion is less for samples prepared with plasma polymerization coating of

Table 29.5 Intensities of ESCA F_{1s}. Peaks of CF_4 Plasma–Treated Nylon 6 Films Immersed in Water, F 1s Intensity (cps $\times 10^4$)

Immersion time (min)	No pretreatment	M-1	M-2	M-3	M-4	M-5
0	8.16	8.08	8.01	7.96	7.75	6.53
5	7.09	6.96	7.24	7.42	7.33	6.90
10	—	—	—	—	—	6.93
15	—	—	—	—	—	6.92
30	6.47	6.50	6.83	7.10	6.88	6.68
60	6.20	6.37	6.72	7.01	6.42	6.65
90	6.59	6.19	6.25	7.01	7.23	6.69
120	6.05	6.27	6.57	6.95	7.20	6.57

methane, and its reduction is dependent on the condition of plasma polymerization of methane, i.e., the higher the W/FM value, the lower the decay.

Results of XPS analysis also follow trends similar to those observed for the contact angles of water described above. Intensities of XPS F 1s peaks are summarized in Table 29.5. The changes due to water immersion observed by XPS are considerably less than those observed by measuring the contact angle of water. These trends may reflect the difference in the nature of these two measurements. XPS observation is not strictly restricted to the top surface (i.e., the depth of electron escape is on order of 2.5–5.0 nm), whereas the contact angle tends to show the character of the atomic level top surface. Consequently, the correlation between advancing contact angle of water and XPS F 1s peak intensity is dependent on the substrate polymer when CF_4 plasma treatment is applied onto polymer films without plasma polymerization coating of methane. In other words, although there is a clear correlation between contact angle of water and XPS F 1s peak intensity, two separate correlation lines can be drawn for nylon 6 and PET.

The surface characteristics of nylon 6 treated with CH_4 plasma are obviously different from those of untreated nylon 6. When the combined character of thickness and mobility (tightness of structure) of plasma polymer of methane reaches a certain level, it is anticipated that the CF_4 plasma treatment of such a surface would have surface dynamic characteristics significantly different from that of nylon 6. Figure 29.4 shows correlations between advancing contact angle of water and XPS F 1s peak intensity for nylon 6 and for nylon 6/M-5.

Changes in profiles of XPS signals due to water immersion are shown in Figure 29.5 for CF_4 plasma–treated film without pretreatment, and Figures 29.6 and 29.7 contain the same information for CF_4 plasma–treated films with CH_4 plasma polymerization coating at different levels of W/FM. The influence of water immersion is seen in the decrease of the C 1s peak at 293 eV and the increase of the C 1s peak at 285 eV and the increase of the lower binding energy O 1s peak (original peak in nylon 6). Figures 29.5 and 29.6 show that (1) the CH_4 plasma (polymerization) covers most of the surface and (2) the changes caused by water immersion described above are reduced remarkably because CF_4 plasma labeling is now on the plasma polymerization coating of CH_4. Comparing Figure 29.6 and Figure 29.7, it can be seen that the effect of water immersion becomes hardly recognizable for sample methane plasma polymerization coating with high W/FM.

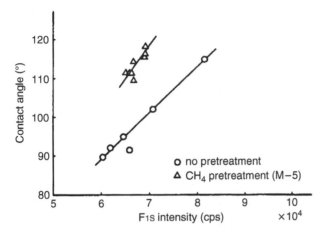

Figure 29.4 The correlation between contact angle of water and ESCA F 1s peak intensity for nylon 6 and nylon 6/M-5; W/FM for CH_4 plasma polymerization 7.0 GJ/kg.

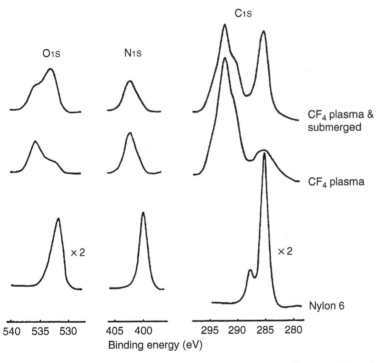

Figure 29.5 Changes in ESCA signals of nylon 6 film on CF_4 plasma treatment and subsequent water immersion (120 min).

3.2. Poly(ethylene terephthalate)

The effect of the thickness of plasma polymer of CH_4 on the contact angles of water is summarized in Table 29.6, and the results of XPS analysis are summarized in Table 29.7. The trend in the decrease of water contact angle on immersion seen for

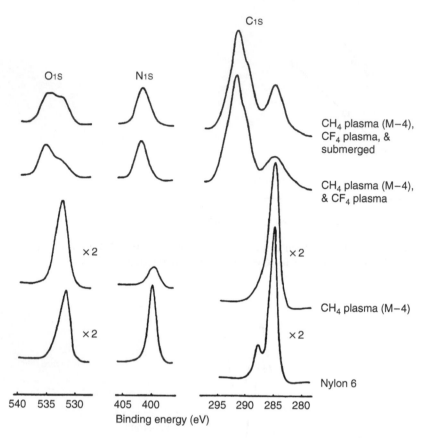

Figure 29.6 Changes in XPS signals of nylon 6 film on CH_4 plasma polymerization, CF_4 plasma treatment, and water immersion (120 min); W/FM for the CH_4 plasma polymerization 5.5 GJ/kg.

PET is similar to that seen for nylon 6. The effect of thickness of plasma polymer of methane, although relatively small, can be seen in the correlation between advancing contact angle and XPS F 1s peak intensity, as shown in Figure 29.8. The correlation lines for the plasma polymerization coating of CH_4 deviate from that for PET (without plasma polymerization coating), and the contact angle of water becomes less sensitive to XPS F 1s peak intensity, depending on the thickness of plasma polymer of methane. A too-thick layer tends to lose its effectiveness.

The general trend observed for samples prepared without plasma polymerization coating is also seen for samples prepared with plasma polymerization coating, namely, changes of XPS F 1s signal intensity as a function of water immersion time indicate that XPS F 1s peak intensities also follow similar changes observed for contact angles of water. On the other hand, plots of XPS F 1s peak intensities for dry films and for water-immersed films, as functions of W/FM values of the plasma polymerization coating, as given in Figure 29.9, show that the XPS F 1s peak intensity decreases with increasing value of W/FM, which indicates that the surface implantation of fluorine-containing moieties by CF_4 plasma becomes progressively more difficult with increasing value of W/FM. This observation is in accordance with

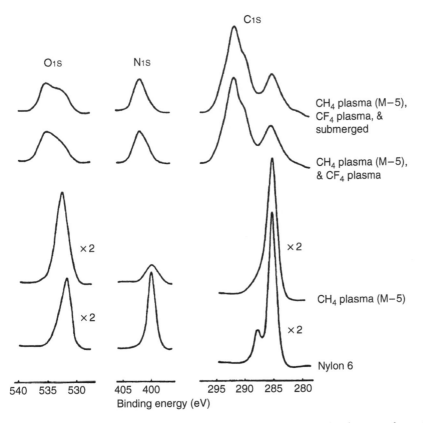

Figure 29.7 Changes in ESCA signals of nylon 6 film on CH_4 plasma polymerization, CF_4 plasma treatment, and water immersion (120 min); W/FM for the CH_4 plasma polymerization 5.5 GJ/kg.

Table 29.6 Contact Angles of Water on CF_4 Plasma–Treated PET Films Immersed in Water

		Contact angle (degrees)					
		Thickness 120 nm			Thickness 60 nm		
Submerged time (min)	PET no pretreatment	PET/ M-1	PET/ M-2	PET/ M-3	PET/ M-4	PET/ M-5	PET/ M-6
0	97.0	98.5	94.3	54.5	97.2	102.7	87.0
5	76.0	81.0	89.5	82.0	80.5	98.3	103.5
15	—	77.0	88.5	84.0	78.5	97.5	105.0
30	72.3	79.5	89.5	80.5	76.5	99.0	94.0
60	72.0	78.0	88.0	80.0	74.5	96.3	86.0
90	71.5	76.5	89.0	80.5	75.5	98.5	89.0
120	71.5	75.8	87.8	80.0	73.8	98.0	90.0
No CF_4 treatment	69.0	69.5	74.0	75.0	72.5	73.8	74.5

Table 29.7 Intensities of ESCA F_{1s}. Peaks of CF Plasma–Treated PET Films Immersed in Water

Submerged time (min)	PET no pretreatment	XPS F 1s Intensity (cps $\times 10^4$)					
		Thickness 120 nm			Thickness 60 nm		
		PET/ M-1	PET/ M-2	PET/ M-3	PET/ M-4	PET/ M-5	PET/ M-6
0	9.31	9.13	8.24	3.45	9.17	8.09	5.31
5	6.98	6.88	6.97	6.87	7.42	6.64	6.22
15	—	7.11	7.16	6.73	7.01	7.04	5.91
30	5.24	6.81	6.00	6.13	6.47	6.06	6.07
60	6.00	6.09	6.07	5.03	6.12	5.80	5.80
90	5.32	6.00	6.04	5.14	6.58	5.71	5.76
120	5.76	5.82	6.06	5.16	6.47	5.77	5.49

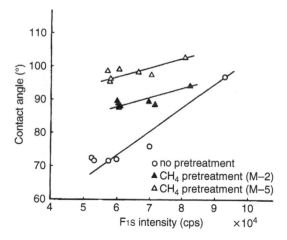

Figure 29.8 The correlation between contact angle of water and ESCA F 1s peak intensity for PET; PET/M2: W/FM 4.9 GJ/kg, 120 nm, PET/M-5: W/FM 5.1 GJ/kg, 60 nm.

the concept that the CF_4 plasma treatment is the surface implantation of fluorine-containing moieties, which proceeds via elimination of HF. The presence of hydrogen atoms on the surface is a key factor for such a reaction. Results indicate that the high W/FM conditions of plasma polymerization of methane, which forms a tighter network, decrease the number of hydrogen atoms on the surface. Changes in profiles of XPS signals of PET films are shown in Figures 29.10–29.12.

Figure 29.13 depicts the influence of conditions of plasma polymerization of methane on the decay rate constant, k in the relationship given by $A_t = A_0 t^{-k}$. The values of k are calculated from the slopes of initial linear portions of log (contact angle) versus log (water immersion time in minutes). Results clearly show that the higher the value of W/FM, the higher the surface dynamic stability of the final surface with CF_4 plasma treatment. The trends shown in the figure seem to suggest that there exists a critical level of energy input that transforms the network of plasma

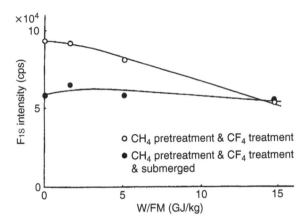

Figure 29.9 ESCA F 1s peak intensities for dry and water-immersed CF_4 plasma treated plasma polymerization coating of CH_4 (on PET) as functions of W/FM values of CH_4 plasma polymerization (thickness $= 60$ nm).

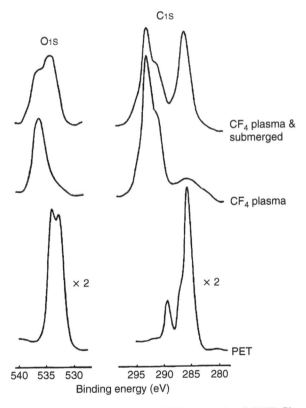

Figure 29.10 Changes in ESCA signals of PET films due to CF_4 plasma labeling and subsequent water immersion for 120 min.

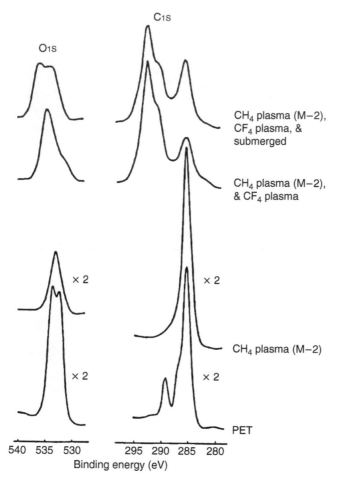

Figure 29.11 Changes in ESCA signals of PET films due to the plasma polymerization coating of CH_4, CF_4 plasma labeling, and subsequent water immersion for 120 min; W/FM for CH_4 plasma polymerization 5.0 GJ/kg, thickness 120 nm.

polymer of methane to an imperturbable state. Plasma polymers of methane formed with energy input level below the critical value (approximately $W/FM = 5$ GJ/kg) are not tight enough and the coatings show some level of surface dynamical perturbability. Thus, the often claimed "tight network of plasma polymer" or "highly cross-linked plasma polymer" cannot be intuitively assumed but can be obtained only under certain conditions, indicating the importance of the *system-dependent* aspect of LCVD.

4. STABILITY OF TOP SURFACE OF LCVD FILM

Because of high packing density of type A plasma polymer (LCVD film), the segmental mobility discussed above section is minimal, and the surface of type A LCVD film is anticipated to be imperturbable. However, there are some observations

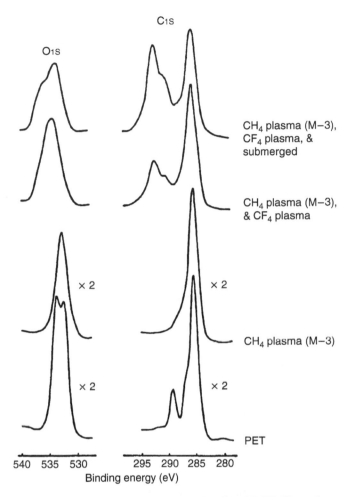

O₁ₛ ... C₁ₛ

CH₄ plasma (M−3),
CF₄ plasma, &
submerged

CH₄ plasma (M−3),
& CF₄ plasma

× 2 ... × 2

CH₄ plasma (M−3)

× 2 ... × 2

PET

540 535 530 ... 295 290 285 280
Binding energy (eV)

Figure 29.12 Changes in ESCA signals of PET films due to the plasma polymerization coating of CH_4, CF_4 plasma labeling, and subsequent water immersion for 90 min; W/FM for CH_4 plasma polymerization 15 GJ/kg, thickness 120 nm.

that the top surface of some films is rather instable. This involves the stability of the top surface of a plasma-polymerized film. It has been observed that the thickness of some deposition decreases when the surface is rinsed with a solvent, while the remaining layer shows no sign of further loss of material.

The growth mechanism of LCVD film, which is described in Figure 5.3, indicates that each step of growth is essentially oligomerization rather than polymerization. Accordingly, the last reactive species that deposit on the top surface at the end of LCVD operation might not be well incorporated into the three-dimensional network of the rest of film. In other words, the top surface region of an anticipated type A LCVD film is composed of type B LCVD film of the same monomer. The extent of this aspect is highly dependent on the nature of monomer as well as the operational conditions of LCVD, and cannot be described quantitatively for general cases. If the top surface is highly populated with oligomeric entities that

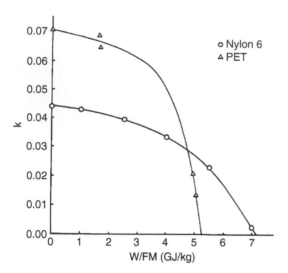

Figure 29.13 Effect of W/FM of plasma polymerization coating of CH_4 on the decay rate constant, k, for the coating deposited on nylon 6 and PET film.

are not well incorporated into the three-dimensional network of plasma polymer, the surface could be highly perturbable or even unstable, which is worse than the perturbable surface.

Although data for the loss of materials from the top surface are not available, the adhesion characteristics of a paint applied on the surface of LCVD film on an aluminum alloy seems to shed light on the nature of the top surface region of cathodically deposited TMS film, of which details are described in Chapter 32. The adhesion of a primer applied on the surface of LCVD film of TMS prepared by flow system reactor is very poor regardless of types of primer. This is probably due to the characteristically high rate of TMS deposition leading to a relatively high concentration of Si-based oligomers at the top surface region. The TMS film prepared by a closed-system reactor yields excellent adhesion of the same primer because the top surface region of LCVD film has a higher concentration of C-based materials. The Ar plasma treatment that is applied immediately after the deposition of TMS film significantly increases the adhesion in both cases. The more dramatic improvement of adhesion is obtained when the TMS LCVD film is treated by hexafluoroethane (HFE) plasma, which is an etching gas for Si-containing materials. (Ar plasma treatment was used to replace HFE plasma treatment, due to the wall contamination problem of F-containing moieties with aluminum oxides, which is described in Chapter 31.)

The role of the second plasma treatment by HFE or Ar seems to be the removal of type B plasma polymer of TMS from the top surface region or possibly converting the type B plasma polymer to type A plasma polymer. Electron spin resonance (ESR) data (described in Chapter 6) indicate that the number of Si-based dangling bonds decreases by these second plasma treatments. The weight loss observed with some plasma polymers and the ESR data for TMS film suggest that type B plasma polymer in the top surface region of an LCVD film could be up to nearly 30% of the deposition, if careful selection and control of operational parameters were not rigorously executed. This assessment reemphasizes the important point that the "imperturbable surface" or "highly cross-linked tight network" is not a trademark of LCVD film but is an attainable goal that can be reached by careful design and execution of the process, which constitutes the core of low-pressure plasma interface engineering.

REFERENCES

1. Nicolson, P.C.; Vogt, J. Biomaterials **2001**, *22*, 3237.
2. Ho, C.-P.; Yasuda, H. J. Biomed. Mater. Res. **1988**, *22*, 919.
3. Lin, Y.; Yasuda, H.; Miyama, M.; Yasuda, T. J. Polym. Sci. **1996**, *A: 34*, 1843.
4. Steennis, F. H. IEEE Trans. Elect. Insul. **1990**, *25*, 989.
5. Ross, T.; Geurts, W.S.M.; Smith, J.J.; van der Maas, J.H.; Lutz, E.T.G. IEEE Trans. Elect. Insul. Int. Sy. Elect. Insul., Toronto, Canada, **June 3–6 1990**.
6. Yasuda, T.; Okuno, T.; Yoshida, K.; Yasuda, H. J. Polym. Sci. Phys. Ed. **1988**, *26*, 2061.

30

Creating Adhesion to Substrate Surface

1. MODIFICATION OF SURFACE ENERGY OF SUBSTRATE

1.1. Plasma Treatments of Low-Density Polyethylene

The majority of commercially available polymers have amphoteric hydrophobicity/ hydrophilicity and surfaces are not hydrophilic, i.e., the contact angle of water is higher than 45 degrees (see Chapters 26 and 27), and water does not spread on the surface. Consequently, when a water-borne coating is applied on the surface, the coating does not spread well on the surface. Such an unprintable or uncoatable surface could be made printable or coatable by increasing the surface energy by various means. Aggressive chemicals could be used to partially oxidize the surface to increase the surface energy; however, the remaining chemicals causes increase of volatile organic content (VOC), which is a growing problem in recent years with respect to the environmental pollution. Plasma treatment, such as oxygen plasma treatment, does not cause a VOC problem and has been widely used in modifying the surface of rather hydrophobic polymers.

The stability of the surface after plasma modification is a major concern for many investigators attempting to improve wettability by plasma surface modification. Hydrophobic recovery, i.e., the decrease in surface hydrophilicity, as indicated by an increase in contact angle with aging time, is a common example of surface instability encountered after plasma modification with reactive plasmas [1–3]. However, this is not a serious problem in terms of attempting to make surface printable, if the subsequent printing could be performed in a reasonable time, i.e., print it before it becomes unprintable again. The problem associated with the surface dynamic instability could be seen in an episode, in which the surface modification and subsequent printing on the surface could not be done within a reasonable time.

In order to increase the moisture resistance of cardboard (wood product) the extrusion coating of polyethylene (PE) is often applied on the surface of the cardboard. In order to make the surface of PE-coated cardboard printable, the corona discharge was applied on the PE surface. The PE coating as well as the corona discharge treatment is applied in line at the site of cardboard manufacturing. PE-coated cardboards with printable surface manufactured in Canada were shipped to a customer in Japan after printability of the treated surface was confirmed.

However, the customer found the surface unprintable, which was done more than a couple of weeks after the corona discharge was applied.

However, the real concern here is the durability of the coating applied on the modified surface. The problem is that the polar groups are introduced on the surface in order to make the surface printable or coatable and create some level of adhesion between the surface and the coating; however, the adhesion based on the polar–polar interaction is vulnerable to water. As soon as water molecules reach the interface, they interact with the surface and the coating more strongly than the existing interaction between them, yielding delamination of the coating.

Thus, making the surface coatable and making the surfaces coatable with the durable coating are two different tasks. The latter task is much more difficult. Oxygen plasma treatment of polymer surface is effective in making the surface coatable, but more often than not, the coating fails when water molecules reach the interface. Many other factors could be responsible for the failure of the coating that is applied on a plasma–treated polymer surface.

Reactive plasmas vary in their ability to chemically etch polymers [4–7], depending on the volatility of the products they form. The formation of nonvolatile oligomers may occur depending on the type of substrate, the reactive gas used, and the discharge conditions. Oxidized oligomers resting on a surface provide an extremely unstable surface because they are not permanently attached. For example, hydrophobic recovery of O_2 plasma–treated polyetheretherketone (PEEK) films after washing with acetone was attributed to the presence of the degradation products [8]. Furthermore, exposure to remote O_2 plasma rather than direct O_2 plasma was found to be more effective in minimizing hydrophobic recovery because the remote O_2 plasma relies on the chemical reactivity of species and the radiation damage is absent.

Inorganic and organic gases such as H_2, O_2, CO_2, H_2O (vapor), N_2, and CF_4 create chemically reactive but not material-depositing luminous gases. Unlike polymer-forming plasma, reactive plasmas do not form solid polymeric deposits, but physically and chemically react with the surface of polymeric materials. These reactions can result in cross-linking, oxidation, or etching of the surface macromolecules. However, chemical reaction of reactive plasmas with the surface of polymeric materials, referred to as chemical etching, is typically the dominant process. Chemical etching is an ablative process that causes oxidation and subsequent chain scission of macromolecules in the outermost surface layer. This ultimately removes surface material in the form of volatile organic products that exit the reactor via the vacuum system.

The chemical structure of macromolecules and the type of reactive gas determine the extent of degradation and thus the amount of material removed, i.e., weight loss. The extent of degradation caused by oxygen plasmas varies according to the type of atoms and their arrangement in the polymer, as described in Chapter 9.

The presence of oligomers on the plasma–treated surface can be shown by the behavior of the surface in the Wilhelmy force measurements [9]. Most relatively stable (surface dynamically) conventional polymers exhibit Wilhelmy force loops that are parallelogram shaped but vary in size and vertical position depending on three factors: geometry of the cross-section of a sample, meniscus shape change, and surface configuration change on the immersion and the emersion process as described in Chapter 26. However, many conventional and plasma-modified polymers are perturbable on changing contacting medium. Consequently, those polymers exhibit deviations from the ideal parallelogram-shaped force loop, which are mainly attributed to surface configuration change. The extent of surface configuration change has been characterized by measuring the extent of intrinsic hysteresis from a Wilhelmy force loop [10–12].

Intrinsic hysteresis is the difference between the first and second immersion lines in a Wilhelmy force loop. For many polymeric surfaces, the second immersion line is elevated above the first immersion line, meaning that the water-immersed surface becomes more hydrophilic. Polymers exhibiting large intrinsic hysteresis values are considered highly perturbable in contrast to imperturbable polymer surfaces. Intrinsic hysteresis is absent for imperturbable polymers, as indicated by the retracing of the first and second immersion lines, $(F/L)_{D,a,1}$ and $(F/L)_{D,a,2}$, respectively. An example of this occurrence in the case of untreated low-density polyethylene (LDPE) is depicted in Figure 30.1a. This is typical force loop behavior for many conventional polymers. Figure 30.1b depicts the same force loop for the O_2 plasma–treated LDPE, in which the second immersion line overshot, i.e., fell below, the first immersion line. The overshooting means that the surface immersed in water became more hydrophobic than the original surface, which contradicts the principle of surface dynamics, i.e., water-immersed polymer surface becomes more hydrophilic than dry surface.

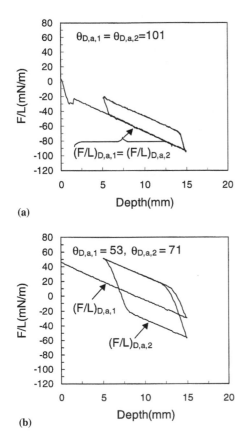

(a)

(b)

Figure 30.1 Comparison of Wilhelmy force loops; (a): LDPE, (b): O_2 plasma–treated LDPE.

Overshooting of immersion lines may be understood by reconsidering possible residual products from the etching process. Since O_2 plasma etches volatile products from the surface of polymers, it is also plausible that nonvolatile hydrophilic oligomers are formed because of chain scission and unzipping of the polymer backbone. These hydrophilic oligomers are only loosely affixed to the surface and thus provide an unstable surface state.

During the first immersion (i.e., advancing), the three-phase contact line moves over the surface laden with oxidized oligomers, which are perceived as part of the surface. However, since the oligomers are only loosely affixed to the surface, they are easily removed by water during the first emersion (receding). This exposes a less oxidized or more hydrophobic surface state that is subsequently rewetted during the second immersion. The increase of hydrophobicity is evident upon calculation of the first and second immersion advancing contact angles, which correspond to 53 degrees and 71 degrees, respectively, in Figure 30.1b.

The correlation between the extent of the overshooting and the plasma susceptibility expressed the weight loss rate as depicted in Figure 30.2. The polymers most susceptible to weight loss degradation from O_2 plasma exposure exhibited the least amount of overshooting. Conversely, polymers less susceptible to weight loss degradation from O_2 plasma exposure exhibited greater overshooting. This correlation is opposite to what we anticipate based on the plasma susceptibility of polymers. However, if we distinguish the degradation of polymer and the ease of removal of degraded species, the correlation found makes good sense. Namely, the polymers most susceptible to weight loss degradation, such as polyoxymethylene (POM), degrade quickly without leaving nonvolatile products, whereas the polymers least susceptible to weight loss degrade slowly yielding nonvolatile products on the surface, which causes the surface dynamic instability.

The short-term increase of the hydrophobicity that occurs within the time scale of wetting during Wilhelmy force measurements should be distinguished from

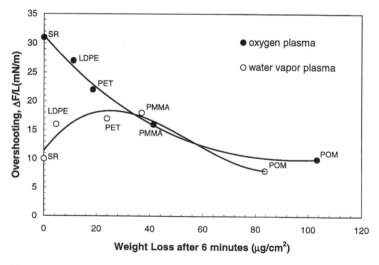

Figure 30.2 Correlation between the overshooting in Wilhelmy force loop and the plasma susceptibility expressed by the weight loss rates.

long-term hydrophobic recovery resulting from surface configuration change, which occurs over a period of days, weeks, or even months under ambient conditions, and recognized often as "hydrophobic recovery." The short-term increase of the hydrophobicity is permanent, whereas long-term hydrophobic recovery may be reversed via immersion in water. The presence of loosely attached hydrophilic oligomers is detrimental in creating water-insensitive adhesion, when a coating is applied on the treated surface.

1.2. Effect of Discharge Conditions on Wetting and Water Film Stability

Wettability, indicated by the first immersion advancing contact angle, $\theta_{D,a,1}$, and water film stability, indicated by the fluid holding time (FHT) (see Chapter 26), varied profoundly for plasma-modified LDPE depending on the glow discharge conditions. The effects of plasma exposure time and gas flow rate on $\theta_{D,a,1}$ and FHT are depicted in Figure 30.3 for O_2 plasma treatment and Figure 30.4 for H_2O plasma treatment. The most obvious result from both O_2 and H_2O vapor plasmas is that wettability and FHT increased with increasing plasma exposure time. However, the most important result of increasing exposure time to both plasmas is the reduction in overshooting. Surface oligomers generated in the early stage are subsequently removed by prolonged exposure, i.e., the degradation of polymers precedes the etching (removal) of the degraded materials.

The effects of system pressure and input power on wettability and FHT immediately following exposure to O_2 and H_2O vapor plasmas are depicted as the dark lines in Figures 30.5 for O_2 plasma treatment and Figure 30.6 for H_2O plasma treatment. Wettability and FHT changed very little with system pressure, which was changed at a fixed flow rate, as compared to input power for both plasmas. The general trends fit the principle of plasma treatment that can be given by the processing parameter $(W/FM)*t$; i.e., the lower the W/FM, the longer treatment is necessary; conversely, with the higher W/FM, the shorter treatment will suffice.

The overshooting was generally smaller with H_2O vapor plasma than O_2 plasma. The lower pressure treatment yields the smaller overshooting, indicating that the removal of fragmented moieties is more efficient in the lower pressure. Thus, plasma gas and the conditions of plasma processing have profound influence on the performance (with respect to the adhesion) of the coated products, in which coating is applied on the plasma–pretreated surface. Plasma treatment to make the surface coatable does not necessarily yield good adhesion of the coating. In order to create good adhesion of a coating that is applied on a plasma–treated polymer surface, it is necessary to avoid the formation of weak boundary due to oligomeric moieties created by plasma treatment of the substrate surface.

2. PARYLENE C FILM

2.1. Improving Adhesion of Parylene C Film on Smooth Surface

Parylene C, or monochloro-substituted poly(para-xylylene), is a polymer that has excellent bulk mechanical properties as well as excellent barrier properties for

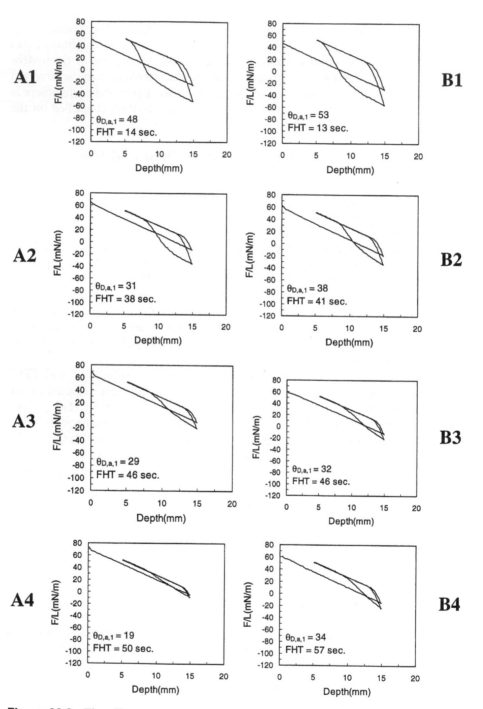

Figure 30.3 The effects of O_2 flow rate and the treatment time on Wilhelmy force loop; Column A: 1 sccm, B: 10 sccm, Row from top; (1): 0.2 min, (2): 1 min, (3): 2 min, (4): 4 min. System pressure and input power were fixed at 50 mtorr and 36 W respectively. The dynamic advancing contact angle of water and fluid holding time, FHT, are shown on each plot.

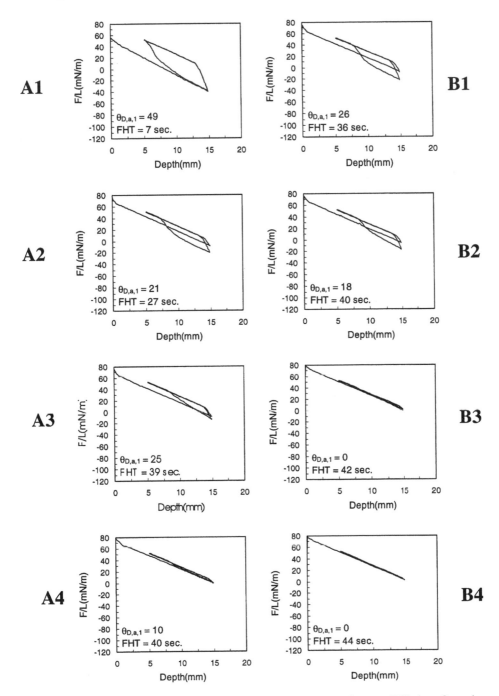

Figure 30.4 The effects of H$_2$O flow rate and the treatment time on Wilhelmy force loop; Column A: 1 sccm, B: 10 sccm, Row from top; (1): 0.2 min, (2): 1 min, (3): 2 min, (4): 4 min. System pressure and input power were fixed at 50 mtorr and 36 W respectively: the dynamic advancing contact angle of water and fluid holding time, FHT, are shown on each plot.

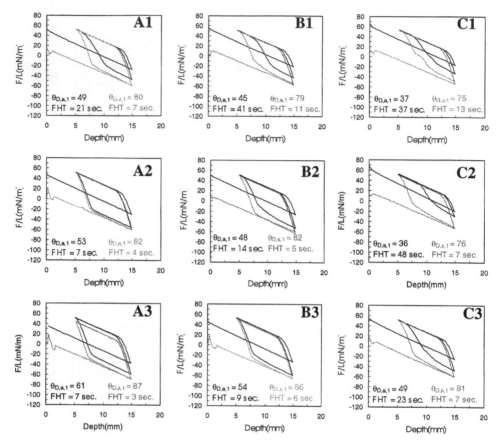

Figure 30.5 The effects of input power; (A): 8 W, (B): 30 W, (C): 63 W, and system pressure; (1): 25 mtorr, (2): 50 mtorr, (3): 100 mtorr, on Wilhelmy force loops of O_2 plasma–treated LDPE: oxygen flow rate and plasma treatment time were fixed at 10 sccm and 0.2 min, respectively, dark-colored force loops were taken just after samples were removed from the reactor, and gray-colored force loops were taken two weeks later after equilibrating with ambient air.

various gases, organic solvents, and water. Parylene C has one major deficiency insofar as the Parylene C coating is concerned, i.e., its adhesion to most substrates with smooth or nonporous surfaces is very poor; there is practically no adhesion at all. The *no adhesion* is a great advantage, on the other hand, in obtaining a freestanding ultrathin film, and Parylene C has been utilized in such applications. This deficiency in no adhesion has restricted its application as protective coatings.

The process of Parylene polymerization is presented schematically in Figure 2.1 for Parylene N (no substituent on phenyl ring). Parylene dimer is heated until it sublimes. This dimer vapor passes through a high-temperature pyrolysis zone (about 650°C) where it cracks and becomes monomer vapor. The monomer polymerizes and polymer is deposited on surfaces in the deposition chamber, of which temperature should be less than 60°C. Parylene C deposition, which is typically carried out at room temperature and completed in vacuum, is a process with no solvents, no

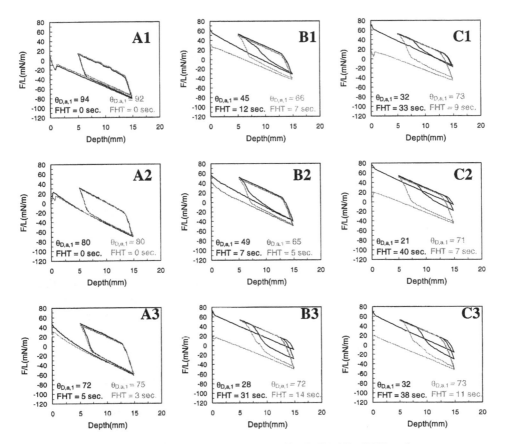

Figure 30.6 The effects of input power; (A): 8 W, (B): 30 W, (C): 63 W, and system pressure; (1): 25 mtorr, (2): 50 mtorr, (3): 100 mtorr, on Wilhelmy force loops of H_2O plasma treated LDPE: oxygen flow rate and plasma treatment time were fixed at 10 sccm and 0.2 min, respectively, dark-colored force loops were taken just after samples were removed from the reactor, and gray-colored force loops were taken two weeks later after equilibrating with ambient air.

curing, and no liquid phase. Its use introduces essentially no concern with regard to the operator's health and safety, air pollution, or waste disposal. As a dry-process, nonsolvent-based coating, Parylene C is not affected by VOC restriction.

Parylene C film could be utilized in system approach interface engineering (SAIE) if the adhesion of Parylene C film to the substrate and the adhesion of a coating, which is applied on the surface of Parylene C, could be improved. The scheme of utilizing Parylene C film in SAIE is depicted in Figure 30.7. Methods to improve its adhesion have been developed, but improvement is limited due to the lack of specific chemical interactions in the interface [13]. Low-temperature plasma deposition has proved to be a very effective process in improving the adhesion properties of materials while maintaining their desirable bulk properties.

A very important factor in the utilization of plasma deposition is that both processes are carried out in vacuum. Enhanced adhesion of Parylene C and

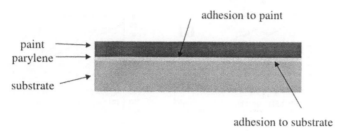

adhesion to paint

paint
parylene

substrate

adhesion to substrate

- Poor adhesion to smooth surfaces
  - No chemical bonding ⇒ Plasma interface modification ⇒ Free radicals ⇒ Chemical bonding
- Not paintable
  - Hydrophobic surface ⇒ Plasma surface treatment ⇒ Hydrophilic surface ⇒ Paintable

Figure 30.7 The plasma treatment effects on the surface and interface of chemically inert Parylene C coating.

Parylene N to smooth surface materials has been reported with the application of plasma depositions [13,14]. It was reported that excellent adhesion of Parylene C coating to a cold-rolled steel surface was achieved using plasma polymer coatings, in turn giving rise to corrosion protection of the metal [15]. Another major deficiency of Parylene C is its poor painting properties when paint is applied on a Parylene C film, due to its extremely hydrophobic surface. Because of this, surface modification of Parylene films is necessary to enhance their adhesion performance with spray primers.

The reactive species in Parylene C deposition that interacts with the substrate surface is para-xylylene, in which two free radicals exist in the para position of a benzene ring. Para-xylylene is relatively stable and reacts only with other free radicals or with other para-xylylene units. In order to create a good adhesion of Parylene C film to a smooth-surface substrate, it is necessary to create free radicals on the substrate surface. With the aid of plasma interface modification, it is possible to achieve strong adhesion of Parylene coatings to such smooth surfaces. Strong adhesion of Parylene C coating to bare 7075-T6 (an aluminum alloy) panels was achieved with the application of plasma polymers [16].

Figure 30.8 shows the scanned picture of Parylene C coating on bare 7075-T6 after the tape test. Figure. 30.8a and b are from the same sample, which passed the dry tape test. However, after boiling in water for 6 h, the tape test pulled off the Parylene C coating leaving the blue color of the trimethylsilane (TMS) plasma polymer visible. The samples in Figure 30.8c and d, all passed the dry tape test. After boiling in water for 8 h, the tape test could not remove the Parylene C coating from the substrate, which indicates that a strong adhesion of the Parylene C coating to the smooth surface was obtained with proper TMS/CH_4 plasma treatments.

The adhesion of Parylene C coatings to three different DC plasma polymer–coated bare 7075-T6 surfaces was examined; the results are summarized in Table 30.1. Plasma polymer of TMS/hexafluoroethane (HFE) showed the best results among the three types of plasma polymers. After exposure to air for around 1 day before deposition of Parylene C, this plasma polymer still showed a very good adhesion to Parylene C coatings. As shown in the table, radio frequency argon

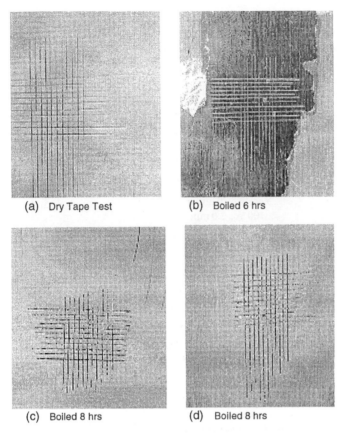

(a)   Dry Tape Test   (b)   Boiled 6 hrs

(c)   Boiled 8 hrs   (d)   Boiled 8 hrs

Figure 30.8   Adhesion improvement of Parylene C to TMS or TMS/CH$_4$ plasma polymer coated 7075-T6 panels with subsequent Ar RF plasma treatment at 100 W and 50 mtorr; (a) and (b) on TMS plasma polymer, treatment time 1.0 min, (c) on TMS/CH$_4$ plasma polymer, treatment time 2.0 min, (d) on TMS/CH$_4$ plasma polymer, treatment time 10.0 min: the blue color (dark) in (b) is the color of TMS film, indicating that Parylene C film was peeled off.

plasma treatment on the TMS/HFE surface in the Parylene reactor was not necessary and such a treatment made the adhesion weaker.

2.2.   Improving Adhesion of Paints on Parylene C Film

Parylene C polymer films have a very hydrophobic surface, which makes it difficult to improve their painting properties. Surface modification of Parylene C films is necessary to enhance their adhesion performance with respect to paints. Because of its energetically milder characteristics and large-scale advantages [17,18], low-temperature cascade arc torch (LPCAT) plasma was used for the surface treatment of Parylene C coating. The water contact angle of Parylene C coatings is about 90 degrees before LPCAT treatment. After argon LPCAT treatment, the surface of Parylene C coatings became more wettable. Changes in the contact angle of the Parylene C surface with LPCAT plasma treatment time are shown in

Table 30.1 Tape Test Results of Parylene C Coatings on Different DC Plasma Polymer
Surfaces

			Tape test ratings	
Plasma polymer type, exposure time in air (h)		RF plasma treatment before Parylene	Dry	Boiled in water for 1, 2, 4, 6, 8 h
TMS/HFE[a]	0.4	No	5	5, 5, 5, 5, 5
	1.2	No	5	5, 5, 5, 5, 5
	23.5	No	5	5, 5, 5, 5, 5
	63	No	5	5, 5, 5, 5, 4
	192	No	4	4, 4, 4, 4, 4
	36	Yes, Ar 20 mtorr, 100 W, 2.0 min	4	4, 4, 4, 4, 4
TMS[b]	0.4	No	5	5, 5, 4, 4, 3
	22.5	No	5	5, 5, 4, 4, 4
TMS/CH$_4$[c]	0.4	No	4	4, 4, 4, 4, 4
	120	No	4	4, 4, 4, 4, 3

[a]Plasma polymer of TMS + plasma polymer of HFE.
[b]Plasma polymer of TMS.
[c]Plasma polymer of TMS + plasma polymer of CH$_4$.

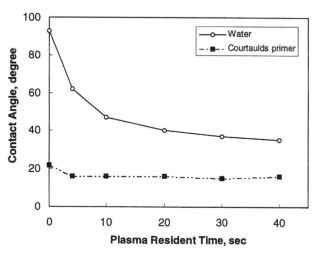

Figure 30.9 Contact angle changes of water and a primer on the Parylene C surfaces with
Ar LTCAT treatment time; Ar flow rate 1000 sccm, arc current 6.0 A.

Figure 30.9. LPCAT treatment under stronger plasma conditions for a longer time
yields the more wettable surface. However, the sessile droplet contact angle of a
paint on Parylene C surface (resident time 0) is low and minimal change occurred
with LTCAT treatment. Thus, the adhesion problem is not due to the wetting
difficulty.

2.2.1. Solvent-Borne Primer on LPCAT-Treated Parylene C Films

After cascade arc torch treatment, Parylene C surfaces were painted with a solvent-borne primer (519X303, Courtauld Aerospace). After the primer coatings were cured, the painted samples underwent the adhesion tape test under dry conditions and after boiling in water for a certain time period.

Figure 30.10 shows the typical scanned pictures of the tape test results of primer coatings on untreated and plasma–treated Parylene C. It is evident that, without any treatment, the painting properties of Parylene C coatings are very poor. The dry tape test removed all the primers from the Parylene C surface. In contrast, as seen from the figure, plasma treatment by argon LPCAT significantly improved the primer adhesion to Parylene C. The samples with longer treatment time passed the cross-cutting tape test even after boiling in water for as long as 8 h. This indicates that very strong adhesion of the primer coating to the Parylene C polymer has been achieved.

Table 30.2 summarizes the tape test results of the Courtauld primer coating on Parylene C surfaces with and without argon LPCAT plasma treatment. The adhesion performance of Parylene C films with respect to the Courtauld primer was improved in varying degrees depending on treatment conditions and treatment time.

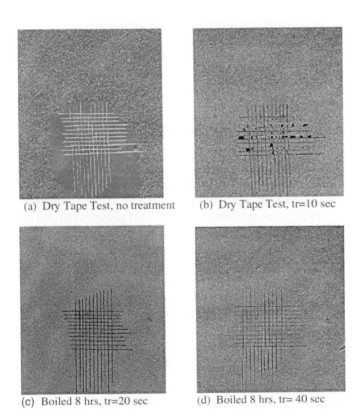

(a) Dry Tape Test, no treatment   (b) Dry Tape Test, tr=10 sec

(c) Boiled 8 hrs, tr=20 sec   (d) Boiled 8 hrs, tr= 40 sec

Figure 30.10 Adhesion Improvement of solvent-borne primer (519X303, Courtaulds Aerospace) to the Parylene C surface after LTCAT plasma treatment at various plasma resident times; Ar flow rate 1000 sccm, arc current 6.0 A.

Table 30.2 Tape Test Results for Solvent-Borne Primer (519X303, Courtauld Aerospace) Applied to Argon LTCAT–Treated Parylene C Surfaces

Plasma parameters			Tape test rating	
Ar flow rate (sccm)	Arc current (A)	Treatment time (s)	Dry test	Boiled in H$_2$O for 1, 2, 4, 6, 8 h
	No treatment		0, no adhesion	—
1000	6.0	4	0	—
		10	2	—
		20	5	5, 5, 5, 5, 5
		30	5	5, 5, 5, 5, 5
		40	5	5, 5, 5, 5, 5
1000	2.0	20	0	—
	4.0	20	5	4, 4, 4, 4, 4
	8.0	20	5	5, 5, 5, 5, 5
500	6.0	20	0	—
1500			5	5, 5, 5, 5, 5
2000			5	5, 5, 5, 5, 5

This indicates that adjusting the cascade arc torch treatment conditions can control the adhesion of the Courtauld primer to Parylene C surfaces. Most of the cases with argon LPCAT plasma treatment passed the tape test after 8 h of boiling in water. The general trend evident is that the stronger plasma conditions, such as higher arc current, higher argon flow rate, and longer treatment time, give rise to better adhesion of primers to Parylene C films.

2.2.2. Water-Borne Primer on LPCAT-Treated Parylene C Films

Plasma treatment of Parylene C films has proved to be very effective in improving the painting properties of Parylene C polymers with respect to a solvent-borne primer as described above. The similar effect of LPCAT plasma treatment on the adhesion of Parylene C polymer to water-borne primer (44-GN-36, Deft Corp.) was also observed. Table 30.3 summarizes the tape test results for the Deft primer coatings on Parylene C surfaces treated by LPCAT under different plasma conditions.

It is evident that the adhesion performance of Parylene C coatings with respect to the Deft primer was also improved to various degrees, depending on the LPCAT treatment conditions. Some of the samples passed the dry tape test but failed the tape test performed after boiling in water for a certain period. Samples with stronger LPCAT treatment conditions passed the tape test even after boiling in water for 8 h. These results indicate that, based on different requirements, the painting properties of Parylene C with respect to the Deft (water-borne) primer can also be controlled simply by adjusting the LPCAT plasma treatment conditions.

2.2.3. Radio Frequency Argon Plasma Treatment of Parylene C Films

Radio frequency argon plasma treatment of Parylene C surfaces is very attractive because two processes—Parylene deposition and plasma treatment—can be carried out in the same reactor, the Parylene reactor. It was found that radio frequency

Table 30.3 Tape Test Results for Water-Borne Primer (44-GN-36, Deft Corp.) Applied to the Argon LTCAT–Treated Parylene C Surfaces

Plasma parameters			Tape test rating	
Ar flow rate (sccm)	Arc current (A)	Treatment time (s)	Dry test	Boiled in H$_2$O for 1, 2, 4, 6, 8 h
	No treatment		0, no adhesion	—
1000	6.0	4	0	–
		10	1	–
		20	5	5, 5, 4, 4, 4
		30	5	5, 5, 5, 5, 5
		40	5	5, 5, 5, 5, 5
1000	2.0	20	0	–
	4.0	20	1	–
	8.0	20	5	5, 5, 5, 5, 5
500	60	20	0	–
1500	6.0		5	5, 5, 5, 5, 5
2000	60		5	5, 5, 5, 5, 5

Table 30.4 Tape Test Results of the Adhesion of Solvent-Borne Primer (519X303, Courtauld Aerospace) to Parylene C Surfaces Treated with Ar Radio Frequency Plasma in the Parylene Reactor at 50 mtorr

		Tape test ratings	
Power input (W)	Treatment time (s)	Dry	Boiled in H$_2$O for 1, 2, 4, 6, 8 h
100	20	5	4, 4, 4, 4, 4
	60	5	5, 5, 5, 4, 4
	120	5	5, 5, 5, 5, 5
	240	5	5, 5, 5, 5, 5
50	120	5	5, 5, 5, 5, 5
200	120	5	5, 5, 5, 5, 5

plasma also could be used for the enhancement of Parylene C painting properties. Results of such treatments are shown in Table 30.4. The difference between radio frequency treatment and LPCAT treatment is that a longer treatment period is necessary for the radio frequency treatment to achieve painting properties similar to those obtained with LPCAT treatment.

3. LPCAT (TREATMENT) AND LCVD (DEPOSITION)

3.1. Surface Treatment of Thermoplastic Olefins

Thermoplastic olefins (TPOs) have found increasing application in the automotive industry due to their competitive advantages over steel, such as low cost, light-weight, good mechanical properties, and easy processing. However, these TPO materials are difficult to paint because of their nonpolar and low surface energy characteristics [19,20]. Many traditional treatment methods are available to improve the paintability of TPOs or to enhance primer adhesion to TPOs. These include wet chemical abrasion, acid etching, flame or corona treatment, and application of adhesion-promoting primers and coatings [21]. However, all of these methods have specific disadvantages that are reflected in the high cost of final products; examples include volatile organic compounds/hazardous air pollution, environmental hazards involved in chemical disposal and overtreatment, and material damage due to process control difficulties [22].

The difficulty associated with TPO is its nonpaintable surface due to the poor wettability with paints. Introduction of polar groups on the surface (i.e., increasing the surface energy) can solve this problem. The introduction of polar groups enhances the wettability and also increases the adhesion of a primer to the modified TPO surface. However, the introduction of polar groups also increases the water sensitivity of the primer/TPO interface. Thus, this approach can yield good dry adhesion of a primer, but it is highly unlikely that such adhesion can withstand the action of water on the interface, i.e., the durability of the painted substrate would be poor.

The main excited species in LPCAT (argon as the carrier gas) are excited argon neutrals as described in Chapter 16. Since the majority of LPCATs use argon as the carrier gas, the term LPCAT is used to describe the process that uses Ar as the carrier gas, unless otherwise noted. This distinguishes LPCAT from other conventional plasma processes in which ions and electrons play significant roles. LPCAT treatment induces trapped free radicals on the exposed polymer surface; some of these introduce cross-links (recombination of two free radicals) in the exposed layer and others react with ambient oxygen when the treated substrate is taken out of the reactor. The reaction of free radicals with O_2 yield peroxides (via peroxy free radical), which further convert to oxygen-containing polar moieties such as carbonyl and carboxylic upon prolonged exposure. Thus, LPCAT treatment introduces oxygen-containing moieties on the treated surface.

LPCAT-O_2 (with oxygen introduced in the expansion chamber of LPCAT) treatment follows the same track of reactions that occurs in argon LPCAT treatment but with an enhanced oxidation, which also causes oxidation and oxidative degradation of the exposed polymer chains. These reactions occur without bombardment of energetic ions and electrons because there is no accelerating electron or ion in LPCAT, and the reactive species in plasma are created by argon metastables. In conventional O_2 plasma treatments, the bombardment of energetic species occur simultaneously and hence could cause more damage to the substrate polymer result than LPCAT-O_2 treatment.

LPCAT-CH_4 treatment deposits plasma polymer of methane on the substrate surface. Since the energetic species are consumed in building up the layer of plasma polymer, the irradiation effects of plasma onto the exposed substrate polymer is much less. The excellent adhesion of a primer to the LPCAT-CH_4–treated TPO was interpreted due to the replacement of the weak boundary layer by the interaction of tight network of plasma polymer of methane [23]. Plasma deposition from organic compounds, such as methane and ethane, can produce a thin layer of plasma polymer coating with tight networks.

The stability of a plasma–treated polymer surface is an extremely important issue in industrial applications of plasma processes. Many researchers have indicated that stabilizing the surface layer via cross-linking or CASING (*c*ross-linking via *a*ctivated *s*pecies of *i*nert *g*ases) using noble gas plasma such as He or Ar could slow aging of polymers [24–26].

It is important to recognize that polypropylene, which is the major constituent of TPO, is a typical degrading-type polymer in the radiation chemistry of polymers, i.e., once a free radical is formed on a polymer chain, the free radical unzips the chain rather than cross-links. CASING effect was first found with polyethylene [24], which is a typical cross-linking-type polymer. The same CASING effect, however, could not be anticipated with the treatment of the degrading-type polymers because the degradation of substrate polymer enhances the extent of weak boundary layer.

It is important to pay attention to the potential role of peroxides created on the surface of plasma–treated, including plasma polymer–coated, TPOs in the formation of durable bonds between the substrate and primer. It has been known for decades that the peroxides formed on the irradiated polymers (by γ-ray, X-ray, electron beams, etc.) can be utilized in graft copolymerization of various monomers. This method is known as the "peroxide method" of radiation copolymerization [27]. The trunk polymer is first irradiated by ionizing radiation in a vacuum or in an inert gas environment. The irradiated polymer is exposed to air or oxygen to convert free radicals to peroxides. Thus created peroxides-containing polymers were used as the initiator of the free radical polymerization of the second monomer. The polymer peroxides are decomposed by heat or by the use of reduction/oxidation accelerator, i.e., peroxides are converted to free radicals.

It is well understood that exposure of a polymer to plasma has the irradiation effects described above. The polymer that is treated with inert gas plasma as well as plasma polymers contains large amount of free radicals, which react with oxygen as described in Chapter 6. Polymer peroxides created by plasma treatment very likely act in a manner similar to those of the peroxide method of grafting during the curing process of a primer. In this case, no monomer is involved, but the reaction of free radicals with polymers in a primer can be handled in a similar manner. The aging test (aging before primer application) addresses the question of how long the treated sample can be kept before the application of primer without losing the capability to create durable bonding.

The most severe damage to the adhesion of paint to a substrate is caused by the action of water, and there is no correlation between the dry adhesion strength and the water-caused deterioration of adhesion. The tape test cannot distinguish the adhesion strength when it reaches the maximal level, noted as grade 5. However, this simple test can be effectively utilized by perturbing the adhesion by exposing samples to different environments. The water boiling test is probably the most severe test of this kind. It is safe to judge that the adhesion is excellent and durable if a scribed coated sample passes the tape test after boiling in water for 8 h.

Figures 30.11–30.16 depict the aging aspects of LPCAT-treated TPOs, and Figures 30.17–30.19 depict the durability aspect of LPCAT treatments investigated with three kinds of commercially available TPOs. In figures, LPCAT-air is abbreviated as air plasma, and likewise argon plasma and methane plasma. The results shown in Figures 30.11–30.16 indicate that the wettability or the contact angle of water on a plasma–treated surface is not the major factor that accounts for

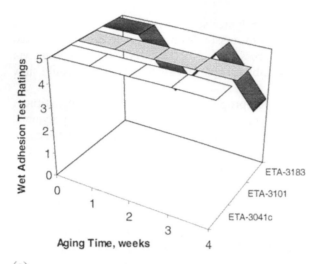

(a)

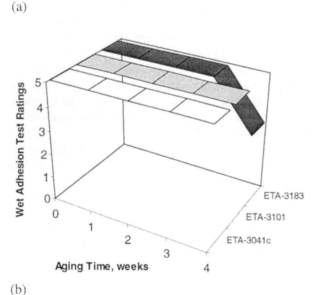

(b)

Figure 30.11 The wet adhesion test results of primer on air-plasma treated TPOs after (a) aging at room temperature and (b) aging at 40°C; #2 & #3 torches, 2.0 A arc current, 1000 sccm argon, 10 sccm air, 270 mtorr pressure, and 36.8 s plasma exposure time.

the adhesion of primer to the treated surfaces. Results also indicate that keeping treated samples in ambient or elevated temperature of 40°C does not affect the adhesion characteristics of treated TPOs. The results for ETA-3183 TPO are more erratic, but methane plasma treatment seems to overcome this problem.

Figures 30.17–30.19 depict the durability of the surface treatments examined with ETA-3183, which is the most difficult one to modify the surface effectively among samples employed. Because the water can reach the interface directly through cross-cuts, this accelerated adhesion test was much more severe when cross-cuts were

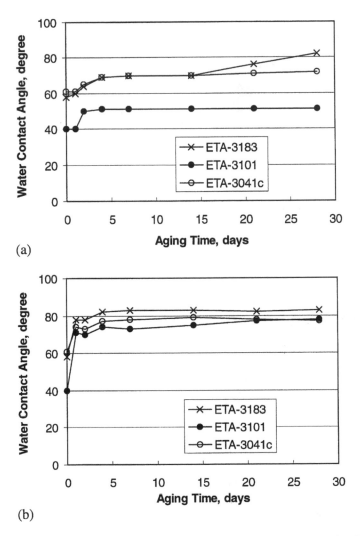

Figure 30.12 Changes in the water contact angles of air-plasma–treated TPOs with increased aging time (a) at room temperature and (b) at 40°C; #2 & #3 torches, 2.0 A arc current, 1000 sccm argon, 10 sccm air, 270 mtorr pressure, and 36.8 s plasma exposure time.

made before immersion than when they were made after immersion. This aspect is evident in the tape test ratings presented in Figure 30.17. To achieve this severity in testing, six cross-cuts were always made on the samples prior to immersion in boiling water.

Figures 30.18 and 30.19 show the changes in the longest water boiling time, which still pass the tape test, of primer–coated TPOs with the arc current and plasma exposure time of LPCAT treatment, respectively. From Figure 30.18 it is evident that durable bonding of the primer to TPOs was obtained with argon and methane LPCAT treatments at a lower arc current. Since the arc current represents energy input in the LPCAT process, plasma treatment conducted at a lower arc current may prevent the overtreatment on the TPO and thus give better adhesion results.

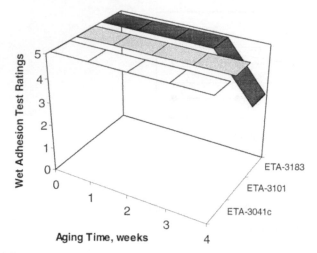

(a)

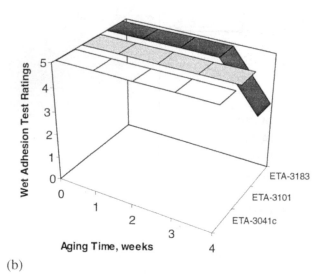

(b)

Figure 30.13 The wet adhesion test results for primer on argon-plasma treated TPOs after (a) aging at room temperature and (b) aging at 40°C; #2 & #3 torches, 2.0 A arc current, 1000 sccm argon, 10 sccm air, and 12.3 s plasma exposure time.

As detailed in Figure 30.19, this overtreatment effect was also observed when LPCAT treatment was applied for longer times at the low current of 2 A.

The overall effect of plasma treatment is a function of the treatment parameter, which can be given as (energy input) × (treatment time), as described in Chapter 8, and there exists a threshold value above which the treatment effect starts to decline [28]. Hence, an excessive energy input at a fixed treatment time or too-long treatment time at a fixed energy input will yield a poor treatment result.

Proper treatment with argon LPCAT could convert the weak boundary layers (WBLs) to a tight and cohesive skin on the TPO surface. However, overtreatment

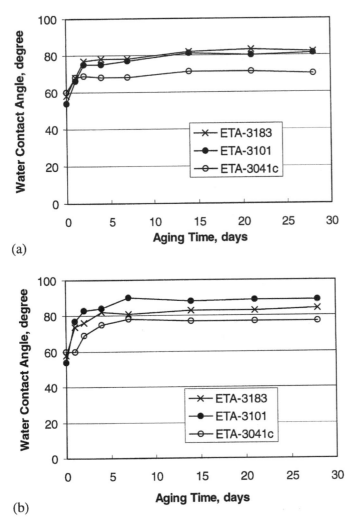

(a)

(b)

Figure 30.14 Changes in the water contact angles of argon-plasma–treated TPOs with increased aging time (a) at room temperature and (b) at 40°C; #2 & #3 torches, 2.0 A arc current, 1000 sccm argon, and 12.3 s plasma exposure time.

with argon LPCAT, such as the treatment time greater than 25 s shown in Figure 30.19, may cause more chain scissions, allowing WBLs to reform on the TPO surface and thus negatively influencing the overall adhesion performance. In the case of methane LPCAT treatment, too-short treatment time could not totally replace the WBL on the TPO surface with tightly cross-linked plasma polymers. A too-long treatment time might produce too-thick plasma coating on the TPO surface, causing the adhesion to become poor at the interface between the plasma polymer and the TPO. Therefore, as seen in Figure 30.19, the "excellent" durable bonding of primer to TPOs was achieved by argon and methane LPCAT treatments only within a relatively narrow range of (power input) × (treatment time).

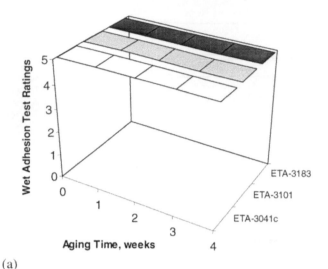

(a)

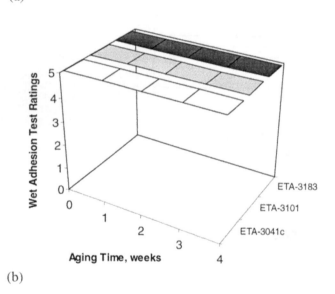

(b)

Figure 30.15 The wet adhesion test results for primer on methane-plasma-treated TPOs after (a) aging at room temperature and (b) aging at 40°C; #2 & #3 torches, 2.0 A arc current, 1000 sccm argon, 10 sccm methane, 270 mtorr pressure, and 1.9 s plasma exposure time.

It is evident in Figures 30.18 and 30.19 that, after the accelerated adhesion test, the primer-coated air LPCAT–treated TPOs showed poor adhesion performance as compared to argon and methane LPCAT–treated TPOs. This is a clear indication that air-plasma treatment can achieve the paintable surface but cannot provide the treated surface that can be painted in a durable manner. The poor durability can be attributed to (1) the water-sensitive nature of adhesion and (2) the excessive degradation of polypropylene that causes weaker boundary layer.

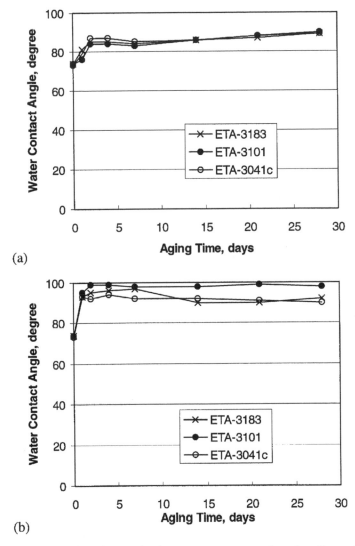

(a)

(b)

Figure 30.16 Changes in the water contact angles of methane-plasma–treated TPOs with increased aging time (a) at room temperature and (b) at 40°C; #2 & #3 torches, 2.0 A arc current, 1000 sccm argon, 10 sccm methane, 270 mtorr pressure, and 1.9 s plasma exposure time.

In the cases of argon and methane LPCAT treatments, both methods improve adhesion of TPOs not only by producing a tightly cross-linked surface layer to replace the original weak boundary layer on the TPO surfaces but also by creating durable bonds between primer and the treated TPO surface. It can be postulated that during the primer curing process the surface-bound peroxides decompose to regenerate free radicals and react with polymers in the primer to form water-insensitive covalent bonds. This is the same principle and reaction scheme known as "peroxide method" in radiation grafting of polymers as described earlier. The results

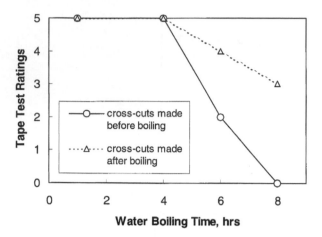

Figure 30.17 Tape adhesion test results at various water boiling times for primer on methane–plasma treated ETA-3183 with six cross-cuts made before and after immersion into boiling water; #2 & #3 torches, 2.0 A arc current, 1000 sccm argon, 10 sccm methane, 270 mtorr pressure, and 1.9 s plasma exposure time.

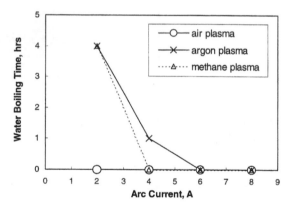

Figure 30.18 The effects of plasma treatment arc current on the water boiling time, which pass the tape test, of primer-coated ETA-3183 specimens; #2 & #3 torches, 1000 sccm argon, 10 sccm air, 270 mtorr pressure, and 12.3 s plasma exposure time (1.9 s for methane plasma).

suggest that such water-insensitive covalent bonds are responsible for the "excellent" primer adhesion obtained with argon and methane LPCAT–treated TPOs.

The ideal surface modification is achieved by modifying only the ultrathin top layer, which generally corresponds to the weak boundary layer, without influencing or altering the bulk characteristics of the material. While conventional plasma treatments tend to yield a modified surface layer that consists of fragmented polymers or oligomers, LPCAT treatment yields a modified ultrathin layer that is reinforced by the induced cross-links. The difference between plasma treatment and LPCAT treatment, and the corresponding results of treatment are schematically depicted in Figures 30.20 and 30.21, respectively.

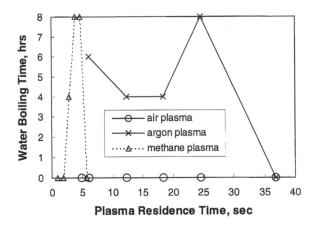

Figure 30.19 The effects of residence time in plasma on the water boiling time, which pass the tape test, of primer-coated ETA-3183 specimens; no. 2 & no. 3 torches, 1000 sccm argon, 2.0 A arc current, 10 sccm air, and 270 mtorr pressure.

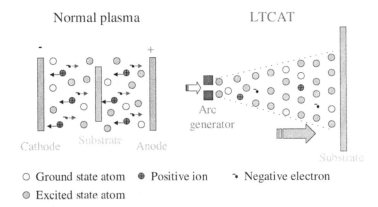

Figure 30.20 Schematic presentation of the difference between plasma treatment (explained by DC glow discharge) and LTCAT treatment.

3.2. Adhesion of Polymer Fibers to Concrete

The incorporation of short discrete fibers in concrete results in improved resistance to fracture, fatigue, and impact loading [29]. Fibers contribute to the toughening of the resultant composite largely through interface stress transfer. Steel and synthetic fibers are among the common fiber types used in fiber-reinforced concrete. The interface characteristics of steel fibers are typically enhanced by providing mechanical anchorages such as hooked or enlarged ends, or by incorporating deformations/indentations along the fiber length. Attempts have been made to enhance bonding between polymeric fibers and cement matrices through mechanical means such as fibrillation and indentation with some success. Surface treatments offer alternative means of enhancing the interface characteristics of such fibers [34–36]. Use of plasma treatment to alter the characteristics of the

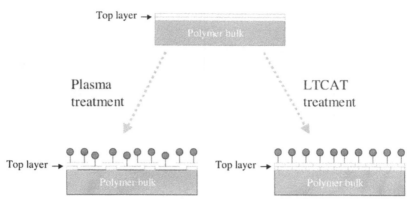

Hydrophilic moiety

Figure 30.21   Schematic presentation of the difference in the structures of modified surface treated by plasma treatment and LTCAT treatment; plasma treatment tends to modify the fragmented polymers while LTCAT treatment tends to strengthen the top layer.

fiber–matrix interface is among the relatively recent and potentially promising techniques to enhance the mechanical performance of fiber-reinforced composite systems [35,36].

Polymeric fibers are popular for reinforcing concrete matrices because of their low density (more number of fibers for a prescribed volume fraction), high tensile strength, ease of dispersion, relative resistance to chemicals, and relatively low cost compared to other kinds of fibers. Polypropylene and polyolefin fibers are typically hydrophobic, resulting in a relatively poor bond with concrete matrices compared to some other types of fibers. Treatment of polypropylene with an aqueous dispersion of colloidal alumina or silica and chlorinated polypropylene enhances the affinity of these fibers toward cement particles. Treatment of polypropylene fibers with a surface-active agent provides better dispersion of the fibers and a stronger bond between cement and fiber. The earlier attempts at surface treatments of polypropylene fibers have had only limited success and have not been commercially attractive.

The effectiveness of polymeric fibers as concrete reinforcement depends on the mechanical bond between the fiber and the cement matrix. A mechanical bond or adhesion with calcium silicate hydrate for polypropylene fiber/concrete has been reported [33]. The mechanical bond of collated fibrillated polypropylene is typically better than that of the monofilament polypropylene fibers because cement matrix penetrating the fibrillated network anchors the network in the matrix. The network structure of fibers leads to bidirectional action of fibers due to the shearing action of aggregate particles during mixing. Fibrillated fibers have a higher tensile strength and modulus of elasticity than films. The fiber geometry and type, fiber volume content, fiber configuration, fiber length, modulus of elasticity, and Poisson ratio of the fiber each have significant effect on the overall interface performance of the fiber in a cement-based matrix.

Li et al. have reported results from tensile and pull-out tests on plasma–treated polyethylene fiber concrete composites ($l = 12.7$ mm, $d = 38$ μm, $V_f = 2\%$) [32]. Among the three treatments (NH_3, CO_2, and Ar) investigated, NH_3 plasma treatment provided the best improvement in bond strength (up to 35% over untreated fiber composites).

Load deflection responses from untreated fibers as well as treated fibers suggest that, from among the various treatments investigated, plasma treatment with argon LPCAT results in the best "overall" performance of the treated-fiber mix [35]. This is true for both fiber types used as well as the fiber volume fractions investigated. Treatments in air and in (methane $+ O_2$) result in smaller residual strengths, in that order, compared to argon LPCAT. The scatter diagram for first crack strength, which also represents the peak load carrying capacity of the specimen, suggests that while argon is the best treatment for the fibrillated polypropylene fiber type, air treatment is marginally superior for the monofilament polypropylene fibers. These observations along with relative values of the contact angles characterizing the wettability of the fiber surface suggest that plasma treatment may be influencing the interface performance more than by merely altering the wettability of the interface. The advantages of argon LPCAT treatment described for TPO surface treatment described earlier are clearly evident in this application also.

Toughness, which is generally measured as the area under the load deflection response at prescribed values of deflection, is greater for all the treatment types compared to untreated fibers (Table 30.5). The Japanese toughness measure T_{JCI} and the ASTM toughness index I_{30}, both computed at relatively large limiting deflections, clearly can be used to highlight the influence of plasma treatment in enhancing the toughness of the fiber-reinforced composite system using treated fibers.

Figure 30.22 shows the energy absorbed per unit cross-sectional area (computed as area under the load deflection curve up to various deflection limits divided by the cross-sectional area of the beam) as a function of specimen deflection for argon-treated and untreated fiber concrete mixes. Clearly, in the fiber-dominated post-cracking regime, the influence of argon LPCAT treatment becomes readily apparent, particularly at large deflections. Toughness enhancements can be achieved through improved fiber-reinforcing parameters (use of higher fiber volume fraction, longer fibers, fibers with higher aspect ratio) and through fiber surface treatments (such as argon LPCAT treatment). The cost effectiveness of these alternate options of toughness enhancement presently favors plasma treatment. Another advantage is that the mixing problems due to balling and segregation often encountered while increasing the fiber volume fraction, fiber aspect ratio, and/or the fiber length are much less, if not completely absent, with argon LPCAT–treated fibers.

4. INCREASING ADHESION OF INORGANIC FILLERS AND FIBERS TO PMMA MATRIX

4.1. Radio Frequency Plasma Treatment of Fillers

The acrylic bone cements have been widely used to successful prosthesis in total joint replacements for the last few decades. While the surgical replacement is very successful, the bone cement is often found as a failed material after long-term use.

Table 30.5 Summary of Results[a] from the Flexural Tests for Fiber-Reinforced Concretes

Mix details	Fiber content and treatment	First crack strength, σ_f (psi)^2	First crack toughness, T_f (lb-in.)^2	I_5	I_{10}	I_{20}	I_{30}	Toughness index, T_{JC1} (lb-in.)^2	Equivalent flexural strength, σ_b (psi)[b]
Plain concrete	0.0%	616	3.68	2.78	3.70	4.25	4.39	16.56	35
Fibrillated	0.25% untreated	500	3.84	2.99	4.32	5.89	7.02	38.94	81
polypropylene	0.25% Ar	697	4.93	3.07	4.49	6.44	8.11	72.87	151
	0.25% CH$_4$+O$_2$	481	2.91	3.41	4.85	6.88	8.52	40.25	87
	0.25% Air	667	3.99	3.07	4.38	6.34	8.05	58.61	122
	0.50% untreated	563	3.58	3.57	5.57	8.90	11.83	82.13	175
	0.50% Ar	687	2.76	4.40	6.59	10.33	14.27	89.48	191
	0.50% CH$_4$+O$_2$	586	4.11	3.86	6.04	9.50	12.76	104.68	203
	0.50% Air	632	3.99	3.78	6.14	10.49	14.46	111.32	215
Monofilament	0.25% untreated	476	2.50	3.30	4.72	6.07	6.81	23.38	52
polyolefin	0.25% Ar	535	2.95	3.23	4.69	6.27	7.27	37.20	74
	0.25% CH$_4$+O$_2$	506	2.78	3.13	4.25	5.33	6.15	32.08	68
	0.25% Air	615	3.96	3.14	4.28	5.48	6.31	44.43	91
	0.50% untreated	474	4.03	2.35	2.99	3.85	4.66	44.37	96
	0.50% Ar	503	3.54	2.85	4.49	6.83	8.88	84.78	173
	0.50% CH$_4$+O$_2$	600	2.73	3.00	4.55	6.95	9.32	65.68	149
	0.50% Air	613	2.69	3.11	4.53	6.60	8.42	57.32	131

[a]Each entry in the table represents average values of 3–4 specimens.
[b]1 MPa = 145 psi, 1 N-m = 0.113 lb-in.

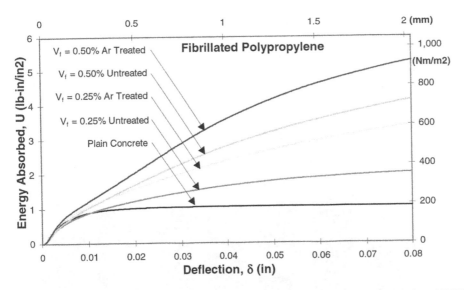

Figure 30.22 Energy absorbed versus deflection for plain concrete, untreated and LPCAT–treated fiber cement composites.

Based on the clinical reports, 10–45% of patients who received the surgical implant with the bone cement need revision surgery mainly due to the cement fracture [36,37]. It is commonly accepted that the fatigue is the major mechanism for the fracture of bone cement [37,38]. The weak resistance of bone cement to the fatigue loading has been addressed to the low molecular weight of matrix, weak interfaces between filler materials and matrix, agglomeration of X-ray-opaque powder, bubbles, and so forth [38–40].

Since the X-ray-opaque filler (e.g., ZrO_2) is one of the basic components of bone cement, the physical properties of bone cement should be treated as a composite material of polymethyl methacrylate (PMMA) and fillers. The physical properties of composite materials depend on the matrices, fillers, and interfaces between them. The most desirable situation may be the combination of good properties of each component material. For this purpose, the role of interface is very important for an efficient stress transfer from the matrix to the fillers [41,42].

The fatigue test results of PMMA bone cement in which the X-ray-opaque powder (ZrO_2) was treated with various kinds of plasma [43]. Figure 30.23 depicts the influence of plasma treatment of the X-ray-opaque powder (zirconium oxide) on the cycles of failure, which represent the fatigue life of specimens at 23 MPa maximal applied stresses under flexural loading. As shown in Figure 30.23, any kind of plasma treatment employed on the X-ray-opaque powder showed improvements in the fatigue properties of bone cement compared with the result of untreated sample. Among the plasma treatments, the hexamethyl disiloxane (HMDSO) plasma–based treatments seemed to have a better effect than the methane plasma–based treatments, possibly due to a better adhesion of silicon-containing monomer onto the surfaces of inorganic materials.

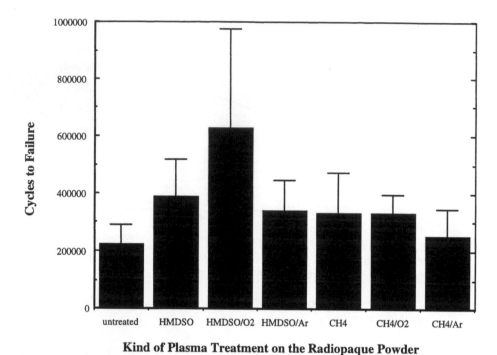

Kind of Plasma Treatment on the Radiopaque Powder

Figure 30.23 Fatigue life of PMMA bone cement with various kinds of plasma–treated X-ray opaque powder at 23 MPa maximum applied stresses under flexural loading.

The best fatigue test result obtained from plasma treatments on X-ray-opaque powder could be seen in the HMDSO plasma treatment following oxygen plasma–posttreated sample. In the HMDSO plasma–based treatments, the oxygen plasma posttreatment showed a significant additional improvement of fatigue properties whereas the argon plasma posttreatment showed negative effect. In the methane plasma–based treatments, oxygen plasma posttreatment showed no additional improvement, and the argon plasma posttreatment also showed the negative effect.

It has been well known that weak interfaces between the inorganic fillers and the organic matrix reduce the mechanical strength of bone cement [38,40,44,45]. The interfacial adhesion strength can be enhanced by plasma treatment, which is generally due to the improved wettability and possibly to the chemical bonds between the filler and the resin [46,47]. Especially in acrylic bone cement, chemical bonds may have an important role in improving the mechanical strength by the plasma treatment.

Since the plasma–treated surfaces contain a high concentration of free radicals as described in Chapter 6, peroxides may be formed in the atmospheric or oxygen environment as a result of oxidation of free radicals. The peroxides on filler surfaces may give free radicals by the reaction with the reducing agent (N,N-dimethyl-p-toluidine) in the bone cement mixture, which can initiate the MMA polymerization. Through this graft polymerization process, the interfacial bond strength between the filler particles and the MMA matrix may be enhanced resulting efficient stress transfer from the matrix to the fillers.

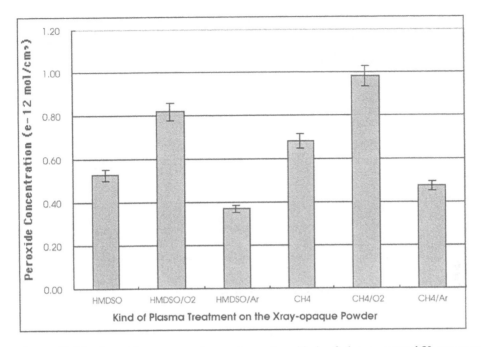

Figure 30.24 Peroxide concentration on the various kinds of plasma–treated X-ray opaque powder surfaces determined by DPPH method.

The peroxide concentration on the plasma–treated X-ray-opaque powder surfaces was estimated from the differential consumption (with respect to the untreated sample) of 2,2-diphenyl-1-pricylhydrazyl (DPPH) as shown in Figure 30.24. (As described in Chapter 6, DPPH consumption is not an accurate indicator of the amount of free radicals. However, DPPH consumption could be used to characterize the surface characteristics of plasma–treated substrate surface.) Regardless of the kinds of plasma-polymerized layer, the oxygen plasma posttreatments showed a larger consumption of DPPH than did the others. The argon plasma posttreatments showed a reduction of DPPH consumption from that for the plasma-polymerized surfaces of the X-ray-opaque powder. This finding seems to be in accordance with the negative effect found in the fatigue cycles for the argon plasma posttreatment.

Figure 30.25 depicts the correlation between the DPPH consumption and the fatigue life. Increasing trends of the fatigue life could be seen as the DPPH consumption increased regardless of the types of plasma treatments. The HMDSO plasma–based treatment on the X-ray-opaque powder showed more pronounced trend in the effects of surface-bound peroxide on the fatigue properties of bone cement.

The increase of fatigue cycle could result from multiple factors, such as wettability of the surface with matrix polymerizing dough, the extent of air bubble entrapment, interfacial voids, and potential formation of chemical bonds between the particle surface and the polymer matrix. What we are seeing is the net cumulative effect of all possible factors. The correlation figure is used to indicate the rough trends to see the contribution of the selected factor.

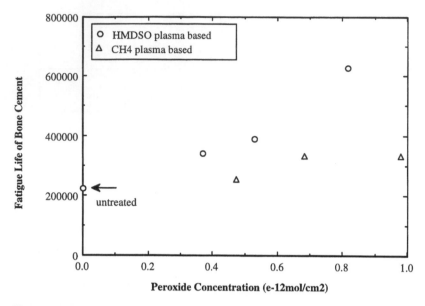

Figure 30.25 Fatigue life of PMMA bone cement as a function of peroxide concentration on the plasma treated X-ray opaque powder surfaces.

If the treated particles that contain peroxides could act as additional initiators of polymerization of MMA, it is anticipated that the effect in the overall polymerization rate of bone cement will be shown. The curing temperature profiles of the bone cement with plasma–treated X-ray-opaque powders were plotted in Figure 30.26. All profiles are identical except the shifting along the time scale, i.e., a shift to the left means that the polymerization is accelerated. The bone cements with untreated or argon plasma posttreated X-ray-opaque powder showed a longer gel time, whereas the bone cements with the oxygen plasma–posttreated X-ray-opaque powder showed a shorter gel time.

The relation between the peroxide concentration on the X-ray-opaque powder and the gel time of bone cement could be seen in Figure 30.27, in which the time at maximal temperature was used instead of gel time or set time to avoid unnecessary error, since the temperature profiles seemed to have identical shape. As the DPPH consumption increased, the gel time decreased (i.e, faster polymerization) regardless of the types of plasma treatment, as seen in the relation between the fatigue properties and the peroxide concentration. The HMDSO plasma–based treatments showed a more pronounced effect than that by the methane plasma–based treatments.

4.2. LPCAT Treatment of Milled Carbon Fibers

In the above section, effect of the interface between the X-ray-opaque filler and PMMA matrix has been investigated. In other words, no additional ingredient was added to the conventional PMMA bone cement recipe. In this section, our attention is focused on effect of the interface between milled carbon fiber and the bone cement matrix. Measurement of fatigue properties such as fatigue failure cycle deals with

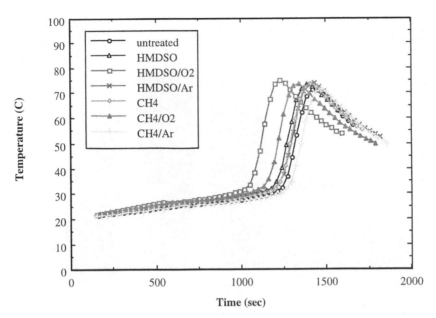

Figure 30.26 Curing temperature profiles of PMMA bone cement with various kinds of plasma–treated X-ray opaque powder.

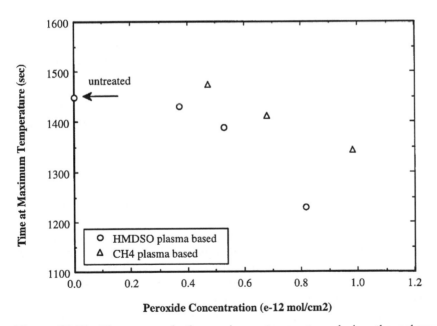

Figure 30.27 Time to reach the maximum temperature during the polymerization of PMMA bone cement as a function of peroxide concentration on the plasma–treated X-ray opaque powder surfaces.

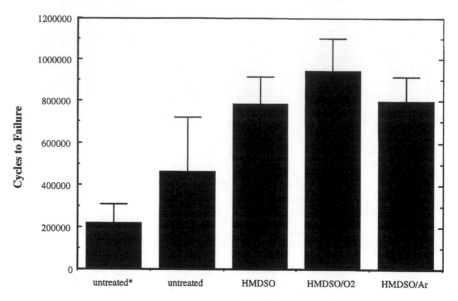

Figure 30.28 Fatigue life of plasma–treated carbon fiber reinforced PMMA bone cement at 23 MPa maximum applied stresses under flexural loading; untreated*: untreated carbon fiber with untreated X-ray opaque powder; all others are with $HMDSO/O_2$ treated X-ray opaque powder; untreated refers to untreated carbon fiber.

characteristic properties of a sample, which are not completely determined by the intrinsic material characteristics. Surface defects, trapped bubbles, and so forth, which influence the fatigue failure cycles, are caused by the processing of a sample. Therefore, the reproducibility of fatigue test results has a characteristically higher standard deviation than that for a material property such as tensile breaking strength. Fatigue life data should be used to find the trend beyond the standard deviation found for the controls.

Figure 30.28 depicts the effect of plasma treatment factors on the fatigue properties of carbon fiber–reinforced bone cements. The fatigue life of plasma–treated carbon fiber–reinforced bone cement showed significant improvements compared with untreated samples. What is shown is the improvement of fatigue failure cycles by the manipulation of interfaces, which is clearly beyond the standard deviation of the control. It should be noted that bone cement specimens shown in the figure contained $HMDSO/O_2$-treated X-ray-opaque powders (ZrO_2) except the first sample marked "untreated*," which had untreated powders. The second column designated "untreated" refers to the treatment of carbon fibers, i.e., untreated carbon fibers and $HMDSO/O_2$-treated X-ray-opaque powders were used. The increases of fatigue failure cycles are far beyond the standard deviation for the control designated by "untreated*." The increase is not in the range of percent, but a nearly fivefold increase is gained.

When the untreated carbon fiber was added into the bone cement, no reinforcing effect could be seen or the addition even deteriorated the fatigue properties. In other words, the addition of (untreated) milled carbon fibers did not

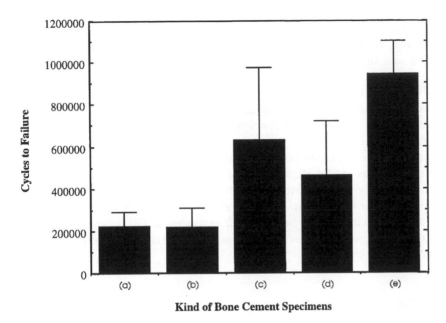

Figure 30.29 Effect of milled carbon fiber reinforcement and plasma treatment on the fatigue life of bone cement; (a) untreated ZrO_2 without fiber, (b) untreated ZrO_2 with untreated fiber, (c) HMDSO/O_2 treated ZrO_2 without fiber, (d) HMDSO/O_2 treated ZrO_2 with untreated fiber, (e) HMDSO/O_2 treated ZrO_2 with HMDSO/O_2 treated fiber.

improve the fatigue failure cycles. It is probably due to the very short fiber length, which is below the critical length. It is also important to note that fiber above the critical length cannot be used in the current mode of bone cement preparation due to the difficulty of mixing and the increase of viscosity. The surface modification of the carbon fibers changed this situation completely. The addition of surface-modified carbon fibers (with HMDSO/O_2) improved the fatigue characteristics of bone cement in a significant manner. Figure 30.29 depicts the contribution of the addition of carbon fibers and of the surface modification of the carbon fibers.

The comparison of (a) and (b) shows the effect of adding untreated milled carbon fibers, which is no improvement. The comparison of (a) and (c) shows the effect of the surface treatment of X-ray-opaque powders on fatigue cycles, which is a significant (nearly threefold) increase. The comparison of (c) and (d) shows the effect of adding untreated carbon fibers on bone cement with treated ZrO_2, which is an appreciable negative effect. The comparison of (d) and (e) shows the effect of the surface treatment of carbon fibers, which is a roughly twofold increase.

From these comparisons it is clear that every quantum improvement in the fatigue cycles can be attributed to the plasma surface modifications of fillers. Among the improvement of fatigue properties by the plasma treatments on the carbon fibers, as shown in Figure 30.28, the oxygen plasma posttreatment showed a better effect that was already seen in the plasma treatment on the X-ray opaque powder. There were reports that the oxygen plasma–treated surface showed a better adhesion to PMMA than the argon plasma–treated surface [46–49].

Carbon fibers treated with $HMDSO/O_2$ plasma showed the highest DPPH consumption per unit surface $(2 \times 10^{-10}\,mol/cm^2)$ among plasma treatments employed, and also showed a significant reduction of the time needed to reach the maximal temperature (1070 s compared to 1170 s for untreated fibers). These are the similar trends found with plasma–treated X-ray-opaque powders. All of these trends strongly indicate that the plasma–treated fillers acting as the additional initiator of MMA polymerization, i.e., PMMA polymers are covalently bonded to the LCVD-coated fillers.

The unique aspect of plasma polymerization technique is that a very significant increase in fatigue cycles can be obtained by using very short fibers. This means that increasing interfacial bonding between the fiber and the matrix can significantly shorten the critical fiber length. Without the surface modification, the milled carbon fibers seem too short to impart any beneficial effect as shown. On the other hand, the use of longer fibers increase the viscosity of the dough mixture and often render it nearly unusable in bone cement.

REFERENCES

1. Yasuda, H.; Sharma, A.K.; Yasuda, T. J. Polym. Sci. **1981**, *19*, 1285.
2. Morra, M.; Occiello, E.; Garbassi, F. J. Adv. Sci. Technol. **1973**, *7*, 1051.
3. Van Der Mei, H.C.; Stokroos, J.M.; Schakenraad, J.M. Busscher, H.J. J. Adv. Sci. Technol. **1991**, *9*, 757.
4. Yasuda, T.; Gazicki, M.; Yasuda, H. J. Appl. Polym. Sci. **1984**, *38*, 201.
5. Yasuda, H.; Lamaze, C.E.; Sakaoku, K. J. Appl. Polym. Sci. **1973**, *17*, 137.
6. Moss, S.J. Polym. Degrad. Stab. **1987**, *17*, 205.
7. Hansen, R.H.; Pascale, J.V.; De Benedictis, T.; Rentzepis, P.M. J. Polym. Sci.: Part A **1965**, *3*, 2205.
8. Inagaki, N.; Tasaka, S.; Horiuchi, T.; Suyama, R. J. Appl. Polym. Sci. **1998**, *68*, 271.
9. Weikart, M.Christopher; Yasuda, K. Hirotsugu. J. Polym. Sci., Part A: Polym. Chem. **2000**, *38*, 3028.
10. Weikart, M.C.; Miyama, M.; Yasuda, H.K. J. Colloid Interface Sci. **1998**, *211*, 28.
11. Miyama, M.; Yang, Y.; Yasuda, T.; Okuno, T.; Yasuda, H. Langmuir **1997**, *13* (20), 5494.
12. Miyama, M.; Yasuda, H.K. Langmuir **1998**, *14* (4), 960.
13. Sharma, A.; Yasuda, H. J. Vac. Sci. Tech. **1982**, *21* (4), 994.
14. Sharma, A.; Yasuda, H. J. Adhesion **1982**, *13*, 201.
15. Yasuda, H.; Chun, B.H.; Cho, D.L.; Lin, T.J.; Yang, D.J.; Antonelli, J.A. Corrosion **1996**, *52* (3), 169.
16. Yu, Qingsong; Deffeyes, Joan; Yasuda, Hirotsugu. Prog. Organic Coatings **2001**, *41*, 247.
17. Fusselman, S.P.; Yasuda, H.K. Plasma Chem. Plasma Proc. **1994**, *14*, 251.
18. Yu Q.S.; Yasuda, H.K. Plasma Chem. Plasma Proc. **1998**, *18* (4), 461.
19. Ryntz, R.A. Polymeric Materials Science and Engineering, Proceedings of ACS Division of Polymeric Materials: Science and Engineering **1990**, *63*, 78.
20. Ryntz, R.A.; Scarlet, K.E.; Henchel, J.A.; Arthur, K.L. Automotive Eng. 37 (May 1993).
21. Kaplan, S.L.; Rose, P.W.; Hansen, W.P.; Sorlien, P.H.; Styrmo, O. Technical Papers, 1993 Regional Technical Conference, Society of Plastics Engineers, p. 84.
22. Strobel, M.; Walzak, M.J.; Hill, J.M.; Lin, A.; Karbashewski, E.; Lyons, C.S. J. Adhesion Sci. Technol. **1995**, *9*, 365.
23. Lin, Y.-S.; Yasuda, H.K. J. Appl. Polym. Sci. **1998**, *67*, 855.
24. Schonborn, H.; Hansen, R.H. J. Appl. Polym. Sci. **1967**, *11*, 1461.

25. Arefei, F.; Andre, V.; Montazer-Rahmati, P. Amouroux, J. Pure Appl. Chem. **1992**, *64*, 715.

26. Arefei-Khonsari, F.; Tatoulian, M.; Shahidzadeh, N. Amouroux, J. In *Plasma Processing of Polymers*; d'Agostino, R.; Faria, P.; Faracassi, F. Eds.; Kluwer Academic: Dordrecht, 1997, p. 165.

27. Chapiro, A. *Radiation Chemistry of Polymeric Systems*; Wiley: New York, 1962.

28. Iriyama, Y.; Yasuda, H. J. Appl. Polym. Sci.; Appl. Polym. Symp. **1988**, *42*, 97.

29. ACI Committee 544, Measurement of Properties of Fiber Reinforced Concrete, ACI Materials Journal, Proceedings, Nov.-Dec. **1988**, *85* (6), 583–593.

30. Fahmy, M.F.; Lovata, N.L. Transport. Res. Rec. **1989**, *1226*, 31.

31. Denes, F.; Feldman, D.; Hua, Z.Q.; Zheng, Z.; Young, R.A. Adhesion Sci. and Technology **1996**, *10* (1), 61.

32. Li, V.; Wu, H.C.; Chan, Y.W. J. Am. Ceram. Soc. **1996**, *79* (3), 700.

33. Rice, E.; Vondran, G.; Kunbargi, H. Materials Research Society Proceedings, Pittsburgh, **1988**, *114*, 145.

34. Hannant, D.J. *Fiber Cements and Fiber Concretes*; Wiley: New York, 1978; pp. 81–98.

35. Zhang, C.; Gopalaratnam, V.S.; Yasuda, H.K. J. Appl. Polym. Sci. 2000, *76*, 1985.

36. Collis, D.K. J. Bone Joint Surg. **1991**, *73-A* (4), 593.

37. Krause, W.; Mathis, R.S. J. Biomed. Mater. Res. **1988**, *22*, 37.

38. Topoleski, L.D.T.; Ducheyne, P.; Cuckler, J.M. J. Biomed. Mater. Res. **1990**, *24*, 135.

39. Haas, S.S.; Brauer, G.M.; Dickson, G. J. Bone Joint Surg. **1975**, *57-A* (3), 380.

40. Kusy, R.P. J. Biomed. Mat. Res. **1978**, *12*, 271.

41. Sheldon, R.P. *Composite Polymeric Materials*; Applied Science: London, 1982; pp. 12–22.

42. Subramanian, S.; Reifsnider, K.L.; Stinchcomb, W.W. Int. J. Fatigue **1995**, *17* (5), 343.

43. Kim, H.Y.; Yasuda, H.K. Appl. Biomater. **1999**, *48*, 135.

44. Pourdeyhimi, B.; Wagner, H.D. J. Biomed. Mater. Res. **1989**, *23*, 63.

45. Dandurand, J.; Delpech, V.; Lebugle, A.; Lamure, A.; Lacabanne, C. J. Biomed. Mater. Res. **1990**, *24*, 1377.

46. Ladizesky, N.H.; Ward, I.M. J. Mat. Sci.: Mater Med. **1995**, *6*, 497.

47. Hild, D.N.; Schwartz, P. J. Mat. Sci.: Mater. Med. **1993**, *4*, 481.

48. Suzuki, M.; Kishida, A.; Iwata, H.; Ikada, Y. Macromolecules, **1986**, *19*, 1804.

49. Hsiue, G.H.; Wang, C.C. J. Polym. Sci.: Polym. Chem. **1993**, *31*, 3327.

31

Corrosion Protection of Aluminum Alloys

1. SAIE APPLIED ON ALUMINUM ALLOYS

An ultrathin layer of plasma polymer of trimethylsilane (TMS) has been utilized in the corrosion protection of aluminum alloys by means of system approach interface engineering (SAIE) [1–4]. SAIE by means of low-temperature plasmas utilizes low-temperature plasma treatment and the deposition of a nanofilm by luminous chemical vapor deposition (LCVD). This approach does not rely on the electro-chemical corrosion–protective agents such as six-valence chromium, and hence the process is totally environmentally benign.

A nanofilm of plasma polymer of TMS (typically 50 nm) is applied on an appropriately prepared surface of an aluminum alloy. Then a corrosion-protective primer coating (typically 30 μm) is applied onto the surface of the plasma nanofilm. Adhesion of a multilayer coating system is the prerequisite for success of the SAIE approach. Therefore, adhesion of the first layer of the nanofilm prepared by plasma polymerization is the most crucial factor in this approach because if this layer delaminated from the substrate surface the rest of coatings could not function at all. It should be emphasized, however, that the issue is not the adhesion of a plasma polymer per se, but rather the adhesion that leads to better corrosion protection by the principle of SAIE, which requires the *water-insensitive adhesion* relevant to the corrosion protection.

The ability of the plasma deposition process to actively interact with the surface of the substrate at the beginning of the process is important for attaining strong adhesion. These films appear to act as barrier films, blocking diffusion of corrosive media into the interface. The tightly adhering film also is important to prevent the lateral penetration of corrosive media from scribes. Without employing electrochemical corrosion inhibitors, the water-insensitive adhesion of a primer with good barrier characteristics is crucially important to provide better corrosion protection than that obtained by the electrochemical corrosion protection agents in the coating systems [5]. Thus, the pretreatment of surface to render good adhesion of plasma polymer is the crucial step. Plasma conditions for sample preparation and the sample identification codes are summarized in Table 28.1.

2. PRETREATMENT OF ALUMINUM ALLOYS SURFACES

The first step of SAIE is the preparation of oxides. The top layer of an aluminum alloy is generally covered with hydrated mixed oxides. Either alkaline cleaning or a combination of alkaline cleaning and deoxidization removes major organic contaminants and this potentially unstable oxide layer. The optimal surface treatment was found to be dependent on the type of alloy used as the substrate. The optimal combination of chemical treatments for one kind of alloy might differ from that for another kind of alloy. For instance, alkaline cleaning and deoxidizing, which works well for an alloy, makes the surface of Alclad worse than the native surface as depicted in Figure 31.1.

2.1. Plasma Sputtering

The oxides on the surface of aluminum alloys are relatively stable (e.g., compared with those of steel) and are difficult to be removed by plasma sputtering. The plasma sputtering was very ineffective and ineffective in the removal of the hydrated layer. These samples, which were pretreated with plasma and subsequently coated with primers, performed poorly in corrosion testing. The extended plasma sputtering was seen to cause some local heating. This heating caused the diffusion of Mg into the interface, which degraded the corrosion resistance. The enrichment of Mg in oxides layer is depicted in Figure 31.2. This is in strong contrast to the case of steel, in which plasma sputtering by $(Ar + H_2)$ effectively removed oxides and provided the foundation for the formation of iron silicides and carbides by the successive plasma polymerization of TMS, yielding excellent adhesion and corrosion protection [6,7].

2.2. Wet Chemical Cleaning

Factors of Al alloy surfaces that influence the adhesion of the TMS nanofilm can be schematically depicted as shown in Figures 31.3–31.5. Figure 31.3 depicts the surface state of as-received Al alloys. The top surface is covered with hydrocarbon film from atmosphere and/or intentionally applied thin film such as that for identification. The major factor is the relatively thick layer of hydrated oxides, which are not stable in

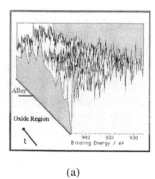
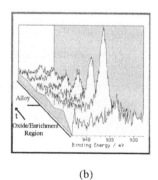
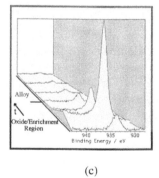

(a)   (b)   (c)

Figure 31.1   Cu 2p 3/2 photoelectron spectra from XPS depth profile runs on [2A] surfaces, a) native surface, b) alkaline cleaned surfaces, and c) alkaline cleaned and deoxidized surface; the arrow shows the direction of sputtering time into the alloy.

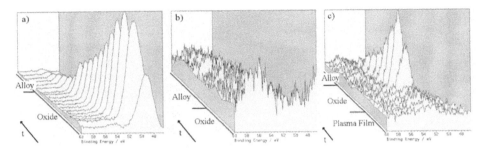

Figure 31.2   Mg 2p spectra from depth profiles showing Mg migration into the oxide layer due to heating during extended plasma pretreatments: the three samples are a) the native acetone cleaned surface, b) alkaline cleaned and deoxidized, and c) alkaline cleaned and deoxidized followed by 10 min of $Ar + H_2$ plasma treatment and 10 min of N_2 plasma treatment prior to deposition of a plasma polymer from $TMS + N_2$, the arrow indicates the evolution as a function of sputtering time, but spacing between spectra is not linear but rather a spectral index, the lines mark the different regions on the samples, as obtained from spectra of the other constituent elements.

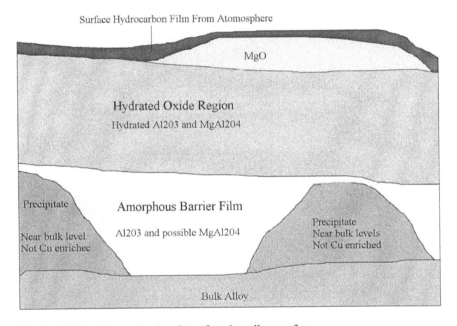

Figure 31.3   Schematic drawing of native alloy surface.

the context of adhesion and corrosion protection. The deposition of the TMS on either of these two top layers did not yield good adhesion. Figure 31.4 depicts the surface state after wet chemical treatments, and Figure 31.5 depicts the final surface state of TMS nanofilm–coated Al alloys.

The reactive species of TMS in plasma react with the stable oxides created by the chemical pretreatment, which is dependent on the type of alloys, and form chemical bonds between the oxides and the depositing plasma polymer. This step is

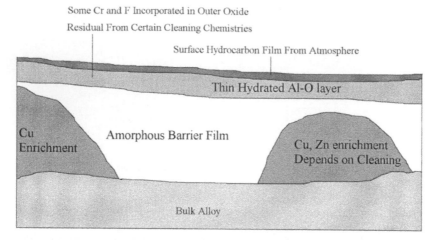

Figure 31.4 Schematic drawing of chemically cleaned alloy surface Mg removed from oxides; precipitate/deposit: $CuAl_2$—2024, $CuAl_2$ and Zn Phase—7075Alk.

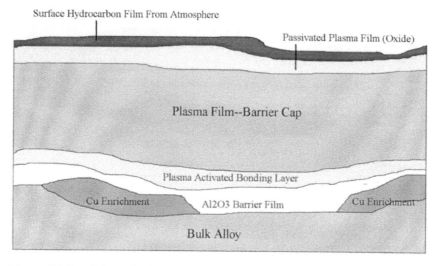

Figure 31.5 Schematic drawing of plasma film on chemically cleaned alloy.

evidenced by the conspicuous shift in XPS signal profiles of elements at the interface as shown in Figure 28.10. Attention should be focused at the plasma polymer/oxide interface, which is indicated by the asterisks.

2.3. Effect of Initial Cleaning Processes

The corrosion resistance of aluminum and Al alloys is largely due to the protective oxide film, which can attain a thickness of about 10 Å within seconds on a freshly exposed aluminum surface [8]. A good corrosion protection system should include protection of the oxide layer and, in addition, should provide a good adhesive base for subsequent paint. The conventional corrosion protection system of aluminum alloys consists of the application of a chromate conversion coating following alkaline cleaning and deoxidization of the surface. The purpose of pretreatments is to remove the contaminants and any defective oxide left after part forming, and thus create a clean surface on which chromium oxide can be grown, which then acts as the corrosion-protective layer and also the adhesive base.

2.3.1. Alclad 2024-T3, [2A]

Alclad is an aluminum alloy clad with a nearly pure aluminum. The base alloy is AA 2024-T3, and the cladding has a specified composition of, in wt %, Cu 0.1, Si + Fe 0.7, Mg 0.05, Mn 0.05, Zn 0.1, Ti 0.03, others 0.3, and Al the balance [9]. This Alclad is designated as 2A in this chapter, i.e., 2 representing 2024 T3 alloy and A representing Alclad. The pure aluminum has an excellent corrosion resistance. The clad is used to protect the base aluminum alloy from corrosion.

Alkaline cleaning is observed to leave a Cu-enriched surface, and alkaline cleaning followed by deoxidization greatly increases the level of enrichment. The extent of this copper enrichment can be seen in the spectra displayed in Figure 31.6. Although the sputter step resolution is somewhat low, it is apparent on observation that the enriched copper layer lies beneath the oxide. The maximal copper signal is at the second data point for both the alkaline-cleaned and the deoxidized samples. By this time the oxygen signal had dropped by more than half on the alkaline-cleaned sample and was less than 10% of the initial value on the deoxidized sample. This elevated copper level persists after the oxygen signal has dropped further in both samples shown. The reasons for the Cu enrichment have not been elucidated.

2.3.2. AA 2024-T3, [2B]

2024 specifications are by composition in wt %: Si 0.5, Fe 0.5, Cu 3.8–4.9, Mn 0.3–0.9, Mg 1.2–1.8, Cr 0.1, Zn 0.25, Ti 0.15, total of others 0.15, and Al the remainder [9]. This alloy is designated as 2B, i.e., 2 representing 2024 and B representing bare (without cladding). XPS surface analysis results indicated that the native surface of [2B] has a high Mg concentration, which was then associated with a decrease in its corrosion resistance. Chemical cleaning of these alloy surfaces by alkaline (Alk) and deoxidization (Dox) procedures leaves an enriched Cu concentration on the surface but reduces the Mg concentration to the bulk level or below it. XPS analysis showed that alkaline-cleaned and deoxidized surfaces have a significant concentration of Cu, while the acetone-cleaned (Ace) surfaces are composed of Mg-Al oxides with no copper. The (Alk) surfaces had very little Mg and the (Dox) surfaces had no measurable Mg, which was in higher concentrations on (Ace) surfaces. The high Mg concentration in the oxide on the (Ace) cleaned surfaces is thought to be composed of Mg oxide channels that allow the electrolyte to penetrate the oxide more easily, which could be why acetone-wiped surfaces coated with E-coat displayed larger corrosion areas.

Figure 31.7 shows the effects of each cleaning method on the thickness of the oxide layer. The oxide layer remaining after alkaline cleaning is thinner than the native oxide, and the oxide layer left on the Dox surface is thinner still.

From Figure 31.8 it is evident that after alkaline cleaning the Mg levels in the oxide have been dramatically reduced. After deoxidization, the Mg level in the oxide diminished to the point that the Mg 2p signal was undetectable, hence no trace for the deoxidized sample is shown in the figure.

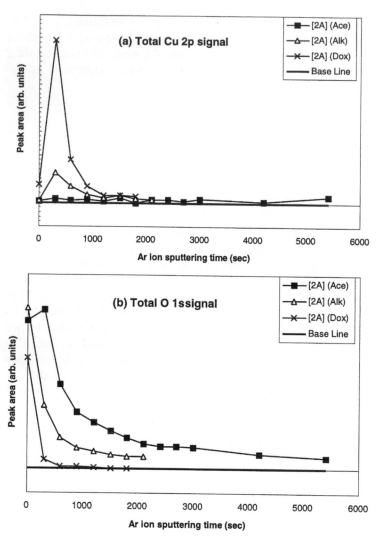

Figure 31.6 XPS depth-profile summary graphs of (a) Cu 2p and (b) O 1s peak area as a function of sputtering time.

Figure 31.9 displays the depth summary generated by the Cu 2p 3/2 spectral area. It is apparent that the copper levels in the near-surface region beneath the oxide are enriched by alkaline cleaning and by alkaline cleaning followed by deoxidization. A copper-enriched surface generally causes accelerated electrochemical corrosion due to galvanic interactions with the bulk alloy. The replacement of the hydrated, mixed oxide layer with a thin oxide layer containing little Mg seems to negate some of the electrochemical activity.

2.3.3. AA 7075-T6, [7B]

2024 specifications are by composition in wt %: Cu 1.6, Mg 2.5, Cr 0.25, Zn 5.6, and Al the remainder [9]. XPS analysis of [7B] surfaces revealed that the chemical

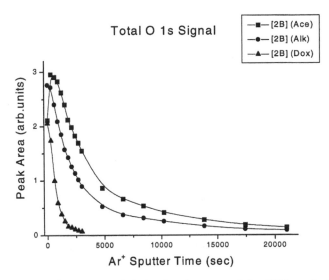

Figure 31.7   Total O 1s signal measured by XPS analysis of chemically pretreated and native (Ace) surfaces of AA 2024-T3 ([2B]).

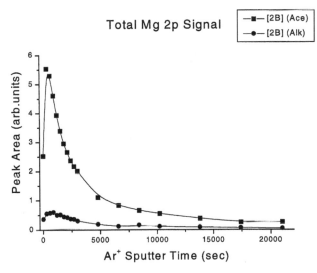

Figure 31.8   Total Mg 2p signal measured by XPS analysis of chemically pretreated and native (Ace) surfaces of AA 2024-T3 ([2B]); the Mg 2p signal from the (Dox) surface is not shown in this figure, because the intensities were below detection limits.

cleaners change the elemental composition of the surface and could have an important role in the corrosion performance of the alloy. XPS depth profiles of the alloy after each of the three cleaning techniques are shown in Figure 31.10. It becomes apparent that the copper levels in the near-surface region beneath the oxide are dramatically enhanced and thus could be a significant factor in local galvanic activity. While the Cu levels increase after each of the chemical cleaning techniques, the enriched magnesium-containing oxide structure seen on the native,

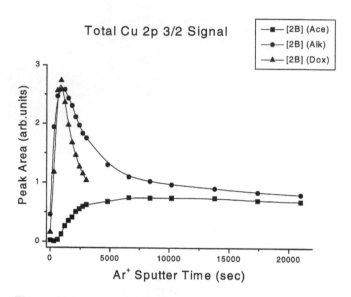

Figure 31.9   Total Cu 2p 3/2 signal measured by XPS analysis of chemically pretreated and native (Ace) surfaces of AA 2024-T3 ([2B]).

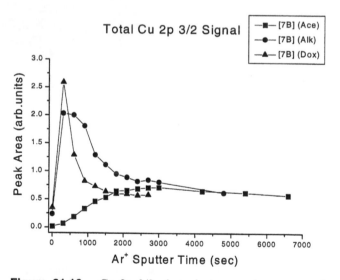

Figure 31.10   Cu 2p 3/2 photoelectron peak area as a function of sputtering time for AA7075-T6 ([7B]) after each of three chemical pretreatments.

as-received surface is removed and replaced with an oxide having no significant Mg content as shown in Figure 31.11.

Figure 31.12 indicates that the chemical cleaners do indeed leave thinner oxides on the surface, with the deoxidizer leaving the thinnest structure. Figure 31.13 indicates that the alkaline cleaner alone leaves a highly zinc-enriched structure on the surface. This zinc enrichment appears to be in the form of an oxide, which is probably hydrated on its topmost surface.

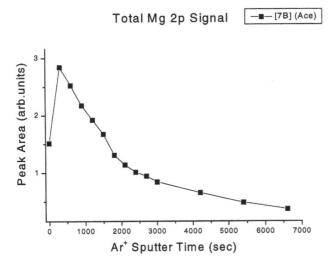

Figure 31.11 Mg 2p photoelectron peak area as a function of sputtering time for AA7075-T6 ([7B]) after each of three chemical pretreatments; no signal was received from (Alk) or (Dox) surfaces.

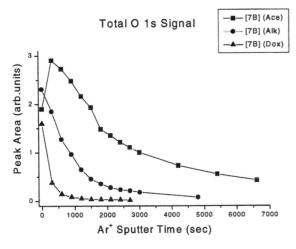

Figure 31.12 O 1s photoelectron peak area as a function of sputtering time for AA7075-T6 ([7B]) after each of three chemical pretreatments.

While a copper-enriched surface has the implication of always causing accelerated electrochemical corrosion, replacing the native, hydroxylated, mixed Al-Mg oxide layer with a thin stable oxide layer seems to allow the plasma films to tightly adhere to the alloy surface. This adhesion, coupled with the barrier properties of the films, appears to provide additional protection of the oxide layer from contact with corrosive agents.

The thickness of the oxide layer remaining after alkaline cleaning is reduced from that of the native oxide. Alkaline cleaning is performed at an elevated

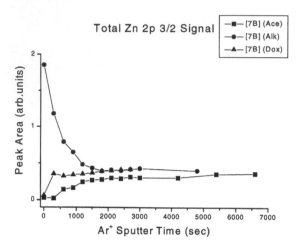

Figure 31.13 Zn 2p 3/2 photoelectron peak area as a function of sputtering time for AA7075-T6 ([7B]) after each of three chemical pretreatments.

temperature (65°C) and may therefore leave a thicker barrier-type oxide, covered by a hydrated oxide layer. Both of these layers have little or no magnesium incorporation after the treatment. Deoxidization after alkaline cleaning substantially reduces the total oxide thickness even further.

Using the sputter rate calculated from a Ta_2O_5 sample, the native oxide is calculated to be just over 20 nm thick. The oxides left after either chemical cleaning method were thin enough to see metallic contributions in both the Al 2p and KLL spectra, with the (Dox) sample having the largest metallic contribution, indicating that the oxides were in the range of just a few nanometers after chemical cleaning.

Deoxidization fully strips the native, magnesium-containing oxide, whereas less aggressive alkaline cleaning was seen to occasionally leave some magnesium-containing oxide complex on sample surfaces. Figure 31.11 only shows the magnesium intensity from a native, acetone-wiped surface because no appreciable Mg 2p signal was measured on either the (Alk) or the (Dox) surfaces that were depth profiled in this set.

The chemical state of aluminum on the surface has a multitude of possible configuration designations. The state in the hydroxylated outer layer corresponds to various mineral phases such as AlO(OH) (boehmite), $Al(OH)_3$, having a modified Auger parameter of 1460.6 on the acetone-cleaned surface and 1461.4 on both the (Alk) and (Dox) surfaces. When capped with a plasma polymer, depth profiles show that the state of the aluminum is seen to be consistent with the many oxides, as well as mixed states with plasma film components.

3. POLARIZATION RESISTANCE

Linear polarization resistance (R_p) is defined as the charge transfer resistance of the solution–metal interface. The linear polarization technique was employed to measure the R_p values of the Al alloy surfaces after different pretreatments. Polarization

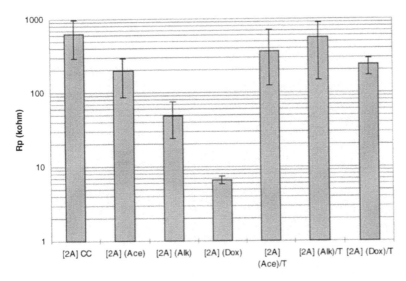

Figure 31.14 Polarization resistance of [2A] panels with different chemical pretreatments and TMS plasma polymer–coated surfaces.

resistance (R_p) was used to evaluate the effect of chemical cleaning and plasma polymer coating on corrosion resistance of aluminum alloys. Each panel was masked with insulating tape so as to expose only a square region of dimension 3×3 cm to the electrolyte aqueous salt solution [aqueous salt solution of 0.5% NaCl + 0.35% $(NH_4)_2SO_4$].

Figure 31.14 shows the R_p values of [2A] with three different chemical pretreatments and with a TMS plasma polymer on each of the three pretreated surfaces, as well as on the control [2A]CC surfaces (chromate conversion–coated 2A). It can be seen that the R_p values of [2A] were decreased to some extent by pretreatment of alkaline cleaning and were drastically reduced by alkaline cleaning plus deoxidization. As observed in the XPS results, the accumulation of Cu elements and removal of oxide layer on [2A] surfaces were presumed responsible for the reduction in corrosion resistance of these chemically pretreated [2A] panels.

In contrast, a significant increase of the R_p values was observed with the application of a thin layer of TMS plasma polymers (about 50 nm) on these chemically treated [2A] surfaces. It was also noted that these TMS plasma polymer–coated [2A] samples have the same level of R_p values as the [2A]CC controls. These results clearly indicate that these plasma polymer coatings have a good corrosion resistance property.

It was also noticed that a simple solvent (acetone) cleaning of [2A] surfaces gave higher R_p values than alkaline cleaning and deoxidization processes. As confirmed by XPS data, it is evident that the [2A] surface, which has a stable aluminum oxide layer with a minimum of alloying elements beneath the oxide layer or penetrating through it, should not be damaged. The acetone cleaning, which removed only surface contamination but did not induce any surface composition change, seems to be a suitable surface pretreatment method for [2A]. With the aid of low-temperature plasma interface engineering, therefore, an excellent corrosion

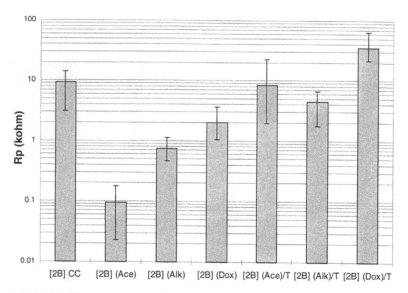

Figure 31.15 Polarization resistance of AA 2024-T3 ([2B]) panels with different chemical pretreatments and plasma polymer coated surfaces.

protection performance was anticipated for these acetone-cleaned [2A] samples. A short plasma sputter cleaning was employed on the acetone-wiped surface to remove tougher organic contaminants and to promote plasma polymer adhesion.

In contrast to Alclad 2024-T3, bare AA 2024-T3 has a significantly lower (three orders of magnitude) R_p value than pure aluminum (Alclad 2024-T3). Alkaline cleaning and alkaline cleaning plus deoxidization increase the R_p value, indicating that chemical treatment of the surface is necessary for this alloy. Figure 31.15 shows the polarization resistance of the [2B] panels with different chemical pretreatment and with a TMS plasma polymer deposited on them. The alkaline-cleaned (Alk) and alkaline-cleaned, deoxidized (Dox) surfaces had higher polarization resistance than acetone-wiped (Ace) surfaces, which implies that these surfaces showed higher corrosion resistance. This supports the XPS finding that chemical cleaning of [2B] eliminated the channels of magnesium-rich, mixed oxides that penetrate the aluminum oxide on the native surface, replacing this native structure with a more stable barrier oxide layer. The native oxide on [2B] had a higher Mg concentration than the bulk composition; this Mg incorporated in the native oxide most likely diffused during solution heat treatment in manufacturing of the alloy panels. Though the Cu concentration on [2B] surfaces increased after alkaline cleaning and after alkaline cleaning followed by deoxidization, the Mg concentration decreased to the level of base alloy elemental concentrations. These more stable oxide layers improved the adhesion of plasma polymers.

Plasma polymer–coated [2B] surfaces showed higher polarization resistance than native and chemically cleaned surfaces. Thus, the corrosion resistance of plasma polymer–coated [2B] was much higher than that of the barrier-type oxides formed after chemical cleaning. Also, as is evident from the higher polarization resistance of the plasma polymers, they are good barriers to water, oxygen, and corrosive species, even under an externally applied potential. The R_p values of the

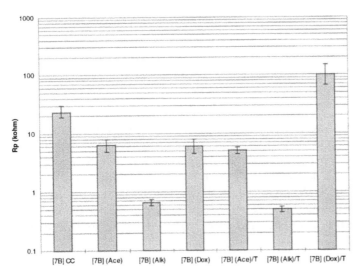

Figure 31.16 Polarization resistance of 3 cm × 3 cm-exposed regions of AA 7075-T6 ([7B]) panels with different chemical pretreatments and plasma polymer coated surfaces.

plasma polymer–coated surfaces of [2B] were nearly equal to or higher than those of the chromate conversion–coated [2B]CC surfaces.

Figure 31.16 shows the polarization resistance of [7B] with pretreatments of acetone wiping, alkaline cleaning, deoxidization, and plasma polymer deposition. Whereas deoxidized (Dox) surfaces show a level of polarization resistance similar to that of acetone-wiped (Ace) surfaces of this alloy, the polarization resistance of alkaline-cleaned (Alk) surfaces is dramatically lower. This decreased R_p value on (Alk) surfaces correlates with the observation of higher Zn concentrations remaining on the surfaces after the (Alk) pretreatment. Although the Cu concentration is slightly higher after (Dox) than (Alk), the Zn enrichment is eliminated after the deoxidization step. Something associated with this higher Zn concentration on the (Alk) surfaces seems to decrease the polarization resistance of these surfaces.

Deoxidized surfaces of [7B] with a plasma polymer coating ([7B] (Dox)/T) showed higher polarization resistance than the chemically deoxidized surfaces without a plasma polymer. This indicates that the added corrosion resistance offered by plasma polymer films is much higher than that of the barrier-type oxides, formed after chemical cleaning, alone. As compared to the chromate conversion–coated surfaces ([7B] CC), the deoxidized and plasma polymer–coated ([7B] (Dox)/T) surfaces showed higher R_p values, suggesting that these surfaces have higher corrosion resistance.

4. CORROSION TEST RESULTS

Two types of corrosion evaluation tests, SO_2 and Prohesion salt spray tests, were employed for the evaluation of corrosion protection characteristics of painted plasma systems. The SO_2 salt spray test was chosen to speed up differentiation of the corrosion protection properties of the different systems investigated. The Prohesion cyclic salt spray test, which is chemically milder than the SO_2 salt spray test, was conducted for a longer period, 2000 h. It is considered a more realistic test, as it better simulates actual service conditions of an aircraft in which both wet and dry periods occur.

4.1. [2A]

Figure 31.17 shows typical scanned images of SO_2 salt spray–tested panels, two controls and two plasma polymer–treated panels. By visual observation, one can easily see that the corrosion performance of the plasma polymer–treated panels, [2A](Ace/O/N)/TN/E and [2A](Dox/AH)/TH/E, is far better than that of the control panels, [2A]CC/E and [2A]CC/A. Figure 31.18 shows typical scanned images of the surfaces of controls and plasma interface–engineered systems of [2A](Ace/O)/TH/E and [2A](Ace/O)/TN)/E after Prohesion salt spray corrosion testing and subsequent

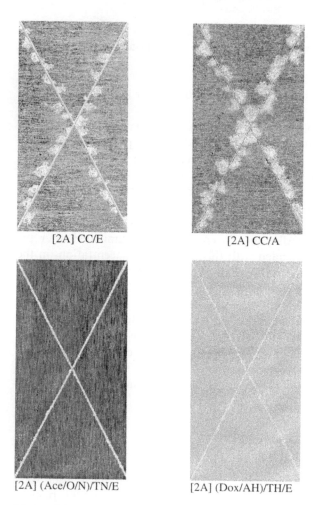

Figure 31.17   Scanned images of SO_2 salt spray–tested (4 weeks) [2A] panels; total scanned area 27 cm^2 and total scribe length within the scanned area 16 cm.

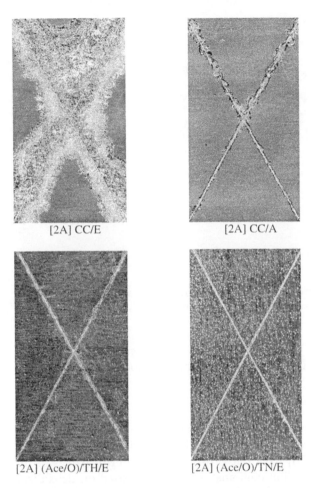

[2A] CC/E   [2A] CC/A

[2A] (Ace/O)/TH/E   [2A] (Ace/O)/TN/E

Figure 31.18 Scanned images of Prohesion salt spray–tested (12 weeks) [2A] panels; total scanned area 27 cm², total scribe length within the scanned area 16 cm: the white spots on [2A] (Ace/O)/TN/E are paint which could not be stripped with Turco paint stripper (not pitting-corrosion sites).

E-coat stripping. Both plasma–treated panels show excellent corrosion protection performance as compared to the control panels. All [2A] panels with different plasma treatments and plasma polymer coatings, which were corrosion tested in both SO_2 and Prohesion salt spray tests, were similarly scanned, and the corrosion width was evaluated by using a scanned image and computer calculation of the corroded area. Figure 31.19 compares the corrosion width obtained by the two methods.

4.2. [2B]

Figures 31.20 and 31.21 show the scanned images of SO_2 and Prohesion salt spray–tested panels of [2B], respectively. Visual observation of these images reveals that panels that were only acetone wiped and E-coated ([2B](Ace)/E) provided poor corrosion resistance. In contrast, the plasma-modified [2B] panels showed excellent corrosion resistance even after 12 weeks of exposure to Prohesion salt spray,

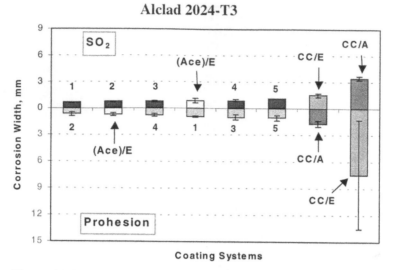

Figure 31.19   Comparison of corrosion width obtained by two salt spray tests; the number indicates the sample in ascending order observed in SO_2 salt spray test.

outperforming both controls of [2B]CC/E and [2B]CC/A. Figure 31.22 shows the comparison of the corrosion width obtained by the two methods.

4.3.   [7B]

Figures 31.23 and 31.24 show typical scanned images of SO_2 and Prohesion salt spray–tested [7B] panels, respectively. Visual observation of these images reveals that the plasma-modified panels of [7B] have outperformed both control panels in the SO_2 salt spray test. These plasma film combinations were prepared on deoxidized [7B] surfaces without any plasma cleaning pretreatment. Figure 31.23 also shows an image of a panel that had simply been deoxidized prior to the application of E-coat, which performed excellently in the SO_2 salt spray test. Figure 31.25 compares the corrosion width obtained by the two methods. The comparisons shown in Figures 31.19, 31.22, and 31.25 indicates that the results obtained by the two methods do not match, partly due to the different duration of tests, and that samples which show good results in one test do not do as well in the other test.

　　　Prohesion salt spray–tested panels in Figure 31.24 show that [7B] (Alk/AH)/T/ E and [7B] (Alk/O)/TH/E systems performed comparably to the controls. Deft primer–coated control panels ([7B] CC/A) displayed extensive pitting corrosion away from the scribe in both tests, indicating that Deft primer may have poor barrier properties. This pitting corrosion away from the scribe was observed on both controls when examined by scanning electron microscopy (SEM).

5.   CHROMATED AND NONCHROMATED SPRAY PAINTS ON PLASMA POLYMER SURFACES

The aluminum alloys that have been investigated in the previous sections, including [2B], [7B], and [2A], are materials used for aircraft skins. Aircraft skins are readily

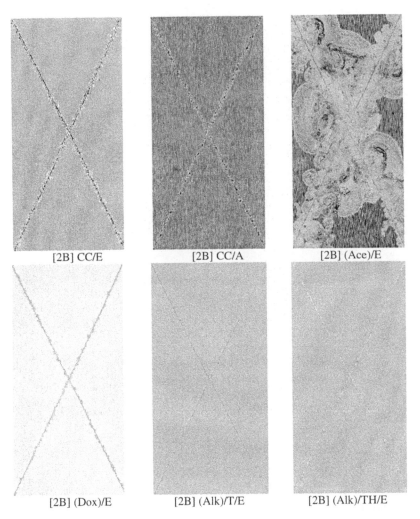

[2B] CC/E   [2B] CC/A   [2B] (Ace)/E

[2B] (Dox)/E   [2B] (Alk)/T/E   [2B] (Alk)/TH/E

Figure 31.20 Scanned images of SO_2 salt spray–tested (4 weeks) panels of AA 2024-T3 ([2B]); total scanned area $27\,cm^2$, total scribe length within the scanned area $16\,cm$.

accessible for inspection and conventional repair, but various internal structural components are neither easily accessible nor easily remedied after the onset of corrosion. Because of the tenacious adhesion of plasma coatings, plasma processes may be more useful in difficult-to-inspect/repair areas, i.e., internal structures of an aircraft. Thus, this method is aimed at the corrosion protection of detailed parts rather than easily accessible aircraft skin. The potential long service life protection offered by these plasma-based systems appears to fit well with the needs of particular detailed parts and internal structural components that cannot be addressed in standard maintenance cycles. Therefore, another two aluminum alloys, i.e., Plate stock AA2124-T851 ([2P]) and Plate stock AA7050-T7451 ([7P]), which are usually used for internal structural parts of aircraft, were introduced and investigated.

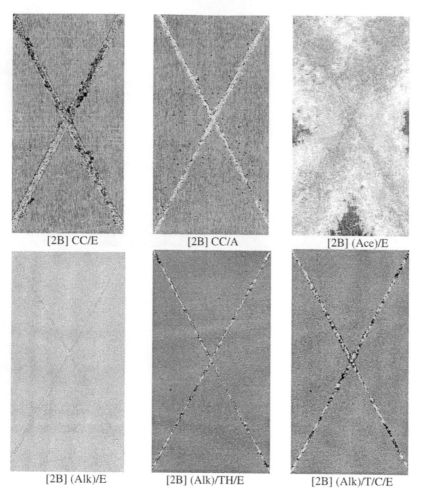

Figure 31.21 Scanned images of Prohesion salt spray–tested (12 weeks) panels of AA 2024-T3 ([2B]); total scanned area 27 cm^2, total scribe length within the scanned area 16 cm.

A PPG cathodic E-coat, designated E2, was used. The chromated spray primers employed were water-borne Deft 44-GN-36 (A) or 44-GN-72 (A1) (Deft Corporation, Irvine, CA) and solvent-borne Courtauld 519X303 (G) (Courtauld Aerospace, Glendale, CA). The nonchromated spray primers were water-borne Dexter 10-PW-22-2 (X) (Dexter Corporation, Waukegan, IL) and water-borne Spraylat EWAE118 (D) (Spraylat Corporation, Chicago, IL).

The plasma deposition step when the anode assembly was removed in flow system is designated by the process code Tfs. In the plasma deposition in a closed reactor system, the reactor chamber was first pumped down to less than 1 mtorr. The reactor chamber was then isolated from the pumping system by closing the main valve located in between. TMS gas, controlled by an MKS mass flow meter (model 247C), was then fed into the reactor. After the system pressure reached a preset point, TMS gas feeding was stopped and DC power was applied to initiate the glow

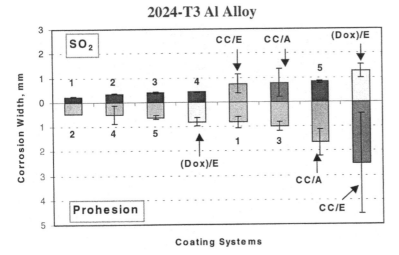

Figure 31.22 Comparison of corrosion width obtained by two salt spray tests; the number indicates the sample in ascending order observed in SO_2 salt spray test.

discharge to start cathodic polymerization. The samples prepared by this process contain the process code Tcs.

The main aim of SAIE is the complete elimination of heavy metals from the coating systems; an approach that primarily relies on tenacious water-insensitive adhesion and good barrier characteristics of a primer has been taken. It should be pointed out that this approach is theoretically incompatible with the approach that utilizes the primers with corrosion inhibitors, e.g., chromated primers. This is because a primer with super barrier characteristics would not allow the migration of inhibitors and would not provide enough water for their electrochemical reaction to form corrosion protection products, as described in Chapter 28. In order to further elucidate the SAIE concept, both *chromated* and *nonchromated* spray primers were employed to generate two types of plasma coating–modified systems, and their corrosion protection behaviors were investigated in this study.

5.1. Chromated Primer Coating Systems

Table 31.1 summarizes the adhesion test results of chromated primers on plasma polymer–coated aluminum alloys. It can be seen that direct application of TMS plasma coatings to Al alloys did not give good primer adhesion. However, an appropriate combination of plasma coatings of TMS followed by hexafluoroethane (C_2F_6, abbreviated as HFE), designated as T/F in Table 31.2, remarkably increased the primer adhesion of Al alloys. The paints applied to T/F plasma–treated Al alloys could not be stripped by the conventional, commercial Turco 5469 paint-stripping solution. The strong adhesion achieved with primers applied to T/F plasma–treated Al alloys was also water insensitive. As listed in Table 31.2, a wet adhesion test, which is the standard tape test performed after boiling the painted specimen, with

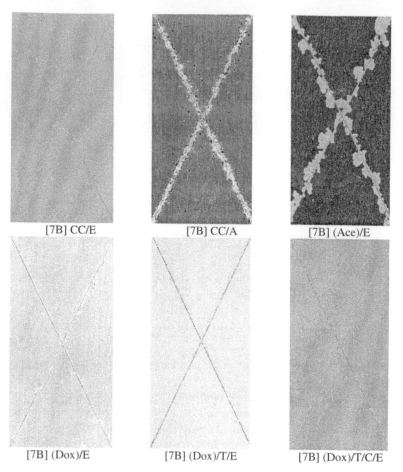

Figure 31.23   Scanned images of SO_2 salt spray–tested (4 weeks) panels of AA 7075-T6 ([7B]); total scanned area $27\,cm^2$, total scribe length within the scanned area $16\,cm$.

prior cross-cuts, in water for up to 8 h, gave the highest tape test ratings of 5. In other words, tenacious and water-insensitive adhesion has been achieved between these primers and the plasma–treated Al alloys.

The poor primer adhesion of the as-deposited TMS plasma coating could be ascribed to its low surface energy (with a water surface contact angle of 120 degrees) and also the possible existence of oligomers on the surface. However, the T/F plasma coating that shows a very good adhesion also has low surface energy, with a water surface contact angle of about 110 degrees. Therefore, the existence of a certain amount of oligomers on the as-deposited TMS plasma-coating surface, which acts as a weak boundary layer for primer adhesion, seems to be the main reason for its poor primer adhesion performance. The reduction of dangling bonds by HFE treatment, described in Chapter 6, strongly supports this interpretation.

The XPS information indicated that the HFE plasma polymerization process deposited a film of roughly 2 nm on the TMS coating, which is itself about 50 nm thick. The plasma polymerization process using HFE to form the thin surface

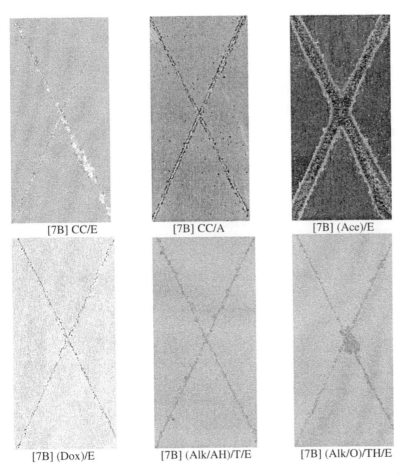

| [7B] CC/E | [7B] CC/A | [7B] (Ace)/E |
| [7B] (Dox)/E | [7B] (Alk/AH)/T/E | [7B] (Alk/O)/TH/E |

Figure 31.24 Scanned images of prohesion salt spray–tested (12 weeks) panels of AA 7075-T6 ([7B]); total scanned area 27 cm^2, total scribe length within the scanned area 16 cm.

plasma coating on top of the deposited plasma polymer of TMS helps to remove and/or fix the oligomeric products that existed on the as-deposited TMS coating surface, and thus significantly improves the primer adhesion performance.

The surface state of a metal is another important factor that will greatly influence the corrosion protection performance of a coating system. As presented in previous sections of this chapter, a proper surface preparation of the Al alloys not only affects the adhesion performance of the coating system but also changes the corrosion protection characteristics of the Al alloy. Figure 31.26 shows the R_p values of several kinds of Al alloys with different chemical pretreatments and plasma T/F–coated surfaces. Deoxidization (Dox) is a very suitable surface preparation for [2B] and [7B] because it increased the corrosion resistance of these Al alloys. As described earlier, either alkaline (Alk) or deoxidization (Dox) degraded the corrosion resistance of Alclad Al alloy of [2A] and gave worse corrosion results in the salt spray tests. Alclad Al alloys of [2A] and [7A] shown here were not treated with these chemical cleaning processes prior to plasma surface treatment. In most cases,

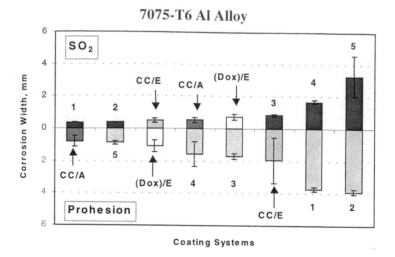

Figure 31.25 Comparison of corrosion width obtained by two salt spray tests; the number indicates the sample in ascending order observed in SO₂ salt spray test.

Table 31.1 Adhesion Test Results of Chromated Spray Primers [Deft 44-GN-36 (A) and Courtauld 519X303 (G)] to Al Alloys Prepared with Chemical Cleanings and Plasma Surface Treatments

		Primer A			Primer G		
Substrate	Surface preparation	Tape test	Boiling 1, 4, 8 h	Turco time	Tape test	Boiling 1, 4, 8 h	Turco time
[2B]	(Alk)	5	3, 3, 3	~5 min	5	5, 5, 5	~30 min
	(Alk)/T	0	—	—	0	—	—
	(Alk)/T/F	5	5, 5, 5	>24 h	5	5, 5, 5	>24 h
[7B]	(Dox)	4	4, 4, 4	~30 min	5	4, 4, 4	~30 min
	(Dox)/T	0	—	—	0	—	—
	(Dox)/T/F	5	5, 5, 5	>24 h	5	5, 5, 5	>24 h
[2A]	(Ace)	3	—	~5 min	5	4, 4, 4	~5 min
	(Ace/O)/T	0	—	—	0	—	—
	(Ace/O)/T/F	5	5, 5, 5	>24 h	5	5, 5, 5	>24 h
[7A]	(Ace)	3	—	~5 min	5	4, 4, 4,	~5 min
	(Ace/O)/T	0	—	—	0	—	—
	(Ace/O)/T/F	5	5, 5, 5	>24 h	5	5, 5, 5	>24 h

Scale: 0–5 indicates poor (0) to excellent (5) performance.

the application of a thin layer of T/F plasma polymers increased the corrosion resistance of the Al alloys.

Figure 31.27 shows the comparison of average corrosion widths of (1) SO₂ salt spray–tested and (2) Prohesion salt spray–tested Al alloy panels and their corresponding control panels. After 4 weeks of SO₂ salt spray test, most of the chromated primer–coated [2B] and [7B] panels, including those with excellent

Table 31.2 Adhesion Test Results of Chromated Primers [Deft 44-GN36 (A) and 44-GN-72 (A1)] to No Anode Assembly Plasma–Treated 7A(Ace/O)

Plasma systems	Plasma coatings	Primer A			Primer A1		
		Tape test	Boiling 1, 4, 8 h	Turco time	Tape test	Boiling 1, 4, 8 h	Turco time
Flow	Tfs	0	—	< 5 min	0	—	~ 6 min
	Tfs/F	5	5, 5, 5	> 24 h	5	5, 5, 5	> 24 h
	Tfs/ (Ar)	5	5, 4, 3	~ 20 min	5	5, 5, 3	~ 1.5 h
Close	Tcs	5	4, 3, 3	~ 5 min	5	3, 3, 3	~ 10 min
	Tcs/(Ar)	5	4, 3, 2	~ 20 min	5	5, 4, 4	> 24 h

Scale: 0–5 indicates poor (0) to excellent performance.

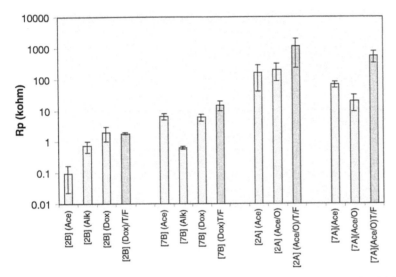

Figure 31.26   Rp values of AA2024-T3 ([2B]), AA7075-T6 ([7B]), Alclad 2024-T3 ([2A]), and Alclad 7075-T6 ([7A]) panels with different chemical pretreatments and plasma polymer–coated surfaces.

adhesion achieved by plasma T/F, showed much larger corrosion widths than their controls ([2B]CC/A and [7B]CC/A). This result evidently proved that the application of chromated primers with tenacious adhesion to plasma T/F–treated [2B] and [7B] could not provide corrosion protection as good as the control systems. In contrast, with the same SO_2 salt spray test as shown in Figure 31.27(a), chromated primer–coated [2A] and [7A] gave better or comparable corrosion results to their controls. These test results reflect the working principle of SAIE, i.e., the combination of the two best processes does not necessarily yield the better results, unless each process is optimized for the new combination.

It can be clearly seen that the tenacious and water-insensitive adhesion achieved by plasma T/F are differentiated in the results of the 12-week Prohesion

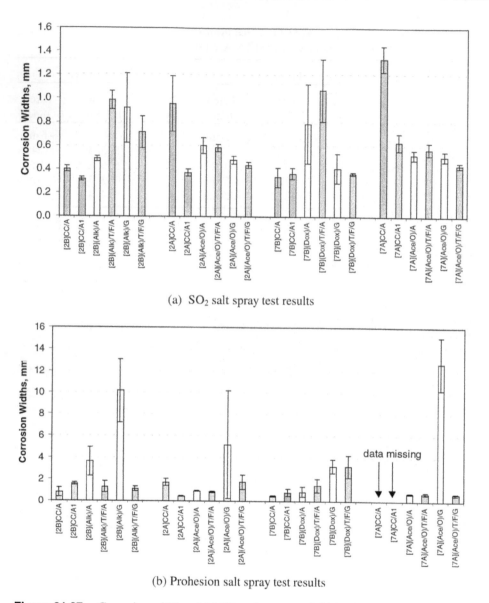

(a) SO₂ salt spray test results

(b) Prohesion salt spray test results

Figure 31.27 Corrosion widths of (a) SO₂ salt spray and (b) Prohesion salt spray–tested Al panels with chromated plasma coating systems prepared by anode magnetron plasmas and their corresponding chromated controls.

salt spray test. These chromated primer coating systems with good adhesion, including those either achieved by chromate conversion coating designated as (CC)/A or plasma T/F coating designated as T/F/A in Figure 31.27b, showed very similar corrosion widths after the test. In contrast, the chromated primer coating systems without good adhesion, which were obtained with direct application of primers to the Al alloy surfaces, showed very large corrosion widths after the test.

After SO_2 and Prohesion salt spray tests, severe pitting corrosion was found on almost all the chromated primer coated [2B] and [7B] panels but not on [2A] and [7A] panels. Figure 31.28 shows the typical scanned images of Prohesion salt spray–tested [2B] and [7B] Al alloy panels. Although a strong and water-insensitive adhesion exists on these, as noted in the adhesion testing results in Table 31.1, the plasma coating systems with chromated primers could not prevent the occurrence of pitting corrosion on [2B] and [7B] surfaces. Chromated primer is good for corrosion protection when combined with chromate conversion coatings, but not in the absence of chromate conversion coatings. Pitting corrosions are due to the poor barrier characteristics of the chromated primers.

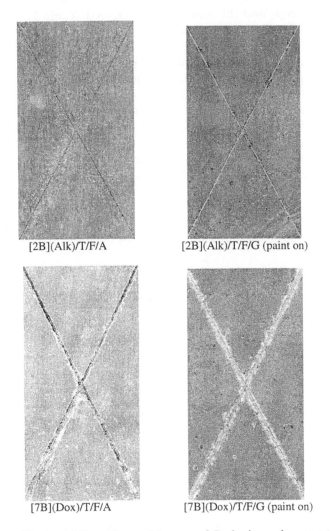

[2B](Alk)/T/F/A   [2B](Alk)/T/F/G (paint on)

[7B](Dox)/T/F/A   [7B](Dox)/T/F/G (paint on)

Figure 31.28 Scanned images of Prohesion salt spray–tested Al panels with chromated plasma coating systems prepared by anode magnetron plasmas: Primer G could not be removed from the whole Al panel due to the strong adhesion, and only the portion with pitting corrosion occurred underneath could be stripped off with paint stripper solution.

5.2. Nonchromated Primer Coating Systems

The complete elimination of heavy metals and other hazardous compounds from the coating systems was the main objective of SAIE. Chromated spray primers with hazardous chromate components do not fit such an objective. Therefore, two kinds of nonchromated and water-borne spray primers, Spraylat EWAE118 (D) and Dexter 10-PW-22-2 (X), were selected to produce chromate-free plasma coating systems for corrosion protection of Al alloys.

DC cathodic polymer polymerization was modified and carried out without using an anode assembly, i.e., with the reactor wall acting as a grounded anode. In order to investigate the consequences of this type of plasma deposition process, DC cathodic polymerization of TMS on Al alloy surfaces was carried out in both flow and closed reactor deposition systems, and the results were compared. Table 31.2 summarizes the adhesion test results of the chromated spray primers to no anode assembly plasma–treated [7A] panels. As compared to Table 31.1, it can be seen that the T/F plasma coatings produced with no anode assembly plasmas gave similar primer adhesion to those T/F plasma polymers obtained with anode magnetron plasmas. As noticed in Table 31.1, some other plasma coating systems, such as plasma Tfs/(Ar) and Tcs/(Ar), also provided excellent primer adhesion.

Since it was observed that fluorine contamination was a possibility and had potentially detrimental effects as described in Chapter 10, the excellent primer adhesion achieved with Tfs/(Ar) and Tcs/(Ar), shown in Table 31.3, has significant importance in the practical application of the plasma technique without any of the potentially deleterious effects of fluorine-based systems. Argon plasma treatments on both flow system TMS (Tfs) and closed system TMS (Tcs) polymers were then investigated as an additional system modification that could provide strong adhesion without the incorporation of fluorine-containing monomers in the quest to produce chromate-free coatings systems.

Table 31.3 summarizes the adhesion test results of nonchromated primers [Spraylat EWAE118 (D) and Dexter 10-PW-22-2(X)] to plasma coatings deposited without an anode assembly, which were produced in both flow and closed system deposition processes. As noted in Table 31.3, closed system TMS plasma polymers (Tcs) showed superior primer adhesion performance to similar ones obtained from a flow system (Tfs). Similar to chromated primers, summarized in

Table 31.3 Adhesion Test Results of Nonchromated Primers [Spraylat EWAE118 (D) and Dexter 10-PW-22-2 (X)] to No Anode Assembly Plasma–Treated 7A(Ace/O)

Plasma systems	Plasma coatings	Primer D			Primer X		
		Tape test	Boiled 1, 4, 8 h	Turco	Tape test	Boiling 1, 4, 8 h	Turco
Flow	Tfs	2	—	—	3	0, —	~10 min
	Tfs/(Ar)	5	5, 3, 3	~15 min	5	5, 5, 5	~13 h
Close	Tcs	5	4, 4, 4	~12 h	5	5, 5, 5	~30 min
	Tcs/(Ar)	5	5, 5, 5	>24 h	5	5, 5, 5	>24 h

Scale: 0–5 indicates poor (0) to excellent (5) performance.

Table 31.1, tenacious and water-insensitive primer adhesion was always achieved with closed system TMS plasma polymers treated with subsequent Ar plasma applications [Tcs/(Ar)].

Figure 31.29 summarizes the corrosion widths along the scribed lines that were calculated from (1) SO_2 salt spray–tested and (2) Prohesion salt spray–tested Al alloy panels and their corresponding control panels. As seen from Fig. 31.29, the corrosion test results showed that the plasma coating systems based on the chromate-free spray primers provided excellent corrosion protection for the Al alloys studied.

All the plasma–coated Al panels, including those prepared in flow and closed reactor systems, showed comparable corrosion widths to, and in many cases much lower corrosion widths than, their corresponding controls. It is also seen in Figure 31.29 that the plasma coating systems prepared in both flow (Tfs) and closed (Tcs) reactor systems showed very similar corrosion test results after the salt spray tests. The application of these chromate-free plasma coating systems to internal structural Al alloys ([2P] and [7P]) gave very small corrosion widths, which were less than 0.2 mm after 4 weeks of SO_2 salt spray testing and less than 0.3 mm after 12 weeks of Prohesion salt spray testing. These results indicated that the plasma coating system combinations explored in this study are very suitable for the corrosion protection of aircraft internal structural Al alloys.

It should be pointed out that the pitting corrosion, which had severe occurrence on chromated primer coated [2B] and [7B] panels shown in Figures 28.3 and 28.4 respectively, was significantly reduced and, in most cases, eliminated with the application of nonchromated primers for the chromate-free plasma coating systems. Figure 31.30 shows the typical scanned images of Prohesion-tested [2B], [7B], [2P], and [7P] Al panels, which were protected with chromate-free plasma coating systems during the tests. A very few small pits were observed on the [2B] surface. There were no pits on [7B], [2P], and [7P] Al surfaces after both SO_2 salt spray and Prohesion salt spray testing. In Figure 31.30, the primers are still in place on the [2P] and [7P] panels because they had such strong adhesion to the plasma treated Al surfaces that the Turco paint-stripping solution could not strip them, even with prolonged application for several days. The results shown in Figures 31.29 and 31.30 demonstrate that the chromate-free plasma coating systems not only provide excellent corrosion protection on a damaged (scribed) Al alloy surface but also prevent the occurrence of pitting corrosion of an undamaged Al alloy surface (away from the scribe lines). Since water-borne primers (primer X and D) was used, the VOC problem, which is often encountered in industrial coating process, can be also avoided with application of the LCVD coating systems.

The mere combination of a chromated primer (with excellent corrosion inhibitors) and tenacious adhesion, which was provided by plasma interface modification, did not provide good corrosion protection of Al alloys. In spite of strong primer adhesion to the Al surface, severe pitting corrosion occurred on these chromated primer–coated [2B] and [7B] panels in both 4 weeks of SO_2 and 12 weeks of Prohesion salt spray tests. In contrast, the chromate-free plasma coating systems, which were produced based on the SAIE approach by combining the superior barrier performance of a primer with tenacious and water-insensitive adhesion, not only provide excellent corrosion protection on damaged (scribed) Al alloy surfaces but also prevent the occurrence of pitting corrosion of undamaged Al alloy surfaces (away from the scribe lines). These results demonstrate that the approach based on

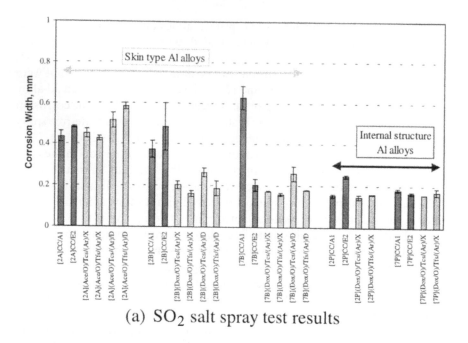

(a) SO₂ salt spray test results

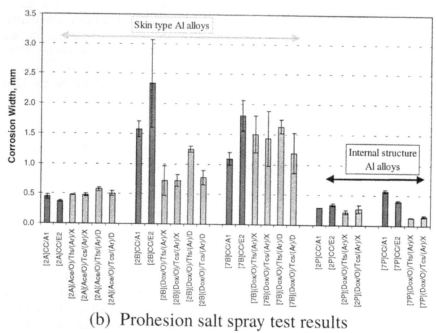

(b) Prohesion salt spray test results

Figure 31.29 Corrosion widths of (a) SO₂ salt spray and (b) Prohesion salt spray–tested Al panels with chromate-free plasma coating systems prepared without using anode assembly and their corresponding chromated controls.

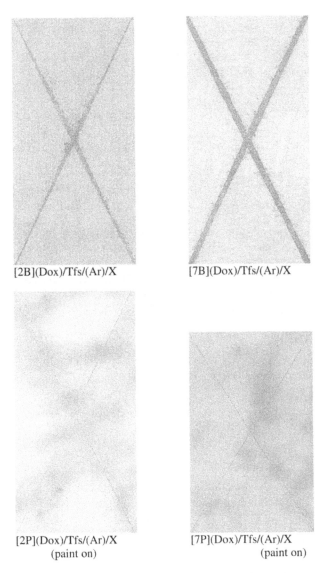

[2B](Dox)/Tfs/(Ar)/X   [7B](Dox)/Tfs/(Ar)/X

[2P](Dox)/Tfs/(Ar)/X   [7P](Dox)/Tfs/(Ar)/X
     (paint on)   (paint on)

Figure 31.30 Scanned images of Prohesion salt spray–tested Al panels coated with nonchromated primers prepared in a flow or closed reactor system without using anode assembly: Primer X could not be removed from plasma–treated [2P] and [7P] surface due to the strong adhesion.

the adhesion and barrier principle and that based on electrochemical corrosion protection are incompatible.

REFERENCES

1. Reddy, C.M.; Yu, Q.S.; Moffitt, C.E.; Wieliczka, D.M.; Johnson, R.; Deffeyes, J.E.; Yasuda, H.K. Corrosion **2000**, *56* (8), 819.

2. Yu, Q.S.; Reddy, C.M.; Moffitt, C.E.; Wieliczka, D.M.; Johnson, R.; Deffeyes, J.E.; Yasuda, H.K. Corrosion **2000**, *56* (9), 887.
3. Moffitt, C.E.; Reddy, C.M.; Yu, Q.S.; Wieliczka, D.M.; Johnson, R.; Deffeyes, J.E.; Yasuda, H.K. Corrosion **2000**, *56* (10), 1032.
4. Yu, Q.S.; Reddy, C.M.; Moffitt, C.E.; Wieliczka, D.M.; Deffeyes, J.E.; Yasuda, H.K. Corrosion **2001**, *57* (9), 802.
5. Yasuda, H.K.; Yu, Q.S.; Reddy, C.M.; Moffitt, C.E.; Wieliczka, D.M.; Deffeyes, J.E. Corrosion **2001**, *57* (8), 670.
6. Wang F. Tinghao; Yasuda, H.; Lin, T.J.; Antonelli, J.A. Prog. in Organic Coatings **1996**, *28*, 291.
7. Wang F. Tinghao; Cho, D.L.; Yasuda, H.; Lin, T.J.; Antonelli, J.A. Prog. Organic Coatings **1997**, *30*, 31.
8. Shreir, L.L.; Jarman, R.A.; Burstein, G.T., Eds. Corrosion 3rd Ed., Buttersworth-Heinemann: Boston, 1994; Vol. 2, Chap. 10.
9. ASM Specialty Handbook. Aluminum and Aluminum Alloys, American Society for Metals, 1993.
10. Yu Iriyama; Yasuda, H. J. Polym. Sci., Polym. Chem. Ed. **1992**, *30*, 1731.
11. Yasuda, H.K.; Yu, Q.S.; Reddy, C.M.; Moffitt, C.E.; Wieliczka, D.M. J. Vac. Sci., Tech. **2001**, *A19* (5), 2074.

32
Corrosion Protection of Ion Vapor–Deposited Aluminum

1. ION VAPOR–DEPOSITED ALUMINUM

Ion vapor deposition (IVD) is a vacuum process that utilizes an ion plating technique to apply a uniform and highly adherent aluminum coating on different metallic materials. IVD Al–coated aluminum alloy aircraft parts have been in service in the aerospace industry for corrosion protection [1,2]. IVD aluminum coatings are used in the aerospace industry as an environmentally friendly replacement for cadmium plating. IVD vacuum equipment is in-place at original equipment makers, military maintenance depots, and a number of coating/plating vendors. Figure 32.1 shows a picture of a typical IVD deposition chamber and associated electronics. IVD of pure aluminum has been in service for about two decades in the aerospace industry where substrate materials have typically been high-strength steels as well as aluminum and titanium alloys.

Historically, IVD was developed in the 1970s for use on fatigue-critical aircraft parts. The aluminum coating provides good fatigue resistance because it is soft and therefore less prone to serve as a crack initiation layer. Its benefits as an environmentally friendly coating have become increasingly appreciated as the use of heavy metals such as cadmium has become more highly regulated. Plating processes, such as cadmium plating, generate hazardous waste when disposal of spent plating solutions is necessary. The IVD process uses high-purity aluminum, and the waste generated is aluminum overspray and is therefore not hazardous. However, for improved corrosion protection, a chromate conversion coating is applied to the surface of aluminum deposited by IVD. The hexavalent chromium in the conversion coat provides additional corrosion protection, but disposal of the spent conversion coat solution generates hazardous waste. In addition, chromated paint is typically required on top of the conversion coating to get acceptable corrosion protection in service. From an environmental standpoint, the ideal process would be one that uses IVD aluminum but does not require the chromate conversion coating and provides good corrosion protection with nonchromated primer paint. This would minimize hazardous waste generated by the process and minimize the potential for worker exposure to harmful heavy metals.

The IVD process is similar to the familiar physical vapor deposition (PVD), with one major difference: during plating, the substrate is held at a high negative

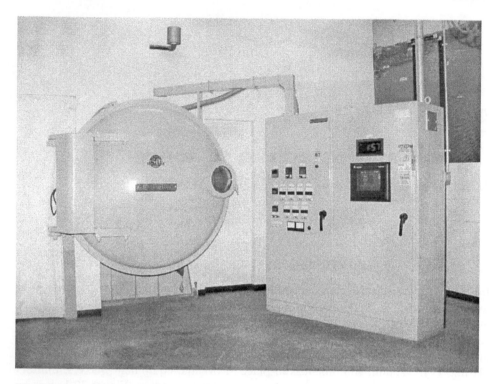

Figure 32.1   IVD deposition equipment in the production line at Boeing, St. Louis, MO.

potential ($\sim$1 kV) with respect to the vacuum chamber and evaporation source [1,2]. This potential produces a DC glow discharge of inert argon gas in the deposition chamber. Numbers of the evaporated aluminum atoms are ionized by this argon glow discharge and accelerated toward the cathode (substrate). Thus, the pure aluminum deposits under continuous bombardment of Ar^+. This produces stronger adhesion and increases the uniformity of the aluminum coating.

As applied, IVD aluminum has an open and columnar surface and therefore low density, as illustrated in Figure 32.2. Because of this open structure, IVD aluminum is susceptible to corrosion as applied. Thus, it is standard practice to use glass bead peening to densify the coating. Following the glass bead peening, chromate conversion coating is required to obtain the desired corrosion resistance and create a good adhesion base for subsequent primer coatings. However, in light of system approach interface engineering (SAIE) for corrosion protection of aluminum described in Chapter 31, the validity of this set of post-IVD processing is questionable, i.e., the combination of two bests does not necessarily produce the better result.

Since low-pressure plasma processes are carried out under vacuum, low-pressure plasma interface engineering can take advantage of the existing IVD vacuum equipment and technology. In principle, plasma interface engineering can be performed in a concurrent mode with IVD. One great benefit of the application of low-pressure plasma to IVD surfaces is the elimination of postdeposition peening and environmentally hazardous chromate conversion coating and

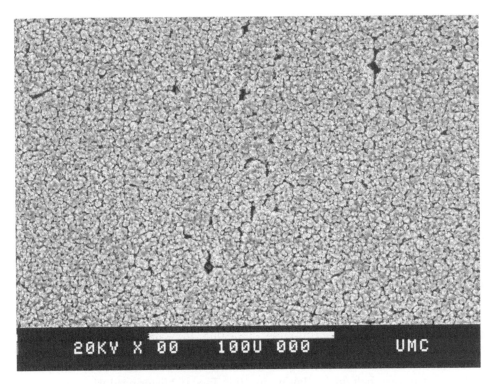

20KV X 00 100U 000 UMC

Figure 32.2 A SEM picture of IVD aluminum coating on 7075-T6 Al substrate.

Table 32.1 Comparison of Ion Vapor Deposition and DC Cathodic Polymerization Processes

Operating parameters	IVD	DC cathodic polymerization
Power supply	DC	DC
Cathode (− charged)	Substrate	Substrate
Anode (grounded)	Chamber and evaporation source	Chamber and anode assembly (if employed)
Applied voltage	Cleaning: −400 to −1000 V Deposition: −400 to −1000 V	Cleaning: −500 to −1000 V Polymerization: −500 to −1000 V
Base pressure	≤ 0.08 mtorr	≤ 1.0 mtorr
System pressure	~10 mtorr	10–100 mtorr
Deposition controlling factor	Evaporation rate	Current density

chromated primer, which saves materials and labor costs as well as environmental remediation cost.

The hybrid process of IVD/cathodic luminous chemical vapor deposition (LCVD) showed great advantage of providing improved corrosion protection by environmentally benign process [3–5]. Table 32.1 shows the comparison of typical operating parameters of IVD and DC cathodic polymerization process. It can be seen that striking similarities exist between DC cathodic polymerization and the industrial IVD process. Since it can utilize the existing IVD vacuum equipment in aerospace industry with no further investment cost, DC cathodic polymerization without anode assembly is a very promising way to apply this technique in practical applications. Therefore, cathodic LCVD without anode assembly was developed for the treatment of IVD Al–coated substrates and compared with results obtained with anode magnetron.

2. CATHODIC LCVD WITH ANODE MAGNETRON

2.1. IVD/Plasma Polymer/E-coat Systems

When E-coat was applied to a thicker class I IVD (about 25 µm thickness), some blisters developed on the E-coat surface during the primer curing process. The blister development on E-coat surface apparently resulted from the air trapping inside the deeper pores of the IVD Al coatings when the E-coat was applied. The trapped air would erupt by the heat and result in the blisters on the E-coat surface during its curing process. It was found that the blisters on E-coat surface could be significantly reduced by appropriately adjusting the curing process with slow heating, which could allow the trapped air to slowly diffuse from the coating system. It was further found that the use of a thinner class II IVD (about 12.5 µm in thickness) was very helpful in eliminating this E-coat blistering during the curing process. Therefore, class II IVD was selected for IVD/plasma polymer/E-coat systems. Processes and sample identification codes for IVD treatments are shown in Table 32.2.

Three kinds of adhesion tests were used to evaluate the interfacial adhesion behaviors of these systems; the results are summarized in Table 32.3. As is evident from the data, excellent water-insensitive adhesion was obtained for all of the plasma interface–engineered IVD/plasma polymer/E-coat systems examined.

In the case of direct application of E-coat to IVD specimens (one of the systems that showed excellent adhesion performance in both the tape test and the accelerated adhesion test), due to the high throw power of the E-coating process, the deep penetration of E-coat into the porous IVD structure creates mechanical interlocking and thus strong adhesion. Because of this development of mechanical interlocking, neither the tape test nor the accelerated adhesion test could distinguish the effect of plasma treatment on adhesion performance.

To determine the significance of plasma treatment, a severe method involving commercial Turco paint stripper was used to evaluate the adhesion properties of these plasma interface–engineered IVD/plasma polymer/E-coat systems. As seen in Table 32.3, the typical Turco delamination time for E-coat on an IVD surface is only about 5 min. In contrast, the application of plasma polymers on IVD surfaces significantly increased the delamination resistance of E-coat in Turco solution. The plasma interface–engineered IVD system [7I](O)T/F/E survived in Turco solution without delaminating for over 24 h. Based on the Turco solution delamination times presented in Table 32.3, it is evident the application of plasma polymers played an important role in improving adhesion, which in turn had significant effects on the corrosion performance of the systems.

Table 32.2 Sample Identification Codes and Associated Plasma Conditions for Sample Preparation

Identification code	Meaning and conditions
[2I]	IVD Al–coated 2024-T3 Al alloy
[2PI]	IVD Al–coated 2124-T851 Al alloy cut from plate
[7PI]	IVD Al–coated 7050-T7451 Al alloy cut from plate
[7A]	Alclad 7075-T6 Al alloy
(Ace)[a]	CH_3COCH_3 wiping with Kimwipes tissue
(O)	O_2 plasma pretreatment (on Al surface: 1 sccm O_2, 100 mtorr, 40 W, 2 min; on TMS polymer surface: 1 sccm oxygen, 50 mtorr, 10 W, 1 min)
(Ar)	Ar plasma treatment (1 sccm argon, 50 mtorr, 10 W, 1 min)
T	TMS plasma polymerization (1 sccm TMS, 50 mtorr, 5 W, 1 min)
TO	Plasma polymerization of TMS and O_2 mixture (1 sccm TMS + 1 sccm O_2, 50 mtorr, 5 W, 1 min)
TN	Plasma polymerization of TMS and N_2 mixture (1 sccm TMS + 1 sccm N_2, 50 mtorr, 5 W, 1 min)
TAr	Plasma polymerization of TMS + Ar mixture in a closed reactor (25 mtorr TMS + Ar based on TMS/Ar ratio, 1000 V, 2 min)
Tfs	Plasma polymerization of TMS in a flow reactor (1 sccm TMS, 50 mtorr, 5 W, 1 min)
Tcs	Plasma polymerization of TMS in a closed reactor (25 mtorr TMS, 1000 V, 2 min)
C	Methane (CH_4) plasma polymerization (1 sccm CH_4, 50 mtorr, 5 W, 1 min)
F	Hexafluoroethane (HFE) plasma polymerization (1 sccm HFE, 50 mtorr, 5 W, 1 min)
T/(Ar)	TMS plasma polymerization succeeded by argon plasma treatment under the following conditions: 1 sccm argon, 50 mtorr, 10 W, 1 min
T/(O)	TMS plasma polymerization succeeded by oxygen plasma treatment under the following conditions: 1 sccm oxygen, 50 mtorr, 10 W, 1 min
CC	Chromate conversion coating (Iridite 14-2)
E	Cathodic E-coat (non-chromated)
E1	Cathodic E-coat PPG ED6650 (nonchromated)
A	Deft primer 44-GN-72 (chromated)
A1	Deft primer 44-GN-72 (chromated)
D	Spraylat primer EWAE118 (nonchromated)
D1	Spraylat primer EWAE118 (nonchromated)
G	Courtaulds primer 519X303 (chromated)
X	Dexter primer 10-PW-22-2 (nonchromated)
/	Process separation mark

[a]Code used in parentheses indicated the surface treatment process; code used without parentheses indicates coating process.

Table 32.3 Adhesion Test Results for E-coat on Plasma–Treated IVD Al–coated 7075-T6 Al ([7I]) Substrates

Coating systems	Dry	Tape test rating	Delamination time in Turco solution
		After boiled in H_2O for 1, 2, 4, 6, 8 h	
[7I]/E	5	5, 5, 5, 5, 5	~5 min
[7I](O)/E	5	5, 5, 5, 5, 5	~5 min
[7I](O)/T/E	5	5, 5, 5, 5, 5	~5 min
[7I](O)/TO/E	5	5, 5, 5, 5, 5	~20 min
[7I](O)/TN/E	5	5, 5, 5, 5, 5	~20 min
[7I](O)/T/F/E	5	5, 5, 5, 5, 5	>24 h

2.2. IVD/Plasma Polymer/Spray Paint Systems

T/F plasma polymer was also selected to improve the adhesion of different spray paints to IVD Al–coated panels. As presented in Table 32.3, T/F plasma polymer [DC plasma–polymerized trimethylsilane (TMS) followed by hexafluoroethane (HFE)] gave rise to such a strong adhesion of E-coat that could not be stripped off after 24-h application of Turco solution. Since the formation of mechanical interlocking between primers and porous IVD surfaces could conceal the role of plasma treatment in enhancing adhesion, bare 7075-T6 aluminum alloy panels with smooth surfaces were first used as substrate to examine the effect of plasma treatment on the adhesion of spray paints.

Table 32.4 displays the adhesion test results for different spray primers applied to T/F plasma–treated bare 7075-T6 alloys. The results demonstrate that this special plasma polymer coating gave rise to excellent water-insensitive adhesion of all three spray primers used in this study: Turco solution could not delaminate any of the primers over a period of 24 h. In addition, up to 6 days aging of T/F plasma coatings in air prior to primer application did not degrade the excellent adhesion performance of the systems.

Table 32.5 shows the adhesion test results for the same spray primers applied to T/F plasma–treated IVD Al–coated 7075-T6 alloys. The DC T/F plasma–treated IVD aluminum panels also gave strong adhesion to subsequent spray paints. The spray paints could not be removed from the IVD Al–coated Al substrates with the Turco solution.

2.3. Corrosion Protection Performance

2.3.1. SO$_2$ Salt Spray Test

Figure 32.3 shows the scanned images of SO_2 salt spray–tested IVD Al–coated 7075-T6 panels: one control, and two E-coated panels. The direct application of E-coat to IVD-coated panels (with no plasma treatment) did not provide corrosion protection as good as that of the chromate conversion–coated control panel: more corrosion creep from the scribed lines was observed on [7I]/E panels than on the [7pI]CC/E

Table 32.4 Adhesion Test Results for Primers Applied to T/F Plasma–Treated Bare 7075-T6 Panels That Had Been Precleaned with Alkaline and Deoxidizer Solutions Prior to Plasma Treatment

Primer	Air exposure time of T/F before primer application	Tape Test Rating		Delamination time in Turco solution (h)
		Dry	After boiled in H_2O for 1, 2, 4, 6, 8 h	
A	~10 min	5	5, 5, 5, 5, 5	> 24
	24 h	5	5, 5, 5, 5, 5	> 24
	48 h	5	5, 5, 5, 5, 5	> 24
	76 h	5	5, 5, 5, 5, 5	> 24
	6 days	5	5, 5, 5, 5, 5	> 24[a]
G	~10 min	5	5, 5, 5, 5, 5	> 24
	24 h	5	5, 5, 5, 5, 5	> 24
	48 h	5	5, 5, 5, 5, 5	> 24
	76 h	5	5, 5, 5, 5, 5	> 24
	6 days	5	5, 5, 5, 5, 5	> 24[a]
D	~10 min	4	4, 4, 4, 4, 4	> 24
	24 h	4	4, 4, 4, 4, 3	> 24
	48 h	4	4, 4, 4, 4, 4	> 24
	76 h	4	4, 4, 4, 4, 4	> 24
	6 days	4	4, 4, 4, 4, 4	> 24[a]

[a] A few blisters begin to develop after 24-h application of Turco solution.

Table 32.5 Adhesion Test Results for Primers Applied to T/F Plasma–treated IVD Al–Coated 7075-T6 Al ([7I]) Panels

Primers	Air exposure time of T/F before primer application	Tape test rating		Delamination time in turco solution (h)
		Dry	After boiling in H_2O for 1, 2, 4, 6, 8 h	
A	2 h	5	5, 5, 5, 5, 5	> 24
	4 days	5	5, 5, 5, 5, 5	> 24
G	2 h	5	5, 5, 5, 5, 5	> 24
	4 days	5	5, 5, 5, 5, 5	> 24
D	2 h	5	5, 5, 5, 5, 5	> 24
	4 days	5	5, 5, 5, 5, 5	> 24

control panels. However, with the aid of plasma interface engineering, [7I](O)/T/F/E clearly outperformed the chromate conversion–coated controls.

All corrosion–tested panels were scanned and the corrosion widths along the scribed lines were calculated as described in the experimental procedures. Figure 32.4

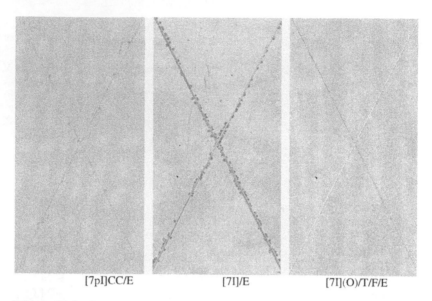

Figure 32.3 Scanned images of SO_2 salt spray–tested (4 weeks) IVD Al–coated 7075-T6 Al panels; the total scanned area $27\,cm^2$, the total scribe length within the scanned area 16 cm.

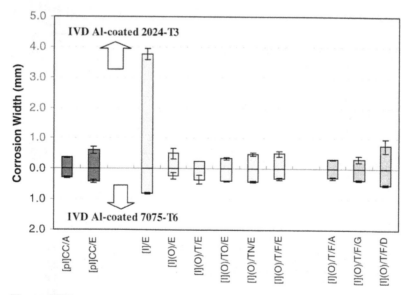

Figure 32.4 Average corrosion widths of SO_2 salt spray–tested IVD aluminum-coated 2024-T3 and 7075-T6 Al panels, including chromated controls, IVD/plasma polymer/E-coat, and IVD/plasma polymer/spray primer coating systems.

presents a comparison of the average corrosion widths of SO_2 salt spray–tested panels of these plasma interface–engineered samples and the controls. The direct application of E-coat to IVD-coated panels without plasma treatment gave large corrosion width values after SO_2 salt spray testing. However, the plasma-tailored

IVD systems on both 7075-T6 and 2024-T3 substrates displayed excellent corrosion protection. Most of the plasma–treated IVD samples outperformed or nearly outperformed the two types of chromate conversion–coated IVD Al alloy controls.

2.3.2. Prohesion Salt Spray Test

Figure 32.5 presents the scanned images of Prohesion salt spray–tested IVD Al–coated 7075-T6 panels. After Prohesion salt spray testing, it was readily evident that plasma interface–engineered samples, like the [7I](O)T/F/E specimen pictured, showed much less corrosion creep along the scribed cross-lines than did the [7pI] CC/E control panels. As compared to the SO_2 salt spray–tested samples, 12 weeks of Prohesion salt spray testing led to more corrosion on all samples.

The corrosion widths of Prohesion salt spray–tested IVD Al–coated Al panels were calculated and are summarized in Figure 32.6. As is evident from the data, after 12 weeks of Prohesion salt spray testing, IVD/plasma polymer/spray paint systems showed better corrosion protection overall than IVD/plasma polymer/E-coat systems. All the IVD/plasma polymer/spray paint systems outperformed the cathodic E-coated controls and showed corrosion test results comparable to those of the Deft primer–coated controls.

In the IVD/plasma polymer/E-coat systems evaluated, oxygen plasma treatment only (without applying any plasma polymer prior to E-coating) did not provide good corrosion protection of IVD Al–coated Al alloys. As a result, very large corrosion width values were obtained for oxygen plasma–treated IVD Al–coated 7075-T6 and 2024-T3 panels shown as [I](O)/E in Figure 32.6.

From the corrosion test results shown in Figures 32.4 and 32.6, it should be noted that chromate-free plasma coating systems (e.g., [I](O)/T/F/E, [I](O)/T/E, [I](O)/TO/E, [I](O)/TN/E, and [I](O)/T/F/D) gave rise to excellent corrosion

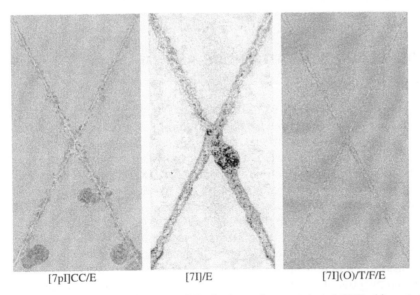

| [7pI]CC/E | [7I]/E | [7I](O)/T/F/E |

Figure 32.5 Scanned images of Prohesion salt spray–tested IVD Al–coated 7075-T6 Al panels; the total scanned area 27 cm², the total scribe length within the scanned area 16 cm.

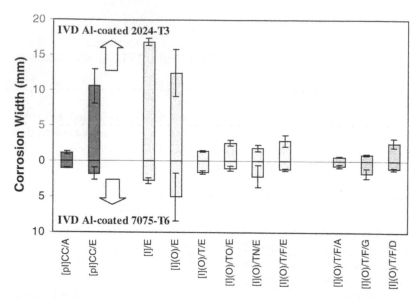

Figure 32.6 Average corrosion widths of Prohesion salt spray–tested IVD aluminum–coated 2024-T3 and 7075-T6 Al panels, including chromated controls, IVD/plasma polymer/E-coat, and IVD/plasma polymer/spray primer coating systems.

protection of IVD Al–coated aluminum alloys. They outperformed the cathodic E-coated controls and showed corrosion test results comparable to those of the Deft primer–coated controls. Thus, without using the conventional chromate conversion coatings, excellent corrosion protection of IVD aluminum alloys can be achieved with chromate-free, plasma interface–engineered coating systems.

3. CATHODIC LCVD WITHOUT ANODE ASSEMBLY

3.1. Nature of Anode

DC cathodic polymerization employed in the previous section was carried out with the substrate as the cathode and grounded anode assembly [3]. Because of limitation of its size and shape of substrate (cathode), the use of anode assembly is impractical for large-scale operation. In contrast, DC cathodic polymerization without using anode assembly will be more compatible to industrial IVD processes. Figure 32.7 shows a comparison of operation parameters of DC plasmas (TMS and oxygen) conducted without anode assembly and with anode magnetron. In comparison with anode magnetron plasmas, for both TMS and oxygen gases, a higher voltage of about 100 V is necessary to sustain the plasmas at a certain power input with no anode assembly operation [4]. Figure 32.8 depicts deposition profiles in no anode assembly plasmas with different operation modes.

The original purpose of using anode magnetron was to eliminate the edge effect of plasma etching and to lower the breakdown voltage. It was found, however, that the edge effect is not a serious problem in plasma deposition and the anode

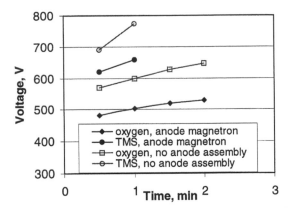

Figure 32.7 Voltage change with discharge time in DC anode magnetron plasmas and no anode assembly operation; 1 sccm TMS, 50 mtorr, 5 W, and 2 sccm oxygen, 100 mtorr, 40 W.

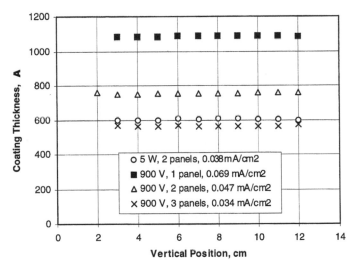

Figure 32.8 TMS deposition profile in no anode assembly plasmas with different operation modes; 1 sccm TMS, 50 mtorr, 1 min.

magnetron tends to yield a peak in the middle of a substrate. Figure 32.9 shows the deposition profile of TMS coatings produced by no anode assembly plasmas and anode magnetron plasmas. With similar plasma conditions, a similar TMS plasma coating thickness was obtained by no anode assembly operation to anode magnetron plasma. It was also noted that no anode assembly operation produced a more uniform TMS coating than anode magnetron plasmas, which showed a mild peak at the center of the substrate. DC cathodic polymerization without anode assembly yields the higher coating thickness than that obtained by with magnetron anode assembly, which is due to collecting more species in gas phase to the cathode simply because there is no other surface to collect them. As Figure 32.8 shows, however,

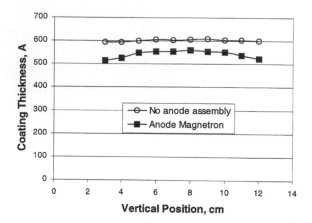

Figure 32.9 TMS deposition profile in no anode assembly and anode magnetron plasmas; 1 sccm TMS, 50 mtorr, 5 W, 1min.

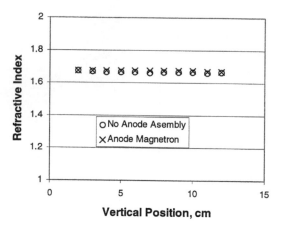

Figure 32.10 Refractive indices of TMS plasma coatings prepared by no anode assembly and anode magnetron plasmas; 1 sccm TMS, 50 mtorr, 5 W, 1 min.

the voltage necessary to sustain discharge increases significantly without anode magnetron, which lowers the breakdown voltage of discharge.

3.2. Coating Properties

Refractive index is one of the simplest methods to describe the quality of plasma polymer coatings. Figure 32.10 shows the refractive indices of TMS plasma polymer coatings produced by no anode assembly plasmas and anode magnetron plasmas. It can be seen that, at similar plasma conditions, DC discharge without anode assembly produced TMS plasma coating with refractive indices identical to those prepared by anode magnetron discharge.

DC polarization technique is very useful in predicting the corrosion protection properties of coatings on metal substrates. Figure 32.11 compares the polarization

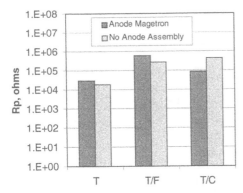

Figure 32.11 Polarization resistance of plasma polymer coated Alclad 7075-T6 alloys in 0.5% NaCl and 0.35% $(NH_4)_2SO_4$ aqueous solution.

resistance (R_p) of several plasma-coated Alclad 7075-T6 panels. It can be seen that DC discharge without anode assembly and that with anode magnetron produced plasma coatings on Alclad 7075-T6 with very similar polarization resistance. These data indicated that DC discharge with substrate alloy as the cathode without anode assembly can be used in LCVD for corrosion protection of aluminum alloy satisfactorily.

3.3. Adhesion Enhancement to Subsequent Spray Primers

The corrosion protection of plasma interface–engineered coating systems relies on the tenacious water-insensitive adhesion and good barrier characteristics of the coatings [3]. DC cathodic polymerization and plasma treatment have been demonstrated as efficient in improving the primer adhesion to metallic substrates.

Table 32.6 compares the adhesion test results of Deft primers (44-GN-36 and 44-GN-72) to plasma coatings prepared by no anode assembly and anode magnetron discharge. It can be seen that plasma polymers prepared by discharge with anode magnetron and without anode assembly provided equivalent adhesion of Deft primers. It is worth noting that, other than plasma T/F, plasma T/C, T/O, and T/Ar also gave good adhesion to Deft 44-GN-72 primer, which could not be stripped off by Turco stripper for over 24 h.

In order to eliminate hazardous heavy metals in the plasma–engineered coating systems, two other spray paint primers of Spraylat EWAE118 and Dexter 10-PW-22-2, which are chromate free, were selected to produce the plasma interface–engineered coating systems. The adhesion performance of Spraylat EWAE118 and Dexter 10-PW-22-2 primers to DC plasma–treated Alclad 7075-T6 aluminum alloys was evaluated and the adhesion test results were summarized in Table 32.7. It can be seen that an excellent adhesion was achieved on T/(Ar)-treated aluminum surfaces for these two primers. They survived the tape test even after water boiling for 8 h. After the application of Turco paint stripper, Spraylat primer on T/(Ar)-treated panels did not show any blisters in the beginning 15 min and Dexter primer stuck firmly to the T/(Ar)-treated substrate for about 13 h. Reasons for the poor primer

Table 32.6 Adhesion Test Results of Chromated Primers (Deft 44-GN-36 and 44-GN-72) to Plasma Coated Alclad 7075-T6 Alloy

7A (Ace/O)	Plasma mode	Deft 44-GN-36			Deft 44-GN-72		
		Dry tape test	Boiling test (8 h)	Turco	Dry tape test	Boiling test (8 h)	Turco
T	AM[a]	0	—	< 5 min	0	—	~6 min
	NA[b]	0	—	< 5 min	0	—	~6 min
T/F	AM	5	4	> 24 h	5	4	> 24 h
	NA	5	4	> 24 h	5	4	> 24 h
T/C	AM	5	4	~15 min	5	4	> 24 h
	NA	5	4	~15 min	5	4	> 24 h
T/(O)	AM	4	3	~15 min	5	4	~4 h
	NA	5	3	~10 min	4	3	> 24 h
T/(Ar)	AM	5	3	~10 min	4	4	> 24 h
	NA	5	3	~20 min	5	3	~1.5 h

[a]Anode magnetron.
[b]No anode assembly.

Table 32.7 Adhesion Test Results of Nonchromated Primers (Spraylat EWDY048 and Dexter 10-PW-22-2) to Plasma-Coated Alclad 7075-T6 Alloy Under IVD Conditions

7A (Ace/O)	Spraylat EWAE118			Dexter 10-PW-22-2		
	Dry tape test	Boiling 1, 4, 8 h	Turco	Dry tape test	Boiling 1, 4, 8 h	Turco
T	2	—	—	3	0, —	~10 min
T/(O)	5	0, —	~5 min	5	5, 5, 5	~30 min
T/(Ar)	5	5, 3, 3	~15 min	5	5, 5, 5	~13 h

adhesion of as-deposited TMS plasma coating, which is shown in Tables 32.6 and 32.7, are discussed in Chapter 31.

3.4. Corrosion Test Results

All the chromate-free plasma coating systems of IVD Al–coated Al alloys performed extremely well in SO_2 salt spray test as depicted in Figure 32.12. Scanned images of SO_2 salt spray–tested plasma coating systems of IVD panels and their chromated controls are shown in Figure 32.12a.

After the SO_2 salt spray test, the chromate-free LCVD coating systems of both IVD Al–coated 2024-T3, plate stock 2124-T851 and 7050-T7451 (not shown in Figure 32.12) showed no corrosion along the scribe lines or away from the scribe. It should be especially noted that these LCVD-tailored IVD coating systems are nonstrippable with conventional paint strippers (Turco 5469). The pattern seen on Prohesion salt spray–tested sample (Fig. 32.12b, right side) is the stain on the remaining paint caused by the Turco stripping test. In contrast, the control panels

[2I]CC/A   [2I](O)/T/(Ar)/X

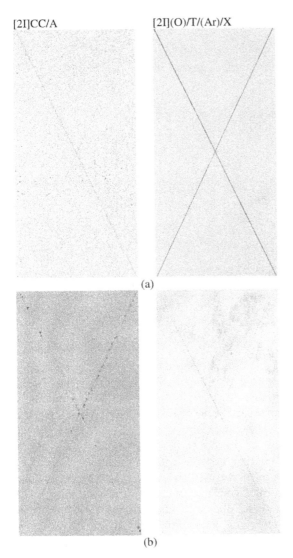

(a)

(b)

Figure 32.12 Scanned images of (a) SO$_2$ salt spray tested and (b) Prohesion salt spray–tested chromate-free plasma coating systems of IVD 2024-T3 ([2I]) and their chromated control panels, left column: some blisters were observed on chromate conversion coatings after the paints were removed.

clearly showed some blisters on the chromate conversion coating, which will very possibly influence the long-term corrosion performance of these coating systems.

After a 12-week Prohesion salt spray test, Deft primer–coated IVD controls performed much better than E-coated controls, although it was hard to distinguish the difference among the two controls with a 4-week SO$_2$ salt spray test. Figure 32.12b presents the typical scanned images of Prohesion salt spray–tested IVD-coated aluminum alloys. By visual observation, one easily can see that clear corrosion creep appeared along the scribed lines for Deft primer coated controls

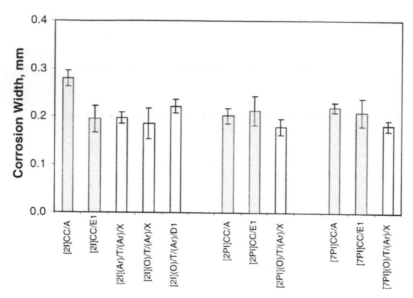

Figure 32.13 Corrosion widths of SO₂ salt spray–tested chromate-free plasma coating systems of IVD panels prepared under IVD conditions and their chromated controls.

([2I]CC/A). In contrast, the chromate-free plasma coating systems of [2I](O)/T/(Ar)/X showed nearly no sign of corrosion on the whole panel.

The corrosion test results were evaluated and the corrosion widths along the scribe lines were calculated according to the procedures described previously. Figure 32.13 shows the comparison of average corrosion widths of SO₂ salt spray–tested IVD panels of chromate-free plasma coating systems and the chromated controls. Almost all the plasma coating systems outperformed their chromated controls. One most important fact that should be pointed out here is the nonstrippable nature of LCVD systems on IVD panels, which would play a significant role in the long-term corrosion protection of Al alloys.

The corrosion widths of Prohesion salt spray–tested alloys are calculated and summarized in Figure 32.14. E-coated IVD controls (CC/E), i.e., the combination coating systems of chromate conversion coating with nonchromated E-coat, showed very large corrosion widths for all the IVD Al–coated aluminum alloys. This combination did not provide good corrosion protection, which could be taken as proof that the two completely different approaches (electrochemical corrosion protection and corrosion protection by barrier adhesion principle) should not be mixed.

Among the three types of aluminum alloys employed in the study, all the chromate-free coating systems except [2I](O)/T/(Ar)/D1 performed extremely well after the Prohesion salt spray test. It can be seen that these LCVD-tailored coating systems showed very small corrosion widths and outperformed their corresponding chromated controls. The only exception is the [2I](O)/T/(Ar)/D1 specimen, which performed well in the SO₂ salt spray test but exhibited severe corrosion in the Prohesion test with large corrosion widths. Since there is no corrosion inhibitor in the LCVD-tailored coating systems, the excellent corrosion performance must result

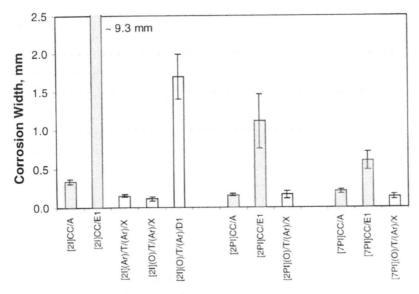

Figure 32.14 Corrosion widths of Prohesion salt spray–tested chromate-free plasma coating systems of IVD panels prepared under IVD conditions and their chromated controls.

from the tenacious adhesion at the primer/IVD interface that was achieved by DC cathodic polymerization and plasma treatment during IVD/LCVD operation.

It was noted (from Figure 32.14) that the [2I](O)/T/(Ar)/D1 specimen, which performed well in the SO$_2$ salt spray test, exhibited much larger corrosion widths along the scribed lines, and much pit corrosion away from the scribed lines was visually observed after Prohesion test. The worse performance of Spraylat primer–coated samples likely resulted from its weaker adhesion to plasma–treated IVD surface than Dexter primer, which has been shown earlier from the adhesion test results. Another possible reason might be due to its inferior barrier properties to Dexter primer because there was no pit corrosion observed for Dexter primer–coated samples but many pits on Spraylat primer–coated panels after the Prohesion test.

4.   CATHODIC LCVD IN CLOSED REACTOR

In the IVD coating process, the reactor is filled with argon gas to a certain pressure and the aluminum evaporation is arranged to face the parts. Before initiating the ensuring plasma for the additional film deposition, the monomer gas has to be filled into the whole reactor to maintain a certain system pressure. If the reactor has a large volume, such as an industrial IVD reactor, large amounts of monomer gases will be necessary to fill the reactor to a certain pressure for CVD operation following IVD processing. In this case, if the plasma deposition is operated in a flowing mode with the aim of achieving a very thin plasma coating, very little monomer will be effectively utilized and most monomer will be wasted and lost through the exhaust.

In contrast, DC cathodic polymerization in a closed system seems to be the most efficient way to operate plasma deposition in a large IVD reactor. In such a closed system, a certain amount of monomers deposit the required thickness of film, which can be calculated based on the total surface area of substrate charged, is introduced into the reactor, and plasma deposition can be started. In this operation mode, there is minimal loss of gases and the most efficient utilization of the monomer can be achieved. It is also possible to keep the argon that is used in IVD operation as an additional process gas in the plasma film deposition, if a closed system operation of TMS polymerization is utilized.

In order to explore the possibility of efficiently operating plasma deposition in an industrial IVD reactor, DC cathodic polymerization of TMS in a closed system mode under conditions similar to the IVD operation was investigated. The corrosion protection properties of the plasma coatings obtained under such operation were also studied on IVD Al–coated Al alloys.

4.1. Closed Vs. Flow System LCVD

In a flow system plasma polymerization, the system pressure is continuously adjusted by controlling the opening of a throttle valve connected to the pumping system. Because of fragmentation of original monomer in a plasma state, the composition of gas phase changes on the inception of the plasma state. The increase in the total number of gas molecules is compensated by the increased pumping rate in a flow system, and a steady-state flow of a consistent composition of gas phase is established at a predetermined system pressure.

In closed system plasma polymerization, a fixed number of monomer molecules are contained in a reactor and glow discharge is initiated. The system pressure in such a system (in a given volume) is proportional to the total number of gas phase molecules. The fragmentation of monomer molecules as well as the ablation of gaseous species from the deposited material increases the pressure, whereas deposition decreases the pressure. Thus, the system pressure change with plasma polymerization time will indicate the change in the overall balance between the plasma fragmentation/ablation and the plasma film deposition.

Figure 4.2 depicts the system pressure change in a closed system reactor when plasma polymerization of TMS is carried out. The system pressure increases continuously while the glow discharge is on but remains at a constant value as soon as the glow discharge is turned off. This indicates that the total number of gas phase species increases with time in spite of the deposition of plasma polymer of TMS. Figure 4.1 depicts the change of gas phase species detected by mass spectrometry during the plasma polymerization of TMS. The results indicate that the deposition of Si containing species takes place in the early stage of plasma polymerization, and the deposition of C-containing species lags behind the deposition of Si-containing species. In the later stage, the main species that constitute the plasma phase is hydrogen.

According to this scheme of plasma polymerization of TMS in a closed system, it is anticipated that the atomic composition of the plasma polymer should continuously change with the plasma polymerization time. Figure 13.21 depicts comparison of XPS cross-section profile of C/Si ratios for plasma polymers deposited in a flow system reactor and that in a closed system reactor. The results clearly show that a closed system plasma polymerization of TMS indeed produces a film with graded composition, i.e., with increasing carbon content from the interface with the substrate to the top surface of film.

Considering that the system pressure continues to increase after most of the polymerizable species are exhausted in the gas phase, plasma polymerization of TMS in a closed system can be visualized as a time-delayed, consecutive application of three fundamental processes. The sequence takes the order: (1) deposition of Si species, (2) deposition of C species, and (3) H_2 plasma treatment of the deposited plasma polymer.

4.2. Plasma Parameter Changes

Figure 32.15 shows the system pressure change with discharge time during plasma polymerization of TMS only and its mixtures with argon gas at different ratios. DC cathodic polymerization of a TMS/Ar mixture has its advantages over that of pure TMS monomers because the argon addition can help to achieve a more stable glow discharge. Also since inert argon gas is always present in the IVD process, DC cathodic polymerization of a TMS/Ar mixture can provide a more compatible process with the industrial IVD system. The result indicated that the presence of Ar increases the overall fragmentation/ablation effect, i.e., the pressure increase is more than the partial pressure of Ar added to the system as can be seen in Figure 32.15.

Figure 32.16 shows the current change with discharge time during plasma polymerization of TMS and its mixtures with argon gas. It was noted that, for TMS monomer only or its mixture with less argon addition (2.5 mtorr and 12.5 mtorr), the DC current to maintain the plasma dropped first after the plasma was ignited at a constant DC potential of 1000 V applied to the substrate. This phenomenon obviously resulted from disappearance of TMS monomers due to polymer

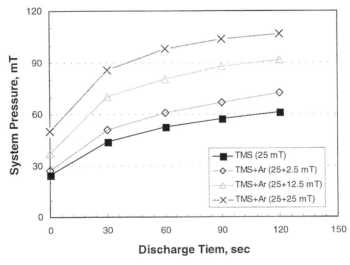

Figure 32.15 System pressure change with discharge time during plasma polymerization of TMS and its mixtures with argon at different ratios; Alclad 7075-T6 substrate, 25 mtorr TMS + Ar (based on TMS/Ar ratios), 2 panels of Alclad 7075-T6, DC 1000 V.

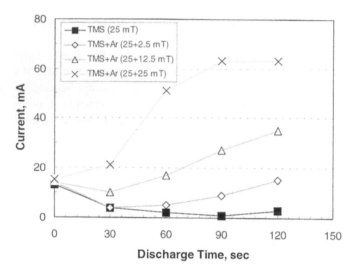

Figure 32.16   DC current change with discharge time during plasma polymerization of TMS and its mixtures with argon at different ratios; Alclad 7075-T6 substrate, 25 mtorr TMS + Ar (based on TMS/Ar ratios), 2 panels of Alclad 7075-T6, DC 1000 V.

deposition and from the change of discharge characteristics of the gases produced from fragmentation of the TMS monomer. Therefore, the current decrease was not observed when a larger amount of argon gases existed in the plasma system, and with 25 mtorr of argon mixed with TMS the discharge current increased with time.

4.3.   Coating Properties

Figure 13.15, shows the dependence of the thickness and refractive index of TMS plasma coating on the plasma polymerization time in a closed reactor system using TMS monomers. It can be seen that the coating thickness increased very fast in the first 90 s. After 90 s, the TMS coating thickness stopped growing with the deposition time, but the refractive index of TMS coating kept increasing with the deposition time. This increase obviously resulted from the continuing bombardment by the reactive species in the plasma.

Figure 32.17 shows the similar plots obtained with the addition of argon into the closed reactor system using TMS. It can be seen that the coating thickness of plasma polymers was solely determined by the amount of TMS monomer filling the reactor system. The addition of argon did not affect the coating thickness but increased the coating quality of TMS plasma polymers that was reflected from the increase of film refractive index.

4.4.   XPS Analysis of TMS Films

Figure 32.18a and b summarizes the XPS results from three TMS plasma films produced in closed reactor system with and without addition of the second gas. Figure 32.19a and b summarizes the XPS results from three TMS plasma films produced in closed reactor system with and without addition of the second gas discharge treatment. These films were deposited on Alclad 7075-T6 panels that were

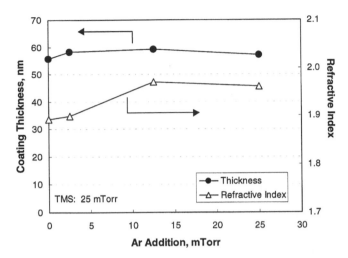

Figure 32.17 Thickness and refractive index changes of closed system TMS plasma coatings with argon pressure in the reactor system; TMS 25 mtorr, 2 panels of Alclad 7075-T6, DC 1000 V, 2 min.

acetone cleaned and then oxygen plasma–treated prior to film deposition. Alclad 7075-T6 panels with smooth surfaces were chosen for such a purpose because they provide similar Al coating surface to IVD Al–coated Al alloys. One film was removed from the reactor right after TMS plasma deposition; the second was treated with argon plasma and the third with oxygen plasma treatment after the TMS deposition.

TMS plasma films produced from closed and flow systems both have a surface chemical structure similar to silicon carbide bonding with Si 2p bonding energy close to 99.5 eV. The second O_2 plasma treatment on these TMS films changed the surface structure to silicon oxide with Si 2p bonding energy shifted to 103 eV. As seen from Figure 32.18b, the Ar plasma treatment on flow system TMS polymers resulted in effects similar to those of O_2 plasma treatment. In contrast, as noted in Figure 32.18a, the Ar plasma treatment on closed system TMS polymers has a surface composed of the intermediary bonding and is thought to be some siliconoxycarbide bonding or Si_2O bonding with various possible silicon suboxides [8–10].

Figure 32.19 shows C/Si ratios formed from the XPS sputter depth profiles of the TMS plasma polymers with and without additional plasma treatment. As deposited, without a second plasma treatment, the closed system TMS plasma film has a surface that is carbon rich (with a C/Si ratio of about 4.7) and low oxygen content (with an O/Si ratio of about 0.7). From Figure 32.19a, it is observed that the as-deposited TMS plasma film shows a gradual structure change from the surface with more carbon (C/Si ratio of about 4.7) to lower carbon (C/Si ratio of about 1.7) in the bulk film. This also manifests itself as a higher C/Si ratio at the surface than the bulk value, which is unique to this film.

The O_2 plasma treatment on the TMS film is seen to significantly reduce the incorporation of carbon in the outer region of the TMS plasma films (including both closed system and flow system TMS polymers) and change the surface from a carbon-rich state to one rich in SiO_2 bonding. This result indicates that the O_2

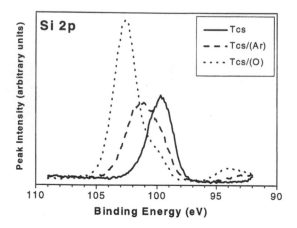

(a) Closed system TMS

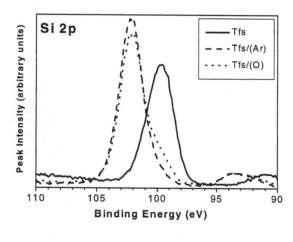

(b) Flow system TMS

Figure 32.18 Si 2p photoelectron spectra from the top surface of TMS plasma coatings produced in (a) closed reactor system and (b) flow reactor system with and without second surface treatment by O_2 or Ar plasmas.

plasma has the effect of removing carbon from the film surface as a volatile compound and restructuring the surface into a silicon–oxide rich layer.

The argon plasma treatment of flow system TMS polymers had a similar ratio in the depth profile as oxygen plasma treatment, shown in Figure 32.19b. In contrast, as noted in Figure 32.19a, the argon plasma modified the surface of closed system TMS film to an intermediary position with a certain amount of carbon loss and silicon enrichment near the surface. The argon plasma–treated closed system TMS film surface has a distinctly different silicon structure, somewhat intermediary between the bulk and true silicon oxide as shown in Figure 32.18. It is not asserted that the Ar treatment actually induces oxygen bonding simply by its own interaction. These samples were exposed to the atmosphere prior to XPS analysis; hence any

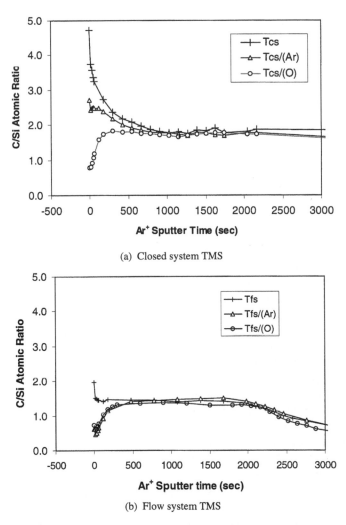

(a) Closed system TMS

(b) Flow system TMS

Figure 32.19 Cross-sectional depth profile of XPS measured C/Si ratios of TMS plasma polymer films prepared in a (a) closed reactor system and (b) flow reactor system with and without second surface treatment by O_2 or Ar plasmas.

active sites formed during the Ar treatment then had ample time to react, forming oxygen bonds upon exposure to atmospheric gases.

4.5. Primer Adhesion

From the XPS analysis results, it was found that TMS plasma coatings prepared in a closed system have a carbon-rich surface, which is similar to TMS polymers followed by methane plasma deposition, i.e., T/C as described in a flow system. The adhesion study also showed that the T/C plasma treatment on aluminum alloy provided excellent adhesion to spray paint primers. The primer adhesion performance of closed system TMS plasma polymers was also examined. Since the porous surface

Table 32.8 Adhesion Performance in Various Adhesion Tests, Indicating the Influence of Second Plasma Treatment Effects of TMS Plasma Films (Under IVD Conditions) on their Adhesion Performance to Chromated Primers (Deft 44-GN-36 and 44-GN-72)[a]

TMS coating	2nd plasma (1 min)	Deft 44-GN-36			Deft 44-GN-72		
		Tape test	Boiling 1, 4, 8 h	Turco	Dry tape test	Boiling 1, 4, 8 h	Turco time
Flow Tfs	—	0	—	< 5 min	0	—	~6 min
	C	5	4	~15 min	5	4	> 24 h
Closed Tcs	—	5	4, 3, 3	~5 min	5	3, 3, 3	~10 min
	(Ar)	5	4, 3, 2	~20 min	5	5, 4, 4	> 24 h

[a]Substrates Alclad 7075-T6 [7A(Ace/(O)].

structure of IVD Al coating may conceal the plasma treatment effect on primer adhesion, Alclad 7075-T6 Al alloy panels were selected as the substrates for adhesion investigation based on the fact that they can provide smooth surfaces while still providing an Al coating similar to that of IVD Al coatings.

Table 32.8 lists the adhesion test results of two chromated spray paint primers applied to these closed system TMS plasma coatings with and without subsequent surface treatment by oxygen or argon plasmas. Due to their similar chemical structure to the flow system T/C plasma polymer, TMS plasma polymers prepared in a closed reactor have much better primer adhesion than those produced in a flow reactor. Therefore, the more organic (carbon-rich) top surface of closed system TMS coatings was considered as the main factor contributing to the superiority of their primer adhesion performance over those prepared in a flow system. Excellent primer adhesion performance was obtained on closed system TMS polymer surface with appropriate second argon plasma treatment, indicating that the intermediate bonding structures observed by the XPS analysis may have an important role in primer adhesion.

In order to produce chromate-free plasma coating systems, the adhesion of closed system TMS coatings to nonchromated primers (Spraylat EWAE118 and Dexter 10-PW-22-2) was also investigated, and the adhesion test results are summarized in Table 32.9. Closed system TMS plasma polymers showed superior primer adhesion performance to those obtained from a flow system. Similar to chromated primers, summarized in Table 32.8, excellent primer adhesion was always achieved with closed system TMS plasma polymers treated with subsequent Ar plasma applications.

It should be pointed out that excellent primer adhesion was also obtained with TMS plasma polymers from a (TMS + Ar) mixture in a closed reactor system. This result indicated that, to achieve equally good primer adhesion, TMS polymerization with subsequent Ar plasma treatment could be replaced by one process of cathodic polymerization of a (TMS + Ar) mixture. Since the addition of argon to TMS can help stabilize the gas discharge, plasma polymerization of a (TMS + Ar) mixture is very important in the practical operation of plasma deposition process in conjunction with the industrial IVD process. Plasma polymerization of a mixture of TMS and argon in a closed system also has the advantage of being more

Table 32.9 Adhesion Performance in Various Adhesion Tests, Indicating the Influence of Second Plasma Treatment Effects of TMS Plasma Films (Under IVD Conditions) on their Adhesion Performance to Nonchromated Primers (Spraylat EWAE118 and Dexter 10-PW-22-2)[a]

TMS coating	2nd Plasma (1 min)	Spraylat EWAE118			Dexter 10-PW-22-2		
		Tape test	Boiled 1,4,8 h	Turco	Tape test	Boiling 1,4,8 h	Turco
Flow Tfs	—	2	—	—	3	0, —	~10 min
	(Ar)	5	5, 3, 3	~15 min	5	5, 5, 5	~13 h
Closed Tcs	—	5	4, 4, 4	~12 h	5	5, 5, 5	~30 min
	(Ar)	5	5, 5, 5	>24 h	5	5, 5, 5	>24 h
Closed TAr (2:1)	—	5	5, 5, 5	>24 h	5	5, 5, 5	~14 h
Closed TAr (1:1)	—	5	5, 5, 5	>24 h	5	5, 5, 5	>24 h
Closed TAr (1:2)	—	5	3, 3, 3	~10 min	5	5, 5, 5	~14 h

[a]Alclad 7075-T6 [7A(Ace/O)].

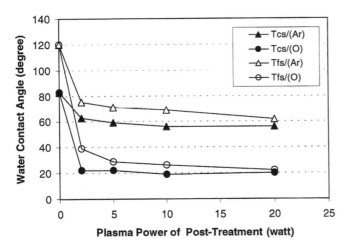

Figure 32.20 Water contact angle change of closed system and flow system TMS polymer surfaces with the power input of argon and oxygen plasma posttreatment.

compatible with the IVD process due to argon coexistence, excellent adhesion performance, and the benefit of one process of combining TMS plasma polymerization and second plasma treatment of the TMS polymers.

In order to study the surface property change of closed system TMS plasma polymers, the water contact angle change was studied as a function of the plasma power input for the second plasma treatment and the results are shown in Figure 32.20. Reflecting the C-rich top surface, the plasma polymer of TMS prepared by the closed system reactor has significantly lower contact angle (about 80 degrees) than that for the sample prepared by a flow system reactor (about 120 degrees), without postdeposition treatment (value from zero power in the graph).

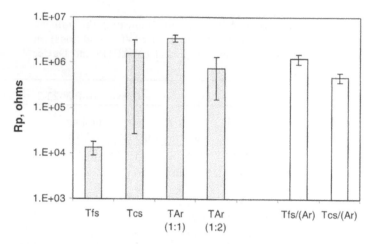

Figure 32.21 Polarization resistance of TMS plasma coated Alcad 7075-T6 (Ace/O) aluminum substrates under IVD conditions (without anode assembly) with or without second plasma treatment; 25 mtorr TMS + Ar (based on TMS/Ar ratios), DC 1000 V, 2 min for closed system TMS; 1 sccm TMS, 50 mtorr, DC 5 W, 1 min for flow system; 2 sccm oxygen or argon, 50 mtorr, 1 min for second plasma treatments.

It can be seen that the second Ar or O_2 plasma treatment lowered the water contact angles on closed system TMS polymer surfaces, but not as much as the change introduced on the flow system sample.

4.6. Polarization Resistance Measurements

Figure 32.21 summarizes results of R_p measurement of TMS plasma–coated Alclad 7075-T6 panels under IVD conditions in both closed and flow reactor systems, with and without subsequent Ar plasma treatments. TMS plasma coatings produced in a closed reactor system showed higher R_p values than those obtained in a flow reactor system. The second Ar discharge treatment of the flow system TMS coating remarkably increased the R_p value to nearly comparable to that for the closed system TMS. The second Ar plasma treatment of the closed system TMS coating, on the other hand, reduced the value of R_p slightly from that for the closed system TMS.

Besides the advantageous features described earlier, DC cathodic plasma polymerization of TMS mixed with argon also provides an opportunity to combine the two processes of TMS deposition and second plasma treatment into a single step. TMS plasma coating thus produced also maintains excellent corrosion protection properties on the aluminum alloy substrates.

4.7. Corrosion Test Results (IVD/LCVD System)

Figure 32.22 shows the comparison of average corrosion widths from both (a) SO_2 salt spray–tested and (b) Prohesion salt spray–tested IVD panels. Both SO_2 and Prohesion test results show that *chromate-free* LCVD coating systems provided excellent corrosion protection on IVD Al–coated Al alloys, having comparable or

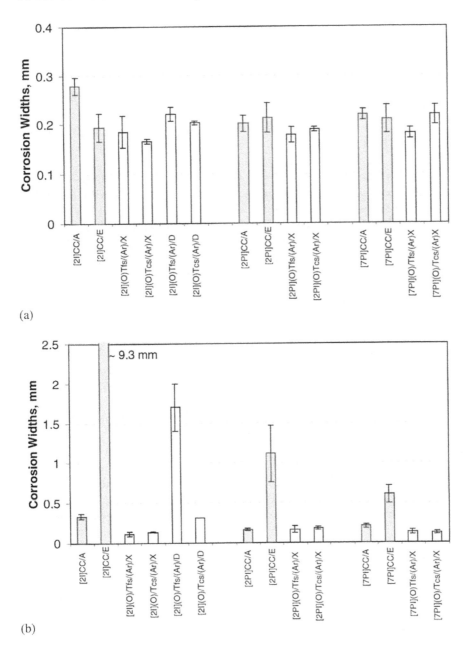

Figure 32.22 Corrosion widths of (a) SO_2 salt spray and (b) Prohesion salt spray–tested IVD Al–coated Al alloy panels protected with chromate-free plasma coating systems and their chromated controls.

lower corrosion widths after the tests than their *chromated controls* of CC/A and CC/E on IVD Al–coated Al panels.

In comparison TMS LCVD coatings in flow system (Tfs), LCVD coating in closed system TMS (Tcs) provided equally good corrosion protection of IVD

Al–coated Al alloys. From Figure 32.22b, it was also noted that the coating system of [2I](O)/Tcs/(Ar)/D obtained with closed system TMS plasma films (Tcs) gave acceptable performance in the Prohesion test, whereas [2I](O)/Tfs/(Ar)/D prepared in the flow system corroded very badly, having much larger corrosion widths.

It should be emphasized that the primers in the plasma coating systems applied to the IVD Al–coated Al alloys could not be removed by the commercial Turco paint stripper solution. This tenacious and water-insensitive adhesion at the primer/IVD interface achieved by TMS cathodic polymerization in a closed reactor system must be responsible for the excellent corrosion protection performance of these plasma coating systems. In other words, excellent corrosion protection of IVD Al–coated Al alloys can be accomplished with chromate-free primer coatings with the aid of tenacious and water-insensitive interface adhesion.

DC cathodic polymerization of TMS in a closed reactor system not only possesses the benefit of efficient operation in a large-scale reactor but also yields high-quality plasma coatings as compared with those obtained in a flow plasma reactor. XPS analysis results indicated that the TMS plasma coatings under such operation have unique chemical structures, which have a more organic (carbon-rich) top surface and then gradually change through the film bulk to a more inorganic (silicon-rich) structure near the aluminum substrate. The organic top surface of the coating provided a compliant adhesion base with subsequent spray paint primers that are usually epoxy-type primers. At the other interface of the plasma coating, the inorganic structure was compatible with IVD aluminum, bonding strongly to the aluminum oxide surface. With further argon plasma treatment of TMS coating surfaces, or direct polymerization of TMS mixed with argon, tenacious adhesion was obtained between chromate-free primers and the TMS plasma film–coated IVD Al substrates.

DC cathodic polymerization of TMS mixed with argon improved the primer adhesion performance of the closed system TMS plasma polymers. Moreover, the addition of a certain amount of argon into the TMS plasma system further increased the plasma coating quality, reflected in the increase in refractive indices. Based on the higher compatibility with the IVD process, the excellent adhesion performance, and the benefit of one process combining TMS plasma polymerization and the postdeposition plasma treatment, DC cathodic polymerization of TMS mixed with argon in a closed system is being considered as a more realistic and favorable approach in practical applications.

REFERENCES

1. Nevill, B.T. 36th Annual Technical Conference Proceedings of the Society of Vacuum Coaters, Albuquerque, NM, 1993, p. 379.
2. Steube, K.E.; McCrary, L.E. J. Vac. Sci. Technol. **1974**, *11* (1), 362.
3. Qingsong Yu; Joan Deffeyes; Hirotsugu Yasuda. Prog. in Organic Coatings **2001**, *42*, 100–109.
4. Qingsong Yu; Joan Deffeyes; Hirotsugu Yasuda. Prog. Organic Coatings **2001**, *43*, 243.
5. Qingsong Yu; Moffitt, C.E.; Wieliczka, D.M.; Deffeyes, J.; Yasuda, H. Prog. Organic Coatings **2002**, *44*, 37.
6. Wrobel, A.M.; Kryszewski, M.; Gazicki, M. J. Macromol. Sci.-Chem. **1983**, *A20*, 583.
7. Gazicki, M.; Wrobel, A.M.; Kryszewski, M. J. Appl. Polym. Sci. **1977**, *21*, 2013.

8.   Himpsel, F.J.; McFeely, F.R.; Taleb-Ibrahimi, A.; Yarmoff, J.A.; Hollinger, G. Phys. Rev. B **1988**, *38*, 6084.

9.   McFeely, F.R.; Zhang, K.Z.; Banaszak Holl, M.M.; Lee, S.; Bender IV, J.E. J. Vac. Sci. Technol. B **1996**, *14*, 2824.

10. Alexander, M.R.; Short, R.D.; Jones, F.R.; Michaeli, W.; Blomfield, C. J. Appl. Surf. Sci. **1999**, *137*, 179.

33

Corrosion Protection of Cold-Rolled Steel and Pure Iron

System approach interface engineering (SAIE) by means of low-pressure plasma in corrosion protection of metals was first applied on cold-rolled steel (CRS) [1–3]. Unlike aluminum alloys described in the previous chapters, oxides on the surface of steel and iron are less stable than those on aluminum alloys, i.e., iron oxides formed on the surface do not have corrosion protecting capability of iron underneath and hence are more vulnerable for corrosion. On the other hand, the oxides on the surface of steel can be removed by plasma treatment, particularly of $(Ar + H_2)$ plasma. Hence, SAIE could be performed in totally dry processes. The improvement of corrosion protection by means of environmentally benign process was the motivation for the approach.

1. PLASMA CLEANING AND IN SITU DEPOSITION OF TMS POLYMER ON COLD-ROLLED STEEL

The pretreatment of the steel substrate surface by oxygen is essentially to remove organic surface contaminants such as hydrocarbons. However, this pretreatment step results in a further buildup of oxides free of carbon on the substrate surface [4]. Argon plasma sputter cleaning will virtually remove the surface absorbents and the oxide layer, leaving the steel surface in a metallic state [1]. Trimethylsilane (TMS) plasma polymers were in situ deposited onto steel substrates with two different surface states. The TMS film structure was then characterized by reflection-absorption infrared spectroscopy (RAIR). Figure 33.1a is the IR spectrum of the TMS plasma polymer deposited on the $(Ar + H_2)$ plasma–cleaned steel surface. Figure 33.1b shows the IR spectrum of the TMS film deposited on the oxygen plasma–cleaned surface.

Figures clearly distinguish the difference between these two TMS depositions on different substrate surface states. Table 33.1 lists the peak assignments for both (a) and (b). In Figure 33.1a, two distinct bands are observed near 3394 and $3190 \, cm^{-1}$. These locations and peak shapes would be consistent with the presence of either isolated O-H bonds or N-H functionality. The strong, sharp peaks at

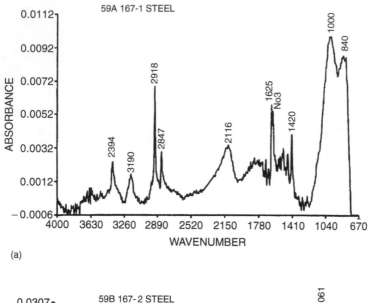

(a)

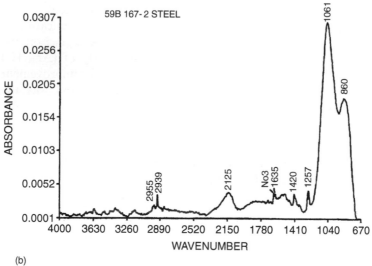

(b)

Figure 33.1 RAIR spectra of the TMS film deposited on (a) $(Ar + H_2)$ plasma pretreated and (b) O_2 plasma pretreated steel surface; adapted from reference [4].

2847 and 2918 cm^{-1} are due to the symmetrical and asymmetrical methylene C-H stretches. The peak expected from the symmetrical methyl C-H stretch at 2960 cm^{-1} is not above the noise level of this spectrum.

These band locations are usually associated with highly ordered, crystalline materials. However, weak bands are observed near 721 and 648 cm^{-1} that are consistent with the location of the symmetrical and asymmetrical S-$(CH_3)_3$ stretching modes (not shown on these spectra). The broad band centered at 2116 cm^{-1} is consistent with the location of the Si-H stretching modes. The strong bands near 1000 and 840 cm^{-1} are most likely associated with Si-C and Si-H

Table 33.1 Peaks Assignment for Figure 33.1a and b

Peak wavenumber (cm^{-1})		
Fig. 33.1a	Fig. 33.1b	IR peak assignment
3394	—	-OH bond
3190	—	-OH bond
2918	2918	CH_2 antisymmetrical stretching
2847	—	CH_2 symmetrical stretching
2116	2128	Si-H stretching
1625	1625	Surface-adsorbed water
1420	1420	Si-CH_2 deformation or Si-CH_3 asymmetrical stretching
—	1257	Si-CH_3 symmetrical deformation
—	1061	Si-O-Si, Si-O-alkyl stretching
1000	—	Si-C stretching or Si-CH_3 rocking

functionality, respectively. Sharp bands were also observed at 1420 and near $1625\,cm^{-1}$. The band near $1625\,cm^{-1}$ is most likely associated with surface adsorbed water. The $1420\,cm^{-1}$ band could be associated with Si-CH_2 deformation or Si-CH_3 asymmetrical stretching modes, although those bands do not usually absorb strongly. This IR spectrum suggests that the TMS polymer film structure is poly(alkylsilane) with a -Si-$(CH_2)_n$. chains or cross-linked network.

A number of differences are observed between the RAIR spectra obtained from the oxygen plasma–treated sample and the $(Ar + H_2)$ plasma–treated sample. Absorptions observed near 3190 and $3394\,cm^{-1}$ in (a) were not present in (b). In addition, the absorbance of the methylene C-H stretching bands are greatly reduced and a band near $2955\,cm^{-1}$ assigned to symmetrical methyl group stretching is now observed. A new, strong band is observed near $1257\,cm^{-1}$ that matches well with the expected location of a strong, sharp IR band associated with the Si-CH_3 symmetrical deformation. A strong band found near $1000\,cm^{-1}$ in (a) has shifted to $1061\,cm^{-1}$ in (b), which is indicative of the presence of Si-O functionality (Si-O-Si or Si-O-alkyl stretching). An increase in Si-O content was not surprising due to the use of oxygen plasma in substrate pretreatment. The band at $860\,cm^{-1}$ in (a) is also stronger, indicating more Si-C structure than in (b). Absorption due to Si-H also occurs in the vicinity of this band.

The TMS plasma polymer deposited on the oxygen-pretreated steel surface has Si-O-Si or Si-O-alkyl chain structure, which is similar to the film deposited from a mixture of TMS and O_2. Oxygen was always found in the plasma film when the steel surface was pretreated with oxygen plasma. The source of oxygen is very likely from the oxide layer. During TMS plasma deposition, oxygen redeposited with the TMS to form the final film. These results were obtained during an in situ experiment, and the treated surface was not exposed to ambient environment before the deposition of plasma polymer of TMS. Therefore, the influence of oxides on the chemical structure of plasma polymer of TMS is quite evident.

2. INTERFACIAL CHEMISTRY CHARACTERIZED BY XPS AND SPUTTERED NEUTRAL MASS SPECTROSCOPY

The superior corrosion performance and strong adhesion of the plasma coating system can be attributed to the coating properties and, more importantly, to the nature of interfacial chemistry. Two techniques were applied to study the surface and interfacial chemistry of the plasma coating system: (1) in situ plasma deposition and XPS analysis and (2) in-depth profiling of sputtered neutral mass spectroscopy (SNMS).

2.1. Surface Structure of Plasma of TMS Films

Figure 33.2 shows XPS spectra of the surfaces of the TMS plasma polymer film deposited on $(Ar + H_2)$ plasma–pretreated steel (a, b, c) and on O_2 plasma–pretreated steel (d, e, f). As shown in the spectra, the surface of the plasma film is functional in nature with functional groups of C-OH, C=O, and Si-OH. Two films basically ended up with the same surface structure. This is also confirmed by XPS analysis of the film during the film aging in air after the film deposition, which indicated that the film surfaces were saturated with a fixed surface structure after a few hours of air exposure [4]. This is due to a well-known phenomenon that the residual free radicals of the plasma polymer surface reacted with oxygen after exposure to air [5]. Curve deconvolution of C 1s peaks showed structures of C-Si, C-C, C-O, and C=O. The analysis clearly shows a silicon carbide type of structure, which is consistent with the IR results. The functional surfaces of TMS films provide bonding sites for the subsequent electrodeposition of primer (E-coat).

2.2. In Situ Plasma Deposition and XPS Analysis

The steel sample was cleaned by $(Ar + H_2)$ plasma, followed by plasma deposition. Transferring the sample from the deposition chamber to the XPS chamber can monitor the surface at any depositing state. As indicated in Figure 33.3, at 5 s deposition, the surface state at this stage reflected the interfacial bonding nature of the plasma film onto the steel. It clearly shows that the C peak comprises peaks at 284.6 eV (C-C) and 283.0 eV, indicative of C-Fe or C-Si bonds. The Si 2p peak also clearly shows a combination of Si-Si (100.6 eV) and Si-Fe (99.7 eV) peaks. Further deposition (15 and 45 s) resulted in establishment of TMS film without influence of the substrate steel, as reflected by C 1s (284.5 eV) and Si 2p (100.80 eV) single peaks.

The early stage of TMS plasma deposition on an oxygen plasma–treated steel substrate shows significantly different results. No split in the Si 2p and C 1s peaks at the early stage of deposition was observed. This confirmed that the plasma/steel surface interactions were very different in the two cases. Interactions with metallic state (rather than metal oxide) were stronger and resulted in stronger interfacial bonding.

2.3. SNMS In-depth Compositional Profiling of Plasma Coating

TMS plasma polymer coated steels were analyzed by SNMS. This analytical technique permitted obtainment of high resolution and quantitative information

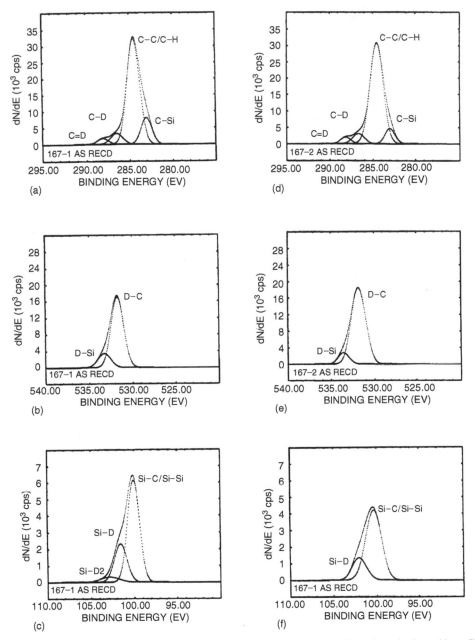

Figure 33.2 XPS spectra of surfaces of TMS plasma polymer films deposited on $(Ar + H_2)$ plasma pretreated steel (a, b, c) and O_2 plasma pretreated steel (d, e, f); adapted from reference [4].

about the plasma coating on steel. More importantly, the transitional region from metal to the coating (*interphase*) can be thoroughly characterized by the in-depth profiling. Figures 33.4 and 33.5 are in-depth compositional profiles of the TMS plasma coating on $(Ar + H_2)$– and O_2 plasma–treated steels, respectively.

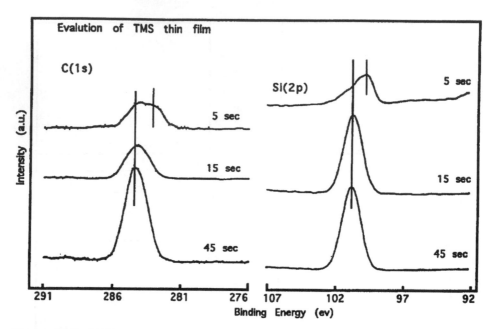

Figure 33.3 XPS spectra of C 1s and Si 2p of TMS plasma polymer at early stages of deposition: deposition and XPS analysis were performed in situ so no interference from air exposure was involved; adapted from reference [4].

Figure 33.4a shows that the TMS plasma film bulk bears the composition of C_7Si_3. In the interphase region, 20–30 nm deep from the film surface, a mixed composition of C, Si, and Fe was detected. A small amount (av. 3 at. %) of oxygen was detected, indicating the residual oxide. More detailed analysis on the atom clusters' in-depth profiling provides rich information about the interphase region. Figure 33.4b shows that iron carbide species (Fe-C) were clearly detected within the interphase region. This strongly supports the results found from the in situ deposition and XPS analysis. The Fe-Si cluster was not within the detectable range. The existence of oxygen implies incomplete removal of the oxide layer. Covalent bonds of Fe-C and possibly Fe-Si were formed during the plasma deposition on the metallic state of the steel surface. These interfacial covalent bonds provide strong adhesion and chemical resistance at the interface between plasma polymer and steel. Iron silicide, a particularly amorphous phase, is known to have high corrosion resistance properties [6].

Figure 33.5a shows the in-depth profiling of the case in which the TMS plasma polymer was deposited on oxygen plasma–treated steel. In the interphase section, a significant amount of oxygen (30%) was detected. The buildup of oxide layer due to oxygen plasma was clearly identified. In-depth mass spectroscopy of the atom cluster in Figure 33.5b shows the existence of Fe-O species. Possible interfacial bridging bonds are Fe-O-Si and Fe-O-C, which provide strong adhesion in the dry condition. However, corrosion test results show that such types of interfacial bonds are not as stable as Fe-C or Fe-Si in the existence of water and ionic species (acidic or basic condition). Fe-C and Fe-Si were not detected in this case.

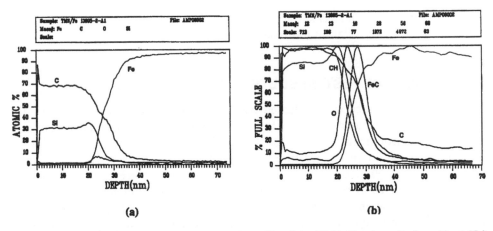

Figure 33.4 SNMS in-depth compositional profile of the TMS film deposited on $(Ar + H_2)$ plasma pretreated steel, (a) atomic compositional profile, (b) profiles of atom clusters.

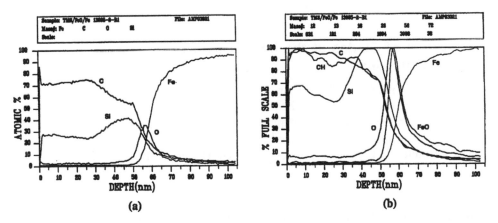

Figure 33.5 SNMS in-depth compositional profile of the TMS film deposited on O_2 plasma pretreated steel, (a) atomic compositional profile, (b) profiles of atom clusters.

Based on the analyses of RAIR, XPS, and SNMS, the structure of the TMS plasma polymer on steel surfaces can be hypothetically constructed. When treated with oxygen plasma, the steel surface was covered with an oxide layer. The TMS plasma polymer was deposited on the oxide surface. Interfacial bonds are consisted with Fe-O-Si and Fe-O-C. Based on the RAIR and SNMS analyses, the structure of the film bulk is a mixture of polysiloxane and polyalkylsilane with Si-O-Si and Si-O-alkyl linkage. Oxygen incorporation from the oxide via plasma/oxide surface interactions was taken into consideration.

When treated with $(Ar + H_2)$ plasma, the steel surface was basically free of oxide. Interactions of plasma with steel surface created an interphase with mixed composition of C, Si, and Fe. Possible bond formation includes C-Si, C-Fe, and Si-Fe in which C-Si and Fe-C bonds were confirmed by RAIR and SNMS analysis. The significance of the removal of oxides is twofolds: (1) formation of more stable covalent bonds between steel surface and TMS film and (2) by doing so, blocks the potential sites for reoxidation (corrosion).

3. CORROSION TEST RESULTS (COLD-ROLLED STEEL)

Corrosion performance was evaluated by the scab corrosion test. In the test, the primer-coated steel panels were scribed and subjected to 20 cycles of exposure to damaging environments as follows: 15 min immersion in 5% NaCl solution, 75 min air dry at room temperature followed, by 22.5 h exposure to 85% relative humidity and a 60°C environment. The tested samples were examined visually for failure such as corrosion, film lifting, peeling, adhesion loss, or blistering. The distance between the scribe line and the unaffected coating was a measure of the corrosion creep. An average of 10 measurements equally spaced along the scribe line was used to describe the corrosion test results. (The computer-assisted quantitative analysis of corrosion width, used in handling data described in previous chapters, was developed later and was not available to evaluate these data.)

3.1. Effect of Oxides at the Interface on Corrosion Test Results

The effect of interface of steel on the corrosion test result is summarized in Table 33.2. The results found for the panels, which were E-coated without phosphate or plasma polymer coatings, are quite astonishing. A scribe creep of 3.0 mm in a 4-week corrosion test was obtained for an oxide-removed CRS surface (without zinc phosphate nor plasma polymer), which is better than the E-coat on the

Table 33.2 Effect of the Presence or Absence of Oxides on the Corrosion Test Result Scribe Creep Width on GM Scab Test

Interfacial system	4-week test (mm)	8-week test (mm)
E-coat/Zn phosphate-Cr/oxides/steel	3.3	5.8
E-coat/Zn phosphate-Cr/oxides/Zn/steel	1.0	1.8
E-coat/oxides/steel	∞^a	∞^a
E-coat/steel[b]	3.0	4.7
E-coat/plasma polymer of TMS/oxides/steel	2.9	5.4
E-coat/plasma polymer of TMS/steel	0.2	0.5

Conditions of plasma treatments:
    Cathodic treatment:
        Oxygen 2 sccm, 20 W, 2 min, 50 mtorr
        $(Ar + H_2)$ 2 sccm Ar + 4 sccm H_2, 80 W, 16 min, 50 mtorr
    Cathodic polymerization:
        TMS 2 sccm TMS + 4 sccm H_2, 80 W, 2 min, 50 mtorr
[a]A scribe creep width more than 7 mm was considered as a total delamination.
[b]E-coat was applied within a minute after the CRS panel was taken from the plasma reactor. The surface contains a small but unknown amount of oxide reformed during this transfer process.

phosphate-chromated CRS. The same E-coat deposited on an oxidized CRS surface was totally delaminated on the scab test. It is important to recognize that the only difference between these two samples is the presence or absence of oxides. These two contrasting results dramatically demonstrated the importance of interface engineering and the potentially damaging role of the cathodic polarization during the process of cathodic E-coat deposition.

Although a good corrosion performance was obtained for the E-coat directly applied on an oxide-removed CRS surface (E-coat/oxide-free CRS), its corrosion performance was not as good as that for the oxide-free metallic surface protected by a plasma polymer of TMS (E-coat/plasma polymer/oxide-free CRS), which was better than the same E-coat deposited on a zinc phosphate–chromated electrogalvanized steel (EGS) (E-coat/phosphate-chromate/EGS), the second row in Table 33.2. The result means that the single step of low-pressure plasma interface engineering that deposits roughly 50 nm of TMS LCVD film can replace galvanizing and zinc phosphate-chromate conversion coating, both of which have significant environmental impacts.

These results pointed out the importance of the cathodic plasma treatment of CRS. In order to examine the effect of the pretreatment process on the corrosion performance, two sets of experiments were carried out in which $(Ar + H_2)$ plasma treatment time (at a fixed wattage of 80 W) and the discharge power (at a fixed treatment time of 12 min) were varied while all other operational parameters were kept constant. The results are shown in Figure 33.6, which clearly shows that the cathodic plasma treatment of CRS is a crucially important factor indicating the importance of the removal of oxides in this interface engineering approach.

Figure 33.7 depicts the influence of plasma pretreatment of CRS surface as well as the hydrophilicity of the plasma polymers on corrosion test results. The left half of the bar graph represents hydrophilic interface and some of top surface are also hydrophilic. The right half of the bar graph represents the water-insensitive interface and nonhydrophilic top surfaces except the plasma polymer of CH_4, which was intentionally kept in air for 10 min before application of E-coat. The figure indicates two important factors, i.e., the removal of oxides from CRS/plasma polymer interface, and nonhydrophilic top surface of plasma coatings, for corrosion protection of CRS by plasma interface engineering, which involves application of cathodic E-coat. While the air exposure of plasma polymer of CH_4 severely deteriorated the corrosion protection of E-coated sample, the same exposure of TMS surface showed no effect. This difference seems to reflect the reactivity of double bonds described in Chapter 7.

3.2. Effect of Lead-Free E-coat

Most commercially available cathodic E-coat paints contain lead in the recipe, and the removal of lead from E-coat generally leads to an appreciably inferior performance on scab test [7]. The effect of lead-free E-coat on the corrosion performance of plasma interface–engineered systems was examined and the results are shown in Table 33.3. The elimination of lead resulted in significant deterioration of the corrosion protection of the phosphated samples. In a strong contrast to the conventional surface preparation, the plasma interface–engineered systems showed

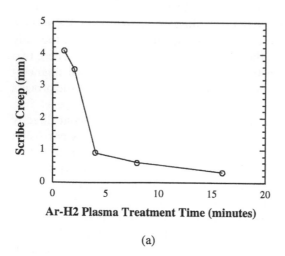

(a)

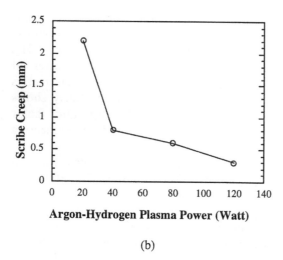

(b)

Figure 33.6 Effect of $(Ar + H_2)$ plasma pretreatment of steel surface on the corrosion performance of E-coat/plasma polymer combined coating system: (a) effect of treatment time, (b) effect of discharge power.

practically no decrease of corrosion performance due to the absence of lead in E-coat.

The effect of the removal of lead was found most severely for the phosphated CRS (3.3 vs. ∞). The effect was much less but significant for the phosphated EGS (1.0 vs. 3.4). The interfacial system with oxide-free CRS showed practically no difference (0.5 vs. 0.6), and a marginal difference (3.2 vs. 3.5) was found for the oxide-containing surface.

According to these results, the role of lead in a cathodic E-coat might be speculated as the buffering of the cathodic polarization during the process of E-coat deposition. In consideration of the use of lead-free E-coat, the plasma interface–engineered system seems to have a distinctive advantage over other conventional systems because the effect of cathodic polarization is minimal.

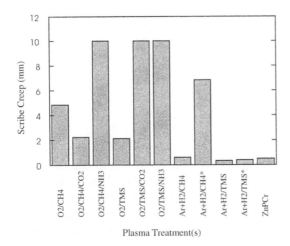

Figure 33.7 Scab test results of cathodic E-coat coated on plasma treated CRS and phosphated CRS (as control); *: E-coat was applied after 10 min air exposure.

Table 33.3 Effect of Removal of Lead from E-coat on the Corrosion Test Result

	Scribe creep width (mm) on 4-week GM scab test	
Interfacial system	Leaded E-coat[a]	Lead-free E-coat[b]
E-coat/Zn phosphate-Cr/oxides/steel	3.3	∞[c]
E-coat/Zn phosphate-Cr/oxides/Zn/steel	1.0	3.4
E-coat/plasma polymer of TMS/oxides/steel	3.2	3.4
E-coat/plasma polymer of TMS/steel	0.5	0.6

The conditions of plasma treatments
  Cathodic treatment:
    Oxygen 2 sccm, 20 W, 2 min, 50 mtorr
    $(Ar + H_2)$ 1 sccm Ar + 4 sccm H_2, 40 W, 12 min, 50 mtorr
  Cathodic polymerization:
    TMS 2 sccm TMS + 4 sccm H_2, 80 W, 2 min, 50 mtorr
[a]Cormax I, DuPont.
[b]An experimental paint.
[c]A scribe creep width more than 7 mm was considered as a total delamination.

All experimental data and discussions presented above seem to confirm the validity of the hypothesis that if reducible elements present in the surface state (including the ultrathin layer of plasma polymer) on which a cathodic E-coat is applied, those elements will be subjected to the cathodic reduction during the process of the cathodic E-coat deposition, and a weak boundary or defective spots would be created in the interface of the E-coat and the substrate (plasma polymer–coated steel). The validity of the hypothesis implies that the best result of a cathodic E-coat cannot be realized unless the adverse effect of the cathodic reduction can be minimized. In this context, the presence or absence of oxides is a very important factor in the use of a cathodic E-coat, besides the fact that the presence of oxides makes the interface more sensitive to water. For the same reason, the chemical stability of the top surface of plasma coatings is equally important. Data shown in Figure 33.7 indicate that hydrophilic component on plasma polymer seriously damages corrosion protection of E-coated CRS.

4. CORROSION PROTECTION OF PURE IRON

SAIE by low-pressure plasma polymerization of TMS was extended to pure iron [8]. Polished pure iron samples (3×3 cm) were plasma pretreated before deposition of TMS plasma polymer. Two to six samples of pure iron were placed on a CRS plate (15×10 cm) maintaining the electrical contact so that each pure iron sample acts as the cathode of DC discharge. A few small pieces of silicon water were also placed on the CRS plate to maintain the electrical contact and were used for the estimation of the thickness of TMS plasma polymer by ellipsometry.

Two CRS plates, with pure iron samples and silicone wafers attached, were placed back to back by metallic clips and hanged in the middle of two anodes (15.5 cm apart) equipped with magnetic enhancement. DC discharge was created between the center cathode and two magnetron anodes. Two different pretreatments of samples, i.e., $(Ar + H_2)$ plasma and O_2 plasma, were applied before plasma polymerization of TMS. Samples thus prepared are summarized in Table 33.4. Appreciable difference in the deposition rate of TMS depending on the type of plasma pretreatment is evident, i.e., the deposition of TMS on O_2 plasma–treated surface is significantly greater than that on $(Ar + H_2)$ plasma–pretreated sample, indicating the participation of oxygen emanating from oxides in plasma polymerization of TMS.

4.1. XPS Analysis of TMS Film

Sputter profiles of the samples O110 and ArH68 (see abbreviation in Table 33.4) are shown in Figure 33.8. Figure 33.8a indicates that $(Ar + H_2)$ plasma treatment applied wasn't enough to remove all oxides, and a significant level of O_2 was found

Table 33.4 Description of Plasma Interface–Engineered Iron Samples

Sample	Pretreatment	TMS treatment time (s)	Coating thickness (nm)
ArH	$Ar + H_2$	None	
ArH22	$Ar + H_2$	10	22.5
ArH35	$Ar + H_2$	40	34.5
ArH57	$Ar + H_2$	90	57.0
ArH68	$Ar + H_2$	120	68.5
O	O_2	None	
O23	O_2	15	23.4
O50	O_2	40	50.0
O92	O_2	90	92.0
O116	O_2	120	116.4

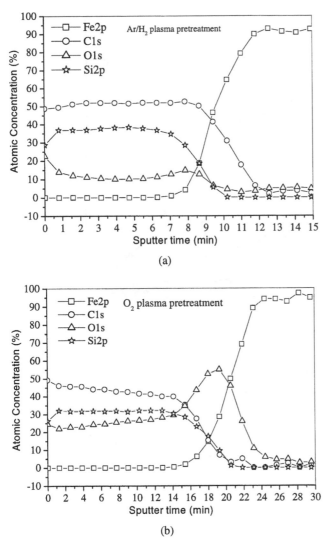

Figure 33.8 XPS depth profile of plasma polymer of TMS deposited on (a) $(Ar + H_2)$ plasma treated pure iron, (b) O_2 plasma treated pure iron.

throughout the film thickness. A clear oxygen peak at the interface can be observed for the O_2 plasma–pretreated surface shown in Figure 33.8b, and oxygen was found throughout the entire thickness of the film with a gradient of concentration. The oxygen concentration slowly decreased with increasing distance to the metal oxide layer. This indicates that the incorporation of oxygen from the iron oxide into the film, which is also observed in the in situ polymerization on CRS described in previous sections. The presence of O_2 throughout the film thickness in both cases differs significantly from the similar profiles made with plasma polymer of TMS deposited on CRS, which were treated similarly. Figure 33.9 depicts the Auger depth profile of TMS films deposited on CRS.

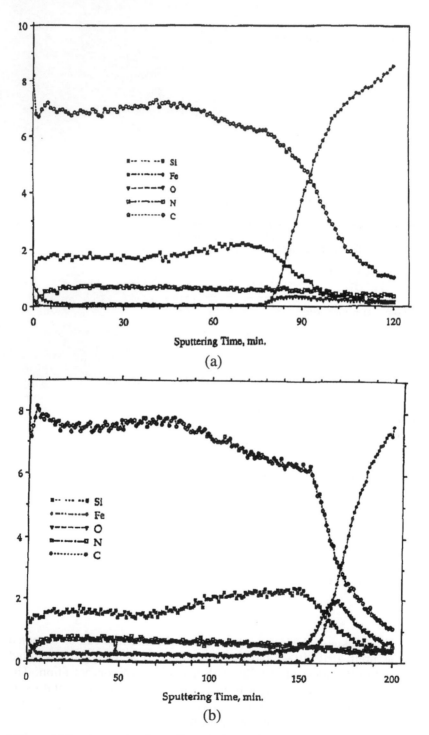

Figure 33.9 Auger depth profile of TMS film on cold-roll steel, (a) $(Ar + H_2)$ plasma pretreated, (b) O_2 plasma pretreated.

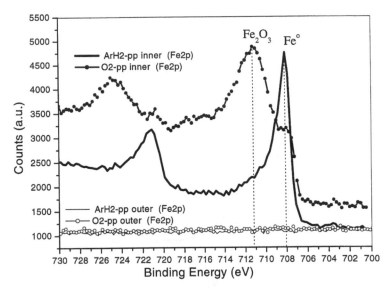

Figure 33.10 Comparison of XPS Fe 2p signals for TMS films deposited on O_2 plasma treated and $(Ar + H_2)$ plasma–treated pure iron.

In the case of CRS, approximately the same $(Ar + H_2)$ plasma treatment nearly completely removed oxides, and no O_2 was found in the TMS film, as seen in Figure 33.9a. A small but clear O_2 peak was found at the interface, but very small O_2 was found in the film as depicted in Figure 33.9b, i.e., no feeding of O from oxides to depositing TMS film was observed beyond the transitional zone.

Figure 33.10 depicts the difference of the state of Fe at the film/metal interface and also at the top surface of the coated samples. Iron is in its metallic form at the interface for $(Ar + H_2)$ plasma–treated sample, whereas iron is in an oxidized state at the interface for O_2 plasma–pretreated sample, as anticipated. The Si 2p peak and oxygen peaks in Figures 33.11 and 33.12, respectively, shows that some of oxygen at the interface of O_2 plasma pretreated sample is in the form of Si-O groups, indicating the reaction of Si-based moieties with O atoms in oxides.

4.2. Morphology of Film Surface

Figures 33.13 shows the topography of the two plasma polymer layers deposited under different conditions on a polished iron surface. Both films show a similar topography as observed by atomic force microscopy, but the film deposited on O_2 plasma–pretreated polished iron showed a little more grainy surface than $(Ar + H_2)$ plasma–pretreated sample. In Figure 33.14 the root mean square value is plotted against the film thickness. The grainy surface (O_2 plasma pretreated), which showed a higher deposition rate, increased the roughness as the thickness increased as expected.

4.3. Electrochemical Impedance Spectroscopy

The electrochemical tests were carried out in a quiescent aerated borate-sulfate buffer solution (0.025 $Na_2B_4O_7 \cdot 10$ $H_2O + 0.5\,M$ Na_2SO_4) prepared from distilled water

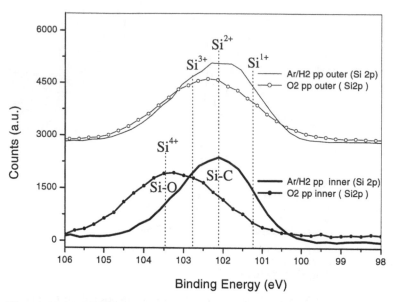

Figure 33.11   Effects of plasma pretreatment on Si 2p signal of plasma polymer of TMS.

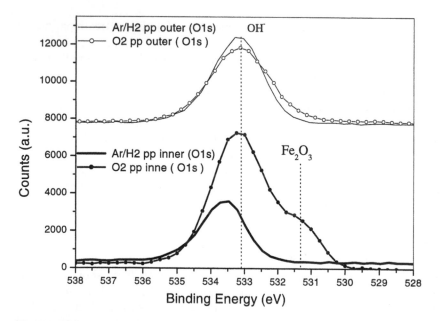

Figure 33.12   Effects of plasma pretreatment on O 1s signal of plasma polymer of TMS.

with pH = 8.9 at room temperature. The time of testing ranged from 5 h to 1 or 2 days depending on the protection properties of the coatings.

Electrochemical Impedence Spectroscopy (EIS) data are shown in Figure 33.15 (Bode plot) and Figure 33.16 (phase angle–frequency plot). All the samples exhibit

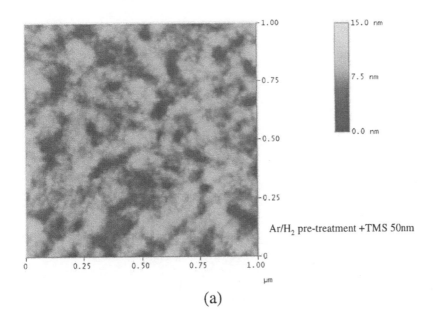

Ar/H$_2$ pre-treatment +TMS 50nm

(a)

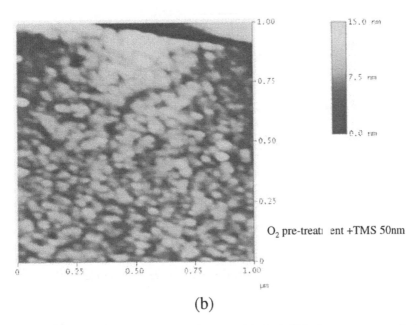

O$_2$ pre-treatment +TMS 50nm

(b)

Figure 33.13 AFM images of TMS films, (a) (Ar + H$_2$) plasma pretreated, (b) O$_2$ plasma pretreated.

predominantly capacitive behavior and only one time constant is observed in the first hours of exposure. This kind of impedance response is characteristic of an insulating polymer layer. The coating capacitance is generally considered to provide information on the degree of water penetration through the coatings and, in

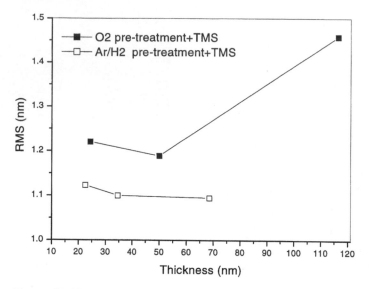

Figure 33.14   Effect of film thickness on the roughness.

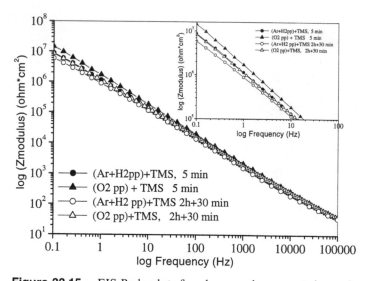

Figure 33.15   EIS Bode plots for plasma polymer coated pure iron samples.

principle, its value is expected to increase with immersion time. Figure 33.17 (top right) presents the evolution of C_p vs. the immersion time for the two systems studied. For both systems, the capacitance values increased with time, and it is interesting to note that characteristic gradients are observed for both conditions. The capacitance values for the $(Ar + H_2)$ plasma–pretreated iron sample showed a slow increase in comparison with the capacitance values obtained for the O_2 plasma–pretreated sample. This indicates that the water uptake is higher for the O_2 plasma–pretreated iron sample. The water uptake, φ, was calculated from the

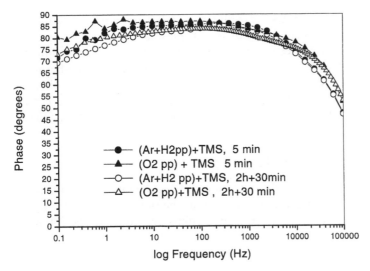

Figure 33.16 Phase angle–frequency plots for plasma polymer–coated pure iron samples.

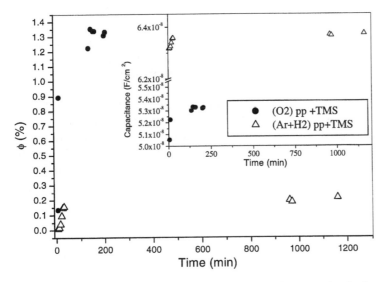

Figure 33.17 Time dependent increase of capacitance and calculated volume fraction of water.

measured coating capacitance C_p using the Brasher–Kingsbury equation, which is given by

$$\varphi = \frac{\log(C_p/C_{p_0})}{\log \varepsilon_w}$$

where C_{p_0} is the coating capacitance extrapolated for $t = 0$ (i.e., the dry film capacitance) and ε_w is the dielectric constant of water ($\varepsilon_w = 78.5$ at T $= 25°$C).

Figure 33.17 (bottom left) shows the evolution of the water volume fraction, φ, in the polymer for both systems studied. For the O_2 plasma–pretreated sample, the water volume fraction increases steeply, denoting a faster ingress of water into the layer in comparison with the $(Ar + H_2)$ plasma–pretreated sample. This higher water uptake for the O_2 plasma–pretreated sample, 1.3% in comparison to 0.2% obtained for the $Ar + H_2$ plasma–pretreated sample in the saturation stage is in agreement with the higher oxygen atomic concentration in the film observed by XPS for the O_2 plasma–pretreated sample.

4.4. Analysis by Scanning Kelvin Probe

The Kelvin probe was used to study the oxidation state at the interface to the metal surface. Grundmeier and Stratmann have shown that the potential measured by the SKP depends on the oxidation state of iron oxide films according to the following equation [9]:

$$\Delta\Psi_{Pol}^{Ref} = -\frac{\Delta\mu_{Fe^{2+}/Fe^{3+}}^{0}}{F} + \Delta\phi_{Pol}^{Ox} - \frac{W_{Ref}}{F} + \chi_{Pol} + \frac{RT}{F}\ln\frac{[Fe^{3+}]}{[Fe^{2+}]}$$

In order to study the effect of the aging after atmospheric exposure on the potential at the inner buried interface of plasma polymer–coated iron, two different plasma–pretreated iron samples were used. For each pretreatment, different TMS plasma polymer thicknesses were studied: 20, 50, and 70 nm for $Ar + H_2$ plasma pretreatment, 20, 55, and 115 nm for O_2 plasma pretreatment. The Scanning Kelvin Probe (SKP) data shown in Figure 33.18 for $(Ar + H_2)$ plasma–pretreated and O_2 plasma–pretreated samples depicts a correlation between the potential at the inner buried interface and the polymer thickness for respective sample. SKP results showed that the plasma polymer effectively inhibits reoxidation of the interface

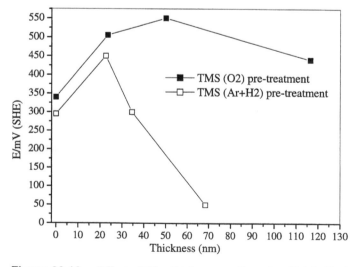

Figure 33.18   Effect of film thickness on Scanning Kelvin Probe potential.

acting as a barrier against the diffusion of oxygen and water molecules during the aging process.

It is clear from the study on pure iron that oxides participate in LCVD of TMS, and characteristics of plasma polymer films differ depending on the extent of oxides present on the surface when LCVD is applied. Oxides on the surface of pure iron are more stable than those on steel and hence more difficult to remove, but this can be effected by plasma pretreatment with $(Ar + H_2)$ mixture. SAIE by LCVD involving removal of oxides provides excellent corrosion protection of pure iron. The key factor of SAIE by LCVD for corrosion protection of metals in general is the handling of oxides, which depends on the characteristic nature of the metal oxide to be handled. Once strong chemical bonds were formed between nanofilm of plasma polymer, either through oxides or direct bonding to the substrate metal, the LCVD film acts as the barrier to corrosive species.

REFERENCES

1. Wang, F. Tinghao; Yasuda, H.; Lin, T. J.; Antonelli, J. A. Prog. Organic Coatings **1996**, *28*, 291.
2. Yasuda, H.; Wang, F. Tinghao; Cho, D. L.; Lin, T. J.; Antonelli, J. A. Prog. Organic Coatings **1997**, *30*, 31.
3. Lin, T.J.; Antonelli, J.; Yang, D.J.; Yasuda, H.K.; Wang, F.T. Prog. Organic Coatings **1997**, *31*, 351–361.
4. Ohuchi, F.S.; Lin, T.J.; Yang, D.J.; Antonelli, J.A. Thin Solid Films **1994**, *10*, 245.
5. Yasuda, H. *Plasma Polymerization*; Academic Press: Orlando, FL, 1985; 72–177.
6. Brusic, V.; MacInnes, R.D.; Aboaf, J. In Proc. 4th Int. Symp. Passivity Met., Electrochem. Soc., R.P. Frankenthal and J. Kruger Eds. Princeton, NJ, 1978, p. 170.
7. Walker, P. J. Oil Col. Chem. Assoc. **1982**, *65*, 415.
8. Data to be published (2003).
9. Grundmeier, G.; Stratmann, M. Appl. Surf. Sci. **1999**, *141*, 43.

34

Membrane Preparation and Modification

1. GENERAL PRINCIPLES IN MEMBRANE APPLICATION OF LUMINOUS CHEMICAL VAPOR DEPOSITION AND LUMINOUS GAS TREATMENT

The membrane in a broad sense is a thin layer that separates two distinctively different phases, i.e., gas/gas, gas/liquid, or liquid/liquid. No characteristic requirement, such as polymer, solid, etc., applies to the nature of materials that function as a membrane. A liquid or a dynamically formed interface could also function as a membrane. Although the selective transport through a membrane is an important feature of membranes, it is not necessarily included in the broad definition of the membrane. The overall transport characteristics of a membrane depends on both the transport characteristics of the bulk phase of membrane and the interfacial characteristics between the bulk phase and the contacting phase or phases, including the concentration polarization at the interface. The term "membrane" is preferentially used for high-throughput membranes, and membranes with very low throughput are often expressed by the term "barrier."

Luminous chemical vapor deposition (LCVD) and luminous gas treatment (LGT), which does not yield the "primary deposition," could be used in the preparation and modification of membrane and barrier [1]. The term "primary deposition" refers to the direct deposition of material from the luminous gas (LCVD) in contrast to "secondary deposition" that results from the deposition of ablated material in LGT. It should be emphasized, however, that both methods are nanofilm technologies and require the substrate membrane on which LCVD nanofilm is deposited or the surface is modified. Accordingly, their use should be limited to special cases where such a nanofilm could be incorporated into membrane or the LGT of surface is warranted.

One of the primary advantages of plasma polymerization is the ability to conformably deposit films with thicknesses in the range from 10 to 100 nm, where conventional polymer coating methods are difficult or impossible. On the other hand, this advantageous feature sets the limitation of the approach in rather stringent requirements for the suitable substrates. For the same reason, applications that require a relatively thick coating (e.g., thickness greater than $1\,\mu m$) are considered inappropriate for the most advantageous use of plasma polymerization coating.

The transport flux across a membrane per unit thickness, which is expressed by permeability coefficient, P, and its dependence on the nature of permeant, which could be represented by the permeability ratio for different permeant A and B, $\alpha = P_A/P_B$, are two important parameters for dealing with membranes. In most cases, the smaller value of P is used in the denominator so that the value of α is greater than unity.

The use of LCVD or LGT in membranes is aimed at increasing the value of α for a set of permeants of interest. The increase of α, however, is achieved at the expense of the permeant throughput, i.e., P_A and P_B both decrease by the application of LCVD layer. Therefore, the use of high-flux membrane or substrate is mandatory. In barrier applications, the selection of good barrier (low values of Ps) is mandatory because the barrier characteristic of a film can be expressed by ℓ/P, where ℓ is the thickness of film, and there is a limit in the increase of barrier characteristics by application of nanofilm.

The permeation of a simple permeant, e.g., O_2, through a polymeric membrane could occur, in principle, by two different mechanisms. One is the transport through pores, and the other is the transport through the free volume of polymer solid. The size of pore and its distribution is the most crucial parameter in the former case (porous membrane), and the value of α is determined by the molecular sizes of permeants A and B. The values of Ps are in the reverse order of the size of permeant, i.e., $P_{N2} > P_{O2} > P_{CO2}$, and P can be dealt as a kinetic parameter.

In the latter case (nonporous membrane), the space in which the transport occurs is not fixed in size and location. The free volume is the volume that is not occupied by the polymer molecules in the solid phase, and its size and location fluctuate with time at a given temperature. Accordingly, the transport through such a membrane is completely different from the transport through fixed pores, and can be expressed by the *solution–diffusion* mechanism. The permeant is first dissolved in the membrane phase, and the dissolved permeant diffuses through the membrane following the chemical potential gradient.

In the solution–diffusion case, P is given by $P = sD$, where s is the solubility coefficient and D is the diffusion coefficient. Accordingly, P is a product of kinetic parameter D and thermodynamic parameter s. In such membranes, the solubility coefficient is a dominant factor that controls the value of P. Since a larger gas generally has the higher solubility coefficient, while the diffusivity is low, the permeability of a larger gas has the higher value of P, and the value of P follows the order of molecular size, i.e., $P_{N2} < P_{O2} < P_{CO2}$.

The permeability ratio, e.g., CO_2/O_2, can be used to distinguish the permeation mechanism through a membrane. When a plasma polymer is deposited on a porous film, the transport mechanism changes from flow to solution–diffusion as the thickness of deposition increases and a homogeneous (nonporous) film covers pores, i.e., a porous membrane is converted to a nonporous membrane. Figure 34.1 depicts the change of gas transfer rate K_0/thickness as a function of thickness of glow discharge polymer of tetramethyldisiloxane deposited on porous polypropylene film. The gas transfer rate drops sharply as the majority of pores are covered, but P_{CO2}/P_{O2} is still less than unity, indicating that the dominant transport mechanism is the flow through the remaining small pores. When a nonporous film covers the total

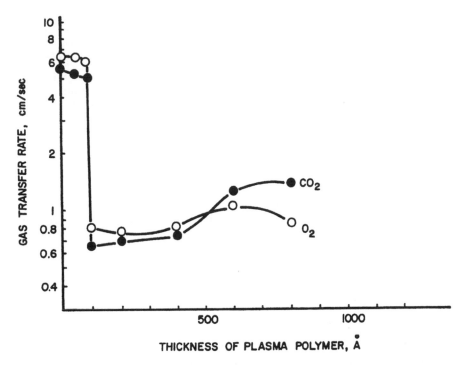

Figure 34.1 Change of transport mechanisms on deposition of plasma polymer of tetramethyldisiloxane on porous polypropylene film.

membrane area the permeability ratio changes to that for a solution–diffusion type of membrane, i.e., P_{CO2}/P_{O2} becomes greater than 1.

The permeability ratio of O_2/N_2 for many polymers is roughly 4, which means that the larger O_2 permeates through a polymer roughly four times faster and the smaller N_2 under identical conditions. Since simple gases such as N_2, O_2, and CO_2 do not specifically interact with polymer molecules, the value of α (among these gases) for many polymers is nearly constant. However, the near-constant rule does not apply if very small gas, such as H_2, and He, is paired because some of the free volume is large enough for such a small gas and behaves as pores for the gas.

The deviation from the near-constant rule also occurs if the gas interacts specifically with polymer molecules. This is seen in the case of water vapor permeability. The molecular size of H_2O is approximately the same as that of O_2; however, the solubility of H_2O is much greater than that of O_2 in orders of magnitudes and varies greatly depending on the nature of polymers. Consequently, there is no near-constant value of α that is found for many gas pairs, and the permeability ratio of H_2O/O_2 for polymers spreads in orders of magnitude.

Type A plasma polymers (see Chapter 8), so far as the transport characteristics are concerned, could be viewed as nanoscale molecular sieves, which are not solution–diffusion membranes. Therefore, the increase of α by reducing the transport rate of the denominator permeant with the minimal reduction in the transport rate of the numerator permeant is the main viable principle for LCVD and LGT modification of membranes.

Thus, plasma polymers (Type A) are, by category, barriers. A nanofilm of such a molecular sieve barrier could be used (1) in improving the selectivity of a high-throughput substrate membrane by depositing a nanoscale sieve or (2) in preparing a composite membrane by depositing on an appropriate porous supporting membrane. The deposition of plasma polymers on silicone polymer films, which have high flux but low perm selectivity, to impart selectivity with minimal reduction in the throughput is a typical example of the first case. If the substrate polymer is a barrier, the deposition of a plasma polymer further improves the barrier characteristics of the substrate, as demonstrated by the inner surface coating of a poly(ethylene terephthalate) (PET) bottle.

The deposition of a plasma polymer on an appropriate porous substrate to form a composite reverse osmosis membrane is a typical example of the second case. In both cases, however, the selection of the substrate membrane is the crucially important factor, particularly in the second case. In the application of nanofilm, the pore size of the substrate membrane must have the uniformity of pore size in nanometer scale, which is an extremely difficult requirement.

Luminous vapor treatment without depositing film (LGT) could be used to modify the surface characteristics of membranes. Type B plasma polymer also could be used for this purpose. General schemes of membrane application of LGT and LCVD are schematically depicted in Figures 34.2 and 34.3, respectively [2]. Since the luminous gas interacts with the substrate material, the selection of the membrane material and the gas to be used in these possible schemes is important, and it should not be considered that any combinations of gas and material could be used in any mode of application.

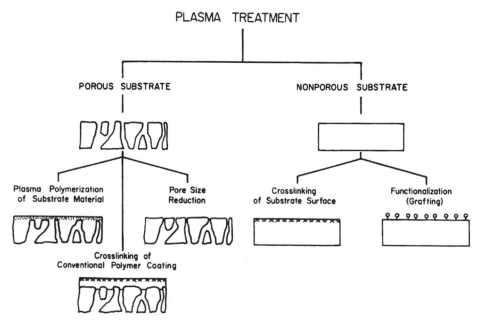

Figure 34.2 Schematic principle of plasma treatment of membranes.

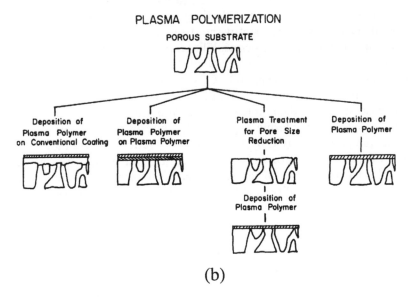

PLASMA POLYMERIZATION
NONPOROUS SUBSTRATE

Deposition of
Plasma Polymer

Deposition of
Plasma Polymer
on Conventional Coating

(a)

PLASMA POLYMERIZATION
POROUS SUBSTRATE

Deposition of
Plasma Polymer
on Conventional Coating

Deposition of
Plasma Polymer
on Plasma Polymer

Plasma Treatment
for Pore Size
Reduction

Deposition of
Plasma Polymer

Deposition of
Plasma Polymer

(b)

Figure 34.3 Modes of plasma polymerization coating, (a) on nonporous membrane, and (b) on porous membrane.

2. LUMINOUS GAS TREATMENT

2.1. Luminous Gas Treatment of Microporous Membrane

LGT could be used to reduce the pore size of a microporous membrane. The modification is by the principle of the LCVD of the ablated substrate material by the action of luminous gas. This method greatly depends on the type membrane material. If the microporous membrane is made of oxygen-containing polymers, the redeposition is largely hampered and the treatment simply ablates the membrane surface. If the membrane is made of polymers that contain elements or ligands that can be easily deposited by LCVD, this method provides a simple process to improve the selectivity of microporous membranes. This type of treatment was used for the preparation of reverse osmosis membranes. Helium plasma was used to modify the surface of porous acrylonitrile copolymer films that showed some degree of the reverse osmosis membrane characteristics [3]. The original membrane is a porous membrane that has no salt rejection capability. By simply treating the surface with He glow discharge, a porous membrane was converted to a reverse osmosis membrane, which indicates the simple luminous gas treatment could modify the transport characteristics of a microporous membrane to a certain extent, if the membrane material meets the requirements for the redeposition of ablated moieties by LGT.

2.2. Luminous Gas Treatment of Nonporous Membrane to Increase the Selectivity

Silicone rubber film has very high gas permeability among conventional commercially available polymers, but the selectivity is poor. If the selectivity is increased by a simple gas plasma treatment without sacrificing the flux, it could be used as a gas separation membrane. The permeability ratios for O_2/N_2 and CO_2/CH_4 gas pairs obtained from single-gas permeability measurements together with the permeation rate of the more permeable gas (O_2, CO_2) are plotted as a function of the argon plasma treatment time in Figures 34.4 and 34.5. The permeability ratios for both gas pairs increase rapidly in the first few minutes of plasma treatment and level off during prolonged periods of plasma exposure. In spite of the decrease of the permeability of the numerator gas (CO_2), the permeability ratio increases indicating that the decreases of the denominator gas (N_2 or CH_4) are greater in these cases.

An argon plasma treatment of a polydimethylsiloxane film caused the reduction of the oxygen permeability coefficients, with the coefficients decreasing with increasing argon plasma treatment time [4]. Such a deduction of gas permeability by Ar plasma treatment was interpreted due to the introduction of cross-linking on the top layer of the surface. Such a concept was presented as the cross-linking by activated species of inert gases ("CASING") [5]. The decrease of gas permeability as a function of cross-linking density is not so clear-cut an observation because the cross-linking density needed to reduce the chain mobility to hinder the transport of small gases is far greater than that needed to see the changes in other mechanical properties of polymers, and if such a high cross-linking had been indeed introduced by Ar plasma treatment is questionable. The deposition of plasma polymers of ablated moieties is more likely the cause of the permeability reduction.

The effect of an argon plasma treatment of natural rubber on the permeation of oxygen and nitrogen was investigated as one facet of a study of the gas separation properties of plasma polymer films deposited on various homogeneous polymer films [6]. The permeability coefficients of the untreated film were 8.79×10^{-8} cm^2/s for nitrogen and 1.79×10^{-7} cm^2/s for oxygen. The nitrogen permeability coefficient decreased to 7.96×10^{-s} cm^2/s after the film was exposed to argon plasma for 30 min;

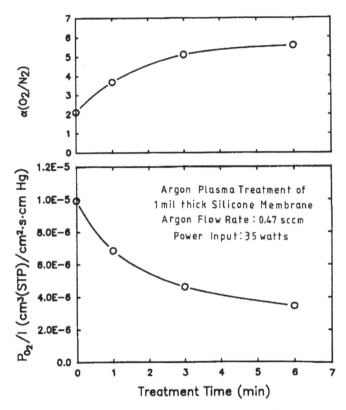

Figure 34.4 Effect of Ar plasma treatment a silicone membrane on the gas flux and the O_2/N_2 selectivity.

the oxygen permeability coefficient increased to $1.96 \times 10^{-7} \, cm^2/s$. An argon plasma treatment of 240 min yielded a film with a nitrogen permeability of $6.80 \times 10^{-8} \, cm^2/s$ and the oxygen permeability coefficient remained at $1.96 \times 10^{-7} \, cm^2/s$.

The diffusion coefficients of the untreated natural rubber film, as determined by diffusion time lag measurements, were $1.04 \times 10^{-6} \, cm^2/s$ for nitrogen and $1.15 \times 10^{-6} \, cm^2/s$ for oxygen. The nitrogen diffusion coefficient decreased to $2.8 \times 10^{-7} \, cm^2/s$ and the oxygen diffusion coefficient decreased to $3.2 \times 10^{-7} \, cm^2/s$ for the 240-min argon plasma–treated film. This indicates that the increase of oxygen permeability coefficients (to sustain O_2 permeability of untreated film) is due to an increase in the solubility of oxygen in the Ar plasma–treated natural rubber film. A plausible reason for the increased oxygen solubility is the formation of polar groups on the Ar plasma–treated natural rubber surface due to the introduction of polar groups caused by the post-plasma reactions of dangling bonds with ambient oxygen, which makes the top surface more specifically interact with oxygen.

2.3. Functionalization of Membrane Surface

The functionalization of membrane surface could be achieved by the grafting of functional groups created by luminous gas. Such a treatment has been used to attach amino groups to polymer surfaces. The luminous gas of ammonia, a mixture of

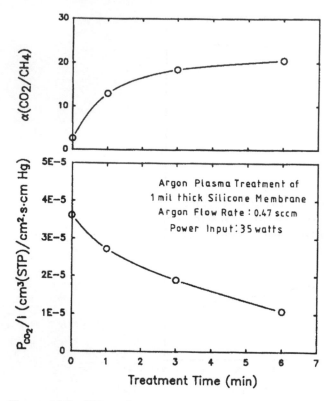

Figure 34.5 Effect of Ar plasma treatment a silicone membrane on the gas flux and the CO_2/CH_4 selectivity.

nitrogen and hydrogen, and organic amines such as arylamine were used to create the amino groups, which then combine various functional substances such as proteins to the surfaces without substantially altering the bulk properties of the polymers. However, it should be bore in mind that the original functional group fragments considerably in the luminous gas phase and the yield of the original functional group, e.g., $-NH_2$ on the treated surface, is relatively low.

Simple gas discharge treatment, such as of Ar, introduces numerous dangling bonds on the surface, which subsequently react with ambient oxygen when the treated surface is taken from the discharge reactor yielding variety of functional groups from peroxide, carbonyl, carboxylic acid, and so forth. These functional groups, including the surface-bound free radicals, could be utilized to functionalize the surface by reacting with appropriate chemicals depending on the stage of the free radicals' cascading reactions. The excellent adhesion of primer coating applied on the surface of plasma polymer–coated metals, described in Chapter 32, is due to the reaction of functional groups discussed here.

3. LCVD ON NONPOROUS MEMBRANE

The deposition of a nanofilm by LCVD on nonporous polymer film involves the interaction of species in the luminous gas phase and the surface of the film. This aspect is clearly demonstrated in the deposition of trimethylsilane (TMS) on O_2 plasma–treated iron vs. on $(Ar + H_2)$ plasma–treated iron. The deposition rate as well as characteristics of LCVD layer depends on the extent and nature of oxides on the surface of iron as described in Chapter 33. Consequently, the permeation properties of the nanofilm depend on the nature of the polymer that constitutes the substrate film, i.e., the same result cannot be anticipated when different substrate films are used. For example, the plasma-polymerized cyanogen bromide coating has a permeability ratio (H_2/CH_4) of 490 when deposited on a poly(phenylene oxide) (PPO) substrate, but when it is deposited on a polysiloxane/polycarbonate copolymer (SC) substrate, the α ratio is 50 [7].

The composite membranes prepared from plasma polymers of the nitrile-type monomers showed the highest permeability ratios of hydrogen to methane, followed by the vinyl-type plasma polymers and then the plasma polymers from the aromatic compounds. The plasma-polymerized composites all showed permeability ratios (H_2/CH_4) greater than unity, compared to a ratio of 0.87 for the substrate material itself. The highest hydrogen/methane permeability ratio was 36.6, attained from the composite prepared from cyanogen bromide, with a reduction in hydrogen permeability from 1.0×10^{-6} cm^2/s to 3.0×10^{-7} cm^2/s. The plasma polymer prepared from benzonitrile under these conditions gave a permeability ratio of 33.0 with only a 21% reduction in the hydrogen permeability.

Plasma polymer films from cyanogen bromide and benzonitrile were also deposited on PPO films. The composite films gave hydrogen-to-methane permeability ratios of 297 and 68 for cyanogen bromide and benzonitrile plasma–polymerized composites, respectively. The PPO substrate film itself has a higher permeability ratio (23.5) than the SC substrate (0.87), although lower hydrogen (6.42×10^{-7}) and methane $(2.73-10^{-8})$ permeability than the SC substrate.

Tables 34.1 and 34.2 show the gas permeation results pertaining to plasma polymers prepared from various nitrile-type monomers that were deposited onto 1-mil thick silicone and silicone-carbonate copolymer sheets [7]. Shown also are the results obtained from the uncoated polymer sheets for comparisons. Both silicone and silicone-carbonate polymer films prior to the plasma polymer deposition show H_2/CH_4 permeability ratios of 0.79 and 0.97, respectively. After being coated with a

Table 34.1 H_2 Permeation Rates and the Permeability Ratio of H_2/CH_4 for Plasma Polymers of Nitrile-Type Monomers Deposited on Silicone Sheet

Monomer	Flow rate (sccm)	Power (W)	Pressure (mtorr)	Thickness (Å)	$P/\ell \ X \ 10^6 \ H_2$	Permeability ratio α (H_2/CH_4)
Methacrylonitrile	1.98	60	39.4	590	7.01	39.5
Methacrylonitrile	2.03	60	39.2	1330	7.48	20.3
Methacrylonitrile	1.97	60	39.5	1610	3.41	19.1
Benzonitrile	0.31	70	19.0	1860	5.63	28.7
Benzonitrile	0.30	70	18.5	2030	6.22	35.4
Blank silicone sheet					13.9	0.79

Units of permeation rate: cm^3 (STP)/cm^2-s-cm Hg. Substrate: 1-mil-thick silicone sheet (MEM 100, General Electric).

Table 34.2 H_2 Permeation Rates and the Permeability Ratio of H_2/CH_4 for Plasma Polymers of Nitrile-Type Monomers Deposited on Silicone-Carbonate Copolymer Sheet

Monomer	Flow rate (sccm)	Power (W)	Pressure (mtorr)	Thickness (Å)	$P/\ell \ X \ 10^6 \ H_2$	Permeability ratio α (H_2/CH_4)
Crotononitrile	1.26	33	29.8	270	4.37	36.3
Crotononitrile	1.27	57	29.7	150	4.45	15.8
Acrylonitrile	1.27	54	29.9	520	3.02	21.7
Butyronitrile	1.26	70	29.4	140	6.22	37.4
Butyronitrile	1.27	70	29.6	190	7.01	35.2
Butyronitrile	1.26	70	30.2	440	6.89	33.1
Blank silicone-carbonate sheet					8.22	0.97

Units of permeation rate: cm^3 (STP)/cm^2-s-cm Hg. Substrate: 1-mil-thick silicone-carbonate sheet (MEM 213, General Electric).

thin layer of plasma polymer of nitrile-type monomers, both films show permeability ratios of H_2/CH_4 as high as above 35 with only a 15–50% reduction in the permeation rate of H_2. Most conventional polymers that have permeability comparable to those presented here show much lower gas selectivity.

4. LCVD ON NONPOROUS/POROUS COMPOSITE MEMBRANE

The first structure depicted in Figure 34.3b is that of a plasma polymer film deposited on the porous substrate coated with conventional polymer. The usefulness of this technique is that a conventional coating that has good casting or coating properties can be used to cover the surface pores of the substrate. A nanofilm of a plasma polymer can then be deposited on the coating to improve the selectivity. This is particularly useful when the surface pores of the substrate are too large to be covered by a plasma polymer film thin enough to take advantage of the nanofilm technology of LCVD. The gas permeability obtained with this type of composite membrane is summarized in Table 34.3 [2]. Although the composite membranes prepared did not show a permeability ratio of H_2/CH_4 as high as those using homogeneous (nonporous) silicon-carbonate sheet as the substrate (Table 34.2), the hydrogen permeability of the composite membranes is one to two orders of magnitude higher than that obtained with thin silicone-carbonate sheets. A substantial increase in the permeation rate obtained with these double-coated membranes, while retaining a reasonable separation factor, is a unique advantage of the application of nanoscale molecular sieve obtainable by LCVD.

5. LCVD ON POROUS MEMBRANE

5.1. Reverse Osmosis Membrane

The final structure shown in Figure 34.3b is the deposition of a plasma polymer film directly onto the surface of a porous substrate, which yields a unique LCVD

Table 34.3 H_2 Permeation Rates and Permeability Ratio of H_2/CH_4 for Plasma Polymers of Butyronitrile Deposited on Silicone-Carbonate–Coated Polysulfone Porous Membranes

No.	Flow rate (sccm)	Power (W)	Pressure (mtorr)	ℓ (Å)	$P/\ell \times 10^4\ H_2{}^a$	$P/\ell \times 10^4\ H_2{}^b$	Permeability ratio^c α (H_2/CH_4)
12	1.25	26	29.6	45	6.18	2.61	15.2 (1.4)
13	1.25	38	29.9	44	9.48	2.01	12.0 (1.2)
14	1.26	55	29.6	50	8.49	3.90	8.0 (1.3)
15	1.26	61	30.7	50	—	3.83	4.1 (—)
16	0.63	35	19.5	50	8.19	5.49	3.2 (1.4)
17	1.25	26	30.2	90	9.76	2.30	14.6 (1.2)
18	1.25	61	30.1	230	7.05	0.32	8.1 (1.3)
19	1.25	72	30.0	—	9.36	0.90	12.4 (1.2)

Units of permeation rate, P/ℓ: cm^3 (STP)/cm^2-s-cm Hg.
^{a}Hydrogen permeation rate of the membrane prior to the plasma polymer coating.
^{b}Hydrogen permeation rate of the membrane after the plasma polymer coating.
^{c}The values in the parentheses are the permeability ratios for the uncoated composite membrane. The location of substrate within a reactor is not the same.

nanofilm/porous composite membrane. This approach was the first application of LCVD film for reverse osmosis membranes [8]. Since the membrane on which LCVD film is deposited is porous and has a great inner surface area, LCVD process is greatly influenced by the nature of the porous substrate membrane and the nature of monomer used and differs significantly from that on nonporous or nonabsorbing substrates. Considering the monomer absorption characteristics of substrate membrane, the following methods of membrane preparation were employed [8–10].

Method I: Flow System. Monomer is fed in from the liquid monomer through an injection tube placed in the center of the reaction tube while the system was pumping.

Method II: Closed System With Sorption of Monomer by a Porous Substrate. Monomer is fed into a closed system at a constant flow rate until the pressure of the system reaches a predetermined level. There is no monomer feed-in and no pumping during the glow discharge.

Method III: Closed System Without Sorption of Monomer by a Porous Substrate. Monomer vapor is stored in a reservoir without exposing the substrate to the monomer. As soon as the monomer vapor is introduced from the reservoir into a reaction tube that contains a porous substrate, glow discharge is initiated.

Method IV: Semiclosed System With Sorption of Monomer by a Substrate. The procedure is similar to method II, except that monomer vapor is continuously fed into a closed system from a liquid monomer source during glow discharge.

Method V: Semiclosed System Without Sorption of Monomer by a Substrate. Monomer vapor is allowed to build up in a reservoir from a liquid monomer source without exposing a substrate to the vapor. As soon as the monomer vapor is introduced into the reaction tube, glow discharge is initiated. Monomer vapor is continuously fed into a closed system from a liquid monomer source during glow discharge.

Method VI: Semiclosed System Connected to a Monomer Flow. The procedure is the same as method V, but as soon as glow discharge is initiated, the system
is connected to a pump. This method is used when the pressure of a reaction tube
rises due to either the decomposition of monomer or the plasma-susceptible
substrate.

Method VII: Semiclosed System Connected to Pump. The procedure is the
same as method II, but as soon as glow discharge is initiated, the system is connected
to a pump.

An important factor that is encountered in polymer deposition onto porous
substrate is the adsorption or sorption characteristics of a porous substrate. This
factor influences the polymer deposition in the following manner. Porous substrates
in general have a large internal surface area and are good adsorbing material.
Consequently, the degassing of the substrate to maintain a desired vacuum level is
dependent on how much and how strongly gases and vapors (particularly water
vapor) are held by a porous substrate. Some of the vapors, particularly water vapor
in a hydrophilic porous substrate, are difficult to degas by pumping. Thus, they will
evolve when glow discharge is initiated, causing an uncontrollable effect due to this
additional water vapor plasma. In this respect, the hydrophilicity of the porous
substrate plays a significant role in determining the reproducibility of polymer
deposition.

When monomer vapor is introduced into the reaction system, some monomers
will be adsorbed or sorbed by a porous substrate. The partition between vapor phase
and sorbed phase is dependent on the adsorbing capability of a porous substrate.
For instance, when a porous glass tube is used as a substrate, nearly 100% of the
monomer fed into a closed system is adsorbed, and it is difficult to establish a steady-
state flow of monomer vapor until the substrate is saturated with the monomer,
which takes several hours at the flow rates generally used in plasma polymerization.

The pressure increase in a reaction chamber that contains a porous substrate
when a monomer vapor is introduced by a given flow rate can be utilized to calculate
the sorption capability of the porous substrate. The pressure buildup curves are
shown in Figure 34.6 for Millipore filter and porous polysulfone film. The pressure
buildup curve with a porous glass tube is too slow to be presented in the same time
scale. From the slope of the linear portion of the pressure buildup curve, the ratio of
monomer sorbed/monomer fed into the system is estimated as 0.636 for the
polysulfone film, 0.926 for Millipore filter, and 0.9987 for the porous glass tube.

The measurement of pressure change in a reaction chamber that contains a
well-degassed substrate in a glow discharge under vacuum will provide information
on the combined effects of plasma susceptibility and desorption characteristics of a
porous substrate. In Figure 34.7, plots of pressure vs. time of plasma treatment are
presented for Millipore filter and porous polysulfone film. The results indicate that
porous polysulfone is very stable against glow discharge and behaves similarly to
glass tube alone. The increase of pressure with a Millipore filter is most likely due to
the degradation of polymers since a Millipore filter disappeared when it was exposed
to prolonged glow discharge.

The pressure changes with nondegradable porous glass are shown in
Figure 34.8 as plots similar to those of Figure 34.7. A porous tube, which was
baked at 600°C and kept in an airtight glass tube container for several weeks, was

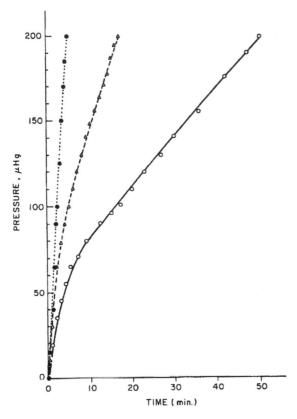

Figure 34.6 Pressure buildup in a chamber containing a substrate when 4-vinyl pyridine is introduced at constant flow rate: (○) Millipore filter; (●) glass slide; (△) porous polysulfone film.

placed in the reaction tube and degassed by vacuum. After the overnight degassing, the pressure of the reaction tube reached below the 0.5 μm Hg level, and the leakage rate (including the degassing rate from the porous glass tube) was confirmed to be below the acceptable level. As seen by curve A in Figure 34.8, this tube showed marked increase of pressure in a glow discharge of the gas phase that is in equilibrium with the porous tube, but with a clear indication of leveling off with exposure time. When the same tube was baked again at 500°C and used immediately after cooling, the pressure was reduced to nearly the same level as that of the plain reaction tube (considering the increase in total surface area with the porous glass tubes in the reaction chamber).

The sorption capability of a porous substrate and the effect of sorbed monomer on the polymerization of monomer vapor by glow discharge can be visualized by the pressure change in the similar system with a monomer (4-vinylpyridine), as shown in Figure 34.9. The monomer was introduced by a constant feed-in rate into a closed system that contains a porous substrate until the pressure of the system reached 200 μm Hg. Consequently, the total amount of monomer introduced and the amount of monomer sorbed by a porous substrate are different depending on the absorbing

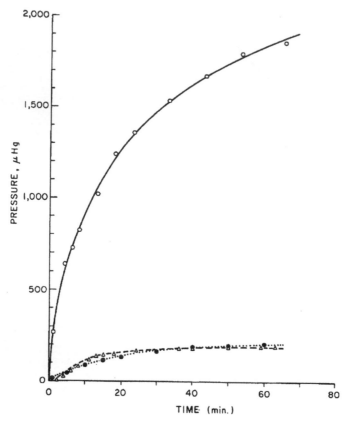

Figure 34.7 Change of pressure in a chamber which contains a substrate in glow discharge (under vacuum): (O) Millipore filter; (●) glass slide; (△) porous polysulfone film.

capability of the substrate. Without a porous substrate, the pressure of the system decreases as soon as a glow discharge is initiated because 4-vinylpyridine is group I monomer that polymerizes with minimal evolution of hydrogen. With a porous substrate that contains a certain amount of sorbed monomer, the pressure of the system increases at the initial stage proportionally to the amount of monomer absorbed, indicating that desorption rate of monomer exceeds the monomer consumption rate owing to polymerization.

The addition of nitrogen or water vapor changes the pattern of monomer adsorption, release of sorbed monomer, and the rate of monomer vapor consumption by polymerization considerably, as seen in the similar plots shown in Figures 34.10–34.12. With the addition of water vapor, the hydrophilicity of the porous substrate plays an important role that can be seen in the difference of pressure change observed when reversing the order of exposure to monomer and to water vapor. Namely, the amount of water vapor sorbed by hydrophilic Millipore filter is greater when water vapor is introduced first, whereas hydrophobic polysulfone sorbs more water when relatively more hydrophilic (than polysulfone) 4-vinylpyridine monomer is introduced first. Results shown in Figures 34.11 and 34.12 indicate that excessive water vapor prevents vapor phase polymerization

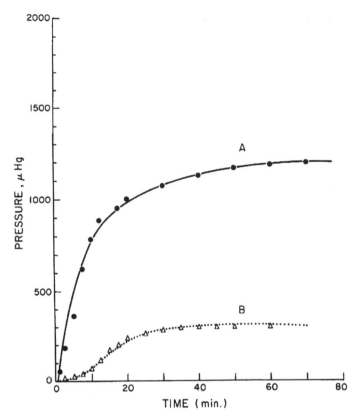

Figure 34.8 Change of pressure in a chamber which contains a porous glass tube in a glow discharge (under vacuum): (A) porous glass tube stored in a dust-free container after the tube is baked at 600°C; (B) same tube used in A, but immediately after the tube is baked at 600°C.

of 4-vinylpyridine. Reverse osmosis characteristics of porous substrates used are summarized in Table 34.4. The use of Millipore filter VC (nominal pore size 100 nm) as a substrate failed to produce good reverse osmosis membranes. It is estimated that a pore size of approximately 30 nm or less is preferred in this application.

The aspect of hole filling by plasma deposition can be demonstrated by the transport characteristics of LCVD-prepared membranes. First, the porosity as porous media calculated from the gas permeability dependence on the applied pressure can be correlated to the salt rejection of the composite membrane as shown in Figure 34.13. The effective porosity ε/q^2, where ε is the porosity and q is the tortuosity factor, is measured in dry state and may not directly correlate to the porosity of the membranes in wet state. The effective porosity of LCVD-prepared membranes was measured before the reverse osmosis experiment. The decrease of porosity (as porous media) is clearly reflected in the increase in salt rejection in reverse osmosis.

It can also be done by analysis of reverse osmosis properties of a membrane as a function of the effective driving pressure according to the relationship developed by Yasuda and Lamaze [11]. Salt rejection R_s is a result of transport depletion of salt in

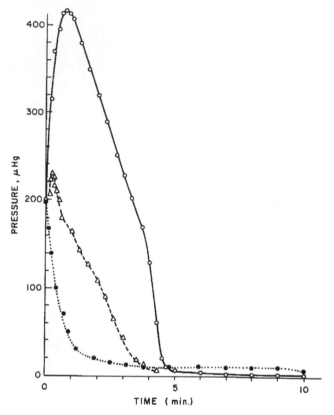

Figure 34.9 Pressure change observed in a chamber containing a substrate pseudo-saturated to an initial pressure of 200 1A Hg of 4-vinylpyridine, during glow discharge: (○) Millipore filter; (●) glass slide; (△) porous polysulfone film.

relation to water flux and can be expressed as

$$R_s = \omega + [P_2RT/P_1V_1(\Delta p - \Delta \pi)]^{-1}$$

$$\omega = RTK_1/P_1V_1 - 1$$

where ω is a parameter to describe the extent of flow of water (not the diffusive flux of water); K_1 is hydraulic permeability; P_1 is diffusive water permeability; V_1 is molar volume of water; P_2 is diffusive salt permeability; ($\Delta p - \Delta \pi$ is effective pressure, p_{eff}; Δp is differential hydrostatic pressure; and $\Delta \pi$ is differential osmotic pressure.

The water transport mechanism changes from the flow mechanism in porous membrane to the diffusive transport in nonporous homogeneous membrane due to the deposition of a homogeneous LCVD layer that fills the pore, i.e., water transport changes from bulk flow to diffusive flow when pores are covered by LCVD film.

It is necessary to minimize the y intercept; $\omega = RTK_1/P_1V_1 = \alpha(K_1/P_1)$, thereby maximizing the diffusive water permeability and minimizing the hydraulic water permeability. Reverse osmosis membranes can be characterized simply by observing salt rejection as a function of effective pressure. Figure 34.14 is a

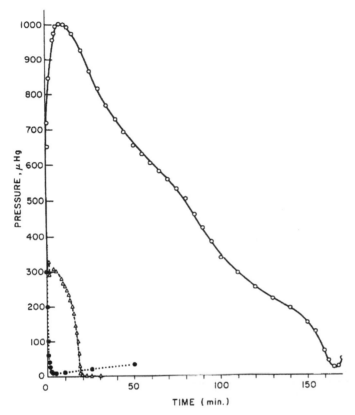

Figure 34.10 Pressure change observed during glow discharge in a chamber containing a substrate pseudo-saturated to $100\,\mu m\,Hg$ of nitrogen and $200\,\mu m\,Hg$ of 4-vinyl-pyridine ($300\,\mu m\,Hg$ total pressure): (O) Millipore filter; (●) glass slide; ($\triangle$) porous polysulfone film.

plot of $1/p_{eff}$ vs. $1/R_s$ of three plasma-polymerized reverse osmosis membranes. The membranes vary only in the treatment time: 4, 6, and 7 min as indicated.

Membranes can be classified according to mode of transport: (1) diffusion membranes for which $\omega = 1$; (2) diffusion–flow membranes, $\omega > 1$; and (3) flow membranes $\omega \gg 1$ [12]. In Figure 34.14, the 7-min line represents a diffusion membrane; all others are best described as diffusion–flow type.

The slope of the plot:

$$\text{Slope} = P_2 RT / P_1 V_1 = \alpha P_2 / P_1$$

should also be minimal for good reverse osmosis membrane. Minimizing salt-diffusive permeability P_2 gives the best rejection and maximizing the diffusive water permeability P_1 gives the best flux. Figure 34.14 shows an improvement in membrane with increased treatment time. The 7-min membrane is the best, having the smallest intercept, $\omega = 1$, and the smallest slope, 20 (psi). Indeed, reverse osmosis tests gave good results for this membrane with 99.0% salt rejection and 4.2 gfd water

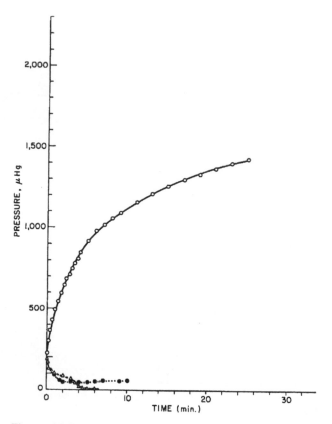

Figure 34.11   Change of pressure observed during glow discharge in a chamber containing a substrate pseudo-saturated first with water (to 20 μm Hg) and then with 200 μm Hg of 4-vinylpyridine (to 220 μm Hg total pressure): (○) Millipore filter; (●) glass slide; (△) porous polysulfone film.

flux with 3.5% NaC at 1500 psi. This result supports the importance of hole filling and of forming homogeneous diffusion-type membranes by LCVD.

In order to increase the diffusive water transport it is necessary to have some hydrophilic moieties in LCVD film. However, it is necessary to maintain the tight network to prevent excessive swelling that allows the bulk flow of water (salt water). Figure 34.15 depicts the influence of adding N_2 gas to the monomer on the salt rejection/water flux correlation. The addition of N_2 to 4-picoline greatly changes the reverse osmosis characteristics of plasma polymers. In Figure 34.15, the change of reverse osmosis characteristics with deposition time is shown in a plot of log water flux vs. salt rejection for 4-picoline (20 μm Hg), 4-picoline (50 μm Hg), 4-picoline (20 μm Hg) + N_2 (60 μm Hg), and 4-picoline (50 μm Hg) + N_2 (50 μm Hg). Without addition of N_2, better reverse osmosis results are obtained by employing low pressure, although this procedure requires a longer deposition time. With addition of N_2, reverse osmosis characteristics are improved at both vapor pressure levels evidenced by the shift of curves toward upper right-hand sides. The improvement is more pronounced at higher vapor pressure (i.e., 50 μm Hg), and better membranes are obtained with considerably shorter deposition time.

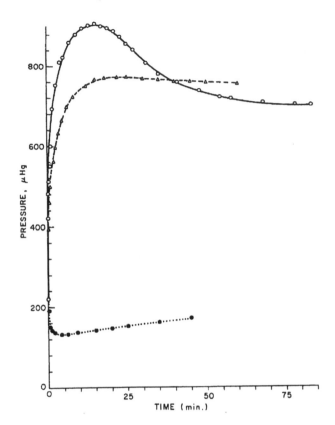

Figure 34.12 Pressure change during glow discharge in a chamber containing a substrate pseudo-saturated first with 4-vinylpyridine (to 200 μm Hg) and then with 20 μm Hg of water (to 220 μm Hg total pressure): (○) Millipore filter; (●) glass slide; (△) porous polysulfone film.

Table 34.4 Reverse Osmosis Characteristics of Porous Substrates[a,b]

Porous substrate	Thickness (μm)	Pore size (Å)	Water flux (gfd)	Salt rejection (%)
Millipore filter	48	250	400–500	0
Polysulfone	13	—	300–400	2–4
Porous glass	1100	50	0.6–1.0	5–10

[a]Water flux and salt rejection were measured with 1.2% NaCl solution at 1200 psi applied pressure.
[b]Figures cited were taken after the initial drop of water flux, which generally occurred in 2–3 h was observed and the flux was stabilized.

Reverse osmosis membranes prepared by LCVD on porous membrane showed unique but very peculiar reverse osmosis membrane performance [10,13]. In general, the reverse osmosis membrane performance declines with time, i.e., salt rejection and particularly water flux decline with time, which is recognized as membrane fouling.

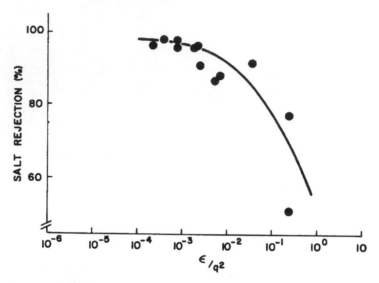

Figure 34.13 Reverse osmosis characteristics of composite membranes prepared by plasma polymerization of benzene/H_2O/N_2 compared with those from acetylene/H_2O/N_2 represented by the solid line; porous polysulfone film as the substrate, 3.5% NaCl at 1500 psi.

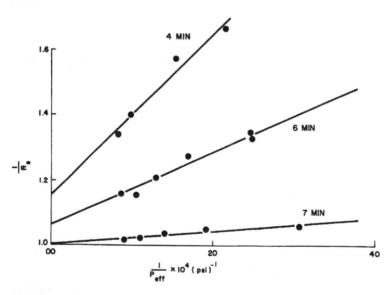

Figure 34.14 The change of $1/R_s$ vs. $1/(\Delta p - \Delta \pi)$ dependence as a function of plasma deposition time.

Figures 34.16 and 34.17 depict the opposite phenomenon of the membrane fouling, i.e., salt rejection and water flux increases with the operation time during the first 3 months, in which period most reverse osmosis membranes show the beginning of the membrane fouling. The reason for this phenomenon has not been elucidated;

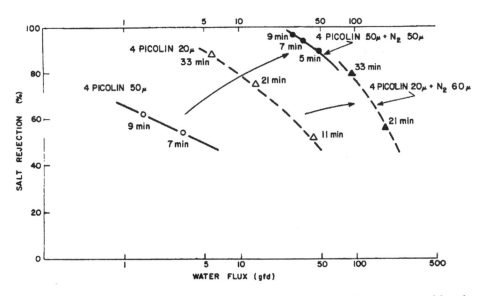

Figure 34.15 Reverse osmosis characteristics of composite membranes prepared by plasma polymerization of 4-picoline with and without N_2 at two vapor pressures; porous polysulfone film as the substrate; 1.2% NaCl at 1200 psi.

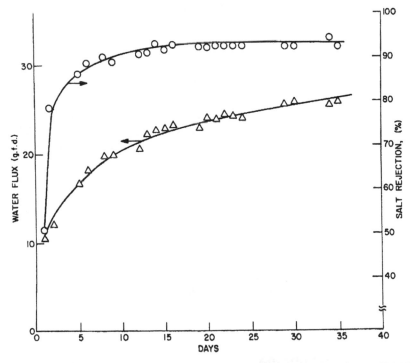

Figure 34.16 Change of salt rejection and water flux with time (1.2% NaCl, 1300 psi); membrane of plasma-polymerized 4-vinylpyridine on Millipore vs filter.

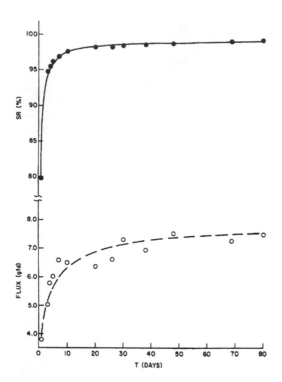

Figure 34.17   Change of salt rejection (SR) and water flux with time (3.5% NaCl, 1500 psi); membrane of plasma-polymerized 4-vinylpyridine on porous polysulfone film.

however, it is likely due to the slow reaction of dangling bonds and short-range rearrangement of hydrophilic moieties within the nanofilm for more uniform distribution.

The effect of LCVD reaction time is depicted in Figure 34.18, which shows the decline of performance after passing the maximum as observed with many LCVD-prepared membranes. In reverse osmosis membrane, the increase of flux as salt rejection declines was not observed. Figure 34.18 plots water flux vs. salt rejection for a series of acetylene/CO/H_2O discharge polymer membranes on cellulose nitrate/cellulose acetate (CNCA) porous membrane. The results bear three important observations: First, good reverse osmosis membranes can be produced by CNCA with acetylene/CO/H_2O plasma. Second, much shorter deposition times are sufficient for this system, reflecting the fact that carbon monoxide is incorporated with ease. For instance, deposition time of only 2 min results in membranes with salt rejection around 94% and with water fluxes of 32 gal/ft^2 day (gfd). Lastly, prolonged plasma reaction times have negative effects on the salt rejection properties of the membrane. This trend is generally observed in every kind of application of LCVD.

5.2. Gas Separation Membrane

All reverse osmosis membranes shown above were with flat sheet porous membranes. However, this was also extended to hollow fiber membranes. The first hollow fibers

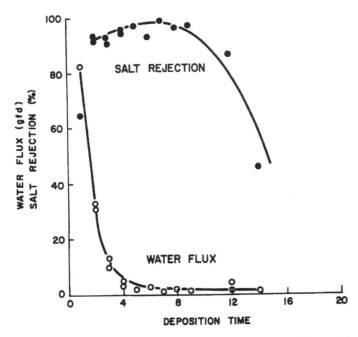

Figure 34.18 Water flux and salt rejection as functions of deposition time, in minutes, for composite membranes of acetylene/CO/H$_2$O (CNCA porous membrane used as substrate).

with porous wall were prepared (by Monsanto on an order from Research Triangle Institute, Research Triangle Park NC) for reverse osmosis membrane preparation by LCVD. This type of composite membrane has received considerable attention for use as gas separation membranes. Figure 34.19 shows the variation of (1) the oxygen enrichment factor and (2) the total (air) flux constant (P_{air}/thickness) when plasma-polymerized composite membranes were exposed to feed air at an applied pressure of 100 psig [2]. The composite membranes were prepared by deposition of plasma-polymerized perfluoro-l-methyldecalin coatings onto porous polysulfone hollow fibers, under three different plasma polymerization conditions. The oxygen enrichment factor increases with increasing coating thickness and passes through a maximum. At the same time the air flux constant decreases and passes through a minimum with increasing coating thickness. There is initially a lag period during which the separation factor remains 1 with increasing coating thickness. Once a certain coating thickness is reached the enrichment factor begins to increase. The value of this thickness, at which the enrichment factor just starts to increase, decreases with increasing W/FM. This coating thickness is a combination of two factors: (1) the surface pores of the porous hollow fiber must be completely covered before any separation occurs, and (2) the tightness within the plasma polymer coating is increasing with increasing coating thickness. The separation factor then increases (and the flux constant decreases) due to the increasing tightness and the increasing coating thickness. When the internal stress within the plasma polymer film exceeds the cohesive strength of the plasma polymer network, cracking occurs causing the decline of the separation factor of the composite membranes.

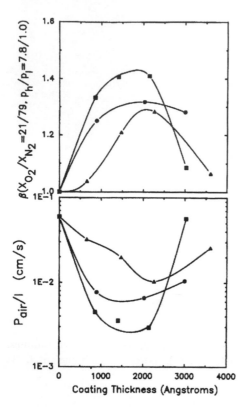

Figure 34.19 Variation of (a) oxygen enrichment factors and (b) air flux constants as a function of plasma polymer coating thickness, perfluro-1-methyldecalin plasma polymer coatings deposited onto porous polysulfone hollow fibers at ▲: $5.35 \times 10^7 \, \text{J/kg}$; ●: $6.64 \times 10^8 \, \text{J/kg}$; and ■: $8.05 \times 10^7 \, \text{J/kg}$.

Figure 34.20 shows the effect of different substrate hollow fibers on the separation properties of composite membranes prepared from plasma polymer films of 1,1,3,3-tetramethyldisiloxane [2]. The membranes were all prepared after the electrodes had been conditioned by operation at the given plasma conditions, so it was considered that the plasma polymer coatings were all nearly identical, and the major source of the differences in the air separation properties were due to the different substrate hollow fibers. The membrane designated HF135 and HF236 were based on porous polysulfone hollow fibers, with the HF135 substrate fiber having smaller surface pore sizes than the HF236 substrate fiber. HF638 was prepared on a porous polypropylene (Celguard) hollow fiber. Both HF236 and HF638 show the lag before the oxygen enrichment factors increase above 1, indicative of the necessity of completely covering the surface pores. The HF135 substrate hollow fiber has an average surface pore size of about 17 nm, and therefore only a thin plasma polymer coating was necessary to completely cover the pores at these plasma polymerization conditions. Once the surface pores are covered, the HF135 and HF236 membranes reach the same level of oxygen enrichment factor.

The air flux constant for the HF236 membrane is slightly higher than that for the HF135 membranes due to a thinner substrate fiber wall and a greater substrate

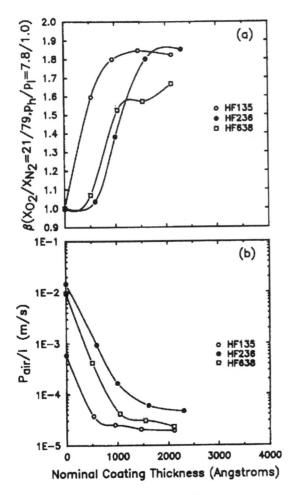

Figure 34.20 The influence of different substrate hollow fibers on (a) air enrichment factor, and (b) air flux of composite membranes prepared by deposition of plasma polymerization coating of 1,1,3,3-tetramethyldisiloxane.

surface porosity. The HF638 membranes never achieve the same level of oxygen enrichment factor as the other two series of membranes, even though the air flux constants for this membrane are intermediate to those of the HF135 and HF236 membranes. This is postulated to be primarily due to polypropylene substrate for HF236 membranes. The polypropylene is more susceptible to interaction with the plasma than the polysulfone for HF135 and HF 236 membranes, and therefore is altered (degraded) to a greater extent.

The separation properties of the membranes vary not only with the thickness of the plasma polymer coatings but also with the conditions of LCVD. One such plot of the variation is shown in Figure 34.21 [14]. The upper portion of this figure shows how the specific conversion parameter, DR/FM, varies as a function of the composite energy input parameter, W/FM. DR is the nominal deposition rate of plasma polymer film on the surface of a quartz crystal thickness monitor located in the plasma reactor, adjacent to the hollow fibers.

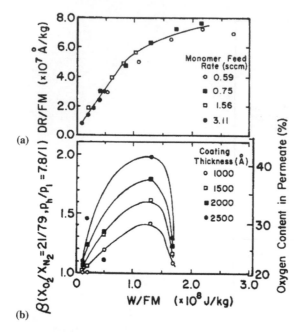

Figure 34.21 (a) Correlation between specific conversion factor DR/FM and composite plasma energy input parameter W/FM for various monomer flow rate, (b) effect of W/FM on the oxygen enrichment factor for composite membranes prepared at various coating thickness of perfloro-2-butyltetrahydrofuran plasma polymer deposited on porous polysulfone hollow fibers.

Figure 34.21 shows a very important correlation between the deposition characteristics and the membrane performance shown as the same function of W/FM. The deposition plot (Fig. 34.21a) typically shows three regions: the power-deficient region at low values of W/FM, the monomer-deficient region at high values of W/FM, and a transition region that is intermediate between the two deficient regions as described in Chapter 8.

In reference to the oxygen enrichment factor, it is observed that the maximal oxygen enrichment, for any thickness of plasma polymer coating, appears at a W/FM value that is in the transition region of the DR/FM vs. W/FM plot. This is the general trend observed for plasma polymerization coating of type A plasma polymers. It is also important to note that such a plot, as shown in Figure 34.21a, is necessary to find the optimal conditions of LCVD. There is an unfortunate trend that the best results are reported as functions of inappropriate experimental parameters without consideration of the domain of LCVD.

6. MACROPOROUS MEMBRANE FOR GAS/LIQUID REACTIONS

Dealing with polymer membranes, "membrane science" is nearly synonymous with the transport mechanisms through the bulk phase of membrane, and the importance of interfacial aspects of membranes is often not recognized or is totally ignored.

In such a narrow concept of a membrane, the macroporous membrane is not considered to be a membrane if the pore size is beyond some arbitrarily set range of preference, e.g., that for microporous membranes. If the pore size of a microporous membrane is greater than the range, the macroporous media is not considered as a membrane. Consequently, a plate with a hole whose diameter is 250 µm is generally not considered as a membrane. However, the macroporous membranes with micrometer- to millimeter-sized holes function well in the transport of gases into liquid phase, which is an important chemical engineering process. The above-mentioned single-hole plate with LCVD surface modification of the plate that surrounds the hole played a historically important role in establishing the principle involved in the sessile bubble formation and detachment, which is described in Chapter 27.

A polymer membrane employed in water aeration device is a typical example of macroporous membranes, i.e., membranes with sub-millimeter holes. A rubber film with numbers of slits functions as a thin layer that separates the gas phase and the liquid phase, which allows the transfer of the gas into the liquid phase in a controlled manner. Such a film is appropriately termed as aeration membrane.

In the domain where the entity that is transported through a membrane is immiscible or not completely soluble in the contacting (exit) phase, such as the case of gas phase air or oxygen in water, the interfacial factor becomes overwhelmingly important over the transport characteristics of the bulk membrane phase, which is empty space. It is important to recognize that the surface of macroporous membrane consists of the solid phase and the gas phase (in the pore diameter exposed to the interface), and the interfacial aspect of the solid surface dominates the behavior of the gas phase that expands out of the pore.

7. MEMBRANE FOR GAS BUBBLING

It is common knowledge that the hole size is the most important factor, but it was not well understood how the interfacial energy for the solid phase could have an equally important role in determining the size of bubble that emerges from the hole, particularly as the size of hole decreases. The chemical reactions that occur between the gas phase reactant and the liquid phase reactant greatly depend on the surface-to-volume ratio of the bubble. Therefore, there is a great need for a device that creates small bubbles in the low gas feeding rate regime.

An obvious approach consists of creating small holes on the solid surface; however, such attempts often end up with large bubbles emerging from small holes because the interfacial energy between the solid surface and the liquid is often neglected, and most readily available membrane materials (polymers) have an amphoteric hydrophobic/hydrophilic surface. It is not easy to carry out experiments that include the size of hole and the interfacial energy as the key parameters, particularly covering a wide range of the interfacial energy. The control of the interfacial energy by LCVD provided the key contribution to elucidate the mechanism of the bubble formation from a hole surrounded by solid surface [15–17].

Oxygenation of water or water suspension such as blood can be done by (1) blowing oxygen gas into the liquid via a porous membrane and (2) bubbleless oxygenation via a gas-permeable (nonporous) membrane. Both the methods have advantages and disadvantages, which seem to be complementary, and the choice is entirely dependent on the natures of applications. One feature could be an advantage in one application but a disadvantage in another application, and vice versa. Some characteristic factors for both processes could be examined by using the cases of blood oxygenation and oxygenation (aeration) of water.

8. OXYGENATION OF BLOOD

Oxygenation of blood can be achieved by bubbling oxygen through blood contained in a vessel, which is part of an overall flow system of blood. This process is practiced in the bubble oxygenator. Oxygen literally bubbles through blood retained in the vessel. The advantage of this approach is that the process is simple and the oxygenation is fast and efficient because an oxygen bubble contacts with many red blood cells before it leaves the blood phase. It should be emphasized that the essential step of blood oxygenation is the transfer of oxygen to red blood cells, which differs significantly from oxygenation of liquid, e.g., water. The disadvantage of this approach is that bubbling of oxygen, particularly with large bubbles at a high flow rate, could cause trauma to red blood cells and hemolysis (breakdown of red blood cells).

Bubbles could be created by multiple capillaries or holes created on a surface. However, the formation of small bubbles on a membrane surface, particularly at a low flow rate, is not as easy as it might be conceived. Furthermore, the basic principle for how to make small bubbles had not been available until the two reference papers were published in 1994 as described in Chapter 27.

Nonporous membrane carries out the bubbleless oxygenation of blood, i.e., no bubble emerges from the membrane surface. The dissolved oxygen is transferred from the membrane phase to the liquid phase, which makes contact with the membrane surface. Characteristic oxygen transfer rate (per unit membrane area) could be high enough to cope with requirements for blood oxygenation processes. The advantage of this approach is that no gas phase oxygen has contact with blood, and consequently the trauma caused by bubbling could be greatly reduced. This process is practiced in the membrane oxygenator.

The disadvantage of this approach is that the concentration polarization at membrane surface becomes the rate-determining step as the membrane transport rate increases. Because the transfer of oxygen to liquid phase precedes the transfer of oxygen to a red blood cell in most cases, it is necessary to maintain a certain flow rate of liquid at the membrane/liquid interface to reduce the concentration polarization at the membrane/blood interface and/or to provide a device to remove oxygenated red blood cells from the membrane surface and bring in unoxygenated red blood cells to the membrane surface in order to take advantage of high oxygen transport rate through the membrane.

These necessary actions increase the trauma to red blood cells and negate the advantage of membrane oxygenation. Another disadvantage is that a unit volume of blood is exposed to a much larger surface area of a foreign body (membrane surface) compared to that in a bubble oxygenator. Considering the pros and cons of these two approaches, blood oxygenation via small bubbles at a low flow rate seems to be a very viable approach that takes advantages of creating many small bubbles at a low flow rate regime by control of surface energy of macroporous membranes.

9. OXYGENATION OF WATER

Oxygenation or aeration of water is essentially the same process as blood oxygenation, except the requirements are much less stringent because oxygen transfer occurs to the liquid phase. Most aeration is carried out by bubbling air into water by pipes or membrane devices. Aeration membranes are essentially sheets of rubber with many holes in the form of slits. The main functions of membrane are (1) providing numbers of holes and (2) providing valve mechanism to prevent flooding of the pipe system when the system is depressurized. Under applied pressure, membrane expands and the slit opens. When the aeration system is depressurized, membrane contracts and the slits close by the hydrostatic pressure of water.

The advantage of a membrane device over direct bubbling from pipes is the reduction of energy necessary to oxygenate water. Many large bubbles leave the water phase with little oxygenation of water. The efficiency increases with the surface-to-volume ratio of a bubble. Thus, the creation of small bubbles in the low flow rate regime should increase the efficiency of oxygenation and reduces the energy consumption significantly.

It has been indicated that the size of a bubble slowly formed at a submerged, horizontal, circular orifice and detached by buoyancy alone is controlled by (1) the size of the orifice and (2) the liquid surface tension [18,19]. Dealing with oxygenation of water, only the hole size remains as the parameter that could be used to control the size of bubble emerging from the surface of an aeration membrane, if one accepted the above principle. However, many efforts to reduce the size of the bubble by creating small holes on the aeration membrane did not succeed. The size of bubble emerging from much smaller holes was practically identical to the normal membrane with larger holes. The true factors that control the size of bubble are described in Chapter 27. In short, however, if the second factor mentioned above, i.e., liquid surface tension, is replaced with "interfacial tension," then the above-mentioned principle is right. However, the difference between liquid surface tension and interfacial tension is huge, and it is not just switching the terminology as described in Chapter 26.

The role of LCVD coating in macroporous membranes is schematically depicted in Figure 34.22. Nanofilm of LCVD film covers the top surface and the wall of holes. This principle was applied to a model macroporous membrane. A circular and clear-edged hole (0.55 mm or 0.25 mm diameter) was created on a stainless steel sheet, one of which is shown in Figure 34.23. Cathodic plasma polymerization of TMS or a mixture of TMS and O_2 coats the surface of the sheet and the edge of the hole. As shown in Figure 27.3, changing the mole ratio of oxygen to TMS can control the surface energy of stainless steel surface in a wide range. In order to prevent the flooding of the gas-feeding chamber, hydrophobic porous membrane was utilized as depicted in Figure 34.24. After a series of bubbling experiments with a membrane (stainless steel sheet with a hole), plasma coating was mechanically removed, and another plasma coating with different surface energy was applied.

Figure 34.22 Schematic cross-sectional view of plasma polymerization coated large pore membrane.

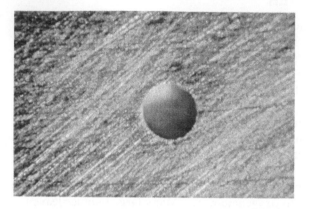

Figure 34.23 Pictorial view of a single hole stainless steel membrane.

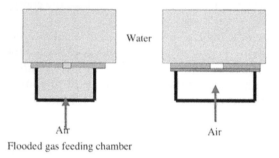

Flooded gas feeding chamber

Goretex (porous Teflon) film to prevent flooding of the gas feeding chamber

Figure 34.24 Prevention of the flooding of gas feeding chamber.

Thus, membranes with exactly the same hole with different surface energies were prepared one at a time.

The bubble volume was calculated from the total volume of air passed through a hole and the number of bubbles, which was obtained by playing the recorded videotape, in a given time. The bubble diameters, surface-to-volume ratio, and so forth are derived from the calculated bubble volume, assuming the spherical bubble, although bubbles in water are far from a sphere as seen in figures presented in Chapter 27.

7.1. Data Obtained by a Model Membrane with Single Hole

The flow rate of air simply increases the frequency of bubble formation and detachment, and has no effect on the size of bubble as depicted in Figure 34.25 for a

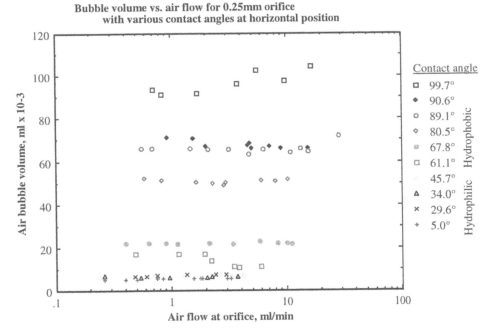

Figure 34.25 Effect of airflow rate on bubble size for 0.25-mm orifice.

0.25-mm hole and Figure 34.26 for a 0.50-mm hole. The effect found with stainless steel membrane was also found when a small hole was made on polytetrafluoroethylene film as shown in Figure 34.27 in the similar plot.

The aeration rate increases as the bubble size (bubble volume) decreases as depicted in Figure 34.28 as plots of dissolved oxygen concentration against the aeration time. Consequently, the efficiency of aeration increases as the bubble surface-to-volume ratio increases as shown in Figure 34.29. It is interesting to note that the oxygenation efficiency is higher with the lower flow rate of air. This is probably because the probability of some bubbles passing through already oxygenated water increases at the higher flow rate. This aspect is important in blood oxygenation in which the transfer of oxygen to red blood cells is the main objective, which could be achieved more efficiently with small bubbles at a low oxygen flow rate.

The fundamental factors of bubble formation and detachment carried out using a single hole placed on a rigid stainless steel plate have important implications for small-bubble oxygenation membranes and eventually for the general gas–liquid reactions. The main requirements for efficient small-bubble oxygenation membrane are (1) small hole size, and (2) hydrophilic surface so that the contact angle of water is less than 45 degrees. The implications of these fundamental requirements for membranes with multiple holes are as follows: The diameter of a spherical bubble that detaches from the orifice is always greater than the diameter of the hole. If too many holes are created on a surface, the probability of merging of developing bubbles increases and it becomes difficult to create many small bubbles. The distance between two holes should be greater than the diameter of a spherical bubble that develops on the membrane surface.

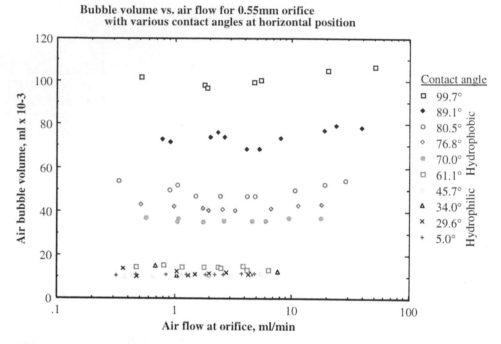

Figure 34.26   Effect of airflow rate on bubble size for 0.50-mm orifice.

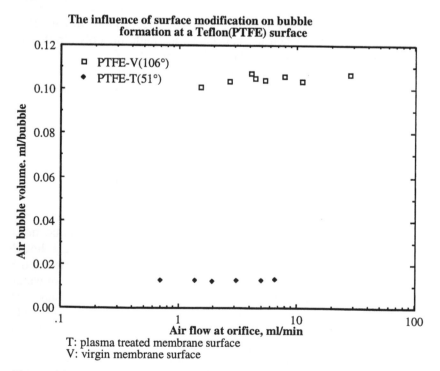

Figure 34.27   Effect of surface modification of Teflon surface on bubble size, number in parenthesis indicate the contact angle of water.

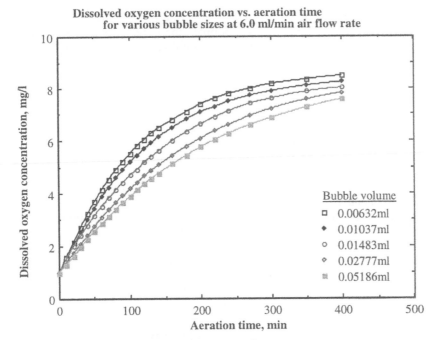

Figure 34.28 Effect of bubble size on aeration rate.

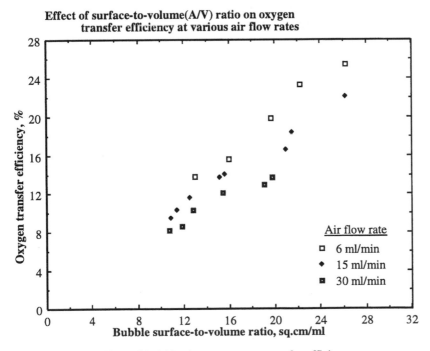

Figure 34.29 Effect of bubble size on oxygen transfer efficiency.

A horizontal surface is best for small-bubble formation. A tilted surface with multiple holes increases the probability of bubbles merging. Creation of sliding bubbles should be avoided. Therefore, a flat horizontal surface is best for creating many small bubbles.

REFERENCES

1. Yasuda, H. J. Membr. Sci. **1984**, *18*, 273.
2. Kramer, P. W.; Yeh, Y.-S.; Yasuda, H. J. Membr. Science **1989**, *46*, 1.
3. Shimomura, T.; Hirakawa, M.; Murase, I.; Sasaki, M.; Sano, T. J. Appl. Polym. Sci., Appl. Polym. Symp. **1984**, *38*, 173.
4. Chang, F.Y.; Shen, M.; Bell, A.T. J. Appl. Polym. Sci. **1973**, *17*, 2915.
5. Hansen, R.H.; Schonhorn, H. Polym. Lett. **1966**, *4*, 203.
6. Kawakami, M.; Yamashita, Y.; Iwamoto, M.; Kagawa, S. J. Membr. Sci. **1984**, *19*, 249.
7. Stancell, A.F.; Spencer, A.T. J. Appl. Polym. Sci. **1972**, *16*, 1505.
8. Yasuda, H.; Lamaze, C.E. J. Appl. Polym. Sci. **1973**, *17*, 201.
9. Yasuda, H.; Marsh, H.C.; Isai, J. J. Appl. Polym. Sci. **1975**, *19*, 2157.
10. Yasuda, H.; Marsh, H.C. J. Appl. Polym. Sci. **1975**, *19*, 2981.
11. Yasuda, H.; Lamaze, C.E. J. Polym. Sci. **1971**, *A-2* (9), 1537.
12. Yasuda, H.; Peterlin, A. J. Appl. Polym. Sci. **1973**, *17*, 433.
13. Buck, K.R.; Davar, V.K. Br. Polym. J. **1970**, *2*, 238.
14. Nomura, H.; Kramer, P.W.; Yasuda, H. Thin Solid Films **1984**, *118*, 187.
15. Lin, J.N.; Banerji, S.K.; Yasuda, H. Langmuir **1994**, *10*, 936.
16. Lin, J.N.; Banerji, S.K.; Yasuda, H. Langmuir **1994**, *10*, 945.
17. Yasuda, H.; Lin, J.N. J. Appl. Polym. Sci. **2003**, *90*, 387.
18. Janczuk, B. J. Colloid Interface Sci., **1983**, *93*, 411.
19. Wark, I.W. J. Phys. Chem. **1933**, *37*, 623.

35

Application of Luminous Chemical Vapor Deposition in Biomaterials

1. SIGNIFICANCE OF IMPERTURBABLE SURFACE IN BIOCOMPATIBILITY

The imperturbability of a surface is a vitally important factor, in the author's view, for the biological compatibility of artificial surfaces. The imperturbability of a surface could be attained by (1) molecular configuration of macromolecule, such as the case of poly(oxyethyle) (POE), or (2) immobile network structure in the top surface, such as the case of the tight network of luminous chemical vapor deposition (LCVD) nanofilm.

The change of surface configuration occurs in order to minimize the free energy difference at the interface, which is recognized as the interfacial tension. It is important to recognize that this interfacial change precedes any change that occurs in the subsurface layer or in the bulk of the material. Furthermore, the change of surface characteristics is not necessarily a consequence of long-range morphological or conformational changes of macromolecules. Short-range motions, such as rotation along the axis of the polymer chain and short-range conformational change, could alter surface configuration.

Adopting the term *surface configuration*, those observations by Langmuir could be rephrased so that the general surface properties could be addressed by the same term.

1. The surface properties of a polymer are determined by the surface configuration rather than the configuration of a macromolecule.
2. Surface configuration changes responding to the conditions under which a surface exists. This is the basic phenomenon that is observed under the topics of *surface dynamics.*
3. Surface configuration can be fixed by chemical reactions. This is the foundation for the creation of an imperturbable surface. LCVD processing is extremely useful in this particular aspect.

With consideration of surface properties as a function of the contacting phase or the surrounding medium, two serious questions arise: (1) "what is a surface?" and (2) "what is an interface?" In recognition of the fact that surface properties depend on the surrounding medium, including air or vacuum, one comes to a conclusion that the surface (exterior of an object) always exists as an interface with a surrounding medium, which could be in gas, liquid, or solid phase. The surrounding medium includes ambient air and vacuum. The interface with ambient air or with vacuum is customarily recognized as the surface, and surface characteristics are expressed by the properties of the interface with air or vacuum.

The surface configuration is a function of the nature of the contacting medium (environment) as well as of temperature. The arrangement of surface atoms under consideration changes by virtue of the change of conformation of macromolecules, but the configuration of macromolecules remains unchanged. The change of interfacial characteristics in real situations are further complicated by the contributions and their changes of surface impurities and low molecular weight components, which tend to accumulate on the surface.

In a strict sense, there are no surface characteristics that can describe behavior of a material in a biological system. Every surface is an interface with a specific contacting medium, and interfacial characteristics cannot be described without specifying the contacting medium. The phenomena recognized by the terms such as *surface dynamics*, *surface reconstruction*, *hydrophobic recovery*, and so forth are reflections of the changes of the reference state because surface characteristics are generally measured with interface with air or vacuum. In other words, how a synthetic polymer behaves in contact with blood or biological tissues is difficult to predict from the interfacial characteristics of the materials with air or vacuum alone. A biological system is very different from a simple contacting phase, e.g., vacuum, air, water, etc., and the aspect of interfacial properties becomes the most important factor for comprehending the biocompatibility of artificial materials.

2. PRINCIPLE OF LCVD COATING FOR BIOMATERIALS

The three unique and important features of type A LCVD nanofilm—imperturbable surface (Chapter 29), nanoscale molecular sieve (Chapter 34), and new surface state of material (Chapter 24)—make LCVD coating an ideal tool in preparation of biomaterials. It should be reiterated that these three features of LCVD films are limited to type A plasma polymers described in Chapter 8, and type B plasma polymers should be excluded in LCVD coatings for biomaterials based on the concept of imperturbable surface. The particularly important aspect is that the LCVD nanofilm becomes the new surface state of the substrate material, i.e., it is not just a coating placed on the surface. The first and second features describe the nature of the new surface state.

So far as the biocompatibility of a man-made material is concerned, the "imperturbable" aspect of the surface describes the way that LCVD coating could be best utilized in biomaterials. The interaction, or more precisely its absence, of a man-made material (not existing in the natural state of biological system) with a biological system is generally expressed by the very vague term "biocompatibility." The materials that have biocompatibility are referred to as biomaterials. However, the true sense of biocompatibility cannot be described without specifying the details of the biological system and how the biomaterial is placed in that system.

The biocompatibility that could be attained by LCVD coatings could be illustrated by the "blood compatibility," which again is a vague term but indicates how well a material could be allowed to contact with blood without causing coagulation. The interaction of a material surface with blood could be described by such terms as thrombogenic surface, which causes the coagulation of blood, and nonthrombogenic surface, which does not cause the coagulation of blood.

The nonthrombogenic surface could be further divided to the antithrombogenic surface, which positively prevents the coagulation of blood, and the athrombogenic surface, which does not interfere with the natural coagulation of blood but does not cause the coagulation either. An example of an antithrombogenic surface is a heparinized surface, on which the anticoagulation agent heparin is incorporated chemically or physically. A heparinized surface prevents the coagulation of blood very effectively and solves acute coagulation problems associated with the use of biomaterials.

While a biologically active surface performs well based on the specific biological reaction, it is a highly perturbable surface tailored for the specific reaction that could, in principle, cause other biological reactions. For instance, a heparinized surface seems to increase hemolysis (breakdown of red blood cells). When the biologically active agents wear out, the surface of the treated material returns to the untreated surface, which required the surface modification to be blood compatible in the first place.

In contrast to an antithrombogenic surface, an athrombogenic surface does not participate in any biological interaction, i.e., does not cause blood coagulation and does not prevent coagulation, and stays neutral in the biological environment. Such an ideal surface cannot be created by the surface modification of existing materials by conventional means of chemical reaction and physical treatment because those means require the surface to be highly perturbable; consequently, the product is also highly perturbable. It is important to recognize that the adsorption of a protein on a material surface, which is a relatively simple interaction of a biological component with the surface, involves the perturbation of surface such as the change of the surface configuration described in Chapter 25.

LCVD coating for biomaterials is the neutral approach illustrated by the athrombogenic surface. The advantage of the neutral approach in biomaterials is not well recognized, simply because finding such a surface in the domains beyond the imperturbable surfaces of type A plasma polymers is extremely difficult.

The surface is a crucially important factor of biomaterial, and without an appropriate biocompatibility the biomaterial could not function. On the other hand, the bulk properties of materials are equally important in the use of biomaterials. An opaque material cannot be used in vision correction, and soft flexible materials cannot be used in bone reinforcement. The probability of finding a material that fulfills all requirements in physical and chemical bulk properties for a biomaterial application and whose surface properties are just right for a specific application is very close to zero, if not absolutely zero. From this point of view, all biomaterials should be surface treated to cope with the biocompatibility. However, if the surface treatment alters the bulk properties, it defeats the purpose. In this sense, tunable LCVD nanofilm coating that causes the minimal effect on the bulk material is the best tool available in the domain of biomaterials.

3. APPLICATIONS OF LCVD COATING ON SILICONE CONTACT LENS

The three key features of LCVD coating ideally suited for biomaterial surface, and the important balance between the bulk properties and the surface properties, could be best illustrated by examples of nanofilms of methane plasma polymer on a contact lens made of polydimethylsiloxane elastomer. Hence, some details of processing factors and their influence on the overall properties of the product are described in the following sections.

Silicone rubber (polydimethylsiloxane) has great advantages as a contact lens material because of its very high oxygen permeability, softness, and excellent mechanical strength and durability. However, practical utilization is hampered by inherent surface characteristics of elastomer, i.e., high tackiness and highly hydrophobic surface properties. The characteristically high permeability of the material to various molecules also accounts for several problems encountered with silicone contact lenses. When a silicone rubber contact lens is placed on the cornea, many lipid-soluble materials in the tear film adhere and eventually penetrate into the bulk phase of the lens, resulting in degradation in optical clarity, which cannot be eliminated with routine lens cleaning. The characteristically high permeability due to the high free volume means, on the other hand, that silicone rubber is one of the poorest barrier materials not only for gases and water vapor but also for lipids and lipid-soluble materials.

Liquid water on one side of a silicone contact lens permeates through the lens by a solution–diffusion mechanism and evaporates on the other side quickly according to the permeability of water, whereas the solubility of water in silicone polymer is low [1]. The high water vapor permeability was speculated to be one of the reasons causing the "suction cup effect" that makes the lens stationary on one spot and tenaciously stick to the cornea; this may damage the corneal epithelium and result in other complications. However, the high permeability per se cannot be the reason for the suction cup effect if the exterior surface is covered by the tear film, i.e., if there is no driving force for water permeation.

The hydrophobicity of the surface prevents the wetting by tear and tends to expose dry surface of a contact lens. Therefore, rapid dehydration of the corneal tissues could occur, which could cause the damage of corneal epithelium. However, this explanation seems to be oversimplified in light of the adsorption of protein, which makes a hydrophobic surface wettable by tear fluid, as described in Chapter 26. Moreover, the highly hydrophobic surface characteristic of silicone rubber tends to encourage the deposition of protein and mucus of the tear on the surface of the lens. Lipids and lipid-soluble materials follow the same track and eventually penetrate into the bulk phase of the contact lens. Because of these undesirable factors, the use of silicone contact lenses of various chemical compositions and with surface treatments has not been successful but rather disastrous because of the interfacial characteristics of silicone contact lens on the cornea, which cannot be offset by these efforts. It indicates that more profound surface modification to cope with the problems rather than mere surface treatment is needed in capitalizing on the advantageous bulk properties of silicone polymers.

The major problems associated with silicone rubber used for contact lens stem from the surface properties. The surface is hydrophobic and hinders the spontaneous spreading of tear on the surface. The surface of silicone rubber has "tackiness" that is characteristics of elastomer, which is probably the main reason why silicone lens tends to stick on the cornea. Plasma polymerization of CH_4 was applied on the surface of silicone contact lens and flat sheet of the same material to investigate the effect of coating on the stickiness of the silicone surface (change of surface state) and the imbibition of lipids and lipid-soluble material by silicone polymer (barrier characteristics of the coating) by means of the absorption of oil-soluble dye as a function of the operational parameter of CH_4 LCVD nanofilm placed on the surface of silicone contact lenses [2,3].

3.1. Deposition Rate and Contact Angle of Water

Figure 35.1 depicts the dependence of the deposition rate on the discharge parameter W/FM. The figure is essentially the same as Figure 8.3 and, if plotted using D.R./FM, the same as Figure 8.4. Such a plot is necessary to identify the domain of LCVD, namely, the energy-deficient domain or monomer-deficient domain. Based on this figure, 2.6 GJ/kg (energy-deficient domain) and 21.8 GJ/kg (monomer-deficient domain) were selected for analysis of coating characteristics.

Figure 35.2 shows the dependence of the contact angle of water on the coated surface on the same operational parameter. The contact angle of water decreases (wettability increases) as the value of W/FM increases until the threshold value is reached at roughly 22 GJ/kg. The increase of wettability is primarily due to the postplasma incorporation of polar functionalities, indicating that the surface concentration of dangling bonds increases with the energy input.

3.2. Dye Penetration Test

Owing to the high free volume, silicone rubber absorbs oil-soluble organic dyes easily, and the oil-soluble dye absorption test is an effective indicator of the uniformity and the barrier characteristics of the coating. When a droplet of

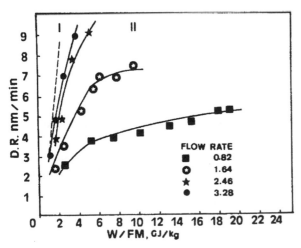

Figure 35.1 The dependence of deposition rate on the energy input parameter W/FM.

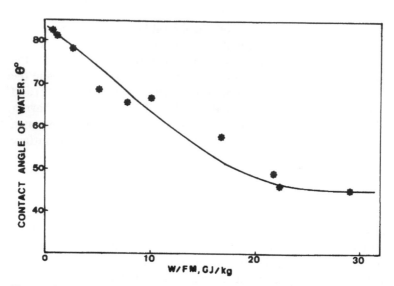

Figure 35.2 The dependence of contact angle of water on plasma coated surface on the energy input parameter W/FM.

oil-soluble Sudan red in butanol is placed on the interior surface of a silicone contact lens (a lens as a cup) for several minutes and rinsed with butanol, the lens shows a bull's eye of the rising sun. This means that the oil-soluble dye penetrated into the bulk of silicone contact lens in the short contact time. When the stained lens is kept for several days, the brilliant red bull's eye is transformed to a lightly stained contact lens as a whole because the dye penetrated from the droplet of dye solution placed on the surface migrates to the entire body of the contact lens. These observations show two important characteristics of the silicone contact lens: (1) high absorption characteristic of oil-soluble material and (2) high diffusivity of the absorbed dye molecules within the bulk phase of the contact lens.

If the coating applied is a good barrier to the dye, no stain is observed. If pinholes exist, spot staining occurs, and if the coating is permeable to the dye molecule, diffused staining occurs, with intensity inversely proportional to the barrier characteristic of the coating. Moreover, if plasma film can prevent the Sudan red dye molecules (MW 380.48) from penetrating into the highly permeable silicone rubber contact lens (no stain), it can be safely assumed that the coating is an effective barrier to the diffusion of larger lipids and lipid-soluble molecules in tears. Thus, the simple dye test serves as a crucially important test that provides information pertinent to the integrity and barrier characteristics of LCVD nanofilm placed on a silicone contact lens.

Table 35.1 summarizes the results of dye tests of plasma polymerization coatings of methane and a mixture of (methane + nitrogen) prepared under various conditions. In this table, the plus sign indicates that the sample passed the test, and the minus sign means that the sample failed the test. Plasma-polymerized methane coating thickness, as read from the thickness monitor, was 12 nm for all coatings shown in the table. It shows that for a pure methane coating at $W/FM = 2.6$ GJ/kg, the various kinds of mechanical resistance of the coating film are better than those prepared under other conditions, i.e., $W/FM = 21.8$ GJ/kg. This implies that at

Table 35.1 Dye Test for Contact Lenses Clampted on Rotating Disk and Coated with Methane Plasma Polymer or Methane Plasma Copolymer at Power $= 75\,W^a$

Flow rate of CH$_4$ (sccm)	W/FM (G/kg)	Gas mixed with CH$_4$ to initiate Plasma	Treatment after CH$_4$ or mixed CH$_4$ Plasma	Dye Test[b]					
				1	2	3	4	5	6
0.29	21.8			+	−		−		
0.29	21.8			+	−		−		
						(half in day)			
1.64	3.8			+	+ −	+ −	−	+ −	+ −
3.28	1.9			+	+ −	+ −	+ −	+ −	+ −
2.46	2.6			+	+	+	+	+	+
2.46	2.6		Immersed in 70°C water for 10 min	+	+	+	+	+	+
2.46	2.6		O$_2$ plasma, $F = 10$ sccm, $W = 12.5$ W, 2 min	+	+	+	+	+	+
2.46	2.6		Wet air passing	+	+	+	+	+	+
2.46	2.6		Wet air plasma $W = 50$ W, $F = 10$ sccm, 2 min, followed by immersion in 25°C water (Using lenses which can pass the 5-min dye test)	+	+ −	+ −	+ −		
2.46	0.8	N$_2$, $F = 3$ sccm		+	+		−	+	+ −
2.46	0.8	N$_2$, $F = 3$ sccm	Immersion in 70°C water for 10 min	+	−	−	−		+ −
2.46	1.5	N$_2$, $F = 1$ sccm	Immersion in 25°C water	+	+ −		−		
2.46	1.5	Air, $F = 1$ sccm	Immersion in 25°C water	+	+		+ −		

[a]The thickness reading was 12 nm. All the coated contact lenses were dipped in dye for 5 min.
[b]Dye test: 1, no mechanical work; 2, rolling once; 3, rolling 3, times; 4, rolling 10 times; 5, slight folding; 6, bending. +, no red stains appeared; −, red cloud appeared; + −, few red stains appeared.

$W/FM = 2.6$ GJ/kg, the adhesion of methane plasma polymer to silicone lenses and the cross-linked network of methane plasma polymer are optimal. The coating prepared at the higher W/FM seems to be too brittle and failed under mechanical stress.

3.3. Effect of Coating Thickness

Tables 35.2 and 35.3 depict the effect of coating thickness on the dye penetration tests for coatings prepared at 2.6 GJ/kg and 21.8 GJ/kg, respectively. Holding a

Table 35.2　Results of Dye Test of Contact Lenses Coated with Methane Plasma Polymer at $W/FM = 2.6\,GJ/kg$

Reading thickness (nm)	Coating thickness (nm)	5-min dye test	
		No mechanical work	Flexing 10 times
10	2.5	+	−
12	3.0	+	+
20	5.0	+	+
25	6.2	+	+
50	12.5	+	+
75	18.8	+	+
100	25.0	+	+
125	31.2	+	+
150	37.5	+	+
175	43.8	+	+
200	50.0	+	−
250	62.5	+	−
300	75.0	+	−
350	87.5	+	−
400	100	+	−
440	110	−	−

All contact lenses were clamped on the edges of empty holes of rotating disk to be coated. All coated lenses were dipped in dye for 5 min. Power = 75 W, flow rate = 2.46 sccm. +, no stain appeared; −, red cloud appeared.

Table 35.3　Results of Dye Test of Contact Lenses Clamped on the Rotating Disk and Coated with Methane Plasma Polymer at $W/FM = 21.8\,GJ/kg$[a]

Reading thickness (nm)	Coating thickness (nm)	5-min dye test	
		No mechanical work	Flexing 10 times
10	3.5	−	−
12	4.2	+	−
20	6.9	+	−
30	10.4	+	+ −
50	17.3	+	+ −
70	24.2	+	+ −
150	51.9	+	−
200	69.2	+	−
250	86.5	+	−
300	103.8	+	−
350	112.1	+	−
400	138.4	+	−

[a]All the coated contact lenses were dipped in dye for 5 min.
Power = 75 W, flow rate = 0.29 sscm. +, not stain appeared; −, red cloud appeared; + −, few red stains appeared.

coated contact lens folded between the thumb and the index finger tightly and rolling the folded contact lens back and forth repeatedly applied flexing of the coatings. The results of the dye tests are related to the internal stress of the coating film and the adhesive strength between the plasma coating and the substrate material. Adhesive strength provides the positive effect to get a tenacious coating on the substrate, whereas internal stress causes cracking of the film. Rolling of coated contact lens 10 times was done to test the mechanical breakdown of the coating. Comparing Tables 35.2 and 35.3, it is clear that at a W/FM of 2.6 GJ/kg, well-covered pinhole-free uniform coating can be obtained from 2.5 to 100 nm, and mechanically resistant film can be obtained from 3 to 44 nm. On the other hand, at a W/FM of 21.8 GJ/kg, well-covered flawless film is obtained from 4.2 to at least 140 nm, but mechanically resistant coating film is never achieved.

When coating thickness is less than the minimal threshold values (2.5 nm at low W/FM, 4.2 nm at high W/FM), it is impossible to get well-covered uniform coating; therefore, the dye penetrates into the contact lenses and stains appear. As the coating thickness gradually increases, the uniform film becomes well developed. Nevertheless, the internal stress simultaneously increases with the coating thickness until it reaches the threshold value and cracking of the film occurs. Therefore, the barrier characteristic is lost and the red stains appear.

When an external flexing force is imposed on the coated substrate, the film of W/FM 21.8 GJ/kg cannot maintain the integrity of the coating. The coating of W/FM 2.6 GJ/kg has a 35-nm durable range, which means that the coating film of low W/FM is more capable of resisting mechanical work than that of high W/FM. Compared with the low-energy film, the high-energy film is accompanied by higher internal stress. The cohesive forces of all the coating films of various thickness at $W/FM = 21.8$ GJ/kg cannot stand the total force of internal stress and external force, and thus cracking results. The cohesive force of the film of $W/FM = 2.6$ GJ/kg can stand the total force until the coating thickness reaches 44 nm. At that point, the cohesive force is overcome by the combination of internal stress and external force. These results clearly demonstrate that so far as the quality of LCVD coating as function of discharge energy is concerned, the principle of "the more the better" definitely does not apply, and the identification of the domain of LCVD is crucial.

3.4. Stickiness of the Surface

The effects of the composite parameter W/FM on the friction coefficient of the contact lens coated by methane plasma polymer at a fixed coating thickness 31.2 nm are shown in Figure 35.3. The friction coefficient, which is the tangential force divided by the normal force, was calculated from the sliding angle of a contact lens placed on a glass plate that was coated with plasma polymer of tetrafluoroethylene (TFE). The friction coefficient of the methane plasma polymerized contact lens is independent of W/FM in the range 2.6–29.1 GJ/kg.

Figure 35.4 depicts the change of frictional coefficient as a function of coating thickness at $W/FM = 2.6$ GJ/kg, which indicates that the application of LCVD coating effectively reduces the frictional coefficient and above thickness roughly 15 nm the frictional coefficient is independent of the coating thickness. The uncoated silicone rubber contact lenses are very sticky, with a falling angle of greater than 90 degrees. But after the silicone rubber contact lens is coated with 2.5 nm of methane

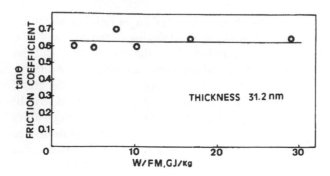

Figure 35.3   Effect of discharge energy input on frictional coefficient of coating.

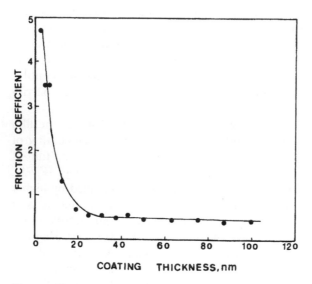

Figure 35.4   Effect of coating thickness on the frictional coefficient of coating of CH_4 plasma polymerization coating.

plasma polymer, the falling angle decreases to 78 degrees; in other words, the friction coefficient is 4.7. When coating thickness is between 2.5 and 25 nm, the friction coefficient decreases very fast, as does the tackiness. Over 25 nm, the friction coefficient is independent of coating thickness and becomes stable around 0.4.

Many factors influence tackiness, such as humidity, surface properties, elasticity, and shape of the tested material. The most plausible explanation for the reduction of friction coefficient by plasma polymerization coating of methane on the contact lens is that the coating changes the surface state of the silicone rubber. The thickness dependence agrees well with that for the necessary thickness to change the surface state described in Chapter 24. Although it cannot be expressed numerically, a short-term coating dramatically changes the feeling by touching the surface with fingers from "tacky" to "slippery."

The dependence of the falling angles of contact lenses on the coating thickness at a W/FM of 2.6 GJ/kg and the results of the dye tests in Table 35.2 showing uniformity and mechanical resistance are combined in Figure 35.5. It shows that the

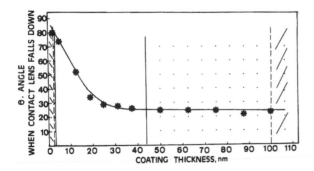

Figure 35.5 The range of coating thickness that satisfy both frictional coefficient and the dye penetration test at W/FM 2.6 GJ/kg.

uniform and pinhole-free film is in the range of coating thickness from 2.5 to 100 nm, whereas the mechanical resistant film is from 3.0 to 43.8 nm thick. This indicates that a plasma polymerization coating of methane, whose thickness is roughly 3.0–40 nm prepared under the condition of 2.6 GJ/kg normalized energy input, satisfies the requirements for obtaining uniform, flawless, highly mechanically durable film with an appropriate friction coefficient (without tackiness). It should be emphasized, however, these conditions are dependent on the nature of the substrate and should not be taken in the generic sense.

3.5. Overall Effects of LCVD Coating

Properties of CH_4 plasma polymerization coated (thickness 5 nm) silicone contact lens are compared with those for uncoated lens in Table 35.4. In the coating process designated as condition C, the energy input at 2.6 GJ/kg was used ($F = 2.46$ sccm, 75 W), and wet oxygen was introduced into the reactor after the polymerization and evacuation of gas. It was intended to enhance the reaction of dangling bonds with oxygen and water to yield a slightly more hydrophilic surface than the normal exposure to ambient air. The post-deposition oxygen plasma treatment of the coating was found to reduce the contact angle of water significantly, but at the expense of surface dynamic stability, and hence not adopted in the sample preparation.

This sample represents the coating at near the low end of coating thickness that satisfies both the dye penetration test and the reduction of surface stickiness. The CH_4 LCVD coating at the thickness of 5 nm effectively removed all problematic features of silicone contact lens with 12% reduction in oxygen permeability, which is still much higher than that of most other materials due to the nanoscale molecular sieve aspect of LCVD film. The true significance of this coating is the change of surface state of silicone lens to the imperturbable surface state of LCVD film.

4. LCVD ON INNER SURFACE OF POLYMER TUBE

LCVD coating of the inner surface of a small-diameter tube, e.g., 3 mm ID and 10 cm long, requires a special equipment that creates a luminous gas phase in the inside volume of the tube. If such a tube is placed in a reactor used for the coating of

Table 35.4 Comparison of Properties of Coated Contact Lenses and Uncoated Lenses
under Condition C

Feature	Uncoated	Coated
Wettability (degree; dyne/cm)		
Contact angle of water, θ	81	60 (26% reduction)
Dispersion component of surface energy	22.54	26.96 (19.6% increase)
Polar component of surface energy	5.63	12.50 (122% increase)
Solid surface energy	28.17	39.46 (40% reduction)
Permeability (cm^3-STP) $(cm)/(cm^2)(s)(cm\ Hg)$		
O_2	577×10^{-10}	508×10^{-10} 12% reduction
N_2	282×10^{-10}	193.3×10^{-10} (31% reduction)
Co_2	2442×10^{-10}	2046×10^{-10} (16% reduction)
Tackiness		
Falling angle of contact lens	90	75
Friction coefficient ($\tan \theta$)		3.73
ESCA	∞	
O 1s/C 1s	0.46	0.271
Si 2p/C 1s	0.55	0.195
Dye test	Stain	No stain
Softness	Good	Good
Transparency	Good	Good

contact lenses described in the previous section, the main coating occurs on the outer surface of the tube and on the inner surface near the opening only because the luminous gas cannot penetrate into the inner volume. As the aspirate ratio (length/ID) increases this situation becomes worse, and an ordinary LCVD reactor cannot do coating of the inner surface of long tubes such as those used in extracorporeal devices.

This problem could be solved if one could crease the luminous gas inside the tube and keep feeding the monomer continuously to compensate for the consumption of monomer due to the deposition. If the luminous gas phase could be kept at a small volume within the tube (typically 2 cm of the quartz tube), and the tube could be moved with respect to the luminous gas phase, then the inside wall of a considerably long (e.g., 16 m) plastic tube could be coated uniformly by LCVD [4].

Figure 35.6 shows pictures of the inner tube coating reactor. The reactor is a dual-vacuum system. The top of a sample tube is connected to the monomer feed-in system and the bottom is connected to a pumping system, which constitutes one vacuum system. The majority of the tube is rolled on a spool in the bottom section (not shown). The outside of the tube (within a quartz tube) is pumped by another pump system. The monomer flows from top to bottom, and the tube is pulled from bottom to top. In this counterflow mode, the deposition occurs on the uncoated portion of the tube. The coating thickness is controlled by the W/FM and the pulling speed, which is controlled by a surface speed control device, of the tube that determines the resident time of a section of surface in the glow.

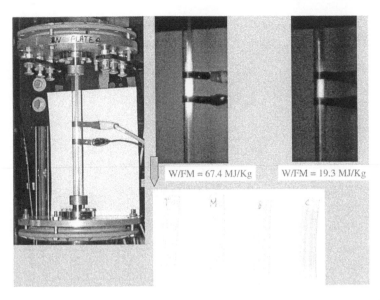

Figure 35.6 Inner tube coating reactor with dual vacuum systems, dye tested coated tube samples are compared with that of uncoated control; T: top, M: middle, B: bottom of 1m long tube, C: uncoated control.

Although monomer is introduced only to the inner tube for the inner surface coating of a tube, the same or a different monomer can be fed into the outside space of the tube for the outer surface coating of a tube, if necessary. The pressure difference between the outside and the inside of the tube determines where the luminous gas phase is created. For the inner surface coating, the pressure in the outside of the tube is kept at the lowest pressure of the pump system. When monomer is fed to both inner tube and outer tube, both sides of the tube can be coated in one operation, by adjusting flow rates and pressure in both spaces to create a luminous gas phase on both sides of the tube. Although most operations were for the inner surface coating, the reactor can be operated in (1) inside only, (2) outside only, (3) inside with coating A and outside with coating A, and (4) inside with A and outside with B.

Figure 35.7 shows a 1-m-long silicone rubber tube coated and dye penetration tested by filling the coated tube with Sudan red solution and rinsed by the solvent. A certain length on the top portion of a tube is above the reaction zone and cannot be coated, and the certain length at the bottom portion of a tube cannot reach the reaction zone because the connection to the pumping system must be kept intact. These uncoated sections show stain by Sudan red, but the coated section does not show stain, indicating that coating is uniform and pinhole free in the entire length of the coated tube. The pressure of the reaction zone drops as the tube moves from the high-pressure side to the low-pressure side and makes it difficult to coat a very long tube, particularly with small ID.

The uniformity of the plasma polymer coating with respect to chemical composition along 10-m coated silicone tubing was examined by XPS. Specimens of about 2 cm in length were cut from the top, middle, and bottom sections of the

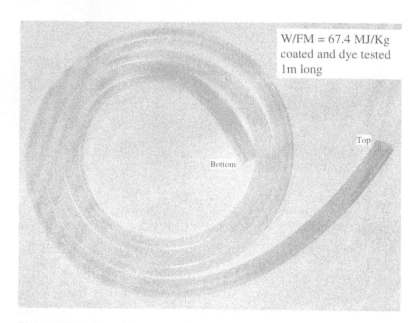

W/FM = 67.4 MJ/Kg
coated and dye tested
1m long

Top

Bottom

Figure 35.7   Pictorial view of coated and dye penetration tested tube, top and bottom part of the whole tube that didn't reach the plasma zone is stained, but no stain occurred in the coated section of the tube.

Table 35.5   Variation of Surface Compositions of Plasma Polymers along a 10-m Coated Tubing

Monomer	Flow rate (sccm)	Power input (W)	Atomic composition (%)			
			C	O	F	Si
Tetrafluoroethylene	0.43	20	T-46.3	18.7	20.7	14.4
			M-47.7	14.2	26.7	11.3
			E-46.8	15.5	26.1	11.6
Hexafluoroethane	0.13	23	T-46.3	14.0	35.8	4.0
			M-46.8	16.1	30.9	6.2
			E-49.8	16.8	26.9	6.6
Hexafluoroethane/H_2	0.32/0.22	9	T-49.0	15.5	29.8	5.7
			M-49.6	17.2	28.2	5.1
			E-49.4	16.0	32.5	2.1

T, Top; M, middle; E, end section of a 10-m coated tubing.

coated tubing. The XPS results obtained from these specimens, pertaining to plasma polymers of tetrafluoroethylene, hexafluoroethane, and hexafluoroethane/H_2, are given in Table 35.5. As can be seen, the variation of surface compositions along the tubing is small and thus confirms the uniformity of the coating layer.

The effect of W/FM parameter on the chemical nature of the glow discharge polymer of propylene was studied by means of XPS. Figure 35.8 depicts the

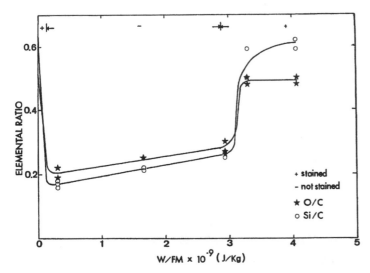

Figure 35.8 XPS elemental ratios, O/C and Si/C, and barrier characteristic manifested by oil-soluble dye test as a function of the normalized power input parameter W/FM.

dependency of the elemental ratios, O/X and Si/C, and the barrier characteristic manifested by the oil-soluble dye test on W/FM. In an intermediate range of W/FM values (from approximately 100 MJ/kg to 2.9 GJ/kg), an effective coating with barrier characteristic is obtained. The increase of the ratios above W/FM 3 GJ/kg is due to the evolution of the Si element from the Silastic substrate as a consequence of the interaction of energetic species in the glow discharge that is further incorporated in the polymerization process (XPS shown Si 2p signal). In many (nearly all) cases, when plasma polymerization coating of a hydrocarbon is applied on silicone rubber, Si is found on the top surface of the plasma polymer layer as seen in this case. This is somewhat similar to the observation that O atoms from oxides of pure iron migrate and participate in LCVD of TMS through the entire thickness of a coating.

How strongly the coating adheres to the underlying silicone substrate in the presence of a flowing fluid can be visualized from a test in which the silicone tubing after being coated with plasma polymer of tetrafluoroethylene was subsequently extracted with n-hexane for 3 h under continuous flow. The surface was examined by ESCA before and after the extraction. Figure 35.9 shows the XPS spectra. The elemental ratios of these surfaces determined from XPS are given in Table 35.6. A significant decrease in O/C and F/C ratios is observed after the extraction. The peaks at high binding energy in the C 1s spectrum of the sample after extraction appear to decrease, indicating a decrease in surface concentrations of the various types of carbon bonded to fluorine moieties. Also, the C1s spectrum exhibits a wide shoulder located in the vicinity of 287 eV. The increased signal intensity in this region could reflect the increasing concentration of carbon bonded to oxygen functionalities, as evidenced by the appearance of a shoulder located toward the higher binding energy in the O 1s spectrum.

One possible explanation for these changes seen in the XPS spectra is that the plasma polymer film prepared under the coating conditions used may contain some low molecular weight species, which are soluble in the extraction solvent. These

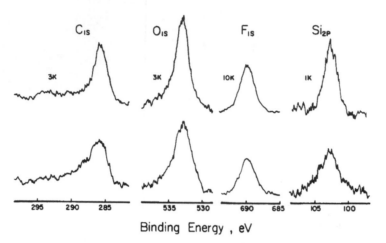

Top: Before Extraction   Bottom: After Extraction

Binding Energy , eV

Figure 35.9 Effect of solvent extraction on XPS spectra of plasma polymer of tetrafluoroethylene deposited on Silastic tube; extracted with *n*-hexane for 3 h under continuous flow.

Table 35.6 XPS Elemental Ratios for Plasma Polymers of Tetrafluoro-ethylene Before and After Extraction with *n*-Hexane

Sample	O/C	Si/C	F/C
Before extraction	0.43	0.34	0.43
After extraction	0.36	0.34	0.33

soluble fractions of the plasma polymer film are removed upon extraction and a change of surface composition results. Even though the tubing swelled considerably during the extraction, only a small change of the ESCA spectra of the surface is observed, indicating a strong adhesion of the overall coating to the substrate. The sample weight loss of some of the tubes coated with plasma polymer of tetrafluoroethylene after being extracted for 2 h was also measured. The percentage weight loss varied from 0.5% to 2% depending on the coating conditions, as compared with a 2.5% weight loss with untreated Silastic tubes. This means that LCVD coating actually reduced the extractable material content of Silastic tubes.

The retardation of the migration of silicone oligomers from the silicone tubing after being coated indicates that the plasma polymer coating applied to the tube substrate is in a form of continuous film with barrier characteristics instead of some other physical forms (e.g., powder). Direct evidence of this aspect can also be seen from the fact that all test specimens prepared had passed the dye penetration test.

It should be recognized that the coated tubes were subjected to a severe extraction condition. And so, due to the mechanical flexing and to the tensile stress resulted from the swelling of the silicone tubing during the extraction, the surface of the plasma polymer coating was stained by the oil-soluble dye after the extraction,

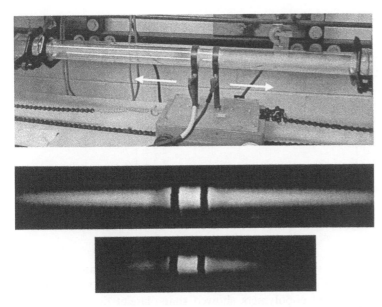

Figure 35.10 Pictorial view of a tube reactor with mobile radio frequency coil.

which may indicate that the coating network has been disintegrated. However, by manipulating the coating conditions, specifically increasing the energy input level for the glow discharge, a coating layer with strong internal structure can be formed, as evidenced by the fact that the polymer surface after the extraction was not stained by the dye.

This moving-tube reactor was also used to coat the inner surface of commercially available Gore-Tex artificial graft (approximately 30 cm long) by placing inside of a silicone rubber tube, in which the Gore-Tex graft can snugly fit, and passing through the reaction zone. Such a relatively short tube (e.g., 30 cm) can be also coated by placing in a glass tube and moving radio frequency coil, instead of moving tube. Figure 35.10 depicts such a moving-coil reactor. The ID of the glass tube should be close to the OD of the sample tube. If the cross-sectional area of the outside of sample become large, the glow occurs on the outside of tube and the inner surface of sample cannot be coated. Since the surrounding influences the coupling by radio frequency coil, it is more difficult to maintain uniform glow by moving the radio frequency coil than the moving tube in stationary radio frequency coil, and possible length much shorter, e.g., less than 60 cm, than the moving-tube mode.

5. LCVD FOR BLOOD COMPATIBILITY

5.1. Smooth Surface

None of the smooth surfaces prepared by the depositing of plasma polymer on the smooth surface of Silastic tubing showed detectable thrombus formation by gamma camera imaging. Therefore, they were evaluated by measurement of relative rates of platelet consumption. Table 35.7 shows the results obtained using the plasma polymers described previously [5].

Table 35.7 Platelet Consumption Rates for Various Plasma Polymers

Surface, plasma polymer of	Platelet consumption (no./cm^2/day $\times 10^{-8}$)
Tetrafluoroethylene Ia	1.9 ± 0.9
Tetrafluoroethylene Ib	3.9 ± 1.3
Tetrafluoroethylene IIa	3.7 ± 1.1
Tetrafluoroethylene IIb	3.2 ± 0.8
(Hexafluoroethane + H$_2$) I	2.5 ± 1.7
(Hexafluoroethane + H$_2$) II	4.7 ± 1.2
Hexafluoroethane	3.4 ± 1.1
Methane I	5.6 ± 1.8
Methane II	1.1 ± 0.4

Rates of platelet destruction varied from 1.1×10^8 to 5.6×10^8 platelets per cm^2 of exposed surface per day. Since studies evaluating polyurethanes as well as acrylic and methacrylic polymers and copolymers showed that platelet destruction rates may exceed 20×10^8 platelets/cm^2-day, the nine plasma polymers evaluated were considered to be considerably less reactive. Since each polymer was evaluated only four or five times with average results in each case near the lower sensitivity limit for this test system (about 1×10^8 platelets/cm^2-day), further statistical interpretations of the data presented in Table 35.7 would be inappropriate. Thus, due to the passive nature of these materials, conclusions could not be drawn regarding the relative importance of specific surface chemical moieties, i.e., all plasma polymers investigated are relatively nonreactive regardless of type of monomer used. This might imply that all type A plasma polymers have the characteristic feature of imperturbable surface regardless of what kind of atoms and moieties are involved, and because of this feature all plasma polymers tested performed better than most conventional polymers.

5.2. Porous Surface

The surface topography of the Gore-Tex (polytetrafluoroethylene) vascular grafts, both as received and following application of a plasma polymer based on hexafluoroethane/H$_2$, is shown in Figure 35.11. The porous texture of the vascular graft was clearly retained. In fact, the two surfaces are not readily differentiated, except that the coated sample appears to have narrower separation of crystalline phases (although it might be variation due to sample). Thus, plasma polymerization when executed properly may provide a unique means to modify surfaces chemically without altering their bulk material properties.

In contrast to the smooth-surface materials, the expanded Teflon (Gore-Tex) grafts showed substantial platelet deposition over a 1-h exposure period as determined by gamma camera imaging [5]. Results are given in Figure 35.12. Thus, after 60 min the untreated grafts (10 cm $\times$ 4 mm ID) accumulated a total of about 10×10^9 platelets, or approximately 8×10^8 platelets per cm^2 of exposed surface. Smooth-walled polytetrafluoroethylene has previously been shown to consume platelets at a rate of 0.07×10^8 platelets/cm^2-h, which is two orders of magnitude less than the

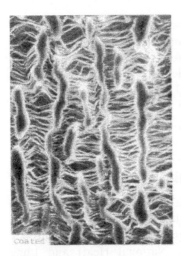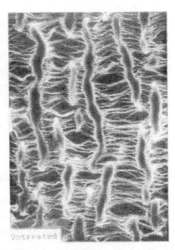

Figure 35.11 SEM picture of standard expanded Teflon (Gore-Tex, 30 μm fibril length); Left: coated with plasma polymer of (hexafluoroethane + H_2), Right: untreated, original magnification; ×500.

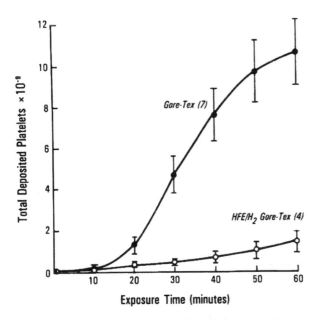

Figure 35.12 The effect of plasma polymerization coating applied on Gore-Tex graft in baboon model.

number of platelets deposited on the (porous) prosthesis over 1 h. Surface texture per se was thus an important variable regulating the extent of acute platelet interactions.

In contrast, identical Teflon grafts treated with a plasma polymer based on (hexafluoroethane + H_2) showed markedly reduced platelet deposition over 1 h as compared with the untreated controls. Platelet deposition averaged $1.4 \pm 0.5 \times 10^9$ platelets per graft (vs. $10.6 \pm 1.6 \times 10^9$ platelets per control graft), and this difference was statistically significant ($p < 0.0001$). Even after 2 h of blood exposure, the treated prostheses had accumulated only $2.4 \pm 0.7 \times 10^9$ platelets per graft. Since the treated and untreated grafts were morphologically identical, this result documents that surface perturbability alone might play a major role in mediating graft thrombus accumulation.

6. BIOCOMPATIBILITY OF IMPERTURBABLE SURFACE

In the early stage of investigation of blood compatibility of synthetic materials in the 1970s, the author carried out the investigation of blood compatibility of plasma polymers. (A series of reports to the Biomaterial Program, Devices and Technology Branch, Division of Heart and Vascular Diseases, National Heart and Lung Institute, National Institutes of Health, which deal with blood compatibility of plasma polymers by various methods of evaluations, are available from National Technical Information Service, 5285 Port Royal Road, Springfield, VA 22151.) Mainly three groups of plasma polymers, based on chemical nature of monomers, were investigated. Those are (1) hydrocarbons, (2) perfluorocarbons, and (3) siloxanes and other Si-containing organic molecules. The evaluation methods used at that time might be not as sophisticated as methods available today, but the methods provided sufficient means to separate good actors and bad actors among various kinds of polymers.

A striking aspect found with plasma polymers was that all plasma polymers, regardless of types of monomer, belonged to the top end of good actors. The common denominator for those plasma polymers was the absence of molecular mobility, which had led to the concept of imperturbable surface being potentially important factor for the good actors in biological interactions. The nature of imperturbable surface has been described in previous chapters.

The imperturbable surface is related to the number of surface configurations that can occur at the surface and its change when the surface has contact with a biological system. The number of surface configuration is governed by the molecular configuration of a polymer molecule. This point can be visualized by comparing the molecular configuration of poly(oxymethylene) and that of POE or poly(ethylene oxide) described in Chapter 23. Top view, side view, bottom view, and end view of stretched short segment of POM and POE are compared side by side in Figure 35.13. This figure compares only a specific conformation of the polymer segments but provides enough difference to indicate the difference in number of surface configurations that could be taken by these two polymers. Because of the symmetry of configuration along the chain axis of POE, the change of surface configuration due to the change of contacting medium is anticipated to be small. This is a similar situation described for the difference of sessile droplet contact angle that indicates the interfacial tension at the solid/air interface and the sessile bubble contact angle that indicates the interfacial tension at the solid/liquid interface for gelatin and agar-agar hydrogels in Chapter 23. The surface of POE is virtually imperturbable by virtue of its molecular configuration, i.e., the surface configuration remains nearly the same regardless of short-range conformational and rotational

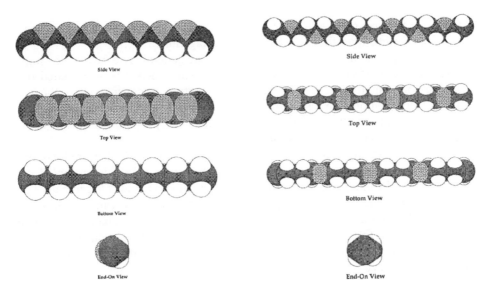

Figure 35.13 Comparison of stretched short segment of POM (left) and POE (right) in top, side, bottom, and end views.

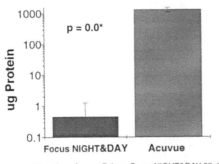

Wear time: Acuvue 7 days, Focus NIGHT&DAY 29 days

Figure 35.14 Effect of LCVD coating on proteins deposition on contact lens surface.

changes. It is known today that the POE surface is one of most biocompatible surfaces.

In contrast to the case of POE, the surface of plasma polymers are imperturbable by virtue of the lack of molecular mobility at the surface. It should be cautioned, however, that the imperturbable surface is limited to type A plasma polymers and does not extend to any plasma polymers. The details of biocompatibility are beyond the scope this book. Only some data that seem to indicate good biocompatibility of surface of plasma polymerization coatings are shown in this chapter.

Figure 35.14 depicts the effect of LCVD coating on the deposition of proteins on the contact lens surface in comparison with a commercially available reference lens [6]. The protein adsorptions were measured after continuous wearing, i.e., 1 week for the control and 1 month for LCVD-coated lens shown at left. The deposition on an LCVD-coated lens (left side) is 3.5 orders of magnitudes less than the reference lens in spite of more than four times longer wearing time.

The adsorption of a protein on a surface, in general, is a relatively simple interaction of the surface with a biological component when it is investigated in a simple one-component system. The protein adsorption is considered as the first important step of more complicated interaction of the surface with a biological system, and numerous efforts were made to reduce the protein adsorption in the avenue of creating biocompatible surfaces.

The protein adsorption becomes much more complicated when multicomponent systems are involved. One surface modification to reduce adsorption of one protein could lead to increased adsorption of other proteins, and it is generally very difficult to reduce the protein adsorption from living biological (multicomponent) system by surface modification. In this context, the remarkable reduction in protein adsorption from a live biological system shown in the figure is an outstanding performance, which is a testimony to the biocompatibility by virtue of an imperturbable surface of LCVD coating.

When a similar plasma polymerization coating was applied on the surface of a memory-expandable stainless steel stent, whose bare surface has notoriously poor blood compatibility, and implanted in a pig without using any drug to suppress blood coagulation, all five coated samples stayed patent, whereas uncoated stents with drug showed partial to total closure [7]. Such a clear-cut result, i.e., 5 out of 5 patency, has been scarcely seen in any animal experiments, which again seems to indicate the superior biocompatibility of imperturbable surfaces created with LCVD coatings.

REFERENCES

1. Refojo, M.F.; Leong, F.L. Contact Intraoc Lens Med. J. **1981**, *7*, 226.
2. Ho, C.-P.; Yasuda, H. J. Biomed. Mater. Res. **1988**, *22*, 919.
3. Ho, C.-P.; Yasuda, H. J. Appl. Polym. Sci. **1989**, *38*, 741.
4. Matsuzawa, Y.; Yasuda, H. Appl. Polym. Symp. **1984**, *18*, 65.
5. Yeh, Y.-S.; Iriyama, Yu; Matsuzawa, Y.; Hanson, S.R.; Yasuda, H. J. Biomed. Mater. Res. **1988**, *22* (9), 795.
6. McKenney, C.; Becker, N.; Thomas, S.; Castillo-Krevolin, C.; Grant, T. American Academy of Optometry, Annual Meeting Optometry and Vision Science **1998**, *75* (12), 276.
7. van der Giessen, W.J.; vanBeusekom, H.M.M.; vanHouten, C.D.; vanWoerkens, L.J.; Verdouw, P.D.; Serruys, P.W. Coronary Artery Dis. **1992**, *3*, 631.

36

Economical Advantages of Luminous Chemical Vapor Deposition

1. ENVIRONMENTALLY BENIGN (GREEN) PROCESSING

The major hurdles to overcome in the development and utilization of luminous chemical vapor deposition (LCVD) processes for industrial scale applications, outside of microelectronic applications, are (1) nonfamiliarity of vacuum processing and the resistance due to psychological fear of it, and (2) the relatively high initial cost of the equipment. In many cases, these two factors are strong enough to make planners shy away from the potential use of low-pressure processing, although the unique characteristics that can be obtained by LCVD have been known for many years.

These two factors are not technical problems, and the second factor is strictly the poor judgment of the economical factor. The economical factor of a processing should be examined by considering the total cost of the process as a whole but not by the initial cost of equipment alone. An important factor that is changing the equation for the economy of chemical processing in recent years is the cost associated with the requirements to keep the environment clean and to minimize the health hazard to workers. In other words, the aspect of "green processing" has become progressively important in determining the economical acceptability of a chemical process.

In dealing with the green aspect of chemical processing, there are two major approaches. One is the front-end approach in that a new nonpolluting or minimally polluting processing is first developed to carry out the processing. Another is the rear-end approach in that a new technique or an additional processing is developed to take care of the pollution created by a conventional chemical process. The first one is a new single process that does not pollute. The second one is two-part process comprising a conventional polluting processing and an added environmental remediation process. Low-pressure LCVD is a front-end approach.

The vacuum LCVD process is probably an ultimate green process because minimal material is used and virtually no effluent yields from the processing. The economical advantage of such a green process could be estimated in two typical cases. One is the case in which LCVD replaces polluting and/or hazardous processes.

Another case is starting a new manufacturing, in which LCVD could be used in competition with other conventional processes.

2.   LCVD TO REPLACE AN EXISTING HAZARDOUS PROCESS

In this case, the obvious advantage is the elimination of polluting and/or hazardous processes. However, the greatest benefit of making a process environmentally benign is intangible. Keeping this great intangible benefit in mind, it is possible to compare tangible advantages in the cost of materials involved and the possible reduction of steps in the overall manufacturing process. As described in Chapter 32, LCVD process can be added by piggybacking on the ion vapor deposition (IVD) process, meaning that there is no need of separate equipment for LCVD. Figure 36.1 compares the steps involved in the current IVD process and IVD/LCVD hybrid process.

      LCVD adds the operation time of an IVD reactor but eliminates two subsequent processes, i.e., glass bead peening and chromate conversion coating, of which processing time is much longer than the added LCVD processing time. Consequently, the addition of LCVD reduces the overall process cycle time as well as the labor cost associated with these two processes and the transferring substrates, which require labor and time. Furthermore, not only is this process environmentally benign, but it also improves the corrosion resistance of a nonchromated primer applied over it. All these are significant but rather intangible advantages.

      The largest cost that can be eliminated is the chromate conversion coating itself and the cost for treatment of spent solution and rinse water. In comparison to these costs, the cost for trimethylsilane (LCVD gas) to be used in the closed system LCVD mode is almost negligible. Therefore, the addition of LCVD process is economically favored, if one considers the cost for overall processing for corrosion protection of IVD-processed metallic objects.

3.   LCVD AS A NEW MANUFACTURING PROCESS

The initial cost of an LCVD reactor, which is substantial, is an extremely important factor in the consideration of a new manufacturing operation. Consequently, the

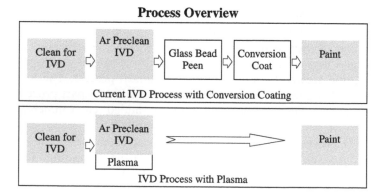

Figure 36.1   Comparison of IVD and IVD/LCVD hybrid processes.

feasibility of LCVD as a manufacturing process largely depends on the commercial value of the product itself and the add-on value attributable to LCVD. Even though LCVD has the advantage of being an ultimate green process, it cannot provide viable manufacturing if the ratio of additional cost to commercially chargeable price is high.

The decisive add-on value, which allows for setting the cost of product well beyond the cost of unprocessed product, and which cannot be attained by other means, is the key element for selecting LCVD processing. Thus, the decisive add-on value contains significant weight of increasing intangible "value." For instance, if a biomedical implant cannot be used because of poor biocompatibility, the feature that the surface could be made biocompatible, and any other conventional means could not achieve the same result, gives tremendous add-on value. This also means, in turn, that LCVD should not be tried to improve a product, of which market value is too low to tolerate an additional cost of processing.

As an example of LCVD process that provides the decisive add-on value, the coating of contact lens is chosen to illustrate the advantage of green processing because a coating of similar thickness can also be obtained by a more or less conventional wet coating process, and advantages and disadvantages could be compared. Furthermore, the role of coating is not just improving surface properties, but is associated with the crucial add-on value that ultimately decides whether or not the lens, including material, design, and other factors, functions satisfactorily in its final application on the human eye.

3.1. Continuous LCVD Coating (Front-End Approach)

Contact lenses are coated with an approximately 20-nm-thick layer of a methane-based plasma polymer by a continuous mode operation, which is shown schematically in Figure 36.2. The total processing time for a contact lens is approximately 40 min, which includes drying wet contact lenses before plasma polymerization coating, evacuation, LCVD, and repressurizing for sample removal. The coating operation could be continuous for approximately 30 days between maintenance breaks. The reactor is capable of coating 30 million contact lenses a year (340 days of operation).

3.2. Layer-By-Layer (LBL) Coating (Rear-End Approach)

A contact lens is coated with approximately 20 nm of coating consisting of alternating layers of an acidic polymer and a basic polymer by consecutive layer-by-layer (LBL) application of dilute solutions of the two types of polymers. LBL coating process is shown schematically in Figure 36.3. Ten baths in a line are considered for the cost comparison. Due to the cross-contamination of the two kinds of polymer solution on the consecutive dipping operation, the solution baths should be cleaned and new sets of dilute solutions used after x number of batches of dipping operation. The calculation of the processing cost shown below is based on 10 batches before the change of solutions. The total capacity of coating process is determined by the number of lines to be employed. The time necessary to complete coating is roughly the same as that for LCVD coating, but the processing is slightly more labor intensive. The most crucial factor, so far as the cost estimate is concerned, is the

Plasma coating process

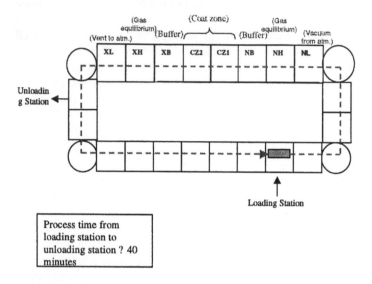

Process time from
loading station to
unloading station ? 40
minutes

(Front-end approach)

Figure 36.2 Schematic diagram of LCVD contact lens coating.

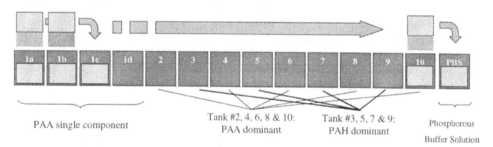

Figure 36.3 Schematic diagram of LBL coating operation.

value of x. If the value of x decreases below 10, the material cost increases proportional to $10/(10 - x)$, i.e., if the solutions must change after five dips, the material cost will increase to twice that cited in the cost estimate.

3.3. Comparison of Cost for Front-End and Rear-End Approaches

The costs of the two coating processes for production of 28 million coated contact lenses are compared in Table 36.1. The cost of LCVD coater is roughly 12 times greater than that for the LBL coating system. However, the plasma coating is an

Table 36.1 Cost Estimate of Materials and Equipment for LBL and LCVD Coatings

Factor	Rear-end approach (LBL coating)		Front-end approach (LCVD coating)	
	Quantity (kg)	Cost (U.S. $)	Quantity (kg)	Cost (U.S. $)
Intial cost of coating		250,000		3 million
Coating materials	7,759,257.4	181,277	1.9	146
Effluent treatment cost		5,971,292		0
Depreciation		25,000		300,000
Total cost		6,177,569		300,146

ultimate green process that uses a minimum of materials and there is virtually no effluent. Only 11 kg of coating materials (LCVD gas) is used to coat 28 million contact lenses. The coating yield is roughly 14%. The cost of coating materials is only less than $300 for 1 year's operation. Since there is no effluent, no cost for the effluent treatment is involved.

The LBL coating process requires the use of very dilute solutions of polymers in order to coat a layer with the thickness in the range of 1–2 nm on each dip-coating operation. This means that a large amount of solvent (water) is needed to make the coating solutions. The yield of nanofilm coating by layer-by-layer application of very dilute solution is roughly 0.05%. The amount of coating materials, of which the majority is water, needed for the production of 28 million contact lenses is roughly 7.76 million kg and its cost is roughly $181,000 in contrast to $290 for the plasma coating.

Without counting the cost of wastewater treatment, the annual cost of LBL coating processes is roughly two-thirds of LCVD. If a set of solutions cannot be used for 10 coating operations, the material cost for the LBL coating without counting wastewater treatment could become greater than that for the plasma coating. With consideration of the cost for wastewater treatment, plasma coating is a much more economical coating process. The cost of wastewater treatment included in Table 36.1 is based on the commercial wastewater treatment cost of $2000 for a 55-gallon drum. With this wastewater treatment cost, the LBL coating operation is more than 20 times more expensive than the LCVD coating on an annual basis.

It is important to recognize that the cost of wastewater treatment for 1 year's operation by LBL coating (recurring cost) is more than twice the initial cost for the LCVD coating system (one time cost subject for the depreciation), which is often conceived as prohibitively expensive for industrial applications. The cost of wastewater treatment could be reduced, to some extent, by installing wastewater treatment facility, but probably not to the extent that would shift the balance considered here. The "green" aspect of processing has become increasingly important in recent years, and the cost for being "not green" might become the major factor in the operational cost as shown in this comparison. This example shows that the cost of environmental remediation in the rear-end approach could be far more expensive than the "expensive" reactor in the front-end approach.

Vacuum processing, in general, could be ultimate green processing in that minimal quantities of materials are consumed and the effluent is contained in the system, which makes the effluent treatment, if it is necessary, easily accomplishable. So far as green processing is concerned, vacuum processing is a typical example of the front-end approach, i.e., creating or choosing processing that does not pollute the environment. The LBL coating process is a typical example of the rear-end approach, i.e., carrying out a process without considering the environmental influence in advance and taking care of the environmental pollution by the subsequent remediation process, which is more difficult and often more costly.

The data presented here clearly show the following important aspects of vacuum processing, which often are unrecognized or ignored. (1) Vacuum processing is an ultimate green processing. (2) The cost of vacuum equipment could be more than offset by the favorable cost of operation that uses the minimal amount of materials and requires no environmental mediation process.

Index